Methods of Multivariate Analysis

WILEY SERIES IN PROBABILITY AND MATHEMATICAL STATISTICS

Established by WALTER A. SHEWHART and SAMUEL S. WILKS

A complete list of the titles in this series appears at the end of this volume

Methods of Multivariate Analysis

ALVIN C. RENCHER

Department of Statistics
Brigham Young University
Provo, Utah

A Wiley-Interscience Publication
JOHN WILEY & SONS, INC.
New York • Chichester • Brisbane • Toronto • Singapore

This text is printed on acid-free paper.

Library of Congress Cataloging-in-Publication Data:

Rencher, Alvin C., 1934–
Methods of multivariate analysis / Alvin C. Rencher.
p. cm. — (Wiley series in probability and mathematical statistics. Applied probability and statistics)
"A Wiley-Interscience publication."
Includes bibliographical references (p. –) and index.
ISBN 0-471-57152-0 (cloth : acid-free paper)
1. Multivariate analysis. I. Title. II. Series.
QA278.R45 1995
519.5′35—dc20 94-23567

Printed in the United States of America

10 9 8 7 6 5 4 3

Contents

Preface

Practitioners and researchers in all applied disciplines often measure several variables on each subject or experimental unit. In some cases, it may be productive to isolate each variable in a system and study it separately. But in most instances, the system is complex, and the variables are intertwined in such a way that when analyzed in isolation they yield little information about the system. All of the variables must be examined simultaneously in order to access the key features of the system that produced them. This can be done readily with multivariate analysis, regardless of how many variables there are or how they are intercorrelated. The multivariate approach enables us to explore the joint performance of the variables and to determine the effect of each variable in the presence of the others.

Multivariate analysis provides both descriptive and inferential procedures—we can search for patterns in the data or test hypotheses about patterns of a priori interest. With multivariate descriptive techniques, we can peer beneath the tangled web of variables on the surface and extract the essence of the system. Multivariate inferential procedures include hypothesis tests that (1) process any number of variables without inflating the Type I error rate and (2) allow for whatever intercorrelations the variables possess. A wide variety of multivariate descriptive and inferential procedures is readily accessible in statistical software packages.

My selection of topics for this volume reflects years of consulting with researchers in many fields of inquiry. A brief overview of multivariate analysis is given in Chapter 1. Chapter 2 reviews the fundamentals of matrix algebra. Chapters 3 and 4 give an introduction to sampling from multivariate populations. Chapters 5, 6, 7, 10, and 11 extend univariate procedures with one dependent variable (such as t-tests, analysis of variance, tests on variances, multiple regression, and multiple correlation) to analogous multivariate techniques involving several dependent variables. A review of each univariate procedure is included. These reviews may provide key insights that the student has missed in previous courses. Chapters 8, 9, 12, and 13 describe multivariate techniques that are not related to univariate procedures. These include finding functions of the variables that discriminate among groups in the samples and finding functions of the variables that reveal the essential dimensionality and characteristic patterns of the system.

To illustrate multivariate applications, I have provided many examples and exercises based on real data sets from a wide variety of disciplines. A practitioner or consultant in multivariate analysis gains insights and acumen from long experience in working with data. It is not expected that a student can achieve this kind of seasoning in a one-semester class. However, the examples provide a good start, and further development is gained by working problems with the data sets. For example, in Chapters 12 and 13, the exercises cover several typical patterns in the covariance or correlation matrix. The student's intuition is expanded by associating these covariance patterns with the resulting configuration of the principal components or factors.

Although this is a methods book, I have included a few derivations. For some readers, an occasional proof provides insights obtainable in no other way. I hope that instructors who do not wish to use proofs will not be deterred by their presence. The proofs can easily be disregarded when reading.

My objective has been to make the book accessible to readers who have taken as few as two statistical methods courses. My classes in multivariate analysis include majors in statistics and majors from other departments. With the applied researcher in mind, I have provided careful intuitive explanations of the concepts and have included many insights heretofore available only in journal articles or in the minds of practitioners.

My overriding goal in preparation of this book has been clarity of exposition. I hope that students and instructors alike will find this multivariate text more comfortable than most. In the final stages of development, I asked my students for written reports on their initial reaction as they read each day's assignment. In general, they found the text to be very readable, but they also made many comments that led to improvements in the manuscript. I will be very grateful if readers will take the time to notify me of errors or of other suggestions they might have for improvements.

I have tried to use standard mathematical and statistical notation as far as possible and to maintain consistency of notation throughout the book. I have refrained from the use of abbreviations and mnemonic devices. These are annoying to those using a book as a reference.

Equations are numbered sequentially throughout a chapter; for example, (3.75) indicates the 75th numbered equation in Chapter 3. Tables and figures are also numbered sequentially throughout a chapter in the form "Table 3.8" or "Figure 3.1." Examples are not numbered sequentially; each example is identified by the same number as the section in which it appears.

When citing references in the text, I have used the standard format involving the year of publication. For a journal article, the year alone suffices, for example, Fisher (1936). But for books, I have included a page number, as in Seber (1984, p. 216).

In Appendix A, tables are provided for many multivariate distributions and tests. These enable the reader to conduct an exact test in many cases for which software packages provide only approximate tests. Appendix B gives answers and hints for the problems.

The diskette included with the text contains (1) all of the data sets and (2) SAS command files for all of the examples in the text. These command files can be adapted for use in working problems or in analyzing data sets encountered in applications. The contents of the diskette are described in Appendix C.

This is the first volume of a two-volume set on multivariate analysis. The second volume, tentatively entitled *An Introduction to Multivariate Statistical Inference and Applications*, is expected to appear in 1996. The two volumes are not necessarily sequential; they can be read independently. I adopted the two-volume format in order to (1) provide broader coverage than would be possible in a single volume and (2) offer the reader a choice of approach.

The second volume includes proofs of many techniques covered in the present volume and also introduces additional topics. The present volume includes many examples and problems using actual data sets, but only a few algebraic problems. The second volume emphasizes derivations of the results and contains fewer examples and problems with real data. The present volume has fewer references to the literature than the other volume, which includes a careful review of the latest developments and a more comprehensive bibliography. In the present volume, I have occasionally referred the reader to "Rencher (1996)" to note that added coverage of a certain subject is available in the second volume.

I am indebted to many individuals in the preparation of this manuscript. My initial exposure to multivariate analysis came in courses taught by Rolf Bargmann at the University of Georgia and D. R. Jensen at Virginia Polytechnic Institute and State University. Additional impetus to probe the subtleties of this field came from research conducted with Bruce Brown at BYU. I wish to thank Bruce Brown, Deane Branstetter, Del Scott, Robert Smidt, and Ingram Olkin for reading various versions of the manuscript and making valuable suggestions. I am grateful to the following students at BYU who helped with computations and typing: Mitchell Tolland, Tawnia Newton, Marianne Matis Mohr, Gregg Littlefield, Suzanne Kimball, Wendy Nielsen, Tiffany Nordgren, David Whiting, Karla Wasden, and Rachel Jones. I dedicate this volume to my wife, LaRue, who supported me patiently during the several years of preparation and revision of the manuscript.

I wish to thank the authors, editors, and owners of copyrights for permission to reproduce the following published figures and tables: Figure 3.8, Tables 3.2, 3.3, 3.4, 5.8, 6.28, and A.4 (Journal of the American Statistical Association); Tables 3.5, 5.3, 8.1 and A.7 (Journal of Quality Technology); Tables 3.7, 6.12, 6.14, 6.29, and A.14 (Elsevier North-Holland Publishing Company); Tables 3.8, 5.5, 6.20, 6.26, 6.27, and 10.1 (Biometrics); Table 3.9 (Genetica); Table 3.10 (American Journal of Clinical Nutrition); Tables A.1 to A.5, A.10, A.12, A.15, 6.16, and 6.21 (Biometrika Trustees); Tables 6.12 and A.6 (John Wiley and Sons); Tables 4.3, 12.2, 12.3, and 12.4 (Applied Statistics);

Tables 5.1 and 5.6 (Psychometrika); Table 5.2 (Psychological Bulletin); Table 5.7 (Springer-Verlag); Table 5.10 (National Academy of Science); Tables 6.24 and A.11 (Journal of Statistical Computation and Simulation); Table 6.3 (Communications in Statistics); Table 6.8 (Routledge Chapman and Hall); Table A.13 (PWS-KENT Publishing Company); Table 6.23 (Colorado State University); Table 7.1 (Institute of Statistical Mathematics); Table 5.9 (C. Y. Cramer); Table A.9 (The Dikewood Corporation); Table 6.6 (H. O. Posten); Table 13.1 (Journal of Pascal, Ada, and Modula); Correlation matrix in Example 13.6 (The Journal of the Acoustical Society of America); Table 3.6 (Diabetologia); Table 6.17 (Quality Progress); Table 6.18 (Scientia Horticulturae); Table 7.2 (R. J. Freund).

ALVIN C. RENCHER

CHAPTER 1

Introduction

1.1 WHY MULTIVARIATE ANALYSIS?

Multivariate analysis consists of a collection of methods that can be used when several measurements are made on each individual or object in one or more samples. We will refer to the measurements as *variables* and to the individuals or objects as *units* (research units, sampling units, or experimental units) or *subjects*. In practice, multivariate data sets are common, although they are not always analyzed as such. But the use of univariate procedures with such data is no longer excusable, given the availability of multivariate techniques and inexpensive computing power to carry them out.

Historically, the bulk of applications of multivariate techniques have been in the behavioral and biological sciences. However, interest in multivariate methods has now spread to numerous other fields of investigation. For example, I have collaborated on multivariate problems with researchers in education, chemistry, physics, geology, engineering, law, business, literature, religion, public broadcasting, nursing, mining, linguistics, biology, psychology, and many other fields. Table 1.1 shows some examples of multivariate observations.

The reader will notice that in some cases all the variables are measured in the same scale (see 1 and 2 in Table 1.1). In other cases, measurements are in different scales (see 3 in Table 1.1). In a few techniques such as profile analysis (Chapters 5 and 6), the variables must be commensurate, that is, similar in scale of measurement; however, most multivariate methods do not require this. Only a few procedures are affected by a change of scale.

Ordinarily the variables are measured simultaneously on each sampling unit. Typically, these variables are correlated. If this were not so, there would be little use for many of the techniques of multivariate analysis. We need to untangle the overlapping information provided by correlated variables and peer beneath the surface to see the underlying structure. Thus the goal of many multivariate approaches is *simplification*. We seek to express "what is going on" in terms of a reduced set of dimensions. In this sense, some multivariate techniques are *exploratory*, and they essentially generate hypotheses rather than test them.

Table 1.1 Examples of Multivariate Data

Units	Variables
1. Students	Several exam scores in a single course
2. Students	Grades in mathematics, history, music, art, physics
3. People	Height, weight, percentage of body fat, resting heart rate
4. Skulls	Length, width, cranial capacity
5. Companies	Expenditures for advertising, labor, raw materials
6. Manufactured items	Various measurements to check on compliance with specifications
7. Applicants for bank loans	Income, education level, length of residence, savings account, current debt load
8. Segments of literature	Sentence length, frequency of usage of certain words and of style characteristics
9. Human hairs	Composition of various elements
10. Birds	Lengths of various bones

On the other hand, if our goal is a formal hypothesis test, we need a technique that will (1) allow several variables to be tested and still preserve the significance level and (2) do this for any intercorrelation structure of the variables. Many such tests are available and are discussed in future chapters.

As the two preceding paragraphs imply, multivariate analysis is concerned generally with two areas, *descriptive* and *inferential* statistics. In the descriptive realm, we often obtain optimal linear combinations of variables. The optimality criterion varies from one technique to another, depending on the goal in each case. Although linear combinations may seem too simple to reveal the underlying structure, we use them for two obvious reasons: (1) mathematical tractability (linear approximations are used throughout all science for the same reason) and (2) they often perform well in practice. These linear functions may also be useful as a follow-up to inferential procedures. When we have a statistically significant test result that compares several groups, for example, we can find the linear combination (or combinations) of variables that led to rejection. Then the contribution of the variables to these linear combinations is of interest.

In the inferential area, many multivariate techniques are extensions of univariate procedures. In such cases we review the univariate procedure before presenting the analogous multivariate approach.

Multivariate inference is especially useful in curbing the researcher's natural tendency to read too much into the data. This is because of the total control provided for experimentwise error rate, that is, no matter how many variables are tested simultaneously, the value of α (the significance level) remains at the level set by the researcher.

Some authors warn against applying the common multivariate techniques to data for which the measurement scale is not interval or ratio. It has been found, however, that many multivariate techniques give reliable results when applied to ordinal data.

For many years the applications lagged behind the theory because the computations were beyond the power of the available desk-top calculators. However, with modern computers, virtually any analysis one desires, no matter how many variables or observations are involved, can be quickly and easily carried out. Perhaps it is not premature to say that multivariate analysis has come of age.

1.2 PREREQUISITES

The mathematical prerequisite for reading this book is matrix algebra. Calculus is not used. But the basic tools of matrix algebra are essential, and the presentation in Chapter 2 is intended to be sufficiently complete so that the reader with no previous experience can master matrix manipulation up to the level required in this book.

The statistical prerequisites are basic operational concepts of the normal distribution, t-tests, confidence intervals, multiple regression, and analysis of variance. These techniques are reviewed as each is extended to the analogous multivariate procedure.

This is a multivariate methods text. Most of the results are given without proof. In a few cases proofs are provided, but the major emphasis is on heuristic explanations. Our goal is an intuitive grasp of multivariate analysis, in the same mode as other statistical methods courses. A few of the problems are algebraic in nature, but the majority involve data sets to be analyzed.

1.3 OBJECTIVES

Having taught a course in multivariate methods for many years and having guided many graduate students in the statistics department as they consulted with clients in other departments who had multivariate research problems, I have formulated three objectives that I hope this book will achieve for the reader.

The first objective is to gain a thorough understanding of the details of various multivariate techniques, their purposes, their assumptions, their limitations, and so on. Many of these techniques are related; yet they differ in some essential ways. These similarities and differences are noted.

The second is to be able to select one or more appropriate techniques for a given multivariate data set. Recognizing the essential nature of a multivariate data set is the first step in a meaningful analysis. This is introduced in Section 1.4.

The third objective is to be able to interpret the results of a computer analysis of a multivariate data set. Reading the manual for a particular program package is not enough to make an intelligent appraisal of the output. Achievement of the first objective and practice on data sets in the text should help achieve the third objective.

1.4 BASIC TYPES OF DATA AND ANALYSIS

We will list four basic types of (continuous) multivariate data and then briefly describe some possible analyses. Some writers would consider this an oversimplification and would prefer elaborate tree diagrams of data structure. However, most data sets can fit into one of these categories, and the simplicity of this structure makes it easier to remember. The four basic data types are as follows:

1. A single sample with several variables measured on each sampling unit (subject or object).

2. A single sample with two sets of variables measured on each unit.

3. Two samples with several variables measured on each unit.

4. Three or more samples with several variables measured on each unit.

Each data type has extensions, and various combinations of the four are possible. A few examples of analyses for each case will now be given:

1. A single sample with several variables measured on each sampling unit:
   a. Test the hypothesis that the means of the variables have specified values.
   b. Test the hypothesis that the variables are uncorrelated and have a common variance.
   c. Find a small set of linear combinations of the original variables that summarizes most of the variation in the data (principal components).
   d. Express the original variables as linear functions of a smaller set of underlying variables that account for the original variables and their intercorrelations (factor analysis).
2. A single sample with two sets of variables measured on each unit:
   a. Determine the number, the size, and the nature of relationships between the two sets of variables. For example, we may wish to relate a set of interest variables to a set of achievement variables. How much overall correlation is there between these two sets (canonical correlation)?
   b. Find a model to predict one set of variables from the other set (multivariate multiple regression).
3. Two samples with several variables measured on each unit:
   a. Compare the means of the variables across the two samples (Hotelling's T^2-test).
   b. Find a linear combination of the variables that best separates the two samples (discriminant analysis).

   c. Find a function of the variables that will accurately allocate the units into the two groups (classification analysis).
4. Three or more samples with several variables measured on each unit:
   a. Compare the means of the variables across the groups (multivariate analysis of variance).
   b. Similar to 3b.
   c. Similar to 3c.

CHAPTER 2

Matrix Algebra

2.1 INTRODUCTION

This chapter introduces the basic elements of matrix algebra used in the remainder of this book. It is essentially a review of the requisite matrix tools and is not intended to be a complete development. However, it is sufficiently self-contained so that those with no previous exposure to the subject should need no other reference. Anyone unfamiliar with matrix algebra should plan to work most of the problems entailing numerical illustrations. It would also be helpful to explore some of the problems involving general matrix manipulation.

With the exception of a few derivations that seemed instructive, most of the results are given without proof. Some additional proofs are requested in the problems. For the remaining proofs, see any general text on matrix theory or one of the specialized matrix texts oriented to statistics, such as Graybill (1969) or Searle (1982).

2.2 NOTATION AND BASIC DEFINITIONS

2.2.1 Matrices, Vectors, and Scalars

A *matrix* is a rectangular or square array of numbers or variables arranged in rows and columns. We use uppercase boldface letters to represent matrices. All entries in matrices will be real numbers or variables representing real numbers. The elements of a matrix are displayed in brackets. For example, the ACT score and GPA for three students can be conveniently listed in the following matrix:

$$\mathbf{A} = \begin{pmatrix} 23 & 3.54 \\ 29 & 3.81 \\ 18 & 2.75 \end{pmatrix}. \tag{2.1}$$

The elements of $\mathbf{A}$ can also be variables, representing possible values of ACT and GPA for three students:

$$\mathbf{A} = \begin{pmatrix} a_{11} & a_{12} \\ a_{21} & a_{22} \\ a_{31} & a_{32} \end{pmatrix}. \tag{2.2}$$

In this double-subscript notation for the elements of a matrix, the first subscript indicates the row; the second identifies the column. The matrix **A** in (2.2) could also be expressed as

$$\mathbf{A} = (a_{ij}), \tag{2.3}$$

where a_{ij} is a general element.

The matrix **A** in (2.1) or (2.2) has three rows and two columns, so it is said to be 3×2. In general, if a matrix **A** has n rows and p columns, it is said to be $n \times p$, or alternatively, we say the *size* of **A** is $n \times p$.

A *vector* is an array with a single column or row. The following could be the test scores of a student in a course in multivariate analysis:

$$\mathbf{x} = \begin{bmatrix} 98 \\ 86 \\ 93 \\ 97 \end{bmatrix}. \tag{2.4}$$

Variable elements in a vector can be identified by a single subscript:

$$\mathbf{x} = \begin{bmatrix} x_1 \\ x_2 \\ x_3 \\ x_4 \end{bmatrix}. \tag{2.5}$$

We will use lowercase letters for column vectors. Row vectors are expressed as

$$\mathbf{x}' = (x_1, x_2, x_3, x_4)$$

or as

$$\mathbf{x}' = (x_1 \quad x_2 \quad x_3 \quad x_4).$$

The notation $\mathbf{x}'$ indicates the *transpose* of **x**, which is defined in Section 2.2.3.

Geometrically, a vector with p elements identifies a point in a p-dimensional space. The elements in the vector are the coordinates of the point. In Section 2.3.3, we define the distance from the origin to the point. In Section 3.12, we define the distance between two vectors.

A single real number is called a *scalar*, to distinguish it from a vector or matrix. Thus 2, −4, and 125 are scalars. A variable representing a scalar will usually be denoted by a lowercase nonbolded letter, such as $a = 5$. A product involving vectors and matrices may reduce to a matrix of size 1×1, which then becomes a scalar.

2.2.2 Equality of Vectors and Matrices

Two matrices are equal if they are the same size and the elements in corresponding positions are equal. Thus if $\mathbf{A} = (a_{ij})$ and $\mathbf{B} = (b_{ij})$, then $\mathbf{A} = \mathbf{B}$ if $a_{ij} = b_{ij}$ for all i and j. For example, let

$$\mathbf{A} = \begin{pmatrix} 3 & -2 & 4 \\ 1 & 3 & 7 \end{pmatrix} \qquad \mathbf{B} = \begin{pmatrix} 3 & 1 \\ -2 & 3 \\ 4 & 7 \end{pmatrix}$$

$$\mathbf{C} = \begin{pmatrix} 3 & -2 & 4 \\ 1 & 3 & 7 \end{pmatrix} \qquad \mathbf{D} = \begin{pmatrix} 3 & -2 & 4 \\ 1 & 3 & 6 \end{pmatrix}.$$

Then $\mathbf{A} = \mathbf{C}$. But even though $\mathbf{A}$ and $\mathbf{B}$ have the same elements, $\mathbf{A} \neq \mathbf{B}$ because the two matrices are not the same size. Likewise, $\mathbf{A} \neq \mathbf{D}$ because $a_{23} \neq b_{23}$. Thus two matrices of the same size are unequal if they differ in a single position.

2.2.3 Transpose and Symmetric Matrices

The *transpose* of a matrix $\mathbf{A}$, denoted by $\mathbf{A}'$, is obtained from $\mathbf{A}$ by interchanging rows and columns. Thus the columns of $\mathbf{A}'$ are the rows of $\mathbf{A}$, and the rows of $\mathbf{A}'$ are the columns of $\mathbf{A}$. The following examples illustrate the transpose of a matrix or vector:

$$\mathbf{A} = \begin{pmatrix} -5 & 2 & 4 \\ 3 & 6 & -2 \end{pmatrix} \qquad \mathbf{A}' = \begin{pmatrix} -5 & 3 \\ 2 & 6 \\ 4 & -2 \end{pmatrix}$$

$$\mathbf{B} = \begin{pmatrix} 2 & -3 \\ 4 & 1 \end{pmatrix} \qquad \mathbf{B}' = \begin{pmatrix} 2 & 4 \\ -3 & 1 \end{pmatrix}$$

$$\mathbf{a} = \begin{pmatrix} 2 \\ -3 \\ 1 \end{pmatrix} \qquad \mathbf{a}' = (2, -3, 1).$$

The transpose operation does not change a scalar, since it has only one row and one column.

If the transpose operator is applied twice to any matrix, the result is the original matrix:

$$(\mathbf{A}')' = \mathbf{A}. \tag{2.6}$$

If the transpose of a matrix is the same as the original matrix, the matrix is said to be *symmetric*. Thus $\mathbf{A}$ is symmetric if $\mathbf{A} = \mathbf{A}'$. For example,

$$\mathbf{A} = \begin{pmatrix} 3 & -2 & 4 \\ -2 & 10 & -7 \\ 4 & -7 & 9 \end{pmatrix} \qquad \mathbf{A}' = \begin{pmatrix} 3 & -2 & 4 \\ -2 & 10 & -7 \\ 4 & -7 & 9 \end{pmatrix}.$$

Clearly, all symmetric matrices are square.

2.2.4 Special Matrices

The *diagonal* of a $p \times p$ square matrix $\mathbf{A}$ consists of the elements $a_{11}, a_{22}, \ldots, a_{pp}$. For example, in the matrix

$$\mathbf{A} = \begin{pmatrix} 5 & -2 & 4 \\ 7 & 9 & 3 \\ -6 & 8 & 1 \end{pmatrix},$$

the elements 5, 9, and 1 comprise the diagonal. If a matrix contains zeros in all off-diagonal positions, it is said to be a *diagonal matrix*. An example of a diagonal matrix is

$$\mathbf{D} = \begin{bmatrix} 10 & 0 & 0 & 0 \\ 0 & -3 & 0 & 0 \\ 0 & 0 & 0 & 0 \\ 0 & 0 & 0 & 7 \end{bmatrix}.$$

This matrix could also be denoted as

$$\mathbf{D} = \text{diag}\,(10, -3, 0, 7). \tag{2.7}$$

A diagonal matrix can be formed from any square matrix by replacing off-diagonal elements by 0s. This is denoted by diag $\mathbf{A}$. Thus for the above matrix

A, we have

$$\text{diag}\,\mathbf{A} = \text{diag}\begin{pmatrix} 5 & -2 & 4 \\ 7 & 9 & 3 \\ -6 & 8 & 1 \end{pmatrix} = \begin{pmatrix} 5 & 0 & 0 \\ 0 & 9 & 0 \\ 0 & 0 & 1 \end{pmatrix}. \tag{2.8}$$

A diagonal matrix with a 1 in each diagonal position is called an *identity* matrix, denoted by **I**. For example, a 3×3 identity matrix is

$$\mathbf{I} = \begin{pmatrix} 1 & 0 & 0 \\ 0 & 1 & 0 \\ 0 & 0 & 1 \end{pmatrix}. \tag{2.9}$$

An *upper triangular matrix* is a square matrix with zeros below the diagonal, for example,

$$\mathbf{T} = \begin{bmatrix} 8 & 3 & 4 & 7 \\ 0 & 0 & -2 & 3 \\ 0 & 0 & 5 & 1 \\ 0 & 0 & 0 & 6 \end{bmatrix}. \tag{2.10}$$

A *lower triangular matrix* is defined similarly.

A vector of 1s will be denoted by **j**:

$$\mathbf{j} = \begin{bmatrix} 1 \\ 1 \\ \vdots \\ 1 \end{bmatrix}. \tag{2.11}$$

A square matrix of 1s is denoted by **J**. For example, a 3×3 matrix **J** is given by

$$\mathbf{J} = \begin{pmatrix} 1 & 1 & 1 \\ 1 & 1 & 1 \\ 1 & 1 & 1 \end{pmatrix}. \tag{2.12}$$

Finally, we denote a vector of zeros by **0** and a matrix of zeros by **O**. For example,

$$\mathbf{0} = \begin{pmatrix} 0 \\ 0 \\ 0 \end{pmatrix} \qquad \mathbf{O} = \begin{pmatrix} 0 & 0 & 0 & 0 \\ 0 & 0 & 0 & 0 \\ 0 & 0 & 0 & 0 \end{pmatrix}. \tag{2.13}$$

2.3 OPERATIONS

2.3.1 Summation and Product Notation

For completeness, we review the standard mathematical notation for sums and products. The sum of a sequence of numbers $a_1, a_2, \ldots, a_n$ is indicated by

$$\sum_{i=1}^{n} a_i = a_1 + a_2 + \cdots + a_n.$$

If the n numbers are all the same, then

$$\sum_{i=1}^{n} a = a + a + \cdots + a = na.$$

The sum of all the numbers in an array with double subscripts, such as

$$\begin{matrix} a_{11} & a_{12} & a_{13} \\ a_{21} & a_{22} & a_{23}, \end{matrix}$$

is indicated by

$$\sum_{i=1}^{2} \sum_{j=1}^{3} a_{ij} = a_{11} + a_{12} + a_{13} + a_{21} + a_{22} + a_{23}.$$

This is sometimes abbreviated to

$$\sum_{i=1}^{2} \sum_{j=1}^{3} a_{ij} = \sum_{i,j} a_{ij}.$$

The product of a sequence of numbers $a_1, a_2, \ldots, a_n$ is indicated by

$$\prod_{i=1}^{n} a_i = (a_1)(a_2) \cdots (a_n).$$

If the n numbers are all equal, the product becomes

$$\prod_{i=1}^{n} a = (a)(a) \cdots (a) = a^n.$$

2.3.2 Addition of Matrices and Vectors

If two matrices (or two vectors) are the same size, their *sum* is found by adding corresponding elements, that is, if $\mathbf{A}$ is $n \times p$ and $\mathbf{B}$ is $n \times p$, then $\mathbf{C} = \mathbf{A} + \mathbf{B}$ is also $n \times p$ and is found as $(c_{ij}) = (a_{ij} + b_{ij})$. For example,

$$\begin{pmatrix} -2 & 5 \\ 3 & 1 \\ 7 & -6 \end{pmatrix} + \begin{pmatrix} 3 & -2 \\ 4 & 5 \\ 10 & -3 \end{pmatrix} = \begin{pmatrix} 1 & 3 \\ 7 & 6 \\ 17 & -9 \end{pmatrix}$$

$$\begin{pmatrix} 1 \\ 3 \\ 7 \end{pmatrix} + \begin{pmatrix} 5 \\ -1 \\ 3 \end{pmatrix} = \begin{pmatrix} 6 \\ 2 \\ 10 \end{pmatrix}.$$

Similarly, the *difference* between two matrices or two vectors of the same size is found by subtracting corresponding elements. Thus $\mathbf{C} = \mathbf{A} - \mathbf{B}$ is found as $(c_{ij}) = (a_{ij} - b_{ij})$. For example,

$$(3 \quad 9 \quad -4) - (5 \quad -4 \quad 2) = (-2 \quad 13 \quad -6).$$

If two matrices are identical, their difference is a zero matrix; that is, $\mathbf{A} = \mathbf{B}$ implies $\mathbf{A} - \mathbf{B} = \mathbf{O}$. For example,

$$\begin{pmatrix} 3 & -2 & 4 \\ 6 & 7 & 5 \end{pmatrix} - \begin{pmatrix} 3 & -2 & 4 \\ 6 & 7 & 5 \end{pmatrix} = \begin{pmatrix} 0 & 0 & 0 \\ 0 & 0 & 0 \end{pmatrix}.$$

Matrix addition is commutative:

$$\mathbf{A} + \mathbf{B} = \mathbf{B} + \mathbf{A}. \tag{2.14}$$

The transpose of the sum (difference) of two matrices is the sum (difference) of the transposes.

$$(\mathbf{A} + \mathbf{B})' = \mathbf{A}' + \mathbf{B}' \tag{2.15}$$

$$(\mathbf{A} - \mathbf{B})' = \mathbf{A}' - \mathbf{B}' \tag{2.16}$$

$$(\mathbf{x} + \mathbf{y})' = \mathbf{x}' + \mathbf{y}' \tag{2.17}$$

$$(\mathbf{x} - \mathbf{y})' = \mathbf{x}' - \mathbf{y}'. \tag{2.18}$$

2.3.3 Multiplication of Matrices and Vectors

In order for the product $\mathbf{AB}$ to be defined, the number of columns in $\mathbf{A}$ must be the same as the number of rows in $\mathbf{B}$, in which case $\mathbf{A}$ and $\mathbf{B}$ are said to be *conformable*. Then the (ij)th element of $\mathbf{C} = \mathbf{AB}$ is

$$c_{ij} = \sum_k a_{ik} b_{kj}. \tag{2.19}$$

Thus c_{ij} is the sum of products of the ith row of **A** and the jth column of **B**. We therefore multiply each row of **A** by each column of **B**, and the size of **AB** consists of the number of rows of **A** and the number of columns of **B**. If **A** is $n \times m$ and **B** is $m \times p$, then $\mathbf{C} = \mathbf{AB}$ is $n \times p$. For example, if

$$\mathbf{A} = \begin{bmatrix} 2 & 1 & 3 \\ 4 & 6 & 5 \\ 7 & 2 & 3 \\ 1 & 3 & 2 \end{bmatrix} \quad \text{and} \quad \mathbf{B} = \begin{bmatrix} 1 & 4 \\ 2 & 6 \\ 3 & 8 \end{bmatrix},$$

then

$$\mathbf{C} = \mathbf{AB} = \begin{bmatrix} 2 \cdot 1 + 1 \cdot 2 + 3 \cdot 3 & 2 \cdot 4 + 1 \cdot 6 + 3 \cdot 8 \\ 4 \cdot 1 + 6 \cdot 2 + 5 \cdot 3 & 4 \cdot 4 + 6 \cdot 6 + 5 \cdot 8 \\ 7 \cdot 1 + 2 \cdot 2 + 3 \cdot 3 & 7 \cdot 4 + 2 \cdot 6 + 3 \cdot 8 \\ 1 \cdot 1 + 3 \cdot 2 + 2 \cdot 3 & 1 \cdot 4 + 3 \cdot 6 + 2 \cdot 8 \end{bmatrix}$$
$$= \begin{bmatrix} 13 & 38 \\ 31 & 92 \\ 20 & 64 \\ 13 & 38 \end{bmatrix}.$$

Note that **A** is 4×3, **B** is 3×2, and **AB** is 4×2. In this case, **AB** is of a different size than either **A** or **B**.

If **A** and **B** are both $n \times n$, then **AB** is also $n \times n$. Clearly, $\mathbf{A}^2$ is defined only if **A** is square.

In some cases **AB** is defined, but **BA** is not defined. In the above example, **BA** cannot be found because **B** is 3×2 and **A** is 4×3 and a row of **B** cannot be multiplied by a column of **A**. Sometimes **AB** and **BA** are both defined but are different in size. For example, if **A** is 2×4 and **B** is 4×2, then **AB** is 2×2 and **BA** is 4×4. If **A** and **B** are square and the same size, then **AB** and **BA** are both defined. But except for a few special cases

$$\mathbf{AB} \neq \mathbf{BA}. \tag{2.20}$$

For example, let

$$\mathbf{A} = \begin{pmatrix} 1 & 3 \\ 2 & 4 \end{pmatrix} \quad \mathbf{B} = \begin{pmatrix} 1 & -2 \\ 3 & 5 \end{pmatrix}.$$

Then

$$\mathbf{AB} = \begin{pmatrix} 10 & 13 \\ 14 & 16 \end{pmatrix} \qquad \mathbf{BA} = \begin{pmatrix} -3 & -5 \\ 13 & 29 \end{pmatrix}.$$

Thus we must be careful to specify the order of multiplication. If we wish to multiply both sides of a matrix equation by a matrix, we must multiply "on the left" or "on the right" and be consistent on both sides of the equation.

Multiplication is distributive over addition or subtraction:

$$\mathbf{A}(\mathbf{B} + \mathbf{C}) = \mathbf{AB} + \mathbf{AC} \tag{2.21}$$

$$\mathbf{A}(\mathbf{B} - \mathbf{C}) = \mathbf{AB} - \mathbf{AC} \tag{2.22}$$

$$(\mathbf{A} + \mathbf{B})\mathbf{C} = \mathbf{AC} + \mathbf{BC} \tag{2.23}$$

$$(\mathbf{A} - \mathbf{B})\mathbf{C} = \mathbf{AC} - \mathbf{BC}. \tag{2.24}$$

Note that, in general, because of (2.20),

$$\mathbf{A}(\mathbf{B} + \mathbf{C}) \neq \mathbf{BA} + \mathbf{CA}. \tag{2.25}$$

Using the distributive law, we can expand products such as $(\mathbf{A} - \mathbf{B})(\mathbf{C} - \mathbf{D})$ to obtain

$$\begin{aligned} (\mathbf{A} - \mathbf{B})(\mathbf{C} - \mathbf{D}) &= (\mathbf{A} - \mathbf{B})\mathbf{C} - (\mathbf{A} - \mathbf{B})\mathbf{D} \quad [\text{by (2.22)}] \\ &= \mathbf{AC} - \mathbf{BC} - \mathbf{AD} + \mathbf{BD} \quad [\text{by (2.24)}]. \end{aligned} \tag{2.26}$$

The transpose of a product is the product of the transposes in reverse order:

$$(\mathbf{AB})' = \mathbf{B}'\mathbf{A}'. \tag{2.27}$$

Note that (2.27) holds as long as $\mathbf{A}$ and $\mathbf{B}$ are conformable. They need not be square.

Multiplication involving vectors follows the same rules as for matrices. Suppose $\mathbf{A}$ is $n \times p$, $\mathbf{a}$ is $p \times 1$, $\mathbf{b}$ is $p \times 1$, and $\mathbf{c}$ is $n \times 1$. Then some possible products are $\mathbf{Ab}$, $\mathbf{c}'\mathbf{A}$, $\mathbf{a}'\mathbf{b}$, $\mathbf{b}'\mathbf{a}$, and $\mathbf{ab}'$. For example, let

$$\mathbf{A} = \begin{pmatrix} 3 & -2 & 4 \\ 1 & 3 & 5 \end{pmatrix} \qquad \mathbf{a} = \begin{pmatrix} 1 \\ -2 \\ 3 \end{pmatrix} \qquad \mathbf{b} = \begin{pmatrix} 2 \\ 3 \\ 4 \end{pmatrix} \qquad \mathbf{c} = \begin{pmatrix} 2 \\ -5 \end{pmatrix}.$$

Then

$$
\begin{aligned}
\mathbf{Ab} &= \begin{pmatrix} 3 & -2 & 4 \\ 1 & 3 & 5 \end{pmatrix} \begin{pmatrix} 2 \\ 3 \\ 4 \end{pmatrix} = \begin{pmatrix} 16 \\ 31 \end{pmatrix} \\
\mathbf{c}'\mathbf{A} &= (2 \quad -5) \begin{pmatrix} 3 & -2 & 4 \\ 1 & 3 & 5 \end{pmatrix} = (1 \quad -19 \quad -17) \\
\mathbf{c}'\mathbf{Ab} &= (2 \quad -5) \begin{pmatrix} 3 & -2 & 4 \\ 1 & 3 & 5 \end{pmatrix} \begin{pmatrix} 2 \\ 3 \\ 4 \end{pmatrix} \\
&= (2 \quad -5) \begin{pmatrix} 16 \\ 31 \end{pmatrix} = -123 \\
\mathbf{a}'\mathbf{b} &= (1 \quad -2 \quad 3) \begin{pmatrix} 2 \\ 3 \\ 4 \end{pmatrix} = 8 \\
\mathbf{b}'\mathbf{a} &= (2 \quad 3 \quad 4) \begin{pmatrix} 1 \\ -2 \\ 3 \end{pmatrix} = 8 \\
\mathbf{ab}' &= \begin{pmatrix} 1 \\ -2 \\ 3 \end{pmatrix} (2 \quad 3 \quad 4) = \begin{pmatrix} 2 & 3 & 4 \\ -4 & -6 & -8 \\ 6 & 9 & 12 \end{pmatrix} \\
\mathbf{ac}' &= \begin{pmatrix} 1 \\ -2 \\ 3 \end{pmatrix} (2 \quad -5) = \begin{pmatrix} 2 & -5 \\ -4 & 10 \\ 6 & -15 \end{pmatrix}.
\end{aligned}
$$

Note that **Ab** is a column vector, **c′A** is a row vector, **c′Ab** is a scalar, and **a′b** = **b′a**. The triple product **c′Ab** was obtained as **c′**(**Ab**). The same result would be obtained if we multiplied in the order (**c′A**)**b**:

$$
(\mathbf{c}'\mathbf{A})\mathbf{b} = (1 \quad -19 \quad -17) \begin{pmatrix} 2 \\ 3 \\ 4 \end{pmatrix} = -123.
$$

This is true in general for a triple product:

$$
\mathbf{ABC} = \mathbf{A}(\mathbf{BC}) = (\mathbf{AB})\mathbf{C}. \tag{2.28}
$$

Thus multiplication of three matrices can be defined in terms of the product of two matrices, since (fortunately) it does not matter which two are multiplied first. Note that **AB** and **BC** must be conformable for multiplication. For example, if **A** is $n \times p$, **B** is $p \times q$, and **C** is $q \times m$, then both multiplications are possible and the product **ABC** is $n \times m$.

A possible exception to the statement that the order of multiplication is arbitrary involves products such as $\mathbf{a}'\mathbf{bc}$. This makes sense only if it is understood that the product $\mathbf{a}'\mathbf{b}$ must be evaluated first to produce a scalar that can then be multiplied by $\mathbf{c}$. It is clear that we cannot take the product $\mathbf{bc}$. To avoid confusion in this case, the triple product should be written $(\mathbf{a}'\mathbf{b})\mathbf{c}$.

We can sometimes factor a sum of triple products on both right and left sides. For example,

$$\mathbf{ABC} + \mathbf{ADC} = \mathbf{A}(\mathbf{B} + \mathbf{D})\mathbf{C}. \tag{2.29}$$

As another illustration, let $\mathbf{X}$ be $n \times p$ and $\mathbf{A}$ be $n \times n$. Then

$$\mathbf{X}'\mathbf{X} - \mathbf{X}'\mathbf{AX} = \mathbf{X}'(\mathbf{X} - \mathbf{AX}) = \mathbf{X}'(\mathbf{I} - \mathbf{A})\mathbf{X}. \tag{2.30}$$

If $\mathbf{a}$ and $\mathbf{b}$ are both $n \times 1$, then

$$\mathbf{a}'\mathbf{b} = a_1b_1 + a_2b_2 + \cdots + a_nb_n \tag{2.31}$$

is a sum of products and is a scalar. On the other hand, $\mathbf{ab}'$ is defined for any size $\mathbf{a}$ and $\mathbf{b}$ and is a matrix, either rectangular or square:

$$\mathbf{ab}' = \begin{bmatrix} a_1 \\ a_2 \\ \vdots \\ a_n \end{bmatrix} (b_1 \quad b_2 \quad \cdots \quad b_p) = \begin{bmatrix} a_1b_1 & a_1b_2 & \cdots & a_1b_p \\ a_2b_1 & a_2b_2 & \cdots & a_2b_p \\ \vdots & \vdots & & \vdots \\ a_nb_1 & a_nb_2 & \cdots & a_nb_p \end{bmatrix}. \tag{2.32}$$

Similarly,

$$\mathbf{a}'\mathbf{a} = a_1^2 + a_2^2 + \cdots + a_n^2 \tag{2.33}$$

$$\mathbf{aa}' = \begin{bmatrix} a_1^2 & a_1a_2 & \cdots & a_1a_n \\ a_2a_1 & a_2^2 & \cdots & a_2a_n \\ \vdots & \vdots & & \vdots \\ a_na_1 & a_na_2 & \cdots & a_n^2 \end{bmatrix}. \tag{2.34}$$

Thus $\mathbf{a}'\mathbf{a}$ is a sum of squares and $\mathbf{a}\mathbf{a}'$ is a square (symmetric) matrix. The products $\mathbf{a}'\mathbf{a}$ and $\mathbf{a}\mathbf{a}'$ are sometimes referred to as the *dot product* and *matrix product*, respectively. The square root of the sum of squares of the elements of $\mathbf{a}$ is the *distance* from the origin to the point $\mathbf{a}$ and is also referred to as the *length* of $\mathbf{a}$:

$$\text{Length of } \mathbf{a} = \sqrt{\mathbf{a}'\mathbf{a}} = \sqrt{\sum_{i=1}^{n} a_i^2}. \tag{2.35}$$

As a special case of (2.34), note that

$$\mathbf{j}\mathbf{j}' = \begin{bmatrix} 1 & 1 & \cdots & 1 \\ 1 & 1 & \cdots & 1 \\ \vdots & \vdots & & \vdots \\ 1 & 1 & \cdots & 1 \end{bmatrix} = \mathbf{J}, \tag{2.36}$$

where $\mathbf{j}$ and $\mathbf{J}$ were defined in (2.11) and (2.12). Other products involving $\mathbf{j}$ are

$$\mathbf{a}'\mathbf{j} = \mathbf{j}'\mathbf{a} = \sum_{i=1}^{n} a_i \tag{2.37}$$

$$\mathbf{A}\mathbf{j} = \begin{bmatrix} \sum_j a_{1j} \\ \sum_j a_{2j} \\ \vdots \\ \sum_j a_{nj} \end{bmatrix}. \tag{2.38}$$

Thus $\mathbf{a}'\mathbf{j}$ is the sum of the elements in $\mathbf{a}$, and $\mathbf{A}\mathbf{j}$ contains the row sums of $\mathbf{A}$.

Since $\mathbf{a}'\mathbf{b}$ is a scalar, it is equal to its transpose:

$$\mathbf{a}'\mathbf{b} = (\mathbf{a}'\mathbf{b})' = \mathbf{b}'(\mathbf{a}')' = \mathbf{b}'\mathbf{a}. \tag{2.39}$$

This allows us to write $(\mathbf{a}'\mathbf{b})^2$ in the form

$$(\mathbf{a}'\mathbf{b})^2 = (\mathbf{a}'\mathbf{b})(\mathbf{a}'\mathbf{b}) = (\mathbf{a}'\mathbf{b})(\mathbf{b}'\mathbf{a}) = \mathbf{a}'(\mathbf{b}\mathbf{b}')\mathbf{a}. \tag{2.40}$$

From (2.18), (2.26), and (2.39) we obtain

$$(\mathbf{x} - \mathbf{y})'(\mathbf{x} - \mathbf{y}) = \mathbf{x}'\mathbf{x} - 2\mathbf{x}'\mathbf{y} + \mathbf{y}'\mathbf{y}. \tag{2.41}$$

If $\mathbf{a}$ and $\mathbf{x}_1, \mathbf{x}_2, \ldots, \mathbf{x}_n$ are all $p \times 1$ and $\mathbf{A}$ is $p \times p$, we obtain the following factoring results as extensions of (2.21) and (2.29):

$$\sum_{i=1}^{n} \mathbf{a}'\mathbf{x}_i = \mathbf{a}' \sum_{i=1}^{n} \mathbf{x}_i \tag{2.42}$$

$$\sum_{i=1}^{n} \mathbf{A}\mathbf{x}_i = \mathbf{A} \sum_{i=1}^{n} \mathbf{x}_i \tag{2.43}$$

$$\sum_{i=1}^{n} (\mathbf{a}'\mathbf{x}_i)^2 = \mathbf{a}' \left(\sum_{i=1}^{n} \mathbf{x}_i \mathbf{x}_i' \right) \mathbf{a} \quad \text{[by (2.40)]} \tag{2.44}$$

$$\sum_{i=1}^{n} \mathbf{A}\mathbf{x}_i (\mathbf{A}\mathbf{x}_i)' = \mathbf{A} \left(\sum_{i=1}^{n} \mathbf{x}_i \mathbf{x}_i' \right) \mathbf{A}'. \tag{2.45}$$

We can express matrix multiplication in terms of row vectors and column vectors. If $\mathbf{a}_i'$ is the ith row of $\mathbf{A}$ and $\mathbf{b}_j$ is the jth column of $\mathbf{B}$, then the (ij)th element of $\mathbf{AB}$ is $\mathbf{a}_i'\mathbf{b}_j$. For example, if $\mathbf{A}$ has three rows and $\mathbf{B}$ has two columns,

$$\mathbf{A} = \begin{pmatrix} \mathbf{a}_1' \\ \mathbf{a}_2' \\ \mathbf{a}_3' \end{pmatrix} \qquad \mathbf{B} = (\mathbf{b}_1, \mathbf{b}_2),$$

then the product $\mathbf{AB}$ can be written as

$$\mathbf{AB} = \begin{pmatrix} \mathbf{a}_1'\mathbf{b}_1 & \mathbf{a}_1'\mathbf{b}_2 \\ \mathbf{a}_2'\mathbf{b}_1 & \mathbf{a}_2'\mathbf{b}_2 \\ \mathbf{a}_3'\mathbf{b}_1 & \mathbf{a}_3'\mathbf{b}_2 \end{pmatrix}.$$

This can be expressed in the form

$$\mathbf{AB} = \begin{pmatrix} \mathbf{a}_1'(\mathbf{b}_1, \mathbf{b}_2) \\ \mathbf{a}_2'(\mathbf{b}_1, \mathbf{b}_2) \\ \mathbf{a}_3'(\mathbf{b}_1, \mathbf{b}_2) \end{pmatrix} = \begin{pmatrix} \mathbf{a}_1'\mathbf{B} \\ \mathbf{a}_2'\mathbf{B} \\ \mathbf{a}_3'\mathbf{B} \end{pmatrix} = \begin{pmatrix} \mathbf{a}_1' \\ \mathbf{a}_2' \\ \mathbf{a}_3' \end{pmatrix} \mathbf{B}. \tag{2.46}$$

Note that the first column of $\mathbf{AB}$ is

$$\begin{pmatrix} \mathbf{a}_1'\mathbf{b}_1 \\ \mathbf{a}_2'\mathbf{b}_1 \\ \mathbf{a}_3'\mathbf{b}_1 \end{pmatrix} = \begin{pmatrix} \mathbf{a}_1' \\ \mathbf{a}_2' \\ \mathbf{a}_3' \end{pmatrix} \mathbf{b}_1 = \mathbf{A}\mathbf{b}_1$$

and likewise the second column is $\mathbf{A}\mathbf{b}_2$. Thus $\mathbf{AB}$ can be written in the form

$$\mathbf{AB} = \mathbf{A}(\mathbf{b}_1, \mathbf{b}_2) = (\mathbf{A}\mathbf{b}_1, \mathbf{A}\mathbf{b}_2).$$

This result holds in general:

$$\mathbf{AB} = \mathbf{A}(\mathbf{b}_1, \mathbf{b}_2, \ldots, \mathbf{b}_p) = (\mathbf{A}\mathbf{b}_1, \mathbf{A}\mathbf{b}_2, \ldots, \mathbf{A}\mathbf{b}_p). \tag{2.47}$$

Any matrix can be multiplied by its transpose. If $\mathbf{A}$ is $n \times p$, then

$\mathbf{AA}'$ is $n \times n$ and is obtained as products of rows of $\mathbf{A}$.

Similarly,

$\mathbf{A}'\mathbf{A}$ is $p \times p$ and is obtained as products of columns of $\mathbf{A}$.

From (2.6) and (2.27), it is clear that both $\mathbf{AA}'$ and $\mathbf{A}'\mathbf{A}$ are symmetric.

In the above illustration for $\mathbf{AB}$ in terms of row and column vectors, the rows of $\mathbf{A}$ were denoted by $\mathbf{a}_i'$ and the columns of $\mathbf{B}$ by $\mathbf{b}_j$. If both rows and columns of a matrix $\mathbf{A}$ are under discussion, as in $\mathbf{AA}'$ and $\mathbf{A}'\mathbf{A}$, we will use the notation $\mathbf{a}_i'$ for rows and $\mathbf{a}_{(j)}$ for columns. To illustrate, if $\mathbf{A}$ is 3×4, we have

$$\mathbf{A} = \begin{pmatrix} a_{11} & a_{12} & a_{13} & a_{14} \\ a_{21} & a_{22} & a_{23} & a_{24} \\ a_{31} & a_{32} & a_{33} & a_{34} \end{pmatrix}$$

$$= \begin{pmatrix} \mathbf{a}_1' \\ \mathbf{a}_2' \\ \mathbf{a}_3' \end{pmatrix} = (\mathbf{a}_{(1)}, \mathbf{a}_{(2)}, \mathbf{a}_{(3)}, \mathbf{a}_{(4)}),$$

where, for example,

$$\mathbf{a}_2' = (a_{21} \quad a_{22} \quad a_{23} \quad a_{24})$$

and

$$\mathbf{a}_{(3)} = \begin{pmatrix} a_{13} \\ a_{23} \\ a_{33} \end{pmatrix}.$$

With this notation for rows and columns of $\mathbf{A}$, we can make more precise the statement above that the elements of $\mathbf{AA}'$ are products of the rows of $\mathbf{A}$ and the elements of $\mathbf{A}'\mathbf{A}$ are products of the columns of $\mathbf{A}$. Thus if we write $\mathbf{A}$ in terms of its rows as

$$\mathbf{A} = \begin{bmatrix} \mathbf{a}_1' \\ \mathbf{a}_2' \\ \vdots \\ \mathbf{a}_n' \end{bmatrix},$$

then by the usual row-by-column definition of matrix multiplication, we have

$$\mathbf{AA}' = \begin{bmatrix} \mathbf{a}_1' \\ \mathbf{a}_2' \\ \vdots \\ \mathbf{a}_n' \end{bmatrix} (\mathbf{a}_1, \mathbf{a}_2, \ldots, \mathbf{a}_n)$$

$$= \begin{bmatrix} \mathbf{a}_1'\mathbf{a}_1 & \mathbf{a}_1'\mathbf{a}_2 & \cdots & \mathbf{a}_1'\mathbf{a}_n \\ \mathbf{a}_2'\mathbf{a}_1 & \mathbf{a}_2'\mathbf{a}_2 & \cdots & \mathbf{a}_2'\mathbf{a}_n \\ \vdots & \vdots & & \vdots \\ \mathbf{a}_n'\mathbf{a}_1 & \mathbf{a}_n'\mathbf{a}_2 & \cdots & \mathbf{a}_n'\mathbf{a}_n \end{bmatrix}. \tag{2.48}$$

Similarly, if we express $\mathbf{A}$ in terms of columns as

$$\mathbf{A} = (\mathbf{a}_{(1)}, \mathbf{a}_{(2)}, \ldots, \mathbf{a}_{(p)}),$$

then

$$\mathbf{A}'\mathbf{A} = \begin{bmatrix} \mathbf{a}'_{(1)} \\ \mathbf{a}'_{(2)} \\ \vdots \\ \mathbf{a}'_{(p)} \end{bmatrix} (\mathbf{a}_{(1)}, \mathbf{a}_{(2)}, \ldots, \mathbf{a}_{(p)})$$

$$= \begin{bmatrix} \mathbf{a}'_{(1)}\mathbf{a}_{(1)} & \mathbf{a}'_{(1)}\mathbf{a}_{(2)} & \cdots & \mathbf{a}'_{(1)}\mathbf{a}_{(p)} \\ \mathbf{a}'_{(2)}\mathbf{a}_{(1)} & \mathbf{a}'_{(2)}\mathbf{a}_{(2)} & \cdots & \mathbf{a}'_{(2)}\mathbf{a}_{(p)} \\ \vdots & \vdots & & \vdots \\ \mathbf{a}'_{(p)}\mathbf{a}_{(1)} & \mathbf{a}'_{(p)}\mathbf{a}_{(2)} & \cdots & \mathbf{a}'_{(p)}\mathbf{a}_{(p)} \end{bmatrix}. \tag{2.49}$$

See problem 2.27 for $\mathbf{A}'\mathbf{A}$ expressed in terms of the rows of $\mathbf{A}$.

Let $\mathbf{A} = (a_{ij})$ be an $n \times n$ matrix and $\mathbf{D}$ be a diagonal matrix, $\mathbf{D} = \text{diag}(d_1, d_2, \ldots, d_n)$. Then, in the product $\mathbf{DA}$, the ith row of $\mathbf{A}$ is multiplied by d_i, and in $\mathbf{AD}$, the jth column of $\mathbf{A}$ is multiplied by d_j. For example, if $n = 3$, we have

$$\mathbf{DA} = \begin{pmatrix} d_1 & 0 & 0 \\ 0 & d_2 & 0 \\ 0 & 0 & d_3 \end{pmatrix} \begin{pmatrix} a_{11} & a_{12} & a_{13} \\ a_{21} & a_{22} & a_{23} \\ a_{31} & a_{32} & a_{33} \end{pmatrix}$$

$$= \begin{pmatrix} d_1a_{11} & d_1a_{12} & d_1a_{13} \\ d_2a_{21} & d_2a_{22} & d_2a_{23} \\ d_3a_{31} & d_3a_{32} & d_3a_{33} \end{pmatrix} \tag{2.50}$$

$$\mathbf{AD} = \begin{pmatrix} a_{11} & a_{12} & a_{13} \\ a_{21} & a_{22} & a_{23} \\ a_{31} & a_{32} & a_{33} \end{pmatrix} \begin{pmatrix} d_1 & 0 & 0 \\ 0 & d_2 & 0 \\ 0 & 0 & d_3 \end{pmatrix}$$

$$= \begin{pmatrix} d_1a_{11} & d_2a_{12} & d_3a_{13} \\ d_1a_{21} & d_2a_{22} & d_3a_{23} \\ d_1a_{31} & d_2a_{32} & d_3a_{33} \end{pmatrix} \tag{2.51}$$

$$\mathbf{DAD} = \begin{pmatrix} d_1^2a_{11} & d_1d_2a_{12} & d_1d_3a_{13} \\ d_2d_1a_{21} & d_2^2a_{22} & d_2d_3a_{23} \\ d_3d_1a_{31} & d_3d_2a_{32} & d_3^2a_{33} \end{pmatrix}. \tag{2.52}$$

In the special case where the diagonal matrix is the identity, we have

$$\mathbf{IA} = \mathbf{AI} = \mathbf{A}. \tag{2.53}$$

If $\mathbf{A}$ is rectangular, (2.53) still holds, but the two identities are of different sizes.

The product of a scalar and a matrix is obtained by multiplying each element of the matrix by the scalar:

$$c\mathbf{A} = (ca_{ij}) = \begin{bmatrix} ca_{11} & ca_{12} & \cdots & ca_{1m} \\ ca_{21} & ca_{22} & \cdots & ca_{2m} \\ \vdots & \vdots & & \vdots \\ ca_{n1} & ca_{n2} & \cdots & ca_{nm} \end{bmatrix}. \tag{2.54}$$

For example,

$$c\mathbf{I} = \begin{bmatrix} c & 0 & \cdots & 0 \\ 0 & c & \cdots & 0 \\ \vdots & \vdots & & \vdots \\ 0 & 0 & \cdots & c \end{bmatrix} \tag{2.55}$$

$$c\mathbf{x} = \begin{bmatrix} cx_1 \\ cx_2 \\ \vdots \\ cx_n \end{bmatrix}. \tag{2.56}$$

Since $ca_{ij} = a_{ij}c$, the product of a scalar and a matrix is commutative:

$$c\mathbf{A} = \mathbf{A}c. \tag{2.57}$$

Multiplication of vectors or matrices by scalars permits the use of linear combinations, such as

$$\sum_{i=1}^{k} a_i\mathbf{x}_i = a_1\mathbf{x}_1 + a_2\mathbf{x}_2 + \cdots + a_k\mathbf{x}_k$$

$$\sum_{i=1}^{k} a_i\mathbf{B}_i = a_1\mathbf{B}_1 + a_2\mathbf{B}_2 + \cdots + a_k\mathbf{B}_k.$$

If $\mathbf{S}$ is a symmetric matrix and $\mathbf{a}$ and $\mathbf{b}$ are vectors, the product

$$\mathbf{a}'\mathbf{S}\mathbf{a} = \sum_i a_i^2 s_{ii} + \sum_{i \neq j} a_i a_j s_{ij} \tag{2.58}$$

is called a *quadratic form*, while

$$\mathbf{a}'\mathbf{S}\mathbf{b} = \sum_{i,j} a_i b_j s_{ij} \tag{2.59}$$

is called a *bilinear form*. Either of these is, of course, a scalar, and can be treated as such. Expressions such as $\mathbf{a}'\mathbf{S}\mathbf{b}/\sqrt{\mathbf{a}'\mathbf{S}\mathbf{a}}$ are permissible (assuming $\mathbf{S}$ is positive definite; see Section 2.7).

2.4 PARTITIONED MATRICES

It is sometimes convenient to partition a matrix into submatrices. A partitioning of a matrix $\mathbf{A}$ into four submatrices could be indicated symbolically as follows:

$$\mathbf{A} = \begin{bmatrix} \mathbf{A}_{11} & \mathbf{A}_{12} \\ \mathbf{A}_{21} & \mathbf{A}_{22} \end{bmatrix}.$$

For example, let the 4×5 matrix $\mathbf{A}$ be partitioned as

$$\mathbf{A} = \left[\begin{array}{ccc|cc} 2 & 1 & 3 & 8 & 4 \\ -3 & 4 & 0 & 2 & 7 \\ 9 & 3 & 6 & 5 & -2 \\ \hline 4 & 8 & 3 & 1 & 6 \end{array}\right] = \begin{pmatrix} \mathbf{A}_{11} & \mathbf{A}_{12} \\ \mathbf{A}_{21} & \mathbf{A}_{22} \end{pmatrix},$$

where

$$\mathbf{A}_{11} = \begin{pmatrix} 2 & 1 & 3 \\ -3 & 4 & 0 \\ 9 & 3 & 6 \end{pmatrix} \qquad \mathbf{A}_{12} = \begin{pmatrix} 8 & 4 \\ 2 & 7 \\ 5 & -2 \end{pmatrix},$$

$$\mathbf{A}_{21} = (4 \quad 8 \quad 3) \qquad \mathbf{A}_{22} = (1 \quad 6).$$

If two matrices $\mathbf{A}$ and $\mathbf{B}$ are conformable and $\mathbf{A}$ and $\mathbf{B}$ are partitioned so that the submatrices are appropriately conformable, then the product $\mathbf{AB}$ can be found by following the usual row-by-column pattern of multiplication on the submatrices as if they were single elements; for example,

$$\mathbf{AB} = \begin{pmatrix} \mathbf{A}_{11} & \mathbf{A}_{12} \\ \mathbf{A}_{21} & \mathbf{A}_{22} \end{pmatrix} \begin{pmatrix} \mathbf{B}_{11} & \mathbf{B}_{12} \\ \mathbf{B}_{21} & \mathbf{B}_{22} \end{pmatrix}$$
$$= \begin{pmatrix} \mathbf{A}_{11}\mathbf{B}_{11} + \mathbf{A}_{12}\mathbf{B}_{21} & \mathbf{A}_{11}\mathbf{B}_{12} + \mathbf{A}_{12}\mathbf{B}_{22} \\ \mathbf{A}_{21}\mathbf{B}_{11} + \mathbf{A}_{22}\mathbf{B}_{21} & \mathbf{A}_{21}\mathbf{B}_{12} + \mathbf{A}_{22}\mathbf{B}_{22} \end{pmatrix}. \qquad (2.60)$$

It can be seen that this formulation is equivalent to the usual row-by-column definition of matrix multiplication. For example, the (1, 1) element of **AB** is the product of the first row of **A** and the first column of **B**. In the (1, 1) element of $\mathbf{A}_{11}\mathbf{B}_{11}$ we have the sum of products of part of the first row of **A** and part of the first column of **B**. In the (1, 1) element of $\mathbf{A}_{12}\mathbf{B}_{21}$ we have the sum of products of the rest of the first row of **A** and the remainder of the first column of **B**.

Multiplication of a matrix and a vector can also be carried out in partitioned form. For example,

$$\mathbf{Ab} = (\mathbf{A}_1, \mathbf{A}_2) \begin{pmatrix} \mathbf{b}_1 \\ \mathbf{b}_2 \end{pmatrix} = \mathbf{A}_1\mathbf{b}_1 + \mathbf{A}_2\mathbf{b}_2, \qquad (2.61)$$

where the partitioning of the columns of **A** corresponds to the partitioning of the elements of **b**. Note that the partitioning of **A** into two sets of columns is indicated by a comma.

The partitioned multiplication in (2.61) can be extended to individual columns of **A** and individual elements of **b**:

$$\mathbf{Ab} = (\mathbf{a}_1, \mathbf{a}_2, \ldots, \mathbf{a}_p) \begin{bmatrix} b_1 \\ b_2 \\ \vdots \\ b_p \end{bmatrix}$$
$$= b_1\mathbf{a}_1 + b_2\mathbf{a}_2 + \cdots + b_p\mathbf{a}_p. \qquad (2.62)$$

Thus **Ab** is expressible as a linear combination of the columns of **A**, the coefficients being elements of **b**. For example, let

$$\mathbf{A} = \begin{pmatrix} 3 & -2 & 1 \\ 2 & 1 & 0 \\ 4 & 3 & 2 \end{pmatrix} \quad \text{and} \quad \mathbf{b} = \begin{pmatrix} 4 \\ 2 \\ 3 \end{pmatrix}.$$

Then

$$\mathbf{Ab} = \begin{pmatrix} 11 \\ 10 \\ 28 \end{pmatrix}.$$

Using a linear combination of columns of **A** as in (2.62), we obtain

$$\begin{aligned} \mathbf{Ab} &= b_1\mathbf{a}_1 + b_2\mathbf{a}_2 + b_3\mathbf{a}_3 \\ &= 4\begin{pmatrix} 3 \\ 2 \\ 4 \end{pmatrix} + 2\begin{pmatrix} -2 \\ 1 \\ 3 \end{pmatrix} + 3\begin{pmatrix} 1 \\ 0 \\ 2 \end{pmatrix} \\ &= \begin{pmatrix} 12 \\ 8 \\ 16 \end{pmatrix} + \begin{pmatrix} -4 \\ 2 \\ 6 \end{pmatrix} + \begin{pmatrix} 3 \\ 0 \\ 6 \end{pmatrix} = \begin{pmatrix} 11 \\ 10 \\ 28 \end{pmatrix}. \end{aligned}$$

We note that the transpose of $\mathbf{A} = (\mathbf{A}_1, \mathbf{A}_2)$ is not equal to $(\mathbf{A}_1', \mathbf{A}_2')$, but rather

$$\mathbf{A}' = (\mathbf{A}_1, \mathbf{A}_2)' = \begin{pmatrix} \mathbf{A}_1' \\ \mathbf{A}_2' \end{pmatrix}. \tag{2.63}$$

2.5 RANK

Before defining the rank of a matrix, we first introduce the notion of linear independence and dependence. A set of vectors $\mathbf{a}_1, \mathbf{a}_2, \ldots, \mathbf{a}_n$ is said to be *linearly dependent* if constants $c_1, c_2, \ldots, c_n$ (not all zero) can be found such that

$$c_1\mathbf{a}_1 + c_2\mathbf{a}_2 + \cdots + c_n\mathbf{a}_n = \mathbf{0}. \tag{2.64}$$

If no constants $c_1, c_2, \ldots, c_n$ can be found satisfying (2.64), the set of vectors is said to be *linearly independent.*

If (2.64) holds, then at least one of the vectors $\mathbf{a}_i$ can be expressed as a linear combination of the other vectors in the set. Thus linear dependence of a set of vectors implies redundancy in the set. Among linearly independent vectors there is no redundancy of this type.

The *rank* of any square or rectangular matrix **A** is defined as

$$\begin{aligned} \text{rank}(\mathbf{A}) &= \text{number of linearly independent rows of } \mathbf{A} \\ &= \text{number of linearly independent columns of } \mathbf{A}. \end{aligned}$$

It can be shown that the number of linearly independent rows of a matrix is always equal to the number of linearly independent columns.

If $\mathbf{A}$ is $n \times p$, the maximum possible rank of $\mathbf{A}$ is the smaller of n and p, in which case $\mathbf{A}$ is said to be of *full rank*. For example,

$$\mathbf{A} = \begin{pmatrix} 1 & -2 & 3 \\ 5 & 2 & 4 \end{pmatrix}$$

has rank 2 because the two rows are linearly independent (neither row is a multiple of the other). However, even though $\mathbf{A}$ is full rank, the columns are linearly dependent because rank 2 implies there are only two linearly independent columns. Thus there exist constants c_1, c_2, and c_3 such that

$$c_1 \begin{pmatrix} 1 \\ 5 \end{pmatrix} + c_2 \begin{pmatrix} -2 \\ 2 \end{pmatrix} + c_3 \begin{pmatrix} 3 \\ 4 \end{pmatrix} = \begin{pmatrix} 0 \\ 0 \end{pmatrix}. \tag{2.65}$$

By (2.62), we can write (2.65) in the form

$$\begin{pmatrix} 1 & -2 & 3 \\ 5 & 2 & 4 \end{pmatrix} \begin{pmatrix} c_1 \\ c_2 \\ c_3 \end{pmatrix} = \begin{pmatrix} 0 \\ 0 \end{pmatrix}$$

or

$$\mathbf{Ac} = \mathbf{0}. \tag{2.66}$$

A solution vector to (2.65) or (2.66) is given by any multiple of $\mathbf{c}' = (14, -11, -12)$. Hence we have the interesting result that a product of a matrix $\mathbf{A}$ and a vector $\mathbf{c}$ is equal to $\mathbf{0}$, even though $\mathbf{A} \neq \mathbf{O}$ and $\mathbf{c} \neq \mathbf{0}$. This is a direct consequence of the linear dependence of the column vectors of $\mathbf{A}$.

Another consequence of the linear dependence of rows or columns of a matrix is the possibility of expressions such as $\mathbf{AB} = \mathbf{CB}$, where $\mathbf{A} \neq \mathbf{C}$. For example, let

$$\mathbf{A} = \begin{pmatrix} 1 & 3 & 2 \\ 2 & 0 & -1 \end{pmatrix} \quad \mathbf{B} = \begin{pmatrix} 1 & 2 \\ 0 & 1 \\ 1 & 0 \end{pmatrix} \quad \mathbf{C} = \begin{pmatrix} 2 & 1 & 1 \\ 5 & -6 & -4 \end{pmatrix}.$$

Then

$$\mathbf{AB} = \mathbf{CB} = \begin{pmatrix} 3 & 5 \\ 1 & 4 \end{pmatrix}.$$

All three matrices **A**, **B**, and **C** are full rank; but being rectangular, they have a rank deficiency in either rows or columns, which permits us to construct $\mathbf{AB} = \mathbf{CB}$ with $\mathbf{A} \neq \mathbf{C}$. Thus in a matrix equation, we cannot, in general, cancel matrices from both sides of the equation.

There are two exceptions to this rule. One involves a nonsingular matrix to be defined in the next section. The other special case occurs when the expression holds for all possible values of the matrix common to both sides of the equation. For example,

$$\text{If } \mathbf{Ax} = \mathbf{Bx} \text{ for all possible values of } \mathbf{x}, \text{ then } \mathbf{A} = \mathbf{B}. \tag{2.67}$$

To see this, let $\mathbf{x} = (1, 0, \ldots, 0)'$. Then the first column of **A** equals the first column of **B**. Now let $\mathbf{x} = (0, 1, 0, \ldots, 0)'$, and the second column of **A** equals the second column of **B**. Continuing in this fashion, we obtain $\mathbf{A} = \mathbf{B}$.

Suppose a rectangular matrix **A** is $n \times p$ of rank p, where $p < n$. We typically shorten this statement to "**A** is $n \times p$ of rank $p < n$."

2.6 INVERSE

If a matrix **A** is square and full rank, then **A** is said to be *nonsingular*, and **A** has a unique *inverse*, denoted by $\mathbf{A}^{-1}$, with the property that

$$\mathbf{AA}^{-1} = \mathbf{A}^{-1}\mathbf{A} = \mathbf{I}. \tag{2.68}$$

For example, let

$$\mathbf{A} = \begin{pmatrix} 3 & 4 \\ 2 & 6 \end{pmatrix}.$$

Then

$$\mathbf{A}^{-1} = \begin{pmatrix} .6 & -.4 \\ -.2 & .3 \end{pmatrix}$$

and

$$\begin{pmatrix} 3 & 4 \\ 2 & 6 \end{pmatrix} \begin{pmatrix} .6 & -.4 \\ -.2 & .3 \end{pmatrix} = \begin{pmatrix} 1 & 0 \\ 0 & 1 \end{pmatrix}.$$

If $\mathbf{A}$ is square and less than full rank, then an inverse does not exist, and $\mathbf{A}$ is said to be *singular*. Note that rectangular matrices do not have inverses as in (2.68), even if they are full rank.

If $\mathbf{A}$ and $\mathbf{B}$ are the same size and nonsingular, then the inverse of their product is the product of their inverses in reverse order,

$$(\mathbf{AB})^{-1} = \mathbf{B}^{-1}\mathbf{A}^{-1}. \tag{2.69}$$

Note that (2.69) holds only for nonsingular matrices. Thus, for example, if $\mathbf{A}$ is $n \times p$ of rank $p < n$, then $\mathbf{A}'\mathbf{A}$ has an inverse, but $(\mathbf{A}'\mathbf{A})^{-1}$ is not equal to $\mathbf{A}^{-1}(\mathbf{A}')^{-1}$ because $\mathbf{A}$ is rectangular and does not have an inverse.

If a matrix is nonsingular, it can be canceled from both sides of an equation, provided it appears on the left (right) on both sides. For example, if $\mathbf{B}$ is nonsingular, then

$$\mathbf{AB} = \mathbf{CB} \quad \text{implies} \quad \mathbf{A} = \mathbf{C},$$

since we can multiply on the right by $\mathbf{B}^{-1}$ to obtain

$$\begin{aligned} \mathbf{ABB}^{-1} &= \mathbf{CBB}^{-1} \\ \mathbf{AI} &= \mathbf{CI} \\ \mathbf{A} &= \mathbf{C}. \end{aligned}$$

Otherwise, if $\mathbf{A}$, $\mathbf{B}$, and $\mathbf{C}$ are rectangular or square and singular, it is easy to construct $\mathbf{AB} = \mathbf{CB}$, with $\mathbf{A} \neq \mathbf{C}$, as illustrated near the end of Section 2.5.

If the symmetric nonsingular matrix $\mathbf{A}$ is partitioned in the form

$$\mathbf{A} = \begin{pmatrix} \mathbf{A}_{11} & \mathbf{a}_{12} \\ \mathbf{a}'_{12} & a_{22} \end{pmatrix},$$

then the inverse is given by

$$\mathbf{A}^{-1} = \frac{1}{b} \begin{pmatrix} b\mathbf{A}_{11}^{-1} + \mathbf{A}_{11}^{-1}\mathbf{a}_{12}\mathbf{a}'_{12}\mathbf{A}_{11}^{-1} & -\mathbf{A}_{11}^{-1}\mathbf{a}_{12} \\ -\mathbf{a}'_{12}\mathbf{A}_{11}^{-1} & 1 \end{pmatrix}, \tag{2.70}$$

where $b = a_{22} - \mathbf{a}'_{12}\mathbf{A}_{11}^{-1}\mathbf{a}_{12}$. A nonsingular matrix of the form $\mathbf{B} + \mathbf{cc}'$, where $\mathbf{B}$ is nonsingular, has as its inverse

$$(\mathbf{B} + \mathbf{cc}')^{-1} = \mathbf{B}^{-1} - \frac{\mathbf{B}^{-1}\mathbf{cc}'\mathbf{B}^{-1}}{1 + \mathbf{c}'\mathbf{B}^{-1}\mathbf{c}}. \tag{2.71}$$

2.7 POSITIVE DEFINITE MATRICES

The symmetric matrix $\mathbf{A}$ is said to be *positive definite* if $\mathbf{x}'\mathbf{A}\mathbf{x} > 0$ for all possible vectors $\mathbf{x}$ (except $\mathbf{x} = \mathbf{0}$). Similarly, $\mathbf{A}$ is *positive semidefinite* if $\mathbf{x}'\mathbf{A}\mathbf{x} \geq 0$ for all $\mathbf{x} \neq \mathbf{0}$.

The diagonal elements a_{ii} of a positive definite matrix are positive. To see this, let $\mathbf{x}' = (0, \ldots, 0, 1, 0, \ldots, 0)$ with a 1 in the ith position. Then $\mathbf{x}'\mathbf{A}\mathbf{x} = a_{ii} > 0$. Similarly, for a positive semidefinite matrix $\mathbf{A}$, $a_{ii} \geq 0$ for all i.

One way to obtain a positive definite matrix is as follows:

$$\text{If } \mathbf{A} = \mathbf{B}'\mathbf{B}, \text{ where } \mathbf{B} \text{ is } n \times p \text{ of rank } p < n, \text{ then } \mathbf{B}'\mathbf{B} \text{ is positive definite.} \tag{2.72}$$

This is easily shown:

$$\mathbf{x}'\mathbf{A}\mathbf{x} = \mathbf{x}'\mathbf{B}'\mathbf{B}\mathbf{x} = (\mathbf{B}\mathbf{x})'(\mathbf{B}\mathbf{x}) = \mathbf{z}'\mathbf{z},$$

where $\mathbf{z} = \mathbf{B}\mathbf{x}$. Thus, $\mathbf{x}'\mathbf{A}\mathbf{x} = \sum_{i=1}^{n} z_i^2$, which is positive ($\mathbf{B}\mathbf{x}$ cannot be $\mathbf{0}$ unless $\mathbf{x} = \mathbf{0}$, because $\mathbf{B}$ is full rank). If $\mathbf{B}$ is less than full rank, then by a similar argument, $\mathbf{B}'\mathbf{B}$ is positive semidefinite.

Note that $\mathbf{A} = \mathbf{B}'\mathbf{B}$ is analogous to $a = b^2$ in real numbers, where the square of any number (including negative ones) is positive.

In another analogy to positive real numbers, a positive definite matrix can be factored into a "square root" in two ways. We give one method below in (2.73) and the other in Section 2.11.7.

A positive definite matrix $\mathbf{A}$ can be factored into

$$\mathbf{A} = \mathbf{T}'\mathbf{T}, \tag{2.73}$$

where $\mathbf{T}$ is a nonsingular upper triangular matrix. One way to obtain $\mathbf{T}$ is the *Cholesky decomposition*, which can be carried out in the following steps.

Let $\mathbf{A} = (a_{ij})$ and $\mathbf{T} = (t_{ij})$ be $n \times n$. Then the elements of $\mathbf{T}$ are found as follows:

$$t_{11} = \sqrt{a_{11}} \qquad t_{1j} = \frac{a_{1j}}{t_{11}} \qquad 2 \le j \le n$$

$$t_{ii} = \sqrt{a_{ii} - \sum_{k=1}^{i-1} t_{ki}^2} \qquad 2 \le i \le n$$

$$t_{ij} = \frac{a_{ij} - \sum_{k=1}^{i-1} t_{ki} t_{kj}}{t_{ii}} \qquad 2 \le i < j \le n$$

$$t_{ij} = 0 \qquad 1 \le j < i \le n$$

For example, let

$$\mathbf{A} = \begin{pmatrix} 3 & 0 & -3 \\ 0 & 6 & 3 \\ -3 & 3 & 6 \end{pmatrix}.$$

Then by the Cholesky method, we obtain

$$\mathbf{T} = \begin{bmatrix} \sqrt{3} & 0 & -\sqrt{3} \\ 0 & \sqrt{6} & \sqrt{1.5} \\ 0 & 0 & \sqrt{1.5} \end{bmatrix},$$

and

$$\mathbf{T}'\mathbf{T} = \begin{bmatrix} \sqrt{3} & 0 & 0 \\ 0 & \sqrt{6} & 0 \\ -\sqrt{3} & \sqrt{1.5} & \sqrt{1.5} \end{bmatrix} \begin{bmatrix} \sqrt{3} & 0 & -\sqrt{3} \\ 0 & \sqrt{6} & \sqrt{1.5} \\ 0 & 0 & \sqrt{1.5} \end{bmatrix}$$
$$= \begin{bmatrix} 3 & 0 & -3 \\ 0 & 6 & 3 \\ -3 & 3 & 6 \end{bmatrix} = \mathbf{A}.$$

2.8 DETERMINANTS

The *determinant* of an $n \times n$ matrix $\mathbf{A}$ is defined as the sum of all $n!$ possible products of n elements such that

1. each product contains one element from every row and every column and
2. the factors in each product are written so that the column subscripts appear in order of magnitude and each product is then preceded by a plus or minus sign according to whether the number of inversions in the row subscripts is even or odd.

An *inversion* occurs whenever a larger number precedes a smaller one. The symbol $n!$ is defined as

$$n! = n(n - 1)(n - 2) \cdots 2 \cdot 1. \tag{2.74}$$

The determinant of **A** is a scalar denoted by $|\mathbf{A}|$. The above definition is not useful in evaluating determinants, except in the case of 2×2 or 3×3 matrices. For larger matrices, other methods are available for manual computation, but determinants are typically evaluated by computer. For a 2×2 matrix, the determinant is found by

$$|\mathbf{A}| = \begin{vmatrix} a_{11} & a_{12} \\ a_{21} & a_{22} \end{vmatrix} = a_{11}a_{22} - a_{21}a_{12}. \tag{2.75}$$

For a 3×3 matrix, the determinant is given by

$$\begin{aligned} |\mathbf{A}| &= a_{11}a_{22}a_{33} + a_{12}a_{23}a_{31} + a_{13}a_{32}a_{21} \\ &\quad - a_{31}a_{22}a_{13} - a_{32}a_{23}a_{11} - a_{33}a_{12}a_{21}. \end{aligned} \tag{2.76}$$

This can be found by the following scheme. The three positive terms are obtained by

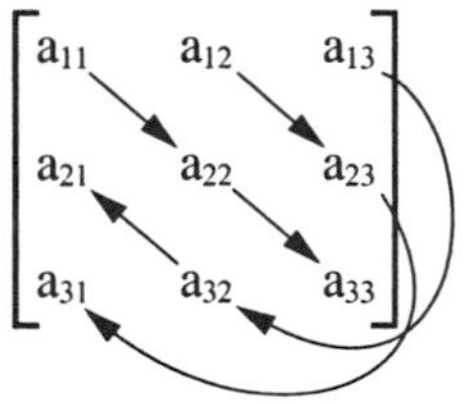

and the three negative terms by

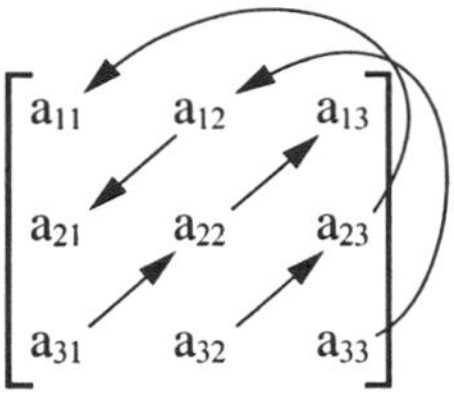

The determinant of a diagonal matrix is the product of the diagonal elements, that is, if $\mathbf{D} = \text{diag}(d_1, d_2, \ldots, d_n)$, then

$$|\mathbf{D}| = \prod_{i=1}^{n} d_i. \tag{2.77}$$

As a special case of (2.77), suppose all diagonal elements are equal, say,

$$\mathbf{D} = \text{diag}(c, c, \ldots, c) = c\mathbf{I}.$$

Then

$$|\mathbf{D}| = |c\mathbf{I}| = \prod_{i=1}^{n} c = c^n. \tag{2.78}$$

The extension of (2.78) to any square matrix $\mathbf{A}$ is

$$|c\mathbf{A}| = c^n|\mathbf{A}|. \tag{2.79}$$

Since the determinant is a scalar, we can carry out operations such as

$$|\mathbf{A}|^2 \qquad |\mathbf{A}|^{1/2} \qquad \frac{1}{|\mathbf{A}|},$$

provided that $|\mathbf{A}| > 0$ for $|\mathbf{A}|^{1/2}$ and that $|\mathbf{A}| \neq 0$ for $1/|\mathbf{A}|$.

If the square matrix $\mathbf{A}$ is singular, its determinant is 0:

$$|\mathbf{A}| = 0 \quad \text{if } \mathbf{A} \text{ is singular.} \tag{2.80}$$

If $\mathbf{A}$ is *near singular*, then there exists a linear combination of the columns that is close to $\mathbf{0}$, and $|\mathbf{A}|$ is also close to 0. If $\mathbf{A}$ is nonsingular, its determinant is nonzero:

$$|\mathbf{A}| \neq 0 \quad \text{if } \mathbf{A} \text{ is nonsingular.} \tag{2.81}$$

If **A** is positive definite, its determinant is positive:

$$|\mathbf{A}| > 0 \quad \text{if } \mathbf{A} \text{ is positive definite.} \tag{2.82}$$

If **A** and **B** are square and the same size, then the determinant of the product is the product of the determinants:

$$|\mathbf{AB}| = |\mathbf{A}||\mathbf{B}|. \tag{2.83}$$

For example, let

$$\mathbf{A} = \begin{pmatrix} 1 & 2 \\ -3 & 5 \end{pmatrix} \quad \text{and} \quad \mathbf{B} = \begin{pmatrix} 4 & 2 \\ 1 & 3 \end{pmatrix}.$$

Then

$$\mathbf{AB} = \begin{pmatrix} 6 & 8 \\ -7 & 9 \end{pmatrix} \qquad |\mathbf{AB}| = 110$$

$$|\mathbf{A}| = 11 \qquad |\mathbf{B}| = 10 \qquad |\mathbf{A}||\mathbf{B}| = 110.$$

The determinant of the inverse of a matrix is the reciprocal of the determinant:

$$|\mathbf{A}^{-1}| = \frac{1}{|\mathbf{A}|} = |\mathbf{A}|^{-1}. \tag{2.84}$$

If a partitioned matrix has the form

$$\mathbf{A} = \begin{pmatrix} \mathbf{A}_{11} & \mathbf{O} \\ \mathbf{O} & \mathbf{A}_{22} \end{pmatrix},$$

where $\mathbf{A}_{11}$ and $\mathbf{A}_{22}$ are square, but not necessarily the same size, then

$$|\mathbf{A}| = \begin{vmatrix} \mathbf{A}_{11} & \mathbf{O} \\ \mathbf{O} & \mathbf{A}_{22} \end{vmatrix} = |\mathbf{A}_{11}||\mathbf{A}_{22}|. \tag{2.85}$$

For a general partitioned matrix,

$$\mathbf{A} = \begin{pmatrix} \mathbf{A}_{11} & \mathbf{A}_{12} \\ \mathbf{A}_{21} & \mathbf{A}_{22} \end{pmatrix},$$

where $\mathbf{A}_{11}$ and $\mathbf{A}_{22}$ are square and nonsingular (not necessarily the same size), the determinant is given by either of the following two expressions:

$$\begin{vmatrix} \mathbf{A}_{11} & \mathbf{A}_{12} \\ \mathbf{A}_{21} & \mathbf{A}_{22} \end{vmatrix} = |\mathbf{A}_{11}||\mathbf{A}_{22} - \mathbf{A}_{21}\mathbf{A}_{11}^{-1}\mathbf{A}_{12}| \tag{2.86}$$

$$= |\mathbf{A}_{22}||\mathbf{A}_{11} - \mathbf{A}_{12}\mathbf{A}_{22}^{-1}\mathbf{A}_{21}|. \tag{2.87}$$

Note the analogy of (2.86) and (2.87) to the case of the determinant of a 2 × 2 matrix as given by (2.75):

$$\begin{vmatrix} a_{11} & a_{12} \\ a_{21} & a_{22} \end{vmatrix} = a_{11}a_{22} - a_{21}a_{12}$$

$$= a_{11}\left(a_{22} - \frac{a_{21}a_{12}}{a_{11}} \right)$$

$$= a_{22}\left(a_{11} - \frac{a_{12}a_{21}}{a_{22}} \right).$$

2.9 TRACE

A simple function of a square matrix is the *trace*, which is denoted by tr($\mathbf{A}$) and defined as the sum of the diagonal elements of $\mathbf{A}$; that is, $\text{tr}(\mathbf{A}) = \sum_{i=1}^{n} a_{ii}$ if $\mathbf{A}$ is $n \times n$. The trace is, of course, a scalar. If $n = 3$, we have

$$\mathbf{A} = \begin{pmatrix} a_{11} & a_{12} & a_{13} \\ a_{21} & a_{22} & a_{23} \\ a_{31} & a_{32} & a_{33} \end{pmatrix},$$

and $\text{tr}(\mathbf{A}) = a_{11} + a_{22} + a_{33}$. For example, suppose

$$\mathbf{A} = \begin{pmatrix} 5 & 4 & 4 \\ 2 & -3 & 1 \\ 3 & 7 & 9 \end{pmatrix}.$$

Then

$$\text{tr}(\mathbf{A}) = 5 + (-3) + 9 = 11.$$

The trace of the sum of two square matrices is the sum of the traces of the two matrices:

$$\text{tr}(\mathbf{A} + \mathbf{B}) = \text{tr}(\mathbf{A}) + \text{tr}(\mathbf{B}). \tag{2.88}$$

An important result for the product of two matrices is

$$\text{tr}(\mathbf{AB}) = \text{tr}(\mathbf{BA}). \tag{2.89}$$

This result holds for any matrices **A** and **B** where **AB** and **BA** are both defined. It is not necessary that **A** and **B** be square or that **AB** equal **BA**. For example, let

$$\mathbf{A} = \begin{pmatrix} 1 & 3 \\ 2 & -1 \\ 4 & 6 \end{pmatrix} \qquad \mathbf{B} = \begin{pmatrix} 3 & -2 & 1 \\ 2 & 4 & 5 \end{pmatrix}.$$

Then

$$\mathbf{AB} = \begin{pmatrix} 9 & 10 & 16 \\ 4 & -8 & -3 \\ 24 & 16 & 34 \end{pmatrix} \qquad \mathbf{BA} = \begin{pmatrix} 3 & 17 \\ 30 & 32 \end{pmatrix}$$

$$\text{tr}(\mathbf{AB}) = 9 - 8 + 34 = 35 \qquad \text{tr}(\mathbf{BA}) = 3 + 32 = 35.$$

2.10 ORTHOGONAL VECTORS AND MATRICES

Two vectors **a** and **b** of the same size are said to be *orthogonal* if

$$\mathbf{a}'\mathbf{b} = a_1 b_1 + a_2 b_2 + \cdots + a_n b_n = 0. \tag{2.90}$$

If $\mathbf{a}'\mathbf{a} = 1$, the vector **a** is said to be *normalized*. The vector **a** can always be normalized by dividing by its length, $\sqrt{\mathbf{a}'\mathbf{a}}$. Thus

$$\mathbf{c} = \frac{\mathbf{a}}{\sqrt{\mathbf{a}'\mathbf{a}}} \tag{2.91}$$

is normalized so that $\mathbf{c}'\mathbf{c} = 1$.

A matrix $\mathbf{C} = (\mathbf{c}_1, \mathbf{c}_2, \ldots, \mathbf{c}_p)$ whose columns are normalized and mutually orthogonal is called an *orthogonal* matrix. Since the elements of $\mathbf{C}'\mathbf{C}$ are products of columns of $\mathbf{C}$, which have the properties $\mathbf{c}_i'\mathbf{c}_i = 1$ for all i and $\mathbf{c}_i'\mathbf{c}_j = 0$ for all $i \neq j$, we have

$$\mathbf{C}'\mathbf{C} = \mathbf{I}. \tag{2.92}$$

If $\mathbf{C}$ satisfies (2.92), it necessarily follows that

$$\mathbf{C}\mathbf{C}' = \mathbf{I} \tag{2.93}$$

holds as well, from which we see that the rows of $\mathbf{C}$ are also normalized and mutually orthogonal. It is clear from (2.92) and (2.93) that $\mathbf{C}^{-1} = \mathbf{C}'$ for an orthogonal matrix $\mathbf{C}$.

We illustrate the creation of an orthogonal matrix by starting with

$$\mathbf{A} = \begin{pmatrix} 1 & 1 & 1 \\ 1 & 1 & -1 \\ 1 & -2 & 0 \end{pmatrix},$$

whose columns are mutually orthogonal. To normalize the three columns, we divide by the respective lengths, $\sqrt{3}$, $\sqrt{6}$, and $\sqrt{2}$, to obtain

$$\mathbf{C} = \begin{bmatrix} 1/\sqrt{3} & 1/\sqrt{6} & 1/\sqrt{2} \\ 1/\sqrt{3} & 1/\sqrt{6} & -1/\sqrt{2} \\ 1/\sqrt{3} & -2/\sqrt{6} & 0 \end{bmatrix}.$$

Note that the rows also became normalized and mutually orthogonal so that $\mathbf{C}$ satisfies both (2.92) and (2.93).

Multiplication by an orthogonal matrix has the effect of rotating axes; that is, if a point $\mathbf{x}$ is transformed to $\mathbf{z} = \mathbf{C}\mathbf{x}$, where $\mathbf{C}$ is orthogonal, then

$$\mathbf{z}'\mathbf{z} = (\mathbf{C}\mathbf{x})'(\mathbf{C}\mathbf{x}) = \mathbf{x}'\mathbf{C}'\mathbf{C}\mathbf{x} = \mathbf{x}'\mathbf{I}\mathbf{x} = \mathbf{x}'\mathbf{x}, \tag{2.94}$$

and the distance to $\mathbf{z}$ is the same as the distance to $\mathbf{x}$.

2.11 EIGENVALUES AND EIGENVECTORS

2.11.1 Definition

For every square matrix **A**, a scalar λ and a nonzero vector **x** can be found such that

$$\mathbf{A}\mathbf{x} = \lambda\mathbf{x}. \tag{2.95}$$

In (2.95), λ is called an *eigenvalue* of **A** and **x** is an *eigenvector*. To find λ and **x**, we write (2.95) as

$$(\mathbf{A} - \lambda\mathbf{I})\mathbf{x} = \mathbf{0}. \tag{2.96}$$

If $|\mathbf{A}-\lambda\mathbf{I}| \neq 0$, then $(\mathbf{A}-\lambda\mathbf{I})$ has an inverse and $\mathbf{x} = \mathbf{0}$ is the only solution. Hence, in order to obtain nontrivial solutions, we set $|\mathbf{A}-\lambda\mathbf{I}| = 0$. We can also see this by noting in (2.66) that $\mathbf{A}\mathbf{c} = \mathbf{0}$ for $\mathbf{c} \neq \mathbf{0}$ requires **A** to be singular; that is, the columns of **A** must be linearly dependent, as in (2.64). Thus in $(\mathbf{A} - \lambda\mathbf{I})\mathbf{x} = \mathbf{0}$, the matrix $\mathbf{A} - \lambda\mathbf{I}$ must be singular in order to find a solution vector **x** that is not **0**.

The equation $|\mathbf{A} - \lambda\mathbf{I}| = 0$ is called the *characteristic equation*. If **A** is $n \times n$, the characteristic equation will have n roots, that is, **A** will have n eigenvalues $\lambda_1, \lambda_2, \ldots, \lambda_n$. The λ's will not necessarily all be distinct or all nonzero. However, if **A** arises from computations on real (continuous) data and is nonsingular, the λ's will all be distinct (with probability 1). After finding $\lambda_1, \lambda_2, \ldots, \lambda_n$, the accompanying eigenvectors $\mathbf{x}_1, \mathbf{x}_2, \ldots, \mathbf{x}_n$ can be found using (2.96).

If we multiply both sides of (2.96) by a scalar k and note by (2.57) that k and $\mathbf{A} - \lambda\mathbf{I}$ commute, we obtain

$$(\mathbf{A} - \lambda\mathbf{I})k\mathbf{x} = k\mathbf{0} = \mathbf{0}. \tag{2.97}$$

Thus if **x** is an eigenvector of **A**, $k\mathbf{x}$ is also an eigenvector, and eigenvectors are unique only up to multiplication by a scalar. Hence we can adjust the length of **x**, but the direction from the origin is unique; that is, the relative values of (ratios of) the components of $\mathbf{x}' = (x_1, x_2, \ldots, x_n)$ are unique. Typically, the eigenvector **x** is scaled so that $\mathbf{x}'\mathbf{x} = 1$.

To illustrate, we will find the eigenvalues and eigenvectors for the matrix

$$\mathbf{A} = \begin{pmatrix} 1 & 2 \\ -1 & 4 \end{pmatrix}.$$

The characteristic equation is

$$|\mathbf{A} - \lambda\mathbf{I}| = \begin{vmatrix} 1 - \lambda & 2 \\ -1 & 4 - \lambda \end{vmatrix} = (1 - \lambda)(4 - \lambda) + 2 = 0$$

or

$$\lambda^2 - 5\lambda + 6 = (\lambda - 3)(\lambda - 2) = 0,$$

from which $\lambda_1 = 3$ and $\lambda_2 = 2$. To find the eigenvector corresponding to $\lambda_1 = 3$, we use (2.96),

$$(\mathbf{A} - \lambda\mathbf{I})\mathbf{x} = \mathbf{0}$$

$$\begin{pmatrix} 1 - 3 & 2 \\ -1 & 4 - 3 \end{pmatrix}\begin{pmatrix} x_1 \\ x_2 \end{pmatrix} = \begin{pmatrix} 0 \\ 0 \end{pmatrix}$$

$$-2x_1 + 2x_2 = 0$$

$$-x_1 + x_2 = 0.$$

As expected, either equation is redundant in the presence of the other, and there remains a single equation with two unknowns: $x_1 = x_2$. The solution vector can be written with an arbitrary constant as

$$\begin{pmatrix} x_1 \\ x_2 \end{pmatrix} = x_1\begin{pmatrix} 1 \\ 1 \end{pmatrix} = c\begin{pmatrix} 1 \\ 1 \end{pmatrix}.$$

If c is set equal to $1/\sqrt{2}$ to normalize the eigenvector, we obtain

$$\mathbf{x}_1 = \begin{pmatrix} 1/\sqrt{2} \\ 1/\sqrt{2} \end{pmatrix}.$$

Similarly, corresponding to $\lambda_2 = 2$, we have

$$\mathbf{x}_2 = \begin{pmatrix} 2/\sqrt{5} \\ 1/\sqrt{5} \end{pmatrix}.$$

2.11.2 tr(A) and |A|

For any square matrix $\mathbf{A}$ with eigenvalues $\lambda_1, \lambda_2, \ldots, \lambda_n$, we have

$$\operatorname{tr}(\mathbf{A}) = \sum_{i=1}^{n} \lambda_i \tag{2.98}$$

$$|\mathbf{A}| = \prod_{i=1}^{n} \lambda_i. \tag{2.99}$$

We illustrate (2.98) and (2.99) using the matrix

$$\mathbf{A} = \begin{pmatrix} 1 & 2 \\ -1 & 4 \end{pmatrix}$$

from the example in Section 2.11.1, for which $\lambda_1 = 3$ and $\lambda_2 = 2$. Using (2.98), we obtain

$$\operatorname{tr}(\mathbf{A}) = \lambda_1 + \lambda_2 = 3 + 2 = 5,$$

and from (2.99), we have

$$|\mathbf{A}| = \lambda_1\lambda_2 = 3(2) = 6.$$

By definition, we obtain

$$\operatorname{tr}(\mathbf{A}) = 1 + 4 = 5 \quad \text{and} \quad |\mathbf{A}| = (1)(4) - (-1)(2) = 6.$$

2.11.3 Positive Definite and Semidefinite Matrices

The eigenvalues and eigenvectors of positive definite and positive semidefinite matrices have the following properties:

1. The eigenvalues of a positive definite matrix are all positive.
2. The eigenvalues of a positive semidefinite matrix are positive or zero, with the number of positive eigenvalues equal to the rank of the matrix.

It is customary to list the eigenvalues of a positive definite matrix in descending order: $\lambda_1 > \lambda_2 > \cdots > \lambda_p$. The eigenvectors $\mathbf{x}_1, \mathbf{x}_2, \ldots, \mathbf{x}_n$ are listed in the same order; $\mathbf{x}_1$ corresponds to λ_1, $\mathbf{x}_2$ corresponds to λ_2, and so on.

2.11.4 Product AB

If **A** and **B** are square and the same size, the eigenvalues of **AB** are the same as those of **BA**, although the eigenvectors are usually different. This result also holds if **AB** and **BA** are both square but of different sizes, as when **A** is $n \times p$ and **B** is $p \times n$. (In this case, the nonzero eigenvalues of **AB** and **BA** will be the same.)

2.11.5 Symmetric Matrix

The eigenvectors of an $n \times n$ symmetric matrix **A** are mutually orthogonal. It follows that if the n eigenvectors of **A** are normalized and inserted as columns of a matrix $\mathbf{C} = (\mathbf{x}_1, \mathbf{x}_2, \ldots, \mathbf{x}_n)$, then **C** is orthogonal.

2.11.6 Spectral Decomposition

It was noted in Section 2.11.5 that if the matrix $\mathbf{C} = (\mathbf{x}_1, \mathbf{x}_2, \ldots, \mathbf{x}_n)$ contains the normalized eigenvectors of a symmetric matrix **A**, then **C** is orthogonal. Therefore, by (2.93), $\mathbf{I} = \mathbf{CC}'$, which, when multiplied by **A**, gives

$$\mathbf{A} = \mathbf{ACC}'.$$

We now substitute $\mathbf{C} = (\mathbf{x}_1, \mathbf{x}_2, \ldots, \mathbf{x}_n)$, where the **x**'s are normalized eigenvectors of **A**:

$$\begin{aligned}
\mathbf{A} &= \mathbf{A}(\mathbf{x}_1, \mathbf{x}_2, \ldots, \mathbf{x}_n)\mathbf{C}' \\
&= (\mathbf{Ax}_1, \mathbf{Ax}_2, \ldots, \mathbf{Ax}_n)\mathbf{C}' \quad \text{[by (2.47)]} \\
&= (\lambda_1\mathbf{x}_1, \lambda_2\mathbf{x}_2, \ldots, \lambda_n\mathbf{x}_n)\mathbf{C}' \quad \text{[by (2.95)]} \\
&= \mathbf{CDC}' \quad \text{[by (2.51)]},
\end{aligned} \tag{2.100}$$

where

$$\mathbf{D} = \begin{bmatrix} \lambda_1 & 0 & \cdots & 0 \\ 0 & \lambda_2 & \cdots & 0 \\ \vdots & \vdots & & \vdots \\ 0 & 0 & \cdots & \lambda_n \end{bmatrix}. \tag{2.101}$$

The relationship in (2.100) between a symmetric matrix **A** and its eigenvalues and eigenvectors is known as the *spectral decomposition* of **A**.

Since **C** is orthogonal and $\mathbf{C}'\mathbf{C} = \mathbf{C}\mathbf{C}' = \mathbf{I}$, we can multiply (2.100) on the left by $\mathbf{C}'$ and on the right by **C** to obtain

$$\mathbf{C}'\mathbf{A}\mathbf{C} = \mathbf{D}. \tag{2.102}$$

Thus a symmetric matrix **A** can be *diagonalized* by an orthogonal matrix containing normalized eigenvectors of **A**, and the resulting diagonal matrix contains eigenvalues of **A**.

2.11.7 Square Root Matrix

If **A** is positive definite, the spectral decomposition of **A** in (2.100) can be modified by taking the square roots of the eigenvalues to produce a *square root matrix*,

$$\mathbf{A}^{1/2} = \mathbf{C}\mathbf{D}^{1/2}\mathbf{C}', \tag{2.103}$$

where

$$\mathbf{D}^{1/2} = \begin{bmatrix} \sqrt{\lambda_1} & 0 & \cdots & 0 \\ 0 & \sqrt{\lambda_2} & \cdots & 0 \\ \vdots & \vdots & & \vdots \\ 0 & 0 & \cdots & \sqrt{\lambda_n} \end{bmatrix}. \tag{2.104}$$

The square root matrix $\mathbf{A}^{1/2}$ is symmetric and serves as the square root of **A**:

$$\mathbf{A}^{1/2}\mathbf{A}^{1/2} = (\mathbf{A}^{1/2})^2 = \mathbf{A}. \tag{2.105}$$

2.11.8 Patterned Matrices

The following result, known as the Perron–Frobenius theorem, is of interest in Chapter 12: If all elements of the positive definite matrix **A** are positive, then all elements of the first eigenvector are positive. (The first eigenvector is the one associated with the first eigenvalue λ_1.)

PROBLEMS

2.1 Let

$$\mathbf{A} = \begin{pmatrix} 4 & 2 & 3 \\ 7 & 5 & 8 \end{pmatrix} \quad \text{and} \quad \mathbf{B} = \begin{pmatrix} 3 & -2 & 4 \\ 6 & 9 & -5 \end{pmatrix}.$$

(a) Find $\mathbf{A} + \mathbf{B}$ and $\mathbf{A} - \mathbf{B}$.

(b) Find $\mathbf{A}'\mathbf{A}$ and $\mathbf{A}\mathbf{A}'$.

2.2 Use the matrices **A** and **B** in problem 2.1:

(a) Find $(\mathbf{A}+\mathbf{B})'$ and $\mathbf{A}'+\mathbf{B}'$ and compare them, thus illustrating (2.15).

(b) Show that $(\mathbf{A}')' = \mathbf{A}$, thus illustrating (2.6).

2.3 Let

$$\mathbf{A} = \begin{pmatrix} 1 & 3 \\ 2 & -1 \end{pmatrix} \qquad \mathbf{B} = \begin{pmatrix} 2 & 0 \\ 1 & 5 \end{pmatrix}.$$

(a) Find **AB** and **BA**.

(b) Find $|\mathbf{AB}|$, $|\mathbf{A}|$, and $|\mathbf{B}|$ and verify that (2.83) holds in this case.

2.4 Use the matrices **A** and **B** in problem 2.3:

(a) Find $\mathbf{A} + \mathbf{B}$ and $\text{tr}(\mathbf{A} + \mathbf{B})$.

(b) Find $\text{tr}(\mathbf{A})$ and $\text{tr}(\mathbf{B})$ and show that (2.88) holds for these matrices.

2.5 Let

$$\mathbf{A} = \begin{pmatrix} 1 & 2 & 3 \\ 2 & -1 & 1 \end{pmatrix} \qquad \mathbf{B} = \begin{pmatrix} 3 & -2 \\ 2 & 0 \\ -1 & 1 \end{pmatrix}.$$

(a) Find **AB** and **BA**.

(b) Compare $\text{tr}(\mathbf{AB})$ and $\text{tr}(\mathbf{BA})$ and confirm that (2.89) holds here.

2.6 Let

$$\mathbf{A} = \begin{pmatrix} 1 & 2 & 3 \\ 2 & 4 & 6 \\ 5 & 10 & 15 \end{pmatrix} \qquad \mathbf{B} = \begin{pmatrix} -1 & 1 & -2 \\ -1 & 1 & -2 \\ 1 & -1 & 2 \end{pmatrix}.$$

(a) Show that $\mathbf{AB} = \mathbf{O}$.
(b) Find a vector $\mathbf{x}$ such that $\mathbf{Ax} = \mathbf{0}$.
(c) Show that $|\mathbf{A}| = 0$.

2.7 Let

$$\mathbf{A} = \begin{pmatrix} 1 & -1 & 4 \\ -1 & 1 & 3 \\ 4 & 3 & 2 \end{pmatrix} \qquad \mathbf{B} = \begin{pmatrix} 3 & -2 & 4 \\ 7 & 1 & 0 \\ 2 & 3 & 5 \end{pmatrix},$$

$$\mathbf{x} = \begin{pmatrix} 1 \\ -1 \\ 2 \end{pmatrix} \qquad \mathbf{y} = \begin{pmatrix} 3 \\ 2 \\ 1 \end{pmatrix}.$$

Find the following:

(a) $\mathbf{Bx}$ **(f)** $\mathbf{x}'\mathbf{y}$
(b) $\mathbf{y}'\mathbf{B}$ **(g)** $\mathbf{xx}'$
(c) $\mathbf{x}'\mathbf{Ax}$ **(h)** $\mathbf{xy}'$
(d) $\mathbf{x}'\mathbf{Ay}$ **(i)** $\mathbf{B}'\mathbf{B}$
(e) $\mathbf{x}'\mathbf{x}$

2.8 Use $\mathbf{x}$, $\mathbf{y}$, and $\mathbf{A}$ as defined in problem 2.7:
(a) Find $\mathbf{x} + \mathbf{y}$ and $\mathbf{x} - \mathbf{y}$.
(b) Find $(\mathbf{x} - \mathbf{y})'\mathbf{A}(\mathbf{x} - \mathbf{y})$.

2.9 Using $\mathbf{B}$ and $\mathbf{x}$ in problem 2.7, find $\mathbf{Bx}$ as a linear combination of columns of $\mathbf{B}$ as in (2.62) and compare with $\mathbf{Bx}$ found in problem 2.7(a).

2.10 Let

$$\mathbf{A} = \begin{pmatrix} 2 & 1 \\ 1 & 3 \end{pmatrix} \qquad \mathbf{B} = \begin{pmatrix} 1 & 4 & 2 \\ 5 & 0 & 3 \end{pmatrix} \qquad \mathbf{I} = \begin{pmatrix} 1 & 0 \\ 0 & 1 \end{pmatrix}.$$

(a) Show that $(\mathbf{AB})' = \mathbf{B}'\mathbf{A}'$, as in (2.27).
(b) Show that $\mathbf{AI} = \mathbf{A}$ and that $\mathbf{IB} = \mathbf{B}$.
(c) Find $|\mathbf{A}|$.

2.11 Let

$$\mathbf{a} = \begin{pmatrix} 1 \\ -3 \\ 2 \end{pmatrix} \quad \text{and} \quad \mathbf{b} = \begin{pmatrix} 2 \\ 1 \\ 3 \end{pmatrix}.$$

(a) Find $\mathbf{a}'\mathbf{b}$ and $(\mathbf{a}'\mathbf{b})^2$.

(b) Find $\mathbf{bb}'$ and $\mathbf{a}'(\mathbf{bb}')\mathbf{a}$.

(c) Compare $(\mathbf{a}'\mathbf{b})^2$ with $\mathbf{a}'(\mathbf{bb}')\mathbf{a}$ and thus illustrate (2.40).

2.12 Let

$$\mathbf{A} = \begin{pmatrix} 1 & 2 & 3 \\ 4 & 5 & 6 \\ 7 & 8 & 9 \end{pmatrix} \qquad \mathbf{D} = \begin{pmatrix} a & 0 & 0 \\ 0 & b & 0 \\ 0 & 0 & c \end{pmatrix}.$$

Find **DA**, **AD**, and **DAD**.

2.13 Let the matrices **A** and **B** be partitioned as follows:

$$\mathbf{A} = \left[\begin{array}{cc|c} 2 & 1 & 2 \\ 3 & 2 & 0 \\ \hline 1 & 0 & 1 \end{array}\right] \qquad \mathbf{B} = \left[\begin{array}{ccc|c} 1 & 1 & 1 & 0 \\ 2 & 1 & 1 & 2 \\ \hline 2 & 3 & 1 & 2 \end{array}\right].$$

(a) Find **AB** as in (2.60) using the indicated partitioning.

(b) Check by finding **AB** in the usual way, ignoring the partitioning.

2.14 Let

$$\mathbf{A} = \begin{pmatrix} 1 & 3 & 2 \\ 2 & 0 & -1 \end{pmatrix} \qquad \mathbf{B} = \begin{pmatrix} 1 & 2 \\ 0 & 1 \\ 1 & 0 \end{pmatrix} \qquad \mathbf{C} = \begin{pmatrix} 2 & 1 & 1 \\ 5 & -6 & -4 \end{pmatrix}.$$

Find **AB** and **CB**. Are they equal? What is the rank of **A**, **B**, and **C**?

2.15 Let

$$\mathbf{A} = \begin{pmatrix} 5 & 4 & 4 \\ 2 & -3 & 1 \\ 3 & 7 & 2 \end{pmatrix} \qquad \mathbf{B} = \begin{pmatrix} 1 & 0 & 1 \\ 0 & 1 & 0 \\ 1 & 2 & 3 \end{pmatrix}.$$

(a) Find $\text{tr}(\mathbf{A})$ and $\text{tr}(\mathbf{B})$.

(b) Find $\mathbf{A} + \mathbf{B}$ and $\text{tr}(\mathbf{A} + \mathbf{B})$. Is $\text{tr}(\mathbf{A} + \mathbf{B}) = \text{tr}(\mathbf{A}) + \text{tr}(\mathbf{B})$?

(c) Find $|\mathbf{A}|$ and $|\mathbf{B}|$.

(d) Find **AB** and $|\mathbf{AB}|$. Is $|\mathbf{AB}| = |\mathbf{A}||\mathbf{B}|$?

2.16 Let

$$\mathbf{A} = \begin{pmatrix} 3 & 4 & 3 \\ 4 & 8 & 6 \\ 3 & 6 & 9 \end{pmatrix}.$$

(a) Show that $|\mathbf{A}| > 0$.

(b) Using the Cholesky decomposition in Section 2.7, find an upper triangular matrix $\mathbf{T}$ such that $\mathbf{A} = \mathbf{T}'\mathbf{T}$.

2.17 Let

$$\mathbf{A} = \begin{pmatrix} 3 & -5 & -1 \\ -5 & 13 & 0 \\ -1 & 0 & 1 \end{pmatrix}.$$

(a) Show that $|\mathbf{A}| > 0$.

(b) Using the Cholesky decomposition in Section 2.7, find an upper triangular matrix $\mathbf{T}$ such that $\mathbf{A} = \mathbf{T}'\mathbf{T}$.

2.18 The columns of the following matrix are mutually orthogonal:

$$\mathbf{A} = \begin{pmatrix} 1 & -1 & 1 \\ 2 & 1 & 0 \\ 1 & -1 & -1 \end{pmatrix}.$$

(a) Normalize the columns of $\mathbf{A}$ by dividing each column by its length; denote the resulting matrix by $\mathbf{C}$.

(b) Show that $\mathbf{C}$ is an orthogonal matrix; that is, $\mathbf{C}'\mathbf{C} = \mathbf{C}\mathbf{C}' = \mathbf{I}$.

2.19 Let

$$\mathbf{A} = \begin{pmatrix} 1 & 1 & -2 \\ -1 & 2 & 1 \\ 0 & 1 & -1 \end{pmatrix}.$$

(a) Find the eigenvalues and associated normalized eigenvectors.

(b) Find $\text{tr}(\mathbf{A})$ and $|\mathbf{A}|$ and show that $\text{tr}(\mathbf{A}) = \sum_{i=1}^{3} \lambda_i$ and $|\mathbf{A}| = \prod_{i=1}^{3} \lambda_i$.

2.20 Let

$$\mathbf{A} = \begin{pmatrix} 3 & 1 & 1 \\ 1 & 0 & 2 \\ 1 & 2 & 0 \end{pmatrix}.$$

(a) The eigenvalues of $\mathbf{A}$ are 1, 4, −2. Find the normalized eigenvectors and use them as columns in an orthogonal matrix $\mathbf{C}$.

(b) Show that $\mathbf{C}'\mathbf{A}\mathbf{C} = \mathbf{D}$, as in (2.102), where $\mathbf{D}$ is diagonal with the eigenvalues of $\mathbf{A}$ on the diagonal.

(c) Show that $\mathbf{A} = \mathbf{C}\mathbf{D}\mathbf{C}'$ as in (2.100).

2.21 For the positive definite matrix

$$\mathbf{A} = \begin{pmatrix} 2 & -1 \\ -1 & 2 \end{pmatrix},$$

calculate the eigenvalues and eigenvectors and find the square root matrix $\mathbf{A}^{1/2}$ as in (2.103). Check by showing $(\mathbf{A}^{1/2})^2 = \mathbf{A}$.

2.22 If $\mathbf{j}$ is a vector of 1s, as defined in (2.11), show that

(a) $\mathbf{j}'\mathbf{a} = \mathbf{a}'\mathbf{j} = \sum_i a_i$ as in (2.37),

(b) $\mathbf{A}\mathbf{j}$ is a column vector whose elements are the row sums of $\mathbf{A}$ as in (2.38), and

(c) $\mathbf{j}'\mathbf{A}$ is a row vector whose elements are the column sums of $\mathbf{A}$.

2.23 Verify (2.41); that is, show that $(\mathbf{x} - \mathbf{y})'(\mathbf{x} - \mathbf{y}) = \mathbf{x}'\mathbf{x} - 2\mathbf{x}'\mathbf{y} + \mathbf{y}'\mathbf{y}$.

2.24 Show that $\mathbf{A}'\mathbf{A}$ is symmetric, where $\mathbf{A}$ is $n \times p$.

2.25 If $\mathbf{a}$ and $\mathbf{x}_1, \mathbf{x}_2, \ldots, \mathbf{x}_n$ are all $p \times 1$ and $\mathbf{A}$ is $p \times p$, show that (2.42)–(2.45) hold:

(a) $\sum_{i=1}^n \mathbf{a}'\mathbf{x}_i = \mathbf{a}' \sum_{i=1}^n \mathbf{x}_i$

(b) $\sum_{i=1}^n \mathbf{A}\mathbf{x}_i = \mathbf{A} \sum_{i=1}^n \mathbf{x}_i$

(c) $\sum_{i=1}^n (\mathbf{a}'\mathbf{x}_i)^2 = \mathbf{a}'(\sum_{i=1}^n \mathbf{x}_i\mathbf{x}_i')\mathbf{a}$

(d) $\sum_{i=1}^n \mathbf{A}\mathbf{x}_i(\mathbf{A}\mathbf{x}_i)' = \mathbf{A}(\sum_{i=1}^n \mathbf{x}_i\mathbf{x}_i')\mathbf{A}'$

2.26 Assume $\mathbf{A} = \begin{pmatrix} \mathbf{a}_1' \\ \mathbf{a}_2' \end{pmatrix}$ is $2 \times p$, $\mathbf{x}$ is $p \times 1$, and $\mathbf{S}$ is $p \times p$.

(a) Show that

$$\mathbf{Ax} = \begin{pmatrix} \mathbf{a}_1'\mathbf{x} \\ \mathbf{a}_2'\mathbf{x} \end{pmatrix}. \tag{2.106}$$

(b) Show that

$$\mathbf{A}'\mathbf{SA} = \begin{pmatrix} \mathbf{a}_1'\mathbf{Sa}_1 & \mathbf{a}_1'\mathbf{Sa}_2 \\ \mathbf{a}_2'\mathbf{Sa}_1 & \mathbf{a}_2'\mathbf{Sa}_2 \end{pmatrix}. \tag{2.107}$$

2.27 If the rows of $\mathbf{A}$ are denoted by $\mathbf{a}_i'$, show that

$$\mathbf{A}'\mathbf{A} = \sum_{i=1}^{n} \mathbf{a}_i\mathbf{a}_i'. \tag{2.108}$$

2.28 Show that the inverse of the partitioned matrix given in (2.70) is correct by multiplying by

$$\begin{pmatrix} \mathbf{A}_{11} & \mathbf{a}_{12} \\ \mathbf{a}_{12}' & a_{22} \end{pmatrix}$$

to obtain an identity.

2.29 Show that the inverse of $\mathbf{B}+\mathbf{cc}'$ given in (2.71) is correct by multiplying by $\mathbf{B} + \mathbf{cc}'$ to obtain an identity.

2.30 Show that $|\mathbf{A}^{-1}| = 1/|\mathbf{A}|$ as in (2.84).

2.31 If $\mathbf{B}$ and $\mathbf{B} + \mathbf{cc}'$ are nonsingular, show that

$$|\mathbf{B} + \mathbf{cc}'| = |\mathbf{B}|(1 + \mathbf{c}'\mathbf{B}^{-1}\mathbf{c}). \tag{2.109}$$

2.32 Show that $\mathbf{CC}' = \mathbf{I}$ in (2.93) follows from $\mathbf{C}'\mathbf{C} = \mathbf{I}$ in (2.92).

2.33 Show that the eigenvalues of $\mathbf{AB}$ are the same as those of $\mathbf{BA}$, as noted in Section 2.11.4.

2.34 If $\mathbf{A}^{1/2}$ is the square root matrix defined in (2.103), show that

(a) $(\mathbf{A}^{1/2})^2 = \mathbf{A}$, as in (2.105),

(b) $|\mathbf{A}^{1/2}|^2 = |\mathbf{A}|$, and

(c) $|\mathbf{A}^{1/2}| = |\mathbf{A}|^{1/2}$.

2.35 Show that

$$(\mathbf{A}')^{-1} = (\mathbf{A}^{-1})'. \qquad (2.110)$$

CHAPTER 3

Characterizing and Displaying Multivariate Data

We review some univariate and bivariate procedures in Sections 3.1, 3.2, and 3.3 and then extend them to vectors of higher dimension in the remainder of the chapter.

3.1 MEAN AND VARIANCE OF A UNIVARIATE RANDOM VARIABLE

Informally, a *random variable* may be defined as a variable whose value depends on the outcome of a chance experiment. Generally, we will consider only *continuous* random variables. Some types of multivariate data are only approximations to this ideal, such as test scores or a seven-point semantic differential (Likert) scale consisting of ordered responses ranging from "strongly disagree" to "strongly agree." Special techniques have been developed for such data, but in many cases, the ordinary methods designed for continuous data work almost as well.

The *density function* $f(y)$ indicates the relative frequency of occurrence of the random variable y. Thus if $f(y_1) > f(y_2)$, then points in the neighborhood of y_1 are more likely to occur than points in the neighborhood of y_2.

The *population mean* of a random variable y is defined (informally) as the mean of all possible values of y and is denoted by μ. The mean is also referred to as the *expected value* of y or $E(y)$. If the density $f(y)$ is known, the mean can be found using methods of calculus, but we will not use these techniques in this text.

If $f(y)$ is unknown, the population mean μ will ordinarily remain unknown unless it has been established from extensive past experience with a stable population. If a large random sample from the population represented by $f(y)$ is available, it is highly probable that the mean of the sample is close to μ.

The *sample mean* of a random sample of n observations $y_1, y_2, \ldots, y_n$ is given by the ordinary arithmetic average

$$\overline{y} = \frac{1}{n} \sum_{i=1}^{n} y_i. \tag{3.1}$$

Generally, $\overline{y}$ will never be equal to μ; by this we mean that the probability is zero that a sample will ever arise in which $\overline{y}$ is exactly equal to μ. However, $\overline{y}$ is considered a good estimator for μ because $E(\overline{y}) = \mu$ and var $(\overline{y}) = \sigma^2/n$, where σ^2 is the variance of y. In other words, $\overline{y}$ is an unbiased estimate of μ and has a smaller variance than a single observation y. The variance σ^2 is defined below. The notation $E(\overline{y})$ indicates the mean of all possible values of $\overline{y}$; that is, conceptually, every possible sample is obtained from the population, the mean of each is found, and the average of all these means is calculated.

If every y in the population is multiplied by a constant a, the expected value is also multiplied by a:

$$E(ay) = aE(y) = a\mu. \tag{3.2}$$

The sample mean has a similar property. If $z_i = ay_i$ for $i = 1, 2, \ldots, n$, then

$$\overline{z} = a\overline{y}. \tag{3.3}$$

The *variance* of the population is defined as var $(y) = \sigma^2 = E(y - \mu)^2$. This is the average squared deviation from the mean and is thus an indication of the extent to which the values of y are spread or scattered. It can be shown that $\sigma^2 = E(y^2) - \mu^2$.

The *sample variance* is defined as

$$s^2 = \frac{\sum_{i=1}^{n} (y_i - \overline{y})^2}{n - 1}, \tag{3.4}$$

which can be shown to be equal to

$$s^2 = \frac{\sum_i y_i^2 - n\overline{y}^2}{n - 1}. \tag{3.5}$$

The sample variance s^2 is generally never equal to the population variance σ^2 (the probability is zero), but is an unbiased estimator for σ^2, that is, $E(s^2) = \sigma^2$. Again the notation $E(s^2)$ implies the mean of all possible sample variances. The square root of either the population or sample variance is called the *standard deviation.*

If each y is multiplied by a constant a, the population variance is multiplied by a^2, or var $(ay) = a^2\sigma^2$. Similarly, if $z_i = ay_i$, $i = 1, 2, \ldots, n$, then the sample variance of z is given by

$$s_z^2 = a^2 s^2. \tag{3.6}$$

3.2 COVARIANCE AND CORRELATION OF BIVARIATE RANDOM VARIABLES

3.2.1 Covariance

If two variables x and y are measured on each research unit (object or subject), we have a *bivariate random variable*. Often x and y will tend to covary; if one is above its mean, the other is more likely to be above its mean, and vice versa. For example, height and weight were observed for a sample of 20 college-age males. The data are given in Table 3.1.

The values of height x and weight y from Table 3.1 are both plotted in the vertical direction in Figure 3.1. The tendency for x and y to stay on the same side of the mean is clear in Figure 3.1. This illustrates positive covariance. With negative covariance the points would tend to deviate simultaneously to opposite sides of the mean.

The *population covariance* is defined as cov $(x, y) = \sigma_{xy} = E[(x-\mu_x)(y-\mu_y)]$, where μ_x and μ_y are the means of x and y, respectively. Thus if x and y are usually both above their means or both below their means, the product $(x - \mu_x)(y - \mu_y)$ will typically be positive, and the average value of the product will be positive. Conversely, if x and y tend to fall on opposite sides of their respective means, the product will usually be negative and the average product will be negative. It can be shown that $\sigma_{xy} = E(xy) - \mu_x\mu_y$.

If the two random variables x and y in a bivariate random variable are added or multiplied, a new random variable is obtained. The mean of $x + y$ or xy is

Table 3.1 Height and Weight for a Sample of 20 College-Age Males

Person	Height, x	Weight, y	Person	Height, x	Weight, y
1	69	153	11	72	140
2	74	175	12	79	265
3	68	155	13	74	185
4	70	135	14	67	112
5	72	172	15	66	140
6	67	150	16	71	150
7	66	115	17	74	165
8	70	137	18	75	185
9	76	200	19	75	210
10	68	130	20	76	220

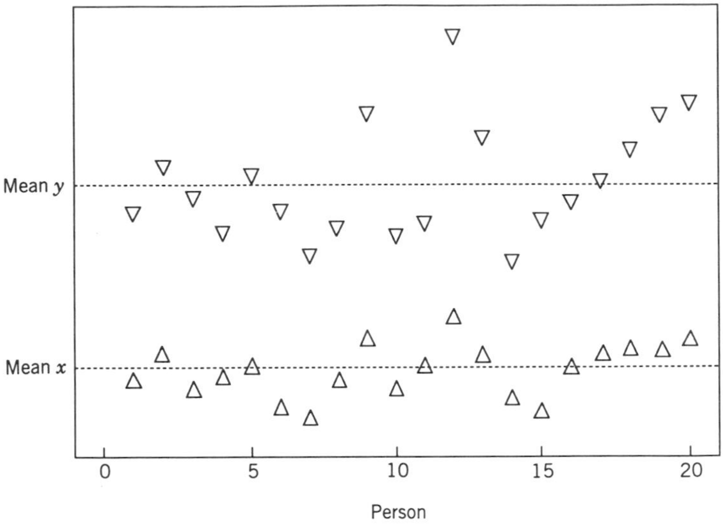

Figure 3.1 Two variables with a tendency to covary.

as follows:

$$E(x + y) = E(x) + E(y) \tag{3.7}$$

$$E(xy) = E(x)E(y) \quad \text{if } x, y \text{ independent.} \tag{3.8}$$

Formally, x and y are independent if their joint density factors into the product of their individual densities: $f(x, y) = g(x)h(y)$. Informally, x and y are independent if the random behavior of either of the variables is not affected by the behavior of the other. Note that (3.7) is true whether or not x and y are independent, but (3.8) holds only for x and y independently distributed.

The notion of independence of x and y is more general than that of zero covariance. The covariance σ_{xy} measures linear relationship only, whereas if two random variables are independent, they are not related either linearly or nonlinearly. Independence implies $\sigma_{xy} = 0$, but $\sigma_{xy} = 0$ does not imply independence. It is easy to show that if x and y are independent, then $\sigma_{xy} = 0$:

$$\begin{aligned} \sigma_{xy} &= E(xy) - \mu_x\mu_y \\ &= E(x)E(y) - \mu_x\mu_y \quad [\text{by } (3.8)] \\ &= \mu_x\mu_y - \mu_x\mu_y = 0. \end{aligned}$$

One way to demonstrate that the converse is not true is to construct examples of bivariate x and y that have zero covariance and yet are related in a nonlinear way (the relationship will have zero slope). This is illustrated in Figure 3.2.

If x and y have a bivariate normal distribution (see Chapter 4), then zero

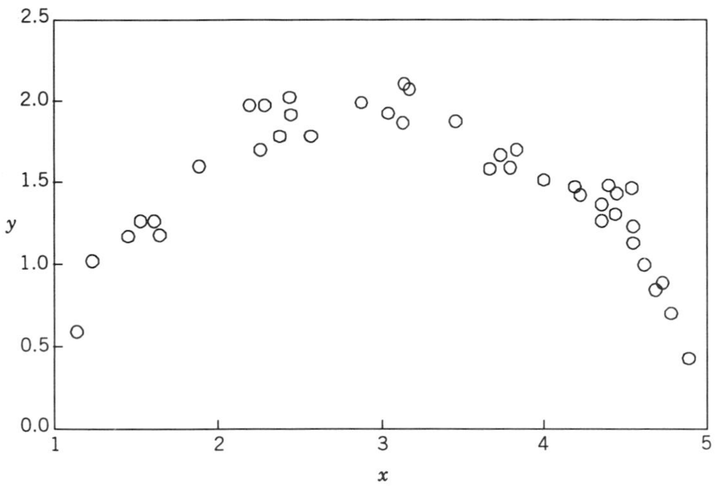

Figure 3.2 A sample from a population where x and y have zero covariance and yet are dependent.

covariance implies independence. This is because (1) the covariance measures only linear relationships and (2) in the bivariate normal case, the mean of y given x (or x given y) is a straight line.

The *sample covariance* is defined as

$$s_{xy} = \frac{\sum_{i=1}^{n}(x_i - \overline{x})(y_i - \overline{y})}{n-1}. \tag{3.9}$$

It can be shown that

$$s_{xy} = \frac{\sum_{i} x_i y_i - n\overline{x}\,\overline{y}}{n-1}. \tag{3.10}$$

Note that s_{xy} is essentially never equal to σ_{xy}, that is, the probability is zero that s_{xy} will equal σ_{xy}. It is true, however, that s_{xy} is an unbiased estimator for σ_{xy}, that is, $E(s_{xy}) = \sigma_{xy}$.

Since $s_{xy} \neq \sigma_{xy}$ in any given sample, this is also true when $\sigma_{xy} = 0$. Thus when the population covariance is zero, no random sample from the population will have zero covariance. The only way a sample from a continuous bivariate distribution will have zero covariance is for the experimenter to choose the values of x and y so that $s_{xy} = 0$. (Such a sample would not be a random sample.) One way to achieve this is to place the values in the form of a grid. This is illustrated in Figure 3.3.

The sample covariance measures only linear relationships. If the points in a bivariate sample follow a curved trend, as, for example, in Figure 3.2, the sample covariance will not measure the strength of the relationship. Note that in the left side of Figure 3.2, y increases as x increases, but y decreases with x on the right side.

To see that s_{xy} measures only linear relationships, note that the slope of a simple linear regression line is

$$b = \frac{\sum_{i=1}^{n}(x_i - \overline{x})(y_i - \overline{y})}{\sum_{i=1}^{n}(x_i - \overline{x})^2} = \frac{s_{xy}}{s_x^2}.$$

Thus s_{xy} is proportional to the slope, which shows only the linear relationship between y and x.

Variables with zero sample covariance can be said to be *orthogonal.* By (2.90), two sets of numbers $a_1, a_2, \ldots, a_n$ and $b_1, b_2, \ldots, b_n$ are orthogonal if $\sum_i a_i b_i = 0$. This is true for $x_i - \overline{x}$ and $y_i - \overline{y}$ when the sample covariance is zero, that is, $\sum_i (x_i - \overline{x})(y_i - \overline{y}) = 0$.

Example 3.2.1. To obtain the sample covariance for the height and weight data in Table 3.1, we first calculate $\overline{x}$, $\overline{y}$, and $\sum_i x_i y_i$, where x is height and y

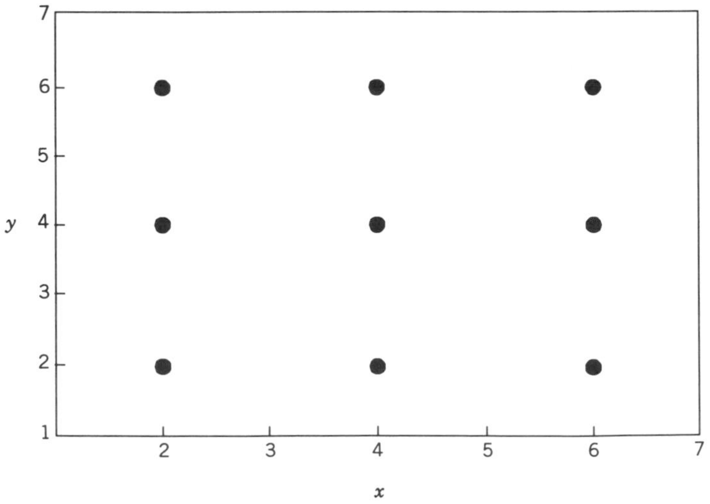

Figure 3.3 A sample of (x, y) values with zero covariance.

is weight:

$$\begin{aligned}\bar{x} &= \frac{69 + 74 + \cdots + 76}{20} = 71.45 \\ \bar{y} &= \frac{153 + 175 + \cdots + 220}{20} = 164.7 \\ \sum_i x_i y_i &= (69)(153) + (74)(175) + \cdots + (76)(220) = 237,805.\end{aligned}$$

Now, by (3.10), we have

$$\begin{aligned}s_{xy} &= \frac{\sum_i x_i y_i - n\bar{x}\,\bar{y}}{n-1} \\ &= \frac{237,805 - (20)(71.45)(164.7)}{19} = 128.88.\end{aligned}$$

By itself, a sample covariance of 128.88 is not very meaningful. We are not sure if this represents a small, moderate, or large amount of relationship between y and x. A method of standardizing the covariance is given in the next section.

3.2.2 Correlation

Since the covariance depends on the scale of measurement of x and y, it is difficult to compare covariances between different pairs of variables. For example, if we change from inches to centimeters, the covariance will change. To circumvent this problem, we can standardize the covariance by dividing by the standard deviations of the two variables. This standardized covariance is called a *correlation*. The *population correlation* of two random variables x and y is

$$\rho_{xy} = \text{corr}(x, y) = \frac{\sigma_{xy}}{\sigma_x \sigma_y} = \frac{E[(x-\mu_x)(y-\mu_y)]}{\sqrt{E(x-\mu_x)^2}\sqrt{E(y-\mu_y)^2}}, \tag{3.11}$$

and the *sample correlation* is

$$r_{xy} = \frac{s_{xy}}{s_x s_y} = \frac{\sum_i (x_i - \bar{x})(y_i - \bar{y})}{\sqrt{\sum_i (x_i - \bar{x})^2 \sum_i (y_i - \bar{y})^2}}. \tag{3.12}$$

Either of these correlations will range between -1 and 1.

We emphasize, as before, that, in practice, r_{xy} will never equal ρ_{xy}. In particular, if $\rho_{xy} = 0$, no sample correlation r_{xy} will be zero unless the sample is contrived to yield such a value.

The sample correlation r_{xy} is related to the cosine of the angle between two vectors. Let θ be the angle between vectors **a** and **b** in Figure 3.4. The vector from the terminal point of **a** to the terminal point of **b** can be represented as $\mathbf{c} = \mathbf{b} - \mathbf{a}$. Then the law of cosines can be stated in vector form as

$$\begin{aligned}\cos\theta &= \frac{\mathbf{a}'\mathbf{a} + \mathbf{b}'\mathbf{b} - (\mathbf{b}-\mathbf{a})'(\mathbf{b}-\mathbf{a})}{2\sqrt{(\mathbf{a}'\mathbf{a})(\mathbf{b}'\mathbf{b})}} \\ &= \frac{\mathbf{a}'\mathbf{a} + \mathbf{b}'\mathbf{b} - (\mathbf{b}'\mathbf{b} + \mathbf{a}'\mathbf{a} - 2\mathbf{a}'\mathbf{b})}{2\sqrt{(\mathbf{a}'\mathbf{a})(\mathbf{b}'\mathbf{b})}} \\ &= \frac{\mathbf{a}'\mathbf{b}}{\sqrt{(\mathbf{a}'\mathbf{a})(\mathbf{b}'\mathbf{b})}}. \end{aligned} \tag{3.13}$$

Since $\cos(90°) = 0$, we see from (3.13) that $\mathbf{a}'\mathbf{b} = 0$ when $\theta = 90°$. Thus **a** and **b** are *perpendicular* when $\mathbf{a}'\mathbf{b} = 0$. By (2.90), two vectors **a** and **b**, such that $\mathbf{a}'\mathbf{b} = 0$, are also said to be *orthogonal*. Hence orthogonal vectors are perpendicular in a geometric sense.

To express the correlation in the form given in (3.13), let the n observation vectors $(x_1, y_1), (x_2, y_2), \ldots, (x_n, y_n)$ in two dimensions be represented as two vectors $\mathbf{x}' = (x_1, x_2, \ldots, x_n)$ and $\mathbf{y}' = (y_1, y_2, \ldots, y_n)$ in n dimensions and let **x** and **y** be centered as $\mathbf{x} - \bar{x}\mathbf{j}$ and $\mathbf{y} - \bar{y}\mathbf{j}$. Then the cosine of the angle θ between

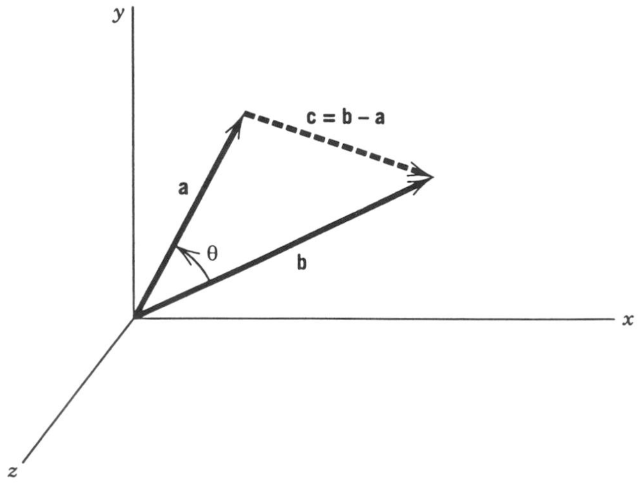

Figure 3.4 Vectors **a** and **b** in 3-space.

them is equal to the sample correlation between x and y:

$$\begin{aligned}\cos\theta &= \frac{(\mathbf{x}-\bar{x}\mathbf{j})'(\mathbf{y}-\bar{y}\mathbf{j})}{\sqrt{[(\mathbf{x}-\bar{x}\mathbf{j})'(\mathbf{x}-\bar{x}\mathbf{j})][(\mathbf{y}-\bar{y}\mathbf{j})'(\mathbf{y}-\bar{y}\mathbf{j})]}} \\ &= \frac{\sum_i (x_i-\bar{x})(y_i-\bar{y})}{\sqrt{\sum_i (x_i-\bar{x})^2 \sum_i (y_i-\bar{y})^2}} \\ &= r_{xy}.\end{aligned} \tag{3.14}$$

Thus if the angle θ between two centered vectors is small so that $\cos\theta$ is near 1, r_{xy} will be close to 1. If the two vectors are perpendicular, $\cos\theta$ and r_{xy} will be zero. If the two vectors have nearly opposite directions, r_{xy} will be close to -1.

Example 3.2.2. To obtain the correlation for the height and weight data of Table 3.1, we first calculate the sample variance of x:

$$\begin{aligned}s_x^2 &= \frac{\sum_i x_i^2 - n\bar{x}^2}{n-1} \\ &= \frac{102,379-(20)(71.45)^2}{19} = 14.576.\end{aligned}$$

Then $s_x = \sqrt{14.576} = 3.8179$ and similarly, $s_y = 37.964$. By (3.12), we have

$$r_{xy} = \frac{s_{xy}}{s_x s_y} = \frac{128.88}{(3.8179)(37.964)} = .889.$$

3.3 SCATTER PLOTS OF BIVARIATE SAMPLES

Figures 3.2 and 3.3 are examples of *scatter plots* of bivariate samples. Earlier, in Figure 3.1, the two variables x and y were plotted separately. The same data (Table 3.1) in a bivariate scatter plot appear in Figure 3.5.

If the origin is shifted to $(\bar{x}, \bar{y})$, as indicated by the dashed lines, it is seen that the first and third quadrants contain most of the points. Scatter plots for correlated data typically show a substantial positive or negative slope.

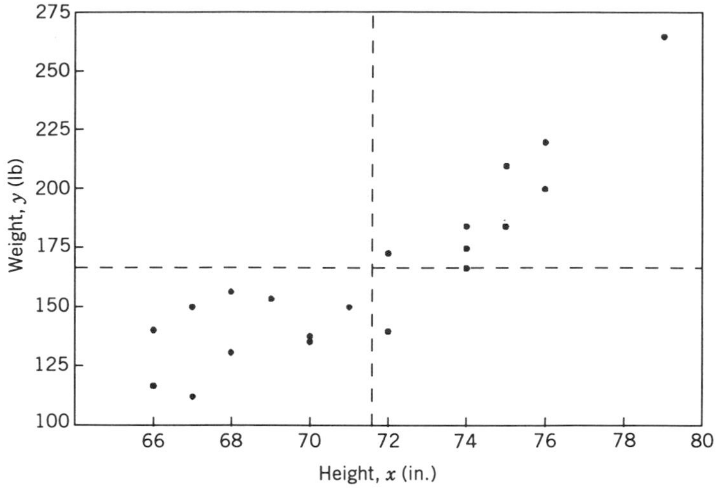

Figure 3.5 Bivariate scatter plot of the data in Figure 3.1.

A hypothetical sample of the uncorrelated variables height and IQ is shown in Figure 3.6. We could change the shape of the cluster by altering the scale on either axis. But because of the independence assumed here, each quadrant is likely to have as many points as any other quadrant. A tall person is as likely to have a high IQ as a low IQ. A person of low IQ is as likely to be short as to be tall.

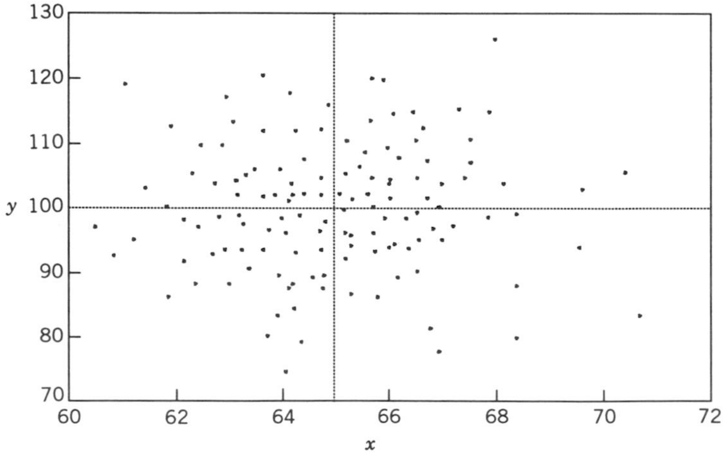

Figure 3.6 A sample of data from a population where x and y are uncorrelated.

3.4 GRAPHICAL DISPLAYS FOR MULTIVARIATE SAMPLES

It is a relatively simple procedure to plot bivariate samples as in Section 3.3. The position of a point shows at once the value of both variables. However, for three or more variables it is a challenge to show graphically the values of all the variables in an observation vector **y**. On a two-dimensional plot, the value of a third variable could be indicated by color or intensity or size of the plotted point. Four dimensions might be represented by starting with a two-dimensional scatter plot and adding two additional dimensions as line segments at right angles, as in Figure 3.7. The "corner point" represents y_1 and y_2; whereas y_3 and y_4 are given by the lengths of the two line segments.

We will now describe various methods proposed for representing several (p) dimensions in a plot of an observation vector.

Profiles represent each point by p vertical bars, with the heights of the bars depicting the values of the variables. Sometimes the profile is outlined by a polygonal line rather than bars.

Stars portray the value of each variable as a point along a ray from the center to the outside of a circle. The points on the rays are usually joined to form a polygon.

Glyphs (Anderson 1960) are circles of fixed size with rays whose lengths

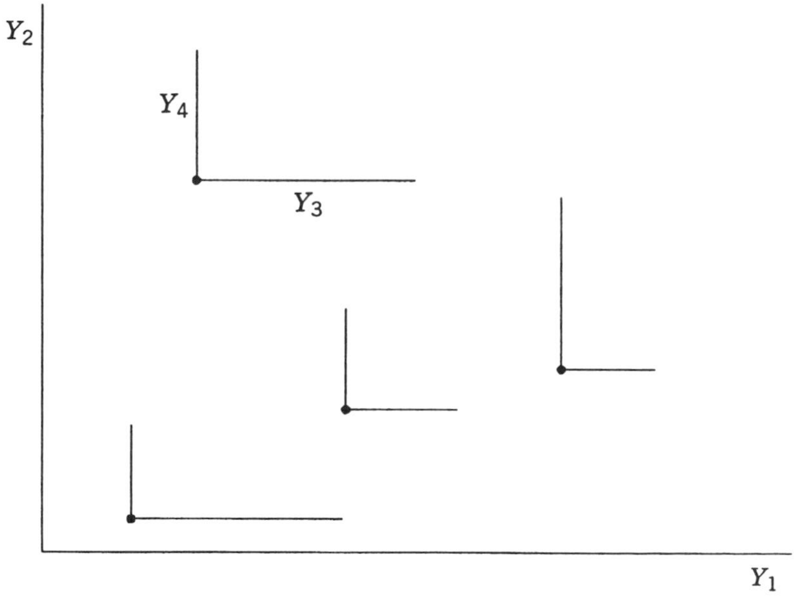

Figure 3.7 Four-dimensional plot.

represent the values of the variables. Anderson suggested using only three lengths of rays, thus rounding the variable values to three levels.

Faces (Chernoff 1973) depict each variable as a feature on a face, such as length of nose, size of eyes, shape of eyes, and so on. Flury and Riedwyl (1981) suggested using asymmetric faces, thus increasing the number of representable variables.

Boxes (Hartigan 1975) show each variable as the length of a dimension of a box. For more than three variables, the dimensions are partitioned into segments.

Among the above five methods, Chambers and Kleiner (1982) prefer the star plots because they "combine a reasonably distinctive appearance with computational simplicity and ease of interpretation." Commenting on the other methods, they state, "Profiles are not so easy to compare as a general shape. Faces are memorable but they are more complex to draw and one must be careful in assigning variables to parameters and in choosing parameter ranges. Faces to some extent disguise the data in the sense that individual data values may not be directly comparable from the plot."

The ith star representing the ith observation vector $\mathbf{y}_i$ has p points corresponding to the p variables. The jth point lies at a distance from the center proportional to y_{ij}, the jth component of $\mathbf{y}_i$. Thus the variables are normalized to the same range, say [0, 1]. By normalizing, we lose information about the mean of each variable unless the variables are comparable so that we can scale all of them the same way. However, the relative magnitudes of the variables can be compared more readily using stars than faces.

Example 3.4. The data in Table 3.2 are from Kleiner and Hartigan (1981). For these data, the above five graphical devices are illustrated in Figure 3.8.

Ehrenberg (1977) made several useful recommendations about how to make tables more informative at a glance. His suggestions are as follows: Round to two digits, order rows and columns by size, keep numbers to be compared

Table 3.2 Percentage of Republican Votes in Residential Elections in Six Southern States for Selected Years

State	1932	1936	1940	1960	1964	1968
Missouri	35	38	48	50	36	45
Maryland	36	37	41	46	35	42
Kentucky	40	40	42	54	36	44
Louisiana	7	11	14	29	57	23
Mississippi	4	3	4	25	87	14
South Carolina	2	1	4	49	59	39

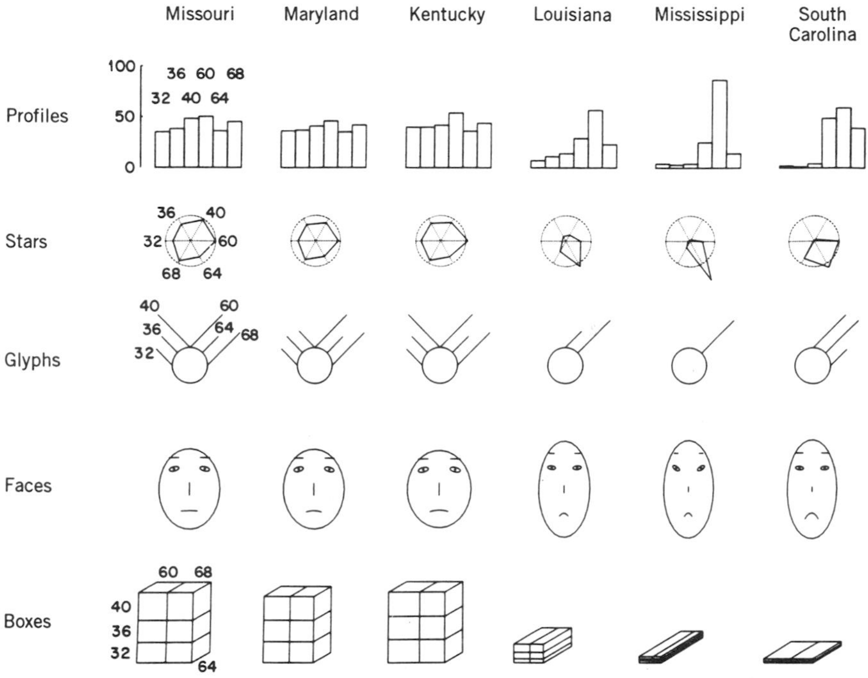

Figure 3.8 Profiles, stars, glyphs, faces, and boxes of percentage of Republican votes in six presidential elections in six southern states. The radius of the circles in the stars is 50%. Assignments of variables to facial features are 1932, shape of face; 1936, length of nose; 1940, curvature of mouth; 1960, width of mouth; 1964, slant of eyes; and 1968, length of eyebrows. (From the Journal of the American Statistical Association, 1981, p. 262.)

close together, and border the table with useful summary statistics. In this spirit, Wainer (1981) augmented Table 3.2 with medians, as in Table 3.3. He further processed this table by iteratively subtracting row-and-column effects to produce Table 3.4. This clarifies structure for both years and states. The unusual residuals are underlined.

Table 3.3 Data of Table 3.2 with Medians

State	1932	1936	1940	1960	1964	1968	Median
Missouri	35	38	48	50	36	45	42
Maryland	36	37	41	46	35	42	39
Kentucky	40	40	42	54	36	44	41
Louisiana	7	11	14	29	57	23	18
Mississippi	4	3	4	25	87	14	9
South Carolina	2	1	4	49	59	39	22
Median	21	24	28	48	46	40	30

Table 3.4 Residuals from Table 3.3 with Row-and-Column Effects

State	1932	1936	1940	1960	1964	1968	State Effect
Kentucky	2	1	1	−1	−3	−6	15
Maryland	0	0	2	−7	0	−6	13
Missouri	0	2	10	−2	0	−2	12
Louisiana	−4	−1	0	1	45	0	−12
South Carolina	1	−1	0	31	157	26	−22
Mississippi	3	1	0	7	85	1	−22
Year effect	−3	−2	0	14	−2	9	26

3.5 MEAN VECTORS

It is a common practice in many texts to use an uppercase letter for a variable name and the corresponding lowercase letter for a particular value or observed value of the random variable, for example, $P(Y > y)$. This notation is convenient in some univariate contexts, but it is often confusing in multivariate analysis, where we use uppercase letters for matrices. In the belief that it is easier to distinguish between a random vector and an observed value than between a vector and a matrix, throughout this text we follow the notation established in Chapter 2. Uppercase boldface letters are used for matrices of random variables or constants, lowercase boldface letters represent vectors of random variables or constants, and lowercase nonbolded letters represent univariate random variables or constants. An attempt will be made to use letters near the beginning of the alphabet for constants and letters near the end for random variables.

Let $\mathbf{y}$ represent a random vector of p variables measured on a sampling unit (subject or object). If there are n individuals in the sample, the n *observation vectors* are denoted by $\mathbf{y}_1, \mathbf{y}_2, \ldots, \mathbf{y}_n$, where

$$\mathbf{y}_i = \begin{bmatrix} y_{i1} \\ y_{i2} \\ \vdots \\ y_{ip} \end{bmatrix}.$$

The *sample mean* vector $\overline{\mathbf{y}}$ can be obtained in two ways:

$$\overline{\mathbf{y}} = \frac{1}{n} \sum_{i=1}^{n} \mathbf{y}_i = \begin{bmatrix} \overline{y}_1 \\ \overline{y}_2 \\ \vdots \\ \overline{y}_p \end{bmatrix}, \tag{3.15}$$

where, for example, $\bar{y}_2 = \sum_i y_{i2}/n$. Thus $\bar{y}_1$ is the mean of the n observations on the first variable, $\bar{y}_2$ is the mean of the second variable, and so on. Hence the sample mean vector can be found either as the average of the n observation vectors or by calculating the average of each of the p variables separately.

All n observation vectors $\mathbf{y}_1, \mathbf{y}_2, \ldots, \mathbf{y}_n$ can be transposed to row vectors and listed in the *data matrix* $\mathbf{Y}$ as follows:

$$\mathbf{Y} = \begin{bmatrix} \mathbf{y}'_1 \\ \mathbf{y}'_2 \\ \vdots \\ \mathbf{y}'_i \\ \vdots \\ \mathbf{y}'_n \end{bmatrix} = (\text{units}) \begin{matrix} \\ 1 \\ 2 \\ \\ i \\ \\ n \end{matrix} \overset{\text{(variables)}}{\begin{matrix} 1 & 2 & & j & & p \\ \left[\begin{matrix} y_{11} & y_{12} & \cdots & y_{1j} & \cdots & y_{1p} \\ y_{21} & y_{22} & \cdots & y_{2j} & \cdots & y_{2p} \\ \vdots & \vdots & & \vdots & & \vdots \\ y_{i1} & y_{i2} & \cdots & y_{ij} & \cdots & y_{ip} \\ \vdots & \vdots & & \vdots & & \vdots \\ y_{n1} & y_{n2} & \cdots & y_{nj} & \cdots & y_{np} \end{matrix} \right] \end{matrix}}. \tag{3.16}$$

Since n is usually greater than p, the data can be more conveniently tabulated by entering the observation vectors as rows rather than columns. Note that the first subscript i corresponds to units (subjects or objects) and the second subscript j refers to variables. This convention will be followed whenever possible.

To calculate $\bar{\mathbf{y}}$ from $\mathbf{Y}$, we need to sum the n entries in each column of $\mathbf{Y}$ and divide by n. This can be indicated in matrix notation using (2.38),

$$\bar{\mathbf{y}} = \frac{1}{n}\mathbf{Y}'\mathbf{j}, \tag{3.17}$$

where $\mathbf{j}$ is a vector of 1s, as defined in (2.11). Note that the rows of $\mathbf{Y}'$ are the columns of $\mathbf{Y}$. For example, the second element of $\mathbf{Y}'\mathbf{j}$ is

$$(y_{12}, y_{22}, \ldots, y_{n2}) \begin{bmatrix} 1 \\ 1 \\ \vdots \\ 1 \end{bmatrix} = \sum_{i=1}^{n} y_{i2}.$$

We now turn to populations. The mean of $\mathbf{y}$ over all possible values in the population is called the *population mean vector* or *expected value* of $\mathbf{y}$. It is defined as a vector of expected values of each variable,

$$E(\mathbf{y}) = E\begin{bmatrix} y_1 \\ y_2 \\ \vdots \\ y_p \end{bmatrix} = \begin{bmatrix} E(y_1) \\ E(y_2) \\ \vdots \\ E(y_p) \end{bmatrix} = \begin{bmatrix} \mu_1 \\ \mu_2 \\ \vdots \\ \mu_p \end{bmatrix} = \boldsymbol{\mu}, \tag{3.18}$$

where μ_j is the population mean of the jth variable.

It can be shown that the expected value of each $\overline{y}_j$ in $\overline{\mathbf{y}}$ is μ_j, that is, $E(\overline{y}_j) = \mu_j$. Thus the expected value of $\overline{\mathbf{y}}$ (over all possible samples) is

$$E(\overline{\mathbf{y}}) = E\begin{bmatrix} \overline{y}_1 \\ \overline{y}_2 \\ \vdots \\ \overline{y}_p \end{bmatrix} = \begin{bmatrix} E(\overline{y}_1) \\ E(\overline{y}_2) \\ \vdots \\ E(\overline{y}_p) \end{bmatrix} = \begin{bmatrix} \mu_1 \\ \mu_2 \\ \vdots \\ \mu_p \end{bmatrix} = \boldsymbol{\mu}. \tag{3.19}$$

Therefore, $\overline{\mathbf{y}}$ is an unbiased estimate of $\boldsymbol{\mu}$. We emphasize again that $\overline{\mathbf{y}}$ is never equal to $\boldsymbol{\mu}$.

Example 3.5. Table 3.5 gives partial data from Kramer and Jensen (1969a). Three variables were measured (in milliequivalents per 100 g) at 10 different locations in the South:

y_1 = available soil calcium
y_2 = exchangeable soil calcium
y_3 = turnip green calcium

Table 3.5 Calcium in Soil and Turnip Greens

Location Number	y_1	y_2	y_3
1	35	3.5	2.80
2	35	4.9	2.70
3	40	30.0	4.38
4	10	2.8	3.21
5	6	2.7	2.73
6	20	2.8	2.81
7	35	4.6	2.88
8	35	10.9	2.90
9	35	8.0	3.28
10	30	1.6	3.20

To find the mean vector $\bar{\mathbf{y}}$, we simply calculate the average of each column and obtain

$$\bar{\mathbf{y}}' = (28.1, 7.18, 3.089).$$

3.6 COVARIANCE MATRICES

The *sample covariance matrix* $\mathbf{S} = (s_{ij})$ is the matrix of sample variances and covariances of the p variables:

$$\mathbf{S} = (s_{ij}) = \begin{bmatrix} s_{11} & s_{12} & \cdots & s_{1p} \\ s_{21} & s_{22} & \cdots & s_{2p} \\ \vdots & \vdots & & \vdots \\ s_{p1} & s_{p2} & \cdots & s_{pp} \end{bmatrix}. \tag{3.20}$$

In $\mathbf{S}$ the sample variances of the p variables are on the diagonal, and all possible pairwise sample covariances appear off the diagonal. The ith row (column) contains the covariances of y_i with the other $p - 1$ variables.

The variance of the ith variable, $s_{ii} = s_i^2$, is calculated as in (3.4) and (3.5):

$$s_{ii} = s_i^2 = \frac{1}{n-1} \sum_{k=1}^{n} (y_{ki} - \bar{y}_i)^2 \tag{3.21}$$

$$= \frac{1}{n-1} \left(\sum_k y_{ki}^2 - n\bar{y}_i^2 \right), \tag{3.22}$$

where $\bar{y}_i$ is the mean of the ith variable, as in (3.15). The covariance of the ith and jth variables, s_{ij}, is calculated as in (3.9) or (3.10):

$$s_{ij} = \frac{1}{n-1} \sum_{k=1}^{n} (y_{ki} - \bar{y}_i)(y_{kj} - \bar{y}_j) \tag{3.23}$$

$$= \frac{1}{n-1} \left(\sum_k y_{ki} y_{kj} - n\bar{y}_i\bar{y}_j \right). \tag{3.24}$$

In this case, we use subscripts i and j for variables and k for observations. Note that in (3.21) the variance s_{ii} is expressed as s_i^2, the square of the standard deviation s_i, and that $\mathbf{S}$ is symmetric because $s_{ij} = s_{ji}$ in (3.23). Other names used for the covariance matrix are *variance matrix, variance–covariance matrix*, and *dispersion matrix.*

By way of notational clarification, we note that in the univariate case, s^2 is the sample variance. But in the multivariate case, we denote the sample covariance matrix as $\mathbf{S}$, not $\mathbf{S}^2$.

The sample covariance matrix $\mathbf{S}$ can be expressed in terms of the observation vectors:

$$\mathbf{S} = \frac{1}{n-1} \sum_{i=1}^{n} (\mathbf{y}_i - \overline{\mathbf{y}})(\mathbf{y}_i - \overline{\mathbf{y}})' \tag{3.25}$$

$$= \frac{1}{n-1} \left(\sum_{i=1}^{n} \mathbf{y}_i \mathbf{y}_i' - n\overline{\mathbf{y}}\,\overline{\mathbf{y}}' \right). \tag{3.26}$$

Since $(\mathbf{y}_i - \overline{\mathbf{y}})' = (y_{i1} - \overline{y}_1, y_{i2} - \overline{y}_2, \ldots, y_{ip} - \overline{y}_p)$, the element in the (1, 1) position of $(\mathbf{y}_i - \overline{\mathbf{y}})(\mathbf{y}_i - \overline{\mathbf{y}})'$ is $(y_{i1} - \overline{y}_1)^2$, and when this is summed over i as in (3.25), the result is the numerator of s_{11} in (3.21). Similarly, the (1, 2) element of $(\mathbf{y}_i - \overline{\mathbf{y}})(\mathbf{y}_i - \overline{\mathbf{y}})'$ is $(y_{i1} - \overline{y}_1)(y_{i2} - \overline{y}_2)$, which sums to the numerator of s_{12} in (3.23). Thus (3.25) is equivalent to (3.21) and (3.23), and likewise (3.26) produces (3.22) and (3.24).

We can also obtain $\mathbf{S}$ directly from the data matrix $\mathbf{Y}$. The first term in the right side of (3.24), $\sum_k y_{ki} y_{kj}$, is the product of the ith and jth columns of $\mathbf{Y}$, while the second term, $n\overline{y}_i\overline{y}_j$, is the (ij)th element of $n\overline{\mathbf{y}}\,\overline{\mathbf{y}}'$. It was noted in (2.49) that $\mathbf{Y}'\mathbf{Y}$ is obtained as products of columns of $\mathbf{Y}$. By (3.17), $\overline{\mathbf{y}} = \mathbf{Y}'\mathbf{j}/n$ and $n\overline{\mathbf{y}} = \mathbf{Y}'\mathbf{j}$; and using (2.36), we have $n\overline{\mathbf{y}}\,\overline{\mathbf{y}}' = \mathbf{Y}'(\mathbf{J}/n)\mathbf{Y}$. Thus $\mathbf{S}$ can be written as

$$\mathbf{S} = \frac{1}{n-1} \left[\mathbf{Y}'\mathbf{Y} - \mathbf{Y}' \left(\frac{1}{n} \mathbf{J} \right) \mathbf{Y} \right]$$

$$= \frac{1}{n-1} \mathbf{Y}' \left(\mathbf{I} - \frac{1}{n} \mathbf{J} \right) \mathbf{Y} \quad \text{[by (2.30)]}. \tag{3.27}$$

Expression (3.27) is a convenient representation of $\mathbf{S}$, since it makes direct use of the data matrix $\mathbf{Y}$. However, the *matrix* $\mathbf{I} - \mathbf{J}/n$ is $n \times n$ and may be unwieldy in computation if n is large.

If $\mathbf{y}$ is a random vector taking on any possible value in a population, the *population covariance matrix* is defined as

$$\boldsymbol{\Sigma} = \operatorname{cov}(\mathbf{y}) = \begin{bmatrix} \sigma_{11} & \sigma_{12} & \cdots & \sigma_{1p} \\ \sigma_{21} & \sigma_{22} & \cdots & \sigma_{2p} \\ \vdots & \vdots & & \vdots \\ \sigma_{p1} & \sigma_{p2} & \cdots & \sigma_{pp} \end{bmatrix}. \tag{3.28}$$

The diagonal elements $\sigma_{ii} = \sigma_i^2$ are the population variances of the y's, and the off-diagonal elements are the population covariances of all possible pairs of y's.

The notation $\boldsymbol{\Sigma}$ for the covariance matrix is widely used and seems natural because $\boldsymbol{\Sigma}$ is the uppercase version of σ. It should not be confused with the same symbol used for summation of a series. The difference should always be apparent from the context. To help further distinguish the two uses, the covariance matrix $\boldsymbol{\Sigma}$ will differ in typeface and in size from the summation symbol $\sum$. Also, whenever they appear together, the summation symbol will have an index of summation, $\sum_{i=1}^{n}$.

The population covariance matrix in (3.28) can also be found as

$$\boldsymbol{\Sigma} = E[(\mathbf{y} - \boldsymbol{\mu})(\mathbf{y} - \boldsymbol{\mu})'], \tag{3.29}$$

which is analogous to (3.25) for the sample covariance matrix.

The matrix $(\mathbf{y} - \boldsymbol{\mu})(\mathbf{y} - \boldsymbol{\mu})'$ is a random matrix. The expected value of a random matrix is defined as the matrix of expected values of the corresponding elements. To see that (3.29) produces population variances and covariances of the p variables as in (3.28), note that

$$\boldsymbol{\Sigma} = E[(\mathbf{y} - \boldsymbol{\mu})(\mathbf{y} - \boldsymbol{\mu})'] = E\begin{bmatrix} y_1 - \mu_1 \\ y_2 - \mu_2 \\ \vdots \\ y_p - \mu_p \end{bmatrix} [y_1 - \mu_1, y_2 - \mu_2, \ldots, y_p - \mu_p]$$

$$= E\begin{bmatrix} (y_1 - \mu_1)^2 & (y_1 - \mu_1)(y_2 - \mu_2) & \cdots & (y_1 - \mu_1)(y_p - \mu_p) \\ (y_2 - \mu_2)(y_1 - \mu_1) & (y_2 - \mu_2)^2 & \cdots & (y_2 - \mu_2)(y_p - \mu_p) \\ \vdots & \vdots & & \vdots \\ (y_p - \mu_p)(y_1 - \mu_1) & (y_p - \mu_p)(y_2 - \mu_2) & \cdots & (y_p - \mu_p)^2 \end{bmatrix}$$

$$= \begin{bmatrix} E(y_1 - \mu_1)^2 & E(y_1 - \mu_1)(y_2 - \mu_2) & \cdots & E(y_1 - \mu_1)(y_p - \mu_p) \\ E(y_2 - \mu_2)(y_1 - \mu_1) & E(y_2 - \mu_2)^2 & \cdots & E(y_2 - \mu_2)(y_p - \mu_p) \\ \vdots & \vdots & & \vdots \\ E(y_p - \mu_p)(y_1 - \mu_1) & E(y_p - \mu_p)(y_2 - \mu_2) & \cdots & E(y_p - \mu_p)^2 \end{bmatrix}$$

$$= \begin{bmatrix} \sigma_{11} & \sigma_{12} & \cdots & \sigma_{1p} \\ \sigma_{21} & \sigma_{22} & \cdots & \sigma_{2p} \\ \vdots & \vdots & & \vdots \\ \sigma_{p1} & \sigma_{p2} & \cdots & \sigma_{pp} \end{bmatrix}.$$

It can be easily shown that $\boldsymbol{\Sigma}$ can be expressed in a form analogous to (3.26):

$$\boldsymbol{\Sigma} = E(\mathbf{yy}') - \boldsymbol{\mu\mu}'. \tag{3.30}$$

Since $E(s_{ij}) = \sigma_{ij}$ for all i, j, the sample covariance matrix $\mathbf{S}$ is an unbiased estimator for $\boldsymbol{\Sigma}$:

$$E(\mathbf{S}) = \boldsymbol{\Sigma}. \tag{3.31}$$

As in the univariate case, we note that it is the average of all possible values of $\mathbf{S}$ that is equal to $\boldsymbol{\Sigma}$. Generally, $\mathbf{S}$ will never be equal to $\boldsymbol{\Sigma}$.

Example 3.6. To calculate the covariance matrix for the calcium data of Table 3.5 using the computational forms (3.22) and (3.24), we need the sum of squares of each column and the sum of products of each pair of columns. We illustrate the computation of s_{13}:

$$\sum_{k=1}^{10} y_{k1}y_{k3} = (35)(2.80) + (35)(2.70) + \cdots + (30)(3.20) = 885.48.$$

From Example 3.5 we have $\bar{y}_1 = 28.1$ and $\bar{y}_3 = 3.089$. By (3.24), we obtain

$$s_{13} = \frac{1}{10-1}[885.48 - 10(28.1)(3.089)] = \frac{17.471}{9} = 1.9412.$$

Continuing in this fashion, we obtain

$$\mathbf{S} = \begin{pmatrix} 140.54 & 49.68 & 1.94 \\ 49.68 & 72.25 & 3.68 \\ 1.94 & 3.68 & .25 \end{pmatrix}.$$

3.7 CORRELATION MATRICES

The sample correlation between the ith and jth variables is defined in (3.12) as

$$r_{ij} = \frac{s_{ij}}{\sqrt{s_{ii}s_{jj}}} = \frac{s_{ij}}{s_i s_j}. \tag{3.32}$$

The *sample correlation matrix* is analogous to the covariance matrix with correlations in place of covariances:

$$\mathbf{R} = (r_{ij}) = \begin{bmatrix} 1 & r_{12} & \cdots & r_{1p} \\ r_{21} & 1 & \cdots & r_{2p} \\ \vdots & \vdots & & \vdots \\ r_{p1} & r_{p2} & \cdots & 1 \end{bmatrix}. \tag{3.33}$$

The second row, for example, contains the correlation of y_2 with each of the y's (including the correlation of y_2 with itself, which is 1). Of course this matrix is symmetric, since $r_{ij} = r_{ji}$.

The correlation matrix can be obtained from the covariance matrix and vice versa. Define

$$\begin{aligned} \mathbf{D}_s &= \text{diag}\,(\sqrt{s_{11}}, \sqrt{s_{22}}, \ldots, \sqrt{s_{pp}}) \\ &= \text{diag}\,(s_1, s_2, \ldots, s_p) \\ &= \begin{bmatrix} s_1 & 0 & \cdots & 0 \\ 0 & s_2 & \cdots & 0 \\ \vdots & \vdots & & \vdots \\ 0 & 0 & \cdots & s_p \end{bmatrix}. \end{aligned} \tag{3.34}$$

Then by (2.52)

$$\mathbf{R} = \mathbf{D}_s^{-1}\mathbf{S}\mathbf{D}_s^{-1} \tag{3.35}$$

and

$$\mathbf{S} = \mathbf{D}_s\mathbf{R}\mathbf{D}_s. \tag{3.36}$$

The *population correlation matrix* analogous to (3.33) is defined as

$$\mathbf{P}_\rho = (\rho_{ij}) = \begin{bmatrix} 1 & \rho_{12} & \cdots & \rho_{1p} \\ \rho_{21} & 1 & \cdots & \rho_{2p} \\ \vdots & \vdots & & \vdots \\ \rho_{p1} & \rho_{p2} & \cdots & 1 \end{bmatrix}, \quad (3.37)$$

where

$$\rho_{ij} = \frac{\sigma_{ij}}{\sigma_i \sigma_j}$$

as in (3.11).

Example 3.7. To obtain the correlation matrix for the calcium data in Table 3.5, we can calculate the individual elements using (3.32) or use the direct matrix operation in (3.35). The diagonal matrix $\mathbf{D}_s$ can be found by taking the square roots of the diagonal elements of **S**,

$$\mathbf{D}_s = \begin{pmatrix} 11.8551 & 0 & 0 \\ 0 & 8.4999 & 0 \\ 0 & 0 & .5001 \end{pmatrix},$$

where we have used the unrounded version of **S** for computation. Then, by (3.35),

$$\mathbf{R} = \mathbf{D}_s^{-1}\mathbf{S}\mathbf{D}_s^{-1} = \begin{pmatrix} 1.000 & .493 & .327 \\ .493 & 1.000 & .865 \\ .327 & .865 & 1.000 \end{pmatrix}.$$

Note that .865 > .493 > .327, which is a different order than the covariances in **S** in Example 3.6. Thus we cannot compare covariances, even within the same matrix **S**.

3.8 MEAN VECTORS AND COVARIANCE MATRICES FOR SUBSETS OF VARIABLES

3.8.1 Two Subsets

Sometimes a researcher is interested in two different kinds of variables, both measured on the same sampling unit. This corresponds to type 2 data in Section 1.4. For example, several classroom behaviors are observed for students, and during the same time period several teacher behaviors are also observed. The researcher wishes to study the relationships between the pupil variables and the teacher variables.

We will denote the two subvectors by $\mathbf{y}$ and $\mathbf{x}$, with p variables in $\mathbf{y}$ and q variables in $\mathbf{x}$. Thus each observation vector in a sample is partitioned as

$$\begin{pmatrix} \mathbf{y}_i \\ \mathbf{x}_i \end{pmatrix} = \begin{bmatrix} y_{i1} \\ \vdots \\ y_{ip} \\ x_{i1} \\ \vdots \\ x_{iq} \end{bmatrix} \qquad i = 1, 2, \ldots, n. \tag{3.38}$$

Note that there are $p + q$ variables in each of n observation vectors. In Chapter 10 we will discuss regression of the y's on the x's, and in Chapter 11 we will define a measure of correlation between the y's and the x's.

For the sample of n observation vectors, the mean vector and covariance matrix have the form

$$\begin{pmatrix} \overline{\mathbf{y}} \\ \overline{\mathbf{x}} \end{pmatrix} = \begin{bmatrix} \overline{y}_1 \\ \vdots \\ \overline{y}_p \\ \overline{x}_1 \\ \vdots \\ \overline{x}_q \end{bmatrix} \tag{3.39}$$

and

$$\mathbf{S} = \begin{pmatrix} \mathbf{S}_{yy} & \mathbf{S}_{yx} \\ \mathbf{S}_{xy} & \mathbf{S}_{xx} \end{pmatrix}, \tag{3.40}$$

where $\mathbf{S}_{yy}$ is $p \times p$, $\mathbf{S}_{yx}$ is $p \times q$, $\mathbf{S}_{xy}$ is $q \times p$, and $\mathbf{S}_{xx}$ is $q \times q$. Note that because of the symmetry of $\mathbf{S}$,

$$\mathbf{S}_{xy} = \mathbf{S}'_{yx}. \tag{3.41}$$

Thus (3.40) could be written

$$\mathbf{S} = \begin{pmatrix} \mathbf{S}_{yy} & \mathbf{S}_{yx} \\ \mathbf{S}'_{yx} & \mathbf{S}_{xx} \end{pmatrix}. \tag{3.42}$$

To illustrate (3.39) and (3.40), let $p = 2$ and $q = 3$. Then

$$\begin{pmatrix} \overline{\mathbf{y}} \\ \overline{\mathbf{x}} \end{pmatrix} = \begin{bmatrix} \overline{y}_1 \\ \overline{y}_2 \\ \hline \overline{x}_1 \\ \overline{x}_2 \\ \overline{x}_3 \end{bmatrix}$$

and

$$\mathbf{S} = \begin{pmatrix} \mathbf{S}_{yy} & \mathbf{S}_{yx} \\ \mathbf{S}_{xy} & \mathbf{S}_{xx} \end{pmatrix} = \left[\begin{array}{cc|ccc} s_{y_1}^2 & s_{y_1y_2} & s_{y_1x_1} & s_{y_1x_2} & s_{y_1x_3} \\ s_{y_2y_1} & s_{y_2}^2 & s_{y_2x_1} & s_{y_2x_2} & s_{y_2x_3} \\ \hline s_{x_1y_1} & s_{x_1y_2} & s_{x_1}^2 & s_{x_1x_2} & s_{x_1x_3} \\ s_{x_2y_1} & s_{x_2y_2} & s_{x_2x_1} & s_{x_2}^2 & s_{x_2x_3} \\ s_{x_3y_1} & s_{x_3y_2} & s_{x_3x_1} & s_{x_3x_2} & s_{x_3}^2 \end{array}\right].$$

The pattern in each of $\mathbf{S}_{yy}$, $\mathbf{S}_{yx}$, $\mathbf{S}_{xy}$, and $\mathbf{S}_{xx}$ is clearly seen in this illustration. For example, the first row of $\mathbf{S}_{yx}$ has the covariances of y_1 with each of x_1, x_2, x_3; the second row exhibits covariances of y_2 with the three x's. On the other hand, $\mathbf{S}_{xy}$ has as its first row the covariances of x_1 with y_1 and y_2, and so on. Thus $\mathbf{S}_{xy} = \mathbf{S}'_{yx}$, and a reordering of $\mathbf{y}$ and $\mathbf{x}$ transposes the covariance matrix for the two sets of variables.

The analogous population results for a partitioned random vector are

$$E\begin{pmatrix} \mathbf{y} \\ \mathbf{x} \end{pmatrix} = \begin{pmatrix} E(\mathbf{y}) \\ E(\mathbf{x}) \end{pmatrix} = \begin{pmatrix} \boldsymbol{\mu}_y \\ \boldsymbol{\mu}_x \end{pmatrix} \tag{3.43}$$

and

$$\text{cov}\begin{pmatrix} \mathbf{y} \\ \mathbf{x} \end{pmatrix} = \boldsymbol{\Sigma} = \begin{pmatrix} \boldsymbol{\Sigma}_{yy} & \boldsymbol{\Sigma}_{yx} \\ \boldsymbol{\Sigma}_{xy} & \boldsymbol{\Sigma}_{xx} \end{pmatrix}, \tag{3.44}$$

where $\boldsymbol{\Sigma}_{xy} = \boldsymbol{\Sigma}'_{yx}$. The submatrix $\boldsymbol{\Sigma}_{yy}$ is a $p \times p$ covariance matrix containing the variances of $y_1, y_2, \ldots, y_p$ on the diagonal and the covariances of each y_i with each y_j off the diagonal. Similarly, $\boldsymbol{\Sigma}_{xx}$ is the $q \times q$ covariance matrix of $x_1, x_2, \ldots, x_q$. The matrix $\boldsymbol{\Sigma}_{yx}$ is $p \times q$ and contains the covariance of each

y_i with each x_j. The covariance matrix for $\mathbf{y}$ with $\mathbf{x}$, $\boldsymbol{\Sigma}_{yx}$, is also denoted by $\text{cov}(\mathbf{y}, \mathbf{x})$, that is,

$$\text{cov}(\mathbf{y}, \mathbf{x}) = \boldsymbol{\Sigma}_{yx}.$$

Note the difference in meaning between $\text{cov}\binom{\mathbf{y}}{\mathbf{x}}$ in (3.44) and $\text{cov}(\mathbf{y}, \mathbf{x}) = \boldsymbol{\Sigma}_{yx}$; $\text{cov}\binom{\mathbf{y}}{\mathbf{x}}$ involves a single vector containing $p + q$ variables, and $\text{cov}(\mathbf{y}, \mathbf{x})$ involves two vectors.

If $\mathbf{x}$ and $\mathbf{y}$ are independent, then $\boldsymbol{\Sigma}_{yx} = \mathbf{O}$. This means that each y_i is independent of each x_j so that $\sigma_{y_i x_j} = 0$ for $i = 1, 2, \ldots, p$; $j = 1, 2, \ldots, q$.

Example 3.8.1. Reaven and Miller (1979; see also Andrews and Herzberg 1985, pp. 215–219) measured five variables in a comparison of normal patients and diabetics. In Table 3.6 we give partial data for normal patients only. The three variables of major interest were

$$\begin{aligned} x_1 &= \text{glucose intolerance} \\ x_2 &= \text{insulin response to oral glucose} \\ x_3 &= \text{insulin resistance} \end{aligned}$$

The two additional variables of minor interest were

$$\begin{aligned} y_1 &= \text{relative weight} \\ y_2 &= \text{fasting plasma glucose} \end{aligned}$$

The mean vector, partitioned as in (3.39), is

$$\begin{pmatrix} \overline{\mathbf{y}} \\ \overline{\mathbf{x}} \end{pmatrix} = \begin{bmatrix} \overline{y}_1 \\ \overline{y}_2 \\ \hline \overline{x}_1 \\ \overline{x}_2 \\ \overline{x}_3 \end{bmatrix} = \begin{bmatrix} .918 \\ 90.41 \\ \hline 340.83 \\ 171.37 \\ 97.78 \end{bmatrix}.$$

The covariance matrix, partitioned as in the illustration following (3.42), is

Table 3.6 Relative Weight, Blood Glucose, and Insulin Levels

Patient Number	y_1	y_2	x_1	x_2	x_3
1	0.81	80	356	124	55
2	0.95	97	289	117	76
3	0.94	105	319	143	105
4	1.04	90	356	199	108
5	1.00	90	323	240	143
6	0.76	86	381	157	165
7	0.91	100	350	221	119
8	1.10	85	301	186	105
9	0.99	97	379	142	98
10	0.78	97	296	131	94
11	0.90	91	353	221	53
12	0.73	87	306	178	66
13	0.96	78	290	136	142
14	0.84	90	371	200	93
15	0.74	86	312	208	68
16	0.98	80	393	202	102
17	1.10	90	364	152	76
18	0.85	99	359	185	37
19	0.83	85	296	116	60
20	0.93	90	345	123	50
21	0.95	90	378	136	47
22	0.74	88	304	134	50
23	0.95	95	347	184	91
24	0.97	90	327	192	124
25	0.72	92	386	279	74
26	1.11	74	365	228	235
27	1.20	98	365	145	158
28	1.13	100	352	172	140
29	1.00	86	325	179	145
30	0.78	98	321	222	99
31	1.00	70	360	134	90
32	1.00	99	336	143	105
33	0.71	75	352	169	32
34	0.76	90	353	263	165
35	0.89	85	373	174	78
36	0.88	99	376	134	80
37	1.17	100	367	182	54
38	0.85	78	335	241	175
39	0.97	106	396	128	80
40	1.00	98	277	222	186
41	1.00	102	378	165	117
42	0.89	90	360	282	160
43	0.98	94	291	94	71
44	0.78	80	269	121	29
45	0.74	93	318	73	42
46	0.91	86	328	106	56

$$\mathbf{S} = \begin{pmatrix} \mathbf{S}_{yy} & \mathbf{S}_{yx} \\ \mathbf{S}_{xy} & \mathbf{S}_{xx} \end{pmatrix}$$

$$= \left[\begin{array}{rr|rrr} .0162 & .2160 & .7872 & -.2138 & 2.189 \\ .2160 & 70.56 & 26.23 & -23.96 & -20.84 \\ \hline .7872 & 26.23 & 1106 & 396.7 & 108.4 \\ -.2138 & -23.96 & 396.7 & 2382 & 1143 \\ 2.189 & -20.84 & 108.4 & 1143 & 2136 \end{array}\right].$$

Notice that $\mathbf{S}_{yy}$ and $\mathbf{S}_{xx}$ are symmetric and that $\mathbf{S}_{xy}$ is the transpose of $\mathbf{S}_{yx}$.

3.8.2 Three or More Subsets

In some cases, three or more subsets of variables are of interest. If the observation vector $\mathbf{y}$ is partitioned as

$$\mathbf{y} = \begin{bmatrix} \mathbf{y}_1 \\ \mathbf{y}_2 \\ \vdots \\ \mathbf{y}_k \end{bmatrix},$$

where $\mathbf{y}_1$ has p_1 variables, $\mathbf{y}_2$ has p_2, and $\mathbf{y}_k$ has p_k, with $p = p_1 + p_2 + \cdots + p_k$, then the sample mean vector and covariance matrix are given by

$$\overline{\mathbf{y}} = \begin{bmatrix} \overline{\mathbf{y}}_1 \\ \overline{\mathbf{y}}_2 \\ \vdots \\ \overline{\mathbf{y}}_k \end{bmatrix} \quad (3.45)$$

and

$$\mathbf{S} = \begin{bmatrix} \mathbf{S}_{11} & \mathbf{S}_{12} & \cdots & \mathbf{S}_{1k} \\ \mathbf{S}_{21} & \mathbf{S}_{22} & \cdots & \mathbf{S}_{2k} \\ \vdots & \vdots & & \vdots \\ \mathbf{S}_{k1} & \mathbf{S}_{k2} & \cdots & \mathbf{S}_{kk} \end{bmatrix}. \quad (3.46)$$

The submatrix $\mathbf{S}_{2k}$, for example, contains the covariances of the variables in $\mathbf{y}_2$ with the variables in $\mathbf{y}_k$.

The corresponding population results are

$$\boldsymbol{\mu} = \begin{bmatrix} \boldsymbol{\mu}_1 \\ \boldsymbol{\mu}_2 \\ \vdots \\ \boldsymbol{\mu}_k \end{bmatrix} \tag{3.47}$$

and

$$\boldsymbol{\Sigma} = \begin{bmatrix} \boldsymbol{\Sigma}_{11} & \boldsymbol{\Sigma}_{12} & \cdots & \boldsymbol{\Sigma}_{1k} \\ \boldsymbol{\Sigma}_{21} & \boldsymbol{\Sigma}_{22} & \cdots & \boldsymbol{\Sigma}_{2k} \\ \vdots & \vdots & & \vdots \\ \boldsymbol{\Sigma}_{k1} & \boldsymbol{\Sigma}_{k2} & \cdots & \boldsymbol{\Sigma}_{kk} \end{bmatrix}. \tag{3.48}$$

3.9 LINEAR COMBINATIONS OF VARIABLES

3.9.1 Sample Properties

We are frequently interested in linear combinations of the variables y_1, $y_2, \ldots, y_p$. For example, two of the types of linear functions we will use are (1) linear combinations that maximize some function and (2) linear combinations that compare variables, for example, $y_1 - y_3$. In this section, we investigate the means, variances, and covariances of linear combinations. These tools are used often in the remainder of the text.

Let $a_1, a_2, \ldots, a_p$ be constants and consider the linear combination of the elements of the vector $\mathbf{y}$,

$$z = a_1 y_1 + a_2 y_2 + \cdots + a_p y_p = \mathbf{a}'\mathbf{y}, \tag{3.49}$$

where $\mathbf{a}' = (a_1, a_2, \ldots, a_p)$. If the same coefficient vector $\mathbf{a}$ is applied to each $\mathbf{y}_i$ in a sample, we have

$$\begin{aligned} z_i &= a_1 y_{i1} + a_2 y_{i2} + \cdots + a_p y_{ip} \\ &= \mathbf{a}'\mathbf{y}_i \qquad i = 1, 2, \ldots, n. \end{aligned} \tag{3.50}$$

The sample mean of z can be found either by averaging the n values $z_1 = \mathbf{a}'\mathbf{y}_1, z_2 = \mathbf{a}'\mathbf{y}_2, \ldots, z_n = \mathbf{a}'\mathbf{y}_n$ or as a linear combination of $\overline{\mathbf{y}}$,

$$\bar{z} = \frac{1}{n} \sum_{i=1}^{n} z_i = \mathbf{a}'\bar{\mathbf{y}}, \tag{3.51}$$

where $\bar{\mathbf{y}}$ is the sample mean vector of $\mathbf{y}_1, \mathbf{y}_2, \ldots, \mathbf{y}_n$. The result in (3.51) is analogous to the univariate result (3.3), $\bar{z} = a\bar{y}$, where $z_i = ay_i, i = 1, 2, \ldots, n$.

Similarly, the sample variance of $z_i = \mathbf{a}'\mathbf{y}_i, i = 1, 2, \ldots, n$, can be found as the sample variance of $z_1, z_2, \ldots, z_n$ or directly from $\mathbf{a}$ and $\mathbf{S}$, where $\mathbf{S}$ is the sample covariance of $\mathbf{y}_1, \mathbf{y}_2, \ldots, \mathbf{y}_n$:

$$s_z^2 = \frac{\sum_{i=1}^{n} (z_i - \bar{z})^2}{n-1} = \mathbf{a}'\mathbf{S}\mathbf{a}. \tag{3.52}$$

Note that $s_z^2 = \mathbf{a}'\mathbf{S}\mathbf{a}$ is the multivariate analogue of the univariate result (3.6), $s_z^2 = a^2 s^2$, where $z_i = ay_i, i = 1, 2, \ldots, n$, and s^2 is the variance of $y_1, y_2, \ldots, y_n$.

Since a variance is always nonnegative, we have $s_z^2 \geq 0$, and therefore $\mathbf{a}'\mathbf{S}\mathbf{a} \geq 0$, for every $\mathbf{a}$. Hence $\mathbf{S}$ is at least positive semidefinite. If the variables are continuous and not linearly related and if $n - 1 > p$ (so that $\mathbf{S}$ is full rank), then $\mathbf{S}$ is positive definite (with probability 1).

If we define another linear combination $w = \mathbf{b}'\mathbf{y} = b_1 y_1 + b_2 y_2 + \cdots + b_p y_p$, where $\mathbf{b}' = (b_1, b_2, \ldots, b_p)$ is a vector of constants different from $\mathbf{a}'$, then the sample covariance of z and w is given by

$$s_{zw} = \mathbf{a}'\mathbf{S}\mathbf{b}. \tag{3.53}$$

The sample correlation between z and w is readily obtained as

$$r_{zw} = \frac{s_{zw}}{\sqrt{s_z^2 s_w^2}} = \frac{\mathbf{a}'\mathbf{S}\mathbf{b}}{\sqrt{(\mathbf{a}'\mathbf{S}\mathbf{a})(\mathbf{b}'\mathbf{S}\mathbf{b})}}. \tag{3.54}$$

We now denote the two constant vectors $\mathbf{a}$ and $\mathbf{b}$ as $\mathbf{a}_1$ and $\mathbf{a}_2$ to facilitate later expansion to more than two such vectors. Let

$$\mathbf{A} = \begin{pmatrix} \mathbf{a}_1' \\ \mathbf{a}_2' \end{pmatrix}$$

and define

$$\mathbf{z} = \begin{pmatrix} \mathbf{a}_1'\mathbf{y} \\ \mathbf{a}_2'\mathbf{y} \end{pmatrix} = \begin{pmatrix} z_1 \\ z_2 \end{pmatrix}.$$

Then we can factor **y** from this expression by (2.106):

$$\mathbf{z} = \begin{pmatrix} \mathbf{a}_1' \\ \mathbf{a}_2' \end{pmatrix} \mathbf{y} = \mathbf{Ay}.$$

If we evaluate the bivariate $\mathbf{z}_i$ for each p-variate $\mathbf{y}_i$ in the sample, we obtain $\mathbf{z}_i = \mathbf{Ay}_i$, $i = 1, 2, \ldots, n$, and the average of $\mathbf{z}$ over the sample can be found from $\overline{\mathbf{y}}$:

$$\overline{\mathbf{z}} = \begin{pmatrix} \overline{z}_1 \\ \overline{z}_2 \end{pmatrix} = \begin{pmatrix} \mathbf{a}_1'\overline{\mathbf{y}} \\ \mathbf{a}_2'\overline{\mathbf{y}} \end{pmatrix} \quad \text{[by (3.51)]} \tag{3.55}$$

$$= \begin{pmatrix} \mathbf{a}_1' \\ \mathbf{a}_2' \end{pmatrix} \overline{\mathbf{y}} = \mathbf{A}\overline{\mathbf{y}} \quad \text{[by (2.106)]}. \tag{3.56}$$

We can use (3.52) and (3.53) to construct the sample covariance matrix for **z**:

$$\mathbf{S}_z = \begin{pmatrix} s_{z_1}^2 & s_{z_1 z_2} \\ s_{z_2 z_1} & s_{z_2}^2 \end{pmatrix}$$

$$= \begin{pmatrix} \mathbf{a}_1'\mathbf{Sa}_1 & \mathbf{a}_1'\mathbf{Sa}_2 \\ \mathbf{a}_2'\mathbf{Sa}_1 & \mathbf{a}_2'\mathbf{Sa}_2 \end{pmatrix}. \tag{3.57}$$

By (2.107), this factors into

$$\mathbf{S}_z = \begin{pmatrix} \mathbf{a}_1' \\ \mathbf{a}_2' \end{pmatrix} \mathbf{S}(\mathbf{a}_1, \mathbf{a}_2) = \mathbf{ASA}'. \tag{3.58}$$

The bivariate results in (3.56) and (3.58) can be readily extended to more than two linear combinations. (See principal components in Chapter 12, for instance, where we transform the y's to fewer dimensions that capture most of the information in the y's.) If we have k linear transformations, they can be expressed as

$$\begin{aligned} z_1 &= a_{11}y_1 + a_{12}y_2 + \cdots + a_{1p}y_p = \mathbf{a}_1'\mathbf{y} \\ z_2 &= a_{21}y_1 + a_{22}y_2 + \cdots + a_{2p}y_p = \mathbf{a}_2'\mathbf{y} \\ &\vdots \\ z_k &= a_{k1}y_1 + a_{k2}y_2 + \cdots + a_{kp}y_p = \mathbf{a}_k'\mathbf{y} \end{aligned}$$

or in matrix notation

$$\mathbf{z} = \begin{bmatrix} z_1 \\ z_2 \\ \vdots \\ z_k \end{bmatrix} = \begin{bmatrix} \mathbf{a}_1'\mathbf{y} \\ \mathbf{a}_2'\mathbf{y} \\ \vdots \\ \mathbf{a}_k'\mathbf{y} \end{bmatrix} = \begin{bmatrix} \mathbf{a}_1' \\ \mathbf{a}_2' \\ \vdots \\ \mathbf{a}_k' \end{bmatrix} \mathbf{y} = \mathbf{Ay},$$

where $\mathbf{z}$ is $k \times 1$, $\mathbf{A}$ is $k \times p$, and $\mathbf{y}$ is $p \times 1$. If $\mathbf{z}_i = \mathbf{Ay}_i$ is evaluated for all $\mathbf{y}_i$, $i = 1, 2, \ldots, n$, then by (3.51) and (2.106), the sample mean of the $\mathbf{z}$'s is

$$\overline{\mathbf{z}} = \begin{bmatrix} \mathbf{a}_1'\overline{\mathbf{y}} \\ \mathbf{a}_2'\overline{\mathbf{y}} \\ \vdots \\ \mathbf{a}_k'\overline{\mathbf{y}} \end{bmatrix} = \begin{bmatrix} \mathbf{a}_1' \\ \mathbf{a}_2' \\ \vdots \\ \mathbf{a}_k' \end{bmatrix} \overline{\mathbf{y}} = \mathbf{A}\overline{\mathbf{y}}. \qquad (3.59)$$

By an extension of (3.57), the sample covariance matrix of the $\mathbf{z}$'s becomes

$$\begin{aligned} \mathbf{S}_z &= \begin{bmatrix} \mathbf{a}_1'\mathbf{Sa}_1 & \mathbf{a}_1'\mathbf{Sa}_2 & \cdots & \mathbf{a}_1'\mathbf{Sa}_k \\ \mathbf{a}_2'\mathbf{Sa}_1 & \mathbf{a}_2'\mathbf{Sa}_2 & \cdots & \mathbf{a}_2'\mathbf{Sa}_k \\ \vdots & \vdots & & \vdots \\ \mathbf{a}_k'\mathbf{Sa}_1 & \mathbf{a}_k'\mathbf{Sa}_2 & \cdots & \mathbf{a}_k'\mathbf{Sa}_k \end{bmatrix} \\ &= \begin{bmatrix} \mathbf{a}_1'(\mathbf{Sa}_1, & \mathbf{Sa}_2, \cdots, \mathbf{Sa}_k) \\ \mathbf{a}_2'(\mathbf{Sa}_1, & \mathbf{Sa}_2, \cdots, \mathbf{Sa}_k) \\ \vdots & \vdots \\ \mathbf{a}_k'(\mathbf{Sa}_1, & \mathbf{Sa}_2, \cdots, \mathbf{Sa}_k) \end{bmatrix} \end{aligned}$$

$$= \begin{bmatrix} \mathbf{a}_1' \\ \mathbf{a}_2' \\ \vdots \\ \mathbf{a}_k' \end{bmatrix} (\mathbf{S}\mathbf{a}_1, \mathbf{S}\mathbf{a}_2, \ldots, \mathbf{S}\mathbf{a}_k) \quad [\text{by } (2.46)]$$

$$= \begin{bmatrix} \mathbf{a}_1' \\ \mathbf{a}_2' \\ \vdots \\ \mathbf{a}_k' \end{bmatrix} \mathbf{S}(\mathbf{a}_1, \mathbf{a}_2, \ldots, \mathbf{a}_k) \quad [\text{by } (2.47)]$$

$$= \mathbf{ASA}'. \tag{3.60}$$

A slightly more general linear transformation is

$$\mathbf{z}_i = \mathbf{A}\mathbf{y}_i + \mathbf{b} \qquad i = 1, 2, \ldots, n. \tag{3.61}$$

The sample mean vector and covariance matrix of $\mathbf{z}$ are given by

$$\bar{\mathbf{z}} = \mathbf{A}\bar{\mathbf{y}} + \mathbf{b} \tag{3.62}$$

$$\mathbf{S}_z = \mathbf{ASA}'. \tag{3.63}$$

Example 3.9.1. Timm (1975, p. 233; 1980, p. 47) reported the results of an experiment where subjects responded to "probe words" at five positions in a sentence. The variables are response times for the ith probe word, y_i, $i = 1, 2, \ldots, 5$. The data are given in Table 3.7.

Table 3.7 Response Times for Five Probe Word Positions

Subject Number	y_1	y_2	y_3	y_4	y_5
1	51	36	50	35	42
2	27	20	26	17	27
3	37	22	41	37	30
4	42	36	32	34	27
5	27	18	33	14	29
6	43	32	43	35	40
7	41	22	36	25	38
8	38	21	31	20	16
9	36	23	27	25	28
10	26	31	31	32	36
11	29	20	25	26	25

These variables are commensurate (same measurement units and similar means and variances), and the researcher may wish to examine some simple linear combinations. Consider the following linear combination for illustrative purposes:

$$\begin{aligned} z &= 3y_1 - 2y_2 + 4y_3 - y_4 + y_5 \\ &= (3, -2, 4, -1, 1)\mathbf{y} = \mathbf{a}'\mathbf{y}. \end{aligned}$$

If z is calculated for each of the 11 observations, we obtain $z_1 = 288, \cdots, z_{11} = 146$ with mean $\bar{z} = 197.0$ and variance $s_z^2 = 2084.0$. These same results can be obtained using (3.51) and (3.52). The sample mean vector and covariance matrix for the data are

$$\bar{\mathbf{y}} = \begin{bmatrix} 36.09 \\ 25.55 \\ 34.09 \\ 27.27 \\ 30.73 \end{bmatrix} \qquad \mathbf{S} = \begin{bmatrix} 65.09 & 33.65 & 47.59 & 36.77 & 25.43 \\ 33.65 & 46.07 & 28.95 & 40.34 & 28.36 \\ 47.59 & 28.95 & 60.69 & 37.37 & 41.13 \\ 36.77 & 40.34 & 37.37 & 62.82 & 31.68 \\ 25.43 & 28.36 & 41.13 & 31.68 & 58.22 \end{bmatrix}.$$

Then, by (3.51),

$$\bar{z} = \mathbf{a}'\bar{\mathbf{y}} = (3, -2, 4, -1, 1) \begin{bmatrix} 36.09 \\ 25.55 \\ 34.09 \\ 27.27 \\ 30.73 \end{bmatrix} = 197.0$$

and by (3.52), $s_z^2 = \mathbf{a}'\mathbf{S}\mathbf{a} = 2084.0$.

We now define a second linear combination:

$$\begin{aligned} w &= y_1 + 3y_2 - y_3 + y_4 - 2y_5 \\ &= (1, 3, -1, 1, -2)\mathbf{y} = \mathbf{b}'\mathbf{y}. \end{aligned}$$

The sample mean and variance of w are $\bar{w} = \mathbf{b}'\bar{\mathbf{y}} = 44.45$ and $s_w^2 = \mathbf{b}'\mathbf{S}\mathbf{b} = 605.67$. The sample covariance of z and w is, by (3.53), $s_{zw} = \mathbf{a}'\mathbf{S}\mathbf{b} = 40.2$.

Using (3.54), we find the sample correlation between z and w to be

$$\begin{aligned} r_{zw} &= \frac{s_{zw}}{\sqrt{s_z^2 s_w^2}} \\ &= \frac{40.2}{\sqrt{(2084)(605.67)}} = .0358. \end{aligned}$$

We now define the three linear functions

$$\begin{aligned} z_1 &= y_1 + y_2 + y_3 + y_4 + y_5 \\ z_2 &= 2y_1 - 3y_2 + y_3 - 2y_4 - y_5 \\ z_3 &= -y_1 - 2y_2 + y_3 - 2y_4 + 3y_5, \end{aligned}$$

which can be written in matrix form as

$$\begin{pmatrix} z_1 \\ z_2 \\ z_3 \end{pmatrix} = \begin{pmatrix} 1 & 1 & 1 & 1 & 1 \\ 2 & -3 & 1 & -2 & -1 \\ -1 & -2 & 1 & -2 & 3 \end{pmatrix} \begin{bmatrix} y_1 \\ y_2 \\ y_3 \\ y_4 \\ y_5 \end{bmatrix}$$

or

$$\mathbf{z} = \mathbf{A}\mathbf{y}.$$

The sample mean vector for $\mathbf{z}$ is given by (3.59) as

$$\bar{\mathbf{z}} = \mathbf{A}\bar{\mathbf{y}} = \begin{pmatrix} 153.73 \\ -55.64 \\ -15.45 \end{pmatrix},$$

and the sample covariance matrix of $\mathbf{z}$ is given by (3.60) as

$$\mathbf{S}_z = \mathbf{ASA}' = \begin{pmatrix} 995.42 & -502.09 & -211.04 \\ -502.09 & 811.45 & 268.08 \\ -211.04 & 268.08 & 702.87 \end{pmatrix}.$$

This can be converted to a correlation matrix by use of (3.35):

$$\mathbf{R}_z = \mathbf{D}_z^{-1}\mathbf{S}_z\mathbf{D}_z^{-1} = \begin{pmatrix} 1.00 & -.56 & -.25 \\ -.56 & 1.00 & .35 \\ -.25 & .35 & 1.00 \end{pmatrix},$$

where

$$\mathbf{D}_z = \begin{pmatrix} 31.55 & 0 & 0 \\ 0 & 28.49 & 0 \\ 0 & 0 & 26.51 \end{pmatrix}$$

is obtained from the square roots of the diagonal elements of $\mathbf{S}_z$.

3.9.2 Population Properties

All the above sample results for linear combinations have population counterparts. Let $z = \mathbf{a}'\mathbf{y}$, where $\mathbf{a}$ is a vector of constants. Then the *population mean* of z is

$$E(z) = E(\mathbf{a}'\mathbf{y}) = \mathbf{a}'E(\mathbf{y}) = \mathbf{a}'\boldsymbol{\mu} \tag{3.64}$$

and the *population variance* is

$$\sigma_z^2 = \text{var}(\mathbf{a}'\mathbf{y}) = \mathbf{a}'\boldsymbol{\Sigma}\mathbf{a}. \tag{3.65}$$

Let $w = \mathbf{b}'\mathbf{y}$, where $\mathbf{b}$ is a vector of constants different from $\mathbf{a}$. The *population covariance* of $z = \mathbf{a}'\mathbf{y}$ and $w = \mathbf{b}'\mathbf{y}$ is

$$\text{cov}(z, w) = \sigma_{zw} = \mathbf{a}'\boldsymbol{\Sigma}\mathbf{b}. \tag{3.66}$$

The *population correlation* of z and w is then

$$\begin{aligned} \rho_{zw} = \text{corr}(\mathbf{a}'\mathbf{y}, \mathbf{b}'\mathbf{y}) &= \frac{\sigma_{zw}}{\sigma_z \sigma_w} \\ &= \frac{\mathbf{a}'\boldsymbol{\Sigma}\mathbf{b}}{\sqrt{(\mathbf{a}'\boldsymbol{\Sigma}\mathbf{a})(\mathbf{b}'\boldsymbol{\Sigma}\mathbf{b})}}. \end{aligned} \tag{3.67}$$

If $\mathbf{Ay}$ represents several linear combinations, the *population mean vector* and *covariance matrix* are given by

$$E(\mathbf{Ay}) = \mathbf{A}E(\mathbf{y}) = \mathbf{A}\boldsymbol{\mu} \tag{3.68}$$

$$\text{cov}(\mathbf{Ay}) = \mathbf{A}\boldsymbol{\Sigma}\mathbf{A}'. \tag{3.69}$$

The general linear transformation $\mathbf{z} = \mathbf{Ay} + \mathbf{b}$ has population mean vector and covariance matrix

$$E(\mathbf{Ay} + \mathbf{b}) = \mathbf{A}E(\mathbf{y}) + \mathbf{b} = \mathbf{A}\boldsymbol{\mu} + \mathbf{b} \tag{3.70}$$

$$\text{cov}(\mathbf{Ay} + \mathbf{b}) = \mathbf{A}\boldsymbol{\Sigma}\mathbf{A}'. \tag{3.71}$$

3.10 MEASURES OF OVERALL VARIABILITY

The covariance matrix contains the variances of the variables and the covariances between pairs of variables and is thus a multifaceted picture of the overall variation in the data. Sometimes it is desirable to have a single numerical value for the overall multivariate scatter. One such measure is the *generalized sample variance*, defined as the determinant of the covariance matrix:

$$\text{Generalized sample variance} = |\mathbf{S}|. \tag{3.72}$$

The generalized sample variance has a geometric interpretation. The extension of an ellipse to more than two dimensions is called a *hyperellipsoid*. A p-dimensional hyperellipsoid $(\mathbf{y} - \overline{\mathbf{y}})'\mathbf{S}^{-1}(\mathbf{y} - \overline{\mathbf{y}}) = a^2$ centered at $\overline{\mathbf{y}}$ and based on $\mathbf{S}^{-1}$ to standardize the distance to the center has axes proportional to the square roots of the eigenvalues of $\mathbf{S}$. It can be shown that the volume of the ellipsoid is proportional to $|\mathbf{S}|^{1/2}$. If the smallest eigenvalue λ_p is zero, there is no axis in that direction, and the ellipsoid lies wholly in a $(p - 1)$-dimensional subspace of p-space. Consequently, the volume in p-space is zero. By (2.99), $|\mathbf{S}| = \lambda_1\lambda_2 \cdots \lambda_p$. Hence, if $\lambda_p = 0$, $|\mathbf{S}| = 0$. A zero eigenvalue indicates a redundancy in the form of a linear relationship among the variables. As will be seen in Section 12.7, the eigenvector corresponding to the zero eigenvalue reveals the form of the linear dependency. The solution to the dilemma is to remove one or more variables.

Another measure of overall variability, the *total sample variance*, is simply the trace of $\mathbf{S}$:

$$\text{Total sample variance} = s_{11} + s_{22} + \cdots + s_{pp} = \text{tr}(\mathbf{S}). \tag{3.73}$$

This measure of overall variation ignores covariance structure altogether but is found useful for comparison purposes in techniques such as principal components (Chapter 12).

In general, for both the above measures, relatively large values reflect a broad scatter about $\overline{\mathbf{y}}$, while lower values indicate closer concentration about $\overline{\mathbf{y}}$. In the case of $|\mathbf{S}|$, however, an extensive scatter may be masked by small eigenvalues that reduce $|\mathbf{S}|$. An extremely small value of $|\mathbf{S}|$ or $|\mathbf{R}|$ may indicate either small scatter or *multicollinearity*, a term indicating near linear relationships in a set of variables. Multicollinearity may be due to high pairwise correlations or to a high multiple correlation between one variable and several of the other variables.

When the variables are highly collinear, then $\mathbf{S}$ (or $\mathbf{R}$) becomes nearly singular and the inverse is unstable, in the sense that large changes in $\mathbf{S}^{-1}$ result from minor changes in $\mathbf{S}$. In this case, $|\mathbf{R}|$ is close to zero. Thus $|\mathbf{R}|$ is a measure of the amount of intercorrelation among the variables. For other measures of intercorrelation, see Rencher (1996, Section 1.7).

3.11 ESTIMATION OF MISSING VALUES

It is not uncommon to find missing measurements in an observation vector, that is, missing values for one or more variables. A small number of rows with missing entries in the data matrix does not constitute a problem; we can simply discard each entire row that has a missing value. However, a small portion of missing data, if widely distributed, can lead to a substantial loss of data with this procedure. For example, in a large data set with $n = 550$ and $p = 85$, only about 1.5% of the $550 \times 85 = 46{,}750$ measurements were missing. However, nearly half of the observation vectors turned out to be incomplete.

The distribution of missing values in a data set is an important consideration. Randomly missing variable values scattered throughout a data matrix are less serious than a pattern of missing values that depends to some extent on the values of the variables.

We discuss two methods of estimating the missing values or "filling the holes" in the data matrix. This is also called *imputation*. Both of the procedures presume that the missing values occur at random. If the selection process for missing values is related to some of the variables, then the techniques may not estimate the missing responses very well.

The first method is very simple: substitute a mean for each missing value, specifically the average of the available data in the column of the data matrix in which the unknown lies. Replacing an observation by its mean reduces the variance and the absolute value of the covariance. Therefore, the sample covariance matrix **S** computed from the data matrix **Y** with means imputed for missing values is biased. However, it is positive definite.

The second technique is a regression approach. Conceptually, the data matrix is partitioned into two parts, one containing all rows with missing entries and the other comprising all the complete rows. Suppose y_{ij} is missing. Then using the data in the submatrix with complete rows, y_j is regressed on the other variables to obtain a prediction equation $\hat{y}_j = b_0 + b_1 y_1 + \cdots + b_{j-1} y_{j-1} + b_{j+1} y_{j+1} + \cdots + b_p y_p$. Then the nonmissing entries in the ith row are entered as independent variables in the regression equation to obtain the predicted value, $\hat{y}_{ij}$. The regression method was first proposed by Buck (1960) and is a special case of the EM algorithm (Dempster, Laird, and Rubin 1977).

The regression method can be improved by iteration. This could proceed in the following way. Estimate all missing entries in the data matrix using regression. After filling in the missing entries, use the full data matrix to obtain new prediction equations. Use these prediction equations to calculate new predicted values $\hat{y}_{ij}$ for missing entries. Use the new data matrix to obtain revised prediction equations and new predicted values $\hat{y}_{ij}$. Continue this process until the predicted values stabilize.

A modification may be needed if the missing entries are so pervasive that it is difficult to find data to estimate the initial regression equations. In this case, the process could be started by using means as in the first method and then beginning the iteration.

The regression approach will ordinarily yield better results than the method of inserting means. However, if the other variables are not very highly correlated with the one to be predicted, the regression technique is essentially equivalent to imputing means. The regression method underestimates the variances and covariances, though to a lesser extent than the method based on means.

Example 3.11. We illustrate the iterated regression method of estimating missing values. Consider the calcium data of Table 3.5 as reproduced below and suppose the entries in parentheses are missing:

Location Number	y_1	y_2	y_3
1	35	(3.5)	2.80
2	35	4.9	(2.70)
3	40	30.0	4.38
4	10	2.8	3.21
5	6	2.7	2.73
6	20	2.8	2.81
7	35	4.6	2.88
8	35	10.9	2.90
9	35	8.0	3.28
10	30	1.6	3.20

We first regress y_2 on y_1 and y_3 for observations 3–10 and obtain $\hat{y}_2 = b_0 + b_1y_1 + b_3y_3$. When this is evaluated for the nonmissing entries in the first row ($y_1 = 35$ and $y_3 = 2.80$), we obtain $\hat{y}_2 = 4.097$. Similarly, we regress y_3 on y_1 and y_2 for observations 3–10 to obtain $\hat{y}_3 = c_0 + c_1y_1 + c_2y_2$. Evaluating this for the nonmissing entries in the second row yields $\hat{y}_3 = 3.011$. We now insert these estimates for the missing values and calculate the regression equations based on all 10 observations. Using the revised equation $\hat{y}_2 = b_0 + b_1y_1 + b_3y_3$, we obtain a new predicted value, $\hat{y}_2 = 3.698$. Similarly, we obtain a revised regression equation for y_3 that gives a new predicted value, $\hat{y}_3 = 2.981$. With these values inserted, we calculate new equations and new predicted values, $\hat{y}_2 = 3.672$ and $\hat{y}_3 = 2.976$. At the third iteration we obtain $\hat{y}_2 = 3.679$ and $\hat{y}_3 = 2.975$. There is very little change in subsequent iterations. These values are closer to the actual values, $y_2 = 3.5$ and $y_3 = 2.70$, than the first regression estimates, $\hat{y}_2 = 4.097$ and $\hat{y}_3 = 3.011$. They are also much better estimates than the means of the second and third columns, $\overline{y}_2 = 7.589$ and $\overline{y}_3 = 3.132$.

3.12 DISTANCE BETWEEN VECTORS

In a univariate setting, the distance between two points is simply the difference (or absolute difference) between their values. For statistical purposes, this difference may not be very informative. For example, we do not want to know how many centimeters apart two means are, but rather how many standard deviations apart they are. Thus we examine the (squared) standardized or statistical distances, such as

$$\frac{(\mu_1 - \mu_2)^2}{\sigma^2} \quad \text{and} \quad \frac{(\overline{y} - \mu)^2}{\sigma_{\overline{y}}^2}.$$

To obtain a useful distance measure in a multivariate setting, we must consider not only the variances of the variables but also their covariances or correlations. The simple Euclidean distance between two vectors, $(\mathbf{y}_1 - \mathbf{y}_2)'(\mathbf{y}_1 - \mathbf{y}_2)$, is not useful because there is no adjustment for the variances or the covariances. For a statistical distance, we standardize by inserting the inverse of the covariance matrix:

$$d^2 = (\mathbf{y}_1 - \mathbf{y}_2)'\mathbf{S}^{-1}(\mathbf{y}_1 - \mathbf{y}_2). \tag{3.74}$$

Other examples are

$$D^2 = (\overline{\mathbf{y}} - \boldsymbol{\mu})'\mathbf{S}^{-1}(\overline{\mathbf{y}} - \boldsymbol{\mu}) \tag{3.75}$$

$$\Delta^2 = (\overline{\mathbf{y}} - \boldsymbol{\mu})'\boldsymbol{\Sigma}^{-1}(\overline{\mathbf{y}} - \boldsymbol{\mu}) \tag{3.76}$$

$$\Delta^2 = (\boldsymbol{\mu}_1 - \boldsymbol{\mu}_2)'\boldsymbol{\Sigma}^{-1}(\boldsymbol{\mu}_1 - \boldsymbol{\mu}_2). \tag{3.77}$$

These (squared) distances between two vectors were first proposed by Mahalanobis (1936) and are often referred to as *Mahalanobis distances*. If a random variable has a larger variance than another, it receives relatively less weight in a Mahalanobis distance. Similarly, two highly correlated variables do not contribute as much as two variables that are less correlated. In essence, then, the use of the inverse of the covariance matrix in a Mahalanobis distance (involving random variables) has the effect of (1) standardizing all variables to the same variance and (2) eliminating correlations. To illustrate this, we use the square root matrix defined in (2.103) to rewrite (3.76) as

$$\begin{aligned}\Delta^2 &= (\overline{\mathbf{y}} - \boldsymbol{\mu})'\boldsymbol{\Sigma}^{-1}(\overline{\mathbf{y}} - \boldsymbol{\mu}) = (\overline{\mathbf{y}} - \boldsymbol{\mu})'(\boldsymbol{\Sigma}^{1/2}\boldsymbol{\Sigma}^{1/2})^{-1}(\overline{\mathbf{y}} - \boldsymbol{\mu}) \\ &= [(\boldsymbol{\Sigma}^{1/2})^{-1}(\overline{\mathbf{y}} - \boldsymbol{\mu})]'[(\boldsymbol{\Sigma}^{1/2})^{-1}(\overline{\mathbf{y}} - \boldsymbol{\mu})] = \mathbf{z}'\mathbf{z},\end{aligned}$$

where $\mathbf{z} = (\boldsymbol{\Sigma}^{1/2})^{-1}(\overline{\mathbf{y}} - \boldsymbol{\mu}) = (\boldsymbol{\Sigma}^{1/2})^{-1}\overline{\mathbf{y}} - (\boldsymbol{\Sigma}^{1/2})^{-1}\boldsymbol{\mu}$. Now, by (3.71) it can be shown that

$$\text{cov}(\mathbf{z}) = \frac{1}{n}\mathbf{I}. \tag{3.78}$$

Hence the transformed variables $z_1, z_2, \ldots, z_p$ are uncorrelated, and each has variance equal to $1/n$. If the appropriate covariance matrix for the random vector were used in a Mahalanolis distance, the variances would reduce to 1. For example, if $\text{cov}(\bar{\mathbf{y}}) = \boldsymbol{\Sigma}/n$ were used above, we would obtain $\text{cov}(\mathbf{z}) = \mathbf{I}$.

PROBLEMS

3.1 If $z_i = ay_i$ for $i = 1, 2, \ldots, n$, show that $\bar{z} = a\bar{y}$ as in (3.3).

3.2 If $z_i = ay_i$ for $i = 1, 2, \ldots, n$, show that $s_z^2 = a^2s^2$ as in (3.6).

3.3 For the data in Figure 3.3, show that $\sum_i (x_i - \bar{x})(y_i - \bar{y}) = 0$.

3.4 Show that $(\mathbf{x} - \bar{x}\mathbf{j})'(\mathbf{y} - \bar{y}\mathbf{j}) = \sum_i (x_i - \bar{x})(y_i - \bar{y})$, thus verifying (3.14).

3.5 For $p = 3$ show that

$$\frac{1}{n-1}\sum_{i=1}^{n}(\mathbf{y}_i - \bar{\mathbf{y}})(\mathbf{y}_i - \bar{\mathbf{y}})' = \begin{pmatrix} s_{11} & s_{12} & s_{13} \\ s_{21} & s_{22} & s_{23} \\ s_{31} & s_{32} & s_{33} \end{pmatrix},$$

which illustrates (3.25).

3.6 Show that $\bar{z} = \mathbf{a}'\bar{\mathbf{y}}$ as in (3.51), where $z_i = \mathbf{a}'\mathbf{y}_i, i = 1, 2, \ldots, n$.

3.7 Show that $s_z^2 = \mathbf{a}'\mathbf{S}\mathbf{a}$ as in (3.52), where $z_i = \mathbf{a}'\mathbf{y}_i, i = 1, 2, \ldots, n$.

3.8 Show that

$$\begin{pmatrix} \mathbf{a}_1' \\ \mathbf{a}_2' \end{pmatrix} \mathbf{S}(\mathbf{a}_1, \mathbf{a}_2) = \begin{pmatrix} \mathbf{a}_1'\mathbf{S}\mathbf{a}_1 & \mathbf{a}_1'\mathbf{S}\mathbf{a}_2 \\ \mathbf{a}_2'\mathbf{S}\mathbf{a}_1 & \mathbf{a}_2'\mathbf{S}\mathbf{a}_2 \end{pmatrix},$$

thus verifying (3.58).

3.9 If the rows of $\mathbf{A}$ are denoted by $\mathbf{a}_i'$, show that

$$\text{tr}(\mathbf{ASA}') = \sum_{i=1}^{k} \mathbf{a}_i'\mathbf{S}\mathbf{a}_i. \tag{3.79}$$

3.10 Use (3.71) to verify (3.78), $\text{cov}(\mathbf{z}) = \mathbf{I}/n$, where $\mathbf{z} = (\boldsymbol{\Sigma}^{1/2})^{-1}(\bar{\mathbf{y}} - \boldsymbol{\mu})$.

3.11 Use the calcium data in Table 3.5:

(a) Calculate **S** using the data matrix **Y** as in (3.27).

(b) Obtain **R** by calculating r_{12}, r_{13}, and r_{23} as in (3.32) and (3.33).

(c) Find **R** using (3.35).

3.12 Use the calcium data in Table 3.5:

(a) Find the generalized sample variance $|\mathbf{S}|$ as in (3.72).

(b) Find the total sample variance tr(**S**) as in (3.73).

3.13 Use the probe word data of Table 3.7:

(a) Find the generalized sample variance $|\mathbf{S}|$ as in (3.72).

(b) Find the total sample variance tr(**S**) as in (3.73).

3.14 For the probe word data in Table 3.7, find **R** using (3.35).

3.15 For the variables in Table 3.5, define $z = 3y_1 - y_2 + 2y_3 = (3, -1, 2)\mathbf{y}$. Find $\bar{z}$ and s_z^2 in two ways:

(a) Evaluate z for each row of Table 3.5 and find $\bar{z}$ and s_z^2 directly from $z_1, z_2, \ldots, z_{10}$ using (3.1) and (3.5).

(b) Use $\bar{z} = \mathbf{a}'\bar{\mathbf{y}}$ and $s_z^2 = \mathbf{a}'\mathbf{S}\mathbf{a}$ as in (3.51) and (3.52).

3.16 For the variables in Table 3.5, define $w = -2y_1 + 3y_2 + y_3$ and define z as in the previous problem. Find r_{zw} in two ways:

(a) Evaluate z and w for each row of Table 3.5 and find r_{zw} from the 10 pairs (z_i, w_i), $i = 1, 2, \ldots, 10$, using (3.10).

(b) Find r_{zw} using (3.54).

3.17 For the variables in Table 3.5, find the correlation between y_1 and $\frac{1}{2}(y_2 + y_3)$.

3.18 Define the following linear combinations for the variables in Table 3.5:

$$\begin{aligned} z_1 &= y_1 + y_2 + y_3 \\ z_2 &= 2y_1 - 3y_2 + 2y_3 \\ z_3 &= -y_1 - 2y_2 - 3y_3 \end{aligned}$$

(a) Find $\bar{\mathbf{z}}$ and $\mathbf{S}_z$ using (3.59) and (3.60).

(b) Find $\mathbf{R}_z$ from $\mathbf{S}_z$ using (3.35).

3.19 For the data in Table 3.8 (Elston and Grizzle 1962), consisting of measurements y_1, y_2, y_3, and y_4 of the ramus bone at four different ages on each of 20 boys, find

Table 3.8 Ramus Bone Length at Four Ages for 20 Boys

	Age			
Individual	8 yr (y_1)	$8\frac{1}{2}$ yr (y_2)	9 yr (y_3)	$9\frac{1}{2}$ yr (y_4)
1	47.8	48.8	49.0	49.7
2	46.4	47.3	47.7	48.4
3	46.3	46.8	47.8	48.5
4	45.1	45.3	46.1	47.2
5	47.6	48.5	48.9	49.3
6	52.5	53.2	53.3	53.7
7	51.2	53.0	54.3	54.5
8	49.8	50.0	50.3	52.7
9	48.1	50.8	52.3	54.4
10	45.0	47.0	47.3	48.3
11	51.2	51.4	51.6	51.9
12	48.5	49.2	53.0	55.5
13	52.1	52.8	53.7	55.0
14	48.2	48.9	49.3	49.8
15	49.6	50.4	51.2	51.8
16	50.7	51.7	52.7	53.3
17	47.2	47.7	48.4	49.5
18	53.3	54.6	55.1	55.3
19	46.2	47.5	48.1	48.4
20	46.3	47.6	51.3	51.8

(a) $\bar{\mathbf{y}}$, $\mathbf{S}$, and $\mathbf{R}$ and

(b) $|\mathbf{S}|$ and tr($\mathbf{S}$).

3.20 For the data in Table 3.8, define $z = y_1 + 2y_2 + y_3 - 3y_4$ and $w = -2y_1 + 3y_2 - y_3 + 2y_4$:

(a) Find $\bar{z}$, $\bar{w}$, s_z^2, and s_w^2 using (3.51) and (3.52).

(b) Find s_{zw} and r_{zw} using (3.53) and (3.54).

3.21 For the data in Table 3.8 define

$$
\begin{aligned}
z_1 &= 2y_1 + 3y_2 - y_3 + 4y_4 \\
z_2 &= -2y_1 - y_2 + 4y_3 - 2y_4 \\
z_3 &= 3y_1 - 2y_2 - y_3 + 3y_4
\end{aligned}
$$

Find $\bar{\mathbf{z}}$, $\mathbf{S}_z$, and $\mathbf{R}_z$, using (3.59), (3.60), and (3.35), respectively.

3.22 The data in Table 3.9 consist of head measurements on first and second sons (Frets 1921). Define y_1 and y_2 as the measurements on the first son and x_1 and x_2 for the second son.

Table 3.9 Measurements on the First and Second Adult Sons in a Sample of 25 families.

First Son		Second Son	
Head Length, y_1	Head Breadth, y_2	Head Length, x_1	Head Breadth, x_2
191	155	179	145
195	149	201	152
181	148	185	149
183	153	188	149
176	144	171	142
208	157	192	152
189	150	190	149
197	159	189	152
188	152	197	159
192	150	187	151
179	158	186	148
183	147	174	147
174	150	185	152
190	159	195	157
188	151	187	158
163	137	161	130
195	155	183	158
186	153	173	148
181	145	182	146
175	140	165	137
192	154	185	152
174	143	178	147
176	139	176	143
197	167	200	158
190	163	187	150

(a) Find the mean vector for all four variables and partition it into $\begin{pmatrix} \bar{\mathbf{y}} \\ \bar{\mathbf{x}} \end{pmatrix}$ as in (3.39).

(b) Find the covariance matrix for all four variables and partition it into

$$\mathbf{S} = \begin{pmatrix} \mathbf{S}_{yy} & \mathbf{S}_{yx} \\ \mathbf{S}_{xy} & \mathbf{S}_{xx} \end{pmatrix}$$

as in (3.40).

3.23 Table 3.10 contains data from O'Sullivan and Mahan (1966; see also Andrews and Herzberg 1985, p. 214) with measurements of blood glucose levels on three occasions for 52 women. The y's represent fasting glucose measurements on the three occasions; the x's are glucose measurements 1 hour after sugar intake. Find the mean vector and covariance matrix for all six variables and partition them into $\begin{pmatrix} \overline{\mathbf{y}} \\ \overline{\mathbf{x}} \end{pmatrix}$, as in (3.39), and

Table 3.10 Blood Glucose Measurements on Three Occasions

Fasting			One Hour After Sugar Intake		
y_1	y_2	y_3	x_1	x_2	x_3
60	69	62	97	69	98
56	53	84	103	78	107
80	69	76	66	99	130
55	80	90	80	85	114
62	75	68	116	130	91
74	64	70	109	101	103
64	71	66	77	102	130
73	70	64	115	110	109
68	67	75	76	85	119
69	82	74	72	133	127
60	67	61	130	134	121
70	74	78	150	158	100
66	74	78	150	131	142
83	70	74	99	98	105
68	66	90	119	85	109
78	63	75	164	98	138
103	77	77	160	117	121
77	68	74	144	71	153
66	77	68	77	82	89
70	70	72	114	93	122
75	65	71	77	70	109
91	74	93	118	115	150
66	75	73	170	147	121
75	82	76	153	132	115
74	71	66	143	105	100

Table 3.10 (*Continued*)

Fasting			One Hour After Sugar Intake		
y_1	y_2	y_3	x_1	x_2	x_3
76	70	64	114	113	129
74	90	86	73	106	116
74	77	80	116	81	77
67	71	69	63	87	70
78	75	80	105	132	80
64	66	71	83	94	133
67	71	69	63	87	70
78	75	80	105	132	80
64	66	71	83	94	133
71	80	76	81	87	86
63	75	73	120	89	59
90	103	74	107	109	101
60	76	61	99	111	98
48	77	75	113	124	97
66	93	97	136	112	122
74	70	76	109	88	105
60	74	71	72	90	71
63	75	66	130	101	90
66	80	86	130	117	144
77	67	74	83	92	107
70	67	100	150	142	146
73	76	81	119	120	119
78	90	77	122	155	149
73	68	80	102	90	122
72	83	68	104	69	96
65	60	70	119	94	89
52	70	76	92	94	100

Note: Measurements are in mg/100 ml.

$$\mathbf{S} = \begin{pmatrix} \mathbf{S}_{yy} & \mathbf{S}_{yx} \\ \mathbf{S}_{xy} & \mathbf{S}_{xx} \end{pmatrix}$$

as in (3.40).

CHAPTER 4

The Multivariate Normal Distribution

4.1 MULTIVARIATE NORMAL DENSITY FUNCTION

Many univariate tests and confidence intervals are based on the univariate normal distribution. Similarly, the vast majority of multivariate procedures have as their underpinning the multivariate normal distribution.

The following are some of the useful features of the multivariate normal distribution: (1) only means, variances, and covariances need be estimated in order to completely describe the distribution; (2) bivariate plots show linear trends; (3) if the variables are uncorrelated, they are independent; (4) linear functions of multivariate normal variables are also normal; (5) as in the univariate case, the convenient form of the density function lends itself to derivation of many properties and test statistics; and (6) even when the data are not multivariate normal, the multivariable normal may serve as a useful approximation, especially in inferences involving sample mean vectors, which are approximately normal by the central limit theorem (see Section 4.3.2).

Since the multivariate normal density is an extension of the univariate normal density and shares many of its features, we review the univariate normal density function in Section 4.1.1. We then describe the multivariate normal density in Sections 4.1.2–4.1.4.

4.1.1 Univariate Normal Density

If a random variable y, with mean μ and variance σ^2, is normally distributed, its density is given by

$$f(y) = \frac{1}{\sqrt{2\pi}\sqrt{\sigma^2}} e^{-(y-\mu)^2/2\sigma^2} \qquad -\infty < y < \infty. \tag{4.1}$$

When y has the density (4.1), we say that y is distributed as $N(\mu, \sigma^2)$, or simply

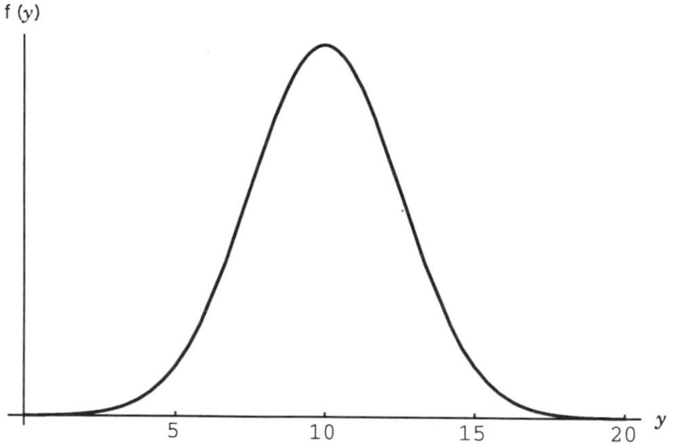

Figure 4.1 The normal density curve.

y is $N(\mu, \sigma^2)$. This function is represented by the familiar bell-shaped curve illustrated in Figure 4.1 for $\mu = 10$ and $\sigma = 2.5$.

4.1.2 Multivariate Normal Density

If **y** has a multivariate normal distribution with mean vector $\boldsymbol{\mu}$ and covariance matrix $\boldsymbol{\Sigma}$, the density is given by

$$g(\mathbf{y}) = \frac{1}{(\sqrt{2\pi})^p |\boldsymbol{\Sigma}|^{1/2}} e^{-(\mathbf{y}-\boldsymbol{\mu})'\boldsymbol{\Sigma}^{-1}(\mathbf{y}-\boldsymbol{\mu})/2}, \tag{4.2}$$

where p is the number of variables. When **y** has the density (4.2), we say that **y** is distributed as $N_p(\boldsymbol{\mu}, \boldsymbol{\Sigma})$, or simply **y** is $N_p(\boldsymbol{\mu}, \boldsymbol{\Sigma})$.

The term $(y - \mu)^2/\sigma^2 = (y - \mu)(\sigma^2)^{-1}(y - \mu)$ in the exponent of the univariate normal density (4.1) measures the squared distance from y to μ in standard deviation units. Similarly, the term $(\mathbf{y} - \boldsymbol{\mu})'\boldsymbol{\Sigma}^{-1}(\mathbf{y} - \boldsymbol{\mu})$ in the exponent of the multivariate normal density (4.2) is the squared generalized distance from **y** to $\boldsymbol{\mu}$, or the Mahalanobis distance,

$$\Delta^2 = (\mathbf{y} - \boldsymbol{\mu})'\boldsymbol{\Sigma}^{-1}(\mathbf{y} - \boldsymbol{\mu}). \tag{4.3}$$

The characteristics of this distance between **y** and $\boldsymbol{\mu}$ were discussed in Section 3.12.

In the coefficient of the exponential function in (4.2), $|\boldsymbol{\Sigma}|^{1/2}$ appears as the analogue of $\sqrt{\sigma^2}$ in (4.1). In the next section, we discuss the effect of $|\boldsymbol{\Sigma}|$ on the density.

4.1.3 Generalized Population Variance

In Section 3.10, we referred to $|\mathbf{S}|$ as a generalized sample variance. Analogously, $|\boldsymbol{\Sigma}|$ is a *generalized population variance.* If σ^2 is small in the univariate normal, the y values are concentrated near the mean. Similarly, a small value of $|\boldsymbol{\Sigma}|$ in the multivariate case indicates that the $\mathbf{y}$'s are concentrated close to $\boldsymbol{\mu}$ in p-space or that there is multicollinearity among the variables. The term *multicollinearity* indicates that the variables are highly intercorrelated, in which case the effective dimensionality is less than p. (See Chapter 12 for a discussion of finding a reduced number of new dimensions that represent the data.) In the presence of multicollinearity, one or more eigenvalues of $\boldsymbol{\Sigma}$ will be near zero and $|\boldsymbol{\Sigma}|$ will be small, since $|\boldsymbol{\Sigma}|$ is the product of the eigenvalues, by (2.99).

Figure 4.2 shows, for the bivariate case, a comparison of a distribution with small $|\boldsymbol{\Sigma}|$ and a distribution with larger $|\boldsymbol{\Sigma}|$. An alternative way to portray the concentration of points in the bivariate normal distribution is with contour plots. Figure 4.3 shows contour plots for the two distributions in Figure 4.2. Each ellipse contains a different proportion of observation vectors $\mathbf{y}$. The contours in Figure 4.3 can be found by setting the density function equal to a constant and solving for $\mathbf{y}$, as illustrated in Figure 4.4. The bivariate normal density surface sliced at a constant height traces an ellipse, which contains a given proportion of the observations.

In both Figures 4.2 and 4.3, small $|\boldsymbol{\Sigma}|$ appears on the left and large $|\boldsymbol{\Sigma}|$ appears on the right. In Figure 4.3*a*, there is a larger correlation between y_1 and y_2. In Figure 4.3*b*, the variances are larger (in the natural directions). In general, for any number of variables p, a decrease in intercorrelations among the variables or an increase in the variances will lead to a larger $|\boldsymbol{\Sigma}|$.

4.1.4 Diversity of Applications of the Multivariate Normal

Nearly all the inferential procedures we discuss in this book are based on the multivariate normal distribution. We acknowledge that a major motivation for the widespread use of the multivariate normal is its mathematical tractability. From the multivariate normal assumption, a host of useful procedures can be derived. Practical alternatives to the multivariate normal are fewer than in the

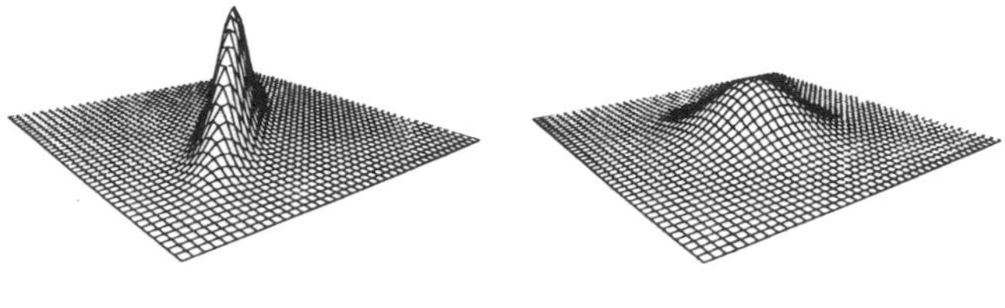

Figure 4.2 Bivariate normal densities.

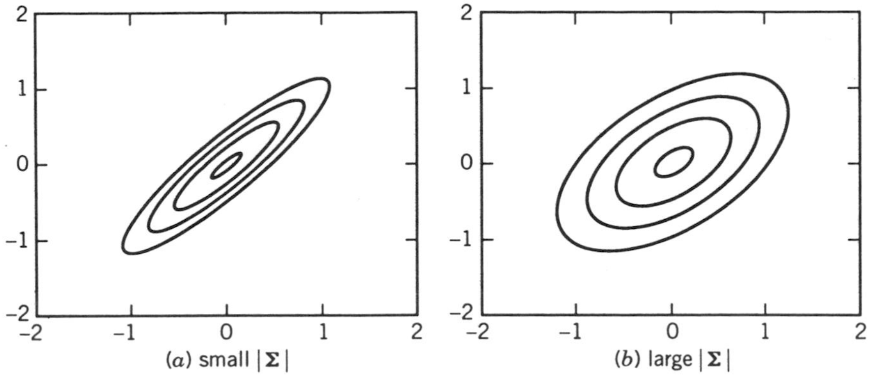

Figure 4.3 Contour plots for the distributions in Figure 4.2.

univariate case. Because it is not as simple to order (or rank) multivariate observation vectors as it is univariate observations, not as many nonparametric procedures are available for multivariate data.

While real data may not often be exactly multivariate normal, the multivariate normal will frequently serve as a useful approximation to the true distribution. Other reasons for our focus on the multivariate normal are the availability of tests and graphical procedures for assessing normality (see Sections 4.4 and 4.5) and the widespread use of procedures based on the multivariate normal in software packages. Fortunately, many of the procedures based on multivariate normality are robust to departures from normality.

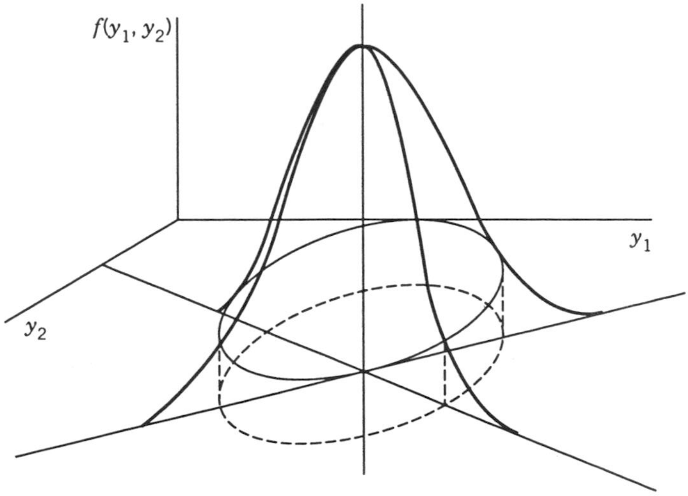

Figure 4.4 Constant density contour for bivariate normal.

4.2 PROPERTIES OF MULTIVARIATE NORMAL RANDOM VARIABLES

We list some of the properties of a random $p \times 1$ vector $\mathbf{y}$ from a multivariate normal distribution $N_p(\boldsymbol{\mu}, \boldsymbol{\Sigma})$:

1. Normality of linear combinations of the variables in $\mathbf{y}$
   a. If $\mathbf{a}$ is a vector of constants, the linear function $\mathbf{a}'\mathbf{y} = a_1y_1 + a_2y_2 + \cdots + a_py_p$ is distributed as $N(\mathbf{a}'\boldsymbol{\mu}, \mathbf{a}'\boldsymbol{\Sigma}\mathbf{a})$. The mean and variance of $\mathbf{a}'\mathbf{y}$ were given previously in (3.64) and (3.65) as $E(\mathbf{a}'\mathbf{y}) = \mathbf{a}'\boldsymbol{\mu}$ and $\text{var}(\mathbf{a}'\mathbf{y}) = \mathbf{a}'\boldsymbol{\Sigma}\mathbf{a}$ for any random vector $\mathbf{y}$. We now have the additional attribute that $\mathbf{a}'\mathbf{y}$ has a (univariate) normal distribution if $\mathbf{y}$ is $N_p(\boldsymbol{\mu}, \boldsymbol{\Sigma})$. This is a fundamental result, since we will often deal with linear combinations.
   b. If $\mathbf{A}$ is a constant $q \times p$ matrix of rank q, where $q \leq p$, then $\mathbf{Ay}$ consists of q linear combinations of the variables in $\mathbf{y}$, with distribution $N_q(\mathbf{A}\boldsymbol{\mu}, \mathbf{A}\boldsymbol{\Sigma}\mathbf{A}')$. Here, again, $E(\mathbf{Ay}) = \mathbf{A}\boldsymbol{\mu}$ and $\text{cov}(\mathbf{Ay}) = \mathbf{A}\boldsymbol{\Sigma}\mathbf{A}'$, in general, as given in (3.68) and (3.69). But we now have the additional feature that the q variables in $\mathbf{Ay}$ have a multivariate normal distribution.
2. Standardized variables
   If $\mathbf{y}$ is $N_p(\boldsymbol{\mu}, \boldsymbol{\Sigma})$, a *standardized vector* $\mathbf{z}$ can be obtained in two ways:

$$\mathbf{z} = (\mathbf{T}')^{-1}(\mathbf{y} - \boldsymbol{\mu}), \tag{4.4}$$

   where $\boldsymbol{\Sigma} = \mathbf{T}'\mathbf{T}$ is factored using the Cholesky procedure in Section 2.7, or

$$\mathbf{z} = (\boldsymbol{\Sigma}^{1/2})^{-1}(\mathbf{y} - \boldsymbol{\mu}), \tag{4.5}$$

   where $\boldsymbol{\Sigma}^{1/2}$ is the symmetric square root matrix of $\boldsymbol{\Sigma}$ defined in (2.103) such that $\boldsymbol{\Sigma} = \boldsymbol{\Sigma}^{1/2}\boldsymbol{\Sigma}^{1/2}$. In either (4.4) or (4.5), it follows from property 1b that $\mathbf{z}$ is distributed as $N_p(\mathbf{0}, \mathbf{I})$; that is, the z's are independently distributed as $N(0, 1)$. Thus in the multivariate case, a standardized vector of random variables has all means equal to 0, all variances equal to 1, *and* all correlations equal to 0.
3. Chi-square distribution
   A *chi-square random variable* with p degrees of freedom is defined as the sum of squares of p independent standard normal random variables. Thus if $\mathbf{z}$ is the standardized vector defined in (4.4) or (4.5), then $\sum_{i=1}^{p} z_i^2 = \mathbf{z}'\mathbf{z}$ has the χ^2-distribution with p degrees of freedom, denoted χ^2_p or $\chi^2(p)$. From either (4.4) or (4.5) we obtain $\mathbf{z}'\mathbf{z} = (\mathbf{y} - \boldsymbol{\mu})'\boldsymbol{\Sigma}^{-1}(\mathbf{y} - \boldsymbol{\mu})$. Hence

$$\text{If } \mathbf{y} \text{ is } N_p(\boldsymbol{\mu}, \boldsymbol{\Sigma}), \text{ then } (\mathbf{y} - \boldsymbol{\mu})'\boldsymbol{\Sigma}^{-1}(\mathbf{y} - \boldsymbol{\mu}) \text{ is } \chi^2_p. \tag{4.6}$$

4. Normality of marginal distributions

a. Any subset of the y's in $\mathbf{y}$ has a multivariate normal distribution, with mean vector consisting of the corresponding subvector of $\boldsymbol{\mu}$ and covariance matrix composed of the corresponding submatrix of $\boldsymbol{\Sigma}$. To illustrate, let $\mathbf{y}_1' = (y_1, y_2, \ldots, y_r)$ denote the subvector containing the first r elements of $\mathbf{y}$ and $\mathbf{y}_2' = (y_{r+1}, \ldots, y_p)$ consist of the remaining $p - r$ elements. Thus $\mathbf{y}$, $\boldsymbol{\mu}$, and $\boldsymbol{\Sigma}$ are partitioned as

$$\mathbf{y} = \begin{pmatrix} \mathbf{y}_1 \\ \mathbf{y}_2 \end{pmatrix} \qquad \boldsymbol{\mu} = \begin{pmatrix} \boldsymbol{\mu}_1 \\ \boldsymbol{\mu}_2 \end{pmatrix} \qquad \boldsymbol{\Sigma} = \begin{pmatrix} \boldsymbol{\Sigma}_{11} & \boldsymbol{\Sigma}_{12} \\ \boldsymbol{\Sigma}_{21} & \boldsymbol{\Sigma}_{22} \end{pmatrix},$$

where $\mathbf{y}_1$ and $\boldsymbol{\mu}_1$ are $r \times 1$ and $\boldsymbol{\Sigma}_{11}$ is $r \times r$. Then $\mathbf{y}_1$ is distributed as $N_r(\boldsymbol{\mu}_1, \boldsymbol{\Sigma}_{11})$. Here, again, $E(\mathbf{y}_1) = \boldsymbol{\mu}_1$ and $\text{cov}(\mathbf{y}_1) = \boldsymbol{\Sigma}_{11}$ hold for any random vector partitioned in this way. But if $\mathbf{y}$ is p-variate normal, then $\mathbf{y}_1$ is r-variate normal.

b. As a special case of the above result, each y_i in $\mathbf{y}$ has the univariate normal distribution $N(\mu_i, \sigma_{ii})$. The converse of this is not true. If the density of each y_i in $\mathbf{y}$ is normal, it does not necessarily follow that $\mathbf{y}$ is multivariate normal.

In the next three properties, let the observation vector be partitioned into two subvectors denoted by $\mathbf{y}$ and $\mathbf{x}$. Or, alternatively, let $\mathbf{x}$ represent some additional variables to be considered along with those in $\mathbf{y}$. Then, as in (3.43) and (3.44),

$$E\begin{pmatrix} \mathbf{y} \\ \mathbf{x} \end{pmatrix} = \begin{pmatrix} \boldsymbol{\mu}_y \\ \boldsymbol{\mu}_x \end{pmatrix} \qquad \text{cov}\begin{pmatrix} \mathbf{y} \\ \mathbf{x} \end{pmatrix} = \begin{pmatrix} \boldsymbol{\Sigma}_{yy} & \boldsymbol{\Sigma}_{yx} \\ \boldsymbol{\Sigma}_{xy} & \boldsymbol{\Sigma}_{xx} \end{pmatrix}.$$

5. Independence
   a. The subvectors $\mathbf{y}$ and $\mathbf{x}$ are independent if $\boldsymbol{\Sigma}_{yx} = \mathbf{O}$.
   b. Two individual variables y_i and y_j are independent if $\sigma_{ij} = 0$.
   Note that this is not true for many nonnormal random variables, as illustrated in Section 3.2.1.
6. Conditional distribution
   If $\mathbf{y}$ and $\mathbf{x}$ are not independent, then $\boldsymbol{\Sigma}_{yx} \neq \mathbf{O}$, and the conditional distribution of $\mathbf{y}$ given $\mathbf{x}$, $f(\mathbf{y}|\mathbf{x})$, is multivariate normal with

$$E(\mathbf{y}|\mathbf{x}) = \boldsymbol{\mu}_y + \boldsymbol{\Sigma}_{yx}\boldsymbol{\Sigma}_{xx}^{-1}(\mathbf{x} - \boldsymbol{\mu}_x) \tag{4.7}$$

and

$$\text{cov}(\mathbf{y}|\mathbf{x}) = \boldsymbol{\Sigma}_{yy} - \boldsymbol{\Sigma}_{yx}\boldsymbol{\Sigma}_{xx}^{-1}\boldsymbol{\Sigma}_{xy}. \tag{4.8}$$

Note that $E(\mathbf{y}|\mathbf{x})$ is a linear function of $\mathbf{x}$, while $\text{cov}(\mathbf{y}|\mathbf{x})$ does not depend on $\mathbf{x}$. The linear trend in (4.7) extends to any pair of variables. Thus to use (4.7) as a check on normality, one can examine bivariate scatter plots of all pairs of variables and look for any nonlinear trends. In (4.7), we have the justification for using the covariance or correlation to measure the relationship between two bivariate normal random variables. As noted in Section 3.2.1, the covariance and correlation are good measures of relationship only for variables with linear trends and are generally unsuitable for nonnormal random variables with a curvilinear relationship. The matrix $\boldsymbol{\Sigma}_{yx}\boldsymbol{\Sigma}_{xx}^{-1}$ in (4.7) is called the *matrix of regression coefficients* because it relates $E(\mathbf{y}|\mathbf{x})$ to $\mathbf{x}$. The sample counterpart of this matrix appears in (10.37).

7. Distribution of the sum of two subvectors
If $\mathbf{y}$ and $\mathbf{x}$ are the same size (both $p \times 1$) and independent, then

$$\mathbf{y} + \mathbf{x} \quad \text{is} \quad N_p(\boldsymbol{\mu}_y + \boldsymbol{\mu}_x, \boldsymbol{\Sigma}_{yy} + \boldsymbol{\Sigma}_{xx}) \tag{4.9}$$

$$\mathbf{y} - \mathbf{x} \quad \text{is} \quad N_p(\boldsymbol{\mu}_y - \boldsymbol{\mu}_x, \boldsymbol{\Sigma}_{yy} + \boldsymbol{\Sigma}_{xx}). \tag{4.10}$$

Here, again, the mean vector and covariance matrix for $\mathbf{y} \pm \mathbf{x}$ hold in general. But if $\mathbf{y}$ and $\mathbf{x}$ are multivariate normal, then $\mathbf{y} \pm \mathbf{x}$ is multivariate normal.

To illustrate property 6, we discuss the conditional distribution for the bivariate normal. Let

$$\mathbf{u} = \begin{pmatrix} y \\ x \end{pmatrix}$$

where

$$E(\mathbf{u}) = \begin{pmatrix} \mu_y \\ \mu_x \end{pmatrix} \qquad \text{cov}(\mathbf{u}) = \boldsymbol{\Sigma} = \begin{pmatrix} \sigma_y^2 & \sigma_{yx} \\ \sigma_{yx} & \sigma_x^2 \end{pmatrix}.$$

By definition $f(y|x) = g(y,x)/h(x)$, where $g(y,x)$ is the joint density of y and x and $h(x)$ is the density of x. Hence

$$g(y,x) = f(y|x)h(x)$$

and because the right side is a product, we seek a function of y and x that is independent of x and whose density can serve as $f(y|x)$. Since linear functions of y and x are normal by property 1a above, we consider $y - \beta x$ and seek the value of β so that $y - \beta x$ and x are independent.

Since $z = y - \beta x$ and x are normal and independent, $\text{cov}(x, z) = 0$. To find cov(x, z), we express x and z as functions of $\mathbf{u}$,

$$x = (0, 1)\begin{pmatrix} y \\ x \end{pmatrix} = (0, 1)\mathbf{u} = \mathbf{a}'\mathbf{u}$$

$$z = y - \beta x = (1, -\beta)\mathbf{u} = \mathbf{b}'\mathbf{u}.$$

Now

$$\begin{aligned} \operatorname{cov}(x, z) &= \operatorname{cov}(\mathbf{a}'\mathbf{u}, \mathbf{b}'\mathbf{u}) \\ &= \mathbf{a}'\mathbf{\Sigma}\mathbf{b} \quad \text{[by (3.66)]} \\ &= (0, 1)\begin{pmatrix} \sigma_y^2 & \sigma_{yx} \\ \sigma_{yx} & \sigma_x^2 \end{pmatrix}\begin{pmatrix} 1 \\ -\beta \end{pmatrix} = (\sigma_{yx}, \sigma_x^2)\begin{pmatrix} 1 \\ -\beta \end{pmatrix} \\ &= \sigma_{yx} - \beta\sigma_x^2. \end{aligned}$$

Since $\operatorname{cov}(x, z) = 0$, we obtain $\beta = \sigma_{yx}/\sigma_x^2$ and $y - \beta x$ becomes

$$y - \frac{\sigma_{yx}}{\sigma_x^2} x.$$

By property 1a above, the density of $y - (\sigma_{yx}/\sigma_x^2)x$ is normal with

$$E\left(y - \frac{\sigma_{yx}}{\sigma_x^2} x \right) = \mu_y - \frac{\sigma_{yx}}{\sigma_x^2}\mu_x$$

and

$$\begin{aligned} \operatorname{var}\left(y - \frac{\sigma_{yx}}{\sigma_x^2} x \right) &= \operatorname{var}(\mathbf{b}'\mathbf{u}) = \mathbf{b}'\mathbf{\Sigma}\mathbf{b} \\ &= \left(1, -\frac{\sigma_{yx}}{\sigma_x^2}\right)\begin{pmatrix} \sigma_y^2 & \sigma_{yx} \\ \sigma_{yx} & \sigma_x^2 \end{pmatrix}\begin{pmatrix} 1 \\ -\dfrac{\sigma_{yx}}{\sigma_x^2} \end{pmatrix} \\ &= \sigma_y^2 - \frac{\sigma_{yx}}{\sigma_x^2}. \end{aligned}$$

For a given value of x, y can be expressed as $y = \beta x + (y - \beta x)$, where βx is a fixed quantity corresponding to the given value of x and $y - \beta x$ is a random deviation. Then $f(y|x)$ is normal, with

$$E(y|x) = \beta x + E(y - \beta x) = \beta x + \mu_y - \beta\mu_x$$
$$= \mu_y + \beta(x - \mu_x) = \mu_y + \frac{\sigma_{yx}}{\sigma_x^2}(x - \mu_x)$$
$$\text{var}(y|x) = \sigma_y^2 - \frac{\sigma_{yx}^2}{\sigma_x^2}.$$

4.3 ESTIMATION IN THE MULTIVARIATE NORMAL

4.3.1 Maximum Likelihood Estimation

When a distribution such as the multivariate normal is assumed to hold for a population, estimates of the parameters are often found by the method of *maximum likelihood*. This technique is conceptually simple: The observation vectors $\mathbf{y}_1, \mathbf{y}_2, \ldots, \mathbf{y}_n$ are considered to be known and values of $\boldsymbol{\mu}$ and $\boldsymbol{\Sigma}$ are sought that maximize the joint density of the **y**'s, called the *likelihood function*. For the multivariate normal, the maximum likelihood estimates of $\boldsymbol{\mu}$ and $\boldsymbol{\Sigma}$ are

$$\hat{\boldsymbol{\mu}} = \overline{\mathbf{y}} \tag{4.11}$$

and

$$\begin{aligned}\hat{\boldsymbol{\Sigma}} &= \frac{1}{n}\sum_{i=1}^{n}(\mathbf{y}_i - \overline{\mathbf{y}})(\mathbf{y}_i - \overline{\mathbf{y}})' \\ &= \frac{1}{n}\mathbf{W} \quad \text{say} \\ &= \frac{n-1}{n}\mathbf{S},\end{aligned} \tag{4.12}$$

where **S** is the sample covariance matrix defined in (3.20) and (3.25). Since $\hat{\boldsymbol{\Sigma}}$ has divisor n instead of $n-1$, it is biased [see (3.31)], and we usually use **S** in place of $\hat{\boldsymbol{\Sigma}}$.

We now give a justification of $\overline{\mathbf{y}}$ as the maximum likelihood estimator of $\boldsymbol{\mu}$. Because the $\mathbf{y}_i$'s constitute a random sample, they are independent, and the joint density is the product of the densities of the **y**'s. The likelihood function is, therefore,

$$
\begin{aligned}
L(\mathbf{y}_1, \mathbf{y}_2, \ldots, \mathbf{y}_n, \boldsymbol{\mu}, \boldsymbol{\Sigma}) &= \prod_{i=1}^{n} f(\mathbf{y}_i, \boldsymbol{\mu}, \boldsymbol{\Sigma}) \\
&= \prod_{i=1}^{n} \frac{1}{(\sqrt{2\pi})^p |\boldsymbol{\Sigma}|^{1/2}} e^{-(\mathbf{y}_i-\boldsymbol{\mu})'\boldsymbol{\Sigma}^{-1}(\mathbf{y}_i-\boldsymbol{\mu})/2} \\
&= \frac{1}{(\sqrt{2\pi})^{np} |\boldsymbol{\Sigma}|^{n/2}} e^{-\sum_{i=1}^{n}(\mathbf{y}_i-\boldsymbol{\mu})'\boldsymbol{\Sigma}^{-1}(\mathbf{y}_i-\boldsymbol{\mu})/2}. \qquad (4.13)
\end{aligned}
$$

To see that $\hat{\boldsymbol{\mu}} = \overline{\mathbf{y}}$ maximizes the likelihood function, we write the exponent of (4.13) in a different form. By adding and subtracting $\overline{\mathbf{y}}$, the exponent in (4.13) becomes

$$
-\frac{1}{2} \sum_{i=1}^{n} (\mathbf{y}_i - \overline{\mathbf{y}} + \overline{\mathbf{y}} - \boldsymbol{\mu})'\boldsymbol{\Sigma}^{-1}(\mathbf{y}_i - \overline{\mathbf{y}} + \overline{\mathbf{y}} - \boldsymbol{\mu}).
$$

When this is expanded in terms of $\mathbf{y}_i - \overline{\mathbf{y}}$ and $\overline{\mathbf{y}} - \boldsymbol{\mu}$, two of the four resulting terms vanish because $\sum_i(\mathbf{y}_i - \overline{\mathbf{y}}) = \mathbf{0}$, and (4.13) becomes

$$
L = \frac{1}{(\sqrt{2\pi})^{np} |\boldsymbol{\Sigma}|^{n/2}} e^{-\sum_{i=1}^{n}(\mathbf{y}_i-\overline{\mathbf{y}})'\boldsymbol{\Sigma}^{-1}(\mathbf{y}_i-\overline{\mathbf{y}})/2 - n(\overline{\mathbf{y}}-\boldsymbol{\mu})'\boldsymbol{\Sigma}^{-1}(\overline{\mathbf{y}}-\boldsymbol{\mu})/2}. \qquad (4.14)
$$

Since $\boldsymbol{\Sigma}^{-1}$ is positive definite, $-n(\overline{\mathbf{y}} - \boldsymbol{\mu})'\boldsymbol{\Sigma}^{-1}(\overline{\mathbf{y}} - \boldsymbol{\mu})/2 \leq 0$ and $0 < e^{-n(\overline{\mathbf{y}}-\boldsymbol{\mu})'\boldsymbol{\Sigma}^{-1}(\overline{\mathbf{y}}-\boldsymbol{\mu})} \leq 1$, with the maximum occurring when the exponent is 0. Therefore, L is maximized when $\hat{\boldsymbol{\mu}} = \overline{\mathbf{y}}$.

The maximum likelihood estimator of the population correlation matrix is the sample correlation matrix, that is,

$$
\hat{\mathbf{P}}_\rho = \mathbf{R}.
$$

Relationships among multinormal variables are linear, as can be seen in (4.7). Thus the estimators **S** and **R** serve well for the multivariate normal because they measure only linear relationships (see Sections 3.2.1 and 4.2). They are not as useful for some nonnormal distributions.

4.3.2 Distribution of $\overline{\mathbf{y}}$ and S

For the distribution of $\overline{\mathbf{y}} = \sum_{i=1}^{n} \mathbf{y}_i/n$, we can distinguish two cases.

(a) When $\overline{\mathbf{y}}$ is based on a random sample $\mathbf{y}_1, \mathbf{y}_2, \ldots, \mathbf{y}_n$ from a multivariate normal distribution $N_p(\boldsymbol{\mu}, \boldsymbol{\Sigma})$, $\overline{\mathbf{y}}$ is $N_p(\boldsymbol{\mu}, \boldsymbol{\Sigma}/n)$.

(b) When $\overline{\mathbf{y}}$ is based on a random sample $\mathbf{y}_1, \mathbf{y}_2, \ldots, \mathbf{y}_n$ from a nonnormal multivariate population with mean vector $\boldsymbol{\mu}$ and covariance matrix $\boldsymbol{\Sigma}$, for large n, $\overline{\mathbf{y}}$ is approximately $N_p(\boldsymbol{\mu}, \boldsymbol{\Sigma}/n)$. More formally, this result is known as the *multivariate central limit theorem*: If $\overline{\mathbf{y}}$ is the mean vector of a random sample $\mathbf{y}_1, \mathbf{y}_2, \ldots, \mathbf{y}_n$ from a population with mean vector $\boldsymbol{\mu}$ and covariance matrix $\boldsymbol{\Sigma}$, then as $n \to \infty$, the distribution of $\sqrt{n}(\overline{\mathbf{y}} - \boldsymbol{\mu})$ approaches $N_p(\mathbf{0}, \boldsymbol{\Sigma})$.

There are p variances in $\mathbf{S}$ and $\binom{p}{2}$ covariances, for a total of

$$p + \binom{p}{2} = p + p(p - 1)/2 = p(p + 1)/2$$

distinct entries. The joint distribution of these $p(p+1)/2$ distinct variables in $\mathbf{W} = (n - 1)\mathbf{S} = \sum_i (\mathbf{y}_i - \overline{\mathbf{y}})(\mathbf{y}_i - \overline{\mathbf{y}})'$ is the Wishart distribution, denoted by $W_p(n - 1, \boldsymbol{\Sigma})$, where $n - 1$ is the degrees of freedom.

The Wishart distribution is the multivariate analogue of the χ^2-distribution, and it has similar uses. As noted in property 3 of Section 4.2, a χ^2 random variable is defined formally as the sum of squares of independent standard normal (univariate) random variables:

$$\sum_{i=1}^{n} z_i^2 = \sum_{i=1}^{n} \frac{(y_i - \mu)^2}{\sigma^2} \quad \text{is} \quad \chi_n^2.$$

If $\overline{y}$ is substituted for μ, then $\sum_i (y_i - \overline{y})^2/\sigma^2 = (n - 1)s^2/\sigma^2$ is χ^2_{n-1}. Similarly, the formal definition of a Wishart random variable is

$$\sum_{i=1}^{n} (\mathbf{y}_i - \boldsymbol{\mu})(\mathbf{y}_i - \boldsymbol{\mu})' \quad \text{is} \quad W_p(n, \boldsymbol{\Sigma}), \tag{4.15}$$

where $\mathbf{y}_1, \mathbf{y}_2, \ldots, \mathbf{y}_n$ are independently distributed as $N_p(\boldsymbol{\mu}, \boldsymbol{\Sigma})$. When $\overline{\mathbf{y}}$ is substituted for $\boldsymbol{\mu}$, the distribution remains Wishart with one less degree of freedom:

$$(n - 1)\mathbf{S} = \sum_{i=1}^{n} (\mathbf{y}_i - \overline{\mathbf{y}})(\mathbf{y}_i - \overline{\mathbf{y}})' \quad \text{is} \quad W_p(n - 1, \boldsymbol{\Sigma}). \tag{4.16}$$

Finally, we note that when sampling from a multivariate normal distribution, $\overline{\mathbf{y}}$ and $\mathbf{S}$ are independent.

4.4 ASSESSING MULTIVARIATE NORMALITY

Many tests and graphical procedures have been suggested for evaluating whether a data set likely originated from a multivariate normal population. One possibility is to check each variable separately for univariate normality. Excellent reviews for both the univariate and multivariate cases have been given by Gnanadesikan (1977, pp. 161–195) and Seber (1984, pp. 141–155). We give a representative sample of univariate and multivariate methods in Sections 4.4.1 and 4.4.2, respectively.

4.4.1 Investigating Univariate Normality

When we have several variables, checking each for univariate normality should not be the sole approach, because (1) the variables are correlated and (2) normality of the individual variables does not guarantee joint normality. On the other hand, it is true that multivariate normality implies individual normality. Hence if even one of the separate variables is not normal, the vector is not multivariate normal. An initial check on the individual variables may therefore be useful.

A basic graphical approach for checking normality is the *Q–Q* plot that compares quantiles of a sample against the population quantiles of the univariate normal. If the points are close to a straight line, there is no indication of departure from normality. Deviation from a straight line indicates nonnormality (at least for a large sample). In fact, the type of nonlinear pattern may reveal the type of departure from normality. Some possibilities are illustrated in Figure 4.5.

Quantiles are similar to the more familiar percentiles, which are expressed in terms of percent; a test score at the 90th percentile, for example, is above 90% of the test scores and below 10% of them. Quantiles are expressed in terms of fractions or proportions. Thus the 90th percentile score becomes the 0.9 quantile score.

The sample quantiles for the *Q–Q* plot are obtained as follows. First we rank the observations $y_1, y_2, \ldots, y_n$ and denote the ordered values by $y_{(1)}, y_{(2)}, \ldots, y_{(n)}$; thus $y_{(1)} \leq y_{(2)} \leq \cdots \leq y_{(n)}$. Then the point $y_{(i)}$ is the i/n sample quantile. For example, if $n = 20$, $y_{(7)}$ is the $\frac{7}{20} = .35$ quantile, because .35 of the sample is less than or equal to $y_{(7)}$. The fraction i/n is often changed to $\left(i - \frac{1}{2}\right)/n$ as a continuity correction. If $n = 20$, $\left(i - \frac{1}{2}\right)/n$ ranges from .025 to .975 and more evenly covers the interval from 0 to 1. With this convention, $y_{(i)}$ is designated as the $\left(i - \frac{1}{2}\right)/n$ sample quantile.

The population quantiles for the *Q–Q* plot are similarly defined corresponding to $\left(i - \frac{1}{2}\right)/n$. If we denote these by $q_1, q_2, \ldots, q_n$, then q_i is the value below which a proportion $\left(i - \frac{1}{2}\right)/n$ of the observations in the population lie, that is, $\left(i - \frac{1}{2}\right)/n$ is the probability of getting an observation less than or equal to q_i. Formally, q_i can be found for the standard normal random variable y with dis-

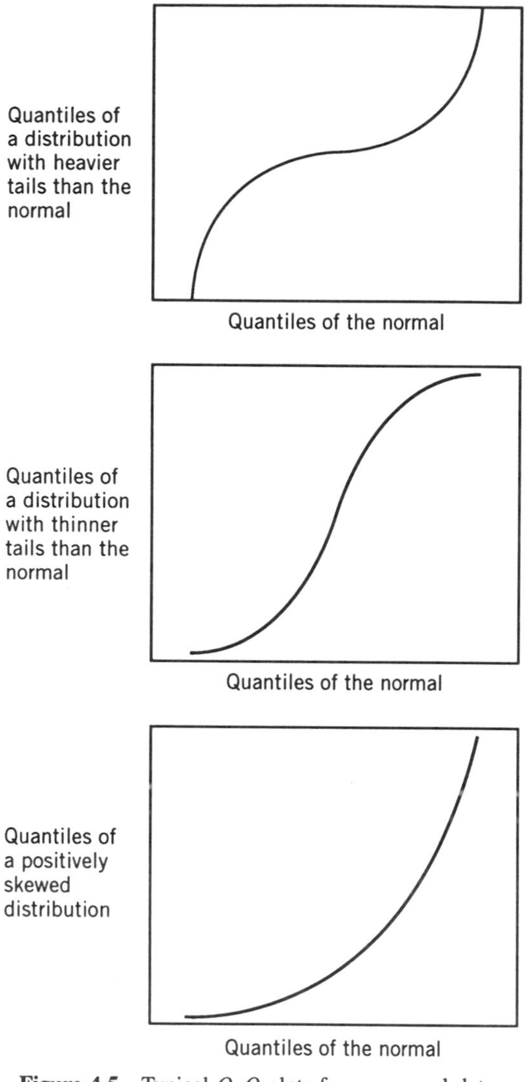

Figure 4.5 Typical *Q–Q* plots for nonnormal data.

tribution $N(0, 1)$ by solving

$$\Phi(q_i) = P(y < q_i) = \frac{i - \frac{1}{2}}{n}, \tag{4.17}$$

which would require numerical integration or tables of the cumulative standard normal distribution, $\Phi(x)$. Another benefit of using $\left(i - \frac{1}{2}\right)/n$ instead of i/n is that $n/n = 1$ would make $q_n = \infty$.

The population need not have the same mean and variance as the sample, since changes in mean and variance merely change the slope and intercept of the plotted line in the *Q–Q* plot. Therefore, we use the standard normal distribution, and the q_i values can easily be found from a table of cumulative standard normal probabilities. We then plot the pairs $(q_i, y_{(i)})$ and examine the resulting *Q–Q* plot for linearity.

Special graph paper, called normal probability paper, is available that eliminates the need to look up the q_i values. We need only plot $\left(i - \frac{1}{2}\right)/n$ in place of q_i, that is, plot the pairs $\left[\left(i - \frac{1}{2}\right)/n, y_{(i)}\right]$ and look for linearity as before. As an even easier alternative, most general-purpose statistical software programs now routinely provide normal probability plots of the pairs $(q_i, y_{(i)})$.

The *Q–Q* plots provide a good visual check on normality and are considered to be adequate for this purpose by many researchers. For those who desire a more objective procedure, several hypothesis tests are available. We give three of these that have good properties and are computationally tractable.

We discuss first a classical approach based on the following measures of skewness and kurtosis:

$$\sqrt{b_1} = \frac{\sqrt{n}\sum_{i=1}^{n}(y_i - \overline{y})^3}{\left[\sum_{i=1}^{n}(y_i - \overline{y})^2\right]^{3/2}} \tag{4.18}$$

and

$$b_2 = \frac{n\sum_{i=1}^{n}(y_i - \overline{y})^4}{\left[\sum_{i=1}^{n}(y_i - \overline{y})^2\right]^{2}}. \tag{4.19}$$

These are sample estimates of the population skewness and kurtosis parameters $\sqrt{\beta_1}$ and β_2, respectively. When the population is normal, $\sqrt{\beta_1} = 0$ and $\beta_2 = 3$. If $\sqrt{\beta_1} < 0$, we have negative skewness; if $\sqrt{\beta_1} > 0$, the skewness is positive. Positive skewness is illustrated in Figure 4.6. If $\beta_2 < 3$, we have negative kurtosis, and if $\beta_2 > 3$, there is positive kurtosis. A distribution with negative kurtosis is characterized by being flatter than the normal distribution, that is, less peaked, with heavier flanks and thinner tails. A distribution with positive kurtosis has a higher peak than the normal, with an excess of values

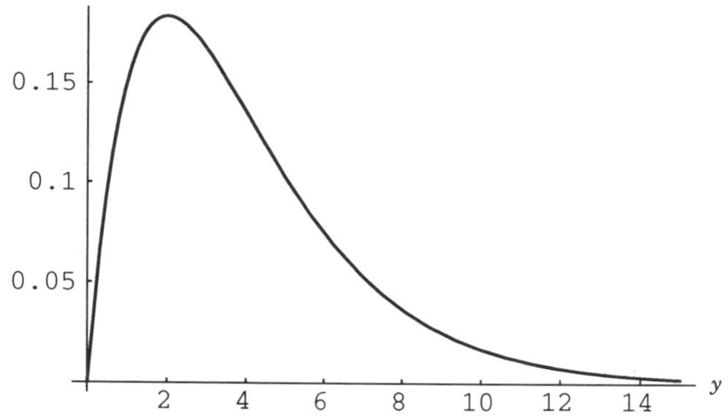

Figure 4.6 A distribution with positive skewness.

near the mean and in the tails but with thinner flanks. Positive and negative kurtosis are illustrated in Figure 4.7.

The test of normality can be carried out using the exact percentage points for $\sqrt{b_1}$ in Table A.1 for $4 \leq n \leq 25$, as given by Mulholland (1977). Alternatively, for $n \geq 8$ the function g as defined by

$$g(\sqrt{b_1}) = \delta \sinh^{-1} \frac{\sqrt{b_1}}{\lambda} \tag{4.20}$$

is approximately $N(0, 1)$, where

$$\sinh^{-1} x = \ln(x + \sqrt{x^2 + 1}). \qquad (4.21))$$

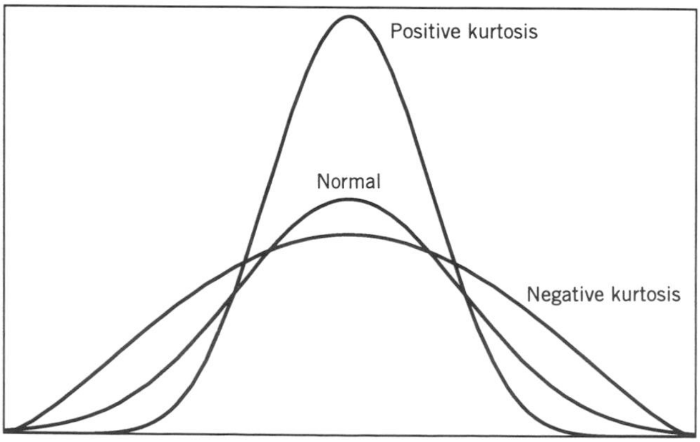

Figure 4.7 Distributions with positive and negative kurtosis compared to the normal.

Table A.2, from D'Agostino and Pearson (1973), gives values for δ and $1/\lambda$. To use b_2 as a test of normality, we can use Table A.3, obtained from D'Agostino and Tietjen (1971), which gives simulated percentiles of b_2 for selected values of n in the range $7 \leq n \leq 50$. Charts of percentiles of b_2 for $20 \leq n \leq 200$ can be found in D'Agostino and Pearson (1973).

Our second test for normality was given by D'Agostino (1971). The observations $y_1, y_2, \ldots, y_n$ are ordered as $y_{(1)} \leq y_{(2)} \leq \cdots \leq y_{(n)}$, and we calculate

$$D = \frac{\sum_{i=1}^{n} \left[i - \frac{1}{2}(n+1) \right] y_{(i)}}{\sqrt{n^3 \sum_{i=1}^{n} (y_i - \overline{y})^2}} \tag{4.22}$$

and

$$Y = \frac{\sqrt{n}[D - (2\sqrt{\pi})^{-1}]}{.02998598}. \tag{4.23}$$

A table of percentiles for Y, given by D'Agostino (1972) for $10 \leq n \leq 250$, is provided in Table A.4.

The final test we report is by Lin and Mudholkar (1980). The test statistic is

$$z = \tanh^{-1} r = \tfrac{1}{2} \ln \frac{1+r}{1-r}, \tag{4.24}$$

where r is the sample correlation of the n pairs (y_i, x_i), $i = 1, 2, \ldots, n$, with x_i defined as

$$x_i = \left[\frac{1}{n} \left(\sum_{j \neq i} y_j^2 - \frac{\left(\sum_{j \neq i} y_j \right)^2}{n-1} \right) \right]^{1/3}. \tag{4.25}$$

If the y's are normal, z is approximately $N(0, 3/n)$. A more accurate upper 100α percentile is given by

$$z_\alpha = \sigma_n \left[u_\alpha + \tfrac{1}{24}(u_\alpha^3 - 3u_\alpha)\gamma_{2n} \right], \tag{4.26}$$

with

$$\sigma_n^2 = \frac{3}{n} - \frac{7.324}{n^2} + \frac{53.005}{n^3} \qquad u_\alpha = \Phi^{-1}(\alpha)$$
$$\gamma_{2n} = -\frac{11.70}{n} + \frac{55.06}{n^2},$$

where Φ is the distribution function of the $N(0, 1)$ distribution; that is, $\Phi(x)$ is the probability of an observation less than or equal to x, as in (4.17). The inverse function Φ^{-1} is essentially a quantile. For example, $u_{.05} = -1.645$ and $u_{.95} = 1.645$.

4.4.2 Investigating Multivariate Normality

Checking for multivariate normality is conceptually not as straightforward as assessing univariate normality, and consequently the state of the art is not as well developed. The complexity of this issue can be illustrated in the context of a goodness-of-fit test for normality. For a goodness-of-fit test in the univariate case, the range covered by a sample $y_1, y_2, \ldots, y_n$ is divided into several intervals, and we count how many y's fall into each interval. These observed frequencies (counts) are compared to the expected frequencies under the assumption that the sample came from a normal distribution with the same mean and variance as the sample. If the n observations $\mathbf{y}_1, \mathbf{y}_2, \ldots, \mathbf{y}_n$ are multivariate, however, the procedure is not so simple. We now have a p-dimensional region that would have to be divided into many more subregions than in the univariate case, and the expected frequencies for these subregions would be less easily obtained. With so many subregions, relatively few would contain observations; many would end up with no observations.

Thus because of the inherent "sparseness" of multivariate data, a goodness-of-fit test would be impractical. The points $\mathbf{y}_1, \mathbf{y}_2, \ldots, \mathbf{y}_n$ are more distant from each other in p-space than in any one of the p individual dimensions. Unless n is very large, a multivariate sample may not provide a very complete picture of the distribution from which it was taken.

As a consequence of the sparseness of the data in p-space, the tests for multivariate normality may not be very powerful. However, some check on the distribution is often desirable. Numerous procedures have been proposed for assessing multivariate normality. We discuss three of these, as well as a program containing three additional tests.

The first procedure is based on the standardized distance from each $\mathbf{y}_i$ to $\overline{\mathbf{y}}$,

$$D_i^2 = (\mathbf{y}_i - \overline{\mathbf{y}})'\mathbf{S}^{-1}(\mathbf{y}_i - \overline{\mathbf{y}}). \tag{4.27}$$

Gnanadesikan and Kettenring (1972) showed that if the $\mathbf{y}_i$'s are multivariate normal, then

$$u_i = \frac{nD_i^2}{(n-1)^2} \tag{4.28}$$

has a beta distribution, which is related to the F. To obtain a Q–Q plot, the values $u_1, u_2, \ldots, u_n$ are ranked to give $u_{(1)} \leq u_{(2)} \leq \cdots \leq u_{(n)}$, and we plot $(v_i, u_{(i)})$, where the quantiles v_i of the beta are given by

$$v_i = \frac{i - \alpha}{n - \alpha - \beta + 1}, \tag{4.29}$$

with α and β defined as

$$\alpha = \frac{p-2}{2p} \tag{4.30}$$

and

$$\beta = \frac{n-p-2}{2(n-p-1)}. \tag{4.31}$$

A nonlinear pattern in the plot would indicate a departure from normality. A formal significance test is also available for $D_{(n)}^2 = \max_i D_i^2$. Table A.6 gives the upper 5 and 1% critical values from Barnett and Lewis (1978).

The second procedure involves scatter plots in two or three dimensions. If p is not too high, the bivariate plots of each pair of variables are often reduced in size and shown on one page, arranged to correspond to the entries in a correlation matrix. In this visual matrix, the eye readily picks out those pairs of variables that show a curved trend, outliers, or other nonnormal appearance. This plot is illustrated in connection with Example 4.5.2. The procedure is based on properties 4 and 6 of Section 4.2, from which we infer that (1) each pair of variables has a bivariate normal distribution and (2) bivariate normal variables follow a straight-line trend.

A popular option in many graphical programs is the ability to dynamically rotate a plot of three variables. While the points are rotating on the screen, a three-dimensional effect is created. The shape of the three-dimensional cloud of points is readily perceived, and we can detect various features of the data. The only drawbacks to this technique are that (1) it is a dynamic display and cannot be printed and (2) if p is very large, the number of subsets of three variables becomes unwieldy, while the number of pairs may still be tractable for plotting. These numbers are compared in Table 4.1, where $\binom{p}{2}$ and $\binom{p}{3}$ represent the number of subsets of sizes 2 and 3, respectively. Thus in many cases, the scatter plots for pairs of variables will continue to be used, even though three-dimensional plotting techniques are available.

The third procedure for assessing multivariate normality is a generalization of the univariate test based on the skewness and kurtosis measures $\sqrt{b_1}$ and b_2 as given by (4.18) and (4.19). The test is due to Mardia (1970). Let $\mathbf{y}$ and $\mathbf{x}$ be independent and identically distributed with mean vector $\boldsymbol{\mu}$ and covariance matrix $\boldsymbol{\Sigma}$. Then skewness and kurtosis for multivariate populations are defined by Mardia as

$$\beta_{1,p} = E[(\mathbf{y} - \boldsymbol{\mu})'\boldsymbol{\Sigma}^{-1}(\mathbf{x} - \boldsymbol{\mu})]^3 \tag{4.32}$$

and

$$\beta_{2,p} = E[(\mathbf{y} - \boldsymbol{\mu})'\boldsymbol{\Sigma}^{-1}(\mathbf{y} - \boldsymbol{\mu})]^2. \tag{4.33}$$

Since third-order central moments for the multivariate normal distribution are zero, $\beta_{1,p} = 0$ when $\mathbf{y}$ is $N(\boldsymbol{\mu}, \boldsymbol{\Sigma})$. It can also be shown that for multivariate normal $\mathbf{y}$,

$$\beta_{2,p} = p(p + 2). \tag{4.34}$$

If we define

$$g_{ij} = (\mathbf{y}_i - \overline{\mathbf{y}})'\hat{\boldsymbol{\Sigma}}^{-1}(\mathbf{y}_j - \overline{\mathbf{y}}), \tag{4.35}$$

where $\hat{\boldsymbol{\Sigma}} = \sum_i(\mathbf{y}_i - \overline{\mathbf{y}})(\mathbf{y}_i - \overline{\mathbf{y}})'/n$ is the maximum likelihood estimator (4.12), then sample estimates of $\beta_{1,p}$ and $\beta_{2,p}$ are given by

Table 4.1 Comparison of Number of Subsets of Sizes 2 and 3

p	$\binom{p}{2}$	$\binom{p}{3}$
6	15	20
8	28	56
10	45	120
12	66	220
15	105	455

$$b_{1,p} = \frac{1}{n^2} \sum_{i=1}^{n} \sum_{j=1}^{n} g_{ij}^3 \tag{4.36}$$

and

$$b_{2,p} = \frac{1}{n} \sum_{i} g_{ii}^2. \tag{4.37}$$

Table A.5 gives percentage points of $b_{1,p}$ and $b_{2,p}$ for $p = 2, 3, 4$, which can be used in testing for multivariate normality (Mardia 1970, 1974). For other values of p or when $n \geq 50$, the following approximate tests are available. For $b_{1,p}$

$$z_1 = \frac{(p+1)(n+1)(n+3)}{6[(n+1)(p+1)-6]} b_{1,p} \tag{4.38}$$

is approximately χ^2 with $\frac{1}{6}p(p+1)(p+2)$ degrees of freedom. Reject if $z_1 \geq \chi^2_{.05}$. With $b_{2,p}$, on the other hand, we wish to reject for large values (distribution too peaked) or small values (distribution too flat). For the upper 2.5% points of $b_{2,p}$ use

$$z_2 = \frac{b_{2,p} - p(p+2)}{\sqrt{8p(p+2)/n}} \tag{4.39}$$

which is approximately $N(0, 1)$. For the lower 2.5% points we have two cases: (a) when $50 \leq n \leq 400$, use

$$z_3 = \frac{b_{2,p} - p(p+2)(n+p+1)/n}{\sqrt{8p(p+2)/(n-1)}} \tag{4.40}$$

which is approximately $N(0, 1)$; (b) when $n \geq 400$, use z_2 as given by (4.39).

Fortran programs for the above tests based on $b_{1,p}$ and $b_{2,p}$ are given by Siotani et al. (1985). They also provide programs for three additional tests for multivariate normality and several tests for univariate normality. Many of these programs are apparently unavailable elsewhere. The three multivariate tests are, briefly, as follows:

1. A test based on the third and fourth central moments

$$E[(y_i - \mu_i)(y_j - \mu_j)(y_k - \mu_k)] \tag{4.41}$$

and

$$E[(y_i - \mu_i)(y_j - \mu_j)(y_k - \mu_k)(y_l - \mu_l)]. \tag{4.42}$$

Under normality, (4.41) is zero and (4.42) is equal to $\sigma_{ij}\sigma_{kl} + \sigma_{ik}\sigma_{jl} + \sigma_{il}\sigma_{jk}$. Estimates of (4.41) and (4.42) are obtained and compared to 0 and $s_{ij}s_{kl} + s_{ik}s_{jl} + s_{il}s_{jk}$, respectively.

2. A multivariate generalization of the Shapiro–Wilk test: Define $z_i = \mathbf{c}'\mathbf{y}_i$, $i = 1, 2, \ldots, n$, where $\mathbf{c}$ is a constant vector, and

$$W(\mathbf{c}) = \frac{\sum_{i=1}^{n} a_i(z_{(i)} - \bar{z})^2}{\sum_{i=1}^{n}(z_i - \bar{z})^2}, \tag{4.43}$$

where $z_{(1)} \leq z_{(2)} \leq \cdots \leq z_{(n)}$ are the ordered values of $z_1, z_2, \ldots, z_n$ and the a_i's are coefficients tabulated in Shapiro and Wilk (1965). The hypothesis of multivariate normality is accepted if

$$\max_{\mathbf{c}}[W(\mathbf{c})] \geq k, \tag{4.44}$$

where k corresponds to the desired significance level, α.

3. A directional normality test: This test is based on an alternative definition of the multivariate normal distribution suggested by property 1a of Section 4.2: If $\mathbf{a}'\mathbf{y}$ is $N(\mathbf{a}'\boldsymbol{\mu}, \mathbf{a}'\boldsymbol{\Sigma}\mathbf{a})$ for all $\mathbf{a}$, then $\mathbf{y}$ is $N_p(\boldsymbol{\mu}, \boldsymbol{\Sigma})$. First the data vectors $\mathbf{y}_1, \mathbf{y}_2, \ldots, \mathbf{y}_n$ are standardized by $\mathbf{z}_i = (\mathbf{S}^{1/2})^{-1}(\mathbf{y}_i - \bar{\mathbf{y}})$, $i = 1, 2, \ldots, n$, where $\mathbf{S}^{1/2}$ is the square root matrix given in (2.103). Then each $\mathbf{z}_i$ is multiplied by a direction vector $\mathbf{d}_\alpha$ to obtain $v_i = \mathbf{d}_\alpha'\mathbf{z}_i$ $i = 1, 2, \ldots, n$. The v's are approximately normal if the $\mathbf{y}$'s are multivariate normal. Several values of $\mathbf{d}_\alpha$ are used to check for normality in different directions. Various univariate normal tests can be applied to the v's.

4.5 OUTLIERS

The detection of outliers has been of concern to statisticians and other scientists for over a century. Many authors have claimed that the researcher can typically expect up to 10% of the observations to have errors in measurement or recording. Occasional stray observations from a different population than the target population are also fairly common. We do not attempt a complete summary of the vast literature covering univariate outliers, but we do review some major concepts and suggested procedures in Section 4.5.1 before moving to the multivariate case in Section 4.5.2. An alternative to detection of outliers is to use robust estimators of $\boldsymbol{\mu}$ and $\boldsymbol{\Sigma}$ (see Rencher 1996, Section 1.10) that are less sensitive to extreme observations than are the standard estimators $\overline{\mathbf{y}}$ and $\mathbf{S}$.

4.5.1 Outliers in Univariate Samples

Excellent surveys of the useful literature on outliers have been given by Beckman and Cook (1983), Hawkins (1980), and Barnett and Lewis (1978). We abstract a few highlights from Beckman and Cook. Many techniques have been proposed for detecting outliers in the residuals from regression, designed experiments, and so on. But we will be concerned only with simple random samples from the normal distribution. Outliers are also known as *discordant observations* or *contaminants*, which imply a discrepancy from what was expected and an origin from a nontarget population, respectively.

There are two principal approaches for dealing with outliers. The first is *identification*, which usually involves deletion of the outlier(s) but may alternatively provide important information about the model or the data. The second method involves *accommodation*, by modifying the method of analysis or the model. Robust methods, in which the influence of outliers is reduced, are the most familiar example of modification of the analysis. An example of a correction to the model is a mixture model that combines two normals with different variances, sometimes used to accommodate contaminants. For example, Marks and Rao (1978) accommodated a particular type of outlier due to patient fatigue by a mixture of two normal distributions.

In small or moderate sized univariate samples, visual methods of identifying outliers are the most frequently used. Tests are also available if a less subjective approach is desired.

Two types of *slippage* models have been proposed to account for outliers. Under the *mean slippage* model, all observations have the same variance, but one or more of the observations arise from a distribution with a different (population) mean. In the *variance slippage* model, one or more of the observations arise from a model with larger (population) variance but the same mean. Thus in the mean slippage model, the bulk of the observations arise from $N(\mu, \sigma^2)$, while the outliers originate from $N(\mu + \theta, \sigma^2)$. For the variance slippage model, the main distribution would again be $N(\mu, \sigma^2)$, with the outliers coming from $N(\mu, a\sigma^2)$ where $a > 1$. These models have led to the development of tests for rejection of outliers. We now briefly discuss some of these tests.

For a single outlier, most tests are based on the maximum studentized residual,

$$\max_i \tau_i = \max_i \left| \frac{y_i - \overline{y}}{s} \right|. \tag{4.45}$$

If the largest or smallest observation is rejected, one could then examine the $n-1$ remaining observations for another possible outlier, and so on. This procedure is called a *consecutive test*. However, if there are two or more outliers, the less extreme ones will often make it difficult to detect the most extreme one, due to inflation of both mean and variance. This effect is called *masking*.

Ferguson (1961) showed that the maximum studentized residual (4.45) is more powerful than most other techniques for detecting intermediate or large shifts in the mean and gave the following guidelines for small shifts:

1. For outliers with small positive shifts in the mean, tests based on sample skewness are best.
2. For outliers with small shifts in the mean in either direction, tests based on the sample kurtosis are best.
3. For outliers with small positive shifts in the variance, tests based on the sample kurtosis are best.

Because of the masking problem in consecutive tests, *block tests* have been proposed for simultaneous rejection of $k > 1$ outliers. These tests work well if k is known, but in practice, it is usually not known. If the value we conjecture for k is too small, we incur the risk of failing to detect any outliers because of masking. If we set k too large, there is a high risk of rejecting more outliers than there really are, an effect known as *swamping*.

4.5.2 Outliers in Multivariate Samples

In the case of multivariate data, the problems in detecting outliers are intensified for several reasons:

1. For $p > 2$ the data cannot be readily plotted to pinpoint the outliers.
2. Multivariate data cannot be ordered as can a univariate sample, where extremes show up readily on either end.
3. An observation vector may have a large recording error in one of its components or smaller errors in several components.
4. A multivariate outlier may reflect slippage in mean, variance, or correlation. This is illustrated in Figure 4.8. Observation 1 causes a small shift in means and variances of both y_1 and y_2 but has little effect on the correlation. Observation 2 has little effect on means and variances, but it reduces the correlation somewhat. Observation 3 has a major effect on means, variances, and correlation.

Of course, as in the univariate case, one approach to outlier identification or accommodation is to use robust methods of estimation. Such methods minimize the influence of outliers in estimation or model fitting. However, an outlier sometimes furnishes valuable information, and the specific pursuit of outliers can be very worthwhile.

We present two methods of multivariate outlier identification, both of which

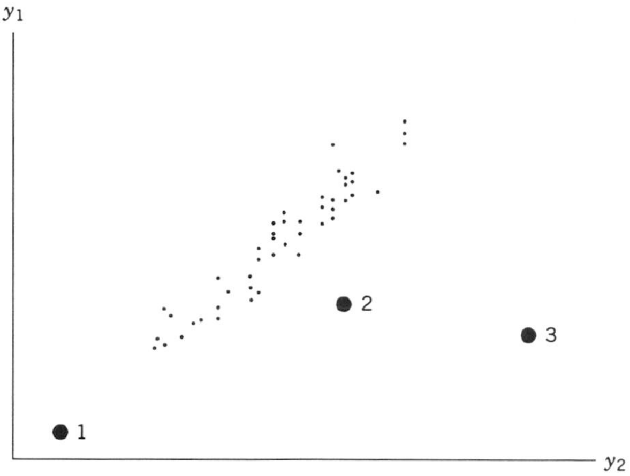

Figures 4.8 Bivariate sample showing three types of outliers.

turn out to be related to methods of assessing multivariate normality. (A third approach based on principal components is given in Section 12.4.) The first method, due to Wilks (1963), is designed for detection of a single outlier. Wilks' statistic is

$$w = \max_i \frac{|(n-2)\mathbf{S}_{-i}|}{|(n-1)\mathbf{S}|}, \tag{4.46}$$

where $\mathbf{S}$ is the usual sample covariance matrix and $\mathbf{S}_{-i}$ is obtained from the same sample with the ith observation deleted. It turns out that w can be expressed in terms of $D^2_{(n)} = \max_i(\mathbf{y}_i - \overline{\mathbf{y}})'\mathbf{S}^{-1}(\mathbf{y}_i - \overline{\mathbf{y}})$ as

$$w = 1 - \frac{nD^2_{(n)}}{(n-1)^2}, \tag{4.47}$$

thus basing a test for an outlier on the distances D_i^2 used in Section 4.4.2 in a graphical procedure for checking multivariate normality. Table A.6 gives the upper 5 and 1% critical values for $D^2_{(n)}$ from Barnett and Lewis (1978).

Yang and Lee (1987) provide an F-test of w as given by (4.47). Define

$$F_i = \frac{n-p-1}{p}\left[\frac{1}{1 - nD_i^2/(n-1)^2} - 1\right], \qquad i = 1, 2, \ldots, n. \tag{4.48}$$

Then the F_i are independently and identically distributed as $F_{p,n-p-1}$, and a test can be constructed in terms of $\max_i F_i$:

$$P\left(\max_i F_i > f\right) = 1 - P(\text{all } F_i \leq f) = 1 - [P(F \leq f)]^n.$$

Therefore, the test can be carried out using an F-table. Note that

$$\max_i F_i = F_{(n)} = \frac{n-p-1}{p}\left(\frac{1}{w} - 1\right), \tag{4.49}$$

where w is given in (4.47).

The second test we discuss is designed for detection of several outliers. Schwager and Margolin (1982) showed that the locally best invariant test for mean slippage is based on Mardia's (1970) sample kurtosis $b_{2,p}$ as defined by (4.35) and (4.37). Essentially this means that among all tests invariant to a class of transformations of the type $\mathbf{z} = \mathbf{Ay} + \mathbf{b}$, where $\mathbf{A}$ is nonsingular, the test using $b_{2,p}$ is most powerful for small shifts in the mean vector. This result holds if the proportion of outliers is no more than 21.13%. With some restrictions on the pattern of the outliers, the permissible fraction of outliers can go as high as $33\frac{1}{3}$%. The hypothesis is H_0: no outliers are present. This hypothesis is rejected for large values of $b_{2,p}$.

A table of critical values of $b_{2,p}$ and some approximate tests were described in Section 4.4.2 following (4.37). Here again we have a test that doubles as a check for multivariate normality and for the presence of outliers. One advantage of this test for outliers is that we do not have to specify the number of outliers and run the attendant risk of masking or swamping. Schwager and Margolin (1982) pointed out that this feature "increases the importance of performing an overall test that is sensitive to a broad range of outlier configurations. There is also empirical evidence that the kurtosis test performs well in situations of practical interest when compared with other inferential outlier procedures."

Sinha (1984) extended the result of Schwager and Margolin to cover the general case of elliptically symmetric distributions. An *elliptically symmetric distribution* is one in which $f(\mathbf{y}) = |\mathbf{\Sigma}|^{-1/2} g[(\mathbf{y} - \boldsymbol{\mu})'\mathbf{\Sigma}^{-1}(\mathbf{y} - \boldsymbol{\mu})]$. By varying the function g, distributions with shorter or longer tails than the normal can be obtained. Of course, the critical value of $b_{2,p}$ would have to be adjusted to correspond to the distribution, but rejection for large values would be a locally best invariant test.

Table 4.2 Values of D_i^2 for the Ramus Bone Data in Table 3.8

Observation Number	D_i^2	Observation Number	D_i^2
1	0.7588	11	2.8301
2	1.2980	12	10.5718
3	1.7591	13	2.5941
4	3.8539	14	0.6594
5	0.8706	15	0.3246
6	2.8106	16	0.8321
7	4.2915	17	1.1083
8	7.9897	18	4.3633
9	11.0301	19	2.1088
10	5.3519	20	10.0931

Example 4.5.2. We use the ramus bone data set of Table 3.8 to illustrate a search for multivariate outliers, while at the same time checking for multivariate normality. An examination of each column of Table 3.8 does not reveal any apparent univariate outliers. We next calculate D_i^2 in (4.27) for each observation vector. The results are given in Table 4.2.

We see that D_9^2, D_{12}^2, and D_{20}^2 seem to stand out as possible outliers. This impression is confirmed when we compute u_i and v_i in (4.28) and (4.29) and plot them in Figure 4.9. The figure shows a distinct departure from linearity due to the three large D_i^2 values noted above and also possibly due to $D_8^2 = 7.99$. In Table A.6, the upper 5% critical value for the maximum value, $D_{(20)}^2$, is given

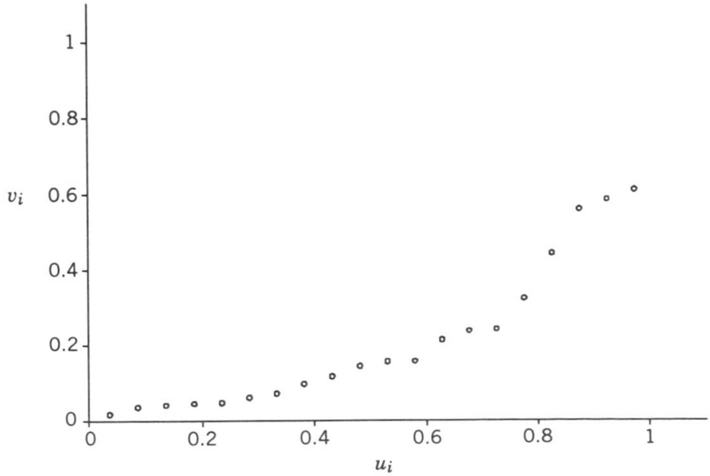

Figure 4.9 *Q–Q* plot of u_i and v_i for the ramus bone data of Table 3.8.

as 11.63. In our case, the largest D_i^2 is $D_9^2 = 11.03$, which does not exceed the critical value. This does not surprise us, since the test was designed to detect a single outlier, and we may have at least three.

We next calculate $b_{1,p}$ and $b_{2,p}$ as given by (4.36) and (4.37):

$$b_{1,p} = 11.338 \qquad b_{2,p} = 28.884.$$

In Table A.5, the upper .01 critical value for $b_{1,p}$ is 9.9; the upper .005 critical value for $b_{2,p}$ is 27.1. Thus both $b_{1,p}$ and $b_{2,p}$ exceed their critical values, and we have significant skewness and kurtosis, apparently caused by the three observations with large values of D_i^2.

The bivariate scatter plots are given in Figure 4.10. The three values are clearly separate from the other observations in the plot of y_1 versus y_4. In Table 3.8, the 9th, 12th, and 20th values of y_4 are not unusual, nor are the 9th, 12th, and 20th values of y_1. However, the increase from y_1 to y_4 is exceptional in each case. If these values are not due to errors in recording the data and if this sample is representative, then we appear to have a mixture of two populations. This should be taken into account in making inferences.

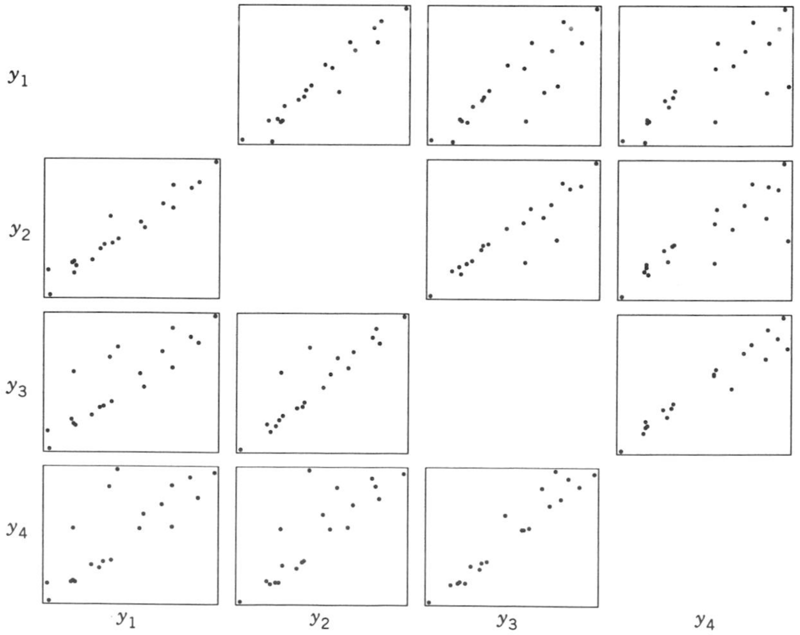

Figure 4.10 Scatter plots for the ramus bone data in Table 3.8.

PROBLEMS

4.1 Consider the two covariance matrices

$$\boldsymbol{\Sigma}_1 = \begin{pmatrix} 14 & 8 & 3 \\ 8 & 5 & 2 \\ 3 & 2 & 1 \end{pmatrix} \qquad \boldsymbol{\Sigma}_2 = \begin{pmatrix} 6 & 6 & 1 \\ 6 & 8 & 2 \\ 1 & 2 & 1 \end{pmatrix}.$$

Show that $|\boldsymbol{\Sigma}_2| > |\boldsymbol{\Sigma}_1|$ and that $\text{tr}(\boldsymbol{\Sigma}_2) < \text{tr}(\boldsymbol{\Sigma}_1)$. Thus the generalized variance of population 2 is greater than the generalized variance of population 1, even though the total variance is less. Comment on why this is true in terms of the correlations and eigenvalues.

4.2 For $\mathbf{z} = (\mathbf{T}')^{-1}(\mathbf{y} - \boldsymbol{\mu})$ as in (4.4), show that $E(\mathbf{z}) = \mathbf{0}$ and $\text{cov}(\mathbf{z}) = \mathbf{I}$.

4.3 Show that the form of the likelihood function in (4.13) follows from the previous expression.

4.4 Show that by adding and subtracting $\overline{\mathbf{y}}$, the exponent of (4.13) has the form given in (4.14), that is,

$$\frac{1}{2}\sum_{i=1}^{n}(\mathbf{y}_i - \overline{\mathbf{y}} + \overline{\mathbf{y}} - \boldsymbol{\mu})'\boldsymbol{\Sigma}^{-1}(\mathbf{y}_i - \overline{\mathbf{y}} + \overline{\mathbf{y}} - \boldsymbol{\mu})$$
$$= \frac{1}{2}\sum_{i=1}^{n}(\mathbf{y}_i - \overline{\mathbf{y}})'\boldsymbol{\Sigma}^{-1}(\mathbf{y}_i - \overline{\mathbf{y}}) + \frac{n}{2}(\overline{\mathbf{y}} - \boldsymbol{\mu})'\boldsymbol{\Sigma}^{-1}(\overline{\mathbf{y}} - \boldsymbol{\mu}).$$

4.5 Show that $\sqrt{b_1}$ and b_2 as given in (4.18) and (4.19) are invariant to the transformation $z_i = ay_i + b$.

4.6 Show that if $\mathbf{y}$ is $N_p(\boldsymbol{\mu}, \boldsymbol{\Sigma})$, then $\beta_{2,p} = p(p+2)$ as in (4.34).

4.7 Show that $b_{1,p}$ and $b_{2,p}$ as given by (4.36) and (4.37) are invariant under the transformation $\mathbf{z}_i = \mathbf{A}\mathbf{y}_i + \mathbf{b}$, where $\mathbf{A}$ is nonsingular. Thus $b_{1,p}$ and $b_{2,p}$ do not depend on the units of measurement; the variables could even be standardized.

4.8 Show that $F_{(n)} = [(n - p - 1)/p](1/w - 1)$ as in (4.49).

4.9 Suppose $\mathbf{y}$ is $N_3(\boldsymbol{\mu}, \boldsymbol{\Sigma})$, where

$$\boldsymbol{\mu} = \begin{pmatrix} 3 \\ 1 \\ 4 \end{pmatrix} \qquad \boldsymbol{\Sigma} = \begin{pmatrix} 6 & 1 & -2 \\ 1 & 13 & 4 \\ -2 & 4 & 4 \end{pmatrix}.$$

(a) Find the distribution of $z = 2y_1 - y_2 + 3y_3$.

(b) Find the joint distribution of $z_1 = y_1 + y_2 + y_3$ and $z_2 = y_1 - y_2 + 2y_3$.

(c) Find the distribution of y_2.

(d) Find the joint distribution of y_1 and y_3.

(e) Find the joint distribution of y_1, y_3, and $\frac{1}{2}(y_1 + y_2)$.

4.10 Suppose $\mathbf{y}$ is $N_3(\boldsymbol{\mu}, \boldsymbol{\Sigma})$ with $\boldsymbol{\mu}$ and $\boldsymbol{\Sigma}$ given in the previous problem.

(a) Find a vector $\mathbf{z}$ such that $\mathbf{z} = (\mathbf{T}')^{-1}(\mathbf{y} - \boldsymbol{\mu})$ is $N_3(\mathbf{0}, \mathbf{I})$ as in (4.4).

(b) Find a vector $\mathbf{z}$ such that $\mathbf{z} = (\boldsymbol{\Sigma}^{1/2})^{-1}(\mathbf{y} - \boldsymbol{\mu})$ is $N_3(\mathbf{0}, \mathbf{I})$ as in (4.5).

(c) What is the distribution of $(\mathbf{y} - \boldsymbol{\mu})'\boldsymbol{\Sigma}^{-1}(\mathbf{y} - \boldsymbol{\mu})$?

4.11 Suppose $\mathbf{y}$ is $N_4(\boldsymbol{\mu}, \boldsymbol{\Sigma})$, where

$$\boldsymbol{\mu} = \begin{bmatrix} -2 \\ 3 \\ -1 \\ 5 \end{bmatrix} \qquad \boldsymbol{\Sigma} = \begin{bmatrix} 11 & -8 & 3 & 9 \\ -8 & 9 & -3 & -6 \\ 3 & -3 & 2 & 3 \\ 9 & -6 & 3 & 9 \end{bmatrix}.$$

(a) Find the distribution of $z = 4y_1 - 2y_2 + y_3 - 3y_4$.

(b) Find the joint distribution of $z_1 = y_1 + y_2 + y_3 + y_4$ and $z_2 = -2y_1 + 3y_2 + y_3 - 2y_4$.

(c) Find the joint distribution of $z_1 = 3y_1 + y_2 - 4y_3 - y_4$, $z_2 = -y_1 - 3y_2 + y_3 - 2y_4$, and $z_3 = 2y_1 + 2y_2 + 4y_3 - 5y_4$.

(d) What is the distribution of y_3?

(e) What is the joint distribution of y_2 and y_4?

(f) Find the joint distribution of y_1, $\frac{1}{2}(y_1 + y_2)$, $\frac{1}{3}(y_1 + y_2 + y_3)$, and $\frac{1}{4}(y_1 + y_2 + y_3 + y_4)$.

4.12 Suppose $\mathbf{y}$ is $N_4(\boldsymbol{\mu}, \boldsymbol{\Sigma})$ with $\boldsymbol{\mu}$ and $\boldsymbol{\Sigma}$ given in the previous problem.

(a) Find a vector $\mathbf{z}$ such that $\mathbf{z} = (\mathbf{T}')^{-1}(\mathbf{y} - \boldsymbol{\mu})$ is $N_4(\mathbf{0}, \mathbf{I})$, as in (4.4).

(b) Find a vector $\mathbf{z}$ such that $\mathbf{z} = (\boldsymbol{\Sigma}^{1/2})^{-1}(\mathbf{y} - \boldsymbol{\mu})$ is $N_4(\mathbf{0}, \mathbf{I})$, as in (4.5).

(c) What is the distribution of $(\mathbf{y} - \boldsymbol{\mu})'\boldsymbol{\Sigma}^{-1}(\mathbf{y} - \boldsymbol{\mu})$?

4.13 Suppose $\mathbf{y}$ is $N_3(\boldsymbol{\mu}, \boldsymbol{\Sigma})$, with

$$\boldsymbol{\mu} = \begin{pmatrix} 2 \\ -3 \\ 4 \end{pmatrix} \qquad \boldsymbol{\Sigma} = \begin{pmatrix} 4 & -3 & 0 \\ -3 & 6 & 0 \\ 0 & 0 & 5 \end{pmatrix}.$$

Which of the following random variables are independent?

(a) y_1 and y_2
(b) y_1 and y_3
(c) y_2 and y_3
(d) (y_1, y_2) and y_3
(e) (y_1, y_3) and y_2

4.14 Suppose $\mathbf{y}$ is $N_4(\boldsymbol{\mu}, \boldsymbol{\Sigma})$, with

$$\boldsymbol{\mu} = \begin{bmatrix} -4 \\ 2 \\ 5 \\ -1 \end{bmatrix} \qquad \boldsymbol{\Sigma} = \begin{bmatrix} 8 & 0 & -1 & 0 \\ 0 & 3 & 0 & 2 \\ -1 & 0 & 5 & 0 \\ 0 & 2 & 0 & 7 \end{bmatrix}.$$

Which of the following random variables are independent?

(a) y_1 and y_2
(b) y_1 and y_3
(c) y_1 and y_4
(d) y_2 and y_3
(e) y_2 and y_4
(f) y_3 and y_4
(g) (y_1, y_2) and y_3
(h) (y_1, y_2) and y_4
(i) (y_1, y_3) and y_4
(j) y_1 and (y_2, y_4)
(k) y_1 and y_2 and y_3
(l) y_1 and y_2 and y_4
(m) (y_1, y_2) and (y_3, y_4)
(n) (y_1, y_3) and (y_2, y_4)

4.15 Assume $\mathbf{y}$ and $\mathbf{x}$ are subvectors, each 2×1, where

$$\begin{pmatrix} \mathbf{y} \\ \mathbf{x} \end{pmatrix} \quad \text{is} \quad N_4(\boldsymbol{\mu}, \boldsymbol{\Sigma})$$

with

$$\boldsymbol{\mu} = \left[\begin{array}{c} 2 \\ -1 \\ \hline 3 \\ 1 \end{array}\right] \qquad \boldsymbol{\Sigma} = \left[\begin{array}{cc|cc} 7 & 3 & -3 & 2 \\ 3 & 6 & 0 & 4 \\ \hline -3 & 0 & 5 & -2 \\ 2 & 4 & -2 & 4 \end{array}\right].$$

(a) Find $E(\mathbf{y}|\mathbf{x})$ by (4.7).
(b) Find $\text{cov}(\mathbf{y}|\mathbf{x})$ by (4.8).

4.16 Suppose $\mathbf{y}$ and $\mathbf{x}$ are subvectors, such that $\mathbf{y}$ is 2×1 and $\mathbf{x}$ is 3×1, with $\boldsymbol{\mu}$ and $\boldsymbol{\Sigma}$ partitioned accordingly:

$$\boldsymbol{\mu} = \begin{bmatrix} 3 \\ -2 \\ \hline 4 \\ -3 \\ 5 \end{bmatrix} \qquad \boldsymbol{\Sigma} = \left[\begin{array}{cc|ccc} 14 & -8 & 15 & 0 & 3 \\ -8 & 18 & 8 & 6 & -2 \\ \hline 15 & 8 & 50 & 8 & 5 \\ 0 & 6 & 8 & 4 & 0 \\ 3 & -2 & 5 & 0 & 1 \end{array}\right].$$

Assume that $\binom{\mathbf{y}}{\mathbf{x}}$ is distributed as $N_5(\boldsymbol{\mu}, \boldsymbol{\Sigma})$.

(a) Find $E(\mathbf{y}|\mathbf{x})$ by (4.7).

(b) Find $\text{cov}(\mathbf{y}|\mathbf{x})$ by (4.8).

4.17 Suppose that $\mathbf{y}_1, \mathbf{y}_2, \ldots, \mathbf{y}_n$ is a random sample from a nonnormal multivariate population with mean $\boldsymbol{\mu}$ and covariance matrix $\boldsymbol{\Sigma}$. If n is large, what is the approximate distribution of each of the following?

(a) $\sqrt{n}(\overline{\mathbf{y}} - \boldsymbol{\mu})$

(b) $\overline{\mathbf{y}}$

4.18 For the ramus bone data treated in Example 4.5.2, check each of the four variables for univariate normality using the following techniques:

(a) Q–Q plots

(b) $\sqrt{b_1}$ and b_2 as given by (4.18) and (4.19)

(c) D'Agostino's test using D and Y given in (4.22) and (4.23)

(d) The test by Lin and Mudholkar using z defined in (4.24)

4.19 For the calcium data in Table 3.5, check for multivariate normality and outliers using the following tests:

(a) Calculate D_i^2 as in (4.27) for each observation.

(b) Compare the largest value of D_i^2 with the critical value in Table A.6.

(c) Compute u_i and v_i in (4.28) and (4.29) and plot them. Is there an indication of nonlinearity or outliers?

(d) Calculate $b_{1,p}$ and $b_{2,p}$ in (4.36) and (4.37) and compare them with critical values in Table A.5.

4.20 For the probe word data in Table 3.7, check each of the five variables for univariate normality and outliers using the following tests:

(a) Q–Q plots

(b) $\sqrt{b_1}$ and b_2 as given by (4.18) and (4.19)

(c) D'Agostino's test using D and Y given in (4.22) and (4.23)

(d) The test by Lin and Mudholkar using z defined in (4.24)

4.21 For the probe word data in Table 3.7, check for multivariate normality and outliers using the following tests:

(a) Calculate D_i^2 as in (4.27) for each observation.

(b) Compare the largest value of D_i^2 with the critical value in Table A.6.

(c) Compute u_i and v_i in (4.28) and (4.29) and plot them. Is there an indication of nonlinearity or outliers?

(d) Calculate $b_{1,p}$ and $b_{2,p}$ in (4.36) and (4.37) and compare them with critical values in Table A.5.

4.22 Six hematology variables were measured on 51 workers (Royston 1983):

y_1 = hemoglobin concentration	y_4 = lymphocyte count
y_2 = packed cell volume	y_5 = neutrophil count
y_3 = white book cell count	y_6 = serum lead concentration

The data are given in Table 4.3. Check each of the six variables for univariate normality using the following tests.

(a) Q–Q plots

(b) $\sqrt{b_1}$ and b_2 as given by (4.18) and (4.19)

(c) D'Agostino's test using D and Y given in (4.22) and (4.23)

(d) The test by Lin and Mudholkar using z defined in (4.24)

4.23 For the hematology data in Table 4.3, check for multivariate normality using the following techniques:

(a) Calculate D_i^2 as in (4.27) for each observation.

(b) Compare the largest value of D_i^2 with the critical value in Table A.6 (extrapolate).

(c) Compute u_i and v_i in (4.28) and (4.29) and plot them. Is there an indication of nonlinearity or outliers?

(d) Calculate $b_{1,p}$ and $b_{2,p}$ in (4.36) and (4.37) and compare them with critical values in Table A.5.

Table 4.3 Hematology Data

Observation Number	y_1	y_2	y_3	y_4	y_5	y_6
1	13.4	39	4100	14	25	17
2	14.6	46	5000	15	30	20
3	13.5	42	4500	19	21	18
4	15.0	46	4600	23	16	18
5	14.6	44	5100	17	31	19
6	14.0	44	4900	20	24	19
7	16.4	49	4300	21	17	18
8	14.8	44	4400	16	26	29
9	15.2	46	4100	27	13	27
10	15.5	48	8400	34	42	36
11	15.2	47	5600	26	27	22
12	16.9	50	5100	28	17	23
13	14.8	44	4700	24	20	23
14	16.2	45	5600	26	25	19
15	14.7	43	4000	23	13	17
16	14.7	42	3400	9	22	13
17	16.5	45	5400	18	32	17
18	15.4	45	6900	28	36	24
19	15.1	45	4600	17	29	17
20	14.2	46	4200	14	25	28
21	15.9	46	5200	8	34	16
22	16.0	47	4700	25	14	18
23	17.4	50	8600	37	39	17
24	14.3	43	5500	20	31	19
25	14.8	44	4200	15	24	29
26	14.9	43	4300	9	32	17
27	15.5	45	5200	16	30	20
28	14.5	43	3900	18	18	25
29	14.4	45	6000	17	37	23
30	14.6	44	4700	23	21	27
31	15.3	45	7900	43	23	23
32	14.9	45	3400	17	15	24
33	15.8	47	6000	23	32	21
34	14.4	44	7700	31	39	23
35	14.7	46	3700	11	23	23
36	14.8	43	5200	25	19	22
37	15.4	45	6000	30	25	18
38	16.2	50	8100	32	38	18
39	15.0	45	4900	17	26	24
40	15.1	47	6000	22	33	16
41	16.0	46	4600	20	22	22
42	15.3	48	5500	20	23	23
43	14.5	41	6200	20	36	21
44	14.2	41	4900	26	20	20
45	15.0	45	7200	40	25	25
46	14.2	46	5800	22	31	22
47	14.9	45	8400	61	17	17
48	16.2	48	3100	12	15	18
49	14.5	45	4000	20	18	20
50	16.4	49	6900	35	22	24
51	14.7	44	7800	38	34	16

CHAPTER 5

Tests on One or Two Mean Vectors

5.1 MULTIVARIATE VERSUS UNIVARIATE TESTS

Hypothesis testing in a multivariate context is more complex than in a univariate setting. The number of parameters may be staggering. The p-variate normal distribution, for example, has p means, p variances, and $\binom{p}{2}$ covariances, where $\binom{p}{2}$ represents the number of pairs among the p variables. The total number of parameters is

$$p + p + \binom{p}{2} = \tfrac{1}{2}p(p + 3).$$

Each parameter corresponds to a hypothesis that could be formulated. Additionally, we might well be interested in testing hypotheses about subsets of these parameters or about functions of them. In some cases, we have the added dilemma of choosing among competing test statistics.

We first discuss the motivation for testing p variables multivariately rather than, or in addition to, univariately. There are at least four arguments for a multivariate approach to hypothesis testing:

1. The use of p univariate tests inflates the Type I error rate, α, whereas the multivariate test preserves the exact α level. For example, if we do $p = 10$ separate univariate tests at the .05 level, the probability of at least one false rejection is greater than .05. If the variables were independent (they rarely are), we would have (under H_0)

$$\begin{aligned} P(\text{at least one rejection}) &= 1 - P(\text{all 10 tests accept}) \\ &= 1 - (.95)^{10} \\ &= .40. \end{aligned}$$

The resulting overall α of .40 is not an acceptable error rate. Typically, the 10 variables are correlated, and the overall α ordinarily lies between .05 and .40.

2. The univariate tests completely ignore the correlations among the variables. In contrast, the multivariate tests make direct use of the covariance matrix.
3. The multivariate test is more powerful in many cases. The *power* of a test is the probability of rejecting H_0 when it is false. Sometimes all p of the univariate tests fail to reach significance, but the multivariate test is significant because small effects on some of the variables combine to jointly indicate significance. For a given sample size, there is a limit to the number of variables a multivariate test can handle without losing power. This is discussed further in Section 5.3.2.
4. Many multivariate tests involving means have as a byproduct the construction of a linear combination of variables that reveals more about how the variables unite to reject the hypothesis.

5.2 TESTS ON $\boldsymbol{\mu}$ WITH $\boldsymbol{\Sigma}$ KNOWN

The test on a mean vector assuming a known $\boldsymbol{\Sigma}$ is introduced to illustrate the issues involved in multivariate testing and to serve as a foundation for the unknown $\boldsymbol{\Sigma}$ case. We first review the univariate case, in which we work with a single variable.

5.2.1 Review of Univariate Test for $H_0\colon \mu = \mu_0$ with σ Known

The hypothesis of interest is that the mean is equal to a given value, μ_0, versus the alternative that it is not equal to μ_0:

$$H_0\colon \mu = \mu_0 \quad \text{vs.} \quad H_1\colon \mu \neq \mu_0.$$

We do not consider one-sided alternative hypotheses because they do not readily generalize to multivariate tests. We assume a random sample of n observations $y_1, y_2, \ldots, y_n$ from a $N(\mu, \sigma^2)$ population with σ^2 known. We calculate $\overline{y} = \sum_i y_i/n$ and compare it to μ_0 using the test statistic

$$z = \frac{\overline{y} - \mu_0}{\sigma_{\overline{y}}} = \frac{\overline{y} - \mu_0}{\sigma/\sqrt{n}}, \tag{5.1}$$

which is distributed as $N(0, 1)$ if H_0 is true. For $\alpha = .05$, reject H_0 if $|z| \geq 1.96$. Equivalently, we can use z^2, which is distributed as χ^2 with one degree of freedom, and reject H_0 if $z^2 \geq (1.96)^2 = 3.84$. If n is large, we are assured by the central limit theorem that z is approximately normal, even if the observations are not from a normal distribution.

We can express z^2 in terms of the square of the standardized distance:

$$z^2 = n\left(\frac{\bar{y} - \mu_0}{\sigma}\right)^2 = n\Delta^2.$$

5.2.2 Multivariate Test for H_0: $\boldsymbol{\mu} = \boldsymbol{\mu}_0$ with $\boldsymbol{\Sigma}$ Known

In the multivariate case we have several variables measured on each sampling unit, and we wish to hypothesize a value for the mean of each variable: H_0: $\boldsymbol{\mu} = \boldsymbol{\mu}_0$ vs. H_1: $\boldsymbol{\mu} \neq \boldsymbol{\mu}_0$. More explicitly, we have

$$H_0: \begin{bmatrix} \mu_1 \\ \mu_2 \\ \vdots \\ \mu_p \end{bmatrix} = \begin{bmatrix} \mu_{01} \\ \mu_{02} \\ \vdots \\ \mu_{0p} \end{bmatrix} \qquad H_1: \begin{bmatrix} \mu_1 \\ \mu_2 \\ \vdots \\ \mu_p \end{bmatrix} \neq \begin{bmatrix} \mu_{01} \\ \mu_{02} \\ \vdots \\ \mu_{0p} \end{bmatrix},$$

where each μ_{0i} is specified from previous experience or is a target value. The vector equality in H_0 implies $\mu_i = \mu_{0i}$ for all $i = 1, 2, \ldots, p$. The vector inequality in H_1 implies at least one $\mu_i \neq \mu_{0i}$. Thus, for example, if $\mu_i = \mu_{0i}$ for all i except 2, for which $\mu_2 \neq \mu_{02}$, then we wish to reject H_0.

To test H_0, we use a random sample of n observation vectors $\mathbf{y}_1, \mathbf{y}_2, \ldots, \mathbf{y}_n$ from $N_p(\boldsymbol{\mu}, \boldsymbol{\Sigma})$, with $\boldsymbol{\Sigma}$ known, and calculate $\bar{\mathbf{y}} = \sum_{i=1}^{n} \mathbf{y}_i/n$. The test statistic is

$$Z^2 = n\Delta^2 = n(\bar{\mathbf{y}} - \boldsymbol{\mu}_0)'\boldsymbol{\Sigma}^{-1}(\bar{\mathbf{y}} - \boldsymbol{\mu}_0), \tag{5.2}$$

where $\Delta^2 = (\bar{\mathbf{y}} - \boldsymbol{\mu}_0)'\boldsymbol{\Sigma}^{-1}(\bar{\mathbf{y}} - \boldsymbol{\mu}_0)$. If H_0 is true, Z^2 is distributed as χ^2_p by (4.6), and we therefore reject H_0 if $Z^2 > \chi^2_{\alpha,p}$. Thus for one variable, z^2 [the square of (5.1)] has a chi-square distribution with one degree of freedom, whereas, for p variables, Z^2 is distributed as chi-square with p degrees of freedom.

If $\boldsymbol{\Sigma}$ is unknown, we could use $\mathbf{S}$ in its place in (5.2) and Z^2 would have an approximate χ^2-distribution. But n would have to be larger than in the analogous univariate situation. The value of n needed for Z^2 to approach the χ^2-distribution depends on p. This will be clarified further in Section 5.3.2.

Example 5.2.2. In Table 3.1, height and weight were given for a sample of 20 college-age males. Let us assume that this sample originated from the bivariate normal $N_2(\boldsymbol{\mu}, \boldsymbol{\Sigma})$, where

$$\boldsymbol{\Sigma} = \begin{pmatrix} 20 & 100 \\ 100 & 1000 \end{pmatrix}.$$

Suppose we wish to test $H_0: \boldsymbol{\mu} = (70, 170)'$. From Example 3.2.1, $\bar{y}_1 = 71.45$ and $\bar{y}_2 = 164.7$. We thus have

$$\begin{aligned} Z^2 &= n(\bar{\mathbf{y}} - \boldsymbol{\mu}_0)'\boldsymbol{\Sigma}^{-1}(\bar{\mathbf{y}} - \boldsymbol{\mu}_0) \\ &= (20)\begin{pmatrix} 71.45 - 70 \\ 164.7 - 170 \end{pmatrix}' \begin{pmatrix} 20 & 100 \\ 100 & 1000 \end{pmatrix}^{-1} \begin{pmatrix} 71.45 - 70 \\ 164.7 - 170 \end{pmatrix} \\ &= (20)(1.45, -5.3)\begin{pmatrix} .1 & -.01 \\ -.01 & .002 \end{pmatrix}\begin{pmatrix} 1.45 \\ -5.3 \end{pmatrix} = 8.4026. \end{aligned}$$

Using $\alpha = .05$, $\chi^2_{.05,2} = 5.99$, and we therefore reject $H_0: \boldsymbol{\mu} = (70, 170)'$ because $Z^2 = 8.4026 > 5.99$.

The rejection region for $\bar{\mathbf{y}} = (\bar{y}_1, \bar{y}_2)'$ is on or outside the ellipse in Figure 5.1; that is, the test statistic $Z^2 = n\Delta^2$ is greater than 5.99 if and only if $\bar{\mathbf{y}}$ is outside the ellipse. If $\bar{\mathbf{y}}$ falls inside the ellipse, H_0 is accepted. Thus distance from $\boldsymbol{\mu}_0$ as well as direction must be taken into account. When the distance is standardized by $\boldsymbol{\Sigma}^{-1}$, all points on the curve are "statistically equidistant" from the center.

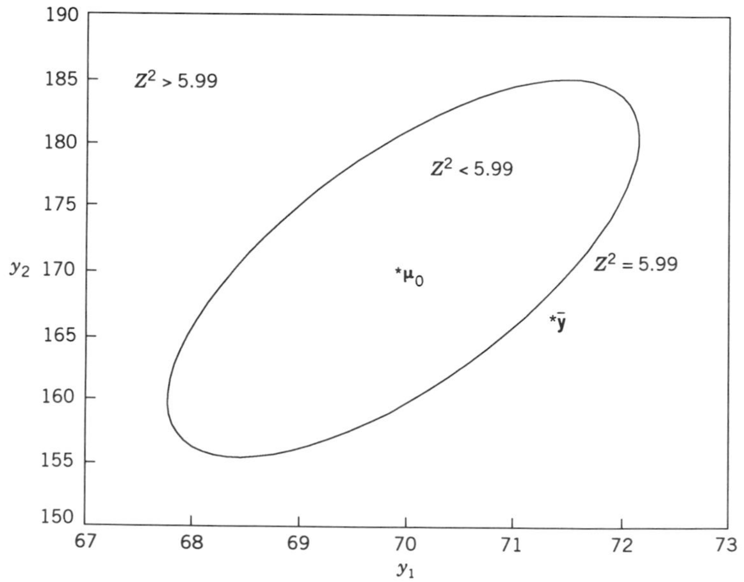

Figure 5.1 Elliptical acceptance region.

Note that the test is sensitive to the covariance structure. If $\text{cov}(y_1, y_2)$ were negative, y_2 would decrease as y_1 increases, and the ellipse would be tilted in the other direction. In this case, $\bar{\mathbf{y}}$ would be in the acceptance region.

Let us now investigate the consequence of testing each variable separately. Using $z_{\alpha/2} = 1.96$ for $\alpha = .05$, we have

$$z_1 = \frac{\bar{y}_1 - \mu_{01}}{\sigma_1/\sqrt{n}} = 1.450 < 1.96$$

and

$$z_2 = \frac{\bar{y}_2 - \mu_{02}}{\sigma_2/\sqrt{n}} = -.7495 > -1.96.$$

Thus both tests accept the hypothesis. In this case neither of the $\bar{y}$'s is far enough from the hypothesized value to cause rejection. But when the positive correlation between y_1 and y_2 is taken into account in the multivariate test, the two evidences against $\boldsymbol{\mu}_0$ combine to cause rejection. This illustrates the third advantage of multivariate tests given in Section 5.1.

Figure 5.2 shows the rectangular acceptance region for the univariate tests superimposed on the elliptical multivariate acceptance region. The rectangle

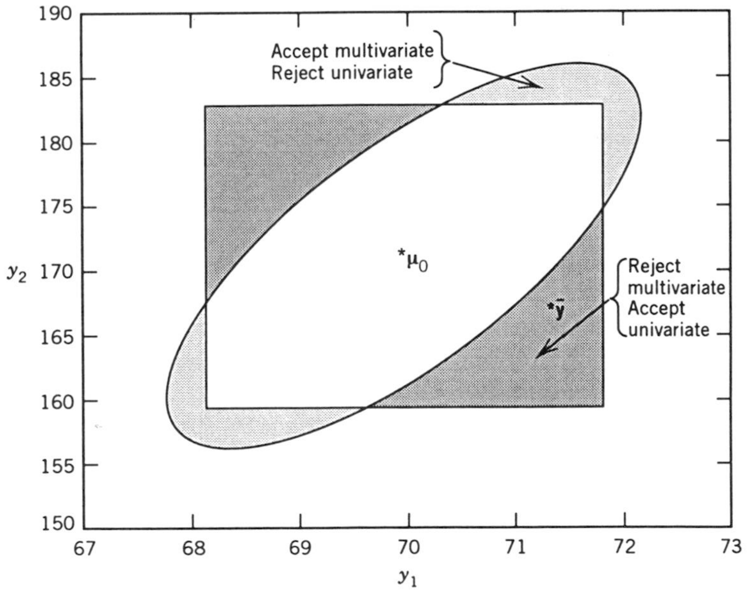

Figure 5.2 Acceptance and rejection regions for univariate and multivariate tests.

was obtained by calculating the two acceptance regions

$$\mu_{01} - 1.96 \frac{\sigma_1}{\sqrt{n}} < \bar{y}_1 < \mu_{01} + 1.96 \frac{\sigma_1}{\sqrt{n}}$$

and

$$\mu_{02} - 1.96 \frac{\sigma_2}{\sqrt{n}} < \bar{y}_2 < \mu_{02} + 1.96 \frac{\sigma_2}{\sqrt{n}}.$$

Points inside the ellipse but outside the rectangle will be rejected in at least one univariate dimension but will be accepted multivariately. This illustrates the inflation of α resulting from univariate tests as discussed in the first reason for multivariate testing in Section 5.1. This phenomenon has been referred to as Rao's paradox. For further discussion see Rao (1966), Healy (1969), and Morrison (1990, p. 174). Points outside the ellipse but inside the rectangle will be rejected multivariately but accepted univariately in both dimensions. In such cases, the multivariate test is more powerful.

Thus in either case represented by the shaded areas, we should use the multivariate test result, not the univariate results. In the one case, the multivariate test is more powerful than the univariate tests; in the other case, the multivariate test preserves α while the univariate tests inflate α. Consequently, when the multivariate and univariate results disagree, our tendency is to trust the multivariate result. In Section 5.5, we discuss various procedures for ascertaining the contribution of the individual variables after the multivariate test has rejected the hypothesis.

5.3 TESTS ON $\boldsymbol{\mu}$ WHEN $\boldsymbol{\Sigma}$ IS UNKNOWN

In Section 5.2, we said little about properties of the tests because the tests discussed were of slight practical consequence due to the assumption that $\boldsymbol{\Sigma}$ is known. We will pay more attention to test properties in the next two sections, first in the one-sample case and then in the two-sample case. The reader may wonder why we include one-sample tests, since we seldom, if ever, have need of a test for $H_0\colon \boldsymbol{\mu} = \boldsymbol{\mu}_0$. However, we will cover this case for two reasons:

1. Many general principles are more easily illustrated in the one-sample framework than in the two-sample case.
2. Some very useful tests can be cast in the one-sample framework. Two examples are (1) $H_0\colon \boldsymbol{\mu}_d = \mathbf{0}$ used in the paired comparison test covered in Section 5.7 and (2) $H_0\colon \mathbf{C}\boldsymbol{\mu} = \mathbf{0}$ used in the profile analysis in Section 5.9, in the analysis of repeated measures in Section 6.9, and in the growth curves in Section 6.10.

5.3.1 Review of Univariate t-Test for $H_0: \mu = \mu_0$ with σ Unknown

We first review the familiar one-sample t-test in the univariate case, with only one variable measured on each sampling unit. We assume that a random sample $y_1, y_2, \ldots, y_n$ is available from $N(\mu, \sigma^2)$. We estimate μ by $\bar{y}$ and σ^2 by s^2, where $\bar{y}$ and s^2 are given by (3.1) and (3.4). To test $H_0: \mu = \mu_0$ vs. $H_1: \mu \neq \mu_0$, we use

$$t = \frac{\bar{y} - \mu_0}{s/\sqrt{n}} = \frac{\sqrt{n}(\bar{y} - \mu_0)}{s}. \tag{5.3}$$

If H_0 is true, t is distributed as t_{n-1}, where $n - 1$ is the degrees of freedom. We reject H_0 if $|\sqrt{n}(\bar{y} - \mu_0)/s| \geq t_{\alpha/2,n-1}$, where $t_{\alpha/2,n-1}$ is a critical value from the t-table.

The first expression in (5.3), $t = (\bar{y} - \mu_0)/(s/\sqrt{n})$, is the *characteristic form* of the t-statistic, which represents a sample standardized distance between $\bar{y}$ and μ_0. In this form, the hypothesized mean is subtracted from $\bar{y}$ and the difference is divided by $s/\sqrt{n}$, the sample standard deviation of $\bar{y}$. We will see an analogous form for the T^2-statistic in the multivariate case in Section 5.3.2.

5.3.2 Hotelling's T^2-Test for $H_0: \boldsymbol{\mu} = \boldsymbol{\mu}_0$ with $\boldsymbol{\Sigma}$ Unknown

We now move to the case in which p variables are measured on each sampling unit. We assume that a random sample $\mathbf{y}_1, \mathbf{y}_2, \ldots, \mathbf{y}_n$ is available from $N_p(\boldsymbol{\mu}, \boldsymbol{\Sigma})$, where $\mathbf{y}_i$ contains the p measurements on the ith sampling unit (subject or object). We estimate $\boldsymbol{\mu}$ by $\bar{\mathbf{y}}$ and $\boldsymbol{\Sigma}$ by $\mathbf{S}$. In order to test $H_0: \boldsymbol{\mu} = \boldsymbol{\mu}_0$ versus $H_1: \boldsymbol{\mu} \neq \boldsymbol{\mu}_0$, we use an extension of the univariate t in (5.3). In squared form, the univariate t can be rewritten as

$$t^2 = \frac{n(\bar{y} - \mu_0)^2}{s^2} = n(\bar{y} - \mu_0)(s^2)^{-1}(\bar{y} - \mu_0).$$

When $\bar{y} - \mu_0$ and s^2 are replaced by $\bar{\mathbf{y}} - \boldsymbol{\mu}_0$ and $\mathbf{S}$, we obtain the test statistic

$$T^2 = n(\bar{\mathbf{y}} - \boldsymbol{\mu}_0)'\mathbf{S}^{-1}(\bar{\mathbf{y}} - \boldsymbol{\mu}_0). \tag{5.4}$$

Alternatively, T^2 can be obtained from Z^2 in (5.2) by replacing $\boldsymbol{\Sigma}$ with $\mathbf{S}$.

The distribution of T^2 was obtained by Hotelling (1931), assuming H_0 is true and sampling is from $N_p(\boldsymbol{\mu}, \boldsymbol{\Sigma})$. The distribution is indexed by two parameters, the dimension p and degrees of freedom $n - 1$. We reject H_0 if $T^2 > T^2_{\alpha,p,n-1}$ and accept H_0 otherwise. Critical values of the T^2-distribution are found in Table A.7, taken from Kramer and Jensen (1969a).

Note that the terminology "accept H_0" is used for expositional convenience to describe our decision when we do not reject the hypothesis. Strictly speaking, we do not accept H_0 in the sense of actually believing it is true. If the sample size were extremely large and we accepted H_0, we could be reasonably certain that the true $\boldsymbol{\mu}$ is close to the hypothesized value $\boldsymbol{\mu}_0$. Otherwise, "accepting H_0" means only that we have failed to reject it.

The T^2-statistic can be expressed as $T^2 = nD^2$, where $D^2 = (\overline{\mathbf{y}} - \boldsymbol{\mu}_0)'\mathbf{S}^{-1}(\overline{\mathbf{y}} - \boldsymbol{\mu}_0)$ is the sample standardized distance. Thus we can view the test from the viewpoint of distance between the observed sample mean vector and the hypothetical mean vector. If the sample mean vector is notably distant from the hypothetical mean vector, we become suspicious of the hypothetical mean vector and wish to reject H_0.

The test statistic is a scalar (univariate) quantity, since $T^2 = n(\overline{\mathbf{y}} - \boldsymbol{\mu}_0)'\mathbf{S}^{-1}(\overline{\mathbf{y}} - \boldsymbol{\mu}_0)$ is a quadratic form. As with the χ^2-distribution of Z^2, the density of T^2 is skewed because the lower limit is zero and there is no upper limit.

The *characteristic form* of the T^2-statistic (5.4) is

$$T^2 = (\overline{\mathbf{y}} - \boldsymbol{\mu}_0)' \left(\frac{\mathbf{S}}{n} \right)^{-1} (\overline{\mathbf{y}} - \boldsymbol{\mu}_0). \tag{5.5}$$

The characteristic form has two features:

1. $\mathbf{S}/n$ is the sample covariance matrix of $\overline{\mathbf{y}}$ and serves as a standardizing matrix in the distance function.
2. $\overline{\mathbf{y}}$ and $\mathbf{S}$ are independent because they are based on a random sample from a multivariate normal distribution (see Section 4.3.2).

In (5.3), the univariate t-statistic is the number of standard deviations $\overline{y}$ is separated from μ_0. In appearance, the T^2-statistic (5.5) is similar, but no such simple interpretation is possible. The distance in (5.5) is in p dimensions; and if we add a variable, the distance increases. (By analogy, the hypotenuse of a right triangle is longer than either of the legs.) Thus we need a test statistic that indicates the significance of the distance from $\overline{\mathbf{y}}$ to $\boldsymbol{\mu}_0$, while allowing for the number of dimensions (see comment 3 below about the T^2-table). Since the resulting T^2-statistic cannot be readily interpreted in terms of the number of standard deviations $\overline{\mathbf{y}}$ is from $\boldsymbol{\mu}_0$, we do not have an intuitive feel for its significance as we do with the univariate t. If $\mathbf{S}$ were diagonal, T^2 would be expressible as the sum of terms like $(\overline{y}_i - \mu_{0i})^2/(s_i^2/n)$, but $\mathbf{S}$ is never diagonal (unless the y's are replaced by principal components; see Chapter 12). To express T^2 as a sum of squared terms, we could factor it as in Section 3.12:

$$T^2 = (\bar{\mathbf{y}} - \boldsymbol{\mu}_0)' \left(\frac{\mathbf{S}}{n} \right)^{-1} (\bar{\mathbf{y}} - \boldsymbol{\mu}_0)$$

$$= \left[\left(\frac{\mathbf{S}^{1/2}}{\sqrt{n}} \right)^{-1} (\bar{\mathbf{y}} - \boldsymbol{\mu}_0) \right]' \left[\left(\frac{\mathbf{S}^{1/2}}{\sqrt{n}} \right)^{-1} (\bar{\mathbf{y}} - \boldsymbol{\mu}_0) \right] = \mathbf{u}'\mathbf{u},$$

where $\mathbf{u} = (\mathbf{S}^{1/2}/\sqrt{n})^{-1}(\bar{\mathbf{y}} - \boldsymbol{\mu}_0)$. Now $\mathbf{u}'\mathbf{u} = \sum_{i=1}^{p} u_i^2$. But u_i, the ith element of $\mathbf{u}$, is a linear combination of all elements of $\bar{\mathbf{y}} - \boldsymbol{\mu}_0$ and is thus not associated exclusively with $\bar{y}_i - \mu_{0i}$. Hence the u_i's or u_i^2's do not provide an intuitive grasp of the significance of T^2; we must compare the calculated value with the table value. However, the T^2-table provides some insights into the behavior of the T^2-distribution. Four of these insights are noted at the end of this section.

If a test leads to rejection of $H_0: \boldsymbol{\mu} = \boldsymbol{\mu}_0$, the question arises as to which variable or variables contributed most to the rejection. This issue is discussed in Section 5.5 for the two-sample T^2-test of $H_0: \boldsymbol{\mu}_1 = \boldsymbol{\mu}_2$, and the results there can be easily adapted to the one-sample test of $H_0: \boldsymbol{\mu} = \boldsymbol{\mu}_0$. For confidence intervals on the individual μ_i, see Rencher (1996, Section 3.4).

The following are some key properties of the T^2-test:

1. We must have $n - 1 > p$. Otherwise, $\mathbf{S}$ is singular and T^2 cannot be computed.
2. In both the one-sample and two-sample cases, the degrees of freedom for the T^2-statistic will be the same as for the analogous univariate t-test, that is, $n - 1$ for one sample and $n_1 + n_2 - 2$ for two samples.
3. The alternative hypothesis is two-sided. Because the space is multidimensional, we do not consider one-sided alternative hypotheses, such as $\boldsymbol{\mu} > \boldsymbol{\mu}_0$. However, even though the alternative hypothesis $H_1: \boldsymbol{\mu} \neq \boldsymbol{\mu}_0$ is essentially two-sided, the critical region is one-tailed (we reject for large values). This is typical of many multivariate tests.
4. In the univariate case, $t_{n-1}^2 = F_{1,n-1}$. The statistic T^2 can also be converted to an F-statistic as follows:

$$\frac{\nu - p + 1}{\nu p} T_{p,\nu}^2 = F_{p,\nu-p+1}. \tag{5.6}$$

Note that the dimension p (number of variables) of the T^2-statistic becomes the first of the two degrees-of-freedom parameters of the F. The degrees of freedom for T^2 is denoted by ν; in this case $\nu = n - 1$. The F transformation is given here in terms of a general ν, since other applications of T^2 will have ν different from $n - 1$.

Equation (5.6) gives an easy way to find critical values for the T^2-test. However, we have included critical values of T^2 in Table A.7 because of the insights they provide into the behavior of the T^2-distribution in particular and multivariate tests in general. The following are some insights that can readily be gleaned from the T^2-tables:

1. The first column contains squares of t-table values; that is, $T^2_{\alpha,1,\nu} = t^2_{\alpha/2,\nu}$. (We use $t^2_{\alpha/2}$ because the univariate test of $H_0\colon \mu = \mu_0$ versus $H_1\colon \mu \neq \mu_0$ is two-tailed.) Thus for $p = 1$, T^2 reduces to t^2. This can easily be seen by comparing (5.4) with the preceding equation.
2. The last row contains χ^2 critical values, that is, $T^2_{p,\infty} = \chi^2_p$. Thus as n increases, $\mathbf{S}$ approaches $\boldsymbol{\Sigma}$, and $T^2 = n(\overline{\mathbf{y}} - \boldsymbol{\mu}_0)'\mathbf{S}^{-1}(\overline{\mathbf{y}} - \boldsymbol{\mu}_0)$ approaches $Z^2 = n(\overline{\mathbf{y}} - \boldsymbol{\mu}_0)'\boldsymbol{\Sigma}^{-1}(\overline{\mathbf{y}} - \boldsymbol{\mu}_0)$ in (5.2), which is distributed as χ^2_p.
3. The values increase along each row; that is, for a fixed ν, the critical value $T^2_{\alpha,p,\nu}$ increases with p. It was noted above that in any given sample, the calculated value of T^2 increases if a variable is added. However, since the critical value also increases, a variable should not be added unless it adds a significant amount to T^2.
4. As p increases, larger values of ν are required for T^2 to approach χ^2. In the univariate case, t in (5.3) is considered a good approximation to the standard normal z in (5.1) when ν is at least 30. In the first column ($p = 1$) of Table A.7, we see $T^2_{.05,1,30} = 4.171$ and $T^2_{.05,1,\infty} = 3.841$ with a ratio of $4.171/3.841 = 1.086$. For $p = 5$, ν must be 100 to obtain the same ratio: $T^2_{.05,5,100}/T^2_{.05,5,\infty} = 1.086$. For $p = 10$, we need $\nu = 200$ to obtain a similar value of the ratio: $T^2_{.05,10,200}/T^2_{.05,10,\infty} = 1.076$. Thus one must be very cautious in stating that T^2 has an approximate χ^2-distribution for large n. The α level (Type I error rate) could be substantially inflated. For example, if $p = 10$ and we assumed $n = 30$ was sufficiently large for a χ^2 approximation to hold, we would reject for $T^2 \geq 18.307$ with a target α level of .05. However, the correct critical value is 34.044, and the misuse of 18.307 would yield an actual α of $P(T^2_{10,29} \geq 18.307) = .314$.

Example 5.3.2. In Table 3.5 we have $n = 10$ observations on $p = 3$ variables: y_1 is available soil calcium, y_2 is exchangeable soil calcium, and y_3 is calcium content in turnip greens. Desirable levels for y_1 and y_2 are 15.0 and 6.0, respectively, and the expected level of y_3 is 2.85. We can, therefore, test the hypothesis

$$H_0\colon \boldsymbol{\mu} = \begin{pmatrix} 15.0 \\ 6.0 \\ 2.85 \end{pmatrix}.$$

In Examples 3.5 and 3.6, $\overline{\mathbf{y}}$ and $\mathbf{S}$ were obtained as

$$\overline{\mathbf{y}} = \begin{pmatrix} 28.1 \\ 7.18 \\ 3.09 \end{pmatrix} \qquad \mathbf{S} = \begin{pmatrix} 140.54 & 49.68 & 1.94 \\ 49.68 & 72.25 & 3.68 \\ 1.94 & 3.68 & .25 \end{pmatrix}.$$

To test H_0, we use (5.4):

$$\begin{aligned} T^2 &= n(\overline{\mathbf{y}} - \boldsymbol{\mu}_0)'\mathbf{S}^{-1}(\overline{\mathbf{y}} - \boldsymbol{\mu}_0) \\ &= 10 \begin{pmatrix} 28.1 - 15.0 \\ 7.18 - 6.0 \\ 3.09 - 2.85 \end{pmatrix}' \begin{pmatrix} 140.54 & 49.68 & 1.94 \\ 49.68 & 72.25 & 3.68 \\ 1.94 & 3.68 & .25 \end{pmatrix}^{-1} \begin{pmatrix} 28.1 - 15.0 \\ 7.18 - 6.0 \\ 3.09 - 2.85 \end{pmatrix} \\ &= 24.559. \end{aligned}$$

From Table A.7, we obtain the critical value $T^2_{.05,3,9} = 16.766$. Since the observed value of T^2 exceeds the critical value, we reject the hypothesis.

5.4 COMPARING TWO MEAN VECTORS

We will first review the univariate two-sample t-test and then proceed with the analogous multivariate test.

5.4.1 Review of Univariate Two-Sample t-test

In the one-variable case we obtain a random sample $y_{11}, y_{12}, \ldots, y_{1n_1}$ from $N(\mu_1, \sigma_1^2)$ and a second random sample $y_{21}, y_{22}, \ldots, y_{2n_2}$ from $N(\mu_2, \sigma_2^2)$. We assume that the two samples are independent and that $\sigma_1^2 = \sigma_2^2 = \sigma^2$, say, with σ^2 unknown. [The assumptions of independence and equal variances are necessary in order for the t-statistic below in (5.7) to have a t-distribution.] From the two samples we calculate $\overline{y}_1$, $\overline{y}_2$, $\text{SS}_1 = \sum_{j=1}^{n_1}(y_{1j} - \overline{y}_1)^2 = (n_1 - 1)s_1^2$, $\text{SS}_2 = \sum_{j=1}^{n_2}(y_{2j} - \overline{y}_2)^2 = (n_2 - 1)s_2^2$, and the pooled variance

$$s_{\text{pl}}^2 = \frac{\text{SS}_1 + \text{SS}_2}{n_1 + n_2 - 2} = \frac{(n_1 - 1)s_1^2 + (n_2 - 1)s_2^2}{n_1 + n_2 - 2},$$

where $n_1 + n_2 - 2$ is the sum of the weights $n_1 - 1$ and $n_2 - 1$ in the numerator. With this denominator, s_{pl}^2 is an unbiased estimator for the common variance, σ^2, that is, $E(s_{\text{pl}}^2) = \sigma^2$.

To test

$$H_0\colon \mu_1 = \mu_2 \quad \text{vs.} \quad H_1\colon \mu_1 \neq \mu_2,$$

we use

$$t = \frac{\bar{y}_1 - \bar{y}_2}{s_{\text{pl}}\sqrt{\dfrac{1}{n_1} + \dfrac{1}{n_2}}}, \tag{5.7}$$

which has a t-distribution with $n_1 + n_2 - 2$ degrees of freedom when H_0 is true. We therefore reject if $|t| \geq t_{\alpha/2, n_1+n_2-2}$.

Note that (5.7) exhibits the *characteristic form* of a t-statistic. In this form, the denominator is the sample standard deviation of the numerator, that is, $s_{\text{pl}}\sqrt{1/n_1 + 1/n_2}$ is an estimate of

$$\sigma_{\bar{y}_1-\bar{y}_2} = \sqrt{\text{var}(\bar{y}_1 - \bar{y}_2)} = \sqrt{\frac{\sigma_1^2}{n_1} + \frac{\sigma_2^2}{n_2}}$$

$$= \sqrt{\frac{\sigma^2}{n_1} + \frac{\sigma^2}{n_2}} = \sigma\sqrt{\frac{1}{n_1} + \frac{1}{n_2}}.$$

5.4.2 Multivariate Two-Sample T^2-Test

We now consider the case where p variables are measured on each sampling unit in two samples. We wish to test

$$H_0: \boldsymbol{\mu}_1 = \boldsymbol{\mu}_2 \quad \text{vs.} \quad H_1: \boldsymbol{\mu}_1 \neq \boldsymbol{\mu}_2.$$

We obtain a random sample $\mathbf{y}_{11}, \mathbf{y}_{12}, \ldots, \mathbf{y}_{1n_1}$ from $N_p(\boldsymbol{\mu}_1, \boldsymbol{\Sigma}_1)$ and a second random sample $\mathbf{y}_{21}, \mathbf{y}_{22}, \ldots, \mathbf{y}_{2n_2}$ from $N_p(\boldsymbol{\mu}_2, \boldsymbol{\Sigma}_2)$. We assume that the two samples are independent and that $\boldsymbol{\Sigma}_1 = \boldsymbol{\Sigma}_2 = \boldsymbol{\Sigma}$, say, with $\boldsymbol{\Sigma}$ unknown. These assumptions are necessary in order for the T^2-statistic in (5.8) below to have a T^2-distribution. A test of $H_0: \boldsymbol{\Sigma}_1 = \boldsymbol{\Sigma}_2$ is given in Section 7.3.2. For an approximate test that can be used when $\boldsymbol{\Sigma}_1 \neq \boldsymbol{\Sigma}_2$, see Rencher (1996, Section 3.9).

The sample mean vectors are $\bar{\mathbf{y}}_1 = \sum_{j=1}^{n_1} \mathbf{y}_{1j}/n_1$ and $\bar{\mathbf{y}}_2 = \sum_{j=1}^{n_2} \mathbf{y}_{2j}/n_2$. Define $\mathbf{W}_1$ and $\mathbf{W}_2$ to be the matrices of sums of squares and cross products for the two samples:

$$\mathbf{W}_1 = \sum_{j=1}^{n_1} (\mathbf{y}_{1j} - \bar{\mathbf{y}}_1)(\mathbf{y}_{1j} - \bar{\mathbf{y}}_1)' = (n_1 - 1)\mathbf{S}_1$$

$$\mathbf{W}_2 = \sum_{j=1}^{n_2} (\mathbf{y}_{2j} - \bar{\mathbf{y}}_2)(\mathbf{y}_{2j} - \bar{\mathbf{y}}_2)' = (n_2 - 1)\mathbf{S}_2.$$

Since $(n_1 - 1)\mathbf{S}_1$ is an unbiased estimate of $(n_1 - 1)\boldsymbol{\Sigma}$ and $(n_2 - 1)\mathbf{S}_2$ is an unbiased estimate of $(n_2 - 1)\boldsymbol{\Sigma}$, we can pool them to obtain an unbiased estimate of the common population covariance matrix, $\boldsymbol{\Sigma}$:

$$\mathbf{S}_{\text{pl}} = \frac{1}{n_1 + n_2 - 2}(\mathbf{W}_1 + \mathbf{W}_2)$$
$$= \frac{1}{n_1 + n_2 - 2}[(n_1 - 1)\mathbf{S}_1 + (n_2 - 1)\mathbf{S}_2].$$

Thus $E(\mathbf{S}_{\text{pl}}) = \boldsymbol{\Sigma}$.

The univariate t^2-statistic (5.7) can be expressed as

$$t^2 = \frac{n_1 n_2}{n_1 + n_2}(\bar{y}_1 - \bar{y}_2)(s_{\text{pl}}^2)^{-1}(\bar{y}_1 - \bar{y}_2).$$

This can be generalized to p variables by substituting $\bar{\mathbf{y}}_1 - \bar{\mathbf{y}}_2$ for $\bar{y}_1 - \bar{y}_2$ and $\mathbf{S}_{\text{pl}}$ for s_{pl}^2 to obtain

$$T^2 = \frac{n_1 n_2}{n_1 + n_2}(\bar{\mathbf{y}}_1 - \bar{\mathbf{y}}_2)'\mathbf{S}_{\text{pl}}^{-1}(\bar{\mathbf{y}}_1 - \bar{\mathbf{y}}_2), \tag{5.8}$$

which is distributed as T^2_{p,n_1+n_2-2} when $H_0\colon \boldsymbol{\mu}_1 = \boldsymbol{\mu}_2$ is true. To carry out the test, we collect the two samples, calculate T^2 by (5.8), and reject H_0 if $T^2 \geq T^2_{\alpha,p,n_1+n_2-2}$. Critical values of T^2 are found in Table A.7. For tables of the power of the T^2-test (probability of rejecting H_0 when it is false) and illustrations of their use, see Rencher (1996, Section 3.10).

The T^2-statistic (5.8) can be expressed in *characteristic form* as

$$T^2 = (\bar{\mathbf{y}}_1 - \bar{\mathbf{y}}_2)'\left[\left(\frac{1}{n_1} + \frac{1}{n_2}\right)\mathbf{S}_{\text{pl}}\right]^{-1}(\bar{\mathbf{y}}_1 - \bar{\mathbf{y}}_2), \tag{5.9}$$

where $(1/n_1 + 1/n_2)\,\mathbf{S}_{\text{pl}}$ is the sample covariance matrix for $\bar{\mathbf{y}}_1 - \bar{\mathbf{y}}_2$ and $\mathbf{S}_{\text{pl}}$ is independent of $\bar{\mathbf{y}}_1 - \bar{\mathbf{y}}_2$ because of sampling from the multivariate normal. For a discussion of robustness of T^2 to departures from the assumptions of multivariate normality and $\boldsymbol{\Sigma}_1 = \boldsymbol{\Sigma}_2$, see Rencher (1996, Section 3.7).

Some key properties of the two-sample T^2-test are given in the following list:

1. It is necessary that $n_1 + n_2 - 2 > p$ for $\mathbf{S}_{\text{pl}}$ to be nonsingular.
2. The T^2-statistic can be expressed in terms of the sample standardized distance between $\bar{\mathbf{y}}_1$ and $\bar{\mathbf{y}}_2$:

$$T^2 = \frac{n_1 n_2}{n_1 + n_2} D^2, \tag{5.10}$$

where

$$D^2 = (\bar{\mathbf{y}}_1 - \bar{\mathbf{y}}_2)' \mathbf{S}_{\text{pl}}^{-1} (\bar{\mathbf{y}}_1 - \bar{\mathbf{y}}_2). \tag{5.11}$$

3. The statistic T^2 is, of course, a scalar. The $3p + p(p-1)/2$ quantities in $\bar{\mathbf{y}}_1, \bar{\mathbf{y}}_2$, and $\mathbf{S}_{\text{pl}}$ have been reduced to a single number. Thus we have a single scale on which T^2 is large if the sample evidence favors $H_1\colon \boldsymbol{\mu}_1 \neq \boldsymbol{\mu}_2$ and small if the evidence supports $H_0\colon \boldsymbol{\mu}_1 = \boldsymbol{\mu}_2$; we reject if the standardized distance between $\bar{\mathbf{y}}_1$ and $\bar{\mathbf{y}}_2$ is large.
4. Since the lower limit of T^2 is zero and there is no upper limit, we would expect the density to be skewed. In fact, as noted in (5.12) below, T^2 is directly related to F, which is a well-known skewed distribution.
5. For degrees of freedom of T^2 we have $n_1 + n_2 - 2$, which is the same as for the corresponding univariate t-statistic (5.7).
6. The alternative hypothesis $H_1\colon \boldsymbol{\mu}_1 \neq \boldsymbol{\mu}_2$ is two sided. A one-sided version of H_1 is not viable. The critical region $T^2 > T^2_\alpha$ is one-tailed, however, as is typical of many multivariate tests.
7. The T^2-statistic can be readily transformed to an F-statistic using (5.6):

$$\frac{n_1 + n_2 - p - 1}{(n_1 + n_2 - 2)p} T^2 = F_{p, n_1 + n_2 - p - 1}, \tag{5.12}$$

where again the dimension p of the T^2-statistic becomes the first degree-of-freedom parameter for the F-statistic.

Example 5.4.2. Four psychological tests were given to 32 men and 32 women. The data are recorded in Table 5.1 (Beall 1945). The variables are

y_1 = pictorial inconsistencies $\quad$ y_3 = tool recognition
y_2 = paper form board $\quad$ y_4 = vocabulary

The mean vectors and covariance matrices of the two samples are

Table 5.1 Four Psychological Test Scores on 32 Males and 32 Females

Males				Females			
y_1	y_2	y_3	y_4	y_1	y_2	y_3	y_4
15	17	24	14	13	14	12	21
17	15	32	26	14	12	14	26
15	14	29	23	12	19	21	21
13	12	10	16	12	13	10	16
20	17	26	28	11	20	16	16
15	21	26	21	12	9	14	18
15	13	26	22	10	13	18	24
13	5	22	22	10	8	13	23
14	7	30	17	12	20	19	23
17	15	30	27	11	10	11	27
17	17	26	20	12	18	25	25
17	20	28	24	14	18	13	26
15	15	29	24	14	10	25	28
18	19	32	28	13	16	8	14
18	18	31	27	14	8	13	25
15	14	26	21	13	16	23	28
18	17	33	26	16	21	26	26
10	14	19	17	14	17	14	14
18	21	30	29	16	16	15	23
18	21	34	26	13	16	23	24
13	17	30	24	2	6	16	21
16	16	16	16	14	16	22	26
11	15	25	23	17	17	22	28
16	13	26	16	16	13	16	14
16	13	23	21	15	14	20	26
18	18	34	24	12	10	12	9
16	15	28	27	14	17	24	23
15	16	29	24	13	15	18	20
18	19	32	23	11	16	18	28
18	16	33	23	7	7	19	18
17	20	21	21	12	15	7	28
19	19	30	28	6	5	6	13

$$\overline{\mathbf{y}}_1 = \begin{bmatrix} 15.97 \\ 15.91 \\ 27.19 \\ 22.75 \end{bmatrix} \qquad \overline{\mathbf{y}}_2 = \begin{bmatrix} 12.34 \\ 13.91 \\ 16.59 \\ 21.94 \end{bmatrix}$$

$$\mathbf{S}_1 = \begin{bmatrix} 5.192 & 4.545 & 6.522 & 5.250 \\ 4.545 & 13.18 & 6.760 & 6.266 \\ 6.522 & 6.760 & 28.67 & 14.47 \\ 5.250 & 6.266 & 14.47 & 16.65 \end{bmatrix}$$

$$\mathbf{S}_2 = \begin{bmatrix} 9.136 & 7.549 & 5.531 & 4.151 \\ 7.549 & 18.60 & 10.73 & 5.446 \\ 5.531 & 10.73 & 30.25 & 13.55 \\ 4.151 & 5.446 & 13.55 & 28.00 \end{bmatrix}.$$

The sample covariance matrices do not appear to indicate a disparity in the population covariance matrices. (A significance test to check this assumption is carried out in Example 7.3.2, and the hypothesis $H_0: \mathbf{\Sigma}_1 = \mathbf{\Sigma}_2$ is not rejected.) The pooled covariance matrix is

$$\mathbf{S}_{\text{pl}} = \frac{1}{32+32-2}[(32-1)\mathbf{S}_1 + (32-1)\mathbf{S}_2]$$

$$= \begin{bmatrix} 7.164 & 6.047 & 6.027 & 4.701 \\ 6.047 & 15.89 & 8.747 & 5.856 \\ 6.027 & 8.747 & 29.46 & 14.01 \\ 4.701 & 5.856 & 14.01 & 22.32 \end{bmatrix}.$$

By (5.8), we obtain

$$T^2 = \frac{n_1 n_2}{n_1 + n_2}(\bar{\mathbf{y}}_1 - \bar{\mathbf{y}}_2)'\mathbf{S}_{\text{pl}}^{-1}(\bar{\mathbf{y}}_1 - \bar{\mathbf{y}}_2) = 96.603.$$

From interpolation in Table A.7, we obtain $T^2_{.01,4,62} = 15.373$, and we therefore reject $H_0: \boldsymbol{\mu}_1 = \boldsymbol{\mu}_2$.

5.4.3 Likelihood Ratio Tests

The maximum likelihood approach to estimation was introduced in Section 4.3.1. As noted there, the likelihood function is the joint density of $\mathbf{y}_1, \mathbf{y}_2, \ldots, \mathbf{y}_n$. The values of the parameters that maximize the likelihood function are the maximum likelihood estimators.

The *likelihood ratio* method of test construction uses the ratio of the maximum value of the likelihood function assuming H_0 is true to the maximum under H_1. The maximum under H_1 is essentially unrestricted. Likelihood ratio tests usually have good power and sometimes have optimum power over a wide class of alternatives.

When applied to multivariate normal samples and $H_0: \boldsymbol{\mu}_1 = \boldsymbol{\mu}_2$, the likelihood ratio approach leads directly to Hotelling's T^2-test in (5.8). Similarly, in the one-sample case, the T^2-statistic (5.4) is the likelihood ratio test. Thus the T^2-test, which we introduced rather informally, is the best test according to certain criteria.

5.5 TESTS ON INDIVIDUAL VARIABLES CONDITIONAL ON REJECTION OF H_0 BY THE T^2-TEST

If the hypothesis $H_0: \boldsymbol{\mu}_1 = \boldsymbol{\mu}_2$ is rejected, the implication is that $\mu_{1i} \neq \mu_{2i}$ for at least one $i = 1, 2, \ldots, p$. But there is no guarantee that $H_0: \mu_{1i} = \mu_{2i}$ will be rejected for some i by a univariate test. However, if we consider a linear combination of the variables, $z = \mathbf{a}'\mathbf{y}$, then there is at least one coefficient vector $\mathbf{a}$ for which

$$t(\mathbf{a}) = \frac{\bar{z}_1 - \bar{z}_2}{\sqrt{(1/n_1 + 1/n_2)\, s_z^2}} \qquad (5.13)$$

will reject the corresponding hypothesis $H_0: \mu_{z_1} = \mu_{z_2}$ or $H_0: \mathbf{a}'\boldsymbol{\mu}_1 = \mathbf{a}'\boldsymbol{\mu}_2$. By (3.51), $\bar{z}_i = \mathbf{a}'\bar{\mathbf{y}}_i$, and from (3.52) the variance estimator s_z^2 is the pooled estimator $\mathbf{a}'\mathbf{S}_{\text{pl}}\mathbf{a}$. Thus (5.13) can be written

$$t(\mathbf{a}) = \frac{\mathbf{a}'\bar{\mathbf{y}}_1 - \mathbf{a}'\bar{\mathbf{y}}_2}{\sqrt{[(n_1 + n_2)/n_1 n_2]\, \mathbf{a}'\mathbf{S}_{\text{pl}}\mathbf{a}}}. \qquad (5.14)$$

Since $t(\mathbf{a})$ can be negative, we work with $t^2(\mathbf{a})$. To find $\mathbf{a}$ for which $z = \mathbf{a}'\mathbf{y}$ will lead to rejection of H_0, we seek the vector $\mathbf{a}$ that maximizes $t^2(\mathbf{a})$. This vector turns out to be

$$\mathbf{a} = \mathbf{S}_{\text{pl}}^{-1}(\bar{\mathbf{y}}_1 - \bar{\mathbf{y}}_2). \qquad (5.15)$$

When $\mathbf{a} = \mathbf{S}_{\text{pl}}^{-1}(\bar{\mathbf{y}}_1 - \bar{\mathbf{y}}_2)$ is used in $z = \mathbf{a}'\mathbf{y}$, then $z = \mathbf{a}'\mathbf{y}$ is called the *discriminant function*. Sometimes the vector $\mathbf{a}$ itself in (5.15) is loosely referred to as the discriminant function.

If $H_0: \boldsymbol{\mu}_1 = \boldsymbol{\mu}_2$ is rejected by T^2 in (5.8), the discriminant function $\mathbf{a}'\mathbf{y}$ will lead to rejection of $H_0: \mathbf{a}'\boldsymbol{\mu}_1 = \mathbf{a}'\boldsymbol{\mu}_2$, using (5.14), with $\mathbf{a} = \mathbf{S}_{\text{pl}}^{-1}(\bar{\mathbf{y}}_1 - \bar{\mathbf{y}}_2)$. We can then examine each a_i in $\mathbf{a}$ for an indication of the contribution of the corresponding y_i to rejection of H_0. The discriminant function will appear again in Section 5.6.2 and in Chapters 8 and 9.

We list these and other procedures that could be used to check each variable following rejection of H_0 by a two-sample T^2-test:

1. Univariate t-tests, one for each variable,

$$t_i = \frac{\bar{y}_{1i} - \bar{y}_{2i}}{\sqrt{[(n_1 + n_2)/n_1 n_2]\, s_{ii}}}, \qquad i = 1, 2, \ldots, p, \qquad (5.16)$$

where s_{ii} is the ith diagonal element of $\mathbf{S}_{\text{pl}}$. Reject $H_0: \mu_{1i} = \mu_{2i}$ if $|t_i| > t_{\alpha/2, n_1+n_2-2}$. For confidence intervals on $\mu_{1i} - \mu_{2i}$, see Rencher (1996, Section 3.6).

2. To adjust the α level resulting from performing the p tests in (5.16), we could use a Bonferroni critical value $t_{\alpha/2p, n_1+n_2-2}$ for (5.16) (Bonferroni 1936). This gives a conservative overall α level, since $t_{\alpha/2p}$ is much greater that $t_{\alpha/2}$. Bonferroni critical values $t_{\alpha/2p,\nu}$ are given in Table A.8, from Bailey (1977).
3. Another critical value that could be used with (5.16) is T_{α,p,n_1+n_2-2}, where T_α is the square root of T^2_α from Table A.7; that is, $T_{\alpha,p,n_1+n_2-2} = \sqrt{T^2_{\alpha,p,n_1+n_2-2}}$. This allows for all p variables to be tested as well as all possible linear combinations, as in (5.14), even linear combinations chosen after seeing the data. Consequently, the use of T_α is even more conservative than $t_{\alpha/2p}$; that is, $T_{\alpha,p,n_1+n_2-2} > t_{\alpha/2p, n_1+n_2-2}$.
4. Partial F- or t-tests [test of each variable adjusted for the other variables; see (5.33) in Section 5.8]
5. Standardized discriminant function coefficients (see Section 8.5)
6. Correlations between the variables and the discriminant function (see Section 8.7.3)
7. Stepwise discriminant analysis (see Section 8.9)

The first three methods are univariate approaches that do not use covariances or correlations among the variables in the computation of the test statistic. The last four methods are multivariate in the sense that the correlation structure is explicitly taken into account in the computation.

Method 6, involving the correlation between each variable and the discriminant function, is recommended in many texts and software packages. However, Rencher (1988) has shown that these correlations are proportional to individual t- or F-tests (see Section 8.7.3). Thus this method is equivalent to method 1 and is a univariate rather than a multivariate approach. Method 7 is often used to identify a subset of important variables or even to rank the variables according to order of entry. But Rencher and Larson (1980) have shown that stepwise methods have a high risk of selecting spurious variables, unless the sample size is very large.

We now consider the univariate procedures 1, 2, and 3. The probability of rejecting one or more of the p univariate tests when H_0 is true is called the *overall* α or *experimentwise error rate*. If we do univariate tests only, with no T^2-test, then the tests based on $t_{\alpha/2p}$ and T_α in procedures 2 and 3 are conservative (overall α too low), and tests based on $t_{\alpha/2}$ in procedure 1 are liberal (overall α too high). However, when these tests are carried out *only* after rejection by the T^2-test, the experimentwise error rates change. Obviously the tests will reject less often (under H_0) if they are carried out only if T^2 rejects. Thus the tests using $t_{\alpha/2p}$ and T_α become even more conservative, and the test using $t_{\alpha/2}$ becomes more acceptable.

Hummel and Sligo (1971) studied the experimentwise error rate for univariate t-tests following rejection of H_0 by the T^2-test. They found that using $t_{\alpha/2}$ for a critical value yields an overall α acceptably close to the nominal .05. In fact, it is slightly conservative, making this the preferred univariate test (within the limits of their study). They also compared this procedure with that of performing univariate tests without a prior T^2-test. For the latter case, the overall α is too high, as expected. Table 5.2 gives an excerpt of Hummel and Sligo's results. The sample size is for each of the two samples; the r^2 in common is for every pair of variables.

Hummel and Sligo therefore recommended performing the multivariate T^2-test followed by univariate t-tests. This procedure appears to have the desired overall α level and will clearly have better power than tests using T_α or $t_{\alpha/2p}$ as a critical value. Table 5.2 also highlights the importance of using univariate t-tests *only if* the multivariate T^2-test is significant. The inflated α's resulting if t-tests are used without regard to the outcome of the T^2-test are clearly evident. Thus among the three univariate procedures (procedures 1, 2, and 3 above), the first appears to be preferred.

Table 5.2 Comparison of Experimentwise Error Rates (Nominal α = 0.05)

Sample Size	Number of Variables	Common r^2 .10	.30	.50	.70
		Univariate Tests Only[a]			
10	3	.145	.112	.114	.077
10	6	.267	.190	.178	.111
10	9	.348	.247	.209	.129
30	3	.115	.119	.117	.085
30	6	.225	.200	.176	.115
30	9	.296	.263	.223	.140
50	3	.138	.124	.102	.083
50	6	.230	.190	.160	.115
50	9	.324	.258	.208	.146
		Multivariate Test Followed by Univariate Tests[b]			
10	3	.044	.029	.035	.022
10	6	.046	.029	.030	.017
10	9	.050	.026	.025	.018
30	3	.037	.044	.029	.025
30	6	.037	.037	.032	.021
30	9	.042	.042	.030	.021
50	3	.038	.041	.033	.028
50	6	.037	.039	.028	.027
50	9	.036	.038	.026	.020

[a]Ignoring multivariate tests.
[b]Carried out only if multivariate test rejects.

Among the multivariate approaches (procedures 4, 5, and 7 above), we prefer the fifth procedure, which compares the (absolute value of) coefficients in the discriminant function to find the effect of each variable in separating the two groups of observations. These coefficients will often tell a different story from the univariate tests because the univariate tests do not take into account the correlations among the variables. A variable will typically have a different effect in the presence of other variables than it has by itself. In the discriminant function $z = \mathbf{a}'\mathbf{y} = a_1y_1 + a_2y_2 + \cdots + a_py_p$, where $\mathbf{a} = \mathbf{S}_{\text{pl}}^{-1}(\overline{\mathbf{y}}_1 - \overline{\mathbf{y}}_2)$, the coefficients $a_1, a_2, \ldots, a_p$ indicate the relative importance of the variables in a multivariate context, something the univariate t-tests cannot do. If the variables are not commensurate (similar in scale and variance), the coefficients should be standardized, as in Section 8.5; this allows for more valid comparisons among the variables. Rencher and Scott (1990) have shown the precise nature of the information in the standardized discriminant function coefficients. For a detailed analysis of the effect of each variable in the presence of the other variables, see Rencher (1993; 1996, Sections 3.3.5 and 3.5.3).

Example 5.5. For the psychological data in Table 5.1, we obtained $\overline{\mathbf{y}}_1, \overline{\mathbf{y}}_2$, and $\mathbf{S}_{\text{pl}}$ in Example 5.4.2. The discriminant function coefficient vector is obtained from (5.15) as

$$\mathbf{a} = \mathbf{S}_{\text{pl}}^{-1}(\overline{\mathbf{y}}_1 - \overline{\mathbf{y}}_2) = \begin{bmatrix} .4856 \\ -.2028 \\ .4654 \\ -.3048 \end{bmatrix}.$$

Thus the linear combination that best separates the two groups is

$$\mathbf{a}'\mathbf{y} = .4856y_1 - .2028y_2 + .4654y_3 - .3048y_4,$$

in which y_1 and y_3 appear to contribute most to separation of the two groups.

5.6 COMPUTATION OF T^2

If one has a program available with matrix manipulation capability, it is a simple matter to compute T^2 using (5.8). However, this approach is somewhat cumbersome for those not accustomed to the use of such a programming language, and many would prefer a more automated procedure. But very few general-purpose statistical programs provide for direct calculation of the two-sample T^2-statistic, perhaps because it is so easy to obtain from other procedures. We will discuss two types of widely available procedures that can be used to compute T^2.

5.6.1 Obtaining T^2 from a MANOVA Program

Multivariate analysis of variance (MANOVA) is discussed in Chapter 6, and the reader may wish to return to the present section after becoming familiar with that material. One-way MANOVA involves a comparison of mean vectors from several samples. Typically, the number of samples is three or more, but the procedure will also accommodate two samples. The two-sample T^2 test is thus a special case of MANOVA.

One can test $H_0\colon \boldsymbol{\mu}_1 = \boldsymbol{\mu}_2$ with the MANOVA program by entering two groups. Four common tests of significance are defined in Section 6.1: Wilks' Λ, the Lawley–Hotelling $U^{(s)}$, Pillai's $V^{(s)}$, and Roy's largest root θ. Without concerning ourselves here with how these are defined or calculated, we show how to use each to obtain T^2:

$$T^2 = (n_1 + n_2 - 2)\,\frac{1-\Lambda}{\Lambda} \tag{5.17}$$

$$T^2 = (n_1 + n_2 - 2)U^{(s)} \tag{5.18}$$

$$T^2 = (n_1 + n_2 - 2)\,\frac{V^{(s)}}{1-V^{(s)}} \tag{5.19}$$

$$T^2 = (n_1 + n_2 - 2)\,\frac{\theta}{1-\theta}. \tag{5.20}$$

These relationships are demonstrated in Section 6.1.7. If the MANOVA program gives eigenvectors of $\mathbf{E}^{-1}\mathbf{H}$ ($\mathbf{E}$ and $\mathbf{H}$ are defined in Section 6.1.2), the eigenvector corresponding to the largest eigenvalue will be equal to (a constant multiple of) the discriminant function $\mathbf{S}_{\text{pl}}^{-1}(\overline{\mathbf{y}}_1 - \overline{\mathbf{y}}_2)$.

5.6.2 Obtaining T^2 from Multiple Regression

In this section, the y's become independent variables in a regression model. For each observation vector $\mathbf{y}_{ij}$ in a two-sample T^2, define a "dummy" group variable as

$$\begin{aligned} w_i &= \frac{n_2}{n_1+n_2} \quad \text{for each of } \mathbf{y}_{11}, \mathbf{y}_{12}, \ldots, \mathbf{y}_{1n_1} \text{ in sample 1} \\ &= -\frac{n_1}{n_1+n_2} \quad \text{for each of } \mathbf{y}_{21}, \mathbf{y}_{22}, \ldots, \mathbf{y}_{2n_2} \text{ in sample 2.} \end{aligned}$$

Then $\overline{w} = 0$ for all $n_1 + n_2$ observations. The prediction equation for the regression of w on the y's can be written as

$$\hat{w}_i = b_0 + b_1 y_{i1} + b_2 y_{i2} + \cdots + b_p y_{ip},$$

where i ranges over all $n_1 + n_2$ observations and the least squares estimate b_0

is [see (10.12)]

$$b_0 = \overline{w} - b_1\overline{y}_1 - b_2\overline{y}_2 - \cdots - b_p\overline{y}_p.$$

Substituting this into the regression equation, we obtain

$$\begin{aligned}\hat{w}_i &= \overline{w} + b_1(y_{i1} - \overline{y}_1) + b_2(y_{i2} - \overline{y}_2) + \cdots + b_p(y_{ip} - \overline{y}_p) \\ &= b_1(y_{i1} - \overline{y}_1) + b_2(y_{i2} - \overline{y}_2) + \cdots + b_p(y_{ip} - \overline{y}_p) \quad (\text{since } \overline{w} = 0).\end{aligned}$$

Let $\mathbf{b}' = (b_1, b_2, \ldots, b_p)$ be the vector of regression coefficients and R^2 the squared multiple correlation. Then we have the following relationships:

$$D^2 = \frac{(n_1 + n_2)(n_1 + n_2 - 2)R^2}{n_1 n_2(1 - R^2)} \tag{5.21}$$

$$T^2 = \frac{n_1 n_2}{n_1 + n_2} D^2 = (n_1 + n_2 - 2)\frac{R^2}{1 - R^2} \tag{5.22}$$

$$\mathbf{a} = \mathbf{S}_{\text{pl}}^{-1}(\overline{\mathbf{y}}_1 - \overline{\mathbf{y}}_2) = \left[\frac{(n_1 + n_2)(n_1 + n_2 - 2)}{n_1 n_2} + D^2\right]\mathbf{b}. \tag{5.23}$$

Thus with ordinary multiple regression, one can easily obtain T^2 and the discriminant function $\mathbf{S}_{\text{pl}}^{-1}(\overline{\mathbf{y}}_1 - \overline{\mathbf{y}}_2)$. We simply define w_i as above, regress the $n_1 + n_2$ w's on the $n_1 + n_2$ $\mathbf{y}_{ij}$'s, and use the resulting R^2 in (5.21) or (5.22). For $\mathbf{b}$, delete the intercept from the regression coefficients for use in (5.23). Actually, since only the relative values of the elements of $\mathbf{a} = \mathbf{S}_{\text{pl}}^{-1}(\overline{\mathbf{y}}_1 - \overline{\mathbf{y}}_2)$ are of interest, it is not necessary to convert from $\mathbf{b}$ to $\mathbf{a}$ in (5.23). We can use $\mathbf{b}$ directly or standardize the values $b_1, b_2, \ldots, b_p$ as in Section 8.5.

Example 5.6.2. We illustrate the regression approach to computation of T^2 using the psychological data in Table 5.1. We set $w = n_2/(n_1 + n_2) = \frac{32}{64} = \frac{1}{2}$ for each observation in group 1 (males) and equal to $-n_1/(n_1 + n_2) = -\frac{1}{2}$ in the second group (females). When w is regressed on the 64 y's, we obtain

$$\begin{bmatrix} b_0 \\ b_1 \\ b_2 \\ b_3 \\ b_4 \end{bmatrix} = \begin{bmatrix} -.729 \\ .049 \\ -.020 \\ .047 \\ -.031 \end{bmatrix} \qquad R^2 = .6091.$$

By (5.22),

$$T^2 = (n_1 + n_2 - 2)\frac{R^2}{1 - R^2} = \frac{62(.6091)}{1 - .6091} = 96.603,$$

as was obtained before in Example 5.4.2. Note that $\mathbf{b}' = (b_1, b_2, b_3, b_4) = (.049, -.020, .047, -.031)$, with the intercept deleted, is proportional to the discriminant function vector **a** from Example 5.5, as we would expect from (5.23).

5.7 PAIRED OBSERVATIONS TEST

As usual, we begin with the univariate case to set the stage for the multivariate presentation.

5.7.1 Univariate Case

Suppose two samples are not independent because there exists a natural pairing between the ith observation y_i in the first sample and the ith observation x_i in the second sample for all i, as, for example, when a treatment is applied twice to the same individual or when subjects are matched according to some criterion, such as IQ or family background. With such pairing, the procedure is often referred to as *paired observations* or *matched pairs*. The two samples thus obtained are correlated, and the two-sample test in (5.7) is not appropriate because the samples must be independent in order for (5.7) to have a t-distribution. We reduce the two samples to one by working with the differences between the paired observations, as in the following layout for two treatments applied to the same subject:

Pair Number	Treatment 1	Treatment 2	Difference $d_i = y_i - x_i$
1	y_1	x_1	d_1
2	y_2	x_2	d_2
⋮	⋮	⋮	⋮
n	y_n	x_n	d_n

To obtain a t-test, it is not sufficient to assume individual normality for each of y and x. To allow for the covariance between y and x, we need the additional assumption that y and x have a bivariate normal distribution with

$$\boldsymbol{\mu} = \begin{pmatrix} \mu_y \\ \mu_x \end{pmatrix} \quad \text{and} \quad \boldsymbol{\Sigma} = \begin{pmatrix} \sigma_y^2 & \sigma_{yx} \\ \sigma_{yx} & \sigma_x^2 \end{pmatrix}.$$

It then follows by property 1a in Section 4.2 that $d_i = y_i - x_i$ is $N(\mu_y - \mu_x, \sigma_d^2)$, where $\sigma_d^2 = \sigma_y^2 - 2\sigma_{yx} + \sigma_x^2$. From $d_1, d_2, \ldots, d_n$ we calculate

$$\overline{d} = \frac{1}{n}\sum_{i=1}^{n} d_i \quad \text{and} \quad s_d^2 = \frac{1}{n-1}\sum_{i=1}^{n}(d_i - \overline{d})^2.$$

To test $H_0\colon \mu_y = \mu_x$, that is, $H_0\colon \mu_d = 0$, we use the one-sample statistic

$$t = \frac{\overline{d}}{s_d/\sqrt{n}}, \tag{5.24}$$

which is distributed as t_{n-1} if H_0 is true. We reject H_0 in favor of $H_1\colon \mu_d \neq 0$ if $|t| > t_{\alpha/2,n-1}$. It is not necessary to assume $\sigma_y^2 = \sigma_x^2$ because there are no restrictions on $\boldsymbol{\Sigma}$.

This test has only $n - 1$ degrees of freedom compared with $2(n - 1)$ for the two independent sample t-test (5.7). In general, the pairing reduces the within-sample variation s_d and thereby increases the power.

If we mistakenly treated the two samples as independent and used (5.7) with $n_1 = n_2 = n$, we would have

$$t = \frac{\overline{y} - \overline{x}}{s_{\text{pl}}\sqrt{2/n}} = \frac{\overline{y} - \overline{x}}{\sqrt{2s_{\text{pl}}^2/n}}.$$

However,

$$E\left(\frac{2s_{\text{pl}}^2}{n}\right) = 2E\left[\frac{(n-1)s_y^2 + (n-1)s_x^2}{(n+n-2)n}\right] = (\sigma_y^2 + \sigma_x^2)/n,$$

whereas $\text{var}(\overline{y} - \overline{x}) = (\sigma_y^2 + \sigma_x^2 - 2\sigma_{yx})/n$. Thus if the test for independent samples (5.7) is used for paired data, it does not have a t-distribution, and in fact underestimates the true average t-value (assuming H_0 is false), since $\sigma_y^2 + \sigma_x^2 > \sigma_y^2 + \sigma_x^2 - 2\sigma_{yx}$, assuming $\sigma_{yx} > 0$, which would be typical in this situation. One could therefore use

$$t = \frac{\overline{y} - \overline{x}}{\sqrt{(s_y^2 + s_x^2 - 2s_{yx})/n}},$$

but $t = \sqrt{n}\,\overline{d}/s_d$ in (5.24) is equal to it and somewhat simpler to use.

5.7.2 Multivariate Case

Here we assume the same natural pairing of sampling units as in the univariate case, but we measure p variables on each sampling unit. Thus $\mathbf{y}_i$ from the first sample is paired with $\mathbf{x}_i$ from the second sample, $i = 1, 2, \ldots, n$. In terms of two treatments applied to each sampling unit, this situation is as follows:

Pair Number	Treatment 1	Treatment 2	Difference $\mathbf{d}_i = \mathbf{y}_i - \mathbf{x}_i$
1	$\mathbf{y}_1$	$\mathbf{x}_1$	$\mathbf{d}_1$
2	$\mathbf{y}_2$	$\mathbf{x}_2$	$\mathbf{d}_2$
$\vdots$	$\vdots$	$\vdots$	$\vdots$
n	$\mathbf{y}_n$	$\mathbf{x}_n$	$\mathbf{d}_n$

To test $H_0: \boldsymbol{\mu}_d = \mathbf{0}$, which is equivalent to $H_0: \boldsymbol{\mu}_y = \boldsymbol{\mu}_x$, calculate

$$\overline{\mathbf{d}} = \frac{1}{n} \sum_{i=1}^{n} \mathbf{d}_i \quad \text{and} \quad \mathbf{S}_d = \frac{1}{n-1} \sum_{i=1}^{n} (\mathbf{d}_i - \overline{\mathbf{d}})(\mathbf{d}_i - \overline{\mathbf{d}})'.$$

We then have

$$T^2 = \overline{\mathbf{d}}' \left(\frac{\mathbf{S}_d}{n} \right)^{-1} \overline{\mathbf{d}} = nD^2 = n\overline{\mathbf{d}}'\mathbf{S}_d^{-1}\overline{\mathbf{d}}, \tag{5.25}$$

where $D^2 = \overline{\mathbf{d}}'\mathbf{S}_d^{-1}\overline{\mathbf{d}}$. This paired comparison T^2-statistic is distributed as $T^2_{p,n-1}$. We reject H_0 if $T^2 > T^2_{\alpha,p,n-1}$.

The cautions expressed in Section 5.7.1 for univariate paired observation data also apply here. If the two samples of multivariate observations are correlated because of a natural pairing of sampling units, the test in (5.25) should be used rather than the two-sample T^2-test in (5.8), which assumes two independent samples. Misuse of (5.8) in place of (5.25) will lead to loss of power.

Since the assumption $\boldsymbol{\Sigma}_y = \boldsymbol{\Sigma}_x$ is not needed for (5.25) to have a T^2-distribution, this test can be used for independent samples when $\boldsymbol{\Sigma}_1 \neq \boldsymbol{\Sigma}_2$ (as long as $n_1 = n_2$). The observations in the two samples would be paired in the order they were obtained or in an arbitrary order. However, in the case of independent samples, the pairing achieves no gain in power to offset the loss of $n - 1$ degrees of freedom.

By analogy with (5.15), the discriminant function for paired observation data becomes

$$\mathbf{a} = \mathbf{S}_d^{-1}\overline{\mathbf{d}}. \tag{5.26}$$

For tests on individual variables, we have

$$t_i = \frac{\bar{d}_i}{\sqrt{\dfrac{s_{d,ii}}{n}}} \qquad i = 1, 2, \ldots, p. \tag{5.27}$$

The critical value for t_i is $t_{\alpha/2p,n-1}$ or $t_{\alpha/2,n-1}$ depending on whether a T^2-test is carried out first (see Section 5.5).

Example 5.7.2. To compare two types of coating for resistance to corrosion, 15 pieces of pipe were coated with each type of coating (Kramer and Jensen 1969b). Two pipes, one with each type of coating, were buried together and left for the same length of time at 15 different locations, providing a natural pairing of the observations. Corrosion for the first type of coating was measured by two variables,

$$y_1 = \text{maximum depth of pit in thousandths of an inch}$$
$$y_2 = \text{number of pits}$$

with x_1 and x_2 defined analogously for the second coating. The data and differences are given in Table 5.3. Thus we have, for example, $\mathbf{y}_1' = (73, 31)$, $\mathbf{x}_1' = (51, 35)$, and $\mathbf{d}_1' = \mathbf{y}_1' - \mathbf{x}_1' = (22, -4)$. For the 15 difference vectors, we

Table 5.3 Depth of Maximum Pits and Number of Pits of Coated Pipes

	Coating 1		Coating 2		Difference	
Location	Depth y_1	Number y_2	Depth x_1	Number x_2	Depth d_1	Number d_2
1	73	31	51	35	22	−4
2	43	19	41	14	2	5
3	47	22	43	19	4	3
4	53	26	41	29	12	−3
5	58	36	47	34	11	2
6	47	30	32	26	15	4
7	52	29	24	19	28	10
8	38	36	43	37	−5	−1
9	61	34	53	24	8	10
10	56	33	52	27	4	6
11	56	19	57	14	−1	5
12	34	19	44	19	−10	0
13	55	26	57	30	−2	−4
14	65	15	40	7	25	8
15	75	18	68	13	7	5

obtain

$$\bar{\mathbf{d}} = \begin{pmatrix} 8.000 \\ 3.067 \end{pmatrix} \qquad \mathbf{S}_d = \begin{pmatrix} 121.571 & 17.071 \\ 17.071 & 21.781 \end{pmatrix}.$$

By (5.25),

$$T^2 = (15)(8.000, 3.067) \begin{pmatrix} 121.571 & 17.071 \\ 17.071 & 21.781 \end{pmatrix}^{-1} \begin{pmatrix} 8.000 \\ 3.067 \end{pmatrix} = 10.819.$$

Since $T^2 = 10.819 > T^2_{.05,2,14} = 8.197$, we reject $H_0\colon \boldsymbol{\mu}_d = \mathbf{0}$ and conclude that the two coatings differ in their effect on corrosion.

5.8 TEST FOR ADDITIONAL INFORMATION

In this section, we are again considering two independent samples. We start with a basic $p \times 1$ vector $\mathbf{y}$ of measurements on each sampling unit and ask whether a $q \times 1$ subvector $\mathbf{x}$ measured in addition to $\mathbf{y}$ will significantly increase the separation of the two samples. It is not necessary that we add new variables. We may be interested in determining whether some of the variables we already have are redundant in the presence of other variables in terms of separating the groups. We have designated the subset of interest by $\mathbf{x}$ for notational convenience.

It is assumed that the two samples are from multivariate normal populations with a common covariance matrix, that is,

$$\begin{pmatrix} \mathbf{y}_{11} \\ \mathbf{x}_{11} \end{pmatrix}, \begin{pmatrix} \mathbf{y}_{12} \\ \mathbf{x}_{12} \end{pmatrix}, \ldots, \begin{pmatrix} \mathbf{y}_{1n_1} \\ \mathbf{x}_{1n_1} \end{pmatrix} \quad \text{are from } N_{p+q}(\boldsymbol{\mu}_1, \boldsymbol{\Sigma})$$

and

$$\begin{pmatrix} \mathbf{y}_{21} \\ \mathbf{x}_{21} \end{pmatrix}, \begin{pmatrix} \mathbf{y}_{22} \\ \mathbf{x}_{22} \end{pmatrix}, \ldots, \begin{pmatrix} \mathbf{y}_{2n_2} \\ \mathbf{x}_{2n_2} \end{pmatrix} \quad \text{are from } N_{p+q}(\boldsymbol{\mu}_2, \boldsymbol{\Sigma}).$$

We partition the sample mean vectors and covariance matrix accordingly:

$$\begin{pmatrix} \overline{\mathbf{y}}_1 \\ \overline{\mathbf{x}}_1 \end{pmatrix} \quad \begin{pmatrix} \overline{\mathbf{y}}_2 \\ \overline{\mathbf{x}}_2 \end{pmatrix} \quad \mathbf{S}_{\text{pl}} = \begin{pmatrix} \mathbf{S}_{yy} & \mathbf{S}_{yx} \\ \mathbf{S}_{xy} & \mathbf{S}_{xx} \end{pmatrix},$$

where $\mathbf{S}_{\text{pl}}$ is the pooled sample covariance matrix from the two samples.

We wish to test the hypothesis that $\mathbf{x}_1$ and $\mathbf{x}_2$ are redundant for separating the two groups, that is, that the extra q variables do not contribute anything significant beyond the information already available in $\mathbf{y}_1$ and $\mathbf{y}_2$ for separating the groups. This is in the spirit of a "full and reduced model" test in regression (see Section 10.2.5b). However, here we are working with a subset of dependent variables as contrasted to the subset of independent variables in the regression setting. Thus both $\mathbf{y}$ and $\mathbf{x}$ are subvectors of dependent variables. In this setting, the independent variables would be grouping variables 1 and 2 corresponding to $\boldsymbol{\mu}_1$ and $\boldsymbol{\mu}_2$.

We are asking not if the x's can significantly separate the two groups by themselves, but whether they provide additional separation beyond the separation already achieved by the y's. If the x's were independent of the y's, we would have $T^2_{p+q} = T^2_p + T^2_q$, but this does not hold, because they are correlated. We must compare T^2_{p+q} for the full set of variables $(y_1, \ldots, y_p, x_1, \ldots, x_q)$ with T^2_p based on the reduced set $(y_1, \ldots, y_p)$.

By definition, the T^2-statistic based on the full set of $p + q$ variables is given by

$$T^2_{p+q} = \frac{n_1 n_2}{n_1 + n_2} \left[\begin{pmatrix} \overline{\mathbf{y}}_1 \\ \overline{\mathbf{x}}_1 \end{pmatrix} - \begin{pmatrix} \overline{\mathbf{y}}_2 \\ \overline{\mathbf{x}}_2 \end{pmatrix} \right]' \mathbf{S}_{\text{pl}}^{-1} \left[\begin{pmatrix} \overline{\mathbf{y}}_1 \\ \overline{\mathbf{x}}_1 \end{pmatrix} - \begin{pmatrix} \overline{\mathbf{y}}_2 \\ \overline{\mathbf{x}}_2 \end{pmatrix} \right], \tag{5.28}$$

while T^2 for the reduced set of p variables is

$$T^2_p = \frac{n_1 n_2}{n_1 + n_2} (\overline{\mathbf{y}}_1 - \overline{\mathbf{y}}_2)' \mathbf{S}_{yy}^{-1} (\overline{\mathbf{y}}_1 - \overline{\mathbf{y}}_2). \tag{5.29}$$

We reject the hypothesis of redundancy of $\mathbf{x}$ if

$$F = \frac{\nu - p - q + 1}{q} \, \frac{T^2_{p+q} - T^2_p}{\nu + T^2_p} \geq F_{\alpha, q, \nu - p - q + 1}, \tag{5.30}$$

or, alternatively, if

$$T^2 = (\nu - p) \frac{T^2_{p+q} - T^2_p}{\nu + T^2_p} \geq T^2_{\alpha, q, \nu - p}, \tag{5.31}$$

where $\nu = n_1 + n_2 - 2$ in both cases. Note that the first degrees-of-freedom parameter in both cases is q, the number of x's.

To prove directly that the statistic defined in (5.30) has an F-distribution, we can use a basic relationship from multiple regression [see (10.27)]:

$$F_{q,\nu-p-q+1} = \frac{(R^2_{p+q} - R^2_p)(\nu - p - q + 1)}{(1 - R^2_{p+q})q}, \qquad (5.32)$$

where R^2_{p+q} is the squared multiple correlation from the full model with $p+q$ independent variables and R^2_p is from the reduced model with p independent variables. If we solve for R^2 in terms of T^2 from (5.22) and substitute this into (5.32), we readily obtain the test statistic in (5.30).

If we are interested in the effect of adding a single x, then $q = 1$, and both (5.30) and (5.31) reduce to

$$t^2 = (\nu - p)\frac{T^2_{p+1} - T^2_p}{\nu + T^2_p} \geq t^2_{\alpha/2,\nu-p} = F_{\alpha,1,\nu-p}. \qquad (5.33)$$

Example 5.8. We use the psychological data of Table 5.1 to illustrate tests on subvectors. We begin by testing the significance of y_3 and y_4, tool recognition and vocabulary, above and beyond y_1 and y_2, pictorial inconsistencies and paper form board. (In the notation of the present section, y_3 and y_4 become x_1 and x_2.) For these subvectors, $p = 2$ and $q = 2$. The value of T^2_{p+q} for all four variables as given by (5.28) was obtained in Example 5.4.2 as 96.6028. For y_1 and y_2, we obtain, by (5.29),

$$\begin{aligned} T^2_p &= \frac{n_1 n_2}{n_1 + n_2}(\bar{\mathbf{y}}_1 - \bar{\mathbf{y}}_2)'\mathbf{S}^{-1}_{yy}(\bar{\mathbf{y}}_1 - \bar{\mathbf{y}}_2) \\ &= \frac{(32)^2}{32+32}\begin{pmatrix} 15.97 - 12.34 \\ 15.91 - 13.91 \end{pmatrix}'\begin{pmatrix} 7.16 & 6.05 \\ 6.05 & 15.89 \end{pmatrix}^{-1}\begin{pmatrix} 15.97 - 12.34 \\ 15.91 - 13.91 \end{pmatrix} \\ &= 31.0126. \end{aligned}$$

By (5.31), the test statistic is

$$T^2 = (\nu - p)\frac{T^2_{p+q} - T^2_p}{\nu + T^2_p} = (62 - 2)\frac{96.6028 - 31.0126}{62 + 31.0126} = 42.311.$$

We reject the hypothesis that $\mathbf{x}$ is redundant, since $42.311 > T^2_{.01,2,60} = 10.137$. We conclude that $\mathbf{x}$ adds a significant amount of separation to $\mathbf{y}$.

To test the effect of each variable adjusted for the other three, we use (5.33). In this case, $p = 3$, $\nu = 62$, and $\nu - p = 59$. The results are given below, where

$T^2_{p+1} = 96.6028$ and T^2_p in each case is based on the three variables, excluding the variable in question. For example, $T^2_p = 89.8718$ for y_2 is based on y_1, y_3, and y_4:

Variable	T^2_p	$(\nu - p)\dfrac{T^2_{p+1} - T^2_p}{\nu + T^2_p}$
y_1	79.7736	7.004
y_2	89.8718	2.615
y_3	32.6253	39.891
y_4	74.1902	9.710

When we compare these four test statistic values with the critical value $t^2_{.025,59} = 4.002$, we see that each variable makes a significant contribution to T^2 except y_2.

5.9 PROFILE ANALYSIS

If $\mathbf{y}$ is $N(\boldsymbol{\mu}, \boldsymbol{\Sigma})$ and the variables in $\mathbf{y}$ are commensurate (measured in the same units and with approximately equal variances), we may wish to compare the means $\mu_1, \mu_2, \ldots, \mu_p$ in $\boldsymbol{\mu}$. This might be of interest when the same research unit is measured at p successive times. Such situations are often referred to as *repeated measures* designs or *growth curves*, which are discussed in some generality in Sections 6.9 and 6.10. In the present section, we discuss one- and two-sample *profile analysis*. Profile analysis for several samples is covered in Section 6.8.

The pattern obtained by plotting $\mu_1, \mu_2, \ldots, \mu_p$ as ordinates and connecting the points is called a *profile*; that is, we draw straight lines connecting the points $(1, \mu_1), (2, \mu_2), \ldots, (p, \mu_p)$. Profile analysis is an analysis of the profile or a comparison of two or more profiles. Profile analysis is often discussed in the context of administering a battery of p psychological or other tests. In growth curve analysis where the variables are measured at time intervals, the responses have a natural order. In profile analysis where the variables arise from test scores, there is ordinarily no natural order. A distinction is not always made between repeated measures of the same variable through time and profile analysis of several different commensurate variables on the same individual.

5.9.1 One-Sample Profile Analysis

We begin with a discussion of the profile of the mean vector $\boldsymbol{\mu}$ from a single sample. A plot of $\boldsymbol{\mu}$ might appear as in Figure 5.3, where we plot $(1, \mu_1), (2, \mu_2), \ldots, (p, \mu_p)$ and connect the points.

In order to compare the means $\mu_1, \mu_2, \ldots, \mu_p$ in $\boldsymbol{\mu}$, the basic hypothesis is

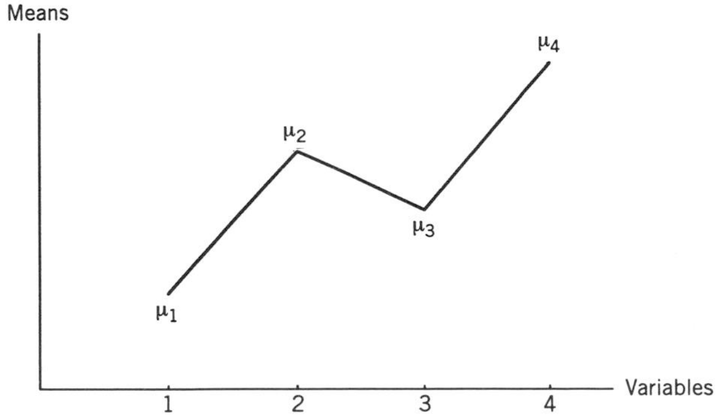

Figure 5.3 Profile of a mean vector.

that the profile is *level* or *flat*:

$$H_0: \mu_1 = \mu_2 = \ldots = \mu_p \quad \text{vs.} \quad H_1: \mu_i \neq \mu_j \quad \text{for some } i \neq j.$$

Equality of the means in H_0 can be expressed as $p - 1$ comparisons,

$$H_0: \begin{bmatrix} \mu_1 - \mu_2 \\ \mu_2 - \mu_3 \\ \vdots \\ \mu_{p-1} - \mu_p \end{bmatrix} = \begin{bmatrix} 0 \\ 0 \\ \vdots \\ 0 \end{bmatrix},$$

or as

$$H_0: \begin{bmatrix} \mu_1 - \mu_2 \\ \mu_1 - \mu_3 \\ \vdots \\ \mu_1 - \mu_p \end{bmatrix} = \begin{bmatrix} 0 \\ 0 \\ \vdots \\ 0 \end{bmatrix},$$

These two expressions can be written in the form $\mathbf{C}_1\boldsymbol{\mu} = \mathbf{0}$ and $\mathbf{C}_2\boldsymbol{\mu} = \mathbf{0}$, where $\mathbf{C}_1$ and $\mathbf{C}_2$ are the $(p-1) \times p$ matrices:

$$\mathbf{C}_1 = \begin{bmatrix} 1 & -1 & 0 & \cdots & 0 \\ 0 & 1 & -1 & \cdots & 0 \\ \vdots & \vdots & \vdots & & \vdots \\ 0 & 0 & 0 & \cdots & -1 \end{bmatrix} \quad \mathbf{C}_2 = \begin{bmatrix} 1 & -1 & 0 & \cdots & 0 \\ 1 & 0 & -1 & \cdots & 0 \\ \vdots & \vdots & \vdots & & \vdots \\ 1 & 0 & 0 & \cdots & -1 \end{bmatrix}.$$

In fact, any $(p-1) \times p$ matrix $\mathbf{C}$ of rank $p - 1$ such that $\mathbf{Cj} = \mathbf{0}$ can be used in

$H_0\colon \mathbf{C}\boldsymbol{\mu} = \mathbf{0}$ to produce $H_0\colon \mu_1 = \mu_2 = \cdots = \mu_p$. If $\mathbf{Cj} = \mathbf{0}$, each row of $\mathbf{C}$ sums to zero by (2.38), and $\mathbf{C}\boldsymbol{\mu}$ is a set of $p-1$ contrasts in the μ's. A linear combination $c_{i1}\mu_1 + c_{i2}\mu_2 + \cdots + c_{ip}\mu_p$ is called a *contrast* in the μ's if the coefficients sum to zero, that is, if $\Sigma_j c_{ij} = 0$. The contrasts in $\mathbf{C}$ must be linearly independent in order to express $H_0\colon \mu_1 = \mu_2 = \cdots = \mu_p$.

From a sample $\mathbf{y}_1, \mathbf{y}_2, \ldots, \mathbf{y}_n$, we obtain estimates $\overline{\mathbf{y}}$ and $\mathbf{S}$ of population parameters $\boldsymbol{\mu}$ and $\boldsymbol{\Sigma}$. To test $H_0\colon \mathbf{C}\boldsymbol{\mu} = \mathbf{0}$, we transform each $\mathbf{y}_i, i = 1, 2, \ldots, n$, to $\mathbf{z}_i = \mathbf{C}\mathbf{y}_i$, which is $(p-1)\times 1$. By (3.59) and (3.60), the sample mean vector and covariance matrix of $\mathbf{z}_i = \mathbf{C}\mathbf{y}_i, i = 1, 2, \ldots, n$, are $\overline{\mathbf{z}} = \mathbf{C}\overline{\mathbf{y}}$ and $\mathbf{S}_z = \mathbf{CSC}'$, respectively. If $\mathbf{y}$ is $N_p(\boldsymbol{\mu}, \boldsymbol{\Sigma})$, then by property 1b in Section 4.2, $\mathbf{z} = \mathbf{Cy}$ is $N_{p-1}(\mathbf{C}\boldsymbol{\mu}, \mathbf{C}\boldsymbol{\Sigma}\mathbf{C}')$. Thus when $H_0\colon \mathbf{C}\boldsymbol{\mu} = \mathbf{0}$ is true, $\mathbf{C}\overline{\mathbf{y}}$ is $N_{p-1}(\mathbf{0}, \mathbf{C}\boldsymbol{\Sigma}\mathbf{C}'/n)$ and

$$T^2 = (\mathbf{C}\overline{\mathbf{y}})'(\mathbf{CSC}'/n)^{-1}(\mathbf{C}\overline{\mathbf{y}}) = n(\mathbf{C}\overline{\mathbf{y}})'(\mathbf{CSC}')^{-1}(\mathbf{C}\overline{\mathbf{y}}) \tag{5.34}$$

is distributed as $T^2_{p-1,n-1}$. We reject $H_0\colon \mathbf{C}\boldsymbol{\mu} = \mathbf{0}$ if $T^2 > T^2_{\alpha,p-1,n-1}$. Note that the dimension $p-1$ corresponds to the number of rows of $\mathbf{C}$. Thus $\overline{\mathbf{z}} = \mathbf{C}\overline{\mathbf{y}}$ is $(p-1)\times 1$ and $\mathbf{S}_z = \mathbf{CSC}'$ is $(p-1)\times(p-1)$.

If the variables have a natural ordering, we could test for a linear trend or polynomial curve in the means by suitably choosing the rows of $\mathbf{C}$. This is discussed in connection with growth curves in Section 6.10. Otherwise, any comparisons of interest can be made as long as they are linearly independent.

5.9.2 Two-Sample Profile Analysis

Suppose two independent groups or samples receive the same set of p tests or measurements. If these tests are comparable, for example, all on a scale of 0 to 100, the variables will often be commensurate.

Rather than testing the hypothesis that $\boldsymbol{\mu}_1 = \boldsymbol{\mu}_2$, we wish to be more specific in comparing the profiles obtained by connecting the points (i, μ_{1i}), $i = 1, 2, \ldots, p$, and (i, μ_{2i}), $i = 1, 2, \ldots, p$. There are three hypotheses of interest in comparing the profiles of two samples. The first of these is "Are the two profiles similar in appearance, or more precisely, are they parallel?" We illustrate this hypothesis in Figure 5.4. If the two profiles are parallel, then one group scored uniformly better than the other group on all p tests.

The parallelism hypothesis can be defined in terms of the slopes. The two profiles are parallel if the two slopes for each segment are the same. If the two profiles are parallel, the two increments for each segment are the same, and it is not necessary to use the actual slopes to express the hypothesis. We can simply compare the increase from one point to the next. This can be expressed as $H_{01}\colon \mu_{1i} - \mu_{1,i-1} = \mu_{2i} - \mu_{2,i-1}$ for $i = 2, 3, \ldots, p$, or

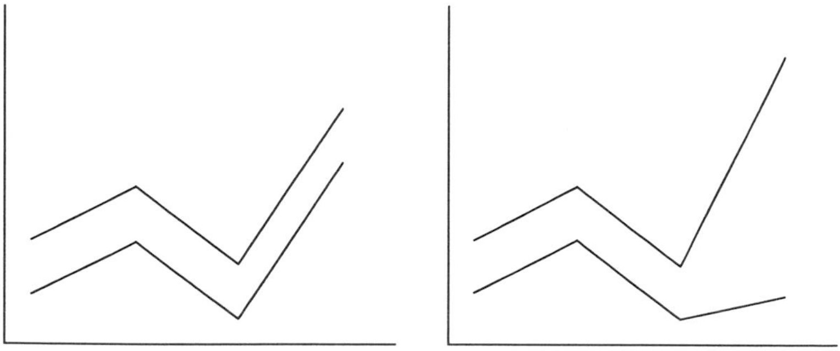

Figure 5.4 Comparison of two profiles under the hypothesis of parallelism.

$$H_{01}: \begin{bmatrix} \mu_{12} - \mu_{11} \\ \mu_{13} - \mu_{12} \\ \vdots \\ \mu_{1p} - \mu_{1,p-1} \end{bmatrix} = \begin{bmatrix} \mu_{22} - \mu_{21} \\ \mu_{23} - \mu_{22} \\ \vdots \\ \mu_{2p} - \mu_{2,p-1} \end{bmatrix},$$

which can be written as $H_{01}: \mathbf{C}\boldsymbol{\mu}_1 = \mathbf{C}\boldsymbol{\mu}_2$, using the contrast matrix

$$\mathbf{C} = \begin{bmatrix} -1 & 1 & 0 & \cdots & 0 \\ 0 & -1 & 1 & \cdots & 0 \\ \vdots & \vdots & \vdots & & \vdots \\ 0 & 0 & 0 & \cdots & 1 \end{bmatrix}.$$

From two samples, $\mathbf{y}_{11}, \mathbf{y}_{12}, \ldots, \mathbf{y}_{1n_1}$ and $\mathbf{y}_{21}, \mathbf{y}_{22}, \ldots, \mathbf{y}_{2n_2}$, we obtain $\overline{\mathbf{y}}_1, \overline{\mathbf{y}}_2$, and $\mathbf{S}_{\text{pl}}$ as estimates of $\boldsymbol{\mu}_1$, $\boldsymbol{\mu}_2$, and $\boldsymbol{\Sigma}$. As in the two-sample T^2 test, we assume that each $\mathbf{y}_{1i}$ in the first sample is $N_p(\boldsymbol{\mu}_1, \boldsymbol{\Sigma})$, and each $\mathbf{y}_{2i}$ in the second sample is $N_p(\boldsymbol{\mu}_2, \boldsymbol{\Sigma})$. If $\mathbf{C}$ is a $(p-1) \times p$ contrast matrix as above, then $\mathbf{C}\mathbf{y}_{1i}$ and $\mathbf{C}\mathbf{y}_{2i}$ are distributed as $N_{p-1}(\mathbf{C}\boldsymbol{\mu}_1, \mathbf{C}\boldsymbol{\Sigma}\mathbf{C}')$ and $N_{p-1}(\mathbf{C}\boldsymbol{\mu}_2, \mathbf{C}\boldsymbol{\Sigma}\mathbf{C}')$, respectively. Under $H_{01}: \mathbf{C}\boldsymbol{\mu}_1 - \mathbf{C}\boldsymbol{\mu}_2 = \mathbf{0}$, $\mathbf{C}\overline{\mathbf{y}}_1 - \mathbf{C}\overline{\mathbf{y}}_2$ is $N_{p-1}[\mathbf{0}, \mathbf{C}\boldsymbol{\Sigma}\mathbf{C}'(1/n_1 + 1/n_2)]$ and

$$\begin{aligned} T^2 &= (\mathbf{C}\overline{\mathbf{y}}_1 - \mathbf{C}\overline{\mathbf{y}}_2)' \left[\left(\frac{1}{n_1} + \frac{1}{n_2} \right) \mathbf{C}\mathbf{S}_{\text{pl}}\mathbf{C}' \right]^{-1} (\mathbf{C}\overline{\mathbf{y}}_1 - \mathbf{C}\overline{\mathbf{y}}_2) \\ &= \frac{n_1 n_2}{n_1 + n_2} (\overline{\mathbf{y}}_1 - \overline{\mathbf{y}}_2)'\mathbf{C}'[\mathbf{C}\mathbf{S}_{\text{pl}}\mathbf{C}']^{-1}\mathbf{C}(\overline{\mathbf{y}}_1 - \overline{\mathbf{y}}_2) \end{aligned} \tag{5.35}$$

is distributed as T^2_{p-1,n_1+n_2-2}. Note that the dimension $p - 1$ is the number of rows of $\mathbf{C}$.

By analogy with the discussion in Section 5.5, if H_{01} is rejected, we can follow up with univariate tests on the individual components of $\mathbf{C}(\overline{\mathbf{y}}_1 - \overline{\mathbf{y}}_2)$. Or we can calculate the discriminant function

$$\mathbf{a} = (\mathbf{C}\mathbf{S}_{\text{pl}}\mathbf{C}')^{-1}\mathbf{C}(\overline{\mathbf{y}}_1 - \overline{\mathbf{y}}_2) \tag{5.36}$$

as an indication of which slope differences contributed most to rejection in the presence of the other components of $\mathbf{C}(\overline{\mathbf{y}}_1 - \overline{\mathbf{y}}_2)$. There should be less need in this case to standardize the components of $\mathbf{a}$ as suggested in Section 5.5, because the variables are assumed to be commensurate. The vector $\mathbf{a}$ is $(p - 1) \times 1$, corresponding to the $p - 1$ segments of the profile. Thus if the second component of $\mathbf{a}$ is largest in absolute value, the divergence in slopes between the two profiles on the second segment contributed most to rejection.

If the data are arranged as in Table 5.4, we see an analogy to a two-way ANOVA model. A plot of the means is often made in a two-way ANOVA; a lack of parallelism indicates interaction between the two factors. Thus the hypothesis H_{01} is analogous to the group by test (variable) interaction hypothesis.

However, the usual ANOVA assumption of independence of observations does not hold here because the variables (tests) are correlated. The ANOVA assumption of independence and homogeneity of variances would require $\text{cov}(\mathbf{y}) = \boldsymbol{\Sigma} = \sigma^2\mathbf{I}$. Thus the test of H_{01} cannot be carried out using a univariate ANOVA approach, since $\boldsymbol{\Sigma} \neq \sigma^2\mathbf{I}$. We therefore proceed with the multivariate approach using T^2.

Table 5.4 Data for Two-Sample Profile Analysis

		Tests (variables)			
		1	2	$\cdots$	p
		Group 1			
$\mathbf{y}'_{11}$	=	$(y_{111}$	y_{112}	$\cdots$	$y_{11p})$
$\mathbf{y}'_{12}$	=	$(y_{121}$	y_{122}	$\cdots$	$y_{12p})$
$\vdots$		$\vdots$	$\vdots$		$\vdots$
$\mathbf{y}'_{1n_1}$	=	$(y_{1n_11}$	y_{1n_12}	$\cdots$	$y_{1n_1p})$
		Group 2			
$\mathbf{y}'_{21}$	=	$(y_{211}$	y_{212}	$\cdots$	$y_{21p})$
$\mathbf{y}'_{22}$	=	$(y_{221}$	y_{222}	$\cdots$	$y_{22p})$
$\vdots$		$\vdots$	$\vdots$		$\vdots$
$\mathbf{y}'_{2n_2}$	=	$(y_{2n_21}$	y_{2n_22}	$\cdots$	$y_{2n_2p})$

Another hypothesis of interest is "Are the two populations or groups at the same *level*?" This hypothesis corresponds to a group (population) main effect in the ANOVA analogy. We can express this hypothesis in terms of the average level of group 1 compared to the average level of group 2:

$$H_{02}: \frac{\mu_{11} + \mu_{12} + \cdots + \mu_{1p}}{p} = \frac{\mu_{21} + \mu_{22} + \cdots + \mu_{2p}}{p}$$

or, equivalently, by (2.37), this can be expressed as

$$H_{02}: \mathbf{j}'\boldsymbol{\mu}_1 = \mathbf{j}'\boldsymbol{\mu}_2.$$

If H_{01} is true, H_{02} can be pictured as in Figure 5.5*a*. If H_{02} is false, then the two profiles differ by a constant (given that H_{01} is true), as in Figure 5.5*b*.

The hypothesis H_{02} can still be true when H_{01} does not hold. Thus the average level of population 1 can equal the average level of population 2 without the two profiles being parallel, as in Figure 5.6. In this case, the "group main effect" is somewhat harder to interpret, as is the case in the analogous two-way ANOVA, where main effects are more difficult to describe in the presence of significant interaction. However, the test may still furnish useful information if a careful description of the results is provided.

To test $H_{02}: \mathbf{j}'(\boldsymbol{\mu}_1 - \boldsymbol{\mu}_2) = 0$, we estimate $\mathbf{j}'(\boldsymbol{\mu}_1 - \boldsymbol{\mu}_2)$ by $\mathbf{j}'(\overline{\mathbf{y}}_1 - \overline{\mathbf{y}}_2)$, which is $N[0, \mathbf{j}'\boldsymbol{\Sigma}\mathbf{j}(1/n_1 + 1/n_2)]$ when H_{02} is true. We can therefore use

$$t = \frac{\mathbf{j}'(\overline{\mathbf{y}}_1 - \overline{\mathbf{y}}_2)}{\sqrt{\mathbf{j}'\mathbf{S}_{\text{pl}}\mathbf{j}(1/n_1 + 1/n_2)}} \qquad (5.37)$$

and reject H_{02} if $|t| \geq t_{\alpha/2, n_1+n_2-2}$.

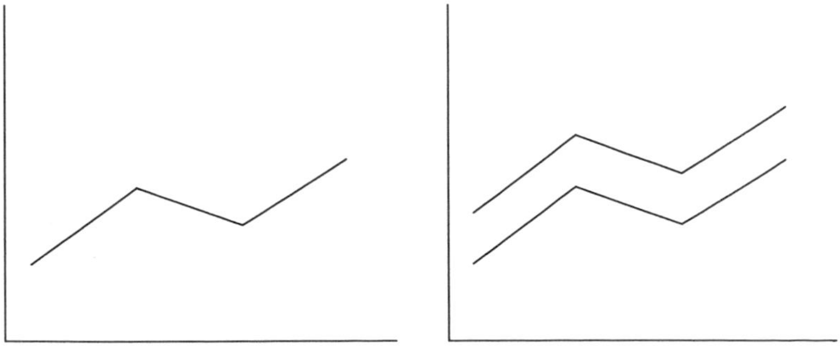

(*a*) Hypothesis is true (*b*) Hypothesis is false

Figure 5.5 Hypothesis H_{02} of equal group effect, assuming parallelism.

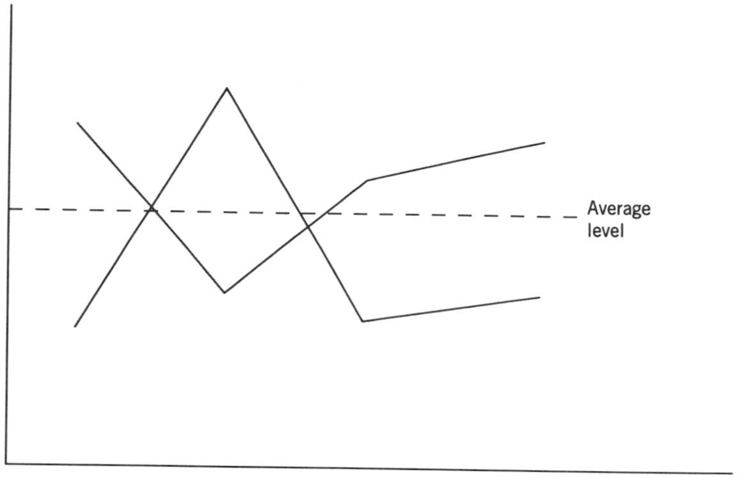

Figure 5.6 Hypothesis H_{02} of equal group effect without parallelism.

The third hypothesis of interest, corresponding to the test (or variable) main effect, is "Are the profiles flat?" Assuming parallelism (assuming H_{01} is true), the "flatness" hypothesis can be pictured as in Figure 5.7. If H_{01} is not true, the test could be carried out separately for each group using the test in Section 5.9.1. If H_{02} is true, the two profiles in Figure 5.7*a*,*b* will be coincident.

To express the third hypothesis in a form suitable for testing, we note from Figure 5.7*a* that the average of the two group means is the same for each test:

$$H_{03}: \tfrac{1}{2}(\mu_{11} + \mu_{21}) = \tfrac{1}{2}(\mu_{12} + \mu_{22}) = \cdots = \tfrac{1}{2}(\mu_{1p} + \mu_{2p})$$

or

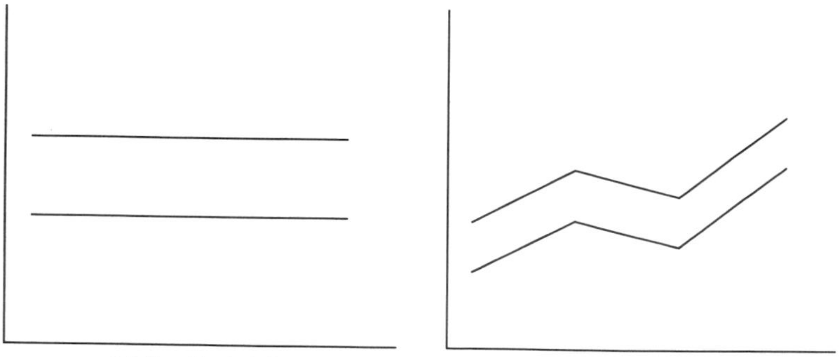

Figure 5.7 Hypothesis H_{03} of equal tests (variables) assuming parallelism.

$$H_{03}\colon \mathbf{C}\,\frac{\boldsymbol{\mu}_1 + \boldsymbol{\mu}_2}{2} = \mathbf{0}, \tag{5.38}$$

where $\mathbf{C}$ is a $(p-1)\times p$ matrix such that $\mathbf{Cj} = \mathbf{0}$. From Figure 5.7*a*, we see that H_{03} could also be expressed as $\mu_{11} = \mu_{12} = \cdots = \mu_{1p}$ and $\mu_{21} = \mu_{22} = \cdots = \mu_{2p}$ or

$$H_{03}\colon \mathbf{C}\boldsymbol{\mu}_1 = \mathbf{0} \quad \text{and} \quad \mathbf{C}\boldsymbol{\mu}_2 = \mathbf{0}.$$

To estimate $\frac{1}{2}(\boldsymbol{\mu}_1 + \boldsymbol{\mu}_2)$, we use the sample grand mean vector based on a weighted average:

$$\overline{\mathbf{y}} = \frac{n_1\overline{\mathbf{y}}_1 + n_2\overline{\mathbf{y}}_2}{n_1 + n_2}.$$

It can easily be shown that under H_{03} (and H_{01}), $E(\mathbf{C}\overline{\mathbf{y}}) = \mathbf{0}$ and $\text{cov}(\overline{\mathbf{y}}) = \boldsymbol{\Sigma}/(n_1+n_2)$. Therefore, $\mathbf{C}\overline{\mathbf{y}}$ is $N_{p-1}[\mathbf{0}, \mathbf{C\Sigma C}'/(n_1+n_2)]$ and

$$\begin{aligned} T^2 &= (\mathbf{C}\overline{\mathbf{y}})'[\mathbf{CS}_{\text{pl}}\mathbf{C}'/(n_1+n_2)]^{-1}(\mathbf{C}\overline{\mathbf{y}}) \\ &= (n_1+n_2)(\mathbf{C}\overline{\mathbf{y}})'(\mathbf{CS}_{\text{pl}}\mathbf{C}')^{-1}\mathbf{C}\overline{\mathbf{y}} \end{aligned} \tag{5.39}$$

is distributed as T^2_{p-1,n_1+n_2-2} when both H_{01} and H_{03} are true. It can be readily shown that H_{03} is unaffected by a difference in the profile levels (unaffected by the status of H_{02}).

Example 5.9.2. The psychological data described in Example 5.4.2, where the four variables are tests, represent the type of data amenable to profile analysis. We use $\overline{\mathbf{y}}_1, \overline{\mathbf{y}}_2$, and $\mathbf{S}_{\text{pl}}$ as given in Example 5.4.2.

The profiles of the two mean vectors $\overline{\mathbf{y}}_1$ and $\overline{\mathbf{y}}_2$ are plotted in Figure 5.8. There appears to be a lack of parallelism.

To test for parallelism, $H_{01}\colon \mathbf{C}\boldsymbol{\mu}_1 = \mathbf{C}\boldsymbol{\mu}_2$, we use the matrix

$$\mathbf{C} = \begin{pmatrix} -1 & 1 & 0 & 0 \\ 0 & -1 & 1 & 0 \\ 0 & 0 & -1 & 1 \end{pmatrix}$$

and obtain

$$\mathbf{C}(\overline{\mathbf{y}}_1 - \overline{\mathbf{y}}_2) = \begin{pmatrix} -1.62 \\ 8.59 \\ -9.78 \end{pmatrix} \qquad \mathbf{CS}_{\text{pl}}\mathbf{C}' = \begin{pmatrix} 10.96 & -7.13 & -1.57 \\ -7.13 & 27.86 & -12.56 \\ -1.57 & -12.56 & 23.76 \end{pmatrix}.$$

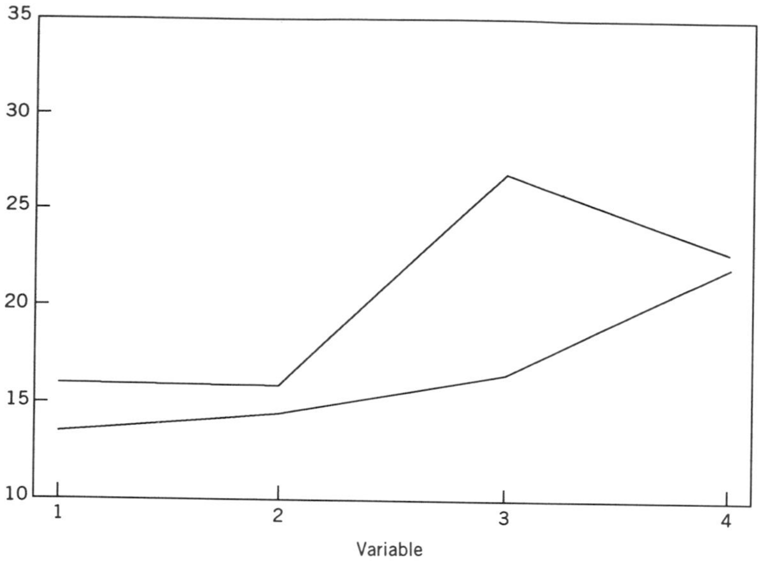

Figure 5.8 Profiles for the psychological data in Table 5.1.

Then, by (5.35),

$$T^2 = \frac{(32)(32)}{32+32}(\bar{\mathbf{y}}_1 - \bar{\mathbf{y}}_2)'\mathbf{C}'(\mathbf{C}\mathbf{S}_{\text{pl}}\mathbf{C}')^{-1}\mathbf{C}(\bar{\mathbf{y}}_1 - \bar{\mathbf{y}}_2) = 75.240.$$

Upon comparison of this value with $T^2_{.01,3,62} = 12.796$ (obtained by interpolation in Table A.7), we reject the hypothesis of parallelism.

In Figure 5.8 the lack of parallelism is most notable in the second and third segments. This can also be seen in the relatively large values of the second and third components of

$$\mathbf{C}(\bar{\mathbf{y}}_1 - \bar{\mathbf{y}}_2) = \begin{pmatrix} -1.62 \\ 8.59 \\ -9.78 \end{pmatrix}.$$

To see which of these made the greatest statistical contribution, we can examine the discriminant function coefficient vector given in (5.36) as

$$\mathbf{a} = (\mathbf{C}\mathbf{S}_{\text{pl}}\mathbf{C}')^{-1}\mathbf{C}(\bar{\mathbf{y}}_1 - \bar{\mathbf{y}}_2) = \begin{pmatrix} -.125 \\ .114 \\ -.359 \end{pmatrix}.$$

Thus the third segment contributed most to rejection in the presence of the other two segments.

To test for equal levels, $H_{02}: \mathbf{j}'\boldsymbol{\mu}_1 = \mathbf{j}'\boldsymbol{\mu}_2$, we use (5.37),

$$t = \frac{\mathbf{j}'(\overline{\mathbf{y}}_1 - \overline{\mathbf{y}}_2)}{\sqrt{\mathbf{j}'\mathbf{S}_{\text{pl}}\mathbf{j}(1/n_1 + 1/n_2)}} = \frac{17.031}{\sqrt{(165.618)(1/32 + 1/32)}} = 5.2936.$$

Comparing this with $t_{.005,62} = 2.658$, we reject the hypothesis of equal levels.

To test the flatness hypothesis, $H_{03}: \frac{1}{2}\mathbf{C}(\boldsymbol{\mu}_1 + \boldsymbol{\mu}_2) = \mathbf{0}$, we first calculate

$$\overline{\mathbf{y}} = \frac{32\overline{\mathbf{y}}_1 + 32\overline{\mathbf{y}}_2}{32 + 32} = \frac{\overline{\mathbf{y}}_1 + \overline{\mathbf{y}}_2}{2} = \begin{bmatrix} 14.16 \\ 14.91 \\ 21.89 \\ 22.34 \end{bmatrix}.$$

Using

$$\mathbf{C} = \begin{pmatrix} 1 & -1 & 0 & 0 \\ 0 & 1 & -1 & 0 \\ 0 & 0 & 1 & -1 \end{pmatrix},$$

we obtain, by (5.39),

$$T^2 = (32 + 32)(\mathbf{C}\overline{\mathbf{y}})'(\mathbf{C}\mathbf{S}_{\text{pl}}\mathbf{C}')^{-1}\mathbf{C}\overline{\mathbf{y}} = 256.689,$$

which exceeds $T^2_{.01,3,62} = 12.796$, so we reject the hypothesis of flatness. However, the assumption of parallelism appears to be unreasonable. A more appropriate approach would be to test each of the two groups separately for flatness using the test of Section 5.9.1. We obtain, by (5.34),

$$T^2 = n_1(\mathbf{C}\overline{\mathbf{y}}_1)'(\mathbf{C}\mathbf{S}_1\mathbf{C}')^{-1}(\mathbf{C}\overline{\mathbf{y}}_1) = 221.126$$

and

$$T^2 = n_2(\mathbf{C}\overline{\mathbf{y}}_2)'(\mathbf{C}\mathbf{S}_2\mathbf{C}')^{-1}(\mathbf{C}\overline{\mathbf{y}}_2) = 103.298.$$

Both of these exceed $T^2_{.01,3,31} = 14.626$, and we have significant lack of flatness.

PROBLEMS

5.1 Show that the characteristic form of T^2 in (5.5) is the same as the original form in (5.4).

5.2 Show that the T^2-statistic in (5.8) can be expressed in the characteristic form given in (5.9).

5.3 Show that $s_d^2 = \sum_{i=1}^n (d_i - \overline{d})^2/(n-1) = s_y^2 + s_x^2 - 2s_{yx}$, as noted at the end of Section 5.7.1.

5.4 Show that $T^2 = n\overline{\mathbf{d}}'\mathbf{S}_d^{-1}\overline{\mathbf{d}}$ in (5.25) has the characteristic form $T^2 = \overline{\mathbf{d}}'(\mathbf{S}_d/n)^{-1}\overline{\mathbf{d}}$.

5.5 Show that the paired observation t-test in (5.24), $t = \overline{d}/(s_d/\sqrt{n})$, has the t_{n-1} distribution.

5.6 Show that the test statistic in (5.30) for additional information in $\mathbf{x}$ above and beyond $\mathbf{y}$ has an F-distribution by solving for R^2 in terms of T^2 from (5.22) and substituting this into (5.32).

5.7 In Section 5.9.2, show that under H_{03} and H_{01}, $E(\mathbf{C}\overline{\mathbf{y}}) = \mathbf{0}$ and $\text{cov}(\overline{\mathbf{y}}) = \boldsymbol{\Sigma}/(n_1+n_2)$, where $\overline{\mathbf{y}} = (n_1\overline{\mathbf{y}}_1 + n_2\overline{\mathbf{y}}_2)/(n_1+n_2)$ and $\boldsymbol{\Sigma}$ is the common covariance matrix of the two populations from which $\overline{\mathbf{y}}_1$ and $\overline{\mathbf{y}}_2$ are sampled.

5.8 Verify that $T^2 = (n_1+n_2)(\mathbf{C}\overline{\mathbf{y}})'(\mathbf{C}\mathbf{S}_{\text{pl}}\mathbf{C}')^{-1}\mathbf{C}\overline{\mathbf{y}}$ in (5.39) has the T^2_{p-1,n_1+n_2-2} distribution.

5.9 Test $H_0\colon \boldsymbol{\mu}' = (6, 11)$ using the data

$$\mathbf{Y} = \begin{bmatrix} 3 & 10 \\ 6 & 12 \\ 5 & 14 \\ 10 & 9 \end{bmatrix}$$

5.10 Use the probe word data in Table 3.7:

(a) Test $H_0\colon \boldsymbol{\mu} = (30, 25, 40, 25, 30)'$.

(b) If H_0 is rejected, test each variable separately, using (5.3).

5.11 For the probe word data in Table 3.7, test $H_0\colon \mu_1 = \mu_2 = \cdots = \mu_5$, using T^2 in (5.34).

5.12 Use the ramus bone data in Table 3.8:

(a) Test H_0: $\boldsymbol{\mu} = (48, 49, 50, 51)'$.

(b) If H_0 is rejected, test each variable separately, using (5.3).

5.13 For the ramus bone data in Table 3.8, test H_0: $\mu_1 = \mu_2 = \mu_3 = \mu_4$, using T^2 in (5.34).

5.14 Four measurements were made on two species of flea beetles (Lubischew 1962). The variables were

y_1 = distance of transverse groove from posterior border of prothorax (μm),

y_2 = length of elytra (in 0.01 mm)

y_3 = length of second antennal joint (μm)

y_4 = length of third antennal joint (μm)

The data are given in Table 5.5.

(a) Test H_0: $\boldsymbol{\mu}_1 = \boldsymbol{\mu}_2$ using T^2.

(b) If the T^2-test in part (a) rejects H_0, carry out a t-test on each variable, as in (5.16).

Table 5.5 Four Measurements on Two Species of Flea Beetles

Haltica oleracea					*Haltica carduorum*				
Experiment Number	y_1	y_2	y_3	y_4	Experiment Number	y_1	y_2	y_3	y_4
1	189	245	137	163	1	181	305	184	209
2	192	260	132	217	2	158	237	133	188
3	217	276	141	192	3	184	300	166	231
4	221	299	142	213	4	171	273	162	213
5	171	239	128	158	5	181	297	163	224
6	192	262	147	173	6	181	308	160	223
7	213	278	136	201	7	177	301	166	221
8	192	255	128	185	8	198	308	141	197
9	170	244	128	192	9	180	286	146	214
10	201	276	146	186	10	177	299	171	192
11	195	242	128	192	11	176	317	166	213
12	205	263	147	192	12	192	312	166	209
13	180	252	121	167	13	176	285	141	200
14	192	283	138	183	14	169	287	162	214
15	200	294	138	188	15	164	265	147	192
16	192	277	150	177	16	181	308	157	204
17	200	287	136	173	17	192	276	154	209
18	181	255	146	183	18	181	278	149	235
19	192	287	141	198	19	175	271	140	192
					20	197	303	170	205

(c) Calculate the discriminant function coefficient vector $\mathbf{a} = \mathbf{S}_{\text{pl}}^{-1}(\overline{\mathbf{y}}_1 - \overline{\mathbf{y}}_2)$.

(d) Show that if the vector **a** found in part (c) is substituted into $t^2(\mathbf{a})$ from (5.14), the result is the same as the value of T^2 found in part (a).

(e) Obtain T^2 using the regression approach in Section 5.6.2.

(f) Test the significance of each variable adjusted for the other three.

(g) Test the significance of y_3 and y_4 adjusted for y_1 and y_2.

5.15 Carry out a profile analysis on the beetles data in Table 5.5.

5.16 Twenty engineer apprentices and 20 pilots were given six tests (Travers 1939). The variables were

y_1 = intelligence
y_2 = form relations
y_3 = dynamometer
y_4 = dotting
y_5 = sensory motor coordination
y_6 = perseveration

The data are given in Table 5.6.

Table 5.6 Comparison of Six Tests on Engineer Apprentices and Pilots

Engineer Apprentices						Pilots					
y_1	y_2	y_3	y_4	y_5	y_6	y_1	y_2	y_3	y_4	y_5	y_6
121	22	74	223	54	254	132	17	77	232	50	249
108	30	80	175	40	300	123	32	79	192	64	315
122	49	87	266	41	223	129	31	96	250	55	319
77	37	66	178	80	209	131	23	67	291	48	310
140	35	71	175	38	261	110	24	96	239	42	268
108	37	57	241	59	245	47	22	87	231	40	217
124	39	52	194	72	242	125	32	87	227	30	324
130	34	89	200	85	242	129	29	102	234	58	300
149	55	91	198	50	277	130	26	104	256	58	270
129	38	72	162	47	268	147	47	82	240	30	322
154	37	87	170	60	244	159	37	80	227	58	317
145	33	88	208	51	228	135	41	83	216	39	306
112	40	60	232	29	279	100	35	83	183	57	242
120	39	73	159	39	233	149	37	94	227	30	240
118	21	83	152	88	233	149	38	78	258	42	271

Table 5.6 (***Continued***)

Engineer Apprentices						Pilots					
y_1	y_2	y_3	y_4	y_5	y_6	y_1	y_2	y_3	y_4	y_5	y_6
141	42	80	195	36	241	153	27	89	283	66	291
135	49	73	152	42	249	136	31	83	257	31	311
151	37	76	223	74	268	97	36	100	252	30	225
97	46	83	164	31	243	141	37	105	250	27	243
109	42	82	188	57	267	164	32	76	187	30	264

(a) Test H_0: $\boldsymbol{\mu}_1 = \boldsymbol{\mu}_2$.

(b) If the T^2-test in part (a) rejects H_0, carry out a t-test for each variable, as in (5.16).

(c) Test each variable adjusted for the other five.

(d) Test the significance of y_4, y_5, and y_6 adjusted for y_1, y_2, and y_3.

5.17 Data were collected in an attempt to find a screening procedure to detect carriers of Duchenne muscular dystrophy, a disease transmitted from female carriers to some of their male offspring (Andrews and Herzberg 1985, pp. 223–228). The following variables were measured on a sample of noncarriers and a sample of carriers:

y_1 = age

y_2 = month in which measurements are taken

y_3 = creatine kinase

y_4 = hemopexin

y_5 = lactate dehydrogenase

y_6 = pyruvate kinase

The data are given in Table 5.7.

(a) Test H_0: $\boldsymbol{\mu}_1 = \boldsymbol{\mu}_2$ using y_3, y_4, y_5, and y_6.

(b) The variables y_3 and y_4 are relatively inexpensive to measure compared to y_5 and y_6. Do y_5 and y_6 contribute an important amount to T^2 above and beyond y_3 and y_4?

(c) The levels of y_3, y_4, y_5, and y_6 may depend on age and season, y_1 and y_2. Do these variables contribute a significant amount to T^2 when adjusted for y_3, y_4, y_5, and y_6?

Table 5.7 Comparison of Carriers and Noncarriers of Muscular Dystrophy

Noncarriers						*Carriers*					
y_1	y_2	y_3	y_4	y_5	y_6	y_1	y_2	y_3	y_4	y_5	y_6
22	6	52.0	83.5	10.9	176	30	10	167.0	89.0	25.6	364
32	8	20.0	77.0	11.0	200	41	10	104.0	81.0	26.8	245
36	7	28.0	86.5	13.2	171	22	8	30.0	108.0	8.8	284
22	11	30.0	104.0	22.6	230	22	8	44.0	104.0	17.4	172
23	1	40.0	83.0	15.2	205	20	10	65.0	87.0	23.8	198
30	5	24.0	78.8	9.6	151	42	9	440.0	107.0	20.2	239
27	8	15.0	87.0	13.5	232	59	8	58.0	88.2	11.0	259
30	11	22.0	91.0	17.5	198	35	9	129.0	93.1	18.3	188
25	10	42.0	65.5	13.3	216	36	6	104.0	87.5	16.7	256
26	2	130.0	80.3	17.1	211	35	2	122.0	88.5	21.6	263
26	3	48.0	85.2	22.7	160	29	4	265.0	83.5	16.1	136
27	7	31.0	86.5	6.9	162	27	4	285.0	79.5	36.4	245
26	10	47.0	53.0	14.6	131	27	9	25.0	91.0	49.1	209
27	3	36.0	56.0	18.2	105	28	4	124.0	92.0	32.2	298
27	7	24.0	57.5	5.6	130	29	8	53.0	76.0	14.0	174
31	4	34.0	92.7	7.9	140	30	2	46.0	71.0	16.9	197
31	9	38.0	96.0	12.6	158	30	7	40.0	85.5	12.7	201
35	10	40.0	104.6	16.1	209	30	8	41.0	90.0	9.7	342
28	4	59.0	88.0	9.9	128	31	6	657.0	104.0	110.0	358
28	8	75.0	81.0	10.1	177	32	2	465.0	86.5	63.7	412
28	9	72.0	66.3	16.4	156	32	5	485.0	83.5	73.0	382
27	7	42.0	77.0	15.3	163	37	2	168.0	82.5	23.3	261
27	3	30.0	80.2	8.1	100	38	6	286.0	109.5	31.9	260
28	6	24.0	87.0	3.5	132	39	1	388.0	91.0	41.6	204
24	9	26.0	84.5	20.7	145	39	9	148.0	105.2	18.8	221
23	8	65.0	75.0	19.9	187	34	6	73.0	105.5	17.0	285
27	3	34.0	86.3	11.8	120	35	4	36.0	92.8	22.0	308
25	2	37.0	73.3	13.0	254	58	8	19.0	100.5	10.9	196
34	3	73.0	57.4	7.4	107	58	2	34.0	98.5	19.9	299
34	7	87.0	76.3	6.0	87	38	1	113.0	97.0	18.8	216
25	7	35.0	71.0	8.8	186	30	8	57.0	105.0	12.9	155
20	7	31.0	61.5	9.9	172	42	8	78.0	118.0	15.5	212
20	5	62.0	81.0	10.2	181	43	11	73.0	104.0	20.6	201
31	6	48.0	79.0	16.8	182	29	3	69.0	111.0	16.0	175
31	7	40.0	82.5	6.4	151						
26	7	55.0	85.5	10.9	216						
26	7	32.0	73.8	8.6	147						
21	11	26.0	79.3	16.4	123						
27	6	25.0	91.0	10.3	135						

5.18 Various aspects of economic cycles were measured for consumers' goods and producers' goods by Tintner (1946). The variables are

$$y_1 = \text{length of cycle}$$

$$y_2 = \text{percentage of rising prices}$$

$$y_3 = \text{cyclical amplitude}$$
$$y_4 = \text{rate of change}$$

The data for several items are given in Table 5.8.

Table 5.8 Cyclical Measurements of Consumer Goods and Producer Goods

Item	y_1	y_2	y_3	y_4	Item	y_1	y_2	y_3	y_4
Consumer Goods					*Producer Goods*				
1	72	50	8	0.5	1	57	57	12.5	0.9
2	66.5	48	15	1.0	2	100	54	17	0.5
3	54	57	14	1.0	3	100	32	16.5	0.7
4	67	60	15	0.9	4	96.5	65	20.5	0.9
5	44	57	14	0.3	5	79	51	18	0.9
6	41	52	18	1.9	6	78.5	53	18	1.2
7	34.5	50	4	0.5	7	48	50	21	1.6
8	34.5	46	8.5	1.0	8	155	44	20.5	1.4
9	24	54	3	1.2	9	84	64	13	0.8
					10	105	35	17	1.8

(a) Test $H_0\colon \boldsymbol{\mu}_1 = \boldsymbol{\mu}_2$ using T^2.

(b) Calculate the discriminant function coefficient vector.

(c) Test for significance of each variable adjusted for the other three.

5.19 Each of 15 students wrote an informal and a formal essay Kramer (1972, p. 100). The variables recorded were the number of words and the number of verbs:

$$y_1 = \text{number of words in the informal essay}$$
$$y_2 = \text{number of verbs in the informal essay}$$
$$x_1 = \text{number of words in the formal essay}$$
$$x_2 = \text{number of verbs in the formal essay}$$

The data are given in Table 5.9. Since each student wrote both types of essays, the observation vectors are paired, and we use the paired comparison test.

(a) Test $H_0\colon \boldsymbol{\mu}_d = \mathbf{0}$.

(b) Find the discriminant function coefficient vector.

(c) Do a univariate t-test on each d_i.

Table 5.9 Number of Words and Number of Verbs

	Informal		Formal			
Student	Words y_1	Verbs y_2	Words x_1	Verbs x_2	$d_1 = y_1 - x_1$	$d_2 = y_2 - x_2$
1	148	20	137	15	+11	+5
2	159	24	164	25	−5	−1
3	144	19	224	27	−80	−8
4	103	18	208	33	−105	−15
5	121	17	178	24	−57	−7
6	89	11	128	20	−39	−9
7	119	17	154	18	−35	−1
8	123	13	158	16	−35	−3
9	76	16	102	21	−26	−5
10	217	29	214	25	+3	+4
11	148	22	209	24	−61	−2
12	151	21	151	16	0	+5
13	83	7	123	13	−40	−6
14	135	20	161	22	−26	−2
15	178	15	175	23	+3	−8

5.20 A number of patients with bronchus cancer were treated with ascorbate and compared with matched patients who received no ascorbate (Cameron and Pauling 1978). The data are given in Table 5.10. The variables measured were

y_1, x_1 = survival time (days) from date of first hospital admission

y_2, x_2 = survival time from date of untreatability

Compare y_1 and y_2 with x_1 and x_2 using a paired comparison T^2-test.

Table 5.10 Survival Times for Bronchus Cancer Patients and Matched Controls

Ascorbate Patients		Matched Controls	
y_1	y_2	x_1	x_2
81	74	72	33
461	423	134	18
20	16	84	20
450	450	98	58
246	87	48	13
166	115	142	49
63	50	113	38
64	50	90	24
155	113	30	18
151	38	260	34

Table 5.10 ***(Continued)***

Ascorbate Patients		Matched Controls	
y_1	y_2	x_1	x_2
166	156	116	20
37	27	87	27
223	218	69	32
138	138	100	27
72	39	315	39
245	231	188	65

5.21 Use the glucose data in Table 3.10:

(a) Test H_0: $\boldsymbol{\mu}_y = \boldsymbol{\mu}_x$ using a paired comparison test.

(b) Test the significance of each variable adjusted for the other two.

CHAPTER 6

Multivariate Analysis of Variance

In this chapter we extend univariate analysis of variance to multivariate analysis of variance, in which we measure more than one variable on each experimental unit. We do not cover multivariate analysis of covariance. For multivariate analysis of covariance, see Rencher (1996, Section 4.10).

6.1 ONE-WAY MODELS

We begin with a review of univariate analysis of variance before covering multivariate analysis of variance with several dependent variables.

6.1.1 Univariate One-way Analysis of Variance (ANOVA)

In the balanced one-way ANOVA, we have a random sample of n observations from each of k normal populations with equal variances, as in the following layout:

	Sample 1, from $N(\mu_1, \sigma^2)$	Sample 2, from $N(\mu_2, \sigma^2)$	$\cdots$	Sample k, from $N(\mu_k, \sigma^2)$
	y_{11}	y_{21}	$\cdots$	y_{k1}
	y_{12}	y_{22}	$\cdots$	y_{k2}
	$\vdots$	$\vdots$		$\vdots$
	y_{1n}	y_{2n}	$\cdots$	y_{kn}
Total	$y_{1.}$	$y_{2.}$	$\cdots$	$y_{k.}$
Mean	$\overline{y}_{1.}$	$\overline{y}_{2.}$	$\cdots$	$\overline{y}_{k.}$
Variance	s_1^2	s_2^2	$\cdots$	s_k^2

The k samples or the populations from which they arise are sometimes referred to as *groups*. The groups may correspond to *treatments* applied by the researcher in an experiment. We have used the "dot" notation for totals and means for each group. Thus, for example,

$$y_{2.} = \sum_{j=1}^{n} y_{2j} \qquad \bar{y}_{2.} = \sum_{j=1}^{n} \frac{y_{2j}}{n}.$$

The k samples are assumed to be independent. The assumptions of independence and common variance are necessary to obtain an F-test.

The model for each observation is

$$\begin{aligned} y_{ij} &= \mu + \alpha_i + \epsilon_{ij} \\ &= \mu_i + \epsilon_{ij} \qquad i = 1, 2, \ldots, k; j = 1, 2, \ldots, k; \end{aligned} \tag{6.1}$$

where $\mu_i = \mu + \alpha_i$ is the mean of the ith population. We wish to compare the sample means $\bar{y}_{i.}, i = 1, 2, \ldots, k$, to see if they are sufficiently different to lead us to believe the population means differ. The hypothesis can be expressed as H_0: $\mu_1 = \mu_2 = \cdots = \mu_k$.

If the hypothesis is true, all y_{ij} are from the same population, $N(\mu, \sigma^2)$, and we can obtain two estimates of σ^2, one based on the sample variances s_1^2, $s_2^2, \ldots, s_k^2$ and the other based on the sample means $\bar{y}_{1.}, \bar{y}_{2.}, \ldots, \bar{y}_{k.}$. The pooled "within-sample" estimator of σ^2 is

$$s_e^2 = \frac{1}{k} \sum_{i=1}^{k} s_i^2 = \frac{\sum_{ij} (y_{ij} - \bar{y}_{i.})^2}{k(n-1)} = \frac{\sum_{ij} y_{ij}^2 - \sum_{i} y_{i.}^2/n}{k(n-1)}. \tag{6.2}$$

Our second estimate of σ^2 (under H_0) is based on the variance of the sample means,

$$s_{\bar{y}}^2 = \frac{\sum_{i} (\bar{y}_{i.} - \bar{y}_{..})^2}{k-1}, \tag{6.3}$$

where $\bar{y}_{..} = \sum_i \bar{y}_{i.}/k$ is the overall mean. If H_0 is true, $s_{\bar{y}}^2$ estimates $\sigma_{\bar{y}}^2 = \sigma^2/n$ [see remarks following (3.1) in Section 3.1], and therefore $E(ns_{\bar{y}}^2) = n(\sigma^2/n) = \sigma^2$, from which the estimate of σ^2 is

$$ns_{\bar{y}}^2 = \frac{n\sum_i (\bar{y}_{i.} - \bar{y}_{..})^2}{k-1} = \frac{\sum_i y_{i.}^2/n - y_{..}^2/kn}{k-1}, \tag{6.4}$$

where $y_{..} = \sum_i y_{i.} = \sum_{ij} y_{ij}$ is the overall total. If H_0 is false, $E(ns_{\bar{y}}^2) = \sigma^2 + n\sum_i \alpha_i^2/(k-1)$, and $ns_{\bar{y}}^2$ will tend to reflect a larger spread in $\bar{y}_{1.}, \bar{y}_{2.}, \ldots, \bar{y}_{k.}$. Since s_e^2 is based on variability within each sample, it estimates σ^2 whether or not H_0 is true; thus $E(s_e^2) = \sigma^2$ in either case.

When sampling from normal distributions, s_e^2, a pooled estimate based on the k values of s_i^2, is independent of $s_{\bar{y}}^2$, which is based on the $\bar{y}_{i.}$'s. We can justify this assertion by noting that $\bar{y}_{i.}$ and s_i^2 are independent in each sample (when sampling from the normal distribution) and that the k samples are independent of each other. Since $ns_{\bar{y}}^2$ and s_e^2 are independent and both estimate σ^2, their ratio forms an F-statistic (see Section 7.3.1):

$$\begin{aligned} F &= \frac{ns_{\bar{y}}^2}{s_e^2} = \frac{\left(\sum_i y_{i.}^2/n - y_{..}^2/kn\right)\Big/(k-1)}{\left(\sum_{ij} y_{ij}^2 - \sum_i y_{i.}^2/n\right)\Big/[k(n-1)]} \\ &= \frac{\text{SSH}/(k-1)}{\text{SSE}/[k(n-1)]} \\ &= \frac{\text{MSH}}{\text{MSE}}, \end{aligned} \tag{6.5}$$

where $\text{SSH} = \sum_i y_{i.}^2/n - y_{..}^2/kn$ and $\text{SSE} = \sum_{ij} y_{ij}^2 - \sum_i y_{i.}^2/n$ are the "between"-sample sum of squares (due to the means) and "within"-sample sum of squares, respectively, and MSH and MSE are the corresponding sample mean squares. The F-statistic (6.5) is distributed as $F_{k-1,k(n-1)}$ when H_0 is true. We reject H_0 if $F > F_\alpha$. The F-statistic (6.5) can be shown to be a simple function of the likelihood ratio.

6.1.2 Multivariate One-Way Analysis of Variance Model (MANOVA)

We often measure several dependent variables instead of just one. In the multivariate case, we assume that k independent random samples of size n are obtained from p-variate normal populations with equal covariance matrices, as in the following layout. (In practice, the observation vectors $\mathbf{y}_{ij}$ would ordinarily be listed in row form, and sample 2 would appear below sample 1, and so on. See, for example, Table 6.2.)

	Sample 1, from $N_p(\boldsymbol{\mu}_1, \boldsymbol{\Sigma})$	Sample 2, from $N_p(\boldsymbol{\mu}_2, \boldsymbol{\Sigma})$	$\cdots$	Sample k, from $N_p(\boldsymbol{\mu}_k, \boldsymbol{\Sigma})$
	$\mathbf{y}_{11}$	$\mathbf{y}_{21}$	$\cdots$	$\mathbf{y}_{k1}$
	$\mathbf{y}_{12}$	$\mathbf{y}_{22}$	$\cdots$	$\mathbf{y}_{k2}$
	$\vdots$	$\vdots$		$\vdots$
	$\mathbf{y}_{1n}$	$\mathbf{y}_{2n}$	$\cdots$	$\mathbf{y}_{kn}$
Total	$\mathbf{y}_{1.}$	$\mathbf{y}_{2.}$	$\cdots$	$\mathbf{y}_{k.}$
Mean	$\overline{\mathbf{y}}_{1.}$	$\overline{\mathbf{y}}_{2.}$	$\cdots$	$\overline{\mathbf{y}}_{k.}$

Totals and means are defined as follows:

Total of the ith sample: $\mathbf{y}_{i.} = \sum_{j=1}^{n} \mathbf{y}_{ij}$.
Overall total: $\mathbf{y}_{..} = \sum_{i=1}^{k} \sum_{j=1}^{n} \mathbf{y}_{ij}$.
Mean of the ith sample: $\overline{\mathbf{y}}_{i.} = \mathbf{y}_{i.}/n$.
Overall mean: $\overline{\mathbf{y}}_{..} = \mathbf{y}_{..}/kn$.

The model for each observation vector is

$$\begin{aligned} \mathbf{y}_{ij} &= \boldsymbol{\mu} + \boldsymbol{\alpha}_i + \boldsymbol{\epsilon}_{ij} \\ &= \boldsymbol{\mu}_i + \boldsymbol{\epsilon}_{ij}. \end{aligned} \tag{6.6}$$

The model for the rth variable ($r = 1, 2, \ldots, p$) in each vector $\mathbf{y}_{ij} = (y_{ij1}, y_{ij2}, \ldots, y_{ijp})'$ is

$$y_{ijr} = \mu_r + \alpha_{ir} + \epsilon_{ijr} = \mu_{ir} + \epsilon_{ijr}.$$

We wish to compare the mean vectors of the k samples for significant differences. The hypothesis is therefore

$$H_0\colon \boldsymbol{\mu}_1 = \boldsymbol{\mu}_2 = \cdots = \boldsymbol{\mu}_k \quad \text{vs.} \quad H_1\colon \text{at least two } \boldsymbol{\mu}\text{'s are unequal.}$$

Equality of the mean vectors implies that the k means are equal for each variable; that is, $\mu_{1r} = \mu_{2r} = \cdots = \mu_{kr}$ for $r = 1, 2, \ldots, p$. If two means differ for just one variable, for example, $\mu_{23} \neq \mu_{43}$, then H_0 is false and we wish to reject it. We can see this by examining the population mean vectors more closely:

$$H_0\colon \begin{bmatrix} \mu_{11} \\ \mu_{12} \\ \vdots \\ \mu_{1p} \end{bmatrix} = \begin{bmatrix} \mu_{21} \\ \mu_{22} \\ \vdots \\ \mu_{2p} \end{bmatrix} = \cdots = \begin{bmatrix} \mu_{k1} \\ \mu_{k2} \\ \vdots \\ \mu_{kp} \end{bmatrix}.$$

Thus H_0 implies p sets of equalities:

$$\begin{array}{ccccccc}
\mu_{11} & = & \mu_{21} & = & \cdots & = & \mu_{k1} \\
\mu_{12} & = & \mu_{22} & = & \cdots & = & \mu_{k2} \\
 & & \vdots & & & & \vdots \\
\mu_{1p} & = & \mu_{2p} & = & \cdots & = & \mu_{kp}.
\end{array}$$

All $p(k-1)$ equalities must hold for H_0 to be true; failure of only one equality will falsify the hypothesis.

In the univariate case, we have "between" and "within" sums of squares SSH and SSE, as in (6.5). In the multivariate case, we have "between" and "within" matrices **H** and **E**, defined analogously as

$$\begin{aligned}
\mathbf{H} &= n\sum_{i=1}^{k}(\overline{\mathbf{y}}_{i.} - \overline{\mathbf{y}}_{..})(\overline{\mathbf{y}}_{i.} - \overline{\mathbf{y}}_{..})' \qquad (6.7)\\
&= \sum_{i=1}^{k}\frac{1}{n}\mathbf{y}_{i.}\mathbf{y}_{i.}' - \frac{1}{kn}\mathbf{y}_{..}\mathbf{y}_{..}'
\end{aligned}$$

and

$$\begin{aligned}
\mathbf{E} &= \sum_{i=1}^{k}\sum_{j=1}^{n}(\mathbf{y}_{ij} - \overline{\mathbf{y}}_{i.})(\mathbf{y}_{ij} - \overline{\mathbf{y}}_{i.})' \qquad (6.8)\\
&= \sum_{ij}\mathbf{y}_{ij}\mathbf{y}_{ij}' - \sum_{i}\frac{1}{n}\mathbf{y}_{i.}\mathbf{y}_{i.}'.
\end{aligned}$$

The $p \times p$ "hypothesis" matrix **H** has "between" sums of squares on the diagonal for each of the p variables. Off-diagonal elements are analogous sums of products for each pair of variables. Of course sums of products can be negative. Assuming there are no linear dependencies in the variables, the rank of **H** is the smaller of p and ν_H, $\min(p, \nu_H)$, where ν_H represents the degrees of freedom for hypothesis; in the one-way case $\nu_H = k - 1$. The $p \times p$ "error" matrix **E** has "within" sums of squares for each variable on the diagonal with analogous sums of products off-diagonal. The rank of **E** is p, unless ν_E is less than p.

Thus **H** has the form

$$\mathbf{H} = \begin{bmatrix} \mathrm{SSH}_{11} & \mathrm{SPH}_{12} & \cdots & \mathrm{SPH}_{1p} \\ \mathrm{SPH}_{12} & \mathrm{SSH}_{22} & \cdots & \mathrm{SPH}_{2p} \\ \vdots & \vdots & & \vdots \\ \mathrm{SPH}_{1p} & \mathrm{SPH}_{2p} & \cdots & \mathrm{SSH}_{pp} \end{bmatrix}, \tag{6.9}$$

where, for example,

$$\mathrm{SSH}_{22} = n \sum_{i=1}^{k} (\bar{y}_{i.2} - \bar{y}_{..2})^2 = \sum_i \frac{y_{i.2}^2}{n} - \frac{y_{..2}^2}{kn}$$

$$\mathrm{SPH}_{12} = n \sum_i (\bar{y}_{i.1} - \bar{y}_{..1})(\bar{y}_{i.2} - \bar{y}_{..2}) = \sum_i \frac{y_{i.1}y_{i.2}}{n} - \frac{y_{..1}y_{..2}}{kn}.$$

In these expressions, the subscript 1 or 2 indicates the first or second variable. Thus, for example, $\bar{y}_{i.2}$ is the second element in $\bar{\mathbf{y}}_{i.}$:

$$\bar{\mathbf{y}}_{i.} = \begin{bmatrix} \bar{y}_{i.1} \\ \bar{y}_{i.2} \\ \vdots \\ \bar{y}_{i.p} \end{bmatrix}.$$

The matrix $\mathbf{E}$ can be expressed in a form similar to (6.9):

$$\mathbf{E} = \begin{bmatrix} \mathrm{SSE}_{11} & \mathrm{SPE}_{12} & \cdots & \mathrm{SPE}_{1p} \\ \mathrm{SPE}_{12} & \mathrm{SSE}_{22} & \cdots & \mathrm{SPE}_{2p} \\ \vdots & \vdots & & \vdots \\ \mathrm{SPE}_{1p} & \mathrm{SPE}_{2p} & \cdots & \mathrm{SSE}_{pp} \end{bmatrix}, \tag{6.10}$$

where, for example,

$$\mathrm{SSE}_{22} = \sum_{i=1}^{k} \sum_{j=1}^{n} (y_{ij2} - \bar{y}_{i.2})^2 = \sum_{ij} y_{ij2}^2 - \sum_i \frac{y_{i.2}^2}{n}$$

$$\mathrm{SPE}_{12} = \sum_{i=1}^{k} \sum_{j=1}^{n} (y_{ij1} - \bar{y}_{i.1})(y_{ij2} - \bar{y}_{i.2}) = \sum_{ij} y_{ij1}y_{ij2} - \sum_i \frac{y_{i.1}y_{i.2}}{n}.$$

Note that the elements of **E** are sums of squares and products, not variances and covariances. To estimate $\mathbf{\Sigma}$, we use $\mathbf{S}_{\text{pl}} = \mathbf{E}/(nk - k)$, so that

$$E\left(\frac{\mathbf{E}}{nk-k}\right) = \mathbf{\Sigma}.$$

6.1.3 Wilks' Test Statistic

The likelihood ratio test of H_0: $\boldsymbol{\mu}_1 = \boldsymbol{\mu}_2 = \cdots = \boldsymbol{\mu}_k$ is given by

$$\Lambda = \frac{|\mathbf{E}|}{|\mathbf{E}+\mathbf{H}|}, \tag{6.11}$$

which is known as Wilks' Λ. It has also been called Wilks' U. We reject H_0 if $\Lambda \leq \Lambda_{\alpha,p,\nu_H,\nu_E}$. Note that rejection is for small values of Λ. Exact critical values $\Lambda_{\alpha,p,\nu_H,\nu_E}$ for Wilks' Λ are found in Table A.9, taken from Wall (1967). The parameters in Wilks' Λ distribution are

$$\begin{aligned} p &= \text{number of variables (dimension)} \\ \nu_H &= \text{degrees of freedom for hypothesis} \\ \nu_E &= \text{degrees of freedom for error} \end{aligned}$$

Wilks' Λ compares the "within" sum of squares and products matrix **E** to the "total" sum of squares and products matrix **E** + **H**. This is similar to the univariate F-test in (6.5) that compares the between sum of squares to the within sum of squares. By using determinants, the test statistic Λ is reduced to a scalar. Thus the multivariate information in **E** and **H** about separation of mean vectors $\overline{\mathbf{y}}_{1.}, \overline{\mathbf{y}}_{2.}, \ldots, \overline{\mathbf{y}}_{k.}$ is channeled into a single scale, on which we can decide if the separation of mean vectors is significant. This is typical of multivariate tests in general.

The mean vectors occupy a space of dimension $s = \min(p, \nu_H)$, and within this space various configurations of these mean vectors are possible. This suggests the possibility that another test statistic may be more powerful than Wilks' Λ. Competing test statistics are discussed in Sections 6.1.4 and 6.1.5.

Some of the properties and characteristics of Wilks' Λ are as follows:

1. In order for the determinants in (6.11) to be positive, it is necessary that $\nu_E \geq p$.
2. For any MANOVA model, the degrees of freedom ν_H and ν_E are always the same as in the analogous univariate case. In the one-way model, for example, $\nu_H = k - 1$ and $\nu_E = k(n - 1)$.

3. The parameters p and ν_H can be interchanged; the distribution of Λ_{p,ν_H,ν_E} is the same as that of $\Lambda_{\nu_H,p,\nu_E+\nu_H-p}$.

4. Wilks' Λ in (6.11) can be expressed in terms of the eigenvalues $\lambda_1, \lambda_2, \ldots, \lambda_s$ of $\mathbf{E}^{-1}\mathbf{H}$, as follows:

$$\Lambda = \prod_{i=1}^{s} \frac{1}{1+\lambda_i}. \tag{6.12}$$

The number of nonzero eigenvalues of $\mathbf{E}^{-1}\mathbf{H}$ is $s = \min(p, \nu_H)$, the rank of $\mathbf{H}$. The matrix $\mathbf{HE}^{-1}$ has the same eigenvalues as $\mathbf{E}^{-1}\mathbf{H}$ (see Section 2.11.4) and can be used in its place to obtain Λ. However, we prefer $\mathbf{E}^{-1}\mathbf{H}$ because we will use its eigenvectors later.

5. The range of Λ is $0 \leq \Lambda \leq 1$, and the test based on Wilks' Λ is an inverse test in the sense that we reject H_0 for small values of Λ. If the sample mean vectors were equal, we would have $\mathbf{H} = \mathbf{O}$ and $\Lambda = |\mathbf{E}|/|\mathbf{E} + \mathbf{O}| = 1$. On the other hand, as the sample mean vectors become more widely spread apart compared to the within-sample variation, $\mathbf{H}$ becomes much "larger" than $\mathbf{E}$, and Λ approaches zero.

6. In Table A.9, the critical values decrease for increasing p. Thus the addition of variables will reduce the power unless the variables contribute to rejection of the hypothesis by causing a significant reduction in Λ.

7. When $\nu_H = 1, 2$ or when $p = 1, 2$, Wilks' Λ transforms to an exact F-statistic. The transformations from Λ to F for these special cases are given in Table 6.1. The hypothesis is rejected when the transformed value of Λ exceeds the upper α-level percentage point of the F-distribution, with degrees of freedom as shown.

Table 6.1 Transformations of Wilks' Λ to Exact Upper Tail F-Tests

Parameters p, ν_H	Statistic Having F-Distribution	Degrees of Freedom
Any p, $\nu_H = 1$	$\frac{1-\Lambda}{\Lambda}\frac{\nu_E - p + 1}{p}$	p, $\nu_E - p + 1$
Any p, $\nu_H = 2$	$\frac{1-\sqrt{\Lambda}}{\sqrt{\Lambda}}\frac{\nu_E - p + 1}{p}$	$2p$, $2(\nu_E - p + 1)$
$p = 1$, any ν_H	$\frac{1-\Lambda}{\Lambda}\frac{\nu_E}{\nu_H}$	ν_H, ν_E
$p = 2$, any ν_H	$\frac{1-\sqrt{\Lambda}}{\sqrt{\Lambda}}\frac{\nu_E - 1}{\nu_H}$	$2\nu_H$, $2(\nu_E - 1)$

8. For values of p and ν_H other than those in Table 6.1, an approximate F-statistic is given by

$$F = \frac{1 - \Lambda^{1/t}}{\Lambda^{1/t}} \frac{\mathrm{df}_2}{\mathrm{df}_1}, \tag{6.13}$$

which has an approximate F-distribution with df_1 and df_2 degrees of freedom, where

$$\mathrm{df}_1 = p\nu_H \qquad \mathrm{df}_2 = wt - \tfrac{1}{2}(p\nu_H - 2)$$

$$w = \nu_E + \nu_H - \tfrac{1}{2}(p + \nu_H + 1) \qquad t = \sqrt{\frac{p^2\nu_H^2 - 4}{p^2 + \nu_H^2 - 5}}.$$

When $p\nu_H = 2$, t is set equal to 1. The approximate F in (6.13) reduces to the exact F values given in Table 6.1, when either ν_H or p is 1 or 2.

A (less accurate) approximate test is given by

$$\chi^2 = -[\nu_E - \tfrac{1}{2}(p - \nu_H + 1)] \ln \Lambda, \tag{6.14}$$

which has an approximate χ^2-distribution with $p\nu_H$ degrees of freedom. Reject H_0 if $\chi^2 > \chi^2_\alpha$. This approximation is accurate to three decimal places when $p^2 + \nu_H^2 \leq \frac{1}{3}f$, where $f = \nu_E - \frac{1}{2}(p - \nu_H + 1)$.

9. If the multivariate test based on Λ rejects H_0, it could be followed by an F-test as in (6.5) on each of the p individual y's. We can formulate a hypothesis comparing the means across the k groups for each variable, namely, H_{0i}: $\mu_{1i} = \mu_{2i} = \cdots = \mu_{ki}, i = 1, 2, \ldots, p$. It does not necessarily follow that any of the F-tests on the p individual variables will reject the corresponding H_{0i}. Conversely, it is possible that one or more of the F's will reject H_{0i} when the Λ-test accepts H_0. In either case, we use the multivariate test result rather than the univariate results. This is similar to the relationship between Z^2-tests and z-tests shown in Figure 5.2. In the three bivariate samples plotted in Figure 6.1, we illustrate the case where Λ rejects H_0: $\boldsymbol{\mu}_1 = \boldsymbol{\mu}_2 = \boldsymbol{\mu}_3$, but the F's accept both of H_{0i}: $\mu_{1i} = \mu_{2i} = \mu_{3i}, i = 1, 2$. There is no significant separation of the three samples in either the y_1 or y_2 direction alone.

10. The Wilks' Λ-test is the likelihood ratio test. Other approaches to test construction lead to different tests. Three such tests are given in Sections 6.1.4 and 6.1.5.

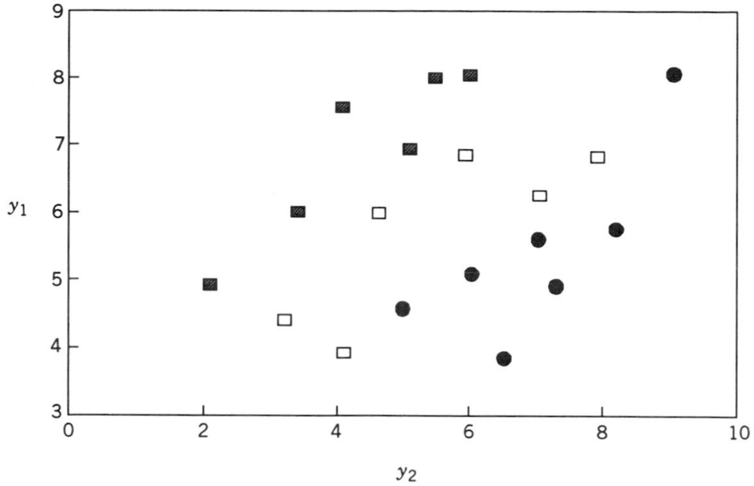

Figure 6.1 Three samples with significant Wilks' Λ but nonsignificant F's.

6.1.4 Roy's Test

Roy's *union–intersection test*, also called *Roy's largest root test*, is given by

$$\theta = \frac{\lambda_1}{1+\lambda_1}, \tag{6.15}$$

where λ_1 is the largest eigenvalue of $\mathbf{E}^{-1}\mathbf{H}$. Critical values for θ are given in Table A.10 (Pearson and Hartley 1972, Pillai 1964, 1965). We reject H_0: $\boldsymbol{\mu}_1 = \boldsymbol{\mu}_2 = \cdots = \boldsymbol{\mu}_k$ if $\theta \geq \theta_{\alpha,s,m,N}$. The parameters s, m, and N are defined as

$$s = \min(\nu_H, p) \qquad m = \tfrac{1}{2}(|\nu_H - p| - 1) \qquad N = \tfrac{1}{2}(\nu_E - p - 1).$$

For $s = 1$, use (6.27) and (6.30) in Section 6.1.7.

The eigenvector $\mathbf{a}$ corresponding to λ_1 is used in the *discriminant function*, $z = \mathbf{a}'\mathbf{y}$. This function maximizes the spread of the transformed means $\bar{z}_{i.} = \mathbf{a}'\bar{\mathbf{y}}_{i.}$ (relative to the within-sample spread). Since this is the function that best separates the means, the coefficients $a_1, a_2, \ldots, a_p$ in the linear combination can be examined for an indication of which variables contribute most to separating the means. The discriminant function is discussed further in Sections 6.1.8 and 6.4 and in Chapter 8.

We do not have a satisfactory F approximation for θ or λ_1, but an upper bound on F that is provided in some software programs is given by

$$F = \frac{(\nu_E - d - 1)\lambda_1}{d} \tag{6.16}$$

with degrees of freedom d and $\nu_E - d - 1$, where $d = \max(p, \nu_H)$. By "upper bound" we mean that the F in (6.16) is greater than or equal to the "true F"; that is, $F \geq F_{d,\nu_E-d-1}$. Therefore, we feel safe if H_0 is accepted by (6.16); but if rejection of H_0 is indicated, we are less sure of our decision.

Many computer programs do not provide eigenvalues of nonsymmetric matrices, such as $\mathbf{E}^{-1}\mathbf{H}$. However, the eigenvalues of $\mathbf{E}^{-1}\mathbf{H}$ are the same as the eigenvalues of the symmetric matrices $(\mathbf{E}^{1/2})^{-1}\mathbf{H}(\mathbf{E}^{1/2})^{-1}$ and $(\mathbf{U}')^{-1}\mathbf{H}\mathbf{U}^{-1}$, where $\mathbf{E}^{1/2}$ is the square root matrix of $\mathbf{E}$ given in (2.103) and $\mathbf{U}'\mathbf{U} = \mathbf{E}$ is the Cholesky factorization of $\mathbf{E}$ (Section 2.7). We demonstrate this for the Cholesky approach. We first multiply the defining relationship $(\mathbf{E}^{-1}\mathbf{H} - \lambda\mathbf{I})\mathbf{a} = \mathbf{0}$ by $\mathbf{E}$ to obtain

$$(\mathbf{H} - \lambda\mathbf{E})\mathbf{a} = \mathbf{0}. \tag{6.17}$$

Then substituting $\mathbf{E} = \mathbf{U}'\mathbf{U}$ into (6.17), multiplying by $(\mathbf{U}')^{-1}$, and inserting $\mathbf{U}^{-1}\mathbf{U} = \mathbf{I}$, we have

$$\begin{aligned}
(\mathbf{H} - \lambda\mathbf{U}'\mathbf{U})\mathbf{a} &= \mathbf{0} \\
(\mathbf{U}')^{-1}(\mathbf{H} - \lambda\mathbf{U}'\mathbf{U})\mathbf{a} &= (\mathbf{U}')^{-1}\mathbf{0} = \mathbf{0} \\
[(\mathbf{U}')^{-1}\mathbf{H} - \lambda\mathbf{U}]\mathbf{U}^{-1}\mathbf{U}\mathbf{a} &= \mathbf{0} \\
[(\mathbf{U}')^{-1}\mathbf{H}\mathbf{U}^{-1} - \lambda\mathbf{I}]\mathbf{U}\mathbf{a} &= \mathbf{0}.
\end{aligned} \tag{6.18}$$

Thus $(\mathbf{U}')^{-1}\mathbf{H}\mathbf{U}^{-1}$ has the same eigenvalues as $\mathbf{E}^{-1}\mathbf{H}$ and has eigenvectors of the form $\mathbf{U}\mathbf{a}$, where $\mathbf{a}$ is an eigenvector of $\mathbf{E}^{-1}\mathbf{H}$.

6.1.5 Pillai and Lawley–Hotelling Tests

There are two additional test statistics for H_0: $\boldsymbol{\mu}_1 = \boldsymbol{\mu}_2 = \cdots = \boldsymbol{\mu}_k$ based on the eigenvalues $\lambda_1, \lambda_2, \ldots, \lambda_s$ of $\mathbf{E}^{-1}\mathbf{H}$. The *Pillai statistic* is given by

$$V^{(s)} = \sum_{i=1}^{s} \frac{\lambda_i}{1 + \lambda_i}. \tag{6.19}$$

We reject H_0 for $V^{(s)} \geq V_\alpha^{(s)}$. The upper percentage points, $V_\alpha^{(s)}$, are given in Table A.11 (Schuurmann et al. 1975), indexed by s, m, and N, which are defined as in Section 6.1.4 for Roy's test. For $s = 1$, use (6.27) and (6.30) in Section 6.1.7. Pillai's test statistic in (6.19) is an extension of Roy's statistic $\theta = \lambda_1/(1 + \lambda_1)$. In many cases the information in the additional terms $\lambda_i/(1 + \lambda_i)$, $i = 2, 3, \ldots, s$, is helpful in rejecting H_0.

For parameter values not included in Table A.11, we can use an approximate F-test:

$$F_1 = \frac{(2N+s+1)V^{(s)}}{(2m+s+1)(s-V^{(s)})}, \tag{6.20}$$

which is approximately distributed as $F_{s(2m+s+1),s(2N+s+1)}$. Two alternative F approximations are given by

$$F_2 = \frac{s(\nu_E - \nu_H + s)V^{(s)}}{p\nu_H(s-V^{(s)})}, \tag{6.21}$$

with $p\nu_H$ and $s(\nu_E - \nu_H + s)$ degrees of freedom, and

$$F_3 = \frac{(\nu_E - p + s)V^{(s)}}{d(s-V^{(s)})}, \tag{6.22}$$

with sd and $s(\nu_E - p + s)$ degrees of freedom, where $d = \max(p, \nu_H)$.

The *Lawley–Hotelling statistic* (Lawley 1938, Hotelling 1951) is defined as

$$U^{(s)} = \sum_{i=1}^{s} \lambda_i, \tag{6.23}$$

and is also known as *Hotelling's generalized T^2-statistic* (see a comment at the end of Section 6.1.7). Table A.12 (Davis 1970a,b, 1980) gives upper percentage points of the test statistic

$$\frac{\nu_E}{\nu_H} U^{(s)}.$$

We reject H_0 for large values of the test statistic. Note that in Table A.12, $p \leq \nu_H$ and $p \leq \nu_E$. If $p > \nu_H$, use $(\nu_H, p, \nu_E + \nu_H - p)$ in place of (p, ν_H, ν_E). (This same pattern in the parameters is found in Wilks' Λ. See property 3 in Section 6.1.3.) If $\nu_H = 1$ and $p > 1$, use the relationship $U^{(1)} = T^2/\nu_E$ (see Section 6.1.7). For other values of the parameters not included in Table A.12, we can use an F approximation:

$$F_1 = \frac{U^{(s)}}{c}, \tag{6.24}$$

which is approximately $F_{a,b}$, where

$$a = p\nu_H \qquad b = 4 + \frac{a+2}{B-1} \qquad c = \frac{a(b-2)}{b(\nu_E - p - 1)}$$

$$B = \frac{(\nu_E + \nu_H - p - 1)(\nu_E - 1)}{(\nu_E - p - 3)(\nu_E - p)}.$$

Alternative F approximations are given by

$$F_2 = \frac{2(sN+1)U^{(s)}}{s^2(2m+s+1)},$$

with $s(2m+s+1)$ and $2(sN+1)$ degrees of freedom, and

$$F_3 = \frac{[s(\nu_E - \nu_H - 1) + 1]U^{(s)}}{sp\nu_H},$$

with $p\nu_H$ and $s(\nu_E - \nu_H - 1)$ degrees of freedom.

6.1.6 Unbalanced One-Way MANOVA

The balanced one-way model can be easily extended to the unbalanced case. The model in (6.6) becomes

$$\mathbf{y}_{ij} = \boldsymbol{\mu} + \boldsymbol{\alpha}_i + \boldsymbol{\epsilon}_{ij} = \boldsymbol{\mu}_i + \boldsymbol{\epsilon}_{ij}, \qquad i = 1, 2, \ldots, k; j = 1, 2, \ldots, n_i.$$

Thus there are n_i observation vectors in the ith group. The mean vectors become $\overline{\mathbf{y}}_{i.} = \sum_{j=1}^{n_i} \mathbf{y}_{ij}/n_i$ and $\overline{\mathbf{y}}_{..} = \sum_{i=1}^{k} \sum_{j=1}^{n_i} \mathbf{y}_{ij}/N$, where $N = \sum_{i=1}^{k} n_i$. Similarly, the total vectors are defined as $\mathbf{y}_{i.} = \sum_{j=1}^{n_i} \mathbf{y}_{ij}$ and $\mathbf{y}_{..} = \sum_{ij} \mathbf{y}_{ij}$. The $\mathbf{H}$ and $\mathbf{E}$ matrices are calculated as

$$\mathbf{H} = \sum_{i=1}^{k} n_i(\overline{\mathbf{y}}_{i.} - \overline{\mathbf{y}}_{..})(\overline{\mathbf{y}}_{i.} - \overline{\mathbf{y}}_{..})' = \sum_{i=1}^{k} \frac{1}{n_i} \mathbf{y}_{i.}\mathbf{y}_{i.}' - \frac{1}{N} \mathbf{y}_{..}\mathbf{y}_{..}' \tag{6.25}$$

$$\mathbf{E} = \sum_{i=1}^{k} \sum_{j=1}^{n_i} (\mathbf{y}_{ij} - \overline{\mathbf{y}}_{i.})(\mathbf{y}_{ij} - \overline{\mathbf{y}}_{i.})' = \sum_{i=1}^{k} \sum_{j=1}^{n_i} \mathbf{y}_{ij}\mathbf{y}_{ij}' - \sum_{i=1}^{k} \frac{1}{n_i} \mathbf{y}_{i.}\mathbf{y}_{i.}'. \tag{6.26}$$

Wilks' Λ and the other tests have the same form as in Sections 6.1.3–6.1.5 using $\mathbf{H}$ and $\mathbf{E}$ from (6.25) and (6.26). In each test we have

$$\nu_H = k - 1 \qquad \nu_E = N - k = \sum_{i=1}^{k} n_i - k.$$

Note that $N = \sum_i n_i$ differs from N used as a parameter in Roy's and Pillai's tests in Sections 6.1.4 and 6.1.5.

6.1.7 Summary of the Four Tests and Relationship to T^2

We compare the four test statistics in terms of the eigenvalues $\lambda_1 > \lambda_2 > \cdots > \lambda_s$ of $\mathbf{E}^{-1}\mathbf{H}$, where $s = \min(\nu_H, p)$:

Pillai	$V^{(s)}$	$= \sum_{i=1}^{s} \dfrac{\lambda_i}{1+\lambda_i}$
Lawley–Hotelling	$U^{(s)}$	$= \sum_{i=1}^{s} \lambda_i$
Wilks' lambda	Λ	$= \prod_{i=1}^{s} \dfrac{1}{1+\lambda_i}$
Roy's largest root	θ	$= \dfrac{\lambda_1}{1+\lambda_1}$

Note that for all four tests we must have $\nu_E \geq p$. As noted in Section 6.1.3 and elsewhere, p is the number of variables, ν_H is the degrees of freedom for the hypothesis, and ν_E is the degrees of freedom for error.

Why do we use four different tests? All four are exact tests; that is, when H_0 is true, each test has probability α of rejecting H_0. However, in a given sample they may lead to different conclusions even when H_0 is true; some may reject H_0 while others accept H_0. This is due to the multidimensional nature of the space in which the mean vectors $\boldsymbol{\mu}_1, \boldsymbol{\mu}_2, \ldots, \boldsymbol{\mu}_k$ lie. A comparison of power and other properties of the tests is given in Section 6.2.

When $\nu_H = 1$, then s is also equal to 1, and there is only one nonzero eigenvalue λ_1. In this case, all four test statistics are functions of each other and give equivalent results. In terms of θ, for example, the other three become

$$U^{(1)} = \lambda_1 = \frac{\theta}{1-\theta} \tag{6.27}$$

$$V^{(1)} = \theta \tag{6.28}$$

$$\Lambda = 1 - \theta. \tag{6.29}$$

In the case of $\nu_H = 1$, all four statistics can be transformed to an exact F using

$$F = \frac{\nu_E - p + 1}{p} U^{(1)}, \tag{6.30}$$

which is distributed as F_{p,ν_E-p+1}.

The equivalence of all four test statistics to Hotelling's T^2 when $\nu_H = 1$ was noted in Section 5.6.1. We now demonstrate the relationship $T^2 = (n_1 + n_2 - 2)U^{(1)}$ in (5.18). For **H** and **E**, we use (6.25) and (6.26), which allow unequal n_i, since we do not require $n_1 = n_2$ in T^2. In this case, with only two groups, $\mathbf{H} = \sum_{i=1}^{2} n_i(\overline{\mathbf{y}}_{i.} - \overline{\mathbf{y}}_{..})(\overline{\mathbf{y}}_{i.} - \overline{\mathbf{y}}_{..})'$ can be expressed as

$$\mathbf{H} = \frac{n_1 n_2}{n_1 + n_2}(\overline{\mathbf{y}}_{1.} - \overline{\mathbf{y}}_{2.})(\overline{\mathbf{y}}_{1.} - \overline{\mathbf{y}}_{2.})' = c(\overline{\mathbf{y}}_{1.} - \overline{\mathbf{y}}_{2.})(\overline{\mathbf{y}}_{1.} - \overline{\mathbf{y}}_{2.})', \tag{6.31}$$

where $c = n_1 n_2/(n_1 + n_2)$. Then $U^{(1)}$ becomes

$$\begin{aligned} U^{(1)} &= \sum_{i=1}^{s} \lambda_i = \text{tr}(\mathbf{E}^{-1}\mathbf{H}) \quad [\text{by (2.98)}] \\ &= \text{tr}[c\mathbf{E}^{-1}(\overline{\mathbf{y}}_{1.} - \overline{\mathbf{y}}_{2.})(\overline{\mathbf{y}}_{1.} - \overline{\mathbf{y}}_{2.})'] \quad [\text{by (6.31)}] \\ &= c\,\text{tr}[(\overline{\mathbf{y}}_{1.} - \overline{\mathbf{y}}_{2.})'\mathbf{E}^{-1}(\overline{\mathbf{y}}_{1.} - \overline{\mathbf{y}}_{2.})] \quad [\text{by (2.89)}] \\ &= \frac{c}{n_1 + n_2 - 2}(\overline{\mathbf{y}}_{1.} - \overline{\mathbf{y}}_{2.})'\left(\frac{\mathbf{E}}{n_1 + n_2 - 2}\right)^{-1}(\overline{\mathbf{y}}_{1.} - \overline{\mathbf{y}}_{2.}) \\ &= \frac{T^2}{n_1 + n_2 - 2}, \end{aligned}$$

since $\mathbf{E}/(n_1 + n_2 - 2) = \mathbf{S}_{\text{pl}}$ (see Section 5.4.2). From this result, (5.17), (5.19), and (5.20) follow immediately using (6.27)–(6.29).

When there are two groups, $\nu_E = n_1 + n_2 - 2$ and $\nu_H = 1$. Thus $T^2/(n_1 + n_2 - 2) = T^2/\nu_E$ and the test statistic $\nu_E U^{(s)}/\nu_H$ reduces to $\nu_E U^{(1)} = T^2$. Because of this relationship, the Lawley–Hotelling statistic $U^{(s)}$ is often called the *generalized* T^2*-statistic*.

Example 6.1.7. In a classical experiment carried out from 1918 to 1934, apple trees of different rootstocks were compared (Andrews and Herzberg 1985, pp. 357–360). The data for eight trees from each of six rootstocks are given in Table 6.2. The variables are

y_1 = trunk girth at 4 years (mm × 100)
y_2 = extension growth at 4 years (m)
y_3 = trunk girth at 15 years (mm × 100)
y_4 = weight of tree above ground at 15 years (lb × 1000)

Table 6.2 Rootstock Data

Rootstock	y_1	y_2	y_3	y_4
1	1.11	2.569	3.58	0.760
1	1.19	2.928	3.75	0.821
1	1.09	2.865	3.93	0.928
1	1.25	3.844	3.94	1.009
1	1.11	3.027	3.60	0.766
1	1.08	2.336	3.51	0.726
1	1.11	3.211	3.98	1.209
1	1.16	3.037	3.62	0.750
2	1.05	2.074	4.09	1.036
2	1.17	2.885	4.06	1.094
2	1.11	3.378	4.87	1.635
2	1.25	3.906	4.98	1.517
2	1.17	2.782	4.38	1.197
2	1.15	3.018	4.65	1.244
2	1.17	3.383	4.69	1.495
2	1.19	3.447	4.40	1.026
3	1.07	2.505	3.76	0.912
3	0.99	2.315	4.44	1.398
3	1.06	2.667	4.38	1.197
3	1.02	2.390	4.67	1.613
3	1.15	3.021	4.48	1.476
3	1.20	3.085	4.78	1.571
3	1.20	3.308	4.57	1.506
3	1.17	3.231	4.56	1.458
4	1.22	2.838	3.89	0.944
4	1.03	2.351	4.05	1.241
4	1.14	3.001	4.05	1.023
4	1.01	2.439	3.92	1.067
4	0.99	2.199	3.27	0.693
4	1.11	3.318	3.95	1.085
4	1.20	3.601	4.27	1.242
4	1.08	3.291	3.85	1.017
5	0.91	1.532	4.04	1.084
5	1.15	2.552	4.16	1.151
5	1.14	3.083	4.79	1.381
5	1.05	2.330	4.42	1.242
5	0.99	2.079	3.47	0.673
5	1.22	3.366	4.41	1.137
5	1.05	2.416	4.64	1.455
5	1.13	3.100	4.57	1.325
6	1.11	2.813	3.76	0.800
6	0.75	0.840	3.14	0.606
6	1.05	2.199	3.75	0.790
6	1.02	2.132	3.99	0.853
6	1.05	1.949	3.34	0.610
6	1.07	2.251	3.21	0.562
6	1.13	3.064	3.63	0.707
6	1.11	2.469	3.95	0.952

The matrices **H** and **E** are given by

$$\mathbf{H} = \begin{bmatrix} .074 & .537 & .332 & .208 \\ .537 & 4.200 & 2.355 & 1.637 \\ .332 & 2.355 & 6.114 & 3.781 \\ .208 & 1.637 & 3.781 & 2.493 \end{bmatrix}$$

$$\mathbf{E} = \begin{bmatrix} .320 & 1.697 & .554 & .217 \\ 1.697 & 12.143 & 4.364 & 2.110 \\ .554 & 4.364 & 4.291 & 2.482 \\ .217 & 2.110 & 2.482 & 1.723 \end{bmatrix}.$$

Their sum is

$$\mathbf{E} + \mathbf{H} = \begin{bmatrix} .394 & 2.234 & .886 & .426 \\ 2.234 & 16.342 & 6.719 & 3.747 \\ .886 & 6.719 & 10.405 & 6.263 \\ .426 & 3.747 & 6.263 & 4.216 \end{bmatrix},$$

and Wilks' Λ is given by (6.11) as

$$\Lambda = \frac{|\mathbf{E}|}{|\mathbf{E}+\mathbf{H}|} = \frac{.6571}{4.2667} = .154.$$

In this case, the parameters of the Wilks' Λ distribution are $p = 4$, $\nu_H = 6 - 1 = 5$, and $\nu_E = 6(8 - 1) = 42$. We reject H_0: $\boldsymbol{\mu}_1 = \boldsymbol{\mu}_2 = \cdots = \boldsymbol{\mu}_6$ because

$$\Lambda = .154 < \Lambda_{.05,4,5,40} = .455.$$

To obtain an approximate F, we first calculate

$$t = \sqrt{\frac{p^2\nu_H{}^2 - 4}{p^2 + \nu_H{}^2 - 5}} = \sqrt{\frac{4^2 5^2 - 4}{4^2 + 5^2 - 5}} = 3.3166$$

$$w = \nu_E + \nu_H - \tfrac{1}{2}(p + \nu_H + 1) = 42 + 5 - \tfrac{1}{2}(4 + 5 + 1) = 42$$

$$\mathrm{df}_1 = p\nu_H = 4(5) = 20$$

$$\mathrm{df}_2 = wt - \tfrac{1}{2}(p\nu_H - 2) = 130.3.$$

Then the approximate F is given by (6.13) as

$$F = \frac{1-\Lambda^{1/t}}{\Lambda^{1/t}} \frac{\text{df}_2}{\text{df}_1} = \frac{1-(.154)^{1/3.3166}}{(.154)^{1/3.3166}} \frac{130.3}{20} = 4.937,$$

which exceeds $F_{.001,20,120} = 2.53$, and we reject H_0.

In this case, the mean vectors represent six points in four-dimensional space. The four eigenvalues of $\mathbf{E}^{-1}\mathbf{H}$ are 1.876, .791, .229, and .026. With these we can calculate the other three test statistics. For Pillai's statistic we have, by (6.19),

$$V^{(s)} = \sum_{i=1}^{4} \frac{\lambda_i}{1+\lambda_i} = 1.305.$$

To find a critical value for $V^{(s)}$ in Table A.11, we need

$$s = \min(\nu_H, p) = 4 \qquad m = \tfrac{1}{2}(|\nu_H - p| - 1) = 0$$
$$N = \tfrac{1}{2}(\nu_E - p - 1) = 18.5.$$

Then $V^{(s)}_{.05} = .645$ (by interpolation). Since $1.305 > .645$, we reject H_0.

For the Lawley–Hotelling statistic we obtain, by (6.23),

$$U^{(s)} = \sum_{i=1}^{s} \lambda_i = 2.921.$$

To make the test, we calculate the test statistic

$$\frac{\nu_E}{\nu_H} U^{(s)} = \tfrac{42}{5}(2.921) = 24.539.$$

The .05 critical value for $\nu_E U^{(s)}/\nu_H$ is given in Table A.12 as 7.6188 (using $\nu_E = 40$), and we therefore reject H_0.

Roy's test statistic is given by (6.15) as

$$\theta = \frac{\lambda_1}{1+\lambda_1} = \frac{1.876}{1+1.876} = .652,$$

which exceeds the .05 critical value .377 obtained (by interpolation) from Table A.10, and we reject H_0.

6.1.8 Measures of Multivariate Association

In multiple regression, a measure of association between the dependent variable y and the independent variables $x_1, x_2, \ldots, x_q$ is given by the *squared multiple correlation*

$$R^2 = \frac{\text{regression sum of squares}}{\text{total sum of squares}}.$$

Similarly, in one-way univariate ANOVA, Fisher's *correlation ratio* η^2 is defined as

$$\eta^2 = \frac{\text{between sum of squares}}{\text{total sum of squares}}.$$

This is a measure of model fit similar to R^2 and gives the proportion of variation in the dependent variable y attributable to differences among the means of the groups. Thus η^2 can be considered to be a measure of association between the dependent variable y and the grouping variable i associated with μ_i or α_i in the model (6.1). It answers the question "How well can we predict y by knowing what group it is from?" In fact, if the grouping variable is represented by $k-1$ *dummy* variables (also called *indicator* or *categorical* variables), then we have a dependent variable related to several independent variables as in multiple regression.

A dummy variable takes on the value 1 for sampling units in a group (sample) and 0 for all other sampling units. (Values other than 0 and 1 could be used.) Thus for k samples (groups), the $k-1$ dummy variables are

$$x_i = \begin{cases} 1 & \text{if sampling unit is in } i\text{th group} \\ 0 & \text{otherwise} \end{cases} \qquad i = 1, 2, \ldots, k-1.$$

Only $k-1$ dummy variables are needed because if $x_1 = x_2 = \cdots = x_{k-1} = 0$, the sampling unit must be from the kth group (see Section 11.6.2 for an illustration). The dependent variable y can be regressed on the $k-1$ dummy variables to produce results equivalent to the usual ANOVA calculations.

In (one-way) MANOVA we need to measure the strength of the association between several dependent variables and several independent (grouping) variables. Various measurements of multivariate association have been proposed. Wilks (1932) suggested a "generalized η^2":

$$\text{MANOVA } \eta^2 = \eta^2_\Lambda = 1 - \Lambda, \tag{6.32}$$

based on the use of $|\mathbf{E}|$ and $|\mathbf{E}+\mathbf{H}|$ as generalizations of sums of squares. We use $1 - \Lambda$ because Λ is small if the spread in the means is large.

We now consider an η^2 based on Roy's statistic, θ. We noted in Section 6.1.4 that the discriminant function is the linear function $z = \mathbf{a}'\mathbf{y}$ that maximizes the spread of the means $\bar{z}_{i.} = \mathbf{a}'\bar{\mathbf{y}}_{i.}$, $i = 1, 2, \ldots, k$, where $\mathbf{a}$ is the eigenvector of $\mathbf{E}^{-1}\mathbf{H}$ corresponding to the largest eigenvalue λ_1. We measure the spread among the means by SSH for z, relative to the within-sample spread SSE for z. The maximum value of this ratio is given by λ_1. Thus

$$\lambda_1 = \frac{\text{SSH}(z)}{\text{SSE}(z)},$$

and by (6.15),

$$\theta = \frac{\lambda_1}{1+\lambda_1} = \frac{\text{SSH}(z)}{\text{SSE}(z)+\text{SSH}(z)}.$$

Hence θ serves directly as a measure of multivariate association:

$$\eta_\theta^2 = \theta = \frac{\lambda_1}{1+\lambda_1}. \tag{6.33}$$

It can be shown that the square root of this quantity,

$$\eta_\theta = \sqrt{\frac{\lambda_1}{1+\lambda_1}}, \tag{6.34}$$

is the maximum correlation between a linear combination of the p dependent variables and a linear combination of the $k-1$ dummy group variables. This type of correlation is often called a *canonical correlation* (see Chapter 11) and is defined for each eigenvalue $\lambda_1, \lambda_2, \ldots, \lambda_s$ as $r_i = \sqrt{\lambda_i/(1+\lambda_i)}$.

Other measures of multivariate association suggested by Cramer and Nicewander (1979) and Muller and Peterson (1984) will now be presented. It is easily shown (Section 11.6.2) that Λ can be expressed as

$$\Lambda = \prod_{i=1}^{s} \frac{1}{1+\lambda_i} = \prod_{i=1}^{s} (1 - r_i^2),$$

where $r_i^2 = \lambda_i/(1+\lambda_i)$ is the ith squared canonical correlation described above.

The *geometric mean* of a set of positive numbers $a_1, a_2, \ldots, a_n$ is defined as $(a_1 a_2 \cdots a_n)^{1/n}$. Thus $\Lambda^{1/s}$ is the geometric mean of the $(1-r_i^2)$'s, and another measure of multivariate association based on Λ is

$$A_\Lambda = 1 - \Lambda^{1/s}. \tag{6.35}$$

In fact, as noted by Muller and Peterson, the F approximation given in (6.13),

$$F = \frac{1 - \Lambda^{1/t}}{\Lambda^{1/t}} \frac{\mathrm{df}_2}{\mathrm{df}_1},$$

is very similar to the univariate F-statistic (10.25) for testing significance in multiple regression,

$$F = \frac{R^2/(\mathrm{df\,model})}{(1 - R^2)/(\mathrm{df\,error})}, \tag{6.36}$$

based on R^2 as defined at the beginning of this section.

Pillai's statistic is easily expressible as the sum of the squared canonical correlations:

$$V^{(s)} = \sum_{i=1}^{s} \frac{\lambda_i}{1 + \lambda_i} = \sum_i r_i^2.$$

The average of the r_i^2 can be used as a measure of multivariate association:

$$A_P = \frac{\sum_{i=1}^{s} r_i^2}{s} = \frac{V^{(s)}}{s}. \tag{6.37}$$

In terms of A_P the F approximation given in (6.21) becomes

$$F_2 = \frac{A_p/p\nu_H}{(1 - A_p)/s(\nu_E - \nu_H + s)}, \tag{6.38}$$

which has an obvious parallel to (6.36).

For the Lawley–Hotelling statistic $U^{(s)}$, a multivariate measure of association can be defined as

$$A_{\mathrm{LH}} = \frac{U^{(s)}/s}{1 + U^{(s)}/s}. \tag{6.39}$$

If $s = 1$, (6.39) reduces to (6.33). In fact, (6.33) is a special case of (6.39) because $U^{(s)}/s = \sum_{i=1}^{s} \lambda_i/s$ is the arithmetic average of the λ_i's. It is easily seen that the F approximation F_3 for the Lawley–Hotelling statistic given at the end of Section 6.1.5 can be expressed in terms of A_{LH} as

$$F_3 = \frac{A_{\text{LH}}/p\nu_H}{(1 - A_{\text{LH}})/[s(\nu_E - \nu_H - 1) + 1]}, \tag{6.40}$$

which is similar to (6.36).

Example 6.1.8. We illustrate some measures of association for the root-stock data in Table 6.2:

$$\begin{aligned}
\eta_\Lambda^2 &= 1 - \Lambda = .846 \\
\eta_\theta^2 &= \theta = .652 \\
A_\Lambda &= 1 - \Lambda^{1/4} = 1 - (.154)^{1/4} = .374 \\
A_{\text{LH}} &= \frac{U^{(s)}/s}{1 + U^{(s)}/s} = \frac{2.921/4}{1 + 2.921/4} = .422 \\
A_P &= V^{(s)}/s = 1.305/4 = .326.
\end{aligned}$$

There is a wide range of values among these measures of association.

6.2 COMPARISON OF THE FOUR MANOVA TEST STATISTICS

When H_0: $\boldsymbol{\mu}_1 = \boldsymbol{\mu}_2 = \cdots = \boldsymbol{\mu}_k$ is true, all the mean vectors are at the same point. Therefore, all four MANOVA test statistics have the same Type I error rate, α, as noted in Section 6.1.7; that is, all have the same probability of rejection when H_0 is true. However, when H_0 is false, the four tests have different probabilities of rejection. We noted in Section 6.1.7 that in a given sample the four tests need not agree, even if H_0 is true. One test could reject H_0 and the others accept H_0, for example.

Historically, Wilks' Λ has played the dominant role in significance tests in MANOVA because it was the first to be derived and has well-known χ^2 and F approximations. It can also be partitioned in certain ways we will find useful later. However, it is not always the most powerful among the four tests. The probability of rejecting H_0 when it is false is known as the *power* of the test.

In univariate ANOVA with $p = 1$, the means $\mu_1, \mu_2, \ldots, \mu_k$ can be uniquely ordered along a line in one dimension, and the usual F-test is uniformly most powerful. In the multivariate case, on the other hand, with $p > 1$, the mean vectors are points in $s = \min(p, \nu_H)$ dimensions. We have four tests, not one of which is uniformly most powerful. The relative powers of the four test statistics depend on the configuration of the mean vectors $\boldsymbol{\mu}_1, \boldsymbol{\mu}_2, \ldots, \boldsymbol{\mu}_k$ in the s-dimensional space. A given test will be more powerful for one configuration of mean vectors than another.

If $\nu_H < p$, then $s = \nu_H$ and the mean vectors lie in an s-dimensional subspace of the p-dimensional space of the observations. The points may, in fact, occupy a subspace of the s dimensions. For example, they may be confined to a line (one dimension) or a plane (two dimensions). This is illustrated in Figure 6.2.

An indication of the pattern of the mean vectors is given by the eigenvalues of $\mathbf{E}^{-1}\mathbf{H}$. If there is one large eigenvalue and the others are small, the mean vectors lie close to a line in space. If there are two large eigenvalues, the mean vectors lie mostly in two dimensions, and so on.

Because Roy's test uses only the largest eigenvalue of $\mathbf{E}^{-1}\mathbf{H}$, it is more powerful than the others if the mean vectors are collinear. The other three tests have greater power than Roy's when the mean vectors are diffuse (spread out in several dimensions).

In terms of power, the tests are ordered $\theta \geq U^{(s)} \geq \Lambda \geq V^{(s)}$ for the collinear case. In the diffuse case and for intermediate structure between collinear and diffuse, the ordering of power is reversed, $V^{(s)} \geq \Lambda \geq U^{(s)} \geq \theta$. The latter ordering also holds for accuracy of the Type I error rate when the population covariance matrices $\boldsymbol{\Sigma}_1, \boldsymbol{\Sigma}_2, \ldots, \boldsymbol{\Sigma}_k$ are not equal. These are comparisons of power. For actual computation of power in a given experimental setting or to find the sample size needed to yield a desired level of power, see Rencher (1996, Section 4.4).

Generally, if group sizes are equal, the tests are sufficiently robust with respect to heterogeneity of covariance matrices so that we need not worry. If the

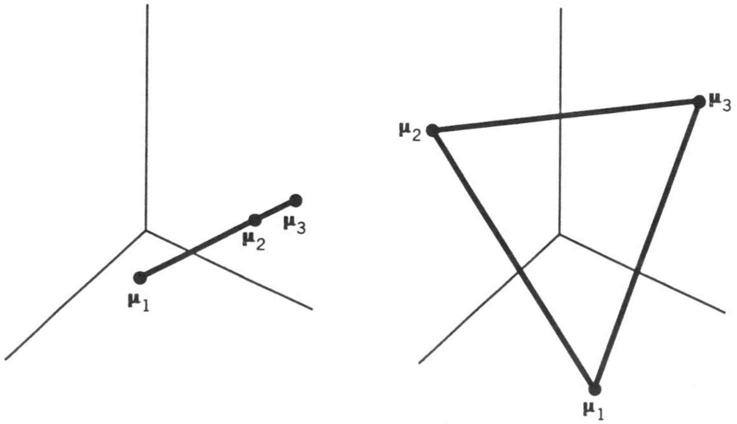

Figure 6.2 Two possible configurations for three mean vectors in 3-space.

n_i are unequal and we have an indication of heterogeneity from, for example, Box's M-test in Section 7.3.2, then the α level of the MANOVA test may be affected as follows. If the larger variances and covariances are associated with the larger samples, the true α level is reduced and the tests become conservative. On the other hand, if the larger variances and covariances come from the smaller samples, α is inflated, and the tests become liberal.

In conclusion, the use of Roy's θ is not recommended in any situation except the collinear case under standard assumptions. In the diffuse case its performance is inferior to that of the other three, both when the assumptions hold and when they do not. If the data come from nonnormal populations exhibiting skewness or positive kurtosis, any of the other three tests perform acceptably well. Among these three, $V^{(s)}$ is superior to the other two when there is heterogeneity of covariance matrices. Indeed $V^{(s)}$ is first in all rankings except those for the collinear case. However, Λ is not far behind, except when there is severe heterogeneity of covariance matrices. It seems likely that Wilks' Λ will continue its dominant role because of its flexibility and historical precedence. In practice, most MANOVA software programs routinely calculate all four test statistics, and they usually reach the same conclusion. In those cases when they differ as to acceptance or rejection of the hypothesis, one can examine the eigenvalues, covariance matrices, and so on, and evaluate the conflicting conclusions in light of the test properties discussed above.

Example 6.2. Wc inspect the eigenvalues of $\mathbf{E}^{-1}\mathbf{H}$ for the rootstock data of Table 6.2 for an indication of the configuration of the six mean vectors in a four-dimensional space. The eigenvalues are 1.876, .791, .229, .026. The first eigenvalue 1.876 constitutes a proportion

$$\frac{1.876}{1.876 + .791 + .229 + .026} = .642$$

of the sum of the eigenvalues. Therefore, the first eigenvalue does not dominate the others and the mean vectors are not collinear. Neither are they completely diffuse, however, since the eigenvalues are far from equal. The first two eigenvalues account for a proportion

$$\frac{1.876 + .791}{1.876 + \cdots + .026} = .913$$

of the sum of the eigenvalues, and thus the six mean vectors lie largely in two dimensions. Since the mean vectors are not collinear, the test statistics Λ, $\mathrm{V}^{(s)}$, and $\mathrm{U}^{(s)}$ will be more appropriate than θ in this case.

6.3 CONTRASTS

As in Sections 6.1.1–6.1.5, we consider only the balanced model where $n_1 = n_2 = \cdots = n_k = n$. We begin with a review of contrasts in the univariate setting before moving to the multivariate case.

6.3.1 Univariate Contrasts

A *contrast* in the population means is defined as a linear combination

$$\delta = c_1\mu_1 + c_2\mu_2 + \cdots + c_k\mu_k, \tag{6.41}$$

where the coefficients satisfy

$$\sum_{i=1}^{k} c_i = 0. \tag{6.42}$$

An unbiased estimator of δ is given by

$$\hat{\delta} = c_1\overline{y}_{1.} + c_2\overline{y}_{2.} + \cdots + c_k\overline{y}_{k.}. \tag{6.43}$$

The sample means $\overline{y}_{i.}$ were defined in Section 6.1.1. Since the $\overline{y}_{i.}$'s are independent with variance σ^2/n, the variance of $\hat{\delta}$ is

$$\text{var}(\hat{\delta}) = \frac{\sigma^2}{n} \sum_{i=1}^{k} c_i^2,$$

which can be estimated by

$$s_{\hat{\delta}}^2 = \frac{\text{MSE}}{n} \sum_{i=1}^{k} c_i^2, \tag{6.44}$$

where MSE was defined following (6.5).

The usual hypothesis to be tested by a contrast is

$$H_0\text{: } \delta = c_1\mu_1 + c_2\mu_2 + \cdots + c_k\mu_k = 0.$$

For example, suppose $k = 4$ and the four groups correspond to four treatments applied in the experiment. Suppose further that a contrast of interest to the researcher is $3\mu_1 - \mu_2 - \mu_3 - \mu_4$. If this contrast is set equal to zero, we have

$$3\mu_1 = \mu_2 + \mu_3 + \mu_4 \quad \text{or} \quad \mu_1 = \tfrac{1}{3}(\mu_2 + \mu_3 + \mu_4),$$

and the experimenter is comparing the first mean with the average of the other three. A contrast is often called a *comparison* among the treatment means.

Assuming normality, H_0 can be tested by

$$t = \frac{\hat{\delta} - 0}{s_{\hat{\delta}}}, \tag{6.45}$$

which is distributed as t_{ν_E}. Alternatively, since $t^2 = F$, we can use

$$F = \frac{\hat{\delta}^2}{s_{\hat{\delta}}^2} = \frac{\left(\sum_i c_i \bar{y}_{i.}\right)^2}{\text{MSE} \sum_i c_i^2/n}$$

$$= \frac{n\left(\sum_i c_i \bar{y}_{i.}\right)^2 \Big/ \sum_i c_i^2}{\text{MSE}}, \tag{6.46}$$

which is F_{1,ν_E}. The numerator of (6.46) is often referred to as the sum of squares for the contrast.

If two contrasts $\delta = \sum_i a_i \mu_i$ and $\gamma = \sum_i b_i \mu_i$ are such that $\sum_i a_i b_i = 0$, the contrasts are said to be *orthogonal*. The two estimated contrasts can be written in the form $\sum_i a_i \bar{y}_{i.} = \mathbf{a}'\bar{\mathbf{y}}$ and $\sum_i b_i \bar{y}_{i.} = \mathbf{b}'\bar{\mathbf{y}}$, where $\bar{\mathbf{y}}' = (\bar{y}_{1.}, \bar{y}_{2.}, \ldots, \bar{y}_{k.})$. Then $\sum_i a_i b_i = \mathbf{a}'\mathbf{b} = 0$, and by the discussion following (3.13), the coefficient vectors $\mathbf{a}$ and $\mathbf{b}$ are perpendicular.

When two contrasts are orthogonal, the two corresponding sums of squares are independent. In fact, for k treatments, we can find $k - 1$ orthogonal contrasts that partition the treatment sum of squares SSH into $k - 1$ independent sums of squares, each with one degree of freedom. In the unbalanced case (Section 6.1.6), orthogonal contrasts such that $\sum_i a_i b_i = 0$ no longer partition SSH into $k - 1$ independent sums of squares. For a discussion of contrasts in the unbalanced case, see Rencher (1996, Sections 4.8.2 and 4.8.3).

6.3.2 Multivariate Contrasts

There are two usages of contrasts in a multivariate setting. We have previously encountered one use in Section 5.9.1, where we considered the hypothesis H_0: $\mathbf{C}\boldsymbol{\mu} = \mathbf{0}$. Each row of $\mathbf{C}$ summed to zero and $\mathbf{C}\boldsymbol{\mu}$ was therefore a set of contrasts comparing the elements $\mu_1, \mu_2, \ldots, \mu_p$ of $\boldsymbol{\mu}$ with each other. In this section, on the other hand, we consider contrasts comparing several mean vectors, not the elements within them.

A contrast among the population mean vectors is defined as

$$\boldsymbol{\delta} = c_1\boldsymbol{\mu}_1 + c_2\boldsymbol{\mu}_2 + \cdots + c_k\boldsymbol{\mu}_k, \tag{6.47}$$

where $\sum_i c_i = 0$. An unbiased estimator of $\boldsymbol{\delta}$ is given by the corresponding contrast in the sample mean vectors:

$$\hat{\boldsymbol{\delta}} = c_1\overline{\mathbf{y}}_{1.} + c_2\overline{\mathbf{y}}_{2.} + \cdots + c_k\overline{\mathbf{y}}_{k.}. \tag{6.48}$$

The sample mean vectors $\overline{\mathbf{y}}_{1.}, \overline{\mathbf{y}}_{2.}, \ldots, \overline{\mathbf{y}}_{k.}$ as defined in Section 6.1.2 were assumed to be independent and to have common covariance matrix, $\text{cov}(\overline{\mathbf{y}}_{i.}) = \boldsymbol{\Sigma}/n$. Thus the covariance matrix for $\hat{\boldsymbol{\delta}}$ is given by

$$\text{cov}(\hat{\boldsymbol{\delta}}) = c_1^2\frac{\boldsymbol{\Sigma}}{n} + c_2^2\frac{\boldsymbol{\Sigma}}{n} + \cdots + c_k^2\frac{\boldsymbol{\Sigma}}{n} = \frac{\boldsymbol{\Sigma}}{n}\sum_{i=1}^{k} c_i^2,$$

which can be estimated by

$$\frac{\mathbf{S}_{\text{pl}}}{n}\sum_{i=1}^{k} c_i^2,$$

where $\mathbf{S}_{\text{pl}} = \mathbf{E}/\nu_E$ is an unbiased estimate of $\boldsymbol{\Sigma}$.

The hypothesis H_0: $c_1\boldsymbol{\mu}_1 + c_2\boldsymbol{\mu}_2 + \cdots + c_k\boldsymbol{\mu}_k = \mathbf{0}$ makes comparisons among the population mean vectors. For example, $\boldsymbol{\mu}_1 - 2\boldsymbol{\mu}_2 + \boldsymbol{\mu}_3 = \mathbf{0}$ is equivalent to

$$\boldsymbol{\mu}_2 = \tfrac{1}{2}(\boldsymbol{\mu}_1 + \boldsymbol{\mu}_3),$$

and we are comparing $\boldsymbol{\mu}_2$ to the average of $\boldsymbol{\mu}_1$ and $\boldsymbol{\mu}_3$. Of course this implies that every element of $\boldsymbol{\mu}_2$ must equal the corresponding element of $\frac{1}{2}(\boldsymbol{\mu}_1 + \boldsymbol{\mu}_3)$:

$$\begin{bmatrix} \mu_{21} \\ \mu_{22} \\ \vdots \\ \mu_{2p} \end{bmatrix} = \begin{bmatrix} \frac{1}{2}(\mu_{11} + \mu_{31}) \\ \frac{1}{2}(\mu_{12} + \mu_{32}) \\ \vdots \\ \frac{1}{2}(\mu_{1p} + \mu_{3p}) \end{bmatrix}.$$

Under appropriate multivariate normality assumptions, H_0: $c_1\boldsymbol{\mu}_1 + c_2\boldsymbol{\mu}_2 + \cdots + c_k\boldsymbol{\mu}_k = \mathbf{0}$ can be tested with

$$T^2 = \hat{\boldsymbol{\delta}}' \left(\frac{\mathbf{S}_{\text{pl}}}{n} \sum_{i=1}^{k} c_i^2 \right)^{-1} \hat{\boldsymbol{\delta}}$$

$$= \frac{n}{\sum_{i=1}^{k} c_i^2} \left(\sum_{i=1}^{k} c_i \overline{\mathbf{y}}_{i.} \right)' \left(\frac{\mathbf{E}}{\nu_E} \right)^{-1} \left(\sum_{i=1}^{k} c_i \overline{\mathbf{y}}_{i.} \right), \qquad (6.49)$$

which is distributed as T^2_{p,ν_E}. In the one-way model under discussion here, $\nu_E = k(n-1)$.

An equivalent test of H_0 can be made with Wilks' Λ. By analogy with the numerator of (6.46), the hypothesis matrix due to the contrast is given by

$$\mathbf{H}_1 = \frac{n}{\sum_{i=1}^{k} c_i^2} \left(\sum_{i=1}^{k} c_i \overline{\mathbf{y}}_{i.} \right) \left(\sum_{i=1}^{k} c_i \overline{\mathbf{y}}_{i.} \right)'. \qquad (6.50)$$

The rank of $\mathbf{H}_1$ is 1 and the test statistic is

$$\Lambda = \frac{|\mathbf{E}|}{|\mathbf{E} + \mathbf{H}_1|}, \qquad (6.51)$$

which is distributed as $\Lambda_{p,1,\nu_E}$. The other three MANOVA test statistics can also be applied here using the single nonzero eigenvalue of $\mathbf{E}^{-1}\mathbf{H}_1$. Because $\nu_H = 1$ in this case, all four MANOVA statistics and T^2 give the same results; that is, all five transform to the same F value using the formulations in Section 6.1.7. If $k - 1$ orthogonal contrasts are used, they partition the $\mathbf{H}$ matrix into $k - 1$ independent $\mathbf{H}_1$ matrices. Each $\mathbf{H}_1$ matrix has one degree of freedom because $\text{rank}(\mathbf{H}_1) = 1$.

Example 6.3.2. We consider the following two orthogonal contrasts for the rootstock data in Table 6.2:

$$\begin{array}{rrrrrr} 2 & -1 & -1 & -1 & -1 & 2 \\ 1 & 0 & 0 & 0 & 0 & -1. \end{array}$$

The first compares $\boldsymbol{\mu}_1$ and $\boldsymbol{\mu}_6$ with the other four mean vectors. The second compares $\boldsymbol{\mu}_1$ vs. $\boldsymbol{\mu}_6$. Thus H_{01}: $2\boldsymbol{\mu}_1 - \boldsymbol{\mu}_2 - \boldsymbol{\mu}_3 - \boldsymbol{\mu}_4 - \boldsymbol{\mu}_5 + 2\boldsymbol{\mu}_6 = \mathbf{0}$ can be written as

$$H_{01}: 2\boldsymbol{\mu}_1 + 2\boldsymbol{\mu}_6 = \boldsymbol{\mu}_2 + \boldsymbol{\mu}_3 + \boldsymbol{\mu}_4 + \boldsymbol{\mu}_5.$$

Dividing both sides by 4 so as to express this in terms of averages, we obtain

$$H_{01}: \tfrac{1}{2}(\boldsymbol{\mu}_1 + \boldsymbol{\mu}_6) = \tfrac{1}{4}(\boldsymbol{\mu}_2 + \boldsymbol{\mu}_3 + \boldsymbol{\mu}_4 + \boldsymbol{\mu}_5).$$

Similarly, the hypothesis for the second contrast can be expressed as

$$H_{02}: \quad \boldsymbol{\mu}_1 = \boldsymbol{\mu}_6.$$

The mean vectors are given by

$\overline{\mathbf{y}}_{1.}$	$\overline{\mathbf{y}}_{2.}$	$\overline{\mathbf{y}}_{3.}$	$\overline{\mathbf{y}}_{4.}$	$\overline{\mathbf{y}}_{5.}$	$\overline{\mathbf{y}}_{6.}$
1.14	1.16	1.11	1.10	1.08	1.04
2.98	3.11	2.82	2.88	2.56	2.21
3.74	4.52	4.46	3.91	4.31	3.60
.87	1.28	1.39	1.04	1.18	.74

For the first contrast, we obtain $\mathbf{H}_1$ from (6.50) as

$$\begin{aligned}
\mathbf{H}_1 &= \frac{n}{\sum_i c_i^2}\left(\sum_i c_i\overline{\mathbf{y}}_{i.}\right)\left(\sum_i c_i\overline{\mathbf{y}}_{i.}\right)' \\
&= \frac{8}{12}(2\overline{\mathbf{y}}_{1.} - \overline{\mathbf{y}}_{2.} - \cdots + 2\overline{\mathbf{y}}_{6.})(2\overline{\mathbf{y}}_{1.} - \overline{\mathbf{y}}_{2.} - \cdots + 2\overline{\mathbf{y}}_{6.})' \\
&= \frac{8}{12}\begin{bmatrix} -.095 \\ -.978 \\ -2.519 \\ -1.680 \end{bmatrix}(-.095, -.978, -2.519, -1.680) \\
&= \begin{bmatrix} .006 & .062 & .160 & .106 \\ .062 & .638 & 1.642 & 1.095 \\ .160 & 1.642 & 4.229 & 2.820 \\ .106 & 1.095 & 2.820 & 1.881 \end{bmatrix}.
\end{aligned}$$

Then

$$\Lambda = \frac{|\mathbf{E}|}{|\mathbf{E} + \mathbf{H}_1|} = \frac{.6571}{1.4824} = .443,$$

which is less than $\Lambda_{.05,4,1,40} = .779$ from Table A.9. We therefore reject H_{01}.

To test the significance of the second contrast, we have

$$\mathbf{H}_1 = \frac{8}{2}(\bar{\mathbf{y}}_{1.} - \bar{\mathbf{y}}_{6.})(\bar{\mathbf{y}}_{1.} - \bar{\mathbf{y}}_{6.})'$$
$$= \frac{8}{2}\begin{bmatrix} .101 \\ .762 \\ .142 \\ .136 \end{bmatrix}(.101, .762, .142, .136)$$
$$= \begin{bmatrix} .041 & .309 & .058 & .055 \\ .309 & 2.326 & .435 & .415 \\ .058 & .435 & .081 & .078 \\ .055 & .415 & .078 & .074 \end{bmatrix}.$$

Then

$$\Lambda = \frac{|\mathbf{E}|}{|\mathbf{E}+\mathbf{H}_1|} = \frac{.6571}{.8757} = .750,$$

which is less than $\Lambda_{.05,4,1,40} = .779$, and we reject H_{02}.

6.4 TESTS ON INDIVIDUAL VARIABLES FOLLOWING REJECTION OF H_0 BY THE OVERALL MANOVA TEST

In Section 6.1, we considered tests of equality of mean vectors, H_0: $\boldsymbol{\mu}_1 = \boldsymbol{\mu}_2 = \cdots = \boldsymbol{\mu}_k$, which implies equality of means for each of the p variables:

$$H_{0r}: \mu_{1r} = \mu_{2r} = \cdots = \mu_{kr} \qquad r = 1, 2, \ldots, p.$$

This hypothesis could be tested for each variable by itself with an ordinary univariate ANOVA F-test, as noted in property 9 in Section 6.1.3. For example, if there are three mean vectors,

$$\boldsymbol{\mu}_1 = \begin{bmatrix} \mu_{11} \\ \mu_{12} \\ \vdots \\ \mu_{1p} \end{bmatrix} \qquad \boldsymbol{\mu}_2 = \begin{bmatrix} \mu_{21} \\ \mu_{22} \\ \vdots \\ \mu_{2p} \end{bmatrix} \qquad \boldsymbol{\mu}_3 = \begin{bmatrix} \mu_{31} \\ \mu_{32} \\ \vdots \\ \mu_{3p} \end{bmatrix},$$

we have H_{01}: $\mu_{11} = \mu_{21} = \mu_{31}, H_{02}$: $\mu_{12} = \mu_{22} = \mu_{32}, \ldots, H_{0p}$: $\mu_{1p} = \mu_{2p} = \mu_{3p}$. Each of these p hypotheses can be tested with a simple ANOVA F-test.

If an F-test is made on each of the p variables regardless of whether the overall MANOVA test of H_0: $\boldsymbol{\mu}_1 = \boldsymbol{\mu}_2 = \boldsymbol{\mu}_3$ rejects H_0, then the overall α level will increase beyond the nominal value because we are making p tests.

As in Section 5.5, we define the *overall* α or *experimentwise error rate* as the probability of rejecting one or more of the p univariate tests when H_0: $\boldsymbol{\mu}_1 = \boldsymbol{\mu}_2 = \boldsymbol{\mu}_3$ is true. We could "protect" against inflation of the experimentwise error rate by performing tests on individual variables *only* if the overall MANOVA test of H_0: $\boldsymbol{\mu}_1 = \boldsymbol{\mu}_2 = \boldsymbol{\mu}_3$ is rejected. In this procedure, the probability of rejection for the tests on individual variables is reduced, and these tests become more conservative. Thus the experimentwise error rate depends on whether or not we do follow-up univariate tests only after rejection of a MANOVA test.

Rencher and Scott (1990) compared the above two procedures for testing the individual variables in a one-way MANOVA model. Since the focus was on α levels, only the case where H_0 was true was considered. Specifically, the two procedures were as follows:

1. A univariate F-test is made on each variable, testing H_{0r}: $\mu_{1r} = \mu_{2r} = \cdots = \mu_{kr}$, $r = 1, 2, \ldots, p$. In this context, the p univariate tests constitute an experiment and one or more rejections are counted as one experimentwise error. No multivariate test is made.
2. The overall MANOVA hypothesis H_0: $\boldsymbol{\mu}_1 = \boldsymbol{\mu}_2 = \cdots = \boldsymbol{\mu}_k$ is tested with Wilks' Λ, and if H_0 is rejected, p univariate F-tests on $H_{01}, H_{02}, \ldots, H_{0p}$ are carried out. Again, one or more rejections among the F-tests are counted as one experimentwise error.

The amount of intercorrelation among the multivariate normal variables was indicated by $\sum_{i=1}^{p}(1/\lambda_i)/p$, where $\lambda_1, \lambda_2, \ldots, \lambda_p$ are the eigenvalues of the population correlation matrix $\mathbf{P}_\rho$. Note that $\sum_i(1/\lambda_i)/p = 1$ for the uncorrelated case and $\sum_i(1/\lambda_i)/p > 1$ for the correlated case. When the variables are highly intercorrelated, one or more of the eigenvalues will be near zero (see Section 4.1.3). Thus the average reciprocal of the eigenvalues is a measure of intercorrelation.

The error rates of these two procedures were investigated for several values of p, n, k, and $\sum_i(1/\lambda_i)/p$. In procedure 1, the probability of rejecting one or more univariate tests when H_0 is true varied from .09 to .31 (α was .05 in each test). Such experimentwise error rates are clearly unacceptable when the nominal value of α is .05. Of course this result is to be expected when several tests are made, each at the .05 level. Nevertheless, this approach is commonly used when the researcher is not familiar with the MANOVA approach or does not have access to appropriate software.

Table 6.3 contains the error rates for procedure 2, univariate F-tests following a rejection by Wilks' Λ. The values range from .022 to .057, comfortably close to the target value of .05. No apparent trends or patterns are seen; the values do not seem to depend on p, k, n, or amount of intercorrelation. Thus when univariate tests are made *only* following a rejection of the overall test, the experimentwise error rate is about right.

Table 6.3 Experimentwise Error Rates for Procedure 2: Univariate *F*-Tests Following Rejection by Wilks' Λ

		$\Sigma_i(1/\lambda_i)/p$							
		1		10		100		1000	
n	p	$k=3$	$k=5$	$k=3$	$k=5$	$k=3$	$k=5$	$k=3$	$k=5$
5	3	.043	.037	.022	.035	.046	.039	.022	.029
5	5	.041	.037	.039	.057	.038	.035	.027	.039
5	7	.030	.042	.035	.045	.039	.037	.026	.048
10	3	.047	.041	.030	.033	.043	.045	.026	.032
10	5	.047	.037	.026	.049	.041	.026	.027	.029
10	7	.034	.054	.037	.047	.047	.040	.040	.044
20	3	.050	.043	.032	.054	.048	.039	.040	.032
20	5	.045	.055	.042	.051	.037	.044	.050	.043
20	7	.055	.051	.029	.040	.033	.051	.039	.033

Based on these results, we recommend making an overall MANOVA test followed by *F*-tests on the individual variables (at the same α level as the MANOVA test) only if the MANOVA test leads to rejection of H_0.

Another procedure that can be used following rejection of the MANOVA test is an examination of the discriminant function coefficients. The discriminant function was defined in Section 6.1.4 as $z = \mathbf{a}'\mathbf{y}$, where $\mathbf{a}$ is the eigenvector associated with the largest eigenvalue λ_1 of $\mathbf{E}^{-1}\mathbf{H}$. Additionally, there are other discriminant functions using eigenvectors corresponding to the other eigenvalues. Since the first discriminant function maximally separates the groups, we can examine its coefficients for the contribution of each variable to group separation. Thus in $z = a_1y_1 + a_2y_2 + \cdots + a_py_p$, if a_2 is larger than the other a_i's, we believe y_2 contributes more than any of the other variables to separation of groups. A method of standardization of the a_i's to adjust for differences in the scale among the variables is given in Section 8.5.

The information in a_i (from $z = \mathbf{a}'\mathbf{y}$) about the contribution of y_i to separation of the groups is fundamentally different from the information provided in a univariate *F*-test that considers y_i alone. The relative size of a_i shows the contribution of y_i in the presence of the other variables, in a manner analogous to a standardized regression coefficient or "beta weight." The individual *F*-tests on $y_1, y_2, \ldots, y_p$ ignore the presence of other variables and thus do not take into account the correlation of each variable with the others. Because we are primarily interested in the collective behavior of the variables, it would seem that the discriminant function coefficients provide more pertinent information than the tests on individual variables. For a detailed analysis of the effect of each variable in the presence of other variables, see Rencher (1993, 1996, Section 4.1.6).

Huberty (1975) compared the standardized coefficients to some correlations that can be shown to be related to individual variable tests (see Section 8.7.3). In a limited simulation, the discriminant coefficients were found to be more valid than the univariate tests in identifying those variables that contribute least to separation of groups. Considerable variation was found from sample to sample in ranking the relative potency of the variables.

Example 6.4. In Example 6.1.7, the hypothesis H_0: $\boldsymbol{\mu}_1 = \boldsymbol{\mu}_2 = \cdots = \boldsymbol{\mu}_6$ was rejected. We can therefore test the four individual variables using the .05 level of significance. For the first variable, y_1 = 4-year trunk girth, we obtain the following ANOVA table:

Source	Sum of Squares	df	Mean Square	F
Rootstocks	.073560	5	.014712	1.93
Error	.319988	42	.007619	
Total	.393548	47		

For $F = 1.93$ the p-value is .1094 and we do not reject H_0. For the other three variables we have

Variable	F	p-Value
y_2 = 4-year extension growth	2.91	.024
y_3 = 15-year trunk girth	11.97	< .0001
y_4 = 15-year weight	12.16	< .0001

Thus for three of the variables, the six means differ significantly. We will examine the standardized discriminant function coefficients for this set of data in Chapter 8 (problem 8.12).

6.5 TWO-WAY CLASSIFICATION

We consider only balanced models, where each cell in the model has the same number of observations, n. For the unbalanced case with unequal cell sizes, see Rencher (1996, Section 4.8).

6.5.1 Review of Univariate Two-Way ANOVA

In the univariate two-way model, we measure one dependent variable y on each experimental unit. The balanced two-way fixed-effects model with factors A and B is

$$y_{ijk} = \mu + \alpha_i + \beta_j + \gamma_{ij} + \epsilon_{ijk} \tag{6.52}$$

$$= \mu_{ij} + \epsilon_{ijk} \tag{6.53}$$

$$i = 1, 2, \ldots, a \qquad j = 1, 2, \ldots, b \qquad k = 1, 2, \ldots, n,$$

where α_i is the effect (on y_{ijk}) of the ith level of A, β_j is the effect of the jth level of B, γ_{ij} is the corresponding interaction effect, and μ_{ij} is the population mean for the ith level of A and the jth level of B. In order to obtain F-tests, we further assume that the ϵ_{ijk} are independently distributed as $N(0, \sigma^2)$.

Let $\overline{\mu}_{i.} = \sum_j \mu_{ij}/b$ be the mean at the ith level of A and define $\overline{\mu}_{.j}$ and $\overline{\mu}_{..}$ similarly. Then if we use side conditions $\sum_i \alpha_i = \sum_j \beta_j = \sum_i \gamma_{ij} = \sum_j \gamma_{ij} = 0$, the effect of the ith level of A can be defined as $\alpha_i = \overline{\mu}_{i.} - \overline{\mu}_{..}$, with similar definitions of β_j and γ_{ij}. That $\sum_i \alpha_i = 0$ if $\alpha_i = \overline{\mu}_{i.} - \overline{\mu}_{..}$ can be shown as follows:

$$\begin{aligned}\sum_{i=1}^{a} \alpha_i &= \sum_{i=1}^{a} (\overline{\mu}_{i.} - \overline{\mu}_{..}) = \sum_{i} \overline{\mu}_{i.} - a\overline{\mu}_{..} \\ &= a\overline{\mu}_{..} - a\overline{\mu}_{..} = 0.\end{aligned}$$

Many texts recommend that the interaction AB be tested first, and if it is found to be significant, then the main effects should not be tested. However, with the side conditions imposed above (side conditions are not necessary in order to obtain tests), the effect of A is defined as the average effect over the levels of B and similarly for the effect of B. With this definition of main effects, the tests for A and B make sense even if AB is significant. Admittedly, interpretation requires more care, and the effect of a factor may vary if the number of levels of the other factor is altered. But in many cases useful information can be gained about the main effects in the presence of interaction.

We illustrate the above statement that $\alpha_i = \overline{\mu}_{i.} - \overline{\mu}_{..}$ represents the effect of the first level of A averaged over the levels of B. Suppose A has two levels and B has three. We represent the means of the six cells as follows:

		B 1	2	3	Mean
A	1	μ_{11}	μ_{12}	μ_{13}	$\overline{\mu}_{1.}$
	2	μ_{21}	μ_{22}	μ_{23}	$\overline{\mu}_{2.}$
	Mean	$\overline{\mu}_{.1}$	$\overline{\mu}_{.2}$	$\overline{\mu}_{.3}$	$\overline{\mu}_{..}$

The means of the rows (corresponding to levels of A) and columns (levels of B) are also given. Then $\alpha_i = \overline{\mu}_{i.} - \overline{\mu}_{..}$ can be expressed as the average of the effect of A at the three levels of B. For example,

$$\begin{aligned}\alpha_1 &= \tfrac{1}{3}[(\mu_{11} - \overline{\mu}_{.1}) + (\mu_{12} - \overline{\mu}_{.2}) + (\mu_{13} - \overline{\mu}_{.3})] \\ &= \tfrac{1}{3}(\mu_{11} + \mu_{12} + \mu_{13}) - \tfrac{1}{3}(\overline{\mu}_{.1} + \overline{\mu}_{.2} + \overline{\mu}_{.3}) = \overline{\mu}_{1.} - \overline{\mu}_{..}.\end{aligned}$$

To estimate α_i, we can use $\hat{\alpha}_i = \overline{y}_{i..} - \overline{y}_{...}$, with similar estimates for β_i and γ_{ij}. The notation $\overline{y}_{i..}$ indicates that y is averaged over j and k to obtain the mean of all nb observations at the ith level of A, namely, $\overline{y}_{i..} = \sum_{jk} y_{ijk}/nb$. The means $\overline{y}_{.j.}, \overline{y}_{ij.}$, and $\overline{y}_{...}$ have analogous definitions.

To construct tests for the significance of factors A and B and the interaction AB, we use the usual sums of squares and degrees of freedom in Table 6.4.

Computational forms of the sums of squares listed in Table 6.4 can be found in many standard (univariate) methods texts. In the balanced model, we have

$$\text{SST} = \text{SSA} + \text{SSB} + \text{SSAB} + \text{SSE},$$

and the four sums of squares on the right are independent. The sums of squares are divided by their corresponding degrees of freedom to obtain mean squares MSA, MSB, MSAB, and MSE. For the fixed effects model, each of MSA, MSB, and MSAB is divided by MSE to obtain an F-test. In the case of factor A, for example, the hypothesis can be expressed as

$$H_{0A}\colon \alpha_1 = \alpha_2 = \cdots = \alpha_a = 0,$$

and the test statistic is $F = \text{MSA}/\text{MSE}$, which is distributed as $F_{a-1,ab(n-1)}$.

In order to define contrasts among the levels of each main effect, we use the model in the form given in (6.53),

$$y_{ijk} = \mu_{ij} + \epsilon_{ijk}.$$

A contrast among the levels of A is given by $\sum_{i=1}^{a} c_i\overline{\mu}_{i.}$, where $\sum_i c_i = 0$. An estimate of the contrast is given by $\sum_i c_i\overline{y}_{i..}$, with variance $\sigma^2 \sum_i c_i^2/nb$, since each $\overline{y}_{i..}$ is based on nb observations. To test H_0: $\sum_i c_i\overline{\mu}_{i.} = 0$, we can use an F-statistic corresponding to (6.46),

Table 6.4 Univariate Two-Way Analysis of Variance

Source	Sum of Squares	df
A	$\text{SSA} = nb\,\Sigma_i(\overline{y}_{i..} - \overline{y}_{...})^2$	$a - 1$
B	$\text{SSB} = na\,\Sigma_j(\overline{y}_{.j.} - \overline{y}_{...})^2$	$b - 1$
AB	$\text{SSAB} = n\,\Sigma_{ij}(\overline{y}_{ij.} - \overline{y}_{i..} - \overline{y}_{.j.} + \overline{y}_{...})^2$	$(a-1)(b-1)$
Error	$\text{SSE} = \Sigma_{ijk}(y_{ijk} - \overline{y}_{ij.})^2$	$ab(n-1)$
Total	$\text{SST} = \Sigma_{ijk}(y_{ijk} - \overline{y}_{...})^2$	$abn - 1$

$$F = \frac{nb\left(\sum_i c_i\bar{y}_{i..}\right)^2 \Big/ \sum_i c_i^2}{\text{MSE}}, \tag{6.54}$$

with 1 and ν_E degrees of freedom. To preserve the experimentwise α, significance tests for more than one contrast could be carried out in the spirit of Section 6.4; that is, contrasts should be chosen prior to seeing the data and tests should be made only if the overall F-test for factor A rejects.

Contrasts $\sum_j c_j \bar{\mu}_{.j}$ among the levels of B are tested in an entirely analogous manner.

6.5.2 Multivariate Two-Way MANOVA

A balanced two-way fixed-effects MANOVA model for p dependent variables can be expressed by substituting vectors in (6.52) and (6.53):

$$\begin{aligned} \mathbf{y}_{ijk} &= \boldsymbol{\mu} + \boldsymbol{\alpha}_i + \boldsymbol{\beta}_j + \boldsymbol{\gamma}_{ij} + \boldsymbol{\epsilon}_{ijk} = \boldsymbol{\mu}_{ij} + \boldsymbol{\epsilon}_{ijk}, \\ i &= 1, 2, \ldots, a \qquad j = 1, 2, \ldots, b \qquad k = 1, 2, \ldots, n, \end{aligned} \tag{6.55}$$

where $\boldsymbol{\alpha}_i$ is the effect of the ith level of A on each of the p variables in $\mathbf{y}_{ijk}$, $\boldsymbol{\beta}_j$ is the effect of the jth level of B, and $\boldsymbol{\gamma}_{ij}$ is the AB interaction effect. We use side conditions $\sum_i \boldsymbol{\alpha}_i = \sum_j \boldsymbol{\beta}_j = \sum_i \boldsymbol{\gamma}_{ij} = \sum_j \boldsymbol{\gamma}_{ij} = \mathbf{0}$ and assume the $\boldsymbol{\epsilon}_{ijk}$ are independently distributed as $N_p(\mathbf{0}, \boldsymbol{\Sigma})$. Under the side condition $\sum_i \boldsymbol{\alpha}_i = \mathbf{0}$, the effect of A is averaged over the levels of B; that is, $\boldsymbol{\alpha}_i = \bar{\boldsymbol{\mu}}_{i.} - \bar{\boldsymbol{\mu}}_{..}$, where $\bar{\boldsymbol{\mu}}_{i.} = \sum_j \boldsymbol{\mu}_{ij}/b$ and $\bar{\boldsymbol{\mu}}_{..} = \sum_{ij} \boldsymbol{\mu}_{ij}/ab$. There are similar definitions for $\boldsymbol{\beta}_j$ and $\boldsymbol{\gamma}_{ij}$.

As in the univariate usage, the mean vector $\bar{\mathbf{y}}_{i..}$ indicates an average over the subscripts replaced by dots, that is, $\bar{\mathbf{y}}_{i..} = \sum_{jk} \mathbf{y}_{ijk}/nb$. The means $\bar{\mathbf{y}}_{.j.}$, $\bar{\mathbf{y}}_{ij.}$, and $\bar{\mathbf{y}}_{...}$ have analogous definitions. The sums of squares and products matrices are given in Table 6.5. Note that the degrees of freedom in Table 6.5 are the same as in the univariate case in Table 6.4. For the two-way model with balanced

Table 6.5 Multivariate Two-Way Analysis of Variance

Source	Sum of Squares and Products Matrix	df
A	$\mathbf{H}_A = nb\,\Sigma_i(\bar{\mathbf{y}}_{i..} - \bar{\mathbf{y}}_{...})(\bar{\mathbf{y}}_{i..} - \bar{\mathbf{y}}_{...})'$	$a - 1$
B	$\mathbf{H}_B = na\,\Sigma_j(\bar{\mathbf{y}}_{.j.} - \bar{\mathbf{y}}_{...})(\bar{\mathbf{y}}_{.j.} - \bar{\mathbf{y}}_{...})'$	$b - 1$
AB	$\mathbf{H}_{AB} = n\,\Sigma_{ij}(\bar{\mathbf{y}}_{ij.} - \bar{\mathbf{y}}_{i..} - \bar{\mathbf{y}}_{.j.} + \bar{\mathbf{y}}_{...})$ $\times(\bar{\mathbf{y}}_{ij.} - \bar{\mathbf{y}}_{i..} - \bar{\mathbf{y}}_{.j.} + \bar{\mathbf{y}}_{...})'$	$(a - 1)(b - 1)$
Error	$\mathbf{E} = \Sigma_{ijk}(\mathbf{y}_{ijk} - \bar{\mathbf{y}}_{ij.})(\mathbf{y}_{ijk} - \bar{\mathbf{y}}_{ij.})'$	$ab(n - 1)$
Total	$\mathbf{T} = \Sigma_{ijk}(\mathbf{y}_{ijk} - \bar{\mathbf{y}}_{...})(\mathbf{y}_{ijk} - \bar{\mathbf{y}}_{...})'$	$abn - 1$

data, the total sum of squares and products matrix is partitioned as

$$\mathbf{T} = \mathbf{H}_A + \mathbf{H}_B + \mathbf{H}_{AB} + \mathbf{E}. \tag{6.56}$$

The structure of any of the hypothesis matrices is similar to that of $\mathbf{H}$ in (6.9). For example, $\mathbf{H}_A$ has on the diagonal the sum of squares for factor A for each of the p variables. The off-diagonal elements of $\mathbf{H}_A$ are corresponding sums of products for all pairs of variables. Thus the rth diagonal element of $\mathbf{H}_A$ corresponding to the rth variable, $r = 1, 2, \ldots, p$, is

$$h_{Arr} = nb \sum_{i=1}^{a} (\bar{y}_{i..r} - \bar{y}_{...r})^2 = \sum_{i=1}^{a} \frac{y_{i..r}^2}{nb} - \frac{y_{...r}^2}{nab},$$

where $\bar{y}_{i..r}$ and $\bar{y}_{...r}$ represent the rth components of $\bar{\mathbf{y}}_{i..}$ and $\bar{\mathbf{y}}_{...}$, respectively, and $y_{i..r}$ and $y_{...r}$ are totals corresponding to $\bar{y}_{i..r}$ and $\bar{y}_{...r}$. The (rs)th off-diagonal element of $\mathbf{H}_A$ is

$$h_{Ars} = nb \sum_{i=1}^{a} (\bar{y}_{i..r} - \bar{y}_{...r})(\bar{y}_{i..s} - \bar{y}_{...s}) = \sum_{i=1}^{a} \frac{y_{i..r}y_{i..s}}{nb} - \frac{y_{...r}y_{...s}}{nab}.$$

For the $\mathbf{E}$ matrix, computational formulas are based on (6.56):

$$\mathbf{E} = \mathbf{T} - \mathbf{H}_A - \mathbf{H}_B - \mathbf{H}_{AB}.$$

Thus

$$e_{rr} = \sum_{ijk} y_{ijkr}^2 - \frac{y_{...r}^2}{nab} - h_{Arr} - h_{Brr} - h_{ABrr}$$

$$e_{rs} = \sum_{ijk} y_{ijkr}y_{ijks} - \frac{y_{...r}y_{...s}}{nab} - h_{Ars} - h_{Brs} - h_{ABrs}.$$

The interaction and main effects can be tested using all four MANOVA test statistics. Thus for Wilks' Λ, we use $\mathbf{E}$ to test each of A, B, and AB:

$$\Lambda_A = \frac{|\mathbf{E}|}{|\mathbf{E} + \mathbf{H}_A|} \quad \text{is} \quad \Lambda_{p,a-1,ab(n-1)}$$

$$\Lambda_B = \frac{|\mathbf{E}|}{|\mathbf{E} + \mathbf{H}_B|} \quad \text{is} \quad \Lambda_{p,b-1,ab(n-1)}$$

$$\Lambda_{AB} = \frac{|\mathbf{E}|}{|\mathbf{E} + \mathbf{H}_{AB}|} \quad \text{is} \quad \Lambda_{p,(a-1)(b-1),ab(n-1)}.$$

In each case, the indicated distribution holds when H_0 is true. To calculate the other three MANOVA test statistics for A, B, and AB, we use the eigenvalues of $\mathbf{E}^{-1}\mathbf{H}_A, \mathbf{E}^{-1}\mathbf{H}_B$, and $\mathbf{E}^{-1}\mathbf{H}_{AB}$.

If the interaction is not significant, interpretation of the main effects is simpler. However, the comments in Section 6.5.1 about testing main effects in the presence of interaction apply to the multivariate model as well. If we define each main effect as the average effect over the levels of the other factor, then main effects can be tested even if the interaction is significant. One must be more careful with the interpretation in case of a significant interaction, but there is information to be gained.

By analogy with the univariate two-way ANOVA in Section 6.5.1, a contrast among the levels of factor A can be defined in terms of the mean vectors, $\sum_{i=1}^{a} c_i \overline{\boldsymbol{\mu}}_{i.}$, where $\sum_i c_i = 0$ and $\overline{\boldsymbol{\mu}}_{i.} = \sum_j \boldsymbol{\mu}_{ij}/b$. Similarly, $\sum_{j=1}^{b} c_j \overline{\boldsymbol{\mu}}_{.j}$ represents a contrast among the levels of B. The hypothesis that these contrasts are $\mathbf{0}$ can be tested by T^2 or any of the four MANOVA test statistics, as in (6.49), (6.50), and (6.51). To test H_0: $\sum_i c_i \overline{\boldsymbol{\mu}}_{i.} = \mathbf{0}$, for example, we can use

$$T^2 = \frac{nb}{\sum_i c_i^2} \left(\sum_i c_i \overline{\mathbf{y}}_{i..} \right)' \left(\frac{\mathbf{E}}{\nu_E} \right)^{-1} \left(\sum_i c_i \overline{\mathbf{y}}_{i..} \right), \tag{6.57}$$

which is distributed as T^2_{p,ν_E} when H_0 is true. Alternatively, with the hypothesis matrix

$$\mathbf{H}_1 = \frac{nb}{\sum_i c_i^2} \left(\sum_i c_i \overline{\mathbf{y}}_{i..} \right) \left(\sum_i c_i \overline{\mathbf{y}}_{i..} \right)', \tag{6.58}$$

we have

$$\Lambda = \frac{|\mathbf{E}|}{|\mathbf{E} + \mathbf{H}_1|},$$

which, under H_0, is $\Lambda_{p,1,\nu_E}$. In the two-way model, $\nu_E = ab(n-1)$. The other three MANOVA test statistics can also be constructed from $\mathbf{E}^{-1}\mathbf{H}_1$. All five test statistics will give equivalent results because $\nu_H = 1$.

If follow-up tests on individual variables are desired, we can infer from Rencher and Scott (1990), as reported in Section 6.4, that if the MANOVA test on factor A or B leads to rejection of H_0, then we can proceed with the univariate F-tests on the individual variables with assurance that the experimentwise error rate will be close to α.

To determine the contribution of each variable in the presence of the others, we can examine the first discriminant function obtained from eigenvectors of

$\mathbf{E}^{-1}\mathbf{H}_A$ or $\mathbf{E}^{-1}\mathbf{H}_B$, as in Section 6.4 for one-way MANOVA. The first discriminant function for $\mathbf{E}^{-1}\mathbf{H}_A$, for example, is $z = \mathbf{a}'\mathbf{y}$, where $\mathbf{a}$ is the eigenvector associated with the largest eigenvalue of $\mathbf{E}^{-1}\mathbf{H}_A$. In $z = a_1y_1 + a_2y_2 + \cdots + a_py_p$, if a_i is larger than the other a's, then y_i contributes more than the other variables to the significance of Λ_A. (In many cases, the a_i's should be standardized as in Section 8.5.) Note that the first discriminant function obtained from $\mathbf{E}^{-1}\mathbf{H}_A$ will not have the same pattern as the first discriminant function from $\mathbf{E}^{-1}\mathbf{H}_B$. This is not surprising since we expect that the relative contribution of the variables to separating the levels of factor A will be different from the relative contribution to separating the levels of B.

A randomized block design or a two-way MANOVA without replication can easily be analyzed in a manner similar to that for the two-way model with replication given here; therefore, no specific details will be given.

Example 6.5.2. Table 6.6 contains data reported by Posten (1962) and analyzed by Kramer and Jensen (1970). The experiment involved a 2×4 design with 4 replications, for a total of 32 observation vectors. The factors were rotational velocity [A_1 (fast) and A_2 (slow)] and lubricants (four types). The experimental units were 32 homogeneous pieces of bar steel. Two variables were measured on each piece of bar steel:

y_1 = ultimate torque
y_2 = ultimate strain

Table 6.6 Two-Way Classification of Measurements on Bar Steel

	A_1		A_2	
Lubricant	y_1	y_2	y_1	y_2
B_1	7.80	90.4	7.12	85.1
	7.10	88.9	7.06	89.0
	7.89	85.9	7.45	75.9
	7.82	88.8	7.45	77.9
B_2	9.00	82.5	8.19	66.0
	8.43	92.4	8.25	74.5
	7.65	82.4	7.45	83.1
	7.70	87.4	7.45	86.4
B_3	7.28	79.6	7.15	81.2
	8.96	95.1	7.15	72.0
	7.75	90.2	7.70	79.9
	7.80	88.0	7.45	71.9
B_4	7.60	94.1	7.06	81.2
	7.00	86.6	7.04	79.9
	7.82	85.9	7.52	86.4
	7.80	88.8	7.70	76.4

We display the totals for each variable for use in computations. The numbers inside the box are cell totals (over the four replications), and the marginal totals are for each level of A and B:

Totals for y_1

	A_1	A_2	
B_1	30.61	29.08	59.69
B_2	32.61	31.34	64.12
B_3	31.79	29.45	61.24
B_4	30.22	29.32	59.54
	125.40	119.19	244.59

Totals for y_2

	A_1	A_2	
B_1	354.0	327.9	681.9
B_2	344.7	310.0	654.7
B_3	352.9	305.0	657.9
B_4	355.4	323.9	679.3
	1407.0	1266.8	2673.8

Using computational forms given for h_{Arr} and h_{Ars} following (6.56), the (1, 1) element of $\mathbf{H}_A$ (corresponding to y_1) is given by

$$h_{A11} = \frac{(125.40)^2 + (119.19)^2}{(4)(4)} - \frac{(244.59)^2}{(4)(4)(2)} = 1.205.$$

For the (2, 2) element of $\mathbf{H}_A$ (corresponding to y_2), we have

$$h_{A22} = \frac{(1407.0)^2 + (1266.8)^2}{16} - \frac{(2673.8)^2}{32} = 614.25.$$

For the (1, 2) element (corresponding to y_1y_2), we obtain

$$h_{A12} = \frac{(125.40)(1407.0) + (119.19)(1266.8)}{16} - \frac{(244.59)(2673.8)}{32}$$
$$= 27.208.$$

Thus

$$\mathbf{H}_A = \begin{pmatrix} 1.205 & 27.208 \\ 27.208 & 614.251 \end{pmatrix}.$$

We obtain $\mathbf{H}_B$ similarly:

$$
\begin{aligned}
h_{B11} &= \frac{(59.69)^2 + \cdots + (59.54)^2}{(4)(2)} - \frac{(244.59)^2}{32} = 1.694 \\
h_{B22} &= \frac{(681.9)^2 + \cdots + (679.3)^2}{8} - \frac{(2673.8)^2}{32} = 74.874 \\
h_{B12} &= \frac{(59.69)(681.9) + \cdots + (59.54)(679.3)}{8} - \frac{(244.59)(2673.8)}{32} \\
&= -9.862 \\
\mathbf{H}_B &= \begin{pmatrix} 1.694 & -9.862 \\ -9.862 & 74.874 \end{pmatrix}.
\end{aligned}
$$

For $\mathbf{H}_{AB}$ we have

$$
\begin{aligned}
h_{AB11} &= \frac{(30.61)^2 + \cdots + (29.32)^2}{4} - \frac{(244.59)^2}{32} - 1.205 - 1.694 = .132 \\
h_{AB22} &= \frac{(354.0)^2 + \cdots + (323.9)^2}{4} - \frac{(2673.8)^2}{32} - 614.25 - 74.874 \\
&= 32.244 \\
h_{AB12} &= \frac{(30.61)(354.0) + \cdots + (29.32)(323.9)}{4} - \frac{(244.59)(2673.8)}{32} \\
&\quad - 27.208 - (-9.862) = 1.585 \\
\mathbf{H}_{AB} &= \begin{pmatrix} .132 & 1.585 \\ 1.585 & 32.244 \end{pmatrix}.
\end{aligned}
$$

The error matrix $\mathbf{E}$ is obtained using the computational forms given for e_{rr} and e_{rs} following (6.56). For example, e_{11} and e_{12} are computed as

$$
\begin{aligned}
e_{11} &= (7.80)^2 + (7.10)^2 + \cdots + (7.70)^2 - \frac{(244.59)^2}{32} - 1.205 \\
&\quad - 1.694 - .132 = 4.897 \\
e_{12} &= (7.80)(90.4) + \cdots + (7.70)(76.4) - \frac{(244.59)(2673.8)}{32} - 27.208 \\
&\quad - (-9.862) - 1.585 = -1.890
\end{aligned}
$$

Proceeding in this fashion, we obtain

$$
\mathbf{E} = \begin{pmatrix} 4.897 & -1.890 \\ -1.890 & 736.390 \end{pmatrix},
$$

with $\nu_E = ab(n - 1) = (2)(4)(4 - 1) = 24$.

To test the main effect of A with Wilks' Λ, we compute

$$\Lambda_A = \frac{|\mathbf{E}|}{|\mathbf{E}+\mathbf{H}_A|} = \frac{3602.2}{7600.2} = .474 < \Lambda_{.05,2,1,24} = .771,$$

and we conclude that velocity has a significant effect on y_1 or y_2 or both.

For the B main effect, we have

$$\Lambda_B = \frac{|\mathbf{E}|}{|\mathbf{E}+\mathbf{H}_B|} = \frac{3602.2}{5208.6} = .6916 > \Lambda_{.05,2,3,24} = .591.$$

We conclude that the effect of lubricants is not significant.

For the AB interaction, we obtain

$$\Lambda_{AB} = \frac{|\mathbf{E}|}{|\mathbf{E}+\mathbf{H}_{AB}|} = \frac{3602.2}{3865.3} = .932 > \Lambda_{.05,2,3,24} = .591.$$

Hence we conclude that the interaction effect is not significant.

We now obtain the other three MANOVA test statistics for each test. For A, the only nonzero eigenvalue of $\mathbf{E}^{-1}\mathbf{H}_A$ is 1.110. Thus

$$V^{(s)} = \frac{\lambda_1}{1+\lambda_1} = .526 \qquad U^{(s)} = \lambda_1 = 1.110$$

$$\theta = \frac{\lambda_1}{1+\lambda_1} = .526.$$

In this case, all three tests give results equivalent to Λ_A because $\nu_H = s = 1$.

For B, $\nu_H = 3$ and $p = s = 2$. The eigenvalues of $\mathbf{E}^{-1}\mathbf{H}_B$ are .418 and .020, and we obtain

$$V^{(s)} = \sum_{i=1}^{s} \frac{\lambda_i}{1+\lambda_i} = .314$$

$$U^{(s)} = \sum_{i=1}^{s} \lambda_i = .438 \qquad \frac{\nu_E}{\nu_H} U^{(s)} = 3.502$$

$$\theta = \frac{\lambda_1}{1+\lambda_1} = .295.$$

With $s = 2$, $m = 0$, and $N = 10.5$, we have $V_{.05}^{(s)} = .439$ and $\theta_{.05} = .364$. The .05 critical value of $\nu_E U^{(s)}/\nu_H$ is 5.1799. Thus $V^{(s)}$, $U^{(s)}$, and θ lead to acceptance of H_0, as does Λ. Of the four tests, θ appears to be closer to rejection. This is because $\lambda_1/(\lambda_1 + \lambda_2) = .418/(.418 + .020) = .954$, indicating that the mean vectors for factor B are essentially collinear, in which case Roy's test is more powerful. If the mean vectors $\overline{\mathbf{y}}_{.1.}, \overline{\mathbf{y}}_{.2.}, \overline{\mathbf{y}}_{.3.}$, and $\overline{\mathbf{y}}_{.4.}$ for the four levels of B were a little further apart, we would have a situation in which the four MANOVA tests do not reach the same conclusion.

For AB, the eigenvalues of $\mathbf{E}^{-1}\mathbf{H}_{AB}$ are .0651 and .0075, from which

$$V^{(s)} = \sum_{i=1}^{s} \frac{\lambda_i}{1+\lambda_i} = .0685 \qquad U^{(s)} = .0726, \frac{\nu_E}{\nu_H} U^{(s)} = .580$$

$$\theta = \frac{\lambda_1}{1+\lambda_1} = .0611.$$

The critical values remain the same as for factor B and all three tests accept H_0, as does Wilks' Λ. With a nonsignificant interaction, interpretation of the main effects is simplified, since we do not have to worry about averaging over the levels of the other factor.

6.6 OTHER MODELS

6.6.1 Higher Order Fixed Effects

A higher order fixed-effects model or factorial experiment presents no new difficulties. As an illustration, consider a three-way classification with three factors A, B, and C and all interactions AB, AC, BC, and ABC. The observation vector $\mathbf{y}$ has p variables as usual. The MANOVA model allowing for main effects and interactions can be written as

$$\mathbf{y}_{ijkl} = \boldsymbol{\mu} + \boldsymbol{\alpha}_i + \boldsymbol{\beta}_j + \boldsymbol{\gamma}_k + \boldsymbol{\delta}_{ij} + \boldsymbol{\eta}_{ik} + \boldsymbol{\tau}_{jk} + \boldsymbol{\phi}_{ijk} + \boldsymbol{\epsilon}_{ijkl}, \tag{6.59}$$

where, for example, $\boldsymbol{\alpha}_i$ is the effect of the ith level of factor A on each of the p variables in $\mathbf{y}_{ijkl}$ and $\boldsymbol{\delta}_{ij}$ is the AB interaction effect on each of the p variables. Similarly, $\boldsymbol{\eta}_{ik}, \boldsymbol{\tau}_{jk}$, and $\boldsymbol{\phi}_{ijk}$ represent the AC, BC, and ABC interactions on each of the p variables.

The matrices of sums of squares and products for main effects, interactions, and error are defined in a fashion similar to that for the matrices detailed for the two-way model in Section 6.5.2. The sum of squares (on the diagonal) for each variable is calculated exactly the same as in a univariate ANOVA for a three-way model. The sums of products (off-diagonal) are obtained analogously. Test construction parallels that for the two-way model, using the matrix for error to test all factors and interactions.

Degrees of freedom for each factor are the same as in the corresponding three-way univariate model. All four MANOVA test statistics can be computed for each test. Contrasts can be defined and tested as in Section 6.5.2. Follow-up procedures on the individual variables (F-tests and discriminant functions) can be used as discussed for the one-way or two-way models in Sections 6.4 and 6.5.2.

6.6.2 Mixed Models

There is a MANOVA counterpart for every univariate ANOVA design. This applies to fixed, random, and mixed models and to experimental structures that are crossed, nested, or a combination. Roebruck (1982) has provided a formal proof that univariate mixed models can be generalized to multivariate mixed models. Schott and Saw (1984) have shown that for the one-way multivariate random effects model, the likelihood ratio approach leads to the same test statistics involving the eigenvalues of $\mathbf{E}^{-1}\mathbf{H}$ as in the fixed-effects model.

In the MANOVA mixed model, the expected mean square matrices have exactly the same pattern as expected mean squares for the corresponding univariate ANOVA model. Thus a table of expected mean squares for the terms in the corresponding univariate model provides direction for choosing the appropriate error matrix to test each term in the MANOVA model. However, if the matrix indicated for "error" has fewer degrees of freedom than p, it will not have an inverse and the test cannot be made.

To illustrate, suppose we have a two-way MANOVA model with A fixed and B random. Then the (univariate) expected mean squares (EMS) and Wilks' Λ-tests are as follows:

Source	EMS	Λ
A	$\sigma^2 + n\sigma_{AB}^2 + nb\sigma_A^{*2}$	$\lvert\mathbf{H}_{AB}\rvert/\lvert\mathbf{H}_{AB}+\mathbf{H}_A\rvert$
B	$\sigma^2 + na\sigma_B^2$	$\lvert\mathbf{E}\rvert/\lvert\mathbf{E}+\mathbf{H}_B\rvert$
AB	$\sigma^2 + n\sigma_{AB}^2$	$\lvert\mathbf{E}\rvert/\lvert\mathbf{E}+\mathbf{H}_{AB}\rvert$
Error	σ^2	

In the expected mean square for factor A, we have used the notation σ_A^{*2} in place of $\sum_{i=1}^{a}\alpha_i^2/(a-1)$. The test for A will be indeterminate (of the form 0/0) if $\nu_{AB} \leq p$, where $\nu_{AB} = (a-1)(b-1)$. When the error matrix is other than $\mathbf{E}$, the degrees of freedom will often fail to exceed p. For example, suppose A has two levels and B has three. Then $\nu_{AB} = 2$, which will ordinarily be less than p. In such a case, we would have little recourse except to compute univariate tests on the p individual variables. However, in this case we would not have the multivariate test to protect against carrying out too many univariate tests and thereby inflating the experimentwise α (see Section 6.4). We would be making the univariate tests every time, not just when the multivariate test leads to rejection of H_0. To protect against inflation of α when making p tests, we could use a Bonferroni correction, as in procedure 2 in Section 5.5. In the case of F-tests, we do not have a table of Bonferroni critical values, as we do for

t-tests, but we can achieve an equivalent result by comparing the p-values for the F-tests against α/p instead of against α.

As another illustration, we consider the analysis for a multivariate split-plot design. For simplicity, we show the associated univariate model in place of the multivariate model. We use the factor names $A, B, AC, \ldots$ to indicate parameters in the model:

$$y_{ijkl} = \mu + A_i + B_{(i)j} + C_k + AC_{ik} + BC_{(i)jk} + \epsilon_{(ijk)l},$$

where A and C are fixed and B is random. Nesting is indicated by bracketed subscripts; for example, B and BC are nested in A. Table 6.7 shows the expected mean squares and corresponding Wilks tests.

Since we use $\mathbf{H}_B$ and $\mathbf{H}_{BC}$, as well as $\mathbf{E}$, to make tests, the following must hold:

$$a(b - 1) \geq p \qquad a(b - 1)(c - 1) \geq p \qquad abc(e - 1) \geq p.$$

To construct the other three MANOVA tests, we use eigenvalues of the following matrices:

Source	Matrix
A	$\mathbf{H}_B^{-1}\mathbf{H}_A$
B	$\mathbf{E}^{-1}\mathbf{H}_B$
C	$\mathbf{H}_{BC}^{-1}\mathbf{H}_C$
AC	$\mathbf{H}_{BC}^{-1}\mathbf{H}_{AC}$
BC	$\mathbf{E}^{-1}\mathbf{H}_{BC}$

With a table of expected mean squares, such as those in Table 6.7, it is a simple matter to determine the error matrix in each case. For a given factor or interaction, such as A, B, or AC, the appropriate error matrix is ordinarily the one whose expected mean square matches that of the given factor except for

Table 6.7 Wilks' Λ Tests for a Typical Split-Plot Design

Source	df	Expected Mean Squares	Wilks' Λ
A	$a - 1$	$\sigma^2 + ce\sigma_B^2 + bce\sigma_A^{*2}$	$\lvert\mathbf{H}_B\rvert/\lvert\mathbf{H}_A + \mathbf{H}_B\rvert$
B	$a(b - 1)$	$\sigma^2 + ce\sigma_B^2$	$\lvert\mathbf{E}\rvert/\lvert\mathbf{H}_B + \mathbf{E}\rvert$
C	$c - 1$	$\sigma^2 + e\sigma_{BC}^2 + abe\sigma_C^{*2}$	$\lvert\mathbf{H}_{BC}\rvert/\lvert\mathbf{H}_C + \mathbf{H}_{BC}\rvert$
AC	$(a - 1)(c - 1)$	$\sigma^2 + e\sigma_{BC}^2 + be\sigma_{AC}^{*2}$	$\lvert\mathbf{H}_{BC}\rvert/\lvert\mathbf{H}_{AC} + \mathbf{H}_{BC}\rvert$
BC	$a(b - 1)(c - 1)$	$\sigma^2 + e\sigma_{BC}^2$	$\lvert\mathbf{E}\rvert/\lvert\mathbf{H}_{BC} + \mathbf{E}\rvert$
Error	$abc(e - 1)$	σ^2	

the last term. For example, factor C, with expected mean square $\sigma^2 + e\sigma_{BC}^2 + abe\sigma_C^{*2}$, is tested by BC, whose expected mean square is $\sigma^2 + e\sigma_{BC}^2$. If H_0: $\sigma_C^{*2} = 0$ is true, they have the same expected mean square.

In some mixed and random models, certain terms have no available error term. When this happens in the univariate case, we can construct an approximate test using Satterthwaites' (1941) or other synthetic mean square approach. In the multivariate case, however, a similar approach is apparently not yet available.

6.7 CHECKING ON THE ASSUMPTIONS

In Section 6.2 we discussed the robustness of the four MANOVA test statistics to nonnormality and heterogeneity of covariance matrices. The MANOVA tests (except for Roy's) are rather robust to these departures from the assumptions, although, in general, as dimensionality increases, robustness decreases.

Even though MANOVA procedures are fairly robust to departures from multivariate normality, we may want to check for gross violations of this assumption. Any of the tests or plots from Section 4.4 could be used. For a two-way design, for example, the tests could be applied separately to the n observations in each individual cell (if n is sufficiently large) or to all the residuals. The residual vectors after fitting the model $\mathbf{y}_{ijk} = \boldsymbol{\mu}_{ij} + \boldsymbol{\epsilon}_{ijk}$ would be

$$\mathbf{y}_{ijk} - \overline{\mathbf{y}}_{ij.} \qquad i = 1, 2, \ldots, a \qquad j = 1, 2, \ldots, b \qquad k = 1, 2, \ldots, n.$$

Outliers can lead to either a Type I or Type II error, and it is advisable to check for them. The tests of Section 4.5 can be run separately for each cell (for sufficiently large n) or for all of the abn residuals, $\mathbf{y}_{ijk} - \overline{\mathbf{y}}_{ij.}$.

A test of the equality of covariance matrices can be made using Box's M-test given in Section 7.3.2. Note the cautions expressed there about the sensitivity of this test to nonnormality and unequal sample sizes.

The assumption of independence of the observation vectors $\mathbf{y}_{ijk}$ is also very important. We are referring, of course, to independence from one observation vector to another. The variables within a vector are assumed to be correlated, as usual. In the univariate case, Barcikowski (1981) showed that a moderate amount of dependence among the observations produces an actual α much greater than the nominal α. This effect is to be expected, since the dependence leads to an underestimate of the variance, so that MSE is reduced and the F-statistic is inflated. We can assume that this effect on error rates carries over to MANOVA.

In univariate ANOVA, a simple measure of dependence among the observations in a one-way model is the so-called *intraclass correlation*:

$$r_c = \frac{\text{MSB} - \text{MSE}}{\text{MSB} + (n-1)\text{MSE}}, \tag{6.60}$$

where MSB and MSE are the between and within mean squares and n is the number of observations per group. This could be calculated for each variable in a MANOVA to check for independence.

In many experimental settings, we do not anticipate a lack of independence. But in certain cases the observations are dependent. For example, if the sampling units are people, they may influence each other as they interact together. In some educational studies, researchers must use entire classrooms as sampling units rather than use individual students. Another example of dependence is furnished by observations that are *serially correlated*, as in a *time series*, for example. Each observation depends to a certain extent on the preceding one, and its random movement is dampened as a result.

6.8 PROFILE ANALYSIS

The two-sample profile analysis of Section 5.9.2 can be extended to k groups. Again we assume that the variables are commensurate, as, for example, when each subject is given a battery of tests. Other assumptions, cautions, and comments expressed in Section 5.9.2 apply here as well.

The basic model is the balanced one-way MANOVA:

$$\mathbf{y}_{ij} = \boldsymbol{\mu}_i + \boldsymbol{\epsilon}_{ij} \qquad i = 1, 2, \ldots, k \qquad j = 1, 2, \ldots, n.$$

To test H_0: $\boldsymbol{\mu}_1 = \boldsymbol{\mu}_2 = \cdots = \boldsymbol{\mu}_k$, we use the usual $\mathbf{H}$ and $\mathbf{E}$ matrices given in (6.7) and (6.8). If the variables are commensurate, we can be more specific and extend H_0 to an examination of the k profiles obtained by plotting each $\boldsymbol{\mu}_i$, as was done with two $\boldsymbol{\mu}_i$'s in Section 5.9.2. We are interested in the same three hypotheses as before:

H_{01}: The k profiles are parallel.

H_{02}: The k profiles are all at the same level.

H_{03}: The k profiles are flat.

The hypothesis of parallelism for two groups was expressed in Section 5.9.2 as H_{01}: $\mathbf{C}\boldsymbol{\mu}_1 = \mathbf{C}\boldsymbol{\mu}_2$, where $\mathbf{C}$ is any $(p-1) \times p$ matrix of rank $p-1$ such that $\mathbf{Cj} = \mathbf{0}$, for example,

$$\mathbf{C} = \begin{bmatrix} 1 & -1 & 0 & \cdots & 0 \\ 0 & 1 & -1 & \cdots & 0 \\ \vdots & \vdots & \vdots & & \vdots \\ 0 & 0 & 0 & \cdots & -1 \end{bmatrix}.$$

For k groups, the analogous hypothesis of parallelism is

$$H_{01}\colon \mathbf{C}\boldsymbol{\mu}_1 = \mathbf{C}\boldsymbol{\mu}_2 = \cdots = \mathbf{C}\boldsymbol{\mu}_k. \tag{6.61}$$

The hypothesis (6.61) is equivalent to the hypothesis H_0: $\boldsymbol{\mu}_{z1} = \boldsymbol{\mu}_{z2} = \cdots = \boldsymbol{\mu}_{zk}$ in a one-way MANOVA on the transformed variables $\mathbf{z}_{ij} = \mathbf{C}\mathbf{y}_{ij}$. By property 1b in Section 4.2, $\mathbf{z}_{ij}$ is distributed as $N_{p-1}(\mathbf{C}\boldsymbol{\mu}_i, \mathbf{C}\boldsymbol{\Sigma}\mathbf{C}')$. Since $\mathbf{C}$ has $p-1$ rows, $\mathbf{C}\mathbf{y}_{ij}$ is $(p-1)\times 1$, $\mathbf{C}\boldsymbol{\mu}_i$ is $(p-1)\times 1$, and $\mathbf{C}\boldsymbol{\Sigma}\mathbf{C}'$ is $(p-1)\times(p-1)$.

By analogy with (3.60), the hypothesis and error matrices for testing H_{01} in (6.61) are

$$\mathbf{H}_z = \mathbf{CHC}' \qquad \mathbf{E}_z = \mathbf{CEC}'.$$

We thus have

$$\Lambda = \frac{|\mathbf{CEC}'|}{|\mathbf{CEC}' + \mathbf{CHC}'|} = \frac{|\mathbf{CEC}'|}{|\mathbf{C}(\mathbf{E}+\mathbf{H})\mathbf{C}'|}, \tag{6.62}$$

which is distributed as $\Lambda_{p-1,\nu_H,\nu_E}$, where $\nu_H = k-1$ and $\nu_E = k(n-1)$. The other three MANOVA test statistics can be obtained from the eigenvalues of $(\mathbf{CEC}')^{-1}(\mathbf{CHC}')$. The test for H_{01} can easily be adjusted for unbalanced data, as in Section 6.1.6. We would calculate $\mathbf{H}$ and $\mathbf{E}$ by (6.25) and (6.26) and use $\nu_E = \sum_i n_i - k$.

The hypothesis that two profiles are at the same level is H_{02}: $\mathbf{j}'\boldsymbol{\mu}_1 = \mathbf{j}'\boldsymbol{\mu}_2$, which generalizes immediately to k profiles at the same level,

$$H_{02}\colon \mathbf{j}'\boldsymbol{\mu}_1 = \mathbf{j}'\boldsymbol{\mu}_2 \cdots = \mathbf{j}'\boldsymbol{\mu}_k. \tag{6.63}$$

For two groups we used a univariate t, as defined in (5.37), to test H_{02}. Similarly, for k groups we can employ an F-test for one-way ANOVA comparing k groups with observations $\mathbf{j}'\mathbf{y}_{ij}$. Alternatively, we can utilize (6.62), with $\mathbf{C} = \mathbf{j}'$,

$$\Lambda = \frac{\mathbf{j}'\mathbf{E}\mathbf{j}}{\mathbf{j}'\mathbf{E}\mathbf{j} + \mathbf{j}'\mathbf{H}\mathbf{j}}, \tag{6.64}$$

which is distributed as Λ_{1,ν_H,ν_E} ($p = 1$ because $\mathbf{j}'\mathbf{y}_{ij}$ is a scalar). This is, of course, equivalent to the F-test on $\mathbf{j}'\mathbf{y}_{ij}$, since by Table 6.1 in Section 6.1.3,

$$F = \frac{1 - \Lambda}{\Lambda} \frac{\nu_E}{\nu_H} \tag{6.65}$$

is distributed as F_{ν_H,ν_E}.

The third hypothesis, that of "flatness," essentially states that the average of the k group means is the same for each variable:

$$H_{03}: \frac{\mu_{11} + \mu_{21} + \cdots + \mu_{k1}}{k} = \frac{\mu_{12} + \mu_{22} + \cdots + \mu_{k2}}{k} = \cdots = \frac{\mu_{1p} + \mu_{2p} + \cdots + \mu_{kp}}{k}$$

or by analogy with (5.38),

$$H_{03}: \frac{\mathbf{C}(\boldsymbol{\mu}_1 + \boldsymbol{\mu}_2 + \cdots + \boldsymbol{\mu}_k)}{k} = \mathbf{0}, \tag{6.66}$$

where $\mathbf{C}$ is defined as in (6.61). The flatness hypothesis can also be stated as "the means of all p variables in each group are the same," or $\mu_{i1} = \mu_{i2} = \cdots = \mu_{ip}, i = 1, 2, \ldots, k$. This can be expressed as H_{03}: $\mathbf{C}\boldsymbol{\mu}_1 = \mathbf{C}\boldsymbol{\mu}_2 = \cdots = \mathbf{C}\boldsymbol{\mu}_k = \mathbf{0}$.

To test H_{03} as given by (6.66), we can extend the T^2-test (5.39). The grand mean vector $(\boldsymbol{\mu}_1 + \boldsymbol{\mu}_2 + \cdots + \boldsymbol{\mu}_k)/k$ can be estimated as in Section 6.1.2 by

$$\overline{\mathbf{y}}_{..} = \sum_{ij} \frac{\mathbf{y}_{ij}}{kn}.$$

Under H_{03} (and H_{01}), $\mathbf{C}\overline{\mathbf{y}}_{..}$ is $N_{p-1}(\mathbf{0}, \mathbf{C\Sigma C}'/kn)$, and H_{03} can be tested by

$$T^2 = kn(\mathbf{C}\overline{\mathbf{y}}_{..})'(\mathbf{CEC}'/\nu_E)^{-1}\mathbf{C}\overline{\mathbf{y}}_{..}, \tag{6.67}$$

where $\mathbf{E}/\nu_E$ is an estimate of $\boldsymbol{\Sigma}$. As in the two-sample case, H_{03} is unaffected by the status of H_{02}. When both H_{01} and H_{03} are true, T^2 in (6.67) is distributed as T^2_{p-1,ν_E}.

Example 6.8. Three vitamin E diet supplements with levels zero, low, and high were compared for their effect on growth of guinea pigs (Crowder and Hand 1990, pp. 21–29). Five guinea pigs received each supplement level and their weight was recorded at the end of weeks 1, 3, 4, 5, 6, and 7. These weights are given in Table 6.8.

The three mean vectors are

$$\begin{aligned}
\bar{\mathbf{y}}'_{1.} &= (466.4, 519.4, 568.8, 561.6, 546.6, 572.0) \\
\bar{\mathbf{y}}'_{2.} &= (494.4, 551.0, 574.2, 567.0, 603.0, 644.0) \\
\bar{\mathbf{y}}'_{3.} &= (497.8, 534.6, 579.8, 571.8, 588.2, 623.2)
\end{aligned}$$

and the overall mean vector is

$$\bar{\mathbf{y}}'_{..} = (486.2, 535.0, 574.3, 566.8, 579.3, 613.1).$$

A profile plot of these means is given in Figure 6.3. There is a high degree of parallelism in the three profiles, with the possible exception of week 6 for group 1.

Table 6.8 Weight of Guinea Pigs under 3 Levels of Vitamin E Supplements

Group	Animal	Week 1	Week 3	Week 4	Week 5	Week 6	Week 7
1	1	455	460	510	504	436	466
1	2	467	565	610	596	542	587
1	3	445	530	580	597	582	619
1	4	485	542	594	583	611	612
1	5	480	500	550	528	562	576
2	6	514	560	565	524	552	597
2	7	440	480	536	484	567	569
2	8	495	570	569	585	576	677
2	9	520	590	610	637	671	702
2	10	503	555	591	605	649	675
3	11	496	560	622	622	632	670
3	12	498	540	589	557	568	609
3	13	478	510	568	555	576	605
3	14	545	565	580	601	633	649
3	15	472	498	540	524	532	583

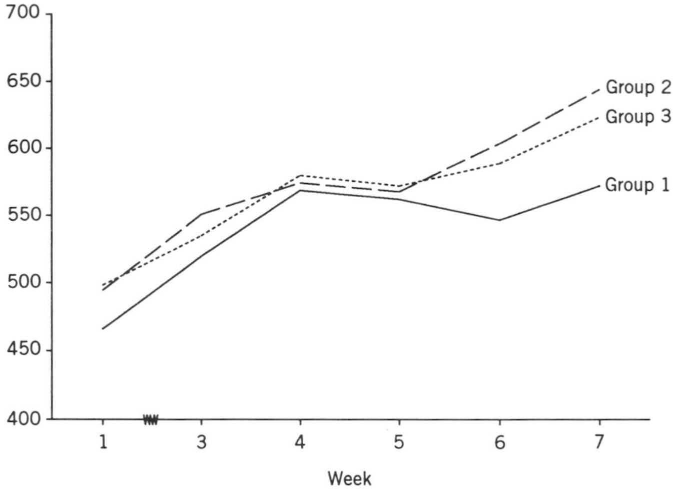

Figure 6.3 Profile of the three groups for the guinea pig data of Table 6.8.

The **H** and **E** matrices are as follows:

$$\mathbf{E} = \begin{bmatrix} 8481.2 & 8538.8 & 4819.8 & 8513.6 & 8710.0 & 8468.2 \\ 8538.8 & 17170.4 & 13293.0 & 19476.4 & 17034.2 & 20035.4 \\ 4819.8 & 13293.0 & 12992.4 & 17077.4 & 17287.8 & 17697.2 \\ 8513.6 & 19476.4 & 17077.4 & 28906.0 & 26226.4 & 28625.2 \\ 8710.0 & 17034.2 & 17287.8 & 26226.4 & 36898.0 & 31505.8 \\ 8468.2 & 20035.4 & 17697.2 & 28625.2 & 31505.8 & 33538.8 \end{bmatrix}$$

$$\mathbf{H} = \begin{bmatrix} 2969.2 & 2177.2 & 859.4 & 813.0 & 4725.2 & 5921.6 \\ 2177.2 & 2497.6 & 410.0 & 411.6 & 4428.8 & 5657.6 \\ 859.4 & 410.0 & 302.5 & 280.4 & 1132.1 & 1392.5 \\ 813.0 & 411.6 & 280.4 & 260.4 & 1096.4 & 1352.0 \\ 4725.2 & 4428.8 & 1132.1 & 1096.4 & 8550.9 & 10830.9 \\ 5921.6 & 5657.6 & 1392.5 & 1352.0 & 10830.9 & 13730.1 \end{bmatrix}$$

Using

$$\mathbf{C} = \begin{bmatrix} 1 & -1 & 0 & 0 & 0 & 0 \\ 0 & 1 & -1 & 0 & 0 & 0 \\ 0 & 0 & 1 & -1 & 0 & 0 \\ 0 & 0 & 0 & 1 & -1 & 0 \\ 0 & 0 & 0 & 0 & 1 & -1 \end{bmatrix}$$

in the test statistic (6.62), we have, as a test for parallelism,

$$\Lambda = \frac{|\mathbf{CEC}'|}{|\mathbf{C(E+H)C}'|} = \frac{3.8238 \times 10^{18}}{2.1355 \times 10^{19}}$$
$$= .1791 > \Lambda_{.05,5,2,12} = .1528.$$

Thus we do not reject the parallelism hypothesis.

To test the hypothesis that the three profiles are at the same level, we use (6.64),

$$\Lambda = \frac{\mathbf{j}'\mathbf{Ej}}{\mathbf{j}'\mathbf{Ej} + \mathbf{j}'\mathbf{Hj}} = \frac{632,605.2}{632,605.2 + 111,288.1}$$
$$= .8504 > \Lambda_{.05,1,2,12} = .6070.$$

Hence we do not reject the "levels" hypothesis. This can also be seen by using (6.65) to transform Λ to F,

$$F = \frac{(1-\Lambda)\nu_E}{\Lambda\nu_H} = \frac{1 - .8504}{.8504}\,\frac{12}{2} = 1.0555,$$

which is clearly nonsignificant ($p = .378$).

To test the "flatness" hypothesis, we use (6.67):

$$T^2 = kn(\mathbf{C}\bar{\mathbf{y}}_{..})'(\mathbf{CEC}'/\nu_E)^{-1}\mathbf{C}\bar{\mathbf{y}}_{..}$$
$$= 15 \begin{bmatrix} -48.80 \\ -39.27 \\ 7.47 \\ -12.47 \\ -33.80 \end{bmatrix}' \begin{bmatrix} 714.5 & -13.2 & 207.5 & -219.9 & 270.2 \\ -13.2 & 298.1 & -174.9 & 221.0 & -216.0 \\ 207.5 & -174.9 & 645.3 & -240.8 & 165.8 \\ -219.9 & 221.0 & -240.8 & 1112.6 & -649.2 \\ 270.2 & -216.0 & 165.8 & -649.2 & 618.8 \end{bmatrix}^{-1} \cdot \begin{bmatrix} -48.80 \\ -39.27 \\ 7.47 \\ -12.47 \\ -33.80 \end{bmatrix}$$
$$= 297.13 > T^2_{.01,5,12} = 49.739.$$

Thus only the "flatness" hypothesis is rejected in this case.

6.9 REPEATED MEASURES DESIGNS

6.9.1 Multivariate versus Univariate Approach

In *repeated measures* designs, the research unit is typically a human or animal subject. Each subject is measured under several treatments or at different points of time. The treatments might be tests, drug levels, various kinds of stimuli, and so on. If the treatments are such that the order of presentation to the various subjects can be varied, then the order should be randomized to avoid an ordering bias. If subjects are measured at successive time points, it may be of interest to determine the degree of polynomial required to fit the curve. This is treated in Section 6.10 as part of an analysis of growth curves.

When comparing means of the treatments applied to each subject, we are examining the *within-subjects* factor. There may also be a *between-subjects* factor if there are several groups of subjects that we wish to compare. In Sections 6.9.2–6.9.6, we consider designs up to a complexity level of two within-subjects factors and two between-subjects factors.

We now discuss univariate and multivariate approaches to hypothesis testing in repeated measures designs. As a framework for this discussion, consider the layout in Table 6.9 for a repeated measures design with one repeated measures (within-subjects) factor, A, and one grouping (between-subjects) factor, B.

This design has often been analyzed as a univariate mixed-model nested design, also called a split-plot design, with subjects nested in factor B (whole-

Table 6.9 Data Layout for k-Groups Repeated Measures Experiment

Factor B (Group)	Subjects	Factor A (Repeated Measures) A_1	A_2	$\cdots$	A_p	
B_1	S_{11}	$(y_{111}$	y_{112}	$\cdots$	$y_{11p})$	$= \mathbf{y}'_{11}$
	S_{12}	$(y_{121}$	y_{122}	$\cdots$	$y_{12p})$	$= \mathbf{y}'_{12}$
	$\vdots$	$\vdots$	$\vdots$		$\vdots$	
	S_{1n}	$(y_{1n1}$	y_{1n2}	$\cdots$	$y_{1np})$	$= \mathbf{y}'_{1n}$
B_2	S_{21}	$(y_{211}$	y_{212}	$\cdots$	$y_{21p})$	$= \mathbf{y}'_{21}$
	S_{22}	$(y_{221}$	y_{222}	$\cdots$	$y_{22p})$	$= \mathbf{y}'_{22}$
	$\vdots$	$\vdots$	$\vdots$		$\vdots$	
	S_{2n}	$(y_{2n1}$	y_{2n2}	$\cdots$	$y_{2np})$	$= \mathbf{y}'_{2n}$
$\vdots$	$\vdots$	$\vdots$	$\vdots$		$\vdots$	
B_k	S_{k1}	$(y_{k11}$	y_{k12}	$\cdots$	$y_{k1p})$	$= \mathbf{y}'_{k1}$
	S_{k2}	$(y_{k21}$	y_{k22}	$\cdots$	$y_{k2p})$	$= \mathbf{y}'_{k2}$
	$\vdots$	$\vdots$	$\vdots$		$\vdots$	
	S_{kn}	$(y_{kn1}$	y_{kn2}	$\cdots$	$y_{knp})$	$= \mathbf{y}'_{kn}$

plot), which is crossed with factor A (repeated measures or split-plot). The univariate model for each y_{ijr} would be

$$y_{ijr} = \mu + B_i + S_{(i)j} + A_r + BA_{ir} + \epsilon_{ijr}, \tag{6.68}$$

where the factor level designations $(B, S, A, \ldots)$ from Table 6.9 are used as parameter values, and the subscript $(i)j$ on S indicates that subjects are nested in factor B. In Table 6.9, the observations y_{ijr} for $r = 1, 2, \ldots, p$ are enclosed in parentheses and denoted by $\mathbf{y}'_{ij}$ to emphasize that these p variables are measured on one subject and thus constitute a vector of correlated variables. The ranges of the subscripts can be seen in Table 6.9: $i = 1, 2, \ldots, k, j = 1, 2, \ldots, n$, and $r = 1, 2, \ldots, p$. With factors A and B fixed and subjects random, the univariate ANOVA is given in Table 6.10.

However, our initial reaction would be to rule out the univariate ANOVA because the y's in each row are correlated and the assumption of independence is critical, as noted in Section 6.7. We will discuss below some assumptions under which the univariate analysis would be appropriate.

In the multivariate approach, the p responses $y_{ij1}, y_{ij2}, \ldots, y_{ijp}$ (repeated measures) for subject S_{ij} constitute a vector $\mathbf{y}_{ij}$, as shown in Table 6.9. The multivariate model for $\mathbf{y}_{ij}$ is a simple one-way MANOVA model,

$$\mathbf{y}_{ij} = \boldsymbol{\mu} + \boldsymbol{\beta}_i + \boldsymbol{\epsilon}_{ij}, \tag{6.69}$$

where $\boldsymbol{\beta}_i$ is a vector of p main effects (corresponding to the p variables in $\mathbf{y}_{ij}$) for factor B and $\boldsymbol{\epsilon}_{ij}$ is an error vector for subject S_{ij}. This model seems to include only factor B, but we show in Section 6.9.3 how to use an approach similar to profile analysis in Section 6.8 to obtain tests on factor A and the BA interaction. The MANOVA assumption that $\text{cov}(\mathbf{y}_{ij}) = \boldsymbol{\Sigma}$ for all i and j allows the p repeated measures to be correlated in any pattern, since $\boldsymbol{\Sigma}$ is completely general. On the other hand, the ANOVA assumptions of homogeneity of variances and independence can be expressed as $\text{cov}(\mathbf{y}_{ij}) = \sigma^2\mathbf{I}$. We would be very surprised if repeated measurements on the same subject were independent.

The univariate ANOVA has been found to be appropriate under less stringent conditions than $\boldsymbol{\Sigma} = \sigma^2\mathbf{I}$. Wilks (1946) showed that the ordinary F-tests of

Table 6.10 Univariate ANOVA for Data Layout in Table 6.9

Source	df	MS	F
B (between)	$k - 1$	MSB	MSB/MSS
S (subjects)	$k(n - 1)$	MSS	
A (within or repeated)	$p - 1$	MSA	MSA/MSE
BA	$(k - 1)(p - 1)$	MSBA	MSBA/MSE
Error (SA interaction)	$k(n - 1)(p - 1)$	MSE	

ANOVA remain valid for a covariance structure of the form

$$\begin{aligned}\text{cov}(\mathbf{y}_{ij}) = \boldsymbol{\Sigma} &= \sigma^2 \begin{bmatrix} 1 & \rho & \rho & \cdots & \rho \\ \rho & 1 & \rho & \cdots & \rho \\ \vdots & \vdots & \vdots & & \vdots \\ \rho & \rho & \rho & \cdots & 1 \end{bmatrix} \\ &= \sigma^2[(1 - \rho)\mathbf{I} + \rho\mathbf{J}], \end{aligned} \tag{6.70}$$

where $\mathbf{J}$ is a square matrix of 1s, as defined in (2.12). The covariance pattern (6.70) is variously known as *uniformity*, *compound symmetry*, or the *intraclass correlation* model. It allows for the variables to be correlated but restricts every variable to have the same variance and every pair of variables to have the same covariance. In a carefully designed experiment with appropriate randomization, this assumption may hold under the hypothesis of no A effect. Alternatively, we could use a test of the hypothesis that $\boldsymbol{\Sigma}$ has the pattern (6.70), as given in Section 7.2.3. If this hypothesis is accepted, one could proceed with the usual ANOVA F-tests.

Bock (1963) and Huynh and Feldt (1970) showed that the most general condition under which univariate F-tests remain valid is that

$$\mathbf{C}\boldsymbol{\Sigma}\mathbf{C}' = \sigma^2\mathbf{I}, \tag{6.71}$$

where $\mathbf{C}$ is a $(p - 1) \times p$ matrix whose rows are *orthonormal* contrasts (orthogonal contrasts that have been normalized to unit length). We can construct $\mathbf{C}$ by choosing any $p - 1$ orthogonal contrasts among the means $\mu_1, \mu_2, \ldots, \mu_p$ of the repeated measures factor and dividing each contrast by $\sqrt{\sum_i c_i^2}$. (This matrix $\mathbf{C}$ is different from $\mathbf{C}$ used in Section 6.8 and in the remainder of Section 6.9, whose rows are contrasts that are not normalized to unit length.) It can be shown that (6.70) is a special case of (6.71). The condition (6.71) is sometimes referred to as *sphericity*, although this term can also refer to the covariance pattern $\boldsymbol{\Sigma} = \sigma^2\mathbf{I}$ on the untransformed $\mathbf{y}_{ij}$ (see Section 7.2.2).

A simple way to test the hypothesis that (6.71) holds is to transform the data by $\mathbf{z}_{ij} = \mathbf{C}\mathbf{y}_{ij}$ and test H_0: $\boldsymbol{\Sigma}_z = \sigma^2\mathbf{I}$, as in Section 7.2.2, using $\mathbf{C}\mathbf{S}_{\text{pl}}\mathbf{C}'$ in place of $\mathbf{S}_{\text{pl}}$, where $\mathbf{S}_{\text{pl}} = \mathbf{E}/\nu_E$.

Thus one procedure for repeated measures designs is to make a preliminary test for (6.70) or (6.71) and, if the hypothesis is accepted, use univariate F-tests, as in Table 6.10. Fehlberg (1980) investigated the use of larger α-values with a preliminary test of structure of the covariance matrix, as in (6.71). He concludes that using $\alpha = .40$ sufficiently controls the problem of falsely accepting sphericity so as to justify the use of a preliminary test.

If the univariate test for the repeated measures factor A is appropriate, it is more powerful because it has more degrees of freedom for error than the corresponding multivariate test. However, even mild departures from (6.71) seriously inflate the Type I error rate of the univariate test for factor A (Box 1954, Davidson 1972, Boik 1981). Because such departures can be easily missed in a preliminary test, Boik (1981) concludes that "on the whole, the ordinary F tests have nothing to recommend them" (p. 248) and emphasized that "there is no justification for employing ordinary univariate F tests for repeated measures treatment contrasts" (p. 254).

Another approach to analysis of repeated measure designs is to adjust the univariate F-test for the amount of departure from sphericity. Box (1954) and Greenhouse and Geisser (1959) showed that when $\mathbf{\Sigma} \neq \sigma^2\mathbf{I}$, an approximate F-test for effects involving the repeated measures is obtained by reducing the degrees of freedom for both numerator and denominator by a factor of

$$\epsilon = \frac{[\text{tr}(\mathbf{\Sigma} - \mathbf{J\Sigma}/p)]^2}{(p-1)\text{tr}(\mathbf{\Sigma} - \mathbf{J\Sigma}/p)^2}, \tag{6.72}$$

where $\mathbf{J}$ is a $p \times p$ matrix of 1s defined in (2.12). For example, in Table 6.10 the F-value for the BA interaction would be compared to F_α with $\epsilon(k-1)(p-1)$ and $\epsilon k(n-1)(p-1)$ degrees of freedom. An estimate $\hat{\epsilon}$ can be obtained by substituting $\hat{\mathbf{\Sigma}} = \mathbf{E}/\nu_E$ in (6.72). Greenhouse and Geisser (1959) showed that ϵ and $\hat{\epsilon}$ vary between $1/(p-1)$ and 1, with ϵ equal to 1 when sphericity holds and ϵ equal to $1/(p-1)$ for general $\mathbf{\Sigma}$. Thus ϵ is a measure of nonsphericity. For a conservative test, Greenhouse and Geisser recommend dividing numerator and denominator degrees of freedom by $p-1$. Huynh and Feldt (1976) provided an improved estimator of ϵ.

The behavior of the approximate univariate F-test with degrees of freedom adjusted by $\hat{\epsilon}$ has been investigated by Collier et al. (1967), Huynh (1978), Davidson (1972), Rogan et al. (1979), and Maxwell and Avery (1982). In these studies, the true α level turned out to be close to the nominal α and the power was close to that of the multivariate test. However, since the ϵ-adjusted F-test is only approximate and has no power advantage over the exact multivariate test, there appears to be no compelling reason to use it. The only case in which we need to fall back on a univariate test is when there are insufficient degrees of freedom to perform a multivariate test, that is, when $p > \nu_E$.

In the remainder of this section, we discuss the multivariate approach to repeated measures. We will cover several models, beginning with the simple one-sample design.

6.9.2 One-Sample Repeated Measures Model

Schematically, a one-sample design with four repeated measures on n subjects would appear as in Table 6.11. There is a superficial resemblance to a univariate randomized block design. However, in the repeated measures design, the observations y_{i1}, y_{i2}, y_{i3}, and y_{i4} are correlated because they are measured on the same subject (experimental unit), whereas in a randomized block design y_{i1}, y_{i2}, y_{i3}, and y_{i4} would be measured on four different experimental units. Thus we have a single sample of n observation vectors $\mathbf{y}_1, \mathbf{y}_2, \ldots, \mathbf{y}_n$.

To test for significance of factor A, we compare the means of the p variables in $\mathbf{y}_i$,

$$E(\mathbf{y}_i) = \boldsymbol{\mu} = \begin{bmatrix} \mu_1 \\ \mu_2 \\ \mu_3 \\ \mu_4 \end{bmatrix}.$$

The hypothesis is H_0: $\mu_1 = \mu_2 = \mu_3 = \mu_4$, which can be reexpressed as H_0: $\mu_1 - \mu_2 = \mu_2 - \mu_3 = \mu_3 - \mu_4 = 0$ or $\mathbf{C}_1 \boldsymbol{\mu} = \mathbf{0}$, where

$$\mathbf{C}_1 = \begin{pmatrix} 1 & -1 & 0 & 0 \\ 0 & 1 & -1 & 0 \\ 0 & 0 & 1 & -1 \end{pmatrix}.$$

To test H_0: $\mathbf{C}_1 \boldsymbol{\mu} = \mathbf{0}$, we calculate $\overline{\mathbf{y}}$ and $\mathbf{S}$ from $\mathbf{y}_1, \mathbf{y}_2, \ldots, \mathbf{y}_n$. Then when H_0 is true, $\mathbf{C}_1\overline{\mathbf{y}}$ is $N_{p-1}(\mathbf{0}, \mathbf{C}_1\boldsymbol{\Sigma}\mathbf{C}_1'/n)$ for p repeated measures on n subjects, and

$$T^2 = n(\mathbf{C}_1\overline{\mathbf{y}})'(\mathbf{C}_1\mathbf{S}\mathbf{C}_1')^{-1}(\mathbf{C}_1\overline{\mathbf{y}}) \qquad (6.73)$$

is distributed as $T^2_{p-1,n-1}$. Note that the dimension is $p - 1$ because $\mathbf{C}_1\overline{\mathbf{y}}$ is $(p - 1) \times 1$.

The multivariate approach involving transformed observations $\mathbf{z}_i = \mathbf{C}_1\mathbf{y}_i$ was first suggested by Hsu (1938) and has been discussed further by Williams (1970) and Morrison (1972). Note that in $\mathbf{C}\,\overline{\mathbf{y}}$ we work with contrasts on the elements $\overline{y}_1, \overline{y}_2, \overline{y}_3$, and $\overline{y}_4$ within the vector

Table 6.11 Data Layout for a Single-Sample Repeated Measures Design

	Factor A (Repeated Measures)				
Subjects	A_1	A_2	A_3	A_4	
S_1	$(y_{11}$	y_{12}	y_{13}	$y_{14})$	$= \mathbf{y}_1'$
S_2	$(y_{21}$	y_{22}	y_{23}	$y_{24})$	$= \mathbf{y}_2'$
S_3	.	.	.	.	.
$\vdots$	$\vdots$	$\vdots$	$\vdots$	$\vdots$	$\vdots$
S_n	$(y_{n1}$	y_{n2}	y_{n3}	$y_{n4})$	$= \mathbf{y}_n'$

$$\bar{\mathbf{y}} = \begin{bmatrix} \bar{y}_1 \\ \bar{y}_2 \\ \bar{y}_3 \\ \bar{y}_4 \end{bmatrix},$$

as opposed to the contrasts involving comparisons of several mean vectors themselves, as, for example, in Section 6.3.2.

The hypothesis H_0: $\mu_1 = \mu_2 = \mu_3 = \mu_4$ can also be expressed as H_0: $\mu_1 - \mu_4 = \mu_2 - \mu_4 = \mu_3 - \mu_4 = 0$, or $\mathbf{C}_2\boldsymbol{\mu} = \mathbf{0}$, where

$$\mathbf{C}_2 = \begin{pmatrix} 1 & 0 & 0 & -1 \\ 0 & 1 & 0 & -1 \\ 0 & 0 & 1 & -1 \end{pmatrix}.$$

The matrix $\mathbf{C}_1$ can be obtained from $\mathbf{C}_2$ by simple row operations, for example, subtracting the second row from the first and the third row from the second. Hence, $\mathbf{C}_1 = \mathbf{A}\mathbf{C}_2$, where

$$\mathbf{A} = \begin{pmatrix} 1 & -1 & 0 \\ 0 & 1 & -1 \\ 0 & 0 & 1 \end{pmatrix}.$$

In fact, H_0 can be expressed as $\mathbf{C}\boldsymbol{\mu} = \mathbf{0}$ for any full-rank $(p-1) \times p$ matrix $\mathbf{C}$ such that $\mathbf{Cj} = \mathbf{0}$, and the same value of T^2 in (6.73) will result. The contrasts in $\mathbf{C}$ can be either linearly independent or orthogonal.

The hypothesis H_0: $\mu_1 = \mu_2 = \cdots = \mu_p = \mu$, say, can also be expressed as

$$H_0\colon \boldsymbol{\mu} = \mu\mathbf{j},$$

where $\mathbf{j}' = (1, 1, \ldots, 1)$. The maximum likelihood estimate of μ is

$$\hat{\mu} = \frac{\bar{\mathbf{y}}'\mathbf{S}^{-1}\mathbf{j}}{\mathbf{j}'\mathbf{S}^{-1}\mathbf{j}}. \tag{6.74}$$

The likelihood ratio test of H_0 is a function of

$$\bar{\mathbf{y}}'\mathbf{S}^{-1}\bar{\mathbf{y}} - \frac{(\bar{\mathbf{y}}'\mathbf{S}^{-1}\mathbf{j})^2}{\mathbf{j}'\mathbf{S}^{-1}\mathbf{j}}.$$

Williams (1970) showed that for any $(p-1) \times p$ matrix $\mathbf{C}$ of rank $p-1$ such that $\mathbf{Cj} = \mathbf{0}$,

$$\bar{\mathbf{y}}'\mathbf{S}^{-1}\bar{\mathbf{y}} - \frac{(\bar{\mathbf{y}}'\mathbf{S}^{-1}\mathbf{j})^2}{\mathbf{j}'\mathbf{S}^{-1}\mathbf{j}} = (\mathbf{C}\bar{\mathbf{y}})'(\mathbf{CSC}')^{-1}(\mathbf{C}\bar{\mathbf{y}})$$

and thus the T^2-test in (6.73) is equivalent to the likelihood ratio test.

Example 6.9.2. The data in Table 6.12 were given by Cochran and Cox (1957, p. 130). As rearranged by Timm (1980), the observations constitute a one-sample repeated measures design with two within-subjects factors. Factor A is a comparison of two tasks; factor B is a comparison of two types of calculators. The measurements are speed of calculation.

To test the hypothesis H_0: $\mu_1 = \mu_2 = \mu_3 = \mu_4$, we use the contrast matrix

$$\mathbf{C} = \begin{pmatrix} 1 & 1 & -1 & -1 \\ 1 & -1 & 1 & -1 \\ 1 & -1 & -1 & 1 \end{pmatrix},$$

where the first row compares the two levels of A, the second row compares the two levels of B, and the third row corresponds to the AB interaction. From the five observation vectors in Table 6.12, we obtain

$$\bar{\mathbf{y}} = \begin{bmatrix} 23.2 \\ 15.6 \\ 20.0 \\ 11.6 \end{bmatrix} \qquad \mathbf{S} = \begin{bmatrix} 51.7 & 29.8 & 9.2 & 7.4 \\ 29.8 & 46.8 & 16.2 & -8.7 \\ 9.2 & 16.2 & 8.5 & -10.5 \\ 7.4 & -8.7 & -10.5 & 24.3 \end{bmatrix}.$$

For the overall test of equality of means, we have, by (6.73),

$$T^2 = n(\mathbf{C}\bar{\mathbf{y}})'(\mathbf{CSC}')^{-1}(\mathbf{C}\bar{\mathbf{y}}) = 29.736 < T^2_{.05,3,4} = 114.986.$$

Since the T^2-test is not significant, we would ordinarily not proceed with tests

Table 6.12 Calculator Speed Data

	A_1		A_2	
Subjects	B_1	B_2	B_1	B_2
S_1	30	21	21	14
S_2	22	13	22	5
S_3	29	13	18	17
S_4	12	7	16	14
S_5	23	24	23	8

based on the individual rows of **C**. We will do so, however, for illustrative purposes.

To test A, B, and AB, we test each row of **C**, in which case $T^2 = n(\mathbf{c}_i'\overline{\mathbf{y}})'(\mathbf{c}_i'\mathbf{S}\mathbf{c}_i)^{-1}\mathbf{c}_i'\overline{\mathbf{y}}$ is the square of the t-statistic

$$t_i = \frac{\sqrt{n}\mathbf{c}_i'\overline{\mathbf{y}}}{\sqrt{\mathbf{c}_i'\mathbf{S}\mathbf{c}_i}} \qquad i = 1,2,3,$$

where $\mathbf{c}_i'$ is the ith row of **C**.

The three results are as follows:

$$\begin{aligned} \text{Factor } A \quad t_1 &= 1.459 < t_{.025,4} = 2.776 \\ \text{Factor } B \quad t_2 &= 5.247 > t_{.005,4} = 4.604 \\ \text{Interaction } AB \quad t_3 &= -.152 \end{aligned}$$

Thus only the main effect for B is significant.

6.9.3 k-Sample Repeated Measures Model

We turn now to the k-sample repeated measures design depicted in Table 6.9. As noted in Section 6.9.1, the multivariate approach to this repeated measures design involves the one-way MANOVA model $\mathbf{y}_{ij} = \boldsymbol{\mu} + \boldsymbol{\beta}_i + \boldsymbol{\epsilon}_{ij} = \boldsymbol{\mu}_i + \boldsymbol{\epsilon}_{ij}$. From the k groups of n observation vectors each, we calculate $\overline{\mathbf{y}}_{1.}, \overline{\mathbf{y}}_{2.}, \ldots, \overline{\mathbf{y}}_{k.}$ and the error matrix **E**.

The layout in Table 6.9 is similar to that of a k-sample profile analysis. To test (the within-subjects) factor A, we need to compare the means of the variables $y_1, y_2, \ldots, y_p$ within **y** averaged across the levels of factor B. We can use a T^2-test analogous to (6.73) for the one-sample case. In the model $\mathbf{y}_{ij} = \boldsymbol{\mu}_i + \boldsymbol{\epsilon}_{ij}$, the mean vectors $\boldsymbol{\mu}_1, \boldsymbol{\mu}_2, \ldots, \boldsymbol{\mu}_k$ correspond to the levels of factor B and are estimated by $\overline{\mathbf{y}}_{1.}, \overline{\mathbf{y}}_{2.}, \ldots, \overline{\mathbf{y}}_{k.}$. To compare the means of $y_1, y_2, \ldots, y_p$ averaged across the levels of B, we use $\overline{\boldsymbol{\mu}}_. = \sum_{i=1}^k \boldsymbol{\mu}_i/k$. Then the hypothesis H_0: $\mu_{.1} = \mu_{.2} = \cdots = \mu_{.p}$ comparing the levels of factor A can be expressed using contrasts:

$$H_0\colon \mathbf{C}\overline{\boldsymbol{\mu}}_. = \mathbf{0}, \qquad (6.75)$$

where **C** is any $(p-1)\times p$ full-rank contrast matrix with $\mathbf{Cj} = \mathbf{0}$. This is equivalent to the "flatness" test of profile analysis. An estimate of $\mathbf{C}\overline{\boldsymbol{\mu}}_.$ is $\mathbf{C}\overline{\mathbf{y}}_{..}$, where $\overline{\mathbf{y}}_{..} = \sum_{i=1}^k \overline{\mathbf{y}}_{i.}/k$ is the grand mean vector. Under H_0, the vector $\mathbf{C}\overline{\mathbf{y}}_{..}$ is distributed as $N_{p-1}(\mathbf{0}, \mathbf{C}\boldsymbol{\Sigma}\mathbf{C}'/N)$, where $N = \sum_i n_i$ for an unbalanced design and $N = kn$ in the balanced case. We can therefore test H_0 with

$$T^2 = N(\mathbf{C}\overline{\mathbf{y}}_{..})'(\mathbf{C}\mathbf{S}_{\text{pl}}\mathbf{C}')^{-1}(\mathbf{C}\overline{\mathbf{y}}_{..}), \tag{6.76}$$

where $\mathbf{S}_{\text{pl}} = \mathbf{E}/\nu_E$. The T^2-statistic in (6.76) is distributed as T^2_{p-1,ν_E} when H_0 is true, where $\nu_E = N - k$. Note that the dimension of T^2 is $p - 1$ because $\mathbf{C}\overline{\mathbf{y}}_{..}$ is $(p - 1) \times 1$.

For the grouping or between-subjects factor B, we wish to compare the means for the k levels of B. The mean response for the ith level of B (averaged over the levels of A) is $\sum_j \mu_{ij}/p = \mathbf{j}'\boldsymbol{\mu}_i/p$. The hypothesis can be expressed as

$$H_0\colon \mathbf{j}'\boldsymbol{\mu}_1 = \mathbf{j}'\boldsymbol{\mu}_2 = \cdots = \mathbf{j}'\boldsymbol{\mu}_k, \tag{6.77}$$

which is analogous to (6.63), the "levels" hypothesis in profile analysis. This is easily tested by calculating a univariate F-statistic for a one-way ANOVA on $z_{ij} = \mathbf{j}'\mathbf{y}_{ij}, i = 1, 2, \ldots, k, j = 1, 2, \ldots, n_i$. There is a z_{ij} corresponding to each subject, S_{ij}. The observation vector for each subject is thus reduced to a single scalar observation, and we have a one-way ANOVA comparing the means $\mathbf{j}'\overline{\mathbf{y}}_{1.}, \mathbf{j}'\overline{\mathbf{y}}_{2.}, \ldots, \mathbf{j}'\overline{\mathbf{y}}_{k.}$.

The AB interaction hypothesis is equivalent to the parallelism hypothesis in profile analysis,

$$H_0\colon \mathbf{C}\boldsymbol{\mu}_1 = \mathbf{C}\boldsymbol{\mu}_2 = \cdots = \mathbf{C}\boldsymbol{\mu}_k. \tag{6.78}$$

In other words, differences or contrasts among the levels of factor A are the same across all levels of factor B. This is easily tested by performing a one-way MANOVA on $\mathbf{z}_{ij} = \mathbf{C}\mathbf{y}_{ij}$ or directly by

$$\Lambda = \frac{|\mathbf{C}\mathbf{E}\mathbf{C}'|}{|\mathbf{C}(\mathbf{E} + \mathbf{H})\mathbf{C}'|}, \tag{6.79}$$

which is distributed as $\Lambda_{p-1,\nu_H,\nu_E}$, with $\nu_H = k - 1$ and $\nu_E = N - k$; that is, $\nu_E = \sum_i (n_i - 1)$ for the unbalanced model or $\nu_E = k(n - 1)$ in the balanced model.

6.9.4 Computation of Repeated Measures Tests

Some statistical software packages have automated repeated measures procedures that are easily implemented. However, if one is unsure as to how the resulting tests correspond to the tests in Section 6.9.3, there are two ways to obtain the tests directly. One approach is to calculate (6.76) and (6.79) outright using a matrix manipulation routine. We would need to have available the **E** and **H** matrices of a one-way MANOVA using a data layout as in Table 6.9.

The second approach uses simple data transformations available in virtually all programs. To test (6.75) for factor A, we would transform each $\mathbf{y}_{ij}$ to $\mathbf{z}_{ij} = \mathbf{C}\mathbf{y}_{ij}$ by using the rows of $\mathbf{C}$. For example, if

$$\mathbf{C} = \begin{pmatrix} 1 & -1 & 0 & 0 \\ 0 & 1 & -1 & 0 \\ 0 & 0 & 1 & -1 \end{pmatrix}$$

then each $\mathbf{y}' = (y_1, y_2, y_3, y_4)$ becomes $\mathbf{z}' = (y_1 - y_2, y_2 - y_3, y_3 - y_4)$. We then test H_0: $\boldsymbol{\mu}_z = \mathbf{0}$ using a one-sample T^2 on all $\mathbf{z}_{ij}$,

$$T^2 = N\bar{\mathbf{z}}'\mathbf{S}_z\bar{\mathbf{z}},$$

where $N = \sum_i n_i$, $\bar{\mathbf{z}} = \sum_{ij} \mathbf{z}_{ij}/N$ and $\mathbf{S}_z$ is the pooled covariance matrix, $\mathbf{S}_z = \mathbf{E}_z/\nu_E$. Reject H_0 if $T^2 \geq T^2_{\alpha,p-1,\nu_E}$.

To test (6.77) for factor B, we sum the components of each observation vector to obtain $z_{ij} = \mathbf{j}'\mathbf{y}_{ij} = y_{ij1} + y_{ij2} + \cdots + y_{ijp}$ and compare the means $\bar{z}_{1.}, \bar{z}_{2.}, \ldots, \bar{z}_{k.}$ by an F-test as in one-way ANOVA.

To test the interaction hypothesis (6.78), we transform each $\mathbf{y}_{ij}$ to $\mathbf{z}_{ij} = \mathbf{C}\mathbf{y}_{ij}$ using the rows of $\mathbf{C}$ as above. Note that $\mathbf{z}_{ij}$ is $(p-1)\times 1$. We then do a one-way MANOVA on $\mathbf{z}_{ij}$ to obtain

$$\Lambda = \frac{|\mathbf{E}_z|}{|\mathbf{E}_z + \mathbf{H}_z|}. \tag{6.80}$$

6.9.5 Repeated Measures with Two Within-Subjects Factors and One Between-Subjects Factor

The repeated measures model with two within-subjects factors and one between-subjects factor corresponds to a one-way MANOVA design in which each observation vector includes measurements on a two-way factorial arrangement of treatments. Thus each subject receives all treatment combinations of two factors. As usual, the sequence of administration of treatment combinations should be randomized for each subject. A design of this type is illustrated in Table 6.13.

Each $\mathbf{y}_{ij}$ in Table 6.13 has nine elements, consisting of responses to the nine treatment combinations $A_1B_1, A_1B_2, \ldots, A_3B_3$. We are interested in the same hypotheses as in a univariate split-plot design but use a multivariate approach to allow for correlated y's. The model for the observation vectors is the one-way MANOVA model

$$\mathbf{y}_{ij} = \boldsymbol{\mu} + \boldsymbol{\gamma}_i + \boldsymbol{\epsilon}_{ij} = \boldsymbol{\mu}_i + \boldsymbol{\epsilon}_{ij},$$

where $\boldsymbol{\gamma}_i$ is the C effect.

Table 6.13 Data Layout for Repeated Measures with Two Within-Subjects Factors and One Between-Subjects Factor

		Within-Subjects Factors								
		A_1			A_2			A_3		
Between-subjects Factor	Subjects	B_1	B_2	B_3	B_1	B_2	B_3	B_1	B_2	B_3
C_1	S_{11}	$(y_{111}$	y_{112}	y_{113}	y_{114}	y_{115}	y_{116}	y_{117}	y_{118}	$y_{119}) = \mathbf{y}'_{11}$
	S_{12}					$\mathbf{y}'_{12}$				
	$\vdots$					$\vdots$				
	S_{1n_1}					$\mathbf{y}'_{1n_1}$				
C_2	S_{21}					$\mathbf{y}'_{21}$				
	S_{22}					$\mathbf{y}'_{22}$				
	$\vdots$					$\vdots$				
	S_{2n_2}					$\mathbf{y}'_{2n_2}$				
C_3	S_{31}					$\mathbf{y}'_{31}$				
	S_{32}					$\mathbf{y}'_{32}$				
	$\vdots$					$\vdots$				
	S_{3n_3}					$\mathbf{y}'_{3n_3}$				

To test factors A, B, and AB, we use contrasts in the y's. An example of contrast matrices would be

$$\mathbf{A} = \begin{pmatrix} 2 & 2 & 2 & -1 & -1 & -1 & -1 & -1 & -1 \\ 0 & 0 & 0 & 1 & 1 & 1 & -1 & -1 & -1 \end{pmatrix} \tag{6.81}$$

$$\mathbf{B} = \begin{pmatrix} 2 & -1 & -1 & 2 & -1 & -1 & 2 & -1 & -1 \\ 0 & 1 & -1 & 0 & 1 & -1 & 0 & 1 & -1 \end{pmatrix} \tag{6.82}$$

$$\mathbf{G} = \begin{bmatrix} 4 & -2 & -2 & -2 & 1 & 1 & -2 & 1 & 1 \\ 0 & 2 & -2 & 0 & -1 & 1 & 0 & -1 & 1 \\ 0 & 0 & 0 & 2 & -1 & -1 & -2 & 1 & 1 \\ 0 & 0 & 0 & 0 & 1 & -1 & 0 & -1 & 1 \end{bmatrix}. \tag{6.83}$$

The rows of **A** are orthogonal contrasts with two comparisons:

$$A_1 \quad \text{vs.} \quad A_2 \text{ and } A_3$$

$$A_2 \quad \text{vs.} \quad A_3.$$

Similarly, **B** compares

$$B_1 \quad \text{vs.} \quad B_2 \text{ and } B_3$$

$$B_2 \quad \text{vs.} \quad B_3.$$

Other orthogonal (or linearly independent) contrasts could be used for A and B. The matrix $\mathbf{G}$ is for the AB interaction and is obtained from products of the rows of $\mathbf{A}$ and the rows of $\mathbf{B}$.

As before, we define $\overline{\mathbf{y}}_{..} = \sum_{ij} \mathbf{y}_{ij}/N$, $\mathbf{S}_{\text{pl}} = \mathbf{E}/\nu_E$, and $N = \sum_i n_i$. If there are k levels of C with mean vectors $\boldsymbol{\mu}_1, \boldsymbol{\mu}_2, \ldots, \boldsymbol{\mu}_k$, then $\overline{\boldsymbol{\mu}}_. = \sum_i \boldsymbol{\mu}_i/k$, and the A main effect corresponding to H_0: $\mathbf{A}\overline{\boldsymbol{\mu}}_. = \mathbf{0}$ can be tested with

$$T^2 = N(\mathbf{A}\overline{\mathbf{y}}_{..})'(\mathbf{A}\mathbf{S}_{\text{pl}}\mathbf{A}')^{-1}(\mathbf{A}\overline{\mathbf{y}}_{..}), \tag{6.84}$$

which is distributed as T^2_{2,ν_E} under H_0, where $\nu_E = \sum_i (n_i - 1)$. The dimension is 2, corresponding to the two rows of $\mathbf{A}$.

Similarly, to test H_0: $\mathbf{B}\overline{\boldsymbol{\mu}}_. = \mathbf{0}$ and H_0: $\mathbf{G}\overline{\boldsymbol{\mu}}_. = \mathbf{0}$ for the B main effect and the AB interaction, we have

$$T^2 = N(\mathbf{B}\overline{\mathbf{y}}_{..})'(\mathbf{B}\mathbf{S}_{\text{pl}}\mathbf{B}')^{-1}(\mathbf{B}\overline{\mathbf{y}}_{..}) \tag{6.85}$$

and

$$T^2 = N(\mathbf{G}\overline{\mathbf{y}}_{..})'(\mathbf{G}\mathbf{S}_{\text{pl}}\mathbf{G}')^{-1}(\mathbf{G}\overline{\mathbf{y}}_{..}), \tag{6.86}$$

which are distributed as T^2_{2,ν_E} and T^2_{4,ν_E}, respectively. In general, if factor A has a levels and factor B has b levels, then $\mathbf{A}$ has $a - 1$ rows, $\mathbf{B}$ has $b - 1$ rows, and $\mathbf{G}$ has $(a - 1)(b - 1)$ rows. The T^2-statistics are then distributed as T^2_{a-1,ν_E}, T^2_{b-1,ν_E}, and $T^2_{(a-1)(b-1),\nu_E}$, respectively.

Factors A, B, and AB can be tested with Wilks' Λ as well as T^2. Define $\mathbf{H}^* = N\overline{\mathbf{y}}_{..}\overline{\mathbf{y}}'_{..}$ from the partitioning $\sum_{ij} \mathbf{y}_{ij}\mathbf{y}'_{ij} = \mathbf{E} + \mathbf{H} + N\overline{\mathbf{y}}_{..}\overline{\mathbf{y}}'_{..}$. This can be used to test H_0: $\overline{\boldsymbol{\mu}}_. = \mathbf{0}$ (not usually a hypothesis of interest) by means of

$$\Lambda = \frac{|\mathbf{E}|}{|\mathbf{E} + \mathbf{H}^*|}, \tag{6.87}$$

which is $\Lambda_{p,1,\nu_E}$ if H_0 is true. Then the hypothesis of interest, H_0: $\mathbf{A}\overline{\boldsymbol{\mu}}_. = \mathbf{0}$, can be tested with

$$\Lambda = \frac{|\mathbf{A}\mathbf{E}\mathbf{A}'|}{|\mathbf{A}(\mathbf{E} + \mathbf{H}^*)\mathbf{A}'|}, \tag{6.88}$$

which is $\Lambda_{a-1,1,\nu_E}$ when H_0 is true, where a is the number of levels of factor A. There are similar expressions for testing factors B and AB. Note that the dimension of Λ in (6.88) is $a - 1$, because $\mathbf{AEA}'$ is $(a-1) \times (a-1)$.

The T^2 and Wilks Λ expressions in (6.84) and (6.87) are related by

$$\Lambda = \frac{\nu_E}{\nu_E + T^2} \tag{6.89}$$

$$T^2 = \nu_E \frac{1-\Lambda}{\Lambda}. \tag{6.90}$$

Factor C is tested exactly as factor B in Section 6.9.3. The hypothesis is

$$H_0\colon \mathbf{j}'\boldsymbol{\mu}_1 = \mathbf{j}'\boldsymbol{\mu}_2 = \cdots = \mathbf{j}'\boldsymbol{\mu}_k$$

as in (6.77), and we perform a univariate F-test on $z_{ij} = \mathbf{j}'\mathbf{y}_{ij}$ in a one-way ANOVA layout.

The AC, BC, and ABC interactions are tested as follows:

AC Interaction. The AC interaction is equivalent to the hypothesis

$$H_0\colon \mathbf{A}\boldsymbol{\mu}_1 = \mathbf{A}\boldsymbol{\mu}_2 = \cdots = \mathbf{A}\boldsymbol{\mu}_k,$$

which states that contrasts in factor A are the same across all levels of factor C. This can be tested by

$$\Lambda = \frac{|\mathbf{AEA}'|}{|\mathbf{A}(\mathbf{E}+\mathbf{H})\mathbf{A}'|},$$

which is distributed as Λ_{2,ν_H,ν_E}, where $a - 1 = 2$ is the number of rows of $\mathbf{A}$ and ν_H and ν_E are from the multivariate one-way model. Alternatively, the test can be carried out by transforming $\mathbf{y}_{ij}$ to $\mathbf{z}_{ij} = \mathbf{A}\mathbf{y}_{ij}$ and doing a one-way MANOVA on $\mathbf{z}_{ij}$.

BC Interaction. The hypothesis

$$H_0\colon \mathbf{B}\boldsymbol{\mu}_1 = \mathbf{B}\boldsymbol{\mu}_2 = \cdots = \mathbf{B}\boldsymbol{\mu}_k$$

is tested by

$$\Lambda = \frac{|\mathbf{BEB}'|}{|\mathbf{B}(\mathbf{E}+\mathbf{H})\mathbf{B}'|},$$

which is Λ_{2,ν_H,ν_E}, where $b - 1 = 2$; H_0 can also be tested by doing MANOVA on $\mathbf{z}_{ij} = \mathbf{B}\mathbf{y}_{ij}$.

ABC Interaction. The hypothesis

$$H_0\colon \mathbf{G}\boldsymbol{\mu}_1 = \mathbf{G}\boldsymbol{\mu}_2 = \cdots = \mathbf{G}\boldsymbol{\mu}_k$$

is tested by

$$\Lambda = \frac{|\mathbf{GEG}'|}{|\mathbf{G}(\mathbf{E}+\mathbf{H})\mathbf{G}'|},$$

which is Λ_{4,ν_H,ν_E}, or by doing MANOVA on $\mathbf{z}_{ij} = \mathbf{G}\mathbf{y}_{ij}$. In this case the dimension is $4 = (a-1)(b-1)$.

The above tests for AC, BC, or ABC can be also carried out with the other three MANOVA test statistics using eigenvalues of the appropriate matrices. For example, for AC we would use $(\mathbf{AEA}')^{-1}(\mathbf{AHA}')$.

Example 6.9.5. The data in Table 6.14 represent a repeated measures design with two within-subjects factors and one between-subjects factor (Timm 1980). Since A and B have three levels each, as in the illustration in this section, we will use the **A**, **B**, and **G** matrices in (6.81), (6.82), and (6.83). The **E** and **H** matrices are 9×9 and will not be shown. The overall mean vector is given by

$$\overline{\mathbf{y}}'_{..} = (46.45, 39.25, 31.70, 38.85, 45.40, 40.15, 34.55, 36.90, 39.15).$$

By (6.84), the test for factor A is

$$\begin{aligned} T^2 &= N(\mathbf{A}\overline{\mathbf{y}}_{..})'(\mathbf{AS}_{\text{pl}}\mathbf{A}')^{-1}(\mathbf{A}\overline{\mathbf{y}}_{..}) \\ &= 20(-.20, 13.80)\begin{pmatrix} 2138.4 & 138.6 \\ 138.6 & 450.4 \end{pmatrix}^{-1}\begin{pmatrix} -.20 \\ 13.80 \end{pmatrix} \\ &= 8.645 > T^2_{.05,2,18} = 7.606. \end{aligned}$$

For factor B, we use (6.85) to obtain

Table 6.14 Data from a Repeated Measures Experiment with Two Within-Subjects Factors and One Between-Subjects Factor

		Within-Subjects Factors								
Between-Subjects Factor	Subjects	A_1			A_2			A_3		
		B_1	B_2	B_3	B_1	B_2	B_3	B_1	B_2	B_3
C_1	S_{11}	20	21	21	32	42	37	32	32	32
	S_{12}	67	48	29	43	56	48	39	40	41
	S_{13}	37	31	25	27	28	30	31	33	34
	S_{14}	42	40	38	37	36	28	19	27	35
	S_{15}	57	45	32	27	21	25	30	29	29
	S_{16}	39	39	38	46	54	43	31	29	28
	S_{17}	43	32	20	33	46	44	42	37	31
	S_{18}	35	34	34	39	43	39	35	39	42
	S_{19}	41	32	23	37	51	39	27	28	30
	$S_{1,10}$	39	32	24	30	35	31	26	29	32
C_2	S_{21}	47	36	25	31	36	29	21	24	27
	S_{22}	53	43	32	40	48	47	46	50	54
	S_{23}	38	35	33	38	42	45	48	48	49
	S_{24}	60	51	41	54	67	60	53	52	50
	S_{25}	37	36	35	40	45	40	34	40	46
	S_{26}	59	48	37	45	52	44	36	44	52
	S_{27}	67	50	33	47	61	46	31	41	50
	S_{28}	43	35	27	32	36	35	33	33	32
	S_{29}	64	59	53	58	62	51	40	42	43
	$S_{2,10}$	41	38	34	41	47	42	37	41	46

$$
\begin{aligned}
T^2 &= N(\mathbf{B}\overline{\mathbf{y}}_{..})'(\mathbf{B}\mathbf{S}_{\text{pl}}\mathbf{B}')^{-1}(\mathbf{B}\overline{\mathbf{y}}_{..}) \\
&= 20(7.15, 10.55)\begin{pmatrix} 305.7 & 94.0 \\ 94.0 & 69.8 \end{pmatrix}^{-1}\begin{pmatrix} 7.15 \\ 10.55 \end{pmatrix} \\
&= 37.438 > T^2_{.01,2,18} = 12.943.
\end{aligned}
$$

By (6.86), the test for the AB interaction is given by

$$
\begin{aligned}
T^2 &= N(\mathbf{G}\overline{\mathbf{y}}_{..})'(\mathbf{G}\mathbf{S}_{\text{pl}}\mathbf{G}')^{-1}(\mathbf{G}\overline{\mathbf{y}}_{..}) \\
&= 61.825 > T^2_{.01,4,18} = 23.487.
\end{aligned}
$$

To test factor C, we carry out a one-way ANOVA on $z_{ij} = \mathbf{j}'\mathbf{y}_{ij}/9$:

Source	Sum of Squares	df	Mean Square	F
Between	3042.22	1	3042.22	8.54
Error	6408.98	18	356.05	

The observed F, 8.54, has a p-value of .0091 and is therefore significant.

The AC interaction is tested by (6.88) as

$$\Lambda = \frac{|\mathbf{AEA}'|}{|\mathbf{A}(\mathbf{E}+\mathbf{H})\mathbf{A}'|} = \frac{3.058 \times 10^8}{3.092 \times 10^8}$$
$$= .9889 > \Lambda_{.05,2,1,18} = .703.$$

For the BC interaction, we have

$$\Lambda = \frac{|\mathbf{BEB}'|}{|\mathbf{B}(\mathbf{E}+\mathbf{H})\mathbf{B}'|} = \frac{4.053 \times 10^6}{4.170 \times 10^6}$$
$$= .9718 > \Lambda_{.05,2,1,18} = .703.$$

For ABC, we obtain

$$\Lambda = \frac{|\mathbf{GEG}'|}{|\mathbf{G}(\mathbf{E}+\mathbf{H})\mathbf{G}'|} = \frac{2.643 \times 10^{12}}{2.927 \times 10^{12}}$$
$$= .9029 > \Lambda_{.05,4,1,18} = .551.$$

In summary, factors A, B, and C and the AB interaction are significant.

6.9.6 Repeated Measures with Two Within-Subjects Factors and Two Between-Subjects Factors

In this case we have a two-way MANOVA design in which each observation vector arises from a two-way factorial arrangement of treatments. This is illustrated in Table 6.15 for a balanced design with three levels of all factors. Each $\mathbf{y}_{ijk}$ has nine elements, consisting of responses to the nine treatment combinations $A_1B_1, A_2B_2, \ldots, A_3B_3$ (see Tables 6.9 and 6.13).

To test A, B, and AB, we can use the same contrast matrices as in (6.81)–(6.83). We define a grand mean vector $\overline{\mathbf{y}}_{\ldots} = \sum \mathbf{y}_{ijr}/N$, where N is the total number of observation vectors; in this illustration, $N = 27$. In general, $N = cdn$, where c and d are the number of levels of factors C and D and n is the number of replications in each cell (in the illustration, $n = 3$). The $\mathbf{E}$ matrix is obtained from the two-way MANOVA and has $\nu_E = cd(n-1)$ degrees of freedom. The test statistics for A, B, and AB are as follows, where $\mathbf{S} = \mathbf{E}/\nu_E$:

Factor A

$$T^2 = N(\mathbf{A}\overline{\mathbf{y}}_{\ldots})'(\mathbf{AS}_{\text{pl}}\mathbf{A}')^{-1}(\mathbf{A}\overline{\mathbf{y}}_{\ldots})$$

Table 6.15 Data Layout for Repeated Measures with Two Within-Subjects Factors and Two Between-Subjects Factors

Between-Subjects Factors			Within-Subjects Factors								
			A_1			A_2			A_3		
C	D	Subject	B_1	B_2	B_3	B_1	B_2	B_3	B_1	B_2	B_3
C_1	D_1	S_{111}					$\mathbf{y}'_{111}$				
		S_{112}					$\mathbf{y}'_{112}$				
		S_{113}					$\mathbf{y}'_{113}$				
	D_2	S_{121}					$\mathbf{y}'_{121}$				
		S_{122}					$\mathbf{y}'_{122}$				
		S_{123}					$\mathbf{y}'_{123}$				
	D_3	S_{131}					$\mathbf{y}'_{131}$				
		S_{132}					$\mathbf{y}'_{132}$				
		S_{133}					$\mathbf{y}'_{133}$				
C_2	D_1	S_{211}					$\mathbf{y}'_{211}$				
		S_{212}					$\mathbf{y}'_{212}$				
		S_{213}					$\mathbf{y}'_{213}$				
	D_2	S_{221}					$\mathbf{y}'_{221}$				
	D_3	⋮					⋮				
C_3	D_1	⋮					⋮				
	D_2	⋮					⋮				
	D_3	S_{333}					$\mathbf{y}'_{333}$				

is distributed as T^2_{a-1,ν_E}.

Factor B

$$T^2 = N(\mathbf{B}\overline{\mathbf{y}}_{...})'(\mathbf{B}\mathbf{S}_{\text{pl}}\mathbf{B}')^{-1}(\mathbf{B}\overline{\mathbf{y}}_{...})$$

is distributed as T^2_{b-1,ν_E}.

AB Interaction

$$T^2 = N(\mathbf{G}\overline{\mathbf{y}}_{...})'(\mathbf{G}\mathbf{S}_{\text{pl}}\mathbf{G}')^{-1}(\mathbf{G}\overline{\mathbf{y}}_{...})$$

is distributed as $T^2_{(a-1)(b-1),\nu_E}$.

To test factors C, D, and CD, we transform to $z_{ijr} = \mathbf{j}'\mathbf{y}_{ijr}$ and do univariate F-tests on a two-way ANOVA design.

To test factors AC, AD, and ACD, we perform a two-way MANOVA on $\mathbf{A}\mathbf{y}_{ijr}$. Then, the C main effect on $\mathbf{A}\mathbf{y}_{ijr}$ tests how the levels of C differ on $\mathbf{A}\mathbf{y}_{ijr}$, which is exactly what we want in the AC interaction. Similarly, the D

main effect on $\mathbf{A}\mathbf{y}_{ijr}$ yields the *AD* interaction, and the *CD* interaction on $\mathbf{A}\mathbf{y}_{ijr}$ gives the *ACD* interaction.

To test factors *BC*, *BD*, and *BCD*, we carry out a two-way MANOVA on $\mathbf{B}\mathbf{y}_{ijr}$. The *C* main effect on $\mathbf{B}\mathbf{y}_{ijr}$ gives the *BC* interaction, the *D* main effect on $\mathbf{B}\mathbf{y}_{ijr}$ yields the *BD* interaction, and the *CD* interaction on $\mathbf{B}\mathbf{y}_{ijr}$ corresponds to the *BCD* interaction.

Finally, to test factors *ABC*, *ABD*, and *ABCD*, we perform a two-way MANOVA on $\mathbf{G}\mathbf{y}_{ijr}$. Then the *C* main effect on $\mathbf{G}\mathbf{y}_{ijr}$ gives the *ABC* interaction, the *D* main effect on $\mathbf{G}\mathbf{y}_{ijr}$ yields the *ABD* interaction, and the *CD* interaction on $\mathbf{G}\mathbf{y}_{ijr}$ corresponds to the *ABCD* interaction.

6.9.7 Additional Topics

Wang (1983) and Timm (1980) give a method for obtaining univariate mixed-model sums of squares from the multivariate **E** and **H** matrices. Crepeau et al. (1985) consider repeated measures experiments with missing data. Federer (1986) discusses the planning of repeated measures designs, emphasizing such aspects as determining the length of treatment period, eliminating carry-over effects, the nature of pre- and posttreatment, the nature of a response to a treatment, treatment sequences, and the choice of a model. Vonesh (1986) discusses sample size requirements to achieve a given power level in repeated measures designs. Patel (1986) presents a model that accommodates both within- and between-subjects covariates in repeated measures designs. Jensen (1982) compares the efficiency and robustness of various procedures.

A *multivariate* or *multiresponse* repeated measurement design will result if more than one variable is measured on each subject at each treatment combination. Such designs are discussed by Timm (1980), Reinsel (1982), Wang (1983), and Thomas (1983). Bock (1975) refers to observations of this type as *doubly multivariate* data.

6.10 GROWTH CURVES

When the subject responds to a treatment or stimulus at successive time periods, the pattern of responses is often referred to as a *growth curve*. As in repeated measures experiments, subjects are usually human or animal. We consider estimation and testing hypotheses about the form of the response curve for a single sample in Section 6.10.1 and extend to growth curves for several samples in Section 6.10.2.

6.10.1 Growth Curve for One Sample

The data layout for a single sample growth curve experiment is analogous to Table 6.11, with the levels of factor *A* representing time periods. Thus we have a sample of n observation vectors $\mathbf{y}_1, \mathbf{y}_2, \ldots, \mathbf{y}_n$, for which we compute $\bar{\mathbf{y}}$ and $\mathbf{S}$. The usual approach is to approximate the shape of the growth curve by a polynomial function of time. If the time points are equally spaced, we can use orthogonal polynomials. This approach will be described first, followed by a method suitable for unequal time intervals.

Orthogonal polynomials are special contrasts that are often used in testing for linear, quadratic, cubic, and higher order trends in quantitative factors. For a complete description and derivation see Guttman (1982, pp. 194–207) or Morrison (1983, pp. 182–188). Here we give only a heuristic introduction to the method.

Suppose we administer a drug and measure a certain reaction at 3-minute intervals. Let $\mu_1, \mu_2, \mu_3, \mu_4$, and μ_5 designate the average responses at 0, 3, 6, 9, and 12 minutes, respectively. To test the hypothesis that there are no trends in the μ_i, we could make a "flatness" test of H_0: $\mu_1 = \mu_2 = \cdots = \mu_5$ using the contrast matrix

$$\mathbf{C} = \begin{bmatrix} -2 & -1 & 0 & 1 & 2 \\ 2 & -1 & -2 & -1 & 2 \\ -1 & 2 & 0 & -2 & 1 \\ 1 & -4 & 6 & -4 & 1 \end{bmatrix}.$$

The four rows of $\mathbf{C}$ are orthogonal polynomials that test for linear, quadratic, cubic, and quartic trends in the means. As noted in Section 6.9.2, any set of orthogonal contrasts in $\mathbf{C}$ will give the same value of T^2 to test H_0: $\mu_1 = \mu_2 = \cdots = \mu_5$. However, in this case we will be interested in using a subset of the rows of $\mathbf{C}$ to determine the shape of the response curve.

Table A.13 (Kleinbaum, Kupper, and Muller 1988) gives orthogonal polynomials up to $p = 10$. The $p-1$ entries for each value of p constitute the matrix $\mathbf{C}$. Some software programs will generate these automatically.

As with all orthogonal contrasts, the rows of $\mathbf{C}$ sum to zero and are mutually orthogonal. It is also apparent that the coefficients in each row increase and decrease in conformity with the desired pattern. Thus the first row, $(-2, -1, 0, 1, 2)$, increases steadily in a straight-line trend. The second row dips down and back up in a quadratic-type bend. The third row moves upward, drops back down, then heads back up in a cubic pattern with two bends. The fourth row bends three times in a quartic curve.

To further illustrate how the orthogonal polynomials pick up trends in the means, consider the three different patterns for $\boldsymbol{\mu}$ depicted in Figure 6.4, where $\boldsymbol{\mu}_a' = (8, 8, 8, 8, 8)$, $\boldsymbol{\mu}_b' = (20, 16, 12, 8, 4)$, and $\boldsymbol{\mu}_c' = (5, 12, 15, 12, 5)$. Let us denote the rows of $\mathbf{C}$ as $\mathbf{c}_1', \mathbf{c}_2', \mathbf{c}_3'$, and $\mathbf{c}_4'$. It is clear that $\mathbf{c}_i'\boldsymbol{\mu}_a = 0$ for $i = 1, 2, 3, 4$; that is, when H_0: $\mu_1 = \cdots = \mu_5$ is true, all four comparisons confirm it. If $\boldsymbol{\mu}$ has the pattern $\boldsymbol{\mu}_b$, only $\mathbf{c}_1'\boldsymbol{\mu}_b$ is nonzero. The other rows are not sensitive to a

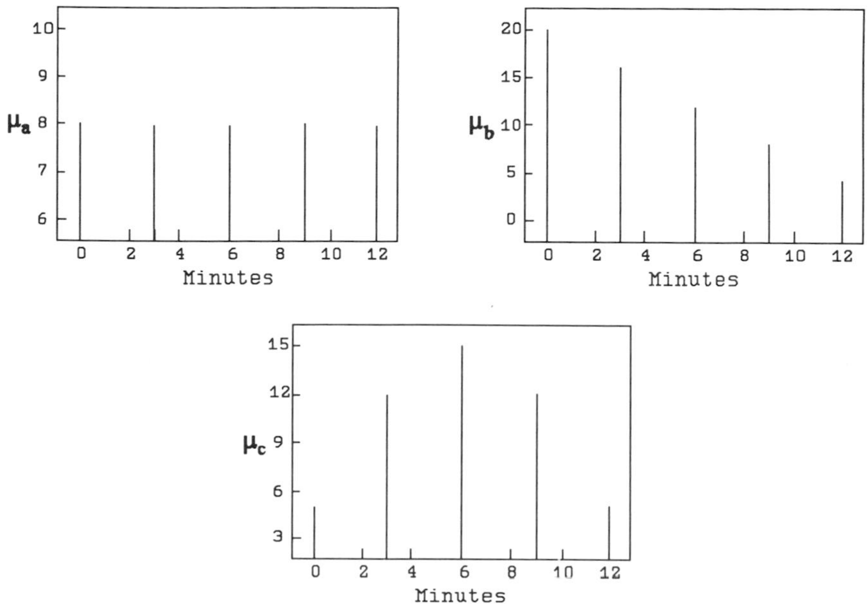

Figure 6.4 Three different patterns for $\boldsymbol{\mu}$.

linear pattern. We illustrate this for $\mathbf{c}_1'$ and $\mathbf{c}_2'$:

$$\mathbf{c}_1'\boldsymbol{\mu}_b = (-2)(20) + (-1)(16) + (0)(12) + (1)(8) + (2)(4) = -44$$
$$\mathbf{c}_2'\boldsymbol{\mu}_b = 2(20) - (16) - 2(12) - 8 + 2(4) = 0.$$

For $\boldsymbol{\mu}_c$, only $\mathbf{c}_2'\boldsymbol{\mu}_c$ is nonzero. For example,

$$\mathbf{c}_1'\boldsymbol{\mu}_c = -2(5) - (12) + 12 + 2(5) = 0$$
$$\mathbf{c}_2'\boldsymbol{\mu}_c = 2(5) - (12) - 2(15) - 12 + 2(5) = -19.$$

Thus the orthogonal polynomials do exactly what we want them to do. Each independently detects the type of curvature it is designed for and ignores other types. Of course real curves generally exhibit a mixture of more than one type of curvature, and in practice more than one orthogonal polynomial contrast may be significant.

To test hypotheses about the shape of the curve, we therefore use the rows of $\mathbf{C}$. Suppose we suspected a priori that there would be a combined linear and quadratic trend. Then we would partition $\mathbf{C}$ as follows:

$$\mathbf{C}_1 = \begin{pmatrix} -2 & -1 & 0 & 1 & 2 \\ 2 & -1 & -2 & -1 & 2 \end{pmatrix}$$
$$\mathbf{C}_2 = \begin{pmatrix} -1 & 2 & 0 & -2 & 1 \\ 1 & -4 & 6 & -4 & 1 \end{pmatrix}.$$

We would test H_0: $\mathbf{C}_1\boldsymbol{\mu} = \mathbf{0}$ by

$$T^2 = n(\mathbf{C}_1\overline{\mathbf{y}})'(\mathbf{C}_1\mathbf{S}\mathbf{C}_1')^{-1}(\mathbf{C}_1\overline{\mathbf{y}}),$$

which is distributed as $T^2_{2,n-1}$, where 2 is the number of rows of $\mathbf{C}_1$, n is the number of subjects in the sample, and $\overline{\mathbf{y}}$ and $\mathbf{S}$ are the mean vector and covariance matrix for the sample. Similarly, H_0: $\mathbf{C}_2\boldsymbol{\mu} = \mathbf{0}$ is tested by

$$T^2 = n(\mathbf{C}_2\overline{\mathbf{y}})'(\mathbf{C}_2\mathbf{S}\mathbf{C}_2')^{-1}(\mathbf{C}_2\overline{\mathbf{y}}),$$

which is $T^2_{2,n-1}$. In this case we might expect the first to reject H_0 and the second to accept H_0.

If we have no a priori expectations as to the shape of the curve, we could proceed as follows. Test the overall hypothesis H_0: $\mathbf{C}\boldsymbol{\mu} = \mathbf{0}$, and if rejected, test each of the rows of $\mathbf{C}$ separately as H_0: $\mathbf{c}_i'\boldsymbol{\mu} = 0$. The respective test statistics are

$$T^2 = n(\mathbf{C}\overline{\mathbf{y}})'(\mathbf{C}\mathbf{S}\mathbf{C}')^{-1}(\mathbf{C}\overline{\mathbf{y}}),$$

which is $T^2_{4,n-1}$, and

$$t_i = \frac{\mathbf{c}_i'\overline{\mathbf{y}}}{\sqrt{\mathbf{c}_i'\mathbf{S}\mathbf{c}_i/n}} \qquad i = 1, 2, 3, 4,$$

each of which is distributed as t_{n-1} (see Example 6.9.2).

In a case where p is large so that $\boldsymbol{\mu}$ has a large number of levels, say 10 or more, we would likely want to stop testing after the first four or five rows of $\mathbf{C}$ and test the remaining rows in one group. However, for larger values of p most tables of orthogonal polynomials give only the first few and omit those corresponding to higher degrees of curvature. We can find a matrix whose rows are orthogonal to the rows of a given matrix as follows. Suppose $p = 11$ so that $\mathbf{C}$ is 10×11 and $\mathbf{C}_1$ contains the first five orthogonal polynomials. Then a matrix $\mathbf{C}_2$, with rows orthogonal to those of $\mathbf{C}_1$, can be obtained by selecting five linearly independent rows of $\mathbf{I} - \mathbf{C}_1'(\mathbf{C}_1\mathbf{C}_1')^{-1}\mathbf{C}_1$, whose rows can easily be shown to be orthogonal to those of $\mathbf{C}_1$. The matrix $\mathbf{I} - \mathbf{C}_1'(\mathbf{C}_1\mathbf{C}_1')^{-1}\mathbf{C}_1$ is not full rank, and some care must be exercised in choosing linearly independent rows. However, if an incorrect choice of $\mathbf{C}_2$ is made, the computer algorithm should indicate this as it attempts to invert $\mathbf{C}_2\mathbf{S}\mathbf{C}_2'$ in

$$T^2 = n(\mathbf{C}_2\overline{\mathbf{y}})'(\mathbf{C}_2\mathbf{S}\mathbf{C}_2')^{-1}(\mathbf{C}_2\overline{\mathbf{y}}).$$

Alternatively, to check for significant curvature beyond the rows of $\mathbf{C}_1$ without finding $\mathbf{C}_2$, we can use the test for additional information in a subset of variables in Section 5.8. We need not find $\mathbf{C}_2$ in order to find the overall T^2, since, as noted in Section 6.9.2, any full rank $(p-1)\times p$ matrix $\mathbf{C}$ such that $\mathbf{C}\mathbf{j} = \mathbf{0}$ will give the same value in the overall T^2-test of H_0: $\mathbf{C}\boldsymbol{\mu} = \mathbf{0}$,

$$T^2 = n(\mathbf{C}\overline{\mathbf{y}})'(\mathbf{C}\mathbf{S}\mathbf{C}')^{-1}(\mathbf{C}\overline{\mathbf{y}}),$$

which is $T^2_{p-1,n-1}$. To obtain T^2, we can conveniently use a simple contrast matrix such as

$$\mathbf{C} = \begin{bmatrix} 1 & -1 & 0 & \cdots & 0 \\ 0 & 1 & -1 & \cdots & 0 \\ \vdots & \vdots & \vdots & & \vdots \\ 0 & 0 & 0 & \cdots & -1 \end{bmatrix}.$$

Let p_1 be the number of orthogonal polynomials of interest in $\mathbf{C}_1$ and p_2 be the number of rows of $\mathbf{C}$ not included in $\mathbf{C}_1$, so that $p_1 + p_2 = p - 1$. Then the test for the p_1 orthogonal polynomials in $\mathbf{C}_1$ is

$$T_1^2 = n(\mathbf{C}_1\overline{\mathbf{y}})'(\mathbf{C}_1\mathbf{S}\mathbf{C}_1')^{-1}(\mathbf{C}_1\overline{\mathbf{y}}),$$

which is $T^2_{p_1,n-1}$. We wish to compare T_1^2 to T^2 to check for significant curvature beyond the rows of $\mathbf{C}_1$. However, the test for additional information in a subset of variables in (5.31) was for the two-sample case. This can be adapted for the one-sample case as follows. The test for any curvature remaining after that accounted for in $\mathbf{C}_1$ is obtained by comparing

$$(n - p_1 - 1)\frac{T^2 - T_1^2}{n - 1 + T_1^2}$$

with $T^2_{\alpha,p_2,n-p_1-1}$.

We now describe an approach that can be used when the time points are not equally spaced. It may also be of interest in the equal-time-increment case because it provides an estimate of the response function.

Suppose we observe the response of the subject at p time points $t_1, t_2, \ldots, t_p$ and that the average response μ at any time point t is a polynomial in t of degree $k < p$:

$$\mu = \beta_0 + \beta_1 t + \beta_2 t^2 + \cdots + \beta_k t^k.$$

This holds for each point t_i and the corresponding average response μ_i. Thus our hypothesis becomes

$$H_0: \begin{bmatrix} \mu_1 = \beta_0 + \beta_1 t_1 + \beta_2 t_1^2 + \cdots + \beta_k t_1^k \\ \mu_2 = \beta_0 + \beta_1 t_2 + \beta_2 t_2^2 + \cdots + \beta_k t_2^k \\ \vdots \\ \mu_p = \beta_0 + \beta_1 t_p + \beta_2 t_p^2 + \cdots + \beta_k t_p^k \end{bmatrix}, \tag{6.91}$$

which, in matrix notation, is

$$H_0: \boldsymbol{\mu} = \mathbf{A}\boldsymbol{\beta}, \tag{6.92}$$

where

$$\mathbf{A} = \begin{bmatrix} 1 & t_1 & t_1^2 & \cdots & t_1^k \\ 1 & t_2 & t_2^2 & \cdots & t_2^k \\ \vdots & \vdots & \vdots & & \vdots \\ 1 & t_p & t_p^2 & \cdots & t_p^k \end{bmatrix} \quad \text{and} \quad \boldsymbol{\beta} = \begin{bmatrix} \beta_0 \\ \beta_1 \\ \vdots \\ \beta_k \end{bmatrix}.$$

In practice, it may be useful to transform the t_i's by subtracting the mean or the smallest value in order to reduce their size for computational purposes.

The following method of testing H_0 is due to Rao (1959, 1973). The model $\boldsymbol{\mu} = \mathbf{A}\boldsymbol{\beta}$ is similar to a regression model $E(\mathbf{y}) = \mathbf{X}\boldsymbol{\beta}$ (see Section 10.2.1). However, in this case, we have $\text{cov}(\mathbf{y}) = \boldsymbol{\Sigma}$ rather than $\sigma^2\mathbf{I}$, as in the standard regression assumption. In place of the usual regression approach of seeking $\hat{\boldsymbol{\beta}}$ to minimize $\text{SSE} = (\mathbf{y} - \mathbf{X}\hat{\boldsymbol{\beta}})'(\mathbf{y} - \mathbf{X}\hat{\boldsymbol{\beta}})$ [see (10.3) and (10.6)], we use a standardized distance as in (3.75), $(\bar{\mathbf{y}} - \mathbf{A}\hat{\boldsymbol{\beta}})'\mathbf{S}^{-1}(\bar{\mathbf{y}} - \mathbf{A}\hat{\boldsymbol{\beta}})$. The value of $\hat{\boldsymbol{\beta}}$ that minimizes $(\bar{\mathbf{y}} - \mathbf{A}\hat{\boldsymbol{\beta}})'\mathbf{S}^{-1}(\bar{\mathbf{y}} - \mathbf{A}\hat{\boldsymbol{\beta}})$ is

$$\hat{\boldsymbol{\beta}} = (\mathbf{A}'\mathbf{S}^{-1}\mathbf{A})^{-1}\mathbf{A}'\mathbf{S}^{-1}\bar{\mathbf{y}}, \tag{6.93}$$

and H_0: $\boldsymbol{\mu} = \mathbf{A}\boldsymbol{\beta}$ can be tested by

$$T^2 = n(\bar{\mathbf{y}} - \mathbf{A}\hat{\boldsymbol{\beta}})'\mathbf{S}^{-1}(\bar{\mathbf{y}} - \mathbf{A}\hat{\boldsymbol{\beta}}), \tag{6.94}$$

which is distributed as $T^2_{p-k-1,n-1}$. The dimension of T^2 is reduced from p to $p - k - 1$ because $k + 1$ parameters have been estimated in $\hat{\boldsymbol{\beta}}$. The T^2-statistic in (6.94) is usually given in the equivalent form

$$T^2 = n(\bar{\mathbf{y}}'\mathbf{S}^{-1}\bar{\mathbf{y}} - \bar{\mathbf{y}}'\mathbf{S}^{-1}\mathbf{A}\hat{\boldsymbol{\beta}}). \tag{6.95}$$

We can estimate the mean response at the ith time point,

$$\begin{aligned}\mu_i &= \beta_0 + \beta_1 t_i + \beta_2 t_i^2 + \cdots + \beta_k t_i^k \\ &= (1, t_i, t_i^2, \ldots, t_i^k)\boldsymbol{\beta} = \mathbf{a}_i'\boldsymbol{\beta},\end{aligned}$$

by

$$\hat{\mu}_i = \mathbf{a}_i'\hat{\boldsymbol{\beta}}. \tag{6.96}$$

Simultaneous confidence intervals for all possible $\mathbf{a}'\boldsymbol{\beta}$ are given by

$$\mathbf{a}'\hat{\boldsymbol{\beta}} \pm \frac{T_\alpha}{\sqrt{n}} \sqrt{\mathbf{a}'(\mathbf{A}'\mathbf{S}^{-1}\mathbf{A})^{-1}\mathbf{a}\left(1 + \frac{T^2}{n-1}\right)}, \tag{6.97}$$

where $T_\alpha = \sqrt{T^2_{\alpha,k+1,n-1}}$ is from Table A.7 and T^2 is given by (6.94) or (6.95). The intervals in (6.97) for $\mathbf{a}'\boldsymbol{\beta}$ include, of course, $\mathbf{a}_i'\boldsymbol{\beta}$ for the p rows of $\mathbf{A}$, that is, confidence intervals for the p time points. If these are all we are interested in, we can shorten the intervals in (6.97) by using a Bonferroni coefficient $t_{\alpha/2p}$ in place of T_α:

$$\mathbf{a}_i'\hat{\boldsymbol{\beta}} \pm \frac{t_{\alpha/2p}}{\sqrt{n}} \sqrt{\mathbf{a}_i'(\mathbf{A}'\mathbf{S}^{-1}\mathbf{A})^{-1}\mathbf{a}_i\left(1 + \frac{T^2}{n-1}\right)}, \tag{6.98}$$

where $t_{\alpha/2p} = t_{\alpha/2p,n-1}$. Bonferroni critical values $t_{\alpha/2p,\nu}$ are given in Table A.8. See procedures 2 and 3 in Section 5.5 for additional comments on the use of $t_{\alpha/2p}$ and T_α.

Example 6.10.1. Potthoff and Roy (1964) reported measurements in a dental study on boys and girls from ages 8 to 14. The data are given in Table 6.16.

To illustrate the methods of this section, we use the data for the boys alone. Later, (Example 6.10.2) we will compare the growth curves of the boys with those of the girls. We first test the overall hypothesis H_0: $\mathbf{C}\boldsymbol{\mu} = \mathbf{0}$, where $\mathbf{C}$

Table 6.16 Dental Measurements

	Girls' Age in Years					Boys' Age in Years			
Subject	8	10	12	14	Subject	8	10	12	14
1	21.0	20.0	21.5	23.0	1	26.0	25.0	29.0	31.0
2	21.0	21.5	24.0	25.5	2	21.5	22.5	23.0	26.5
3	20.5	24.0	24.5	26.0	3	23.0	22.5	24.0	27.5
4	23.5	24.5	25.0	26.5	4	25.5	27.5	26.5	27.0
5	21.5	23.0	22.5	23.5	5	20.0	23.5	22.5	26.0
6	20.0	21.0	21.0	22.5	6	24.5	25.5	27.0	28.5
7	21.5	22.5	23.0	25.0	7	22.0	22.0	24.5	26.5
8	23.0	23.0	23.5	24.0	8	24.0	21.5	24.5	25.5
9	20.0	21.0	22.0	21.5	9	23.0	20.5	31.0	26.0
10	16.5	19.0	19.0	19.5	10	27.5	28.0	31.0	31.5
11	24.5	25.0	28.0	28.0	11	23.0	23.0	23.5	25.0
					12	21.5	23.5	24.0	28.0
					13	17.0	24.5	26.0	29.5
					14	22.5	25.5	25.5	26.0
					15	23.0	24.5	26.0	30.0
					16	22.0	21.5	23.5	25.0

contains orthogonal polynomials for linear, quadratic, and cubic effects:

$$\mathbf{C} = \begin{pmatrix} -3 & -1 & 1 & 3 \\ 1 & -1 & -1 & 1 \\ -1 & 3 & -3 & 1 \end{pmatrix}.$$

From the 16 observation vectors we obtain

$$\overline{\mathbf{y}} = \begin{bmatrix} 22.88 \\ 23.81 \\ 25.72 \\ 27.47 \end{bmatrix} \qquad \mathbf{S} = \begin{bmatrix} 6.02 & 2.29 & 3.63 & 1.61 \\ 2.29 & 4.56 & 2.19 & 2.81 \\ 3.63 & 2.19 & 7.03 & 3.24 \\ 1.61 & 2.81 & 3.24 & 4.35 \end{bmatrix}.$$

To test H_0: $\mathbf{C}\boldsymbol{\mu} = \mathbf{0}$, we use

$$T^2 = n(\mathbf{C}\overline{\mathbf{y}})'(\mathbf{CSC}')^{-1}(\mathbf{C}\overline{\mathbf{y}}) = 77.957,$$

which exceeds $T^2_{.01,3,15} = 19.867$. We now test each row of $\mathbf{C}$ to determine the shape of the growth curve. For the linear effect, we use the first row, $\mathbf{c}_1'$:

$$t_1 = \frac{\mathbf{c}_1'\overline{\mathbf{y}}}{\sqrt{\mathbf{c}_1'\mathbf{S}\mathbf{c}_1/n}} = 7.722 > t_{.005,15} = 2.947.$$

The test of significance of the quadratic component using the second row yields

$$t_2 = \frac{\mathbf{c}_2'\overline{\mathbf{y}}}{\sqrt{\mathbf{c}_2'\mathbf{S}\mathbf{c}_2/n}} = 1.370 < t_{.025,15} = 2.131.$$

To test for a cubic trend, we use the third row of **C**:

$$t_3 = \frac{\mathbf{c}_3'\overline{\mathbf{y}}}{\sqrt{\mathbf{c}_3'\mathbf{S}\mathbf{c}_3/n}} = -.511 > -t_{.025,15} = -2.131.$$

Thus only the linear trend is needed to describe the growth curve.

To model the curve for each variable, we use (6.91),

$$\mu_i = \beta_0 + \beta_1 t_i$$

for μ_i and $t_i, i = 1, 2, 3, 4$, or

$$\boldsymbol{\mu} = \mathbf{A}\boldsymbol{\beta},$$

where

$$\mathbf{A} = \begin{bmatrix} 1 & -3 \\ 1 & -1 \\ 1 & 1 \\ 1 & 3 \end{bmatrix} \qquad \boldsymbol{\beta} = \begin{pmatrix} \beta_0 \\ \beta_1 \end{pmatrix}.$$

The values in the second column of **A** are obtained as $t = \text{age} - 11$. By (6.93), we obtain

$$\hat{\boldsymbol{\beta}} = (\mathbf{A}'\mathbf{S}^{-1}\mathbf{A})^{-1}\mathbf{A}'\mathbf{S}^{-1}\overline{\mathbf{y}} = \begin{pmatrix} 25.002 \\ .834 \end{pmatrix},$$

and our prediction equation is

$$\begin{aligned} \hat{\mu} &= 25.002 + .834t = 25.002 + .834(\text{age} - 11) \\ &= 15.828 + .834(\text{age}). \end{aligned}$$

6.10.2 Growth Curves for Several Samples

For the case of several samples or groups, the data layout would be similar to that in Table 6.9, where the p levels of factor A represent time points. If the time points are equally spaced, we can use orthogonal polynomials in the $(p-1)\times p$ contrast matrix $\mathbf{C}$ and express the hypothesis in the form H_0: $\mathbf{C}\boldsymbol{\mu}_{.} = \mathbf{0}$, where $\boldsymbol{\mu}_{.} = \sum_{i=1}^{k} \boldsymbol{\mu}_i/k$. As in Section 6.9.3, let us denote the sample mean vectors for the k groups as $\bar{\mathbf{y}}_{1.}, \bar{\mathbf{y}}_{2.}, \ldots, \bar{\mathbf{y}}_{k.}$, with grand mean $\bar{\mathbf{y}}_{..}$ and pooled covariance matrix $\mathbf{S}_{\text{pl}} = \mathbf{E}/\nu_E$. The overall test of H_0: $\mathbf{C}\boldsymbol{\mu}_{.} = \mathbf{0}$, no differences in means $\mu_1, \mu_2, \ldots, \mu_p$ averaged across groups, is

$$T^2 = N(\mathbf{C}\bar{\mathbf{y}}_{..})'(\mathbf{C}\mathbf{S}_{\text{pl}}\mathbf{C}')^{-1}(\mathbf{C}\bar{\mathbf{y}}_{..}),$$

which is T^2_{p-1,ν_E} as in (6.76), where $N = \sum_i n_i$ if unbalanced or $N = kn$ if balanced. The corresponding degrees of freedom for error is $\nu_E = N - k$ or $\nu_E = k(n-1)$. A test that the average (over groups) growth curve has a particular form can be tested with $\mathbf{C}_1$, containing a subset of the rows of $\mathbf{C}$,

$$T^2 = N(\mathbf{C}_1\bar{\mathbf{y}}_{..})'(\mathbf{C}_1\mathbf{S}_{\text{pl}}\mathbf{C}_1')^{-1}(\mathbf{C}_1\bar{\mathbf{y}}_{..}),$$

which is distributed as T^2_{r,ν_E}, where r is the number of rows in $\mathbf{C}_1$.

The growth curves for the groups can be compared by the interaction or parallelism test using either $\mathbf{C}$ or $\mathbf{C}_1$. We do a one-way MANOVA on $\mathbf{C}\mathbf{y}_{ij}$ or $\mathbf{C}_1\mathbf{y}_{ij}$, or equivalently calculate by (6.79),

$$\Lambda = \frac{|\mathbf{CEC}'|}{|\mathbf{C}(\mathbf{E}+\mathbf{H})\mathbf{C}'|} \quad \text{or} \quad \Lambda = \frac{|\mathbf{C}_1\mathbf{E}\mathbf{C}_1'|}{|\mathbf{C}_1(\mathbf{E}+\mathbf{H})\mathbf{C}_1'|},$$

which are distributed as $\Lambda_{p-1,k-1,\nu_E}$ and $\Lambda_{r,k-1,\nu_E}$, respectively.

Example 6.10.2. In Example 6.10.1, we found a linear trend for the growth curve for dental measurements of boys in Table 6.16. We now consider the growth curves for the combined group and also compare the girls' and boys' groups.

The two sample sizes are unequal and we use (6.26) to calculate the $\mathbf{E}$ matrix for the two groups,

$$\mathbf{E} = \begin{bmatrix} 135.39 & 67.88 & 97.76 & 67.76 \\ 67.88 & 103.76 & 72.86 & 82.71 \\ 97.76 & 72.86 & 161.39 & 103.27 \\ 67.76 & 82.71 & 103.27 & 124.64 \end{bmatrix},$$

from which we obtain $\mathbf{S}_{\text{pl}} = \mathbf{E}/\nu_E$. Using the $\mathbf{C}$ matrix of Example 6.10.1, we can test the overall hypothesis of equal means for the combined samples,

$$\begin{aligned} T^2 &= N(\mathbf{C}\overline{\mathbf{y}}_{..})'(\mathbf{C}\mathbf{S}_{\text{pl}}\mathbf{C}')^{-1}(\mathbf{C}\overline{\mathbf{y}}_{..}) \\ &= 118.136 > T^2_{.01,3,25} = 15.538. \end{aligned}$$

To test for a linear trend, we use the first row of $\mathbf{C}$ in

$$\begin{aligned} T^2 &= N(\mathbf{c}_1'\overline{\mathbf{y}}_{..})'(\mathbf{c}_1'\mathbf{S}_{\text{pl}}\mathbf{c}_1)^{-1}(\mathbf{c}_1'\overline{\mathbf{y}}_{..}) \\ &= 99.260 > T^2_{.01,1,25} = 7.770. \end{aligned}$$

This is, of course, the square of a t-statistic, but in the T^2 form it can be readily compared with the T^2 above using all three rows of $\mathbf{C}$. The linear trend is seen to dominate the relationship among the means.

We now compare the growth curves of the two groups. Using $\mathbf{C}$, we obtain

$$\begin{aligned} \Lambda &= \frac{|\mathbf{CEC}'|}{|\mathbf{C}(\mathbf{E}+\mathbf{H})\mathbf{C}'|} = \frac{1.3996 \times 10^8}{1.9025 \times 10^8} \\ &= .736 > \Lambda_{.05,3,1,25} = .717. \end{aligned}$$

For the linear trend, we have

$$\begin{aligned} \Lambda &= \frac{|\mathbf{c}_1'\mathbf{E}\mathbf{c}_1|}{|\mathbf{c}_1'(\mathbf{E}+\mathbf{H})\mathbf{c}_1|} = \frac{1184.2}{1427.9} \\ &= .829 < \Lambda_{.05,1,1,25} = .855. \end{aligned}$$

Thus the overall comparison does not reach significance, but the more specific comparison of linear trends does give a significant result.

6.10.3 Additional Topics

Jackson and Bryce (1981) presented methods of analyzing growth curves based on univariate linear models. Snee (1972) and Snee et al. (1979) proposed the use of eigenvalues and eigenvectors of a matrix derived from residuals after fitting the model. If one of the eigenvalues is dominant, certain simplifications result. Bryce (1980) discussed a similar simplification for the two-group case. Geisser (1980) and Fearn (1975, 1977) gave the Bayesian approach to growth curves, including estimation and prediction. Zerbe (1979a,b) provided a randomization test requiring fewer assumptions than normal-based tests.

6.11 TESTS ON A SUBVECTOR

6.11.1 Test for Additional Information

In Section 5.8, we considered tests of significance of the additional information in a subvector when comparing two groups. We now extend to several groups and will use similar notation.

Let **y** be a $p \times 1$ vector of measurements and **x** be a $q \times 1$ vector measured in addition to **y**. We are interested in determining whether **x** makes a significant contribution to the test of H_0: $\boldsymbol{\mu}_1 = \boldsymbol{\mu}_2 = \cdots = \boldsymbol{\mu}_k$ above and beyond **y**. Another way to phrase the question is: Can the separation of groups achieved by **x** be predicted from the separation achieved by **y**? It is not necessary, of course, that **x** represent new variables. It may be that $\binom{\mathbf{y}}{\mathbf{x}}$ is a partitioning of the present variables, and we wish to know if the variables in **x** can be deleted because they do not contribute to rejecting H_0.

We consider here only the one-way MANOVA, but the results could be extended to higher order designs, where various possibilities arise. In a two-way context, for example, it may happen that **x** contributes nothing to the A main effect but does contribute significantly to the B main effect.

It is assumed that we have k samples,

$$\begin{pmatrix} \mathbf{y}_{ij} \\ \mathbf{x}_{ij} \end{pmatrix} \qquad i = 1, 2, \ldots, k, \; j = 1, 2, \ldots, n,$$

from which we calculate

$$\mathbf{E} = \begin{pmatrix} \mathbf{E}_{yy} & \mathbf{E}_{yx} \\ \mathbf{E}_{xy} & \mathbf{E}_{xx} \end{pmatrix}$$

and

$$\mathbf{H} = \begin{pmatrix} \mathbf{H}_{yy} & \mathbf{H}_{yx} \\ \mathbf{H}_{xy} & \mathbf{H}_{xx} \end{pmatrix},$$

where **E** and **H** are $(p+q) \times (p+q)$ and $\mathbf{E}_{yy}$ and $\mathbf{H}_{yy}$ are $p \times p$.

Then

$$\Lambda(\mathbf{y}, \mathbf{x}) = \frac{|\mathbf{E}|}{|\mathbf{E} + \mathbf{H}|} \tag{6.99}$$

is distributed as $\Lambda_{p+q, \nu_H, \nu_E}$ and tests the significance of group separation using the full vector $\binom{\mathbf{y}}{\mathbf{x}}$. In the balanced one-way model, the degrees of freedom are

$\nu_H = k - 1$ and $\nu_E = k(n - 1)$. To test group separation using the reduced vector **y**, we can compute

$$\Lambda(\mathbf{y}) = \frac{|\mathbf{E}_{yy}|}{|\mathbf{E}_{yy} + \mathbf{H}_{yy}|}, \tag{6.100}$$

which is distributed as Λ_{p,ν_H,ν_E}.

To test the hypothesis that the extra variables in **x** do not contribute anything significant to separating the groups beyond the information already available in **y**, we calculate

$$\Lambda(\mathbf{x}|\mathbf{y}) = \frac{\Lambda(\mathbf{y}, \mathbf{x})}{\Lambda(\mathbf{y})}, \tag{6.101}$$

which is distributed as $\Lambda_{q,\nu_H,\nu_E-p}$. Note that the dimension is q, the number of x's. The error degrees of freedom of $\Lambda(\mathbf{x}|\mathbf{y})$ is $\nu_E - p$ because it has been adjusted for the p y's. Thus to test for the contribution of additional variables to separation of groups, we take the ratio of Wilks' Λ for the full set of variables in (6.99) to Wilks' Λ for the reduced set in (6.100). If the addition of **x** makes $\Lambda(\mathbf{y}, \mathbf{x})$ sufficiently smaller than $\Lambda(\mathbf{y})$, then $\Lambda(\mathbf{x}|\mathbf{y})$ will be small enough to reject. Thus (6.101) is a full-versus-reduced model test comparing $\Lambda(\mathbf{y}, \mathbf{x})$ and $\Lambda(\mathbf{y})$ to see if **x** makes a difference.

If we are interested in the effect of adding a single x, then $q = 1$, and (6.101) becomes

$$\Lambda(x|y_1, \ldots, y_p) = \frac{\Lambda(y_1, \ldots, y_p, x)}{\Lambda(y_1, \ldots, y_p)}, \tag{6.102}$$

which is distributed as $\Lambda_{1,\nu_H,\nu_E-p}$. Thus we are inquiring whether x reduces the overall Λ by a significant amount. With a dimension of 1, the Λ-statistic in (6.102) has an exact F transformation from Table 6.1,

$$F = \frac{1 - \Lambda}{\Lambda} \frac{\nu_E - p}{\nu_H}, \tag{6.103}$$

which is distributed as F_{ν_H,ν_E-p}. The statistic (6.102) is often referred to as a *partial Λ-statistic*, and correspondingly, (6.103) is called a *partial F-statistic*.

In (6.102) and (6.103), we have a test of the significance of a variable in the presence of the other variables. For a breakdown of precisely how the contribution of a variable depends on the other variables, see Rencher (1993).

We can rewrite (6.102) as

$$\Lambda(y_1,\ldots,y_p,x) = \Lambda(x|y_1,\ldots,y_p)\Lambda(y_1,\ldots,y_p) \leq \Lambda(y_1,\ldots,y_p), \qquad (6.104)$$

which shows that Wilks' Λ can only decrease with an additional variable.

Example 6.11.1. We use the rootstock data of Table 6.2 to illustrate tests on subvectors. From Example 6.1.7, we have, for all four variables,

$$\Lambda(y_1,y_2,y_3,y_4) = .1540.$$

To test the significance of y_3 and y_4 adjusted for y_1 and y_2, we compute Wilks' Λ for y_1 and y_2,

$$\Lambda(y_1,y_2) = .6990.$$

Then by (6.101),

$$\Lambda(y_3,y_4|y_1,y_2) = \frac{\Lambda(y_1,y_2,y_3,y_4)}{\Lambda(y_1,y_2)} = \frac{.1540}{.6990} = .2203,$$

which is less than the critical value $\Lambda_{.05,2,5,40} = .639$.

Similarly, the test for y_4 adjusted for y_1, y_2, and y_3 is given by (6.102) as

$$\begin{aligned}\Lambda(y_4|y_1,y_2,y_3) &= \frac{\Lambda(y_1,y_2,y_3,y_4)}{\Lambda(y_1,y_2,y_3)} = \frac{.1540}{.2460} \\ &= .6261 < \Lambda_{.05,1,5,39} = .759.\end{aligned}$$

For each of the other variables, we have a similar test:

$$y_3: \quad \Lambda(y_3|y_1,y_2,y_4) = \frac{.1540}{.2741} = .5618 < \Lambda_{.05,1,5,39} = .759$$

$$y_2: \quad \Lambda(y_2|y_1,y_3,y_4) = \frac{.1540}{.1922} = .8014 > \Lambda_{.05,1,5,39} = .759$$

$$y_1: \quad \Lambda(y_1|y_2,y_3,y_4) = \frac{.1540}{.1599} = .9630 > \Lambda_{.05,1,5,39} = .759.$$

Thus the two variables y_3 and y_4, either individually or together, contribute a significant amount to separation of the six groups.

6.11.2 Stepwise Selection of Variables

If there are no variables that we have a priori interest in testing for significance, we can do a data-directed search for the variables that best separate the groups. Such a strategy is often called *stepwise discriminant analysis*, although it could more aptly be called stepwise MANOVA. The procedure appears in many software packages.

We first describe an approach that is usually called *forward selection.* At the first step calculate $\Lambda(y_i)$ for each individual variable and choose the one with minimum $\Lambda(y_i)$ (or maximum associated F). At the second step calculate $\Lambda(y_i|y_1)$ for each of the $p - 1$ variables not entered at the first step, where y_1 indicates the first variable entered. For the second variable we choose the one with minimum $\Lambda(y_i|y_1)$ (or maximum associated partial F), that is, the variable that adds the maximum separation to the one entered at step 1. Denote the variable entered at step 2 by y_2. At the third step calculate $\Lambda(y_i|y_1, y_2)$ for each of the $p-2$ remaining variables and choose the one that minimizes $\Lambda(y_i|y_1, y_2)$ (or maximizes the associated partial F). Continue this process until the F falls below some predetermined threshold value, say F_{in}.

A *stepwise* procedure would follow a similar sequence, except that at each step the variables already selected would be reexamined to see if each still contributes a significant amount. The variable with smallest partial F will be removed if the partial F is less than a second threshold value, F_{out}. If $F_{\text{out}} = F_{\text{in}}$, there is a very small possibility that the procedure will cycle continuously without stopping. This possibility can be eliminated by using a value of F_{out} slightly less than F_{in}. For an illustration of the stepwise procedure, see Example 8.9.

PROBLEMS

6.1 Verify the computational forms given in (6.2) and (6.4); that is, show that

(a) $\sum_{ij}(y_{ij} - \bar{y}_{i.})^2 = \sum_{ij} y_{ij}^2 - \sum_i y_{i.}^2/n$

(b) $n\sum_i(\bar{y}_{i.} - \bar{y}_{..})^2 = \sum_i y_{i.}^2/n - y_{..}^2/kn$

6.2 Show that the eigenvalues of $\mathbf{E}^{-1}\mathbf{H}$ are the same as those of $(\mathbf{E}^{1/2})^{-1}\mathbf{H}(\mathbf{E}^{1/2})^{-1}$, as noted in Section 6.1.4, where $\mathbf{E}^{1/2}$ is the square root matrix defined in (2.103).

6.3 Show that if there is only one nonzero eigenvalue λ_1, then $U^{(1)}$, $V^{(1)}$, and Λ can be expressed in terms of θ, as in (6.27)–(6.29).

6.4 Show that (5.17), (5.19), and (5.20), which relate T^2 to Λ, $V^{(s)}$, and θ, follow from (6.27)–(6.29) and the relationship obtained in Section 6.1.7, $U^{(1)} = T^2/(n_1 + n_2 - 2)$.

6.5 Verify the computational forms of $\mathbf{H}$ and $\mathbf{E}$ shown following (6.25) and (6.26); that is, show that

(a) $\sum_{i=1}^{k} n_i(\overline{\mathbf{y}}_{i.} - \overline{\mathbf{y}}_{..})(\overline{\mathbf{y}}_{i.} - \overline{\mathbf{y}}_{..})' = \sum_{i=1}^{k} \mathbf{y}_{i.}\mathbf{y}_{i.}'/n_i - \mathbf{y}_{..}\mathbf{y}_{..}'/N$

(b) $\sum_{i=1}^{k} \sum_{j=1}^{n_i}(\mathbf{y}_{ij} - \overline{\mathbf{y}}_{i.})(\mathbf{y}_{ij} - \overline{\mathbf{y}}_{i.})' = \sum_{i=1}^{k} \sum_{j=1}^{n_i} \mathbf{y}_{ij}\mathbf{y}_{ij}' - \sum_{i=1}^{k} \mathbf{y}_{i.}\mathbf{y}_{i.}'/n_i$

6.6 Show that for two groups, $\mathbf{H} = \sum_{i=1}^{2} n_i(\overline{\mathbf{y}}_{i.}-\overline{\mathbf{y}}_{..})(\overline{\mathbf{y}}_{i.}-\overline{\mathbf{y}}_{..})'$ can be expressed as $\mathbf{H} = [n_1n_2/(n_1 + n_2)](\overline{\mathbf{y}}_{1.} - \overline{\mathbf{y}}_{2.})(\overline{\mathbf{y}}_{1.} - \overline{\mathbf{y}}_{2.})'$, thus verifying (6.31). Note that

$$\overline{\mathbf{y}}_{..} = \frac{n_1\overline{\mathbf{y}}_{1.} + n_2\overline{\mathbf{y}}_{2.}}{n_1 + n_2}.$$

6.7 Show that θ can be expressed as $\theta = \text{SSH}(z)/[\text{SSE}(z) + \text{SSH}(z)]$, as in Section 6.1.8.

6.8 Show that

$$\prod_{i=1}^{s} \frac{1}{1+\lambda_i} = \prod_{i=1}^{s} (1 - r_i^2),$$

as in Section 6.1.8, where $r_i^2 = \lambda_i/(1+\lambda_i)$.

6.9 Show that the F approximation based on A_P in (6.38) reduces to (6.21) if $A_P = V^{(s)}/s$, as in (6.37).

6.10 Show that if $s = 1$, A_{LH} in (6.39) reduces to (6.33).

6.11 Show that the F approximation denoted by F_3 at the end of Section 6.1.5 is equivalent to (6.40).

6.12 Show that $\text{cov}(\hat{\boldsymbol{\delta}}) = \frac{\boldsymbol{\Sigma}}{n} \sum_{i=1}^{k} c_i^2$, as noted following (6.48).

6.13 If $\mathbf{z}_{ij} = \mathbf{C}\mathbf{y}_{ij}$, where $\mathbf{C}$ is $(p-1)\times p$, show that $\mathbf{H}_z = \mathbf{CHC}'$ and $\mathbf{E}_z = \mathbf{CEC}'$, as used in (6.62).

6.14 Why do $\mathbf{C}$ and $\mathbf{C}'$ not "cancel out" of Wilks' Λ in (6.62)?

6.15 Show that under H_{03} and H_{01}, $\mathbf{C}\overline{\mathbf{y}}_{..}$ is $N_{p-1}(\mathbf{0}, \mathbf{C}\boldsymbol{\Sigma}\mathbf{C}'/kn)$, as noted preceding (6.67).

6.16 Show that $T^2 = kn(\mathbf{C}\overline{\mathbf{y}}_{..})'(\mathbf{CEC}'/\nu_E)^{-1}\mathbf{C}\overline{\mathbf{y}}_{..}$ in (6.67) is distributed as T^2_{p-1,ν_E}.

6.17 Give a justification of the Wilks' Λ test of H_0: $\overline{\boldsymbol{\mu}}_{.} = \mathbf{0}$

6.18 Show that the T^2- and Λ-tests in (6.84) and (6.88) are re
$\Lambda = \nu_E/(\nu_E + T^2)$.

6.19 Obtain T^2 in terms of Λ in (6.90) from (6.89).

6.20 Show that the rows of $\mathbf{C}_1$ are orthogonal to those of $\mathbf{A} = \mathbf{I} - \mathbf{C}_1'(\mathbf{C}_1\mathbf{C}_1')^{-1}\mathbf{C}_1$, as in Section 6.10.1.

6.21 Show that $\hat{\boldsymbol{\beta}}$ in (6.93) minimizes $(\overline{\mathbf{y}} - \mathbf{A}\hat{\boldsymbol{\beta}})'\mathbf{S}^{-1}(\overline{\mathbf{y}} - \mathbf{A}\hat{\boldsymbol{\beta}})$.

6.22 Show that T^2 in (6.95) is equivalent to T^2 in (6.94).

6.23 Baten, Tack, and Baeder (1958) compared judges' scores on fish prepared by three methods. Twelve fish were cooked by each method, and several judges tasted fish samples and rated each on four variables: y_1 = aroma, y_2 = flavor, y_3 = texture, and y_4 = moisture. The data are in Table 6.17. Each entry is an average score for the judges on that fish.

Table 6.17 Judges' Scores on Fish Prepared by Three Methods

Method 1				Method 2				Method 3			
y_1	y_2	y_3	y_4	y_1	y_2	y_3	y_4	y_1	y_2	y_3	y_4
5.4	6.0	6.3	6.7	5.0	5.3	5.3	6.5	4.8	5.0	6.5	7.0
5.2	6.2	6.0	5.8	4.8	4.9	4.2	5.6	5.4	5.0	6.0	6.4
6.1	5.9	6.0	7.0	3.9	4.0	4.4	5.0	4.9	5.1	5.9	6.5
4.8	5.0	4.9	5.0	4.0	5.1	4.8	5.8	5.7	5.2	6.4	6.4
5.0	5.7	5.0	6.5	5.6	5.4	5.1	6.2	4.2	4.6	5.3	6.3
5.7	6.1	6.0	6.6	6.0	5.5	5.7	6.0	6.0	5.3	5.8	6.4
6.0	6.0	5.8	6.0	5.2	4.8	5.4	6.0	5.1	5.2	6.2	6.5
4.0	5.0	4.0	5.0	5.3	5.1	5.8	6.4	4.8	4.6	5.7	5.7
5.7	5.4	4.9	5.0	5.9	6.1	5.7	6.0	5.3	5.4	6.8	6.6
5.6	5.2	5.4	5.8	6.1	6.0	6.1	6.2	4.6	4.4	5.7	5.6
5.8	6.1	5.2	6.4	6.2	5.7	5.9	6.0	4.5	4.0	5.0	5.9
5.3	5.9	5.8	6.0	5.1	4.9	5.3	4.8	4.4	4.2	5.6	5.5

Source: Baten, Tack, and Baeder (1958, p. 8).

(a) Compare the three methods using all four MANOVA tests.

(b) Compute the following measures of multivariate association from Section 6.1.8: $\eta_\Lambda^2, \eta_\theta^2, A_\Lambda, A_{\text{LH}}, A_P$.

(c) Based on the eigenvalues, is the essential dimensionality of the space containing the mean vectors equal to 1 or 2?

(d) Using contrasts, test the following two comparisons: 1 and 2 vs. 3 and 1 vs. 2.

(e) If any of the four tests in (a) is significant, run an ANOVA F-test on each y_i and examine the discriminant function $z = \mathbf{a}'\mathbf{y}$ (Section 6.4).

(f) Test the significance of y_3 and y_4 adjusted for y_1 and y_2.

(g) Test the significance of each variable adjusted for the other three.

6.24 Table 6.18 from Keuls et al. (1984) gives data from a two-way (fixed-effects) MANOVA on snap beans showing the results of four variables: y_1 = yield earliness, y_2 = specific leaf area (SLA) earliness, y_3 = total yield, and y_4 = average SLA. The factors are sowing date (S) and variety (V).

Table 6.18 Snapbean Data

S	V		y_1	y_2	y_3	y_4	S	V		y_1	y_2	y_3	y_4
1	1	1	59.3	4.5	38.4	295	3	1	1	68.1	3.4	42.2	280
		2	60.3	4.5	38.6	302			2	68.0	2.9	42.4	284
		3	60.9	5.3	37.2	318			3	68.5	3.3	41.5	286
		4	60.6	5.8	38.1	345			4	68.6	3.1	41.9	284
		5	60.4	6.0	38.8	325			5	68.6	3.3	42.1	268
1	2	1	59.3	6.7	37.9	275	3	2	1	64.0	3.6	40.9	233
		2	59.4	4.8	36.6	290			2	63.4	3.9	41.4	248
		3	60.0	5.1	38.7	295			3	63.5	3.7	41.6	244
		4	58.9	5.8	37.5	296			4	63.4	3.7	41.4	266
		5	59.5	4.8	37.0	330			5	63.5	4.1	41.1	244
1	3	1	59.4	5.1	38.7	299	3	3	1	68.0	3.7	42.3	293
		2	60.2	5.3	37.0	315			2	68.7	3.5	41.6	284
		3	60.7	6.4	37.4	304			3	68.7	3.8	40.7	277
		4	60.5	7.1	37.0	302			4	68.4	3.5	42.0	299
		5	60.1	7.8	36.9	308			5	68.6	3.4	42.4	285
2	1	1	63.7	5.4	39.5	271	4	1	1	69.8	1.4	48.4	265
		2	64.1	5.4	39.2	284			2	69.5	1.3	47.8	247
		3	63.4	5.4	39.0	281			3	69.5	1.3	46.9	231
		4	63.2	5.3	39.0	291			4	69.9	1.3	47.5	268
		5	63.2	5.0	39.0	270			5	70.3	1.1	47.1	247
2	2	1	60.6	6.8	38.1	248	4	2	1	66.6	1.8	45.7	205
		2	61.0	6.5	38.6	264			2	66.5	1.7	46.8	239
		3	60.7	6.8	38.8	257			3	67.1	1.7	46.3	230
		4	60.6	7.1	38.6	260			4	65.8	1.8	46.3	235
		5	60.3	6.0	38.5	261			5	65.6	1.9	46.1	220
2	3	1	63.8	5.7	40.5	282	4	3	1	70.1	1.7	48.1	253
		2	63.2	6.1	40.2	284			2	72.3	0.7	47.8	249
		3	63.3	6.0	40.0	291			3	69.7	1.5	46.7	226
		4	63.2	5.9	40.0	299			4	69.9	1.3	47.1	248
		5	63.1	5.4	39.7	295			5	69.8	1.4	46.7	236

(a) Test for main effects and interaction using all four MANOVA statistics.

(b) In previous experiments, the second variety gave higher yields. Compare variety 2 with 1 and 3 by means of a test on a contrast.

(c) Test linear, quadratic, and cubic contrasts for sowing date. Interpretation of these for mean vectors is not as straightforward as for univariate means.

(d) If any of the tests in part (a) rejects H_0, carry out ANOVA F-tests on the four variables.

(e) Test the significance of y_3 and y_4 adjusted for y_1 and y_2 in main effects and interaction.

(f) Test the significance of each variable adjusted for the other three in main effects and interaction.

6.25 The barsteel data in Table 6.6 was analyzed in Example 6.5.2 as a two-way fixed-effects design. Consider lubricants to be random so that we have a mixed model. Test for main effects and interaction.

6.26 In Table 6.19, we have a comparison of four reagents (Burdick 1979). The first reagent is the one presently in use and the other three are less expensive reagents that we wish to compare with the first. All four reagents are used with a blood sample from each patient. The three vari-

Table 6.19 Blood Data

	Reagent 1			Reagent 2			Reagent 3			Reagent 4		
Subject	y_1	y_2	y_3	y_1	y_2	y_3	y_1	y_2	y_3	y_1	y_2	y_3
1	8.0	3.96	12.5	8.0	3.93	12.7	7.9	3.86	13.0	7.9	3.87	13.2
2	4.0	5.37	16.9	4.2	5.35	17.2	4.1	5.39	17.2	4.0	5.35	17.3
3	6.3	5.47	17.1	6.3	5.39	17.5	6.0	5.39	17.2	6.1	5.41	17.4
4	9.4	5.16	16.2	9.4	5.16	16.7	9.4	5.17	16.7	9.1	5.16	16.7
5	8.2	5.16	17.0	8.0	5.13	17.5	8.1	5.10	17.4	7.8	5.12	17.5
6	11.0	4.67	14.3	10.7	4.60	14.7	10.6	4.52	14.6	10.5	4.58	14.7
7	6.8	5.20	16.2	6.8	5.16	16.7	6.9	5.13	16.8	6.7	5.19	16.8
8	9.0	4.65	14.7	9.0	4.57	15.0	8.9	4.58	15.0	8.6	4.55	15.1
9	6.1	5.22	16.3	6.0	5.16	16.9	6.1	5.14	16.9	6.0	5.21	16.9
10	6.4	5.13	15.9	6.4	5.11	16.4	6.4	5.11	16.4	6.3	5.07	16.3
11	5.6	4.47	13.3	5.5	4.45	13.6	5.3	4.46	13.6	5.3	4.44	13.7
12	8.2	5.22	16.0	8.2	5.14	16.5	8.0	5.14	16.5	7.8	5.16	16.5
13	5.7	5.10	14.9	5.6	5.05	15.3	5.5	5.02	15.4	5.4	5.05	15.5
14	9.8	5.25	16.1	9.8	5.15	16.6	8.1	5.10	13.8	9.4	5.16	16.6
15	5.9	5.28	15.8	5.8	5.25	16.4	5.7	5.26	16.4	5.6	5.29	16.2
16	6.6	4.65	12.8	6.4	4.59	13.2	6.3	4.58	13.1	6.4	4.57	13.2
17	5.7	4.42	14.5	5.5	4.31	14.9	5.5	4.30	14.9	5.4	4.32	14.8
18	6.7	4.38	13.1	6.5	4.32	13.4	6.5	4.32	13.6	6.5	4.31	13.5
19	6.8	4.67	15.6	6.6	4.57	15.8	6.5	4.55	16.0	6.5	4.56	15.9
20	9.6	5.64	17.0	9.5	5.58	17.5	9.3	5.50	17.4	9.2	5.46	17.5

ables measured for each reagent are y_1 = white blood count, y_2 = red blood count, and y_3 = hemoglobin count.

(a) Analyze as a randomized block design with subjects as blocks.

(b) Compare the first reagent with the other three using a contrast.

6.27 The data in Table 6.20 from Box (1950) show the amount of fabric wear y_1, y_2, and y_3 in three successive periods: (1) the first 1000 revolutions, (2) the second 1000 revolutions, and (3) the third 1000 revolutions of the abrasive wheel. There were three factors: type of abrasive surface, type of filler, and proportion of filler. There were two replications. Carry out a three-way MANOVA, testing for main effects and interactions. (Ignore the repeated measures aspects of the data.)

Table 6.20 Wear of Coated Fabrics in Three Periods (mg)

		Proportion of Filler								
		P_1 (25%)			P_2 (50%)			P_3 (75%)		
Surface Treatment	Filler	y_1	y_2	y_3	y_1	y_2	y_3	y_1	y_2	y_3
T_0	F_1	194	192	141	233	217	171	265	252	207
		208	188	165	241	222	201	269	283	191
	F_2	239	127	90	224	123	79	243	117	100
		187	105	85	243	123	110	226	125	75
T_1	F_1	155	169	151	198	187	176	235	225	166
		173	152	141	177	196	167	229	270	183
	F_2	137	82	77	129	94	78	155	76	91
		160	82	83	98	89	48	132	105	67

6.28 The fabric wear data in Table 6.20 can be considered to be a growth curve model, with the three periods (y_1, y_2, and y_3) representing repeated measurements on the same specimen. We thus have one within-subjects factor, to which we should assign polynomial contrasts $(-1, 0, 1)$ and $(-1, 2, -1)$, and a three-way between-subjects classification. Test for period and the interaction of period with the between-subjects factors and interactions.

6.29 Carry out a profile analysis on the fish data in Table 6.17, testing for parallelism, equal levels, and flatness.

6.30 Rao (1948) measured the weight of cork borings taken from the north (N), east (E), south (S), and west (W) directions of 28 trees. The data are given in Table 6.21. It is of interest to compare the bark thickness in the four directions. This can be done by analyzing the data as a one-sample repeated measures design. Since the primary comparison of interest is

Table 6.21 Weights of Cork Borings (cg) in Four Directions for 28 Trees

Tree	N	E	S	W	Tree	N	E	S	W
1	72	66	76	77	15	91	79	100	75
2	60	53	66	63	16	56	68	47	50
3	56	57	64	58	17	79	65	70	61
4	41	29	36	38	18	81	80	68	58
5	32	32	35	36	19	78	55	67	60
6	30	35	34	26	20	46	38	37	38
7	39	39	31	27	21	39	35	34	37
8	42	43	31	25	22	32	30	30	32
9	37	40	31	25	23	60	50	67	54
10	33	29	27	36	24	35	37	48	39
11	32	30	34	28	25	39	36	39	31
12	63	45	74	63	26	50	34	37	40
13	54	46	60	52	27	43	37	39	50
14	47	51	52	43	28	48	54	57	43

north and south vs. east and west, use the contrast matrix

$$\mathbf{C} = \begin{pmatrix} 1 & -1 & 1 & -1 \\ 1 & 0 & -1 & 0 \\ 0 & 1 & 0 & -1 \end{pmatrix}.$$

(a) Test H_0: $\mu_N = \mu_E = \mu_S = \mu_W$ using the entire matrix $\mathbf{C}$.

(b) If the test in (a) rejects H_0, test each row of $\mathbf{C}$

6.31 Analyze the glucose data in Table 3.10 as a one-sample repeated measures design with two within-subjects factors. Factor A is a comparison of fasting test vs. 1 hour posttest. The three levels of factor B are y_1 (and x_1), y_2 (and x_2), and y_3 (and x_3).

6.32 Table 6.22 gives survival times for cancer patients (Cameron and Pauling 1978; see also Andrews and Herzberg 1985, pp. 203–206). The factors in this two-way design are gender (1 = male, 2 = female) and type of cancer (1 = stomach, 2 = bronchus, 3 = colon, 4 = rectum, 5 = bladder, 6 = kidney). The variables (repeated measures) are y_1 = survival time (days) of patient treated with ascorbate measured from date of first hospital attendance, y_2 = mean survival time for the patient's 10 matched controls (untreated with ascorbate), y_3 = survival time after ascorbate treatment ceased, and y_4 = mean survival time after all treatment ceased for the patient's 10 matched controls. Analyze as a repeated measures design with one within-subjects factor (y_1, y_2, y_3, y_4) and a two-way (unbalanced) design between subjects. Since the two-way classification of subjects is unbalanced, you will need to use a program that allows for this or delete some observations to achieve a balanced design.

Table 6.22 Survival Times for Cancer Patients

Type of Cancer	Gender	Age	y_1	y_2	y_3	y_4
1	2	61	124	264	124	38
1	1	69	42	62	12	18
1	2	62	25	149	19	36
1	2	66	45	18	45	12
1	1	63	412	180	257	64
1	1	79	51	142	23	20
1	1	76	1112	35	128	13
1	1	54	46	299	46	51
1	1	62	103	85	90	10
1	1	46	146	361	123	52
1	1	57	340	269	310	28
1	2	59	396	130	359	55
2	1	74	81	72	74	33
2	1	74	461	134	423	18
2	1	66	20	84	16	20
2	1	52	450	98	450	58
2	2	48	246	48	87	13
2	2	64	166	142	115	49
2	1	70	63	113	50	38
2	1	77	64	90	50	24
2	1	71	155	30	113	18
2	1	39	151	260	38	34
2	1	70	166	116	156	20
2	1	70	37	87	27	27
2	1	55	223	69	218	32
2	1	74	138	100	138	27
2	1	69	72	315	39	39
2	1	73	245	188	231	65
3	2	76	248	292	135	18
3	2	58	377	492	50	30
3	1	49	189	462	189	65
3	1	69	1843	235	1267	17
3	2	70	180	294	155	57
3	2	68	537	144	534	16
3	1	50	519	643	502	25
3	2	74	455	301	126	21
3	1	66	406	148	90	17
3	2	76	365	641	365	42
3	2	56	942	272	911	40
3	2	74	372	37	366	28
3	1	58	163	199	156	31
3	2	60	101	154	99	28
3	1	77	20	649	20	33
3	1	38	283	162	274	80
4	2	56	185	422	62	38
4	2	75	479	82	226	10
4	2	57	875	551	437	62
4	1	56	115	140	85	13

Table 6.22 (*Continued*)

Type of Cancer	Gender	Age	y_1	y_2	y_3	y_4
4	1	68	362	106	122	36
4	1	54	241	645	198	80
4	1	59	2175	407	759	64
5	1	93	4288	464	260	29
5	2	70	3658	694	305	22
5	2	77	51	221	37	21
5	2	72	278	490	109	16
5	1	44	548	433	37	11
6	2	71	205	332	88	91
6	2	63	538	377	96	47
6	2	51	203	147	190	35
6	1	53	296	500	64	34
6	1	57	870	299	260	19
6	1	73	331	585	326	37
6	1	69	1685	1056	46	15

6.33 Analyze the ramus bone data of Table 3.8 as a one-sample growth curve design.

(a) Using a matrix **C** of orthogonal polynomial contrasts, test the hypothesis of overall equality of means, H_0: $\mathbf{C}\boldsymbol{\mu} = \mathbf{0}$.

(b) If thc overall hypothesis in (a) rejects H_0, find the degree of growth curve by testing each row of **C**.

6.34 Table 6.23 contains the weights of 13 male mice measured every 3 days from birth to weaning. The data set was reported and analyzed by

Table 6.23 Weights of 13 Male Mice Measured at Successive Intervals of 3 Days over 21 Days from Birth to Weaning

Mouse	Day 3	Day 6	Day 9	Day 12	Day 15	Day 18	Day 21
1	.190	.388	.621	.823	1.078	1.132	1.191
2	.218	.393	.568	.729	.839	.852	1.004
3	.211	.394	.549	.700	.783	.870	.925
4	.209	.419	.645	.850	1.001	1.026	1.069
5	.193	.362	.520	.530	.641	.640	.751
6	.201	.361	.502	.530	.657	.762	.888
7	.202	.370	.498	.650	.795	.858	.910
8	.190	.350	.510	.666	.819	.879	.929
9	.219	.399	.578	.699	.709	.822	.953
10	.225	.400	.545	.690	.796	.825	.836
11	.224	.381	.577	.756	.869	.929	.999
12	.187	.329	.441	.525	.589	.621	.796
13	.278	.471	.606	.770	.888	1.001	1.105

Williams and Izenman (1981) and by Izenman and Williams (1989) and has been further analyzed by Rao (1984, 1987) and by Lee (1988). Analyze as a one-sample growth curve design.

(a) Using a matrix **C** of orthogonal polynomial contrasts, test the hypothesis of overall equality of means, H_0: $\mathbf{C}\boldsymbol{\mu} = \mathbf{0}$.

(b) If the overall hypothesis in (a) rejects H_0, find the degree of growth curve by testing each row of **C**.

6.35 In Table 6.24, we have measurements of proportions of albumin at four time points on three groups of trout (Beauchamp and Hoel 1974).

Table 6.24 Measurements of Trout

	Time Point			
Group	1	2	3	4
1	.257	.288	.328	.358
1	.266	.282	.315	.464
1	.256	.303	.293	.261
1	.272	.456	.288	.261
2	.312	.300	.273	.253
2	.253	.220	.314	.261
2	.239	.261	.279	.224
2	.254	.243	.304	.254
3	.272	.279	.259	.295
3	.246	.292	.279	.302
3	.262	.311	.263	.264
3	.292	.261	.314	.244

(a) Using a matrix **C** of orthogonal contrasts, test the hypothesis of overall equality of means, H_0: $\mathbf{C}\boldsymbol{\mu}_. = \mathbf{0}$, for the combined samples, as in Section 6.10.2.

(b) If the overall hypothesis is rejected, find the degree of growth curve for the combined samples by testing each row of **C**.

(c) Compare the groups using the entire matrix **C**.

(d) Compare the groups using the row or rows of **C** found to be significant in (b).

6.36 Table 6.25 contains weight gains for three groups of rats (Box 1950). The variables are y_i = gain in ith week, $i = 1, 2, 3, 4$.
The groups are 1 = controls, 2 = thyroxin added to drinking water, and 3 = thiouracil added to drinking water.

(a) Using a matrix **C** of orthogonal contrasts, test the hypothesis of overall equality of means, H_0: $\mathbf{C}\boldsymbol{\mu}_. = \mathbf{0}$, for the combined samples, as in Section 6.10.2.

Table 6.25 Weekly Gains in Weight for 27 Rats

Group 1					Group 2					Group 3				
Rat	y_1	y_2	y_3	y_4	Rat	y_1	y_2	y_3	y_4	Rat	y_1	y_2	y_3	y_4
1	29	28	25	33	11	26	36	35	35	18	25	23	11	9
2	33	30	23	31	12	17	19	20	28	19	21	21	10	11
3	25	34	33	41	13	19	33	43	38	20	26	21	6	27
4	18	33	29	35	14	26	31	32	29	21	29	12	11	11
5	25	23	17	30	15	15	25	23	24	22	24	26	22	17
6	24	32	29	22	16	21	24	19	24	23	24	17	8	19
7	20	23	16	31	17	18	35	33	33	24	22	17	8	5
8	28	21	18	24						25	11	24	21	24
9	18	23	22	28						26	15	17	12	17
10	25	28	29	30						27	19	17	15	18

(b) If the overall hypothesis is rejected, find the degree of growth curve for the combined samples by testing each row of **C**.

(c) Compare the groups using the entire matrix **C**.

(d) Compare the groups using the row or rows of **C** found to be significant in (b).

6.37 Table 6.26 contains measurements of coronary sinus potassium at 2-minute intervals after coronary occlusion on four groups of dogs (Grizzle and Allen 1969).

Table 6.26 Coronary Sinus Potassium Measured at 2-minute Intervals on Dogs

	Time						
Group	1	3	5	7	9	11	13
1	4.0	4.0	4.1	3.6	3.6	3.8	3.1
1	4.2	4.3	3.7	3.7	4.8	5.0	5.2
1	4.3	4.2	4.3	4.3	4.5	5.8	5.4
1	4.2	4.4	4.6	4.9	5.3	5.6	4.9
1	4.6	4.4	5.3	5.6	5.9	5.9	5.3
1	3.1	3.6	4.9	5.2	5.3	4.2	4.1
1	3.7	3.9	3.9	4.8	5.2	5.4	4.2
1	4.3	4.2	4.4	5.2	5.6	5.4	4.7
1	4.6	4.6	4.4	4.6	5.4	5.9	5.6
2	3.4	3.4	3.5	3.1	3.1	3.7	3.3
2	3.0	3.2	3.0	3.0	3.1	3.2	3.1
2	3.0	3.1	3.2	3.0	3.3	3.0	3.0
2	3.1	3.2	3.2	3.2	3.3	3.1	3.1
2	3.8	3.9	4.0	2.9	3.5	3.5	3.4
2	3.0	3.6	3.2	3.1	3.0	3.0	3.0
2	3.3	3.3	3.3	3.4	3.6	3.1	3.1

Table 6.26 (*Continued*)

	Time						
Group	1	3	5	7	9	11	13
2	4.2	4.0	4.2	4.1	4.2	4.0	4.0
2	4.1	4.2	4.3	4.3	4.2	4.0	4.2
2	4.5	4.4	4.3	4.5	5.3	4.4	4.4
3	3.2	3.3	3.8	3.8	4.4	4.2	3.7
3	3.3	3.4	3.4	3.7	3.7	3.6	3.7
3	3.1	3.3	3.2	3.1	3.2	3.1	3.1
3	3.6	3.4	3.5	4.6	4.9	5.2	4.4
3	4.5	4.5	5.4	5.7	4.9	4.0	4.0
3	3.7	4.0	4.4	4.2	4.6	4.8	5.4
3	3.5	3.9	5.8	5.4	4.9	5.3	5.6
3	3.9	4.0	4.1	5.0	5.4	4.4	3.9
4	3.1	3.5	3.5	3.2	3.0	3.0	3.2
4	3.3	3.2	3.6	3.7	3.7	4.2	4.4
4	3.5	3.9	4.7	4.3	3.9	3.4	3.5
4	3.4	3.4	3.5	3.3	3.4	3.2	3.4
4	3.7	3.8	4.2	4.3	3.6	3.8	3.7
4	4.0	4.6	4.8	4.9	5.4	5.6	4.8
4	4.2	3.9	4.5	4.7	3.9	3.8	3.7
4	4.1	4.1	3.7	4.0	4.1	4.6	4.7
4	3.5	3.6	3.6	4.2	4.8	4.9	5.0

The groups are 1 = control dogs, 2 = dogs with extrinsic cardiac denervation 3 weeks prior to coronary occlusion, 3 = dogs with extrinsic cardiac denervation immediately prior to coronary occlusion, and 4 = dogs with bilateral thoracic sympathectomy and stellectomy 3 weeks prior to coronary occlusion.

(a) Using a matrix **C** of orthogonal contrasts, test the hypothesis of overall equality of means, H_0: $\mathbf{C}\boldsymbol{\mu}_{.} = \mathbf{0}$, for the combined samples, as in Section 6.10.2.

(b) If the overall hypothesis is rejected, find the degree of growth curve for the combined samples by testing each row of **C**.

(c) Compare the groups using the entire matrix **C**.

(d) Compare the groups using the row or rows of **C** found to be significant in (b).

6.38 Table 6.27 contains blood pressure measurements at intervals after inducing a heart attack for four groups of rats: group 1 is the controls and groups 2–4 have been exposed to halothane concentrations of .25%, .50%, 1.0%, respectively (Crepeau et al. 1985).

(a) Find the degree of growth curve for the combined sample using the methods in (6.91)–(6.96).

Table 6.27 Blood Pressure Data

	Number of Minutes after Ligation					
Group	1	5	10	15	30	60
1	112.5	100.5	102.5	102.5	107.5	107.5
1	92.5	102.5	105.0	100.0	110.0	117.5
1	132.5	125.0	115.0	112.5	110.0	110.0
1	102.5	107.5	107.5	102.5	90.0	112.5
1	110.0	130.0	115.0	105.0	112.5	110.0
1	97.5	97.5	80.0	82.5	82.5	102.5
1	90.0	70.0	85.0	85.0	92.5	97.5
2	115.0	115.0	107.5	107.5	112.5	107.5
2	125.0	125.0	120.0	120.0	117.5	125.0
2	95.0	90.0	95.0	90.0	100.0	107.5
2	87.5	65.5	85.0	90.0	105.0	90.0
2	90.0	87.5	97.5	95.0	100.0	95.0
2	97.5	92.5	57.5	55.0	90.0	97.5
2	107.5	107.5	145.0	110.0	105.0	112.5
2	102.5	130.0	85.0	80.0	127.5	97.5
3	107.5	107.5	102.5	102.5	102.5	97.5
3	97.5	108.5	94.5	102.5	102.5	107.5
3	100.0	105.0	105.0	105.0	110.0	110.0
3	95.0	95.0	90.0	100.0	100.0	100.0
3	85.0	92.5	92.5	92.5	90.0	110.0
3	82.5	77.5	75.0	65.5	65.0	72.5
3	62.5	75.0	115.0	110.0	100.0	100.0
4	70.0	67.5	67.5	77.5	77.5	77.5
4	45.0	37.5	45.0	45.0	47.5	45.0
4	52.5	22.5	90.0	65.0	60.0	65.5
4	100.0	100.0	100.0	100.0	97.5	92.5
4	115.0	110.0	100.0	110.0	105.0	105.0
4	97.5	97.5	97.5	105.0	95.0	92.5
4	95.0	125.0	130.0	125.0	115.0	117.5
4	72.5	87.5	65.0	57.5	92.5	82.5
4	105.0	105.0	105.0	105.0	102.5	100.0

(b) Repeat (a) for group 1.

(c) Repeat (a) for groups 2–4 combined.

6.39 Table 6.28 from Zerbe (1979a) compares 13 control and 20 obese patients on a glucose tolerance test using plasma inorganic phosphate. Delete the observations corresponding to $\frac{1}{2}$ and $1\frac{1}{2}$ hours so that the time points are equally spaced.

(a) For the control group, use orthogonal polynomials to find the degree of growth curve.

(b) Same as (a) for the obese group.

(c) Find the degree of growth curve for the combined groups, and compare the growth curves of the two groups.

Table 6.28 Plasma Inorganic Phosphate (mg/dl)

	Hours after Glucose Challenge							
Patient	0	$\frac{1}{2}$	1	$1\frac{1}{2}$	2	3	4	5
				Control				
1	4.3	3.3	3.0	2.6	2.2	2.5	3.4	4.4[a]
2	3.7	2.6	2.6	1.9	2.9	3.2	3.1	3.9
3	4.0	4.1	3.1	2.3	2.9	3.1	3.9	4.0
4	3.6	3.0	2.2	2.8	2.9	3.9	3.8	4.0
5	4.1	3.8	2.1	3.0	3.6	3.4	3.6	3.7
6	3.8	2.2	2.0	2.6	3.8	3.6	3.0	3.5
7	3.8	3.0	2.4	2.5	3.1	3.4	3.5	3.7
8	4.4	3.9	2.8	2.1	3.6	3.8	4.0	3.9
9	5.0	4.0	3.4	3.4	3.3	3.6	4.0	4.3
10	3.7	3.1	2.9	2.2	1.5	2.3	2.7	2.8
11	3.7	2.6	2.6	2.3	2.9	2.2	3.1	3.9
12	4.4	3.7	3.1	3.2	3.7	4.3	3.9	4.8
13	4.7	3.1	3.2	3.3	3.2	4.2	3.7	4.3
				Obese				
1	4.3	3.3	3.0	2.6	2.2	2.5	2.4	3.4[a]
2	5.0	4.9	4.1	3.7	3.7	4.1	4.7	4.9
3	4.6	4.4	3.9	3.9	3.7	4.2	4.8	5.0
4	4.3	3.9	3.1	3.1	3.1	3.1	3.6	4.0
5	3.1	3.1	3.3	2.6	2.6	1.9	2.3	2.7
6	4.8	5.0	2.9	2.8	2.2	3.1	3.5	3.6
7	3.7	3.1	3.3	2.8	2.9	3.6	4.3	4.4
8	5.4	4.7	3.9	4.1	2.8	3.7	3.5	3.7
9	3.0	2.5	2.3	2.2	2.1	2.6	3.2	3.5
10	4.9	5.0	4.1	3.7	3.7	4.1	4.7	4.9
11	4.8	4.3	4.7	4.6	4.7	3.7	3.6	3.9
12	4.4	4.2	4.2	3.4	3.5	3.4	3.9	4.0
13	4.9	4.3	4.0	4.0	3.3	4.1	4.2	4.3
14	5.1	4.1	4.6	4.1	3.4	4.2	4.4	4.9
15	4.8	4.6	4.6	4.4	4.1	4.0	3.8	3.8
16	4.2	3.5	3.8	3.6	3.3	3.1	3.5	3.9
17	6.6	6.1	5.2	4.1	4.3	3.8	4.2	4.8
18	3.6	3.4	3.1	2.8	2.1	2.4	2.5	3.5
19	4.5	4.0	3.7	3.3	2.4	2.3	3.1	3.3
20	4.6	4.4	3.8	3.8	3.8	3.6	3.8	3.8

[a]The similarity in the data for patient 1 in the control group and patient 1 in the obese group is coincidental.

6.40 Consider the complete data from Table 6.28 including the observations corresponding to $\frac{1}{2}$ and $1\frac{1}{2}$ hours. Use the methods in (6.91)–(6.96) for unequally spaced time points to analyze each group separately and the combined groups.

6.41 Table 6.29 contains mandible measurements. There were two groups of subjects. Each subject was measured at three time points y_1, y_2, and y_3 for each of three types of activator treatment (Timm 1980). Analyze as a repeated measures design with two within-subjects factors and one between-subjects factor. Use linear and quadratic contrasts for time (growth curve).

Table 6.29 Mandible Measurements

		Activator Treatment								
		1			2			3		
Group	Subject	y_1	y_2	y_3	y_1	y_2	y_3	y_1	y_2	y_3
1	1	117.0	117.5	118.5	59.0	59.0	60.0	10.5	16.5	16.5
	2	109.0	110.5	111.0	60.0	61.5	61.5	30.5	30.5	30.5
	3	117.0	120.0	120.5	60.0	61.5	62.0	23.5	23.5	23.5
	4	122.0	126.0	127.0	67.5	70.5	71.5	33.0	32.0	32.5
	5	116.0	118.5	119.5	61.5	62.5	63.5	24.5	24.5	24.5
	6	123.0	126.0	127.0	65.5	61.5	67.5	22.0	22.0	22.0
	7	130.5	132.0	134.5	68.5	69.5	71.0	33.0	32.5	32.0
	8	126.5	128.5	130.5	69.0	71.0	73.0	20.0	20.0	20.0
	9	113.0	116.5	118.0	58.0	59.0	60.5	25.0	25.0	24.5
2	1	128.0	129.0	131.5	67.0	67.5	69.0	24.0	24.0	24.0
	2	116.5	120.0	121.5	63.5	65.0	66.0	28.5	29.5	29.5
	3	121.5	125.5	127.0	64.5	67.5	69.0	26.5	27.0	27.0
	4	109.5	112.0	114.0	54.0	55.5	57.0	18.0	18.5	19.0
	5	133.0	136.0	137.5	72.0	73.5	75.5	34.5	34.5	34.5
	6	120.0	124.5	126.0	62.5	65.0	66.0	26.0	26.0	26.0
	7	129.5	133.5	134.5	65.0	68.0	69.0	18.5	18.5	18.5
	8	122.0	124.0	125.5	64.5	65.5	66.0	18.5	18.5	18.5
	9	125.0	127.0	128.0	65.5	66.5	67.0	21.5	21.5	21.6

CHAPTER 7

Tests on Covariance Matrices

7.1 INTRODUCTION

We now consider tests of hypotheses involving the variance–covariance structure. These tests are often carried out to check assumptions pertaining to other tests. In Sections 7.2–7.4, we cover three basic types of hypotheses: (1) the covariance matrix has a particular structure, (2) two or more covariance matrices are equal, and (3) certain elements of the covariance matrix are zero, thus implying independence of the corresponding (multivariate normal) random variables. In most cases we use the likelihood ratio approach. The resulting test statistics often involve the ratio of the determinants of the sample covariance matrix under the null hypothesis and under the alternative hypothesis.

7.2 TESTING A SPECIFIED PATTERN FOR $\boldsymbol{\Sigma}$

In this section, the discussion is in terms of a sample covariance matrix $\mathbf{S}$ from a single sample. However, the tests can be applied to a sample covariance matrix $\mathbf{S}_{\text{pl}} = \mathbf{E}/\nu_E$ obtained by pooling across several samples. To allow for either possibility, the degrees-of-freedom parameter has been indicated by ν. For a single sample, $\nu = n - 1$; for a pooled covariance matrix, $\nu = \sum_{i=1}^{k}(n_i - 1) = \sum_{i=1}^{k} n_i - k$.

7.2.1 Testing H_0: $\boldsymbol{\Sigma} = \boldsymbol{\Sigma}_0$

We begin with the basic hypothesis H_0: $\boldsymbol{\Sigma} = \boldsymbol{\Sigma}_0$ versus H_1: $\boldsymbol{\Sigma} \neq \boldsymbol{\Sigma}_0$. The hypothesized covariance matrix $\boldsymbol{\Sigma}_0$ is a target value for $\boldsymbol{\Sigma}$ or a nominal value from previous experience. Note that $\boldsymbol{\Sigma}_0$ is completely specified in H_0, whereas $\boldsymbol{\mu}$ is not specified.

To test H_0, we obtain a random sample of n observation vectors $\mathbf{y}_1, \mathbf{y}_2, \ldots, \mathbf{y}_n$ from $N_p(\boldsymbol{\mu}, \boldsymbol{\Sigma})$ and calculate $\mathbf{S}$. To see if $\mathbf{S}$ is significantly different from $\boldsymbol{\Sigma}_0$, we use the following test statistic, which is a modification of the likelihood ratio (Section 5.4.3):

$$u = \nu[\ln|\boldsymbol{\Sigma}_0| - \ln|\mathbf{S}| + \operatorname{tr}(\mathbf{S}\boldsymbol{\Sigma}_0^{-1}) - p], \tag{7.1}$$

where ν represents the degrees of freedom of $\mathbf{S}$, ln is the natural logarithm (base e), and tr is the trace of a matrix (Section 2.9). Note that if $\mathbf{S} = \boldsymbol{\Sigma}_0$, then $u = 0$, and that u increases with the "distance" between $\mathbf{S}$ and $\boldsymbol{\Sigma}_0$ [see (7.4) and the comment following].

When ν is large, the statistic u in (7.1) is approximately distributed as $\chi^2[\frac{1}{2}p(p+1)]$ if H_0 is true. For moderate size ν,

$$u' = \left[1 - \frac{1}{6\nu - 1}\left(2p + 1 - \frac{2}{p+1}\right)\right]u \tag{7.2}$$

is a better approximation to the $\chi^2[\frac{1}{2}p(p+1)]$ distribution. We reject H_0 if u or u' is greater than $\chi^2[\alpha, \frac{1}{2}p(p+1)]$. Note that the degrees of freedom for the χ^2-statistic, $\frac{1}{2}p(p+1)$, is the number of distinct parameters in $\boldsymbol{\Sigma}$.

We can express u in terms of the eigenvalues $\lambda_1, \lambda_2, \ldots, \lambda_p$ of $\mathbf{S}\boldsymbol{\Sigma}_0^{-1}$ by noting that $\operatorname{tr}(\mathbf{S}\boldsymbol{\Sigma}_0^{-1})$ and $\ln|\boldsymbol{\Sigma}_0| - \ln|\mathbf{S}|$ become

$$\begin{aligned}
\operatorname{tr}(\mathbf{S}\boldsymbol{\Sigma}_0^{-1}) &= \sum_{i=1}^{p} \lambda_i \qquad \text{[by (2.98)]} \\
\ln|\boldsymbol{\Sigma}_0| - \ln|\mathbf{S}| &= -\ln|\boldsymbol{\Sigma}_0|^{-1} - \ln|\mathbf{S}| \\
&= -\ln|\mathbf{S}\boldsymbol{\Sigma}_0^{-1}| \qquad \text{[by (2.83) and (2.84)]} \\
&= -\ln\left(\prod_{i=1}^{p} \lambda_i\right) \qquad \text{[by (2.99)]},
\end{aligned} \tag{7.3}$$

from which (7.1) can be written as

$$\begin{aligned}
u &= \nu\left[-\ln\left(\prod_{i=1}^{p} \lambda_i\right) + \sum_{i=1}^{p} \lambda_i - p\right] \\
&= \nu\left[\sum_{i=1}^{p} (\lambda_i - \ln\lambda_i) - p\right].
\end{aligned} \tag{7.4}$$

A plot of $y = x - \ln x$ will show that $x - \ln x \geq 1$ for all $x > 0$, with equality holding only for $x = 1$. Thus $\sum_{i=1}^{p}(\lambda_i - \ln\lambda_i) > p$ and $u > 0$.

The hypothesis that the variables are independent and have unit variance,

$$H_0: \boldsymbol{\Sigma} = \mathbf{I},$$

can be tested by simply setting $\boldsymbol{\Sigma}_0 = \mathbf{I}$ in (7.1).

7.2.2 Testing Sphericity

The hypothesis that the variables $y_1, y_2, \ldots, y_p$ in $\mathbf{y}$ are independent and have the same variance can be expressed as H_0: $\boldsymbol{\Sigma} = \sigma^2\mathbf{I}$ versus H_1: $\boldsymbol{\Sigma} \neq \sigma^2\mathbf{I}$, where σ^2 is the unknown common variance. This hypothesis is of interest in repeated measures (see Section 6.9.1). Under H_0, the ellipsoid $(\mathbf{y}-\boldsymbol{\mu})'\boldsymbol{\Sigma}^{-1}(\mathbf{y}-\boldsymbol{\mu}) = c^2$ reduces to $(\mathbf{y}-\boldsymbol{\mu})'(\mathbf{y}-\boldsymbol{\mu}) = \sigma^2c^2$, the equation of a sphere; hence the term *sphericity* is applied to the covariance structure $\boldsymbol{\Sigma} = \sigma^2\mathbf{I}$. An alternative sphericity hypothesis is H_0: $\mathbf{C}\boldsymbol{\Sigma}\mathbf{C}' = \sigma^2\mathbf{I}$, where $\mathbf{C}$ is any full-rank $(p-1) \times p$ matrix of orthonormal contrasts. This version of H_0 is also of interest in repeated measures (see Section 6.9.1).

For a random sample $\mathbf{y}_1, \mathbf{y}_2, \ldots, \mathbf{y}_n$ from $N(\boldsymbol{\mu}, \boldsymbol{\Sigma})$, the likelihood ratio for testing H_0: $\boldsymbol{\Sigma} = \sigma^2\mathbf{I}$ is

$$\text{LR} = \left[\frac{|\mathbf{S}|}{(\text{tr}\,\mathbf{S}/p)^p}\right]^{n/2}. \tag{7.5}$$

In some cases that we have considered previously, the likelihood ratio is a simple function of a test statistic such as F, T^2, Wilks' Λ, and so on. However, LR in (7.5) does not reduce to a standard statistic, and we resort to an approximation for its distribution. It has been shown that for a general likelihood ratio statistic LR,

$$-2\ln \text{LR} \quad \text{is approximately } \chi^2_\nu \tag{7.6}$$

for large n, where ν is the total number of parameters minus the number estimated under the restrictions imposed by H_0.

For the likelihood ratio statistic in (7.5), we obtain

$$-2\ln \text{LR} = -n\ln\left[\frac{|\mathbf{S}|}{(\text{tr}\,\mathbf{S}/p)^p}\right] = -n\ln u,$$

where

$$u = (\text{LR})^{2/n} = \frac{p^p|\mathbf{S}|}{(\text{tr}\,\mathbf{S})^p}. \tag{7.7}$$

By (2.98) and (2.99), u becomes

$$u = \frac{p^p \prod_{i=1}^{p} \lambda_i}{\left(\sum_{i=1}^{p} \lambda_i \right)^p}, \tag{7.8}$$

where $\lambda_1, \lambda_2, \ldots, \lambda_p$ are the eigenvalues of $\mathbf{S}$. An improvement over $-n \ln u$ is given by

$$u' = -\left(\nu - \frac{2p^2 + p + 2}{6p} \right) \ln u, \tag{7.9}$$

which has an approximate χ^2-distribution with $\frac{1}{2}p(p+1) - 1$ degrees of freedom. We reject H_0 if $u' \geq \chi^2[\alpha, \frac{1}{2}p(p+1) - 1]$. As noted above, the degrees of freedom in the χ^2 approximation is equal to the total number of parameters minus the number of parameters estimated under H_0. The number of parameters in $\boldsymbol{\Sigma}$ is $p + \binom{p}{2} = \frac{1}{2}p(p+1)$ and the loss of one degree of freedom is due to estimation of σ^2.

We see from (7.8) that if the sample λ_i's are all equal, $u = 1$ and $u' = 0$. Hence, this statistic also tests the hypothesis of equality of the population eigenvalues.

To test H_0: $\mathbf{C\Sigma C'} = \sigma^2\mathbf{I}$, use $\mathbf{CSC'}$ in place of $\mathbf{S}$ in (7.7) and use $p - 1$ in place of p in (7.7)–(7.9) and in the degrees of freedom for χ^2.

The likelihood ratio (7.5) was first given by Mauchly (1940), and his name is often associated with this test. Nagarsenker and Pillai (1973) gave the exact distribution of u and provided a table for $p = 4, 5, \ldots, 10$. Venables (1976) showed that u can be obtained by a union–intersection approach.

Example 7.2.2. We use the probe word data in Table 3.7 to illustrate tests of sphericity. The five variables appear to be commensurate, and the hypothesis H_0: $\mu_1 = \mu_2 = \cdots = \mu_5$ may be of interest. We would expect the variables to be correlated, and H_0 would ordinarily be tested using a multivariate approach, as in Sections 5.9.1 and 6.9.2. However, if $\boldsymbol{\Sigma} = \sigma^2\mathbf{I}$ or $\mathbf{C\Sigma C'} = \sigma^2\mathbf{I}$, then the hypothesis H_0: $\mu_1 = \mu_2 = \cdots = \mu_5$ can be tested with a univariate ANOVA F-test (see Section 6.9.1).

We first test H_0: $\boldsymbol{\Sigma} = \sigma^2\mathbf{I}$. The sample covariance matrix $\mathbf{S}$ was obtained in Example 3.9.1. By (7.7),

$$u = \frac{p^p|\mathbf{S}|}{(\operatorname{tr}\mathbf{S})^p} = \frac{5^5(27,236,586)}{(292.891)^5} = .0395.$$

Then by (7.9), with $n = 11$ and $p = 5$, we have

$$u' = -\left(n - 1 - \frac{2p^2 + p + 2}{6p} \right) \ln u = 26.177.$$

The approximate χ^2-test has $\frac{1}{2}p(p+1)-1 = 14$ degrees of freedom. We therefore compare $u' = 26.177$ with $\chi^2_{.05,14} = 23.68$ and reject H_0: $\mathbf{\Sigma} = \sigma^2\mathbf{I}$.

To test H_0: $\mathbf{C\Sigma C}' = \sigma^2\mathbf{I}$, we use the following matrix of orthonormalized contrasts:

$$\mathbf{C} = \begin{bmatrix} 4/\sqrt{20} & -1/\sqrt{20} & -1/\sqrt{20} & -1/\sqrt{20} & -1/\sqrt{20} \\ 0 & 3/\sqrt{12} & -1/\sqrt{12} & -1/\sqrt{12} & -1/\sqrt{12} \\ 0 & 0 & 2/\sqrt{6} & -1/\sqrt{6} & -1/\sqrt{6} \\ 0 & 0 & 0 & 1/\sqrt{2} & -1/\sqrt{2} \end{bmatrix}.$$

Then using $\mathbf{CSC}'$ in place of $\mathbf{S}$ and with $p - 1 = 4$ for the four rows of $\mathbf{C}$, we obtain

$$u = \frac{(p-1)^{p-1}|\mathbf{CSC}'|}{[\mathrm{tr}\,(\mathbf{CSC}')]^{p-1}} = \frac{4^4(144039.8)}{(93.6)^4} = .480$$

$$u' = 6.170.$$

For degrees of freedom, we now have $\frac{1}{2}(4)(5) - 1 = 9$, and the critical value is $\chi^2_{.05,9} = 16.92$. Hence, we do not reject H_0: $\mathbf{C\Sigma C}' = \sigma^2\mathbf{I}$, and a univariate F-test of H_0: $\mu_1 = \mu_2 = \cdots = \mu_5$ may be justified.

7.2.3 Testing H_0: $\mathbf{\Sigma} = \sigma^2[(1 - \rho)\mathbf{I} + \rho\mathbf{J}]$

In Section 6.9.1, it was noted that univariate ANOVA remains valid if

$$\mathbf{\Sigma} = \sigma^2 \begin{bmatrix} 1 & \rho & \rho & \cdots & \rho \\ \rho & 1 & \rho & \cdots & \rho \\ \vdots & \vdots & \vdots & & \vdots \\ \rho & \rho & \rho & \cdots & 1 \end{bmatrix} \tag{7.10}$$

$$= \sigma^2[(1 - \rho)\mathbf{I} + \rho\mathbf{J}], \tag{7.11}$$

where $\mathbf{J}$ is a square matrix of 1s, as defined in (2.12), and ρ is the population correlation between any two variables. This pattern of equal variances and equal covariances in $\mathbf{\Sigma}$ is variously referred to as *uniformity*, *compound symmetry*, or the *intraclass correlation model*.

We now consider the hypothesis that (7.10) holds:

$$H_0\colon \boldsymbol{\Sigma} = \begin{bmatrix} \sigma^2 & \sigma^2\rho & \cdots & \sigma^2\rho \\ \sigma^2\rho & \sigma^2 & \cdots & \sigma^2\rho \\ \vdots & \vdots & & \vdots \\ \sigma^2\rho & \sigma^2\rho & \cdots & \sigma^2 \end{bmatrix}.$$

From a sample we obtain the sample covariance matrix $\mathbf{S}$. Estimates of σ^2 and $\sigma^2\rho$ under H_0 are given by

$$s^2 = \frac{1}{p}\sum_{i=1}^{p} s_{ii} \quad \text{and} \quad s^2 r = \frac{1}{p(p-1)}\sum_{i \neq j} s_{ij}, \tag{7.12}$$

respectively, where s_{ii} and s_{ij} are from $\mathbf{S}$. Thus s^2 is an average of the variances on the diagonal of $\mathbf{S}$ and $s^2 r$ is an average of the off-diagonal covariances in $\mathbf{S}$. An estimate of ρ can be obtained as $r = s^2 r/s^2$. Using s^2 and $s^2 r$ in (7.12), the estimate of $\boldsymbol{\Sigma}$ under H_0 is then

$$\mathbf{S}_0 = \begin{bmatrix} s^2 & s^2 r & \cdots & s^2 r \\ s^2 r & s^2 & \cdots & s^2 r \\ \vdots & \vdots & & \vdots \\ s^2 r & s^2 r & \cdots & s^2 \end{bmatrix}.$$

To compare $\mathbf{S}$ and $\mathbf{S}_0$, we use the following function of the likelihood ratio:

$$u = \frac{|\mathbf{S}|}{|\mathbf{S}_0|}, \tag{7.13}$$

which can be expressed in the alternative form

$$u = \frac{|\mathbf{S}|}{(s^2)^p (1-r)^{p-1}[1+(p-1)r]}. \tag{7.14}$$

By analogy with (7.9), the test statistic is given by

$$u' = -\left[\nu - \frac{p(p+1)^2(2p-3)}{6(p-1)(p^2+p-4)}\right] \ln u, \tag{7.15}$$

which is approximately $\chi^2[\frac{1}{2}p(p+1)-2]$, where ν is the degrees of freedom of $\mathbf{S}$. We therefore reject H_0 if $u' > \chi^2[\alpha, \frac{1}{2}p(p+1)-2]$. Note that two degrees of freedom are lost due to estimation of σ^2 and ρ.

An alternative approximate test that is more precise when p is large and ν

is relatively small is given by

$$F = \frac{-(\gamma_2 - \gamma_2 c_1 - \gamma_1)\nu}{\gamma_1 \gamma_2} \ln u,$$

where

$$c_1 = \frac{p(p+1)^2(2p-3)}{6\nu(p-1)(p^2+p-4)} \qquad c_2 = \frac{p(p^2-1)(p+2)}{6\nu^2(p^2+p-4)}$$

$$\gamma_1 = \frac{1}{2}p(p+1) - 2 \qquad \gamma_2 = \frac{\gamma_1 + 2}{c_2 - c_1^2}.$$

We reject if $F > F_{\alpha, \gamma_1, \gamma_2}$.

Example 7.2.3. To illustrate this test, we use the cork data of Table 6.21. The main interest was in comparing average thickness, and hence weight, in the four directions. A standard ANOVA approach to this repeated measures design would be valid if (7.10) holds. To check this assumption, we test H_0: $\mathbf{\Sigma} = \sigma^2[(1-\rho)\mathbf{I} + \rho\mathbf{J}]$. The sample covariance matrix is given by

$$\mathbf{S} = \begin{bmatrix} 290.41 & 223.75 & 288.44 & 226.27 \\ 223.75 & 219.93 & 229.06 & 171.37 \\ 288.44 & 229.06 & 350.00 & 259.54 \\ 226.27 & 171.37 & 259.54 & 226.00 \end{bmatrix},$$

from which we obtain

$$|\mathbf{S}| = 25{,}617{,}563.28 \qquad s^2 = \frac{1}{p}\sum_{i=1}^{p} s_{ii} = 271.586$$

$$s^2 r = \frac{1}{p(p-1)}\sum_{i \neq j} s_{ij} = 233.072 \qquad r = \frac{s^2 r}{s^2} = \frac{233.072}{271.586} = .858.$$

From (7.14) and (7.15), we now have

$$u = \frac{25{,}617{,}563.28}{(271.586)^4(1-.858)^3[1+(3)(.858)]} = .461$$

$$u' = -\left[\nu - \frac{p(p+1)^2(2p-3)}{6(p-1)(p^2+p-4)}\right] \ln u$$
$$= \left[27 - \frac{(4)(25)(5)}{(6)(3)(16)}\right] 0.774 = 19.511.$$

Since $19.511 > \chi^2_{.05,8} = 15.5$, we reject H_0 and conclude that $\boldsymbol{\Sigma}$ does not have the pattern in (7.10).

7.3 TESTS COMPARING COVARIANCE MATRICES

An assumption for T^2 or MANOVA tests comparing two or more mean vectors is that the corresponding population covariance matrices are equal: $\boldsymbol{\Sigma}_1 = \boldsymbol{\Sigma}_2 = \cdots = \boldsymbol{\Sigma}_k$. Under this assumption, the sample covariance matrices $\mathbf{S}_1, \mathbf{S}_2, \ldots, \mathbf{S}_k$ reflect a common population $\boldsymbol{\Sigma}$ and are therefore pooled to obtain an estimate of $\boldsymbol{\Sigma}$. If $\boldsymbol{\Sigma}_1 = \boldsymbol{\Sigma}_2 = \cdots = \boldsymbol{\Sigma}_k$ is not true, large differences in $\mathbf{S}_1, \mathbf{S}_2, \ldots, \mathbf{S}_k$ may possibly lead to rejection of H_0: $\boldsymbol{\mu}_1 = \boldsymbol{\mu}_2 = \cdots = \boldsymbol{\mu}_k$. However, the T^2 and MANOVA tests are fairly robust to heterogeneity of covariance matrices as long as the sample sizes are large and equal. For other cases it is useful to have available a test of equality of covariance matrices. We begin with a review of the univariate case.

7.3.1 Univariate Tests of Equality of Variances

The two-sample univariate hypothesis H_0: $\sigma_1^2 = \sigma_2^2$ versus H_1: $\sigma_1^2 \neq \sigma_2^2$ is tested with

$$F = \frac{s_1^2}{s_2^2},$$

where s_1^2 and s_2^2 are the variances of the two samples. If H_0 is true, F is distributed as F_{ν_1,ν_2}, where ν_1 and ν_2 are the degrees of freedom of s_1^2 and s_2^2, respectively. Note that s_1^2 and s_2^2 must be independent, which will hold if the two samples are independent.

For the several-sample case, various procedures have been proposed. We present Bartlett's (1937) test of homogeneity of variances because it has been extended to the multivariate case. To test

$$H_0: \sigma_1^2 = \sigma_2^2 = \cdots = \sigma_k^2,$$

we calculate

$$c = 1 + \frac{1}{3(k-1)}\left[\sum_{i=1}^{k}\frac{1}{\nu_i} - \frac{1}{\sum_i \nu_i}\right], \qquad s^2 = \frac{\sum_{i=1}^{k}\nu_i s_i^2}{\sum_{i=1}^{k}\nu_i},$$

$$\text{and} \quad m = \left(\sum_{i=1}^{k}\nu_i\right)\ln s^2 - \sum_{i=1}^{k}\nu_i \ln s_i^2,$$

where $s_1^2, s_2^2, \ldots, s_k^2$ are independent sample variances with $\nu_1, \nu_2, \ldots, \nu_k$ degrees of freedom, respectively. Then

$$\frac{m}{c} \quad \text{is approximately} \quad \chi^2_{k-1}.$$

We reject H_0 if $m/c > \chi^2_{\alpha,k-1}$.

For an F approximation, we use c and m above and also calculate

$$a_1 = k - 1 \qquad a_2 = \frac{k+1}{(c-1)^2} \qquad b = \frac{a_2}{2 - c + 2/a_2}.$$

Then

$$F = \frac{a_2 m}{a_1(b-m)} \quad \text{is approximately} \quad F_{a_1,a_2}.$$

We reject H_0 if $F > F_\alpha$.

Note that an assumption for either form of the above test is independence of $s_1^2, s_2^2, \ldots, s_k^2$, which will hold for random samples from k distinct populations. This test would be inappropriate for comparing $s_{11}, s_{22}, \ldots, s_{pp}$ from the diagonal of $\mathbf{S}$, since the s_{ii} are correlated.

7.3.2 Multivariate Tests of Equality of Covariance Matrices

For k multivariate populations, the hypothesis of equality of covariance matrices is

$$H_0\colon \boldsymbol{\Sigma}_1 = \boldsymbol{\Sigma}_2 = \cdots = \boldsymbol{\Sigma}_k. \tag{7.16}$$

The test of $H_0\colon \boldsymbol{\Sigma}_1 = \boldsymbol{\Sigma}_2$ for two groups is easily obtained as a special case by setting $k = 2$. We assume independent samples of size $n_1, n_2, \ldots, n_k$ from multivariate normal distributions, as in an unbalanced one-way MANOVA, for example. To make the test, we calculate

$$M = \frac{|\mathbf{S}_1|^{\nu_1/2}|\mathbf{S}_2|^{\nu_2/2}\cdots|\mathbf{S}_k|^{\nu_k/2}}{|\mathbf{S}_{\text{pl}}|^{\sum_i \nu_i/2}}, \tag{7.17}$$

in which $\nu_i = n_i - 1$, $\mathbf{S}_i$ is the covariance matrix of the ith sample, and $\mathbf{S}_{\text{pl}}$ is the pooled sample covariance matrix

$$\mathbf{S}_{\text{pl}} = \frac{\sum_{i=1}^{k} \nu_i \mathbf{S}_i}{\sum_{i=1}^{k} \nu_i} = \frac{\mathbf{E}}{\nu_E}, \tag{7.18}$$

where $\mathbf{E}$ is given by (6.26) and $\nu_E = \sum_i \nu_i = \sum_i n_i - k$. It is clear that we must have every $\nu_i > p$; otherwise M would be zero because a determinant in its numerator would be zero. We can easily modify (7.17) and (7.18) to compare covariance matrices for the cells of a higher order design using $\nu_{ij} = n_{ij} - 1$.

The statistic M is a modification of the likelihood ratio and varies between 0 and 1, with values near 1 favoring the hypothesis and values near 0 leading to rejection. It is not immediately obvious that M in (7.17) behaves in this traditional way, and we offer the following heuristic argument. First we note that (7.17) can be expressed as

$$M = \left(\frac{|\mathbf{S}_1|}{|\mathbf{S}_{\text{pl}}|}\right)^{\nu_1/2} \left(\frac{|\mathbf{S}_2|}{|\mathbf{S}_{\text{pl}}|}\right)^{\nu_2/2} \cdots \left(\frac{|\mathbf{S}_k|}{|\mathbf{S}_{\text{pl}}|}\right)^{\nu_k/2}. \tag{7.19}$$

If $\mathbf{S}_1 = \mathbf{S}_2 = \cdots = \mathbf{S}_k = \mathbf{S}_{\text{pl}}$, then $M = 1$. As the disparity among $\mathbf{S}_1, \mathbf{S}_2, \ldots, \mathbf{S}_k$ increases, M approaches zero. To see this, note that the determinant of the pooled covariance matrix, $|\mathbf{S}_{\text{pl}}|$, lies somewhere near the "middle" of the $|\mathbf{S}_i|$'s and that as a set of variables $z_1, z_2, \ldots, z_n$ increases in spread, $z_{(1)}/\bar{z}$ reduces the product more than $z_{(n)}/\bar{z}$ increases it, where $z_{(1)}$ and $z_{(n)}$ are the minimum and maximum values, respectively. We illustrate this with the two sets of numbers 4, 5, 6 and 1, 5, 9, which have the same mean but different spread. If we assume $\nu_1 = \nu_2 = \nu_3 = \nu$, then for the first set,

$$M_1 = \left[\left(\frac{4}{5}\right)\left(\frac{5}{5}\right)\left(\frac{6}{5}\right)\right]^{\nu/2} = [(.8)(1)(1.2)]^{\nu/2} = (.96)^{\nu/2}$$

and for the second set,

$$M_2 = \left[\left(\frac{1}{5}\right)\left(\frac{5}{5}\right)\left(\frac{9}{5}\right)\right]^{\nu/2} = [(.2)(1)(1.8)]^{\nu/2} = (.36)^{\nu/2}.$$

In M_2, the smallest value, .2, reduces the product proportionally more than the largest value, 1.8, increases it. Another illustration is found in problem 7.8.

Box (1949, 1950) gave χ^2 and F approximations for the distribution of M. Either of these approximate tests is referred to as *Box's M-test.* For the χ^2 approximation, calculate

$$c_1 = \left[\sum_{i=1}^{k} \frac{1}{\nu_i} - \frac{1}{\sum_{i=1}^{k} \nu_i}\right]\left[\frac{2p^2 + 3p - 1}{6(p+1)(k-1)}\right]. \tag{7.20}$$

Then

$$u = -2(1 - c_1) \ln M \quad \text{is approximately} \quad \chi^2\left[\tfrac{1}{2}(k - 1)p(p + 1)\right], \tag{7.21}$$

where M is defined in (7.17). We reject H_0 if $u > \chi^2_\alpha$. If $\nu_1 = \nu_2 = \cdots = \nu_k = \nu$, then c_1 simplifies to

$$c_1 = \frac{(k + 1)(2p^2 + 3p - 1)}{6k\nu(p + 1)}. \tag{7.22}$$

For computational purposes, it may be preferable to use the following form of $\ln M$:

$$\ln M = \frac{1}{2}\sum_{i=1}^{k} \nu_i \ln |\mathbf{S}_i| - \frac{1}{2}\left(\sum_{i=1}^{k} \nu_i\right) \ln |\mathbf{S}_{\text{pl}}|. \tag{7.23}$$

To justify the degrees of freedom of the χ^2 approximation, note first that the total number of parameters estimated under H_1 is $k[\frac{1}{2}p(p + 1)]$, whereas under H_0 we estimate only $\boldsymbol{\Sigma}$, which has $p + \binom{p}{2} = \frac{1}{2}p(p + 1)$ parameters. The difference is $(k-1)\left[\frac{1}{2}p(p + 1)\right]$. Note that the quantity $k[\frac{1}{2}p(p+1)]$ arises from the assumption that all $\boldsymbol{\Sigma}_i$, $i = 1, 2, \ldots, k$, are different. Technically, H_1 can be stated as $\boldsymbol{\Sigma}_i \neq \boldsymbol{\Sigma}_j$ for some $i \neq j$. However, the most general case is all $\boldsymbol{\Sigma}_i$ different, and the distribution of M is derived accordingly.

For the F approximation, we use c_1 from (7.20) and calculate, additionally,

$$c_2 = \frac{(p-1)(p+2)}{6(k-1)} \left[\sum_{i=1}^{k} \frac{1}{\nu_i^2} - \frac{1}{\left(\sum_i \nu_i\right)^2} \right] \tag{7.24}$$

$$a_1 = \tfrac{1}{2}(k-1)p(p+1) \qquad a_2 = \frac{a_1+2}{|c_2 - c_1^2|}$$

$$b_1 = \frac{1 - c_1 - a_1/a_2}{a_1} \qquad b_2 = \frac{1 - c_1 - 2/a_2}{a_2}.$$

If $c_2 > c_1^2$,

$$F = -2b_1 \ln M \quad \text{is approximately} \quad F_{a_1,a_2}. \tag{7.25}$$

If $c_2 < c_1^2$,

$$F = -\frac{a_2 b_2 \ln M}{a_1(1 + 2b_2 \ln M)} \quad \text{is approximately} \quad F_{a_1,a_2}. \tag{7.26}$$

In either case, we reject H_0 if $F > F_\alpha$. If $\nu_1 = \nu_2 = \cdots = \nu_k = \nu$, then c_1 simplifies as in (7.22) and c_2 simplifies to

$$c_2 = \frac{(p-1)(p+2)(k^2+k+1)}{6k^2\nu^2}. \tag{7.27}$$

Exact upper percentage points of $-2 \ln M = \nu(k \ln |\mathbf{S}_{\text{pl}}| - \Sigma_i \ln |\mathbf{S}_i|)$ for the case of equal ν_i are given in Table A.14 for $p = 2, 3, 4, 5$ (Lee et al. 1977).

Box's M-test is calculated routinely in many computer programs for MANOVA. However, we may not wish to use it on a regular basis to check the heterogeneity assumption before performing T^2 or MANOVA tests. Olson (1974) showed that the M-test with equal ν_i may detect some forms of heterogeneity that have only minor effects on the MANOVA tests. The test is also sensitive to some forms of nonnormality. For example, it is sensitive to kurtosis for which the MANOVA tests are rather robust. Thus the M-test may signal covariance heterogeneity in some cases where it is not damaging to the MANOVA tests. Hence we may not wish to automatically rule out standard MANOVA tests if the M-test leads to rejection of H_0. Olson showed that the skewness and kurtosis statistics $b_{1,p}$ and $b_{2,p}$ (see Section 4.4.2) have similar shortcomings.

Example 7.3.2. We test the hypothesis H_0: $\mathbf{\Sigma}_1 = \mathbf{\Sigma}_2$ for the psychological data of Table 5.1. The covariance matrices $\mathbf{S}_1$, $\mathbf{S}_2$, and $\mathbf{S}_{\text{pl}}$ were given in Example 5.4.2. Using these, we obtain, by (7.23),

$$\begin{aligned} \ln M &= \tfrac{1}{2}\left[\nu_1 \ln |\mathbf{S}_1| + \nu_2 \ln |\mathbf{S}_2|\right] - \tfrac{1}{2}(\nu_1 + \nu_2) \ln |\mathbf{S}_{\text{pl}}| \\ &= \tfrac{1}{2}\left[(31)\ln(7917.7) + (31)\ln(58019.2)\right] \\ &\quad - \tfrac{1}{2}(31 + 31)\ln(26977.7) = -7.1324. \end{aligned}$$

For the χ^2 approximation, we compute, by (7.22) and (7.21),

$$c_1 = \frac{(2+1)[2(4^2) + 3(4) - 1]}{6(2)(31)(4+1)} = .06935$$

$$u = -2(1 - c_1)\ln M = 13.275 < \chi^2_{.05,10} = 18.307.$$

For an approximate F-test, we first calculate the following:

$$\begin{aligned} c_2 &= \frac{(4-1)(4+2)}{6(2-1)}\left[\frac{1}{31^2} + \frac{1}{31^2} - \frac{1}{(31+31)^2}\right] = .005463 \\ a_1 &= \frac{1}{2}(2 - 1)(4)(4 + 1) = 10 \\ a_2 &= \frac{10+2}{|.005463 - .06935^2|} = 18377.7 \\ b_1 &= \frac{1 - .06935 - 10/18377.7}{10} = .0930 \\ b_2 &= \frac{1 - .06935 - 2/18377.7}{18377.7} = 5.0634 \times 10^{-5}. \end{aligned}$$

Since $c_2 = .005463 > c_1^2 = .00481$, we use (7.25) to obtain

$$F = -2b_1 \ln M = 1.327 < F_{.05,10,\infty} = 1.83.$$

For an exact test, we compare

$$-2\ln M = 14.265$$

with 19.74, its critical value from Table A.14. Thus all three tests accept H_0.

7.4 TESTS OF INDEPENDENCE

7.4.1 Independence of Two Subvectors

Suppose the observation vector is partitioned into two subvectors of interest, which we label **y** and **x**, as in Section 3.8.1, where **y** is $p \times 1$ and **x** is $q \times 1$. The corresponding partitioning of the population covariance matrix is, by (3.44),

$$\boldsymbol{\Sigma} = \begin{pmatrix} \boldsymbol{\Sigma}_{yy} & \boldsymbol{\Sigma}_{yx} \\ \boldsymbol{\Sigma}_{xy} & \boldsymbol{\Sigma}_{xx} \end{pmatrix},$$

with analogous partitioning of **S** and **R** as in (3.40):

$$\mathbf{S} = \begin{pmatrix} \mathbf{S}_{yy} & \mathbf{S}_{yx} \\ \mathbf{S}_{xy} & \mathbf{S}_{xx} \end{pmatrix} \qquad \mathbf{R} = \begin{pmatrix} \mathbf{R}_{yy} & \mathbf{R}_{yx} \\ \mathbf{R}_{xy} & \mathbf{R}_{xx} \end{pmatrix}.$$

The hypothesis of independence of **y** and **x** can be expressed as

$$H_0\colon \boldsymbol{\Sigma} = \begin{pmatrix} \boldsymbol{\Sigma}_{yy} & \mathbf{O} \\ \mathbf{O} & \boldsymbol{\Sigma}_{xx} \end{pmatrix} \quad \text{or} \quad H_0\colon \boldsymbol{\Sigma}_{yx} = \mathbf{O}.$$

Thus independence of **y** and **x** means that every variable in **y** is independent of every variable in **x**. Note that there is no restriction on $\boldsymbol{\Sigma}_{yy}$ or $\boldsymbol{\Sigma}_{xx}$.

The likelihood ratio test statistic for $H_0\colon \boldsymbol{\Sigma}_{yx} = \mathbf{O}$ is given by

$$\Lambda = \frac{|\mathbf{S}|}{|\mathbf{S}_{yy}||\mathbf{S}_{xx}|} = \frac{|\mathbf{R}|}{|\mathbf{R}_{yy}||\mathbf{R}_{xx}|}, \tag{7.28}$$

which is distributed as $\Lambda_{p,q,n-1-q}$. We reject H_0 if $\Lambda \le \Lambda_\alpha$. Critical values for Wilks' Λ are given in Table A.9. The test statistic in (7.28) is equivalent (when H_0 is true) to the Λ-statistic (10.44) in Section 10.5.1 for testing the significance of the regression of **y** on **x**.

By the symmetry of

$$\frac{|\mathbf{S}|}{|\mathbf{S}_{yy}||\mathbf{S}_{xx}|} = \frac{|\mathbf{S}|}{|\mathbf{S}_{xx}||\mathbf{S}_{yy}|},$$

Λ in (7.28) is also distributed as $\Lambda_{q,p,n-1-p}$. This is equivalent to property 3 in Section 6.1.3.

Note that $|\mathbf{S}_{yy}||\mathbf{S}_{xx}|$ in (7.28) is an estimate of $|\boldsymbol{\Sigma}_{yy}||\boldsymbol{\Sigma}_{xx}|$, which by (2.85) is the determinant of $\boldsymbol{\Sigma}$ when $\boldsymbol{\Sigma}_{yx} = \mathbf{O}$. Thus Wilks' Λ compares an estimate of $\boldsymbol{\Sigma}$ without restrictions to an estimate of $\boldsymbol{\Sigma}$ under H_0: $\boldsymbol{\Sigma}_{yx} = \mathbf{O}$. We can see intuitively that $|\mathbf{S}| < |\mathbf{S}_{yy}||\mathbf{S}_{xx}|$ by noting from (2.86) that $|\mathbf{S}| = |\mathbf{S}_{xx}||\mathbf{S}_{yy} - \mathbf{S}_{yx}\mathbf{S}_{xx}^{-1}\mathbf{S}_{xy}|$, and since $\mathbf{S}_{yx}\mathbf{S}_{xx}^{-1}\mathbf{S}_{xy}$ is positive definite, $|\mathbf{S}_{yy} - \mathbf{S}_{yx}\mathbf{S}_{xx}^{-1}\mathbf{S}_{xy}| < |\mathbf{S}_{yy}|$.

Wilks' Λ in (7.28) can be expressed in terms of eigenvalues:

$$\Lambda = \prod_{i=1}^{s} (1 - r_i^2), \tag{7.29}$$

where $s = \min(p, q)$ and the r_i^2's are the nonzero eigenvalues of $\mathbf{S}_{xx}^{-1}\mathbf{S}_{xy}\mathbf{S}_{yy}^{-1}\mathbf{S}_{yx}$. We could also use $\mathbf{S}_{yy}^{-1}\mathbf{S}_{yx}\mathbf{S}_{xx}^{-1}\mathbf{S}_{xy}$, since the (nonzero) eigenvalues of $\mathbf{S}_{yy}^{-1}\mathbf{S}_{yx}\mathbf{S}_{xx}^{-1}\mathbf{S}_{xy}$ are the same as those of $\mathbf{S}_{xx}^{-1}\mathbf{S}_{xy}\mathbf{S}_{yy}^{-1}\mathbf{S}_{yx}$ (these two matrices are of the form **AB** and **BA**; see Section 2.11.4). The number of nonzero eigenvalues is $s = \min(p, q)$, since s is the rank of both $\mathbf{S}_{yy}^{-1}\mathbf{S}_{yx}\mathbf{S}_{xx}^{-1}\mathbf{S}_{xy}$ and $\mathbf{S}_{xx}^{-1}\mathbf{S}_{xy}\mathbf{S}_{yy}^{-1}\mathbf{S}_{yx}$. The eigenvalues are designated r_i^2 because they are the squared *canonical correlations* between $\mathbf{y}$ and $\mathbf{x}$ (see Chapter 11). In the special case when $p = 1$, (7.29) becomes

$$\Lambda = 1 - r_1^2 = 1 - R^2,$$

where R^2 is the square of the multiple correlation between y and $(x_1, x_2, \ldots, x_q)$.

The other test statistics, $U^{(s)}$, $V^{(s)}$, and Roy's, can also be defined in terms of the r_i^2's (see Section 11.4.1).

Example 7.4.1. Consider the diabetes data in Table 3.6. There is a natural partitioning in the variables, with y_1 and y_2 of minor interest and x_1, x_2, and x_3 of major interest. We test independence of the y's and the x's, that is, H_0: $\boldsymbol{\Sigma}_{yx} = \mathbf{O}$. From Example 3.8.1, the partitioned covariance matrix is

$$\mathbf{S} = \begin{pmatrix} \mathbf{S}_{yy} & \mathbf{S}_{yx} \\ \mathbf{S}_{xy} & \mathbf{S}_{xx} \end{pmatrix} = \left[\begin{array}{cc|ccc} .0162 & .2160 & .7872 & -.2138 & 2.189 \\ .2160 & 70.56 & 26.23 & -23.96 & -20.84 \\ \hline .7872 & 26.23 & 1106 & 396.7 & 108.4 \\ -.2138 & -23.96 & 396.7 & 2382 & 1143 \\ 2.189 & -20.84 & 108.4 & 1143 & 2136 \end{array}\right].$$

To make the test, we compute

$$\Lambda = \frac{|\mathbf{S}|}{|\mathbf{S}_{yy}||\mathbf{S}_{xx}|} = \frac{3.108 \times 10^9}{(1.095)(3.920 \times 10^9)} = .724 < \Lambda_{.05,2,3,40} = .730.$$

Thus we reject the hypothesis of independence.

7.4.2 Independence of Several Subvectors

Let there be k sets of variates so that $\mathbf{y}$ and $\boldsymbol{\Sigma}$ are partitioned as

$$\mathbf{y} = \begin{bmatrix} \mathbf{y}_1 \\ \mathbf{y}_2 \\ \vdots \\ \mathbf{y}_k \end{bmatrix} \quad \text{and} \quad \boldsymbol{\Sigma} = \begin{bmatrix} \boldsymbol{\Sigma}_{11} & \boldsymbol{\Sigma}_{12} & \cdots & \boldsymbol{\Sigma}_{1k} \\ \boldsymbol{\Sigma}_{21} & \boldsymbol{\Sigma}_{22} & \cdots & \boldsymbol{\Sigma}_{2k} \\ \vdots & \vdots & & \vdots \\ \boldsymbol{\Sigma}_{k1} & \boldsymbol{\Sigma}_{k2} & \cdots & \boldsymbol{\Sigma}_{kk} \end{bmatrix},$$

with p_i variables in $\mathbf{y}_i$, where $p_1 + p_2 + \cdots + p_k = p$. Note that $\mathbf{y}_1, \mathbf{y}_2, \ldots, \mathbf{y}_k$ represents a partitioning of $\mathbf{y}$, not a random sample of independent vectors. The hypothesis that the subvectors $\mathbf{y}_1, \mathbf{y}_2, \ldots, \mathbf{y}_k$ are mutually independent can be expressed as H_0: $\boldsymbol{\Sigma}_{ij} = \mathbf{O}$ for all $i \neq j$, or

$$H_0: \boldsymbol{\Sigma} = \begin{bmatrix} \boldsymbol{\Sigma}_{11} & \mathbf{O} & \cdots & \mathbf{O} \\ \mathbf{O} & \boldsymbol{\Sigma}_{22} & \cdots & \mathbf{O} \\ \vdots & \vdots & & \vdots \\ \mathbf{O} & \mathbf{O} & \cdots & \boldsymbol{\Sigma}_{kk} \end{bmatrix}. \tag{7.30}$$

The likelihood ratio statistic is

$$u = \frac{|\mathbf{S}|}{|\mathbf{S}_{11}||\mathbf{S}_{22}|\cdots|\mathbf{S}_{kk}|} \tag{7.31}$$

$$= \frac{|\mathbf{R}|}{|\mathbf{R}_{11}||\mathbf{R}_{22}|\cdots|\mathbf{R}_{kk}|}, \tag{7.32}$$

where $\mathbf{S}$ and $\mathbf{R}$ are obtained from a random sample of n observations and are partitioned as $\boldsymbol{\Sigma}$ above, conforming to $\mathbf{y}_1, \mathbf{y}_2, \ldots, \mathbf{y}_k$. Note that the denominator is the determinant of $\mathbf{S}$ restricted by H_0; that is, with $\mathbf{S}_{ij} = \mathbf{O}$ for all $i \neq j$. In this case, the statistic u does not have Wilks' Λ-distribution as it does in (7.28) when $k = 2$, but a good χ^2 approximation to its distribution is given by

$$u' = -\nu c \ln u, \tag{7.33}$$

where

$$c = 1 - \frac{1}{12f\nu}(2a_3 + 3a_2) \tag{7.34}$$

$$f = \frac{1}{2}a_2 \qquad a_2 = p^2 - \sum_{i=1}^{k} p_i^2 \qquad a_3 = p^3 - \sum_{i=1}^{k} p_i^3$$

and ν is the degrees of freedom of **S** or **R**. We reject the independence hypothesis if $u' > \chi^2_{\alpha,f}$.

The degrees of freedom, $\frac{1}{2}a_2$, arises from the following consideration. The number of parameters in $\boldsymbol{\Sigma}$ unrestricted by the hypothesis is $\frac{1}{2}p(p+1)$. Under the hypothesis (7.30), the number of parameters in each $\boldsymbol{\Sigma}_{ii}$ is $\frac{1}{2}p_i(p_i+1)$, for a total of $\frac{1}{2}\sum_{i=1}^{k} p_i(p_i+1)$. The difference is

$$\begin{aligned} f &= \frac{1}{2}p(p+1) - \frac{1}{2}\sum_{i=1}^{k} p_i(p_i+1) = \frac{1}{2}\left(p^2 + p - \sum_i p_i^2 - \sum_i p_i\right) \\ &= \frac{1}{2}\left(p^2 + p - \sum_i p_i^2 - p\right) = \frac{1}{2}\left(p^2 - \sum_i p_i^2\right) = \frac{a_2}{2}. \end{aligned}$$

Example 7.4.2 For 30 brands of Japanese Seishu wine, Siotani et al. (1963) studied the relationship between

$$\begin{aligned} y_1 &= \text{taste} \\ y_2 &= \text{odor} \end{aligned}$$

and

$$\begin{aligned} x_1 &= \text{pH} & x_5 &= \text{direct reducing sugar} \\ x_2 &= \text{acidity 1} & x_6 &= \text{total sugar} \\ x_3 &= \text{acidity 2} & x_7 &= \text{alcohol} \\ x_4 &= \text{sake meter} & x_8 &= \text{formyl-nitrogen.} \end{aligned}$$

The data are in Table 7.1.

Table 7.1 Seishu Measurements

y_1	y_2	x_1	x_2	x_3	x_4	x_5	x_6	x_7	x_8
1.0	.8	4.05	1.68	.85	3.0	3.97	5.00	16.90	122.0
.1	.2	3.81	1.39	.30	.6	3.62	4.52	15.80	62.0
.5	.0	4.20	1.63	.92	−2.3	3.48	4.46	15.80	139.0
.7	.7	4.35	1.43	.97	−1.6	3.45	3.98	15.40	150.0
−.1	−1.1	4.35	1.53	.87	−2.0	3.67	4.22	15.40	138.0
.4	.5	4.05	1.84	.95	−2.5	3.61	5.00	16.78	123.0
.2	−.3	4.20	1.61	1.09	−1.7	3.25	4.15	15.81	172.0
.3	−.1	4.32	1.43	.93	−5.0	4.16	5.45	16.78	144.0
.7	.4	4.21	1.74	.95	−1.5	3.40	4.25	16.62	153.0
.5	−.1	4.17	1.72	.92	−1.2	3.62	4.31	16.70	121.0
−.1	.1	4.45	1.78	1.19	−2.0	3.09	3.92	16.50	176.0
.5	−.5	4.45	1.48	.86	−2.0	3.32	4.09	15.40	128.0
.5	.8	4.25	1.53	.83	−3.0	3.48	4.54	15.55	126.0
.6	.2	4.25	1.49	.86	2.0	3.13	3.45	15.60	128.0
.0	−.5	4.05	1.48	.30	.0	3.67	4.52	15.38	99.0
−.2	−.2	4.22	1.64	.90	−2.2	3.59	4.49	16.37	122.8
.0	−.2	4.10	1.55	.85	1.8	3.02	3.62	15.31	114.0
.2	.2	4.28	1.52	.75	−4.8	3.64	4.93	15.77	125.0
−.1	−.2	4.32	1.54	.83	−2.0	3.17	4.62	16.60	119.0
.6	.1	4.12	1.68	.84	−2.1	3.72	4.83	16.93	111.0
.8	.5	4.30	1.50	.92	−1.5	2.98	3.92	15.10	68.0
.5	.2	4.55	1.50	1.14	.9	2.60	3.45	15.70	197.0
.4	.7	4.15	1.62	.78	−7.0	4.11	5.55	15.50	106.0
.6	−.3	4.15	1.32	.31	.8	3.56	4.42	15.40	49.5
−.7	−.3	4.25	1.77	1.12	.5	2.84	4.15	16.65	164.0
−.2	.0	3.95	1.36	.25	1.0	3.67	4.52	15.98	29.5
.3	−.1	4.35	1.42	.96	−2.5	3.40	4.12	15.30	131.0
.1	.4	4.15	1.17	1.06	−4.5	3.89	5.00	16.79	168.2
.4	.5	4.16	1.61	.91	−2.1	3.93	4.35	15.70	118.0
−.6	−.3	3.85	1.32	.30	.7	3.61	4.29	15.71	48.0

We test independence of the following four subsets of variables:

$$(y_1, y_2), (x_1, x_2, x_3), (x_4, x_5, x_6), (x_7, x_8).$$

The sample covariance matrix is

$$\mathbf{S} = \begin{bmatrix} \mathbf{S}_{11} & \mathbf{S}_{12} & \mathbf{S}_{13} & \mathbf{S}_{14} \\ \mathbf{S}_{21} & \mathbf{S}_{22} & \mathbf{S}_{23} & \mathbf{S}_{24} \\ \mathbf{S}_{31} & \mathbf{S}_{32} & \mathbf{S}_{33} & \mathbf{S}_{34} \\ \mathbf{S}_{41} & \mathbf{S}_{42} & \mathbf{S}_{43} & \mathbf{S}_{44} \end{bmatrix}$$

$$
= \left[\begin{array}{rr|rrr|rrr|rr}
.16 & .08 & .01 & .006 & .02 & -.03 & .02 & .01 & -.02 & .28 \\
.08 & .22 & -.01 & .003 & .01 & -.07 & .03 & .06 & .04 & -1.83 \\
\hline
.01 & -.01 & .03 & .004 & .03 & -.11 & -.03 & -.03 & -.01 & 4.73 \\
.006 & .003 & .004 & .024 & .020 & -.009 & -.009 & .0004 & .038 & 1.76 \\
.02 & .01 & .03 & .020 & .07 & -.18 & -.03 & -.03 & .05 & 8.97 \\
\hline
.03 & -.07 & -.11 & -.009 & -.18 & 4.67 & -.33 & -.63 & -.14 & -21.15 \\
.02 & .03 & -.03 & -.009 & -.03 & -.33 & .13 & .15 & .05 & -5.06 \\
.01 & .06 & -.03 & .0004 & -.03 & -.63 & .15 & .26 & .13 & -4.93 \\
\hline
-.02 & .04 & -.01 & .038 & .04 & -.14 & .05 & .13 & .35 & 3.43 \\
.28 & -1.83 & 4.73 & 1.76 & 8.97 & -21.15 & -5.06 & -4.93 & 3.43 & 1948
\end{array}\right]
$$

where $\mathbf{S}_{11}$ is 2×2, $\mathbf{S}_{22}$ is 3×3, $\mathbf{S}_{33}$ is 3×3, and $\mathbf{S}_{44}$ is 2×2. We first obtain

$$
\begin{aligned}
u &= \frac{|\mathbf{S}|}{|\mathbf{S}_{11}||\mathbf{S}_{22}||\mathbf{S}_{33}||\mathbf{S}_{44}|} \\
&= \frac{2.925 \times 10^{-7}}{(.0271)(.0000158)(.0344)(672.6)} = .02937.
\end{aligned}
$$

For the χ^2 approximation, we calculate

$$
\begin{aligned}
a_2 &= p^2 - \sum_{i=1}^{4} p_i^2 = 10^2 - (2^2 + 3^2 + 3^2 + 2^2) = 74 \\
a_3 &= p^3 - \sum_{i=1}^{4} p_i^3 = 930 \qquad f = \frac{1}{2} a_2 = 37 \\
c &= 1 - \frac{1}{12 f \nu}(2a_3 + 3a_2) = 1 - \frac{2(930) + 3(74)}{12(37)(29)} = .838.
\end{aligned}
$$

Then,

$$
u' = -\nu c \ln u = -(29)(.838) \ln(.02937) = 85.763,
$$

which exceeds $\chi^2_{.001,37} = 69.35$, and we reject the hypothesis of independence of the four subsets.

7.4.3 Test for Independence of All Variables

If all $p_i = 1$ in the hypothesis (7.30) in Section 7.4.2, we have a special case in which all the variables are mutually independent,

$$H_0: \mathbf{\Sigma} = \begin{bmatrix} \sigma_{11} & 0 & \cdots & 0 \\ 0 & \sigma_{22} & \cdots & 0 \\ \vdots & \vdots & & \vdots \\ 0 & 0 & \cdots & \sigma_{pp} \end{bmatrix}.$$

There is no restriction on the σ_{ii}. With $\sigma_{ij} = 0$ for all $i \neq j$, the corresponding ρ_{ij} are also 0, and an equivalent form of the hypothesis is H_0: $\mathbf{P}_\rho = \mathbf{I}$, where $\mathbf{P}_\rho$ is the population correlation matrix defined in (3.37).

Since all $p_i = 1$, the statistics (7.31) and (7.32) reduce to

$$u = \frac{|\mathbf{S}|}{s_{11}s_{22}\cdots s_{pp}} = |\mathbf{R}|, \tag{7.35}$$

and (7.33) becomes

$$u' = -\left[\nu - \tfrac{1}{6}(2p+5)\right] \ln u, \tag{7.36}$$

which has an approximate χ^2_f-distribution, where ν is the degrees of freedom of $\mathbf{S}$ or $\mathbf{R}$ and $f = \frac{1}{2}p(p-1)$ is the degrees of freedom of χ^2. We reject H_0 if $u' > \chi^2_{\alpha,f}$. Exact percentage points of u' for selected values of n and p are given in Table A.15 (Mathai and Katiyar 1979). Percentage points for the limiting χ^2-distribution are also given for comparison.

Note that $|\mathbf{R}|$ in (7.35) varies between 0 and 1. If the variables were uncorrelated, we would have $\mathbf{R} = \mathbf{I}$ and $|\mathbf{R}| = 1$. On the other hand, if two or more variables were linearly related, $\mathbf{R}$ would not be full rank and we would have $|\mathbf{R}| = 0$. If the variables were highly correlated, $|\mathbf{R}|$ would be close to 0; if the correlations were all small, $|\mathbf{R}|$ would be close to 1.

Example 7.4.3. To test the hypothesis H_0: $\sigma_{ij} = 0$, $i \neq j$, for the probe word data from Table 3.7, we calculate

$$\mathbf{R} = \begin{bmatrix} 1.000 & .614 & .757 & .575 & .413 \\ .614 & 1.000 & .547 & .750 & .548 \\ .757 & .547 & 1.000 & .605 & .692 \\ .575 & .750 & .605 & 1.000 & .524 \\ .413 & .548 & .692 & .524 & 1.000 \end{bmatrix}.$$

Then by (7.35) and (7.36),

$$u = |\mathbf{R}| = .0409$$
$$u' = -[n - 1 - \tfrac{1}{6}(2p+5)] \ln u = 23.97.$$

The exact .01 critical value for u' from Table A.15 is 23.75, and we therefore reject H_0. The approximate χ^2 critical value for u' is $\chi^2_{.01,10} = 23.21$, with which we also reject H_0.

PROBLEMS

7.1 Show that if $\mathbf{S} = \mathbf{\Sigma}_0$ in (7.1), then $u = 0$.

7.2 Verify (7.3); that is, show that $\ln|\mathbf{\Sigma}_0| - \ln|\mathbf{S}| = -\ln|\mathbf{S}\mathbf{\Sigma}_0^{-1}|$.

7.3 Verify (7.4); that is, show that $-\ln(\prod_{i=1}^p \lambda_i) + \sum_{i=1}^p \lambda_i = \sum_{i=1}^p (\lambda_i - \ln \lambda_i)$.

7.4 Verify that the likelihood ratio for H_0: $\mathbf{\Sigma} = \sigma^2\mathbf{I}$ is given by (7.5), LR = $[|\mathbf{S}|/(\text{tr}\,\mathbf{S}/p)^p]^{n/2}$.

7.5 In Section 7.2.2., show that $u = 1$ and $u' = 0$ if all the λ_i's are equal.

7.6 Show that the covariance matrix in (7.10) can be written in the form $\sigma^2[(1-\rho)\mathbf{I} + \rho\mathbf{J}]$, as given in (7.11).

7.7 Show that M in (7.17) can be expressed in the form given in (7.19).

7.8 **(a)** Calculate M as given in (7.19) for

$$\mathbf{S}_1 = \begin{pmatrix} 2 & 1 \\ 1 & 4 \end{pmatrix} \qquad \mathbf{S}_2 = \begin{pmatrix} 4 & 3 \\ 3 & 6 \end{pmatrix}.$$

Assume $\nu_1 = \nu_2 = 5$.

(b) Calculate M for

$$\mathbf{S}_1 = \begin{pmatrix} 2 & 1 \\ 1 & 4 \end{pmatrix} \qquad \mathbf{S}_2 = \begin{pmatrix} 10 & 15 \\ 15 & 30 \end{pmatrix}.$$

Assume $\nu_1 = \nu_2 = 5$.

In (b), $\mathbf{S}_1$ and $\mathbf{S}_2$ differ more than in (a) and M is accordingly much smaller. This illustrates the comments following (7.19).

7.9 Show that the two forms of u in (7.31) and (7.32) reduce to (7.35) when all $p_i = 1$.

7.10 Show that when all $p_i = 1$, c in (7.34) reduces to $1 - (2p+5)/6\nu$, so that (7.33) becomes (7.36).

7.11 In Example 5.2.2, we assumed that for the height and weight data of Table 3.1, the population covariance matrix is

$$\boldsymbol{\Sigma} = \begin{pmatrix} 20 & 100 \\ 100 & 1000 \end{pmatrix}.$$

Test this as a hypothesis using (7.2).

7.12 Test H_0: $\boldsymbol{\Sigma} = \sigma^2\mathbf{I}$ and H_0: $\mathbf{C}\boldsymbol{\Sigma}\mathbf{C}' = \sigma^2\mathbf{I}$ for the calculator speed data of Table 6.12.

7.13 Test H_0: $\boldsymbol{\Sigma} = \sigma^2\mathbf{I}$ and H_0: $\mathbf{C}\boldsymbol{\Sigma}\mathbf{C}' = \sigma^2\mathbf{I}$ for the ramus bone data of Table 3.8.

7.14 Test H_0: $\boldsymbol{\Sigma} = \sigma^2\mathbf{I}$ and H_0: $\mathbf{C}\boldsymbol{\Sigma}\mathbf{C}' = \sigma^2\mathbf{I}$ for the cork data of Table 6.21.

7.15 Test H_0: $\boldsymbol{\Sigma} = \sigma^2[(1-\rho)\mathbf{I}+\rho\mathbf{J}]$ for the probe word data in Table 3.7. Use both χ^2 and F approximations.

7.16 Test H_0: $\boldsymbol{\Sigma} = \sigma^2[(1-\rho)\mathbf{I}+\rho\mathbf{J}]$ for the calculator speed data in Table 6.12. Use both χ^2 and F approximations.

7.17 Test H_0: $\boldsymbol{\Sigma} = \sigma^2[(1-\rho)\mathbf{I}+\rho\mathbf{J}]$ for the ramus bone data in Table 3.8. Use both χ^2 and F approximations.

7.18 Test H_0: $\boldsymbol{\Sigma}_1 = \boldsymbol{\Sigma}_2$ for the beetles data of Table 5.5. Use an exact critical value from Table A.17 as well as χ^2 and F approximations.

7.19 Test H_0: $\boldsymbol{\Sigma}_1 = \boldsymbol{\Sigma}_2$ for the engineer data of Table 5.6. Use an exact critical value from Table A.17 as well as χ^2 and F approximations.

7.20 Test H_0: $\boldsymbol{\Sigma}_1 = \boldsymbol{\Sigma}_2$ for the dystrophy data of Table 5.7. Use an exact critical value from Table A.17 as well as χ^2 and F approximations.

7.21 Test H_0: $\boldsymbol{\Sigma}_1 = \boldsymbol{\Sigma}_2$ for the cyclical data of Table 5.8. Use an exact critical value from Table A.17 as well as χ^2 and F approximations.

7.22 Test H_0: $\boldsymbol{\Sigma}_1 = \boldsymbol{\Sigma}_2 = \boldsymbol{\Sigma}_3$ for the fish data of Table 6.17. Use both χ^2 and F approximations.

7.23 Test H_0: $\boldsymbol{\Sigma}_1 = \boldsymbol{\Sigma}_2 = \cdots = \boldsymbol{\Sigma}_6$ for the rootstock data in Table 6.2. Use both χ^2 and F approximations.

7.24 Test H_0: $\boldsymbol{\Sigma}_{11} = \boldsymbol{\Sigma}_{12} = \cdots = \boldsymbol{\Sigma}_{43}$ for the snapbean data in Table 6.18. Use the χ^2 and F approximations.

7.25 Test independence of (y_1, y_2) and (x_1, x_2) for the sons data in Table 3.9.

7.26 Test independence of (y_1, y_2, y_3) and (x_1, x_2, x_3) for the glucose data in Table 3.10.

7.27 Test independence of (y_1, y_2) and $(x_1, x_2, \ldots, x_8)$ for the seishu data of Table 7.1.

7.28 The data in Table 7.2 relate temperature, humidity, and evaporation (courtesy of R. J. Freund). The variables are

y_1 = maximum daily air temperature
y_2 = minimum daily air temperature
y_3 = integrated area under daily air temperature curve, i.e., a measure of average air temperature
y_4 = maximum daily soil temperature
y_5 = minimum daily soil temperature
y_6 = integrated area under soil temperature curve
y_7 = maximum daily relative humidity
y_8 = minimum daily relative humidity
y_9 = integrated area under daily humidity curve
y_{10} = total wind, measured in miles per day
y_{11} = evaporation

Test independence of the following five groups of variables: (y_1, y_2, y_3), (y_4, y_5, y_6), (y_7, y_8, y_9), y_{10}, and y_{11}.

7.29 Test the independence of all the variables for the calcium data of Table 3.5.

7.30 Test the independence of all the variables for the calculator data of Table 6.12.

7.31 Test the independence of all the variables for the ramus bone data of Table 3.8.

7.32 Test the independence of all the variables for the cork data of Table 6.21.

Table 7.2 Temperature, Humidity, and Evaporation

y_1	y_2	y_3	y_4	y_5	y_6	y_7	y_8	y_9	y_{10}	y_{11}
84	65	147	85	59	151	95	40	398	273	30
84	65	149	86	61	159	94	28	345	140	34
79	66	142	83	64	152	94	41	368	318	33
81	67	147	83	65	158	94	50	406	282	26
84	68	167	88	69	180	93	46	379	311	41
74	66	131	77	67	147	96	73	478	446	4
73	66	131	78	69	159	96	72	462	294	5
75	67	134	84	68	159	95	70	464	313	20
84	68	161	89	71	195	95	63	430	455	31
86	72	169	91	76	206	93	56	406	604	36
88	73	176	91	76	206	94	55	393	610	43
90	74	187	94	76	211	94	51	385	520	47
88	72	171	94	75	211	96	54	405	663	45
58	72	171	92	70	201	95	51	392	467	45
81	69	154	87	68	167	95	61	448	184	11
79	68	149	83	68	162	95	59	436	177	10
84	69	160	87	66	173	95	42	392	173	30
84	70	160	87	68	177	94	44	392	76	29
84	70	168	88	70	169	95	48	396	72	23
77	67	147	83	66	170	97	60	431	183	16
87	67	166	92	67	196	96	44	379	76	37
89	69	171	92	72	199	94	48	393	230	50
89	72	180	94	72	204	95	48	394	193	36
93	72	186	92	73	201	94	47	386	400	54
93	74	188	93	72	206	95	47	389	339	44
94	75	199	94	72	208	96	45	370	172	41
93	74	193	95	73	214	95	50	396	238	45
93	74	196	95	70	210	96	45	380	118	42
96	75	198	95	71	207	93	40	365	93	50
95	76	202	95	69	202	93	39	357	269	48
84	73	173	96	69	173	94	58	418	128	17
91	71	170	91	69	168	94	44	420	423	20
88	72	179	89	70	189	93	50	399	415	15
89	72	179	95	71	210	98	46	389	300	42
91	72	182	96	73	208	95	43	384	193	44
92	74	196	97	75	215	96	46	389	195	41
94	75	192	96	69	198	95	36	380	215	49
96	75	195	95	67	196	97	24	354	185	53
93	76	198	94	75	211	93	43	364	466	53
88	74	188	92	73	198	95	52	405	399	21
88	74	178	90	74	197	95	61	447	232	1
91	72	175	94	70	205	94	42	380	275	44
92	72	190	95	71	209	96	44	379	166	44
92	73	189	96	72	208	93	42	372	189	46
94	75	194	95	71	208	93	43	373	164	47
96	76	202	96	71	208	94	40	368	139	50

CHAPTER 8

Discriminant Analysis: Description of Group Separation

8.1 INTRODUCTION

We use the term *group* to represent either a population or a sample from the population. There are two major objectives in separation of groups:

1. Description of group separation, in which linear functions (discriminant functions) of the variables are used to describe or elucidate the differences between two or more groups. The goals of discriminant analysis include identifying the relative contribution of the p variables to separation of the groups and finding the optimal plane on which the points can be projected to best illustrate the configuration of the groups.
2. Prediction or allocation, in which linear or quadratic functions (classification functions) of the variables are employed to assign an individual sampling unit to one of the groups. The measured values (in the observation vector) for an individual or object are evaluated by the classification functions to see to which group the individual most likely belongs.

For consistency we will use the term *discriminant analysis* only in connection with objective 1. We will refer to all aspects of objective 2 as *classification analysis*, which is the subject of Chapter 9. Unfortunately, there is no general agreement with regard to usage of the terms discriminant analysis and discriminant functions. Many writers, perhaps the majority, use the term discriminant analysis in connection with the second objective, prediction. The linear functions contributing to the first objective, description of group separation, are often referred to as canonical variates or discriminant coordinates. To avoid confusion, we prefer to reserve the term *canonical* for canonical correlation analysis in Chapter 11.

Discriminant functions are linear combinations of variables that best separate groups. They were introduced in Section 5.5 for two groups and in Sections 6.1.4 and 6.4 for several groups. In those sections, interest was centered on follow-up to Hotelling's T^2-tests and MANOVA tests. In this chapter, we further develop these useful multivariate tools.

8.2 THE DISCRIMINANT FUNCTION FOR TWO GROUPS

We assume that the two populations to be compared have the same covariance matrix $\boldsymbol{\Sigma}$ but distinct mean vectors $\boldsymbol{\mu}_1$ and $\boldsymbol{\mu}_2$. We work with samples $\mathbf{y}_{11}, \mathbf{y}_{12}, \ldots, \mathbf{y}_{1n_1}$ and $\mathbf{y}_{21}, \mathbf{y}_{22}, \ldots, \mathbf{y}_{2n_2}$ from the two populations. As usual, each vector $\mathbf{y}_{ij}$ consists of measurements on p variables. The discriminant function is the linear combination of these p variables that maximizes the distance between the two (transformed) group mean vectors. A linear combination $z = \mathbf{a}'\mathbf{y}$ transforms each observation vector to a scalar:

$$\begin{aligned} z_{1j} &= \mathbf{a}'\mathbf{y}_{1j} = a_1 y_{1j1} + a_2 y_{1j2} + \cdots + a_p y_{1jp}, & j &= 1, 2, \ldots, n_1 \\ z_{2j} &= \mathbf{a}'\mathbf{y}_{2j} = a_1 y_{2j1} + a_2 y_{2j2} + \cdots + a_p y_{2jp}, & j &= 1, 2, \ldots, n_2. \end{aligned}$$

Hence the $n_1 + n_2$ observation vectors in the two samples,

$$\begin{matrix} \mathbf{y}_{11} & \mathbf{y}_{21} \\ \mathbf{y}_{12} & \mathbf{y}_{22} \\ \vdots & \vdots \\ \mathbf{y}_{1n_1} & \mathbf{y}_{2n_2}, \end{matrix}$$

are transferred to scalars,

$$\begin{matrix} z_{11} & z_{21} \\ z_{12} & z_{22} \\ \vdots & \vdots \\ z_{1n_1} & z_{2n_2}. \end{matrix}$$

We find the means $\bar{z}_1 = \sum_{j=1}^{n_1} z_{1j}/n_1 = \mathbf{a}'\bar{\mathbf{y}}_1$ and $\bar{z}_2 = \mathbf{a}'\bar{\mathbf{y}}_2$ by (3.51) and wish to find the vector $\mathbf{a}$ that maximizes the standardized difference $(\bar{z}_1 - \bar{z}_2)/s_z$. Since $(\bar{z}_1 - \bar{z}_2)/s_z$ can be negative, we use the squared distance $(\bar{z}_1 - \bar{z}_2)^2/s_z^2$, which, by (3.51) and (3.52), can be expressed as

$$\frac{(\bar{z}_1 - \bar{z}_2)^2}{s_z^2} = \frac{[\mathbf{a}'(\bar{\mathbf{y}}_1 - \bar{\mathbf{y}}_2)]^2}{\mathbf{a}'\mathbf{S}_{\text{pl}}\mathbf{a}}. \tag{8.1}$$

The maximum of (8.1) occurs when

$$\mathbf{a} = \mathbf{S}_{\text{pl}}^{-1}(\overline{\mathbf{y}}_1 - \overline{\mathbf{y}}_2), \tag{8.2}$$

or when $\mathbf{a}$ is any multiple of $\mathbf{S}_{\text{pl}}^{-1}(\overline{\mathbf{y}}_1 - \overline{\mathbf{y}}_2)$. Since any multiple of $\mathbf{a} = \mathbf{S}_{\text{pl}}^{-1}(\overline{\mathbf{y}}_1 - \overline{\mathbf{y}}_2)$ will maximize (8.1), the maximizing vector $\mathbf{a}$ is not unique. However, its "direction" is unique; that is, the relative values or ratios of $a_1, a_2, \ldots, a_p$ are unique. Note that in order for $\mathbf{S}_{\text{pl}}^{-1}$ to exist, we must have $n_1 + n_2 - 2 > p$.

The optimum direction given by $\mathbf{a} = \mathbf{S}_{\text{pl}}^{-1}(\overline{\mathbf{y}}_1 - \overline{\mathbf{y}}_2)$ is effectively parallel to the line joining $\overline{\mathbf{y}}_1$ and $\overline{\mathbf{y}}_2$, because the squared distance $(\overline{z}_1 - \overline{z}_2)^2/s_z^2$ is equivalent to the standardized distance between $\overline{\mathbf{y}}_1$ and $\overline{\mathbf{y}}_2$. This can be seen by substituting (8.2) into (8.1) to obtain

$$\frac{(\overline{z}_1 - \overline{z}_2)^2}{s_z^2} = (\overline{\mathbf{y}}_1 - \overline{\mathbf{y}}_2)'\mathbf{S}_{\text{pl}}^{-1}(\overline{\mathbf{y}}_1 - \overline{\mathbf{y}}_2) \tag{8.3}$$

for $z = \mathbf{a}'\mathbf{y}$ with $\mathbf{a} = \mathbf{S}_{\text{pl}}^{-1}(\overline{\mathbf{y}}_1 - \overline{\mathbf{y}}_2)$. Thus $(\overline{z}_1 - \overline{z}_2)^2/s_z^2 = \mathbf{a}'(\overline{\mathbf{y}}_1 - \overline{\mathbf{y}}_2)$, and any other direction than that represented by $\mathbf{a} = \mathbf{S}_{\text{pl}}^{-1}(\overline{\mathbf{y}}_1 - \overline{\mathbf{y}}_2)$ would yield a smaller difference between $\mathbf{a}'\overline{\mathbf{y}}_1$ and $\mathbf{a}'\overline{\mathbf{y}}_2$.

Figure 8.1 illustrates the separation of two bivariate normal ($p = 2$) groups along the single dimension represented by the discriminant function $z = \mathbf{a}'\mathbf{y}$, where $\mathbf{a}$ is given by (8.2). In this illustration the population covariance matrices are equal. When the linear combination $z_{ij} = \mathbf{a}'\mathbf{y}_{ij} = a_1 y_{ij1} + a_2 y_{ij2}$ is applied to the points $\mathbf{y}_{ij}$, it projects them onto the line of optimum separation of the two groups. Since the two variables y_1 and y_2 are bivariate normal, a linear combi-

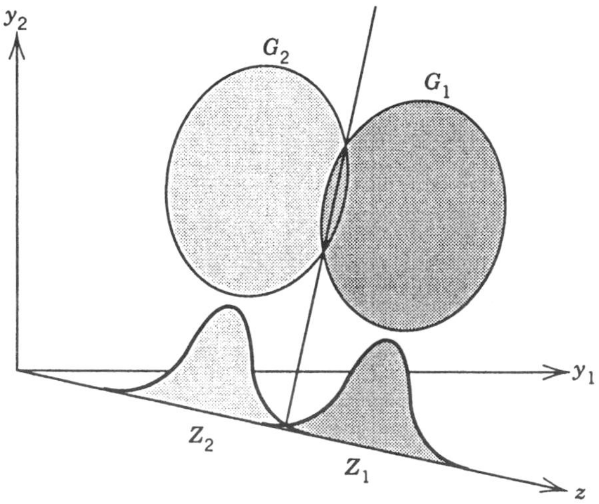

FIGURE 8.1 Two-group discriminant analysis.

nation is univariate normal (see property 1a in Section 4.2). We have therefore indicated this by two univariate normal densities along the line representing z.

The point where the line joining the points of intersection of the two ellipses intersects the discriminant function line z is the point of maximum separation (minimum overlap) of points projected onto the line. If the two populations are multivariate normal with common covariance matrix $\mathbf{\Sigma}$, as illustrated in Figure 8.1, it can be shown that the amount of overlap between them is a function of the distance between mean vectors, $\Delta^2 = (\boldsymbol{\mu}_1 - \boldsymbol{\mu}_2)'\mathbf{\Sigma}^{-1}(\boldsymbol{\mu}_1 - \boldsymbol{\mu}_2)$. It can also be shown that all possible group separation is expressed in a single new dimension.

In Figure 8.2, we illustrate the optimum separation achieved by the discriminant function. Projection in another direction denoted by z' gives a smaller standardized distance between the transformed means $\bar{z}'_1$ and $\bar{z}'_2$ and also a larger overlap between the projected points. Note that the spread of the projected points is greater in the z' dimension, so that $s_{z'}$ is greater and $(\bar{z}'_1 - \bar{z}'_2)/s_{z'}$ is less.

Example 8.2. Samples of steel produced at two different rolling temperatures are compared in Table 8.1 (Kramer and Jensen 1969a). The variables are y_1 = yield point and y_2 = ultimate strength. From the data, we calculate

$$\bar{\mathbf{y}}_1 = \begin{pmatrix} 36.4 \\ 62.6 \end{pmatrix} \qquad \bar{\mathbf{y}}_2 = \begin{pmatrix} 39.0 \\ 60.4 \end{pmatrix} \qquad \mathbf{S}_{\text{pl}} = \begin{pmatrix} 7.92 & 5.68 \\ 5.68 & 6.29 \end{pmatrix}.$$

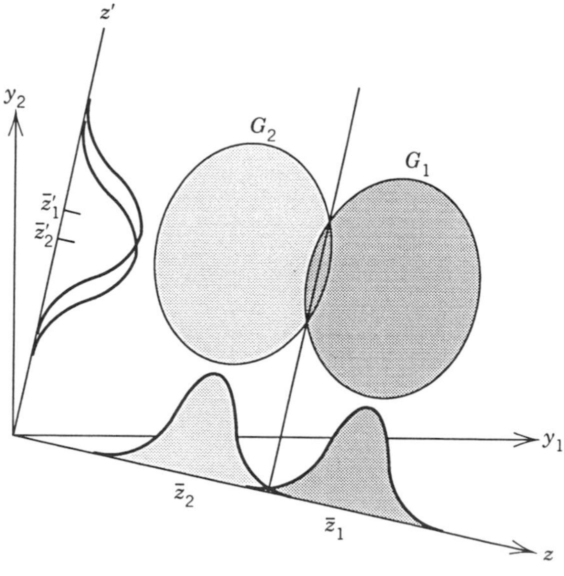

FIGURE 8.2 Separation achieved by the discriminant function.

Table 8.1 Yield Point and Ultimate Strength of Steel Produced at Two Rolling Temperatures

Temperature 1		Temperature 2	
y_1	y_2	y_1	y_2
33	60	35	57
36	61	36	59
35	64	38	59
38	63	39	61
40	65	41	63
		43	65
		41	59

A plot of the data appears in Figure 8.3. We see that if the points were projected on either the y_1 or y_2 axis, there would be considerable overlap. In fact, when the two groups are compared by means of a t-statistic for each variable separately, both t's are nonsignificant:

$$t_1 = \frac{\overline{y}_{11} - \overline{y}_{21}}{\sqrt{s_{11}(1/n_1 + 1/n_2)}} = -1.58$$

$$t_2 = \frac{\overline{y}_{12} - \overline{y}_{22}}{\sqrt{s_{22}(1/n_1 + 1/n_2)}} = 1.48.$$

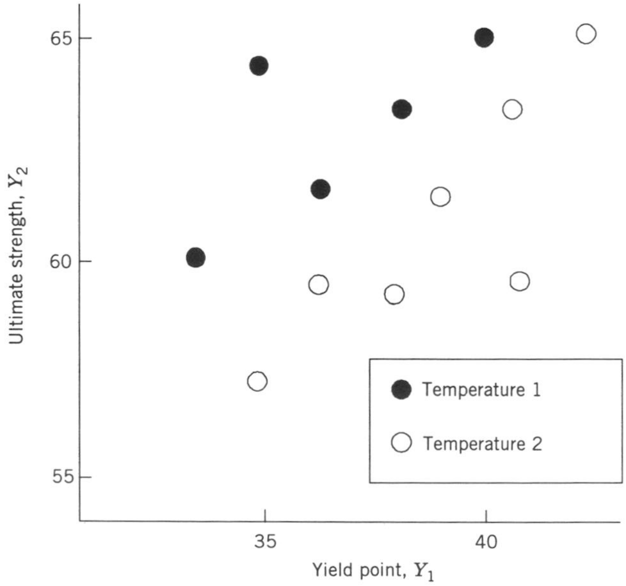

FIGURE 8.3 Ultimate strength and yield point for steel rolled at two temperatures.

Table 8.2 Discriminant Function $z = -1.633y_1 + 1.819y_2$ Evaluated for Data in Table 8.1

Temperature 1	Temperature 2
55.24	46.52
52.17	48.53
59.24	45.26
52.54	47.27
52.91	47.64
	48.01
	40.37

However, it is clear in Figure 8.3 that the two groups can be separated. If they are projected in an appropriate direction as in Figure 8.1, there will be no overlap. The single dimension onto which the points would be projected is the discriminant function

$$z = \mathbf{a}'\mathbf{y} = a_1y_1 + a_2y_2 = -1.633y_1 + 1.819y_2,$$

where **a** is obtained as

$$\mathbf{a} = \mathbf{S}_{\text{pl}}^{-1}(\overline{\mathbf{y}}_1 - \overline{\mathbf{y}}_2) = \begin{pmatrix} -1.633 \\ 1.819 \end{pmatrix}.$$

The values of the projected points are found by calculating z for each observation vector $\mathbf{y}$ in the two groups. The results are given in Table 8.2, where the separation provided by the discriminant function is clearly evident.

8.3 RELATIONSHIP BETWEEN TWO-GROUP DISCRIMINANT ANALYSIS AND MULTIPLE REGRESSION

The mutual connection between multiple regression and two-group discriminant analysis was introduced as a computational device in Section 5.6.2. The roles of independent and dependent variables are, of course, reversed in the two models. The dependent variables (y's) of discriminant analysis become the independent variables in regression.

Let w be a grouping variable (identifying groups 1 and 2) such that $\overline{w} = 0$ and define $\mathbf{b} = (b_1, b_2, \ldots, b_p)'$ as the vector of regression coefficients when w is fit to the y's. Then by (5.23), $\mathbf{b}$ is proportional to the discriminant function coefficient vector $\mathbf{a} = \mathbf{S}_{\text{pl}}^{-1}(\overline{\mathbf{y}}_1 - \overline{\mathbf{y}}_2)$:

$$\mathbf{b} = \frac{n_1 n_2}{(n_1 + n_2)(n_1 + n_2 - 2) + n_1 n_2 D^2} \mathbf{a}, \tag{8.4}$$

where $D^2 = (\bar{\mathbf{y}}_1 - \bar{\mathbf{y}}_2)' \mathbf{S}_{\text{pl}}^{-1} (\bar{\mathbf{y}}_1 - \bar{\mathbf{y}}_2)$. From (5.21) the multiple correlation coefficient R is related to D^2 by

$$R^2 = (\bar{\mathbf{y}}_1 - \bar{\mathbf{y}}_2)' \mathbf{b} = \frac{n_1 n_2 D^2}{(n_1 + n_2)(n_1 + n_2 - 2) + n_1 n_2 D^2}. \tag{8.5}$$

The test statistic (5.30) for the hypothesis that q of the $p + q$ variables are redundant for separating the groups can also be obtained in terms of regression by (5.32) as

$$F = \frac{n_1 + n_2 - p - q - 1}{q} \frac{R_{p+q}^2 - R_p^2}{1 - R_{p+q}^2}, \tag{8.6}$$

where R_{p+q}^2 and R_p^2 are from regressions with $p+q$ and p variables, respectively.

The link between two-group discriminant analysis and multiple regression was first noted by Fisher (1936). Flury and Riedwyl (1985) give further insights into the relationship.

Example 8.3. In Example 5.6.2, the psychological data of Table 5.1 was used in an illustration of the regression approach to computation of $\mathbf{a}$ and T^2. We use the same data to obtain $\mathbf{b}$ and R^2 from $\mathbf{a}$ and D^2.

From the results of Examples 5.4.2 and 5.5, we have

$$D^2 = (\bar{\mathbf{y}}_1 - \bar{\mathbf{y}}_2)' \mathbf{S}_{\text{pl}}^{-1} (\bar{\mathbf{y}}_1 - \bar{\mathbf{y}}_2) = 6.0377$$

$$\mathbf{a} = \begin{bmatrix} .4856 \\ -.2028 \\ .4654 \\ -.3048 \end{bmatrix}.$$

To find $\mathbf{b}$ from $\mathbf{a}$ and D^2, we use (8.4):

$$\mathbf{b} = \frac{(32)(32)}{(64)(62) + (32)(32)(6.0377)} \mathbf{a} = \begin{bmatrix} .049 \\ -.020 \\ .047 \\ -.031 \end{bmatrix}.$$

To find R^2, we use (8.5):

$$R^2 = (\bar{\mathbf{y}}_1 - \bar{\mathbf{y}}_2)'\mathbf{b} = \begin{bmatrix} 3.625 \\ 2.000 \\ 10.594 \\ .812 \end{bmatrix}' \begin{bmatrix} .049 \\ -.020 \\ .047 \\ -.031 \end{bmatrix} = .609.$$

8.4 DISCRIMINANT ANALYSIS FOR SEVERAL GROUPS

8.4.1 Discriminant Functions

As with discriminant analysis for two groups, in discriminant analysis for several groups, we are concerned with finding linear combinations of variables that best separate groups of multivariate observations. Discriminant analysis for several groups may serve any one of various purposes:

1. Examine group separation in a two-dimensional plot. When there are more than two groups, it requires more than one discriminant function to describe group separation. If the points in the p-dimensional space are projected onto a 2-dimensional space represented by the first two discriminant functions, we obtain the best possible view of how the groups are separated.
2. Find a subset of the original variables that separates the groups almost as well as the original set. This topic was introduced in Section 6.11.2.
3. Rank the variables in terms of their relative contribution to group separation. This use for discriminant functions has been mentioned in Sections 5.5, 6.1.4, 6.1.8, and 6.4. In Section 8.5, we standardize the discriminant function coefficients so that a more valid comparison of the variables can be made.
4. Interpret the new dimensions represented by the discriminant functions.
5. Follow up to fixed-effects MANOVA.

Purposes 3 and 4 are closely related. Any of the first four can be used to accomplish purpose 5. Methods of achieving these five goals of discriminant analysis are discussed in subsequent sections. In the present section we review discriminant functions for the several-group case and discuss attendant assumptions. For alternative approaches to estimation of discriminant functions that may be useful in the presence of multicollinearity or outliers, see Rencher (1996, Section 5.11).

For k groups (samples) with n_i observations in the ith group, we transform each observation vector $\mathbf{y}_{ij}$ to obtain $z_{ij} = \mathbf{a}'\mathbf{y}_{ij}$, $i = 1, 2, \ldots, k$; $j = 1, 2, \ldots, n_i$, and find the means $\bar{z}_i = \mathbf{a}'\bar{\mathbf{y}}_i$. As in the two-group case, we seek the vector $\mathbf{a}$ that maximally separates $\bar{z}_1, \bar{z}_2, \ldots, \bar{z}_k$. To express separation among $\bar{z}_1, \bar{z}_2, \ldots,$ and $\bar{z}_k$, we extend the separation criterion (8.1) to the k-group case. Since $\mathbf{a}'(\bar{\mathbf{y}}_1 - \bar{\mathbf{y}}_2) = (\bar{\mathbf{y}}_1 - \bar{\mathbf{y}}_2)'\mathbf{a}$, we can express (8.1) in the form

$$\frac{(\bar{z}_1 - \bar{z}_2)^2}{s_z^2} = \frac{[\mathbf{a}'(\bar{\mathbf{y}}_1 - \bar{\mathbf{y}}_2)]^2}{\mathbf{a}'\mathbf{S}_{\text{pl}}\mathbf{a}} = \frac{\mathbf{a}'(\bar{\mathbf{y}}_1 - \bar{\mathbf{y}}_2)(\bar{\mathbf{y}}_1 - \bar{\mathbf{y}}_2)'\mathbf{a}}{\mathbf{a}'\mathbf{S}_{\text{pl}}\mathbf{a}}. \tag{8.7}$$

To extend (8.7) to k groups, we use the $\mathbf{H}$ matrix from MANOVA in place of $(\bar{\mathbf{y}}_1 - \bar{\mathbf{y}}_2)(\bar{\mathbf{y}}_1 - \bar{\mathbf{y}}_2)'$ [see (6.31)] and $\mathbf{E}$ in place of $\mathbf{S}_{\text{pl}}$ to obtain

$$\lambda = \frac{\mathbf{a}'\mathbf{H}\mathbf{a}}{\mathbf{a}'\mathbf{E}\mathbf{a}}. \tag{8.8}$$

This can be written in the form

$$\mathbf{a}'\mathbf{H}\mathbf{a} = \lambda\mathbf{a}'\mathbf{E}\mathbf{a}$$

or

$$\mathbf{a}'(\mathbf{H}\mathbf{a} - \lambda\mathbf{E}\mathbf{a}) = 0. \tag{8.9}$$

We examine values of λ and $\mathbf{a}$ that are solutions of (8.9) to determine the value of $\mathbf{a}$ that results in maximum λ. The solution $\mathbf{a}' = \mathbf{0}'$ is not permissible because it gives $\lambda = 0/0$ in (8.8). Other possible solutions are found from

$$\mathbf{H}\mathbf{a} - \lambda\mathbf{E}\mathbf{a} = \mathbf{0}, \tag{8.10}$$

which can be written in the form

$$(\mathbf{E}^{-1}\mathbf{H} - \lambda\mathbf{I})\mathbf{a} = \mathbf{0}. \tag{8.11}$$

The solutions of (8.11) are the eigenvalues $\lambda_1, \lambda_2, \ldots, \lambda_s$ and associated eigenvectors $\mathbf{a}_1, \mathbf{a}_2, \ldots, \mathbf{a}_s$ of $\mathbf{E}^{-1}\mathbf{H}$. As in previous discussions of eigenvalues, we consider them to be ranked $\lambda_1 > \lambda_2 > \cdots > \lambda_s$. The number of (nonzero) eigenvalues s is the rank of $\mathbf{H}$, which can be found as the smaller of $k - 1$ or p. Thus the largest eigenvalue λ_1 is the maximum value of $\lambda = \mathbf{a}'\mathbf{H}\mathbf{a}/\mathbf{a}'\mathbf{E}\mathbf{a}$ in (8.8), and the coefficient vector that produces the maximum is the corresponding eigenvector $\mathbf{a}_1$. (This can be confirmed using calculus.) Hence the discriminant function that maximally separates the means is $z_1 = \mathbf{a}_1'\mathbf{y}$.

From the s eigenvectors $\mathbf{a}_1, \mathbf{a}_2, \ldots, \mathbf{a}_s$ of $\mathbf{E}^{-1}\mathbf{H}$ corresponding to $\lambda_1, \lambda_2, \ldots, \lambda_s$, we obtain s discriminant functions $z_1 = \mathbf{a}_1'\mathbf{y}, z_2 = \mathbf{a}_2'\mathbf{y}, \ldots, z_s = \mathbf{a}_s'\mathbf{y}$. These discriminant functions are uncorrelated. They show the dimensions or directions of differences among $\bar{\mathbf{y}}_1, \bar{\mathbf{y}}_2, \ldots, \bar{\mathbf{y}}_k$, where $\bar{\mathbf{y}}_i = \Sigma_{j=1}^{n_i} \mathbf{y}_{ij}/n_i$.

The relative importance of each discriminant function can be assessed by considering its eigenvalue as a proportion of the total:

$$\frac{\lambda_i}{\sum_j \lambda_j}. \tag{8.12}$$

By this criterion, two or three discriminant functions will often suffice to describe the group differences. The discriminant functions associated with small eigenvalues can be neglected. A test of significance for each discriminant function is also available (see Section 8.6).

The matrix $\mathbf{E}^{-1}\mathbf{H}$ is not symmetric. Many algorithms for computation of eigenvalues and eigenvectors accept only symmetric matrices. In Section 6.1.4, it was shown that the eigenvalues of the symmetric matrix $(\mathbf{U}^{-1})'\mathbf{H}\mathbf{U}^{-1}$ are the same as those of $\mathbf{E}^{-1}\mathbf{H}$, where $\mathbf{E} = \mathbf{U}'\mathbf{U}$ is the Cholesky factorization of $\mathbf{E}$. However, an adjustment is needed for the eigenvectors. If $\mathbf{b}$ is an eigenvector of $(\mathbf{U}^{-1})'\mathbf{H}\mathbf{U}^{-1}$, then $\mathbf{a} = \mathbf{U}^{-1}\mathbf{b}$ is an eigenvector of $\mathbf{E}^{-1}\mathbf{H}$.

The discussion above was presented in terms of unequal sample sizes $n_1, n_2, \ldots, n_k$. In applications, this situation is common and can be handled with no difficulty. Ideally, the smallest n_i should exceed the number of variables, p. This is not required mathematically, but will lead to more stable discriminant functions.

Example 8.4.1. The data in Table 8.3 were collected by G. R. Bryce and R. M. Barker (Brigham Young University) as part of a preliminary study of a possible link between football helmet design and neck injuries.

Six head measurements were made on each subject. There were 30 subjects in each of three groups: high school football players (group 1), college football players (group 2), and non–football players (group 3). The six variables are

WDIM = head width at widest dimension
CIRCUM = head circumference
FBEYE = front-to-back measurement at eye level
EYEHD = eye-to-top-of-head measurement
EARHD = ear-to-top-of-head measurement
JAW = jaw width

The eigenvalues of $\mathbf{E}^{-1}\mathbf{H}$ are $\lambda_1 = 1.9178$ and $\lambda_2 = .1159$. The corresponding eigenvectors are

Table 8.3 Head Measurements for Three Groups

Group	WDIM	CIRCUM	FBEYE	EYEHD	EARHD	JAW
1	13.5	57.2	19.5	12.5	14.0	11.0
1	15.5	58.4	21.0	12.0	16.0	12.0
1	14.5	55.9	19.0	10.0	13.0	12.0
1	15.5	58.4	20.0	13.5	15.0	12.0
1	14.5	58.4	20.0	13.0	15.5	12.0
1	14.0	61.0	21.0	12.0	14.0	13.0
1	15.0	58.4	19.5	13.5	15.5	13.0
1	15.0	58.4	21.0	13.0	14.0	13.0
1	15.5	59.7	20.5	13.5	14.5	12.5
1	15.5	59.7	20.5	13.0	15.0	13.0
1	15.0	57.2	19.0	14.0	14.5	11.5
1	15.5	59.7	21.0	13.0	16.0	12.5
1	16.0	57.2	19.0	14.0	14.5	12.0
1	15.5	62.2	21.5	14.0	16.0	12.0
1	15.5	57.2	19.5	13.5	15.0	12.0
1	14.0	61.0	20.0	15.0	15.0	12.0
1	14.5	58.4	20.0	12.0	14.5	12.0
1	15.0	56.9	19.0	13.0	14.0	12.5
1	15.5	59.7	20.0	12.5	14.0	12.5
1	15.0	57.2	19.5	12.0	14.0	11.0
1	15.0	56.9	19.0	12.0	13.0	12.0
1	15.5	56.9	19.5	14.5	14.5	13.0
1	17.5	63.5	21.5	14.0	15.5	13.5
1	15.5	57.2	19.0	13.0	15.5	12.5
1	15.5	61.0	20.5	12.0	13.0	12.5
1	15.5	61.0	21.0	14.5	15.5	12.5
1	15.5	63.5	21.8	14.5	16.5	13.5
1	14.5	58.4	20.5	13.0	16.0	10.5
1	15.5	56.9	20.0	13.5	14.0	12.0
1	16.0	61.0	20.0	12.5	14.5	12.5
2	15.5	60.0	21.1	10.3	13.4	12.4
2	15.4	59.7	20.0	12.8	14.5	11.3
2	15.1	59.7	20.2	11.4	14.1	12.1
2	14.3	56.9	18.9	11.0	13.4	11.0
2	14.8	58.0	20.1	9.6	11.1	11.7
2	15.2	57.5	18.5	9.9	12.8	11.4
2	15.4	58.0	20.8	10.2	12.8	11.9
2	16.3	58.0	20.1	8.8	13.0	12.9
2	15.5	57.0	19.6	10.5	13.9	11.8
2	15.0	56.5	19.6	10.4	14.5	12.0
2	15.5	57.2	20.0	11.2	13.4	12.4
2	15.5	56.5	19.8	9.2	12.8	12.2
2	15.7	57.5	19.8	11.8	12.6	12.5
2	14.4	57.0	20.4	10.2	12.7	12.3
2	14.9	54.8	18.5	11.2	13.8	11.3

Table 8.3 ***(Continued)***

Group	WDIM	CIRCUM	FBEYE	EYEHD	EARHD	JAW
2	16.5	59.8	20.2	9.4	14.3	12.2
2	15.5	56.1	18.8	9.8	13.8	12.6
2	15.3	55.0	19.0	10.1	14.2	11.6
2	14.5	55.6	19.3	12.0	12.6	11.6
2	15.5	56.5	20.0	9.9	13.4	11.5
2	15.2	55.0	19.3	9.9	14.4	11.9
2	15.3	56.5	19.3	9.1	12.8	11.7
2	15.3	56.8	20.2	8.6	14.2	11.5
2	15.8	55.5	19.2	8.2	13.0	12.6
2	14.8	57.0	20.2	9.8	13.8	10.5
2	15.2	56.9	19.1	9.6	13.0	11.2
2	15.9	58.8	21.0	8.6	13.5	11.8
2	15.5	57.3	20.1	9.6	14.1	12.3
2	16.5	58.0	19.5	9.0	13.9	13.3
2	17.3	62.6	21.5	10.3	13.8	12.8
3	14.9	56.5	20.4	7.4	13.0	12.0
3	15.4	57.5	19.5	10.5	13.8	11.5
3	15.3	55.4	19.2	9.7	13.3	11.5
3	14.6	56.0	19.8	8.5	12.0	11.5
3	16.2	56.5	19.5	11.5	14.5	11.8
3	14.6	58.0	19.9	13.0	13.4	11.5
3	15.9	56.7	18.7	10.8	12.8	12.6
3	14.7	55.8	18.7	11.1	13.9	11.2
3	15.5	58.5	19.4	11.5	13.4	11.9
3	16.1	60.0	20.3	10.6	13.7	12.2
3	15.2	57.8	19.9	10.4	13.5	11.4
3	15.1	56.0	19.4	10.0	13.1	10.9
3	15.9	59.8	20.5	12.0	13.6	11.5
3	16.1	57.7	19.7	10.2	13.6	11.5
3	15.7	58.7	20.7	11.3	13.6	11.3
3	15.3	56.9	19.6	10.5	13.5	12.1
3	15.3	56.9	19.5	9.9	14.0	12.1
3	15.2	58.0	20.6	11.0	15.1	11.7
3	16.6	59.3	19.9	12.1	14.6	12.1
3	15.5	58.2	19.7	11.7	13.8	12.1
3	15.8	57.5	18.9	11.8	14.7	11.8
3	16.0	57.2	19.8	10.8	13.9	12.0
3	15.4	57.0	19.8	11.3	14.0	11.4
3	16.0	59.2	20.8	10.4	13.8	12.2
3	15.4	57.6	19.6	10.2	13.9	11.7
3	15.8	60.3	20.8	12.4	13.4	12.1
3	15.4	55.0	18.8	10.7	14.2	10.8
3	15.5	58.4	19.8	13.1	14.5	11.7
3	15.7	59.0	20.4	12.1	13.0	12.7
3	17.3	61.7	20.7	11.9	13.3	13.3

$$\mathbf{a}_1 = \begin{bmatrix} -.948 \\ .004 \\ .006 \\ .647 \\ .504 \\ .829 \end{bmatrix} \qquad \mathbf{a}_2 = \begin{bmatrix} -1.407 \\ .001 \\ .029 \\ -.540 \\ .384 \\ 1.529 \end{bmatrix}.$$

The first eigenvalue accounts for a high proportion of the total:

$$\frac{\lambda_1}{\lambda_1 + \lambda_2} = \frac{1.9178}{1.9178 + .1159} = .94.$$

Thus the mean vectors lie largely in one dimension, and one discriminant function suffices to describe most of the separation among the three groups.

8.4.2 A Measure of Association for Discriminant Functions

Measures of association between the dependent variables $y_1, y_2, \ldots, y_p$ and the independent grouping variable i associated with $\boldsymbol{\mu}_1, \boldsymbol{\mu}_2, \ldots, \boldsymbol{\mu}_k$ were presented in Section 6.1.8. These measures attempt to answer the following question: How well do the variables separate the groups? It was noted that Roy's statistic itself serves as an R^2-like measure of association, since it is the ratio of between to total sum of squares for the first discriminant function, $z_1 = \mathbf{a}_1'\mathbf{y}$:

$$\eta_\theta^2 = \theta = \frac{\lambda_1}{1 + \lambda_1} = \frac{\text{SSH}(z_1)}{\text{SSE}(z_1) + \text{SSH}(z_1)}.$$

Another interpretation of η_θ^2 is the maximum squared correlation between the first discriminant function and the best linear combination of the $k - 1$ (dummy) group membership variables. This was noted following (6.34) in Section 6.1.8. Dummy variables were defined in the first two paragraphs of Section 6.1.8. The maximum correlation is called the canonical correlation (see Chapter 11). The squared canonical correlation can be calculated for each discriminant function:

$$r_i^2 = \frac{\lambda_i}{1 + \lambda_i}, \qquad i = 1, 2, \ldots, s. \tag{8.13}$$

The average squared canonical correlation was used as a measure of association in (6.37).

Example 8.4.2. For the football data of Table 8.3, we obtain the squared canonical correlation between each of the two discriminant functions and the grouping variables,

$$r_1^2 = \frac{\lambda_1}{1+\lambda_1} = \frac{1.9178}{1+1.9178} = .657$$

$$r_2^2 = \frac{\lambda_2}{1+\lambda_2} = \frac{.1159}{1+.1159} = .104.$$

8.5 STANDARDIZED DISCRIMINANT FUNCTIONS

In Section 5.5, it was noted that in the two-group case the relative contribution of the y's to separation of the two groups can best be assessed by comparing the coefficients a_r, $r = 1, 2, \ldots, p$, in the discriminant function

$$z = \mathbf{a}'\mathbf{y} = a_1 y_1 + a_2 y_2 + \cdots + a_p y_p.$$

Similar comments appeared in Sections 6.1.4, 6.1.8, and 6.4 concerning the use of discriminant functions to assess contribution of the y's to separation of several groups. However, such comparisons are informative only if the y's are commensurate, that is, measured on the same scale and with comparable variances. If the y's are not commensurate, we need coefficients a_r^* that are applicable to standardized variables. For the jth observation vector $\mathbf{y}_{ij}$ in the ith group, we can express the discriminant function in terms of standardized variables as

$$z_{ij} = a_1^* \frac{y_{ij1} - \bar{y}_{i1}}{s_1} + a_2^* \frac{y_{ij2} - \bar{y}_{i2}}{s_2} + \cdots + a_p^* \frac{y_{ijp} - \bar{y}_{ip}}{s_p}$$
$$i = 1, 2, \quad j = 1, 2, \ldots, n_i,$$

where $\bar{\mathbf{y}}_i' = (\bar{y}_{i1}, \bar{y}_{i2}, \ldots, \bar{y}_{ip})$ is the mean vector for the ith group, $i = 1, 2$, and s_r is the within-sample standard deviation of the rth variable, obtained as the square root of the rth diagonal element of $\mathbf{S}_{\text{pl}}$. Clearly, these standardized coefficients must be of the form

$$a_r^* = s_r a_r \qquad r = 1, 2, \ldots, p. \tag{8.14}$$

In vector form, this becomes

$$\mathbf{a}^* = (\text{diag}\, \mathbf{S}_{\text{pl}})^{1/2} \mathbf{a}. \tag{8.15}$$

For the several-group case, we can standardize the discriminant functions in an analogous fashion. If we denote the rth coefficient in the mth discriminant function by a_{mr}, then the standardized form is $a_{mr}^* = s_r a_{mr}$, where s_r is the within-group standard deviation obtained from the diagonal of $\mathbf{S}_{\text{pl}} = \mathbf{E}/\nu_E$.

Alternatively, since the mth eigenvector is unique only up to multiplication by a scalar, we can simplify the standardization by using

$$a_{mr}^* = \sqrt{e_{rr}}a_{mr} \qquad r = 1, 2, \ldots, p,$$

where e_{rr} is the rth diagonal element of $\mathbf{E}$. For further discussion of the use of standardized discriminant function coefficients to gauge the relative contribution of the variables to group separation, see Section 8.7.1 [see also Rencher and Scott (1990) and Rencher (1996), Section 5.4].

Example 8.5. In Example 8.4.1, we obtained the discriminant function coefficient vectors $\mathbf{a}_1$ and $\mathbf{a}_2$ for the football data of Table 8.3. Since $\lambda_1/(\lambda_1 + \lambda_2) = .94$, we concentrate on $\mathbf{a}_1$. To standardize $\mathbf{a}_1$, we need the within-sample standard deviations of the variables. The pooled covariance matrix is given by

$$\mathbf{S}_{\text{pl}} = \frac{\mathbf{E}}{87} = \begin{bmatrix} .428 & .578 & .158 & .084 & .125 & .228 \\ .578 & 3.161 & 1.020 & .653 & .340 & .505 \\ .158 & 1.020 & .546 & .077 & .129 & .159 \\ .084 & .653 & .077 & 1.232 & .315 & .024 \\ .125 & .340 & .129 & .315 & .618 & .009 \\ .228 & .505 & .159 & .024 & .009 & .376 \end{bmatrix}.$$

Using the square roots of the diagonal elements of $\mathbf{S}_{\text{pl}}$, we obtain

$$\mathbf{a}_1^* = \begin{bmatrix} \sqrt{.428}(-.948) \\ \sqrt{3.161}(.004) \\ \vdots \\ \sqrt{.376}(.829) \end{bmatrix} = \begin{bmatrix} -.621 \\ .007 \\ .005 \\ .719 \\ .397 \\ .508 \end{bmatrix}.$$

Thus the fourth, first, sixth, and fifth variables contribute most to separating the groups, in that order. The second and third variables are not useful (in the presence of the others) in distinguishing groups.

8.6 TESTS OF SIGNIFICANCE

In order to test hypotheses, we will need the assumption of multivariate normality. This was not explicitly required for the development of discriminant functions.

8.6.1 Tests for the Two-Group Case

By (8.3) we see that the separation of transformed means, $(\bar{z}_1 - \bar{z}_2)^2/s_z^2$, achieved by the discriminant function $z = \mathbf{a}'\mathbf{y}$ is equivalent to the standardized distance between the mean vectors $\bar{\mathbf{y}}_1$ and $\bar{\mathbf{y}}_2$. This standardized distance is proportional to the two-group T^2 (5.8) in Section 5.4.2. Hence the discriminant function coefficient vector $\mathbf{a}$ is significantly different from $\mathbf{0}$ if T^2 is significant.

To test the significance of a subset of the discriminant function coefficients, we can use the test of the corresponding subset of y's given in Section 5.8. To test the hypothesis that the population discriminant function has a specified form $(\mathbf{a}_0'\mathbf{y})$, see Rencher (1996, Section 5.5.1).

8.6.2 Tests for the Several-Group Case

In Section 8.4.1 we noted that the discriminant criterion $\lambda = \mathbf{a}'\mathbf{H}\mathbf{a}/\mathbf{a}'\mathbf{E}\mathbf{a}$ is maximized by λ_1, the largest eigenvalue of $\mathbf{E}^{-1}\mathbf{H}$, and that the remaining eigenvalues $\lambda_2, \ldots, \lambda_s$ correspond to other discriminant dimensions. These eigenvalues are the same as those in the Wilks Λ-test for significant differences among mean vectors, as in (6.12),

$$\Lambda_1 = \prod_{i=1}^{s} \frac{1}{1+\lambda_i},$$

which is distributed as $\Lambda_{p,k-1,N-k}$, where $N = \sum_i n_i$ for an unbalanced design or $N = kn$ in the balanced case. In this form, it is seen that Λ_1 is small if one or more λ_i's are large. Thus Wilks' Λ tests for significance of the eigenvalues and thereby for the discriminant functions. The s eigenvalues represent s dimensions of separation of the mean vectors $\bar{\mathbf{y}}_1, \bar{\mathbf{y}}_2, \ldots, \bar{\mathbf{y}}_k$. We are interested in which, if any, of these dimensions are significant. In the context of discriminant functions, Wilks' Λ is more useful than the other three MANOVA test statistics, because it can be used on a subset of eigenvalues, as we see below.

We can use the χ^2 approximation for Λ_1 given in (6.14), with $\nu_E = N - k = \sum_i n_i - k$ and $\nu_H = k - 1$:

$$\begin{aligned} V_1 &= -[\nu_E - \tfrac{1}{2}(p - \nu_H + 1)] \ln \Lambda_1 \\ &= -[N - 1 - \tfrac{1}{2}(p + k)] \ln \prod_{i=1}^{s} \frac{1}{1+\lambda_i} \\ &= [N - 1 - \tfrac{1}{2}(p + k)] \sum_{i=1}^{s} \ln(1 + \lambda_i), \end{aligned} \tag{8.16}$$

which is approximately χ^2 with $p(k - 1)$ degrees of freedom. The test statistic

Λ_1 and its approximation (8.16) test the significance of all of $\lambda_1, \lambda_2, \ldots, \lambda_s$. If the test leads to rejection of H_0, we conclude that at least one of the λ's is significantly different from zero. Since λ_1 is the largest, we are only sure of its significance, along with that of $z_1 = \mathbf{a}_1'\mathbf{y}$.

To test the significance of $\lambda_2, \lambda_3, \ldots, \lambda_s$, we delete λ_1 from Wilks' Λ and the associated χ^2 approximation to obtain

$$\Lambda_2 = \prod_{i=2}^{s} \frac{1}{1+\lambda_i}$$

$$V_2 = -[N - 1 - \tfrac{1}{2}(p + k)] \ln \Lambda_2$$

$$= [N - 1 - \tfrac{1}{2}(p + k)] \sum_{i=2}^{s} \ln(1 + \lambda_i), \tag{8.17}$$

which is approximately χ^2 with $(p-1)(k-2)$ degrees of freedom. If this test leads to rejection of H_0, we conclude that at least λ_2 is significant along with the associated discriminant function $z_2 = \mathbf{a}_2'\mathbf{y}$. We can continue in this fashion, testing each λ_i in turn until a test fails to reject H_0. (To compensate for making several tests, an adjustment to the α-level of each test could be made as in procedure 2, Section 5.5.) The test statistic at the mth step is

$$\Lambda_m = \prod_{i=m}^{s} \frac{1}{1+\lambda_i},$$

which is distributed as $\Lambda_{p-m+1, k-m, N-k-m+1}$. The statistic

$$V_m = -[N - 1 - \tfrac{1}{2}(p + k)] \ln \Lambda_m$$

$$= [N - 1 - \tfrac{1}{2}(p + k)] \sum_{i=m}^{s} \ln(1 + \lambda_i) \tag{8.18}$$

has an approximate χ^2-distribution with $(p - m + 1)(k - m)$ degrees of freedom. In some cases, more λ's will be statistically significant than the researcher considers to be of practical importance. If $\lambda_i / \sum_j \lambda_j$ is small, the associated discriminant function may not be of interest, even if it is significant.

We can also use an F approximation for each Λ_i. For

$$\Lambda_1 = \prod_{i=1}^{s} \frac{1}{1+\lambda_i},$$

we use (6.13), with $\nu_E = N - k$ and $\nu_H = k - 1$:

$$F = \frac{1 - \Lambda_1^{1/t}}{\Lambda_1^{1/t}} \frac{\text{df}_2}{\text{df}_1},$$

where

$$t = \sqrt{\frac{p^2(k-1)^2 - 4}{p^2 + (k-1)^2 - 5}} \qquad w = N - 1 - \tfrac{1}{2}(p + k)$$

$$\text{df}_1 = p(k - 1) \qquad \text{df}_2 = wt - \tfrac{1}{2}[p(k - 1) - 2].$$

For

$$\Lambda_m = \prod_{i=m}^{s} \frac{1}{1 + \lambda_i} \qquad m = 2, 3, \ldots, s,$$

we use

$$F = \frac{1 - \Lambda_m^{1/t}}{\Lambda_m^{1/t}} \frac{\text{df}_2}{\text{df}_1}$$

with $p - m + 1$ and $k - m$ in place of p and $k - 1$:

$$t = \sqrt{\frac{(p - m + 1)^2(k - m)^2 - 4}{(p - m + 1)^2 + (k - m)^2 - 5}}$$

$$w = N - 1 - \tfrac{1}{2}(p + k)$$

$$\text{df}_1 = (p - m + 1)(k - m)$$

$$\text{df}_2 = wt - \tfrac{1}{2}[(p - m + 1)(k - m) - 2].$$

Example 8.6.2. We test the significance of the two discriminant functions obtained in Example 8.4.1 for the football data. For the overall test we have

$$\Lambda_1 = \prod_{i=1}^{2} \frac{1}{1 + \lambda_i} = \frac{1}{1 + 1.9178} \frac{1}{1 + .1159} = .307.$$

The χ^2 approximation is

$$
\begin{aligned}
V_1 &= -[N - 1 - \tfrac{1}{2}(p + k)] \ln \Lambda_1 \\
&= -[90 - 1 - \tfrac{1}{2}(6 + 3)] \ln(.307) = 99.75,
\end{aligned}
$$

which exceeds the critical value $\chi^2_{.01,12} = 26.217$. Thus at least the first discriminant function is significant. To test the second discriminant function, we have, by (8.17),

$$
\begin{aligned}
V_2 &= -[N - 1 - \tfrac{1}{2}(p + k)] \ln \Lambda_2 \\
&= -[90 - 1 - \tfrac{1}{2}(6 + 3)] \ln \frac{1}{1 + .1159} = 9.27 < \chi^2_{.05,5} = 11.070.
\end{aligned}
$$

For the F approximation for Λ_1, we obtain

$$
\begin{aligned}
t &= \sqrt{\frac{p^2(k-1)^2 - 4}{p^2 + (k-1)^2 - 5}} = \sqrt{\frac{6^2 2^2 - 4}{6^2 + 2^2 - 5}} = 2 \\
w &= N - 1 - \tfrac{1}{2}(p + k) = 90 - 1 - \tfrac{1}{2}(6 + 3) = 84.5 \\
\text{df}_1 &= p(k - 1) = 6(2) = 12 \\
\text{df}_2 &= wt - \tfrac{1}{2}[p(k - 1) - 2] = (84.5)(2) - \tfrac{1}{2}[6(2) - 2] = 164 \\
F &= \frac{1 - \Lambda_1^{1/2}}{\Lambda_1^{1/2}} \frac{\text{df}_2}{\text{df}_1} = \frac{1 - .307^{1/2}}{.307^{1/2}} \frac{164}{12} = 10.994.
\end{aligned}
$$

The p-value for $F = 10.994$ is less than .0001. For the F approximation for Λ_2, we reduce p and k by 1 and obtain

$$
\begin{aligned}
t &= \sqrt{\frac{5^2 1^2 - 4}{5^2 + 1^2 - 5}} = 1 \qquad w = 90 - 1 - \tfrac{1}{2}(6 + 3) = 84.5 \\
\text{df}_1 &= 5(1) = 5 \qquad \text{df}_2 = 84.5(1) - \tfrac{1}{2}[5(1) - 2] = 83 \\
F &= \frac{1 - \Lambda_2}{\Lambda_2} \cdot \frac{\text{df}_2}{\text{df}_1} = \frac{1 - .896}{.896} \frac{83}{5} = 1.924.
\end{aligned}
$$

The p-value for $F = 1.924$ is .099. Thus only the first discriminant function significantly separates groups.

8.7 INTERPRETATION OF DISCRIMINANT FUNCTIONS

There is a close correspondence between interpreting discriminant functions and determining the contribution of each variable, and we shall not always make a distinction. In interpretation, the signs of the coefficients are taken into account; in ascertaining the contribution, the signs are ignored, and the coefficients are ranked in absolute value. We discuss this distinction further in Section 8.7.1.

In the next three sections, we cover three common approaches to assessing the contribution of each variable to separating the groups. Specifically, we wish to learn what the variables contribute to a discriminant function in the presence of each other, allowing for the fact that they are correlated. The three methods are (1) examine the standardized discriminant function coefficients, (2) calculate a partial F-test for each variable, and (3) calculate a correlation between each variable and the discriminant function. The third method is the most widely recommended, but we note in Section 8.7.3 that it is the least useful.

8.7.1 Standardized Coefficients

To offset differing scales among the variables, the discriminant function coefficients can be standardized using (8.14) or (8.15). The standardized coefficients have been adjusted so that they apply to standardized variables. For the observations in the first group, for example, we have

$$z_{1j} = a_1^* \frac{y_{1j1} - \overline{y}_{11}}{s_1} + a_2^* \frac{y_{1j2} - \overline{y}_{12}}{s_2} + \cdots + a_p^* \frac{y_{1jp} - \overline{y}_{1p}}{s_p}$$

$$j = 1, 2, \ldots, n_1.$$

The standardized variables $(y_{ijr} - \overline{y}_{ir})/s_r$ are scale free, and the standardized coefficients $a_1^*, a_2^*, \ldots, a_p^*$ therefore correctly reflect the joint contribution of the variables to the discriminant function z as it maximally separates the groups. The discriminant function coefficient vector is an eigenvector of $\mathbf{E}^{-1}\mathbf{H}$, and as such, it takes into account the sample correlations among the variables. Thus the coefficients indicate the influence of each variable in the presence of the others.

As noted in Section 8.5, this standardization is carried out for each of the s discriminant functions. Typically, each will have a different interpretation; that is, the pattern of the coefficients a_r^* will vary from one function to another.

The absolute values of the coefficients can be used to rank the variables in order of their contribution to separating the groups. If we wish to go further and interpret or "name" a discriminant function, the signs can be taken into account. Thus, for example, $z_1 = .8y_1 - .9y_2 + .5y_3$ has a different meaning than does $z_2 = .8y_1 + .9y_2 + .5y_3$, since z_1 depends on the difference between y_1 and y_2, while z_2 is related to the sum of y_1 and y_2.

The discriminant function is subject to the same limitations as other linear combinations such as regression; for example, (1) the coefficient for a variable may change notably if other variables are added or deleted and (2) the coefficients may not be stable from sample to sample unless the sample size is large relative to the number of variables. With regard to limitation 1, we note that the coefficients reflect the contribution of the variable in the presence of the particular variables at hand. This is, in fact, what we want them to do. As to limitation 2, the processing of a substantial number of variables is not "free." More stable estimates will be obtained from 50 observations on 2 variables than from 50 observations on 20 variables. In other words, if N/p is too small, the variables that rank high in one sample may emerge as less important in another sample.

8.7.2 Partial *F*-Values

For any variable y_k we can calculate a partial F-test showing the significance of y_k after adjusting for the other variables, that is, the separation provided by y_k in addition to that due to the other variables. After computing the partial F for each variable, the variables can then be ranked.

In the case of two groups, the partial F is given by (5.33) as

$$F = (\nu - p + 1)\frac{T_p^2 - T_{p-1}^2}{\nu + T_{p-1}^2}, \tag{8.19}$$

where T_p^2 is the two-sample Hotelling T^2 with all p variables, T_{p-1}^2 is the T^2-statistic with all variables except y_k, and $\nu = n_1 + n_2 - 2$. The F-statistic in (8.19) is distributed as $F_{1,\nu-p+1}$.

For the several-group case, the partial Λ for y_k adjusted for the other $p-1$ variables is given by (6.102) as

$$\Lambda(y_k|y_1, \ldots, y_{k-1}, y_{k+1}, \ldots, y_p) = \frac{\Lambda_p}{\Lambda_{p-1}}, \tag{8.20}$$

where Λ_p is Wilks' Λ for all p variables and Λ_{p-1} involves all variables except y_k. The corresponding partial F is given by (6.103) as

$$F = \frac{1-\Lambda}{\Lambda}\,\frac{\nu_E - p + 1}{\nu_H}, \tag{8.21}$$

where Λ is defined in (8.20), $\nu_E = N - k$, and $\nu_H = k - 1$. The partial Λ-statistic in (8.20) is distributed as $\Lambda_{1,\nu_H,\nu_E-p+1}$, and the partial F in (8.21) is distributed as F_{ν_H,ν_E-p+1}.

The partial F-values are not associated with a single dimension of group separation as are the standardized discriminant function coefficients. For example, y_2 will have a different contribution in each of the s discriminant functions, but the partial F for y_2 constitutes an overall index of the contribution of y_2 to group separation taking into account all dimensions. However, the partial F-values will often rank the variables in the same order as the standardized coefficients for the first discriminant function, especially if $\lambda_1/\sum_j \lambda_j$ is very large so that the first function accounts for most of the available separation.

A partial index of association for y_i similar to the overall measure for all y's, η^2_Λ in (6.32), can be defined by

$$R_i^2 = 1 - \Lambda_i \qquad i = 1, 2, \ldots, p, \tag{8.22}$$

where Λ_i is the partial Λ in (8.20) for y_i. This partial R^2 is a measure of association between the grouping variables and y_i after adjusting for the other $p-1$ y's.

8.7.3 Correlations between Variables and Discriminant Functions

Many textbooks and research papers assert that the best measure of variable importance is the correlation between each variable and a discriminant function. It is claimed that these correlations are more informative than standardized coefficients with respect to the joint contribution of the variables to the discriminant functions. The correlations are often referred to as loadings or structure coefficients and are routinely provided in most major programs. However, Rencher (1988, 1992b, 1996, Section 5.7) has shown that the correlations in question show the contribution of each variable not in a multivariate context but rather in a univariate one. The correlations actually reproduce the t or F for each variable and only show how each variable by itself separates the groups, ignoring the presence of the other variables. But this is precisely what we are trying to avoid in a multivariate analysis. Thus these correlations are useless in gauging the importance of a given variable in the context of others because they provide no information about how the variables contribute jointly to separation of the groups. Consequently, they become misleading if used for interpretation of discriminant functions.

Upon reflection, we could have anticipated this failure of the correlations to perform as envisioned by their proponents. The objection to standardized coefficients is based on the argument that they are "unstable" because they change if some variables are deleted and others added. However, we actually want them to behave this way, so as to reflect the mutual influence of the variables on each other. In a multivariate analysis, interest is centered in the joint performance of the set of variables at hand. To ask for the contribution of each variable independent of all other variables is equivalent to requesting a univariate index that ignores the other variables. We should not be surprised when that is exactly what we get with the above correlations. There is no middle ground between the univariate and multivariate realms; we cannot have all the advantages of both.

8.7.4 Rotation

Rotation of the discriminant function coefficients is sometimes recommended. This is a procedure (see Section 13.5) that attempts to produce a pattern with (absolute values of) coefficients closer to 0 or 1. Discriminant functions with such coefficients might be easier to interpret, but they have two deficiencies: they no longer maximize group separation and they are correlated.

Accordingly, for interpretation of discriminant functions we recommend standardized coefficients rather than correlations or rotated coefficients.

8.8 SCATTER PLOTS

One advantage of the dimension reduction effected by discriminant analysis is the potential for plotting. It was noted in Section 6.2 that the number of large eigenvalues of $\mathbf{E}^{-1}\mathbf{H}$ reflects the dimensionality of the space occupied by the mean vectors. In many data sets, the first two discriminant functions account for most of $\lambda_1 + \lambda_2 + \cdots + \lambda_s$, and consequently the pattern of the mean vectors can be effectively portrayed in a two-dimensional plot. If the essential dimensionality is greater than 2, there may be some distortion in intergroup configuration in a two-dimensional plot.

To plot the first two discriminant functions for the individual observation vectors, simply calculate $z_{1ij} = \mathbf{a}_1'\mathbf{y}_{ij}$ and $z_{2ij} = \mathbf{a}_2'\mathbf{y}_{ij}$ for $i = 1, 2, \ldots, k$; $j = 1, 2, \ldots, n_i$, and plot a scatter plot of the $N = \sum_i n_i$ values of

$$\mathbf{z}_{ij} = \begin{pmatrix} z_{1ij} \\ z_{2ij} \end{pmatrix} = \begin{pmatrix} \mathbf{a}_1'\mathbf{y}_{ij} \\ \mathbf{a}_2'\mathbf{y}_{ij} \end{pmatrix} = \begin{pmatrix} \mathbf{a}_1' \\ \mathbf{a}_2' \end{pmatrix} \mathbf{y}_{ij} = \mathbf{A}\mathbf{y}_{ij}. \tag{8.23}$$

The transformed mean vectors,

$$\overline{\mathbf{z}}_i = \begin{pmatrix} \overline{z}_{1i} \\ \overline{z}_{2i} \end{pmatrix} = \begin{pmatrix} \mathbf{a}_1' \\ \mathbf{a}_2' \end{pmatrix} \overline{\mathbf{y}}_i = \mathbf{A}\overline{\mathbf{y}}_i, \tag{8.24}$$

should be plotted along with the individual values, $\mathbf{z}_{ij}$. In some cases, a plot would show only the transformed mean vectors $\overline{\mathbf{z}}_1, \overline{\mathbf{z}}_2, \ldots, \overline{\mathbf{z}}_k$. For confidence regions, see Rencher (1996, Section 5.8).

We note that the eigenvalues of $\mathbf{E}^{-1}\mathbf{H}$ reveal the dimensionality of the mean vectors, not of the individual points. The dimensionality of the individual observations is p, although the essential dimensionality may be less; that is, because the variables are correlated, the points may have less scatter in some dimen-

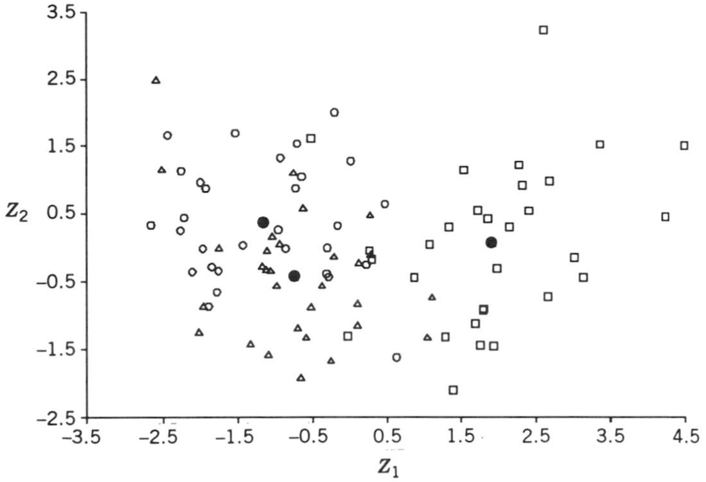

Figure 8.4 Scatter plot of discriminant function values for the football data of Table 8.3.

sions. (The dimensionality of the observation vectors is the concern of principal components; see Chapter 12.) If $s = 2$, for example, so that the mean vectors occupy only two dimensions, the individual observation vectors ordinarily lie in more than two dimensions, and their inclusion in a plot constitutes a projection onto the two-dimensional plane of the mean vectors.

It was noted in Section 8.4.1 that the discriminant functions are uncorrelated. However, they are not orthogonal. Thus the angle between $\mathbf{a}_1$ and $\mathbf{a}_2$ as given by (3.13) is not 90°. In practice, however, the usual procedure is to plot discriminant functions on a rectangular coordinate system. The resulting distortion is generally not serious.

Example 8.8. Figure 8.4 contains a scatter plot of (z_1, z_2) for the observations in the football data of Table 8.3. Each observation in group 1 is denoted by a square, observations in group 2 are denoted by circles, and observations in group 3 are indicated by triangles. We see that the first discriminant function z_1 (the horizontal direction) effectively separates group 1 from groups 2 and 3, while the second discriminant function z_2 (the vertical direction) is less successful in separating group 2 from group 3.

The group mean vectors are indicated by solid circles. They are almost collinear as we would expect since $\lambda_1 = 1.92$ dominates $\lambda_2 = .12$.

8.9 STEPWISE SELECTION OF VARIABLES

In many applications, a large number of dependent variables is available and the experimenter would like to discard those that are redundant (in the presence of the other variables) for separating the groups. Our discussion is limited to procedures that delete or add variables one at a time using stepwise methods. We emphasize that we are selecting *dependent* variables (y's) and the basic model (one-way MANOVA) does not change. In subset selection in regression, on the other hand, we select *independent* variables with a consequent alteration of the model.

A *forward selection* method was discussed in Section 6.11.2. In this approach, the variable entered at each step is the one that maximizes the partial F-statistic based on Wilks' Λ. We thus obtain the maximal additional separation of groups above and beyond the separation already attained by the other variables. Since we choose the variable with maximum partial F at each step, the proportion of these maximum F's that exceed F_α is greater than α. This bias is discussed in Rencher and Larson (1980) and Rencher (1996, Section 5.10).

Backward elimination is a similar operation in which the variable that contributes least is deleted at each step, as indicated by the partial F.

Stepwise selection is a combination of the forward and backward approaches. Variables are selected one at a time, and at each step, the variables are reexamined to see if any variable that entered earlier has become redundant in the presence of recently added variables. The procedure stops when the largest partial F among the variables available for entry fails to exceed a preset threshold value. The stepwise procedure has long been popular with practitioners. More detail about the steps in this procedure was given in Section 6.11.2.

Any of the above procedures is commonly referred to as *stepwise discriminant analysis*. However, as noted in Section 6.11.2, we are actually doing stepwise MANOVA. No discriminant functions are calculated in the selection process. After the subset selection is completed, we can calculate discriminant functions for the selected variables. We could also use the variables in a classification analysis as described in Chapter 9.

Example 8.9. We use the football data of Table 8.3 to illustrate the stepwise procedure outlined in this section and in Section 6.11.2. At the first step, we carry out a univariate F (using ordinary ANOVA) for each variable by itself to determine which variable best separates the three groups:

Variable	F	p-Value
WDIM	2.550	.0839
CIRCUM	6.231	.0030
FBEYE	1.668	.1947
EYEHD	58.162	.0001
EARHD	22.427	.0001
JAW	4.511	.0137

Thus EYEHD is the first variable to "enter." The Wilks Λ value equivalent to $F = 58.162$ is $\Lambda(y_1) = .4279$ (see Table 6.1 with $p = 1$). At the second step we calculate a partial Λ and accompanying partial F using (8.20) and (8.21):

$$\Lambda(y_i|y_1) = \frac{\Lambda(y_1, y_i)}{\Lambda(y_1)}$$
$$F = \frac{1 - \Lambda(y_i|y_1)}{\Lambda(y_i|y_1)} \frac{\nu_E - 1}{\nu_H},$$

where y_1 indicates the variable selected at step 1 and y_i represents each of the five variables to be examined at step 2. The results are

Variable	Partial Λ	Partial F	p-Value
WDIM	.9355	2.964	.0569
CIRCUM	.9997	.012	.9881
FBEYE	.9946	.235	.7911
EARHD	.9525	2.143	.1235
JAW	.9540	2.072	.1322

Due to having the largest partial F, WDIM would enter at this step. With a p-value of .0569, this may be questionable, but we will continue the procedure for illustrative purposes. We next check to see if EYEHD is still significant now that WDIM has entered. The partial Λ and F for EYEHD adjusted for WDIM are $\Lambda = .424$ and $F = 58.47$. Thus EYEHD stays "in." The overall Wilks' Λ for EYEHD and WDIM is $\Lambda(y_1, y_2) = .4003$.

At step 3 we check each of the four remaining variables for possible entry using

$$\Lambda(y_i|y_1, y_2) = \frac{\Lambda(y_1, y_2, y_i)}{\Lambda(y_1, y_2)}$$
$$F = \frac{1 - \Lambda(y_i|y_1, y_2)}{\Lambda(y_i|y_1, y_2)} \frac{\nu_E - 2}{\nu_H},$$

where y_1 = EYEHD, y_2 = WDIM, and y_i represents each of the other four variables. The results are

Variable	Partial Λ	Partial F	p-Value
CIRCUM	.9774	.981	.3793
FBEYE	.9748	1.098	.3381
EARHD	.9292	3.239	.0441
JAW	.8451	7.791	.0008

The indicated variable for entry at this step is JAW. To determine whether one of the first two should be removed after JAW has entered, we calculate the partial Λ and F for each, adjusted for the other two:

Variable	Partial Λ	Partial F	p-Value
WDIM	.8287	8.787	.0003
EYEHD	.4634	49.211	.0001

Thus both previously entered variables remain in the model. The overall Wilks Λ for EYEHD, WDIM, and JAW is $\Lambda(y_1, y_2, y_3) = .3383$.

At step 4 there are three candidate variables for entry. The partial Λ- and F-statistics are

$$\Lambda(y_i|y_1, y_2, y_3) = \frac{\Lambda(y_1, y_2, y_3, y_i)}{\Lambda(y_1, y_2, y_3)}$$

$$F = \frac{1 - \Lambda(y_i|y_1, y_2, y_3)}{\Lambda(y_i|y_1, y_2, y_3)} \frac{\nu_E - 3}{\nu_H},$$

where y_1, y_2, and y_3 are the three variables already entered and y_i represents each of the other three remaining variables. The results are

Variable	Partial Λ	Partial F	p-Value
CIRCUM	.9987	.055	.9462
FBEYE	.9955	.189	.8282
EARHD	.9080	4.257	.0173

Hence EARHD enters at this step, and we check to see if any of the three previously entered variables has now become redundant. The partial Λ and partial F for each of these three are

Variable	Partial Λ	Partial F	p-Value
WDIM	.7889	11.237	.0001
EYEHD	.6719	20.508	.0001
JAW	.8258	8.861	.0003

Consequently, all three variables are retained. The overall Wilks' Λ for all four variables is now $\Lambda(y_1, y_2, y_3, y_4) = .3072$.

At step 5, the partial Λ- and F-values are

Variable	Partial Λ	Partial F	p-Value
CIRCUM	.9999	.003	.9971
FBEYE	.9999	.004	.9965

Thus no more variables will enter.

We summarize the selection process as follows:

Step	Variable Entered	Overall Λ	p-Value	Partial Λ	Partial F	p-Value
1	EYEHD	.4279	.0001	.4279	58.162	.0001
2	WDIM	.4003	.0001	.9355	2.964	.0569
3	JAW	.3383	.0001	.8451	7.791	.0008
4	EARHD	.3072	.0001	.9080	4.257	.0173

PROBLEMS

8.1 Show that if $\mathbf{a} = \mathbf{S}_{\text{pl}}^{-1}(\bar{\mathbf{y}}_1 - \bar{\mathbf{y}}_2)$ is substituted into $[\mathbf{a}'(\bar{\mathbf{y}}_1 - \bar{\mathbf{y}}_2)]^2/\mathbf{a}'\mathbf{S}_{\text{pl}}\mathbf{a}$, the result is (8.3).

8.2 Verify (8.4) for the relationship between **b** and **a**.

8.3 Verify the relationship between R^2 and D^2 shown in (8.5).

8.4 Show that $[\mathbf{a}'(\bar{\mathbf{y}}_1 - \bar{\mathbf{y}}_2)]^2 = \mathbf{a}'(\bar{\mathbf{y}}_1 - \bar{\mathbf{y}}_2)(\bar{\mathbf{y}}_1 - \bar{\mathbf{y}}_2)'\mathbf{a}$ as in (8.7).

8.5 Show that $\mathbf{Ha} - \lambda\mathbf{Ea} = \mathbf{0}$ can be written in the form $(\mathbf{E}^{-1}\mathbf{H} - \lambda\mathbf{I})\mathbf{a} = \mathbf{0}$, as in (8.11).

8.6 Verify (8.14) by substituting $a_r^* = s_r a_r$ into the expression preceding (8.14) to obtain $z_{ij} = a_1 y_{ij1} + a_2 y_{ij2} + \cdots + a_p y_{ijp} - \mathbf{a}'\bar{\mathbf{y}}_i$.

8.7 For the psychological data in Table 5.1, the discriminant function coefficient vector was given in Example 5.5.

(a) Find the standardized coefficients.

(b) Calculate t-tests for the individual variables.

(c) Compare the results of (a) and (b) as to the contribution of the variables to separation of the two groups.

(d) Find the partial F for each variable, as in (8.19), and compare with the standardized coefficients.

8.8 Using the beetle data of Table 5.5, do the following:

(a) Find the discriminant function coefficient vector.

(b) Find the standardized coefficients.

(c) Calculate t-tests for individual variables.

(d) Compare the results of (b) and (c) as to the contribution of each variable to separation of the groups.

(e) Find the partial F for each variable, as in (8.19). Do the partial F's rank the variables in the same order of importance as the standardized coefficients?

8.9 Using the dystrophy data of Table 5.7, do the following:

(a) Find the discriminant function coefficient vector.

(b) Find the standardized coefficients.

(c) Calculate t-tests for individual variables.

(d) Compare the results of (b) and (c) as to the contribution of each variable to separation of the groups.

(e) Find the partial F for each variable, as in (8.19). Do the partial F's rank the variables in the same order of importance as the standardized coefficients?

8.10 For the cyclical data of Table 5.8, do the following:

(a) Find the discriminant function coefficient vector.

(b) Find the standardized coefficients.

(c) Calculate t-tests for individual variables.

(d) Compare the results of (b) and (c) as to the contribution of each variable to separation of the groups.

(e) Find the partial F for each variable, as in (8.19). Do the partial F's rank the variables in the same order of importance as the standardized coefficients?

8.11 Using the fish data in Table 6.17, do the following:

(a) Find the eigenvectors of $\mathbf{E}^{-1}\mathbf{H}$.

(b) Carry out tests of significance for the discriminant functions and find the relative importance of each as in (8.12), $\lambda_i / \sum_j \lambda_j$. Do these two procedures agree as to the number of important discriminant functions?

(c) Find the standardized coefficients and comment on the contribution of the variables to separation of groups.

(d) Find the partial F for each variable, as in (8.21). Do they rank the variables in the same order as the standardized coefficients for the first discriminant function?

(e) Plot the first two discriminant functions for each observation and for the mean vectors.

8.12 For the rootstock data of Table 6.2, do the following:

(a) Find the eigenvectors of $\mathbf{E}^{-1}\mathbf{H}$.

(b) Carry out tests of significance for the discriminant functions and find the relative importance of each as in (8.12), $\lambda_i / \sum_j \lambda_j$. Do these two procedures agree as to the number of important discriminant functions?

(c) Find the standardized coefficients and comment on the contribution of the variables to separation of groups.

(d) Find the partial F for each variable, as in (8.21). Do they rank the variables in the same order as the standardized coefficients for the first discriminant function?

(e) Plot the first two discriminant functions for each observation and for the mean vectors.

8.13 Carry out a stepwise selection of variables on the rootstock data of Table 6.2.

8.14 Carry out a stepwise selection of variables on the engineer data of Table 5.6.

8.15 Carry out a stepwise selection of variables on the fish data of Table 6.17.

CHAPTER 9

Classification Analysis: Allocation of Observations to Groups

9.1 INTRODUCTION

The *descriptive* aspect of discriminant analysis, in which group separation is characterized by means of discriminant functions, was covered in Chapter 8. We turn now to *allocation* of observations to groups, which is the *predictive* aspect of discriminant analysis. We prefer to call this *classification analysis* to clearly distinguish it from the descriptive aspect. However, classification is often referred to simply as discriminant analysis. In engineering and computer science, classification is usually called *pattern recognition.* Some writers use the term classification analysis to describe *cluster analysis*, in which the observations are clustered according to variable values rather than into predefined groups.

In classification, a sampling unit (subject or object) whose group membership is unknown is assigned to a group on the basis of the vector of p measured values, $\mathbf{y}$, associated with the unit. To classify the unit, we must have available a previously obtained sample of observation vectors from each group. Then one approach is to compare $\mathbf{y}$ with the mean vectors $\overline{\mathbf{y}}_1, \overline{\mathbf{y}}_2, \ldots, \overline{\mathbf{y}}_k$ of the k samples and assign the unit to the group whose $\overline{\mathbf{y}}_i$ is closest to $\mathbf{y}$. Thus if we think of the measured values in $\mathbf{y}$ as a *profile* of the unit, then we are comparing this profile to the profile of each group. We assign the unit to the group whose profile $\mathbf{y}$ most resembles.

In this chapter, the term "groups" may refer to either the k samples or the k populations from which they were taken. It should be clear from the context which of the two uses is intended in every case.

We give some examples to illustrate the classification technique:

1. A university admissions committee wants to classify applicants as likely to succeed or likely to fail. The variables available are high school grades in various subject areas, standardized test scores, rating of high school, number of advanced placement courses, etc.

2. A psychiatrist gives a battery of diagnostic tests in order to assign a patient to the appropriate mental illness category.
3. A college student takes aptitude and interest tests in order to determine which vocational area his or her profile best matches.
4. African, or "killer," bees cannot be distinguished visually from ordinary domestic honey bees. Ten variables based on chromatograph peaks can be used to readily identify them (Lavine and Carlson 1987).
5. The Air Force wishes to classify each applicant into the training program where he or she has the most potential.
6. Twelve of the *Federalist Papers* were claimed by both Madison and Hamilton. Can we identify authorship by measuring frequencies of word usage (Mosteller and Wallace 1984)?
7. Variables such as availability of fingerprints, availability of eye witnesses, and time until police arrive can be used to classify burglaries into solvable and unsolvable.
8. One approach to speech recognition by computer consists of an attempt to identify phonemes based on the energy levels in speech waves.
9. A number of variables are measured at five weather stations. Based on these variables, we wish to predict the ceiling at a particular airport in 2 hours. The ceiling categories are closed, low instrument, high instrument, low open, and high open (Lachenbruch 1975, p. 2).

9.2 CLASSIFICATION INTO TWO GROUPS

In the case of two populations, we have a sampling unit to be classified, but we do not know to which of the two populations the subject or object belongs. The information we have available consists of the p variables in $\mathbf{y}$ measured on the sampling unit. In the first illustration in Section 9.1, for example, we have an applicant with high school grades and various test scores recorded in $\mathbf{y}$. We do not know if the applicant will succeed or fail at the university, but we have data on previous students who have matriculated at the university for whom it is now known whether they succeeded or failed. By comparing $\mathbf{y}$ with $\overline{\mathbf{y}}_1$ for those who succeeded and $\overline{\mathbf{y}}_2$ for those who failed, we attempt to predict in which group the student will end up.

When there are two populations, we can use a classification procedure due to Fisher (1936). For Fisher's procedure, the principal assumption is that the two populations have the same covariance matrix. Normality is not required. We obtain a sample from each of the two populations and compute $\overline{\mathbf{y}}_1$, $\overline{\mathbf{y}}_2$, and $\mathbf{S}_{\text{pl}}$. A simple procedure for classification can be based on the discriminant function (see Sections 5.5, 5.6, 8.2, and 8.5 for discussion)

$$z = \mathbf{a}'\mathbf{y} = (\overline{\mathbf{y}}_1 - \overline{\mathbf{y}}_2)'\mathbf{S}_{\text{pl}}^{-1}\mathbf{y}, \tag{9.1}$$

where $\mathbf{y}$ is the vector of measurements on a new sampling unit that we wish to classify into one of the two groups (populations). For convenience we speak of "classifying $\mathbf{y}$" rather than the subject or object associated with $\mathbf{y}$.

Denote the two groups by G_1 and G_2. We can evaluate (9.1) for each observation $\mathbf{y}_{1j}$ from the first sample and obtain $z_{11}, z_{12}, \ldots, z_{1n_1}$, from which, by (3.51), $\overline{z}_1 = \sum_{j=1}^{n_1} z_{1j}/n_1 = \mathbf{a}'\overline{\mathbf{y}}_1 = (\overline{\mathbf{y}}_1 - \overline{\mathbf{y}}_2)'\mathbf{S}_{\text{pl}}^{-1}\overline{\mathbf{y}}_1$. Similarly, $\overline{z}_2 = \mathbf{a}'\overline{\mathbf{y}}_2$. Fisher's (1936) *linear classification procedure* assigns $\mathbf{y}$ to G_1 if $z = \mathbf{a}'\mathbf{y}$ is closer to $\overline{z}_1$ than to $\overline{z}_2$ and assigns $\mathbf{y}$ to G_2 if z is closer to $\overline{z}_2$. This is illustrated in Figure 9.1.

For the configuration in Figure 9.1, we see that z is closer to $\overline{z}_1$ if

$$z > \tfrac{1}{2}(\overline{z}_1 + \overline{z}_2). \tag{9.2}$$

This is true in general because $\overline{z}_1$ is always greater than $\overline{z}_2$, which can easily be shown as follows:

$$\overline{z}_1 - \overline{z}_2 = \mathbf{a}'(\overline{\mathbf{y}}_1 - \overline{\mathbf{y}}_2) = (\overline{\mathbf{y}}_1 - \overline{\mathbf{y}}_2)'\mathbf{S}_{\text{pl}}^{-1}(\overline{\mathbf{y}}_1 - \overline{\mathbf{y}}_2) > 0, \tag{9.3}$$

because $\mathbf{S}_{\text{pl}}^{-1}$ is positive definite. Thus $\overline{z}_1 > \overline{z}_2$. [If $\mathbf{a}$ were of the form $\mathbf{a}' = (\overline{\mathbf{y}}_2 - \overline{\mathbf{y}}_1)'\mathbf{S}_{\text{pl}}^{-1}$, then $\overline{z}_2 - \overline{z}_1$ would be positive.] Since $\frac{1}{2}(\overline{z}_1 + \overline{z}_2)$ is the midpoint,

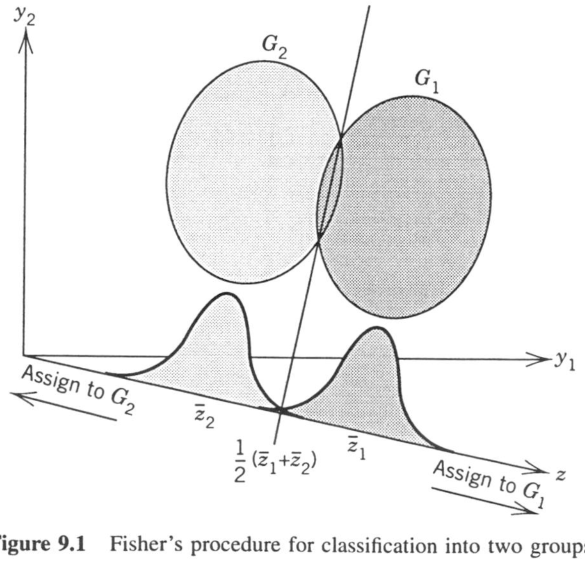

Figure 9.1 Fisher's procedure for classification into two groups.

$z > \frac{1}{2}(\bar{z}_1 + \bar{z}_2)$ implies z is closer to $\bar{z}_1$. Note that the distance from $\bar{z}_1$ to $\bar{z}_2$ is the same as that from $\bar{\mathbf{y}}_1$ to $\bar{\mathbf{y}}_2$.

To express the classification rule in terms of $\mathbf{y}$, we first write $\frac{1}{2}(\bar{z}_1 + \bar{z}_2)$ in the form

$$\tfrac{1}{2}(\bar{z}_1 + \bar{z}_2) = \tfrac{1}{2}(\bar{\mathbf{y}}_1 - \bar{\mathbf{y}}_2)'\mathbf{S}_{\text{pl}}^{-1}(\bar{\mathbf{y}}_1 + \bar{\mathbf{y}}_2). \tag{9.4}$$

Then the classification rule becomes: Assign $\mathbf{y}$ to G_1 if

$$\mathbf{a}'\mathbf{y} = (\bar{\mathbf{y}}_1 - \bar{\mathbf{y}}_2)'\mathbf{S}_{\text{pl}}^{-1}\mathbf{y} > \tfrac{1}{2}(\bar{\mathbf{y}}_1 - \bar{\mathbf{y}}_2)'\mathbf{S}_{\text{pl}}^{-1}(\bar{\mathbf{y}}_1 + \bar{\mathbf{y}}_2), \tag{9.5}$$

and assign $\mathbf{y}$ to G_2 if

$$\mathbf{a}'\mathbf{y} = (\bar{\mathbf{y}}_1 - \bar{\mathbf{y}}_2)'\mathbf{S}_{\text{pl}}^{-1}\mathbf{y} < \tfrac{1}{2}(\bar{\mathbf{y}}_1 - \bar{\mathbf{y}}_2)'\mathbf{S}_{\text{pl}}^{-1}(\bar{\mathbf{y}}_1 + \bar{\mathbf{y}}_2). \tag{9.6}$$

This *linear classification rule* employs the same discriminant function used in Section 8.2 in connection with descriptive separation of groups. Thus in the two-group case the discriminant function serves as a linear classification function. However, in the several-group case in Section 9.3, we use classification functions that are different from the descriptive discriminant functions in Section 8.4.

Fisher's (1936) approach using (9.5) and (9.6) is essentially nonparametric because no distributional assumptions were made. However, if the two populations are normal with equal covariance matrices, then this method is (asymptotically) optimal; that is, the probability of misclassification is minimized, as noted below.

If *prior probabilities* p_1 and p_2 are known for the two populations, the classification rule can be modified. By prior probabilities, we mean that p_1 is the proportion of observations in G_1 and p_2 is the proportion in G_2, where $p_2 = 1 - p_1$. For example, suppose that at a certain university 70% of entering freshmen ultimately graduate. Then $p_1 = .7$ and $p_2 = .3$.

In order to use the prior probabilities, the density functions for the two populations, $f(\mathbf{y}|G_1)$ and $f(\mathbf{y}|G_2)$, must also be known. Then the optimal classification rule that minimizes the probability of misclassification is (Welch 1939): Assign $\mathbf{y}$ to G_1 if

$$p_1 f(\mathbf{y}|G_1) > p_2 f(\mathbf{y}|G_2) \tag{9.7}$$

and to G_2 otherwise. Note that $f(\mathbf{y}|G_1)$ is a convenient notation for the density when sampling from the population represented by G_1. It does not represent a conditional distribution in the usual sense.

Assuming that the two densities are multivariate normal with equal covariance matrices, namely, $f(\mathbf{y}|G_1) = N_p(\boldsymbol{\mu}_1, \boldsymbol{\Sigma})$ and $f(\mathbf{y}|G_2) = N_p(\boldsymbol{\mu}_2, \boldsymbol{\Sigma})$, then application of (9.7) yields (with estimates in place of $\boldsymbol{\mu}_1$, $\boldsymbol{\mu}_2$, and $\boldsymbol{\Sigma}$): Assign $\mathbf{y}$ to G_1 if

$$\mathbf{a}'\mathbf{y} = (\overline{\mathbf{y}}_1 - \overline{\mathbf{y}}_2)'\mathbf{S}_{\text{pl}}^{-1}\mathbf{y} > \tfrac{1}{2}(\overline{\mathbf{y}}_1 - \overline{\mathbf{y}}_2)'\mathbf{S}_{\text{pl}}^{-1}(\overline{\mathbf{y}}_1 + \overline{\mathbf{y}}_2) + \ln\frac{p_2}{p_1} \tag{9.8}$$

and to G_2 otherwise. Because we have substituted estimates for the parameters, the rule in (9.8) is no longer optimal as in (9.7). However, it is *asymptotically optimal* (approaches optimality as the sample size increases).

If $p_1 = p_2$, the normal based classification rule in (9.8) becomes the same as Fisher's procedure given in (9.5). Thus Fisher's rule, which is not based on a normality assumption, has optimal properties when the data come from multivariate normal populations with $\boldsymbol{\Sigma}_1 = \boldsymbol{\Sigma}_2$ and $p_1 = p_2$. (For the case when $\boldsymbol{\Sigma}_1 \neq \boldsymbol{\Sigma}_2$, see Rencher 1995, Section 6.2.2.) Hence, even though Fisher's method is nonparametric, it works better for normally distributed populations or other populations with linear trends. For example, suppose two populations have 95% contours, as in Figure 9.2. If the points are projected in any direction onto a straight line, there will be almost total overlap. A linear discriminant procedure will not successfully separate the two populations.

Example 9.2. For the psychological data of Table 5.1, $\overline{\mathbf{y}}_1, \overline{\mathbf{y}}_2$, and $\mathbf{S}_{\text{pl}}$ were obtained in Example 5.4.2. The discriminant function coefficients were obtained in Example 5.5 as $\mathbf{a}' = (.4856, -.2028, .4654, -.3048)$. For G_1 (the male group), we find

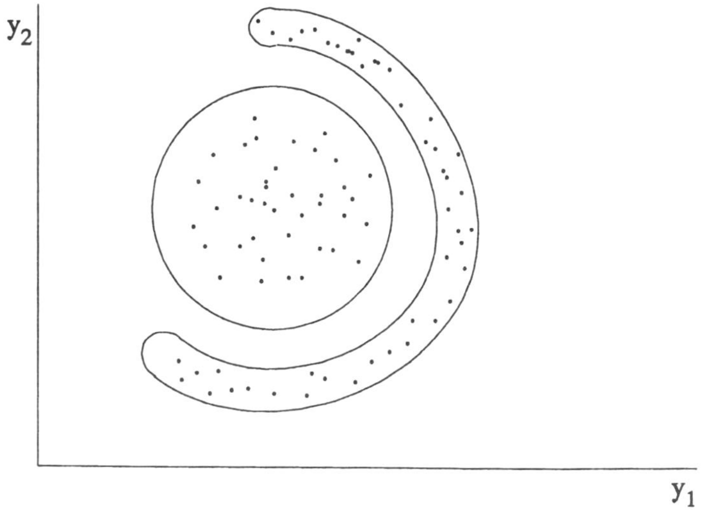

Figure 9.2 Two populations with nonlinear separation.

$$\bar{z}_1 = \mathbf{a}'\bar{\mathbf{y}}_1 = .4856(15.97) - .2028(15.91) + .4654(27.19) - .3048(22.75)$$
$$= 10.2482.$$

Similarly, for G_2 (the female group), $\bar{z}_2 = \mathbf{a}'\bar{\mathbf{y}}_2 = 4.2050$. Thus we assign an observation vector $\mathbf{y}$ to G_1 if

$$z = \mathbf{a}'\mathbf{y} > \tfrac{1}{2}(\bar{z}_1 + \bar{z}_2) = 7.2266$$

and assign $\mathbf{y}$ to G_2 if $z < 7.2266$.

There are no new observations available, so we will illustrate the procedure by classifying two of the observations in G_1. For $\mathbf{y}'_{11} = (15, 17, 24, 14)$, the first observation in G_1, we have $z_{11} = \mathbf{a}'\mathbf{y}_{11} = .4856(15) - .2028(17) + .4654(24) - .3048(14) = 10.7388$, which is greater than 7.2266, and $\mathbf{y}_{11}$ would be correctly classified as belonging to G_1. For $\mathbf{y}'_{14} = (13, 12, 10, 16)$, the fourth observation in G_1, we find $z_{14} = 3.6568$, which would misclassify $\mathbf{y}_{14}$ into G_2.

9.3 CLASSIFICATION INTO SEVERAL GROUPS

In this section we discuss classification rules for several groups. As in the two-group case, we use a sample from each of the k groups to establish a profile to be compared with a vector $\mathbf{y}$ whose group membership is unknown. We classify $\mathbf{y}$ into the group whose profile it most resembles. One measure of resemblance is distance from $\mathbf{y}$ to each of the sample mean vectors $\bar{\mathbf{y}}_1, \bar{\mathbf{y}}_2, \ldots, \bar{\mathbf{y}}_k$.

9.3.1 Equal Population Covariance Matrices: Linear Classification Functions

If $\boldsymbol{\Sigma}_1 = \boldsymbol{\Sigma}_2 = \cdots = \boldsymbol{\Sigma}_k$, we can estimate the common population covariance matrix by a pooled sample covariance matrix

$$\mathbf{S}_{\text{pl}} = \frac{1}{N-k} \sum_{i=1}^{k} (n_i - 1)\mathbf{S}_i = \frac{\mathbf{E}}{N-k},$$

where n_i and $\mathbf{S}_i$ are the sample size and covariance matrix of the ith group, $\mathbf{E}$ is the error matrix from one-way MANOVA, and $N = \Sigma_i n_i$. We compare $\mathbf{y}$ to each $\bar{\mathbf{y}}_i, i = 1, 2, \ldots, k$, by the distance function

$$D_i^2(\mathbf{y}) = (\mathbf{y} - \bar{\mathbf{y}}_i)'\mathbf{S}_{\text{pl}}^{-1}(\mathbf{y} - \bar{\mathbf{y}}_i) \tag{9.9}$$

and assign $\mathbf{y}$ to the group for which $D_i^2(\mathbf{y})$ is smallest.

We can obtain a linear classification rule by expanding (9.9):

$$\begin{aligned} D_i^2(\mathbf{y}) &= \mathbf{y}'\mathbf{S}_{\text{pl}}^{-1}\mathbf{y} - \mathbf{y}'\mathbf{S}_{\text{pl}}^{-1}\overline{\mathbf{y}}_i - \overline{\mathbf{y}}_i'\mathbf{S}_{\text{pl}}^{-1}\mathbf{y} + \overline{\mathbf{y}}_i'\mathbf{S}_{\text{pl}}^{-1}\overline{\mathbf{y}}_i \\ &= \mathbf{y}'\mathbf{S}_{\text{pl}}^{-1}\mathbf{y} - 2\overline{\mathbf{y}}_i'\mathbf{S}_{\text{pl}}^{-1}\mathbf{y} + \overline{\mathbf{y}}_i'\mathbf{S}_{\text{pl}}^{-1}\overline{\mathbf{y}}_i. \end{aligned}$$

The first term on the right can be neglected since it is not a function of i and, consequently, does not change from group to group. The second term is a linear function of $\mathbf{y}$, and the third does not involve $\mathbf{y}$. We thus delete $\mathbf{y}'\mathbf{S}_{\text{pl}}^{-1}\mathbf{y}$ and obtain a *linear classification function*. We will emphasize this by denoting it as $L_i(\mathbf{y})$. If we multiply by $-\frac{1}{2}$ so as to agree with the rule based on the normal distribution and prior probabilities given in (9.11) below, our linear classification rule becomes: Assign $\mathbf{y}$ to the group for which

$$L_i(\mathbf{y}) = \overline{\mathbf{y}}_i'\mathbf{S}_{\text{pl}}^{-1}\mathbf{y} - \tfrac{1}{2}\overline{\mathbf{y}}_i'\mathbf{S}_{\text{pl}}^{-1}\overline{\mathbf{y}}_i \tag{9.10}$$

is a maximum (we reversed the sign when multiplying by $-\frac{1}{2}$). To highlight the linearity of (9.10) as a function of $\mathbf{y}$, we can express it as

$$L_i(\mathbf{y}) = \mathbf{c}_i'\mathbf{y} + c_{i0} = c_{i1}y_1 + c_{i2}y_2 + \cdots + c_{ip}y_p + c_{i0},$$

where $\mathbf{c}_i' = \overline{\mathbf{y}}_i'\mathbf{S}_{\text{pl}}^{-1}$ and $c_{i0} = -\frac{1}{2}\overline{\mathbf{y}}_i'\mathbf{S}_{\text{pl}}^{-1}\overline{\mathbf{y}}_i$. To carry out this procedure, we calculate $\mathbf{c}_i$ and c_{i0} for each of the k groups, evaluate $L_i(\mathbf{y})$, $i = 1, 2, \ldots, k$, and allocate $\mathbf{y}$ to the group for which $L_i(\mathbf{y})$ is largest. This will be the same group for which $D_i^2(\mathbf{y})$ in (9.9) is smallest, that is, the group whose mean vector $\overline{\mathbf{y}}_i$ is closest to $\mathbf{y}$.

For the case of several groups, the optimal rule in (9.7) extends to: Assign $\mathbf{y}$ to the group for which $p_i f(\mathbf{y}|G_i)$ is maximum. With this rule, the probability of misclassification is minimized. If we assume normality with equal covariance matrices and with prior probabilities of group membership, $p_1, p_2, \ldots, p_k$, then $f(\mathbf{y}|G_i) = N_p(\boldsymbol{\mu}_i, \boldsymbol{\Sigma})$, and this rule becomes (with estimates in place of parameters): Calculate

$$L_i'(\mathbf{y}) = \ln p_i + \overline{\mathbf{y}}_i'\mathbf{S}_{\text{pl}}^{-1}\mathbf{y} - \tfrac{1}{2}\overline{\mathbf{y}}_i'\mathbf{S}_{\text{pl}}^{-1}\overline{\mathbf{y}}_i \tag{9.11}$$

for $i = 1, 2, \ldots, k$ and assign $\mathbf{y}$ to the group with maximum value of $L_i'(\mathbf{y})$. It is clear that if $p_1 = p_2 = \cdots = p_k$, then (9.11), which optimizes the classification rate for the normal distribution, reduces to (9.10), which was based on the heuristic approach of minimizing the distance of $\mathbf{y}$ to $\overline{\mathbf{y}}_i$.

The linear functions $L_i(\mathbf{y})$ defined in (9.10) are called *linear classification functions*, although many writers refer to them as *linear discriminant scores*. They are different from the linear discriminant functions in Sections 6.1.4, 6.4, and 8.4.1, whose coefficients are eigenvectors of $\mathbf{E}^{-1}\mathbf{H}$. In fact, there will be k classification functions and $s = \min(p, k-1)$ discriminant functions, where k is the number of groups and p is the number of variables. In many cases we do not need all s discriminant functions to effectively describe group differences, whereas all k classification functions must be used in assigning observations to groups.

Example 9.3.1. For the football data of Table 8.3, the mean vectors for the three groups are as follows:

$$\begin{aligned} \overline{\mathbf{y}}_1' &= (15.2, 58.9, 20.1, 13.1, 14.7, 12.3) \\ \overline{\mathbf{y}}_2' &= (15.4, 57.4, 19.8, 10.1, 13.5, 11.9) \\ \overline{\mathbf{y}}_3' &= (15.6, 57.8, 19.8, 10.9, 13.7, 11.8). \end{aligned}$$

Using these values of $\overline{\mathbf{y}}_i$ and the pooled covariance matrix $\mathbf{S}_{\text{pl}}$, given in Example 8.5, the linear classification functions (9.10) become

$$\begin{aligned} L_1(\mathbf{y}) &= 7.6y_1 + 13.3y_2 + 4.2y_3 - 1.2y_4 + 14.6y_5 + 8.2y_6 - 641.1 \\ L_2(\mathbf{y}) &= 10.2y_1 + 13.3y_2 + 4.2y_3 - 3.4y_4 + 13.2y_5 + 6.1y_6 - 608.0 \\ L_3(\mathbf{y}) &= 10.9y_1 + 13.3y_2 + 4.1y_3 - 2.7y_4 + 13.1y_5 + 5.2y_6 - 614.6. \end{aligned}$$

We note that y_2 and y_3 have essentially the same coefficients in all three functions and hence do not contribute to classification of $\mathbf{y}$. These same two variables were eliminated in the stepwise discriminant analysis in Example 8.9.

We illustrate the use of these linear functions for the first and third observations in group 1. For the first observation ($\mathbf{y}_{11}$), we obtain

$$\begin{aligned} L_1(\mathbf{y}_{11}) &= 7.6(13.5) + 13.3(57.2) + 4.2(19.5) - 1.2(12.5) + 14.6(14.0) \\ &\quad + 8.2(11.0) - 641.1 = 582.124 \\ L_2(\mathbf{y}_{11}) &= 10.2(13.5) + 13.3(57.2) + 4.2(19.5) - 3.4(12.5) + 13.2(14.0) \\ &\quad + 6.1(11.0) - 608.0 = 578.099 \\ L_3(\mathbf{y}_{11}) &= 10.9(13.5) + 13.3(57.2) + 4.1(19.5) - 2.7(12.5) + 13.1(14.0) \\ &\quad + 5.2(11.0) - 614.6 = 578.760. \end{aligned}$$

We classify $\mathbf{y}_{11}$ into group 1 since $L_1(\mathbf{y}_{11}) = 582.1$ exceeds $L_2(\mathbf{y}_{11})$ and $L_3(\mathbf{y}_{11})$.

For the third observation in group 1 ($\mathbf{y}_{13}$), we obtain

$$L_1(\mathbf{y}_{13}) = 567.054 \qquad L_2(\mathbf{y}_{13}) = 570.290 \qquad L_3(\mathbf{y}_{13}) = 569.137.$$

This observation is misclassified into group 2 since $L_2(\mathbf{y}_{13}) = 570.290$ exceeds $L_1(\mathbf{y}_{13})$ and $L_3(\mathbf{y}_{13})$.

9.3.2 Unequal Population Covariance Matrices: Quadratic Classification Functions

Linear classification procedures are sensitive to heterogeneity of covariance matrices. Observations tend to be classified too frequently into groups whose covariance matrices have larger variances on the diagonal. Thus the population covariance matrices should not be assumed to be equal if there is reason to suspect otherwise.

If $\boldsymbol{\Sigma}_1 = \boldsymbol{\Sigma}_2 = \cdots = \boldsymbol{\Sigma}_k$ does not hold, the classification rules can easily be altered to preserve optimality of classification rates. In place of (9.9), we can use

$$D_i^2(\mathbf{y}) = (\mathbf{y} - \overline{\mathbf{y}}_i)'\mathbf{S}_i^{-1}(\mathbf{y} - \overline{\mathbf{y}}_i), \tag{9.12}$$

where $\mathbf{S}_i$ is the sample covariance matrix for the ith group. As before, we would assign $\mathbf{y}$ to the group for which $D_i^2(\mathbf{y})$ is smallest. With $\mathbf{S}_i$ in place of $\mathbf{S}_{\text{pl}}$, (9.12) cannot be reduced to a linear function of $\mathbf{y}$ as in (9.10) but remains a quadratic function. Hence rules based on $\mathbf{S}_i$ are called *quadratic classification rules.*

If we assume normality with unequal covariance matrices and with prior probabilities $p_1, p_2, \ldots, p_k$, then $f(\mathbf{y}|G_i) = N_p(\boldsymbol{\mu}_i, \boldsymbol{\Sigma}_i)$, and the quadratic classification function based on $p_i f(\mathbf{y}|G_i)$ becomes

$$Q_i(\mathbf{y}) = \ln p_i - \tfrac{1}{2}\ln|\mathbf{S}_i| - \tfrac{1}{2}(\mathbf{y} - \overline{\mathbf{y}}_i)'\mathbf{S}_i^{-1}(\mathbf{y} - \overline{\mathbf{y}}_i). \tag{9.13}$$

Since we seek to maximize $p_i f(\mathbf{y}|G_i)$ [note that all three terms of $Q_i(\mathbf{y})$ are negative], the classification rule is: Assign $\mathbf{y}$ to the group for which $Q_i(\mathbf{y})$ is largest. If $p_1 = p_2 = \cdots = p_k$ or if the p_i's are unknown, the term $\ln p_i$ is deleted.

In order to use a quadratic classification rule based on $\mathbf{S}_i$, each n_i must be greater than p so that $\mathbf{S}_i^{-1}$ will exist. This restriction does not apply to linear classification rules based on $\mathbf{S}_{\text{pl}}$. Note the distinction between p, the number of variables, and p_i, the prior probability for the ith group.

9.4 ESTIMATING MISCLASSIFICATION RATES

In Chapter 8, we assessed the effectiveness of the discriminant functions in group separation by the use of significance tests or by examining $\lambda_i/\Sigma_j \lambda_j$. To judge the ability of classification procedures to predict group membership, we usually use the probability of misclassification, which is known as the *error rate*. We could also use its complement, the *correct classification rate*.

A simple estimate of the error rate can be obtained by trying out the classification procedure on the same data set that has been used to compute the classification functions. This method is commonly referred to as *resubstitution*. Each observation vector $\mathbf{y}_{ij}$ is submitted to the classification functions and assigned to a group. We then count the number of correct classifications and the number of misclassifications. The proportion of misclassifications resulting from resubstitution is called the *apparent error rate*. The results can be conveniently displayed in a *classification table* or *confusion matrix*, such as Table 9.1.

Among the n_1 observations in G_1, n_{11} are correctly classified into G_1 and n_{12} are misclassified into G_2, where $n_1 = n_{11} + n_{12}$. Similarly, of the n_2 observations in G_2, n_{21} are misclassified into G_1 and n_{22} are correctly classified into G_2, where $n_2 = n_{21} + n_{22}$. Thus

$$\begin{aligned} \text{Apparent error rate} &= \frac{n_{12} + n_{21}}{n_1 + n_2} \\ &= \frac{n_{12} + n_{21}}{n_{11} + n_{12} + n_{21} + n_{22}}. \end{aligned} \tag{9.14}$$

Similarly, we can define

$$\text{Apparent correct classification rate} = \frac{n_{11} + n_{22}}{n_1 + n_2}. \tag{9.15}$$

Clearly,

$$\text{Apparent error rate} = 1 - \text{apparent correct classification rate}.$$

The method of resubstitution can be readily extended to the case of several groups.

The apparent error rate is easily obtained and is routinely provided by most classification software programs. It is an estimate of the probability that our classification functions based on the present sample will misclassify a future observation. This probability is called the *actual error rate*. Unfortunately, the apparent error rate underestimates the actual error rate because the data set used to compute the classification functions is also used to evaluate them. The classification functions are optimized for the particular sample and may be capitalizing on chance to some degree, especially for small samples. For other estimates

Table 9.1 Classification Table for Two Groups

Actual Group	Number of Observations	Predicted Group 1	Predicted Group 2
1	n_1	n_{11}	n_{12}
2	n_2	n_{21}	n_{22}

Table 9.2 Classification Table for the Psychological Data of Table 5.1

Actual Group	Number of Observations	Predicted Group 1	Predicted Group 2
Male	32	28	4
Female	32	4	28

of error rates, see Rencher (1996, Section 6.4). In Section 9.5 we consider some approaches to reducing the bias in the apparent error rate.

Example 9.4(a). We use the psychological data of Table 5.1 to illustrate the resubstitution method of estimating the error rate for two groups. To compute the apparent error rate, we classify each of the 64 observations using the linear classification procedure obtained in Example 9.2: Classify as G_1 if $\mathbf{a}'\mathbf{y} > 7.229$ and as G_2 otherwise. The resulting classification table is given in Table 9.2. By (9.14),

$$\text{Apparent error rate} = \frac{n_{12} + n_{21}}{n_1 + n_2} = \frac{4 + 4}{32 + 32} = .125.$$

Example 9.4(b). We use the football data of Table 8.3 to illustrate the use of the resubstitution method of estimating the error rate for the case of more than two groups. The sample covariance matrices for the three groups are almost significantly different, and we will use both linear and quadratic classification functions.

The linear classification functions $L_i(\mathbf{y})$ from (9.10) were given in Example 9.3.1 for the football data. Using these, we classify each of the 90 observations. The results are shown in Table 9.3.

An examination of this data set in Example 8.8 showed that groups 2 and

Table 9.3 Classification Table for the Football Data of Table 8.3 Using Linear Classification Functions

Actual Group	Number of Observations	Predicted Group 1	Predicted Group 2	Predicted Group 3
1	30	26	1	3
2	30	1	20	9
3	30	2	8	20

$$\text{Apparent correct classification rate} = \frac{26 + 20 + 20}{90} = .733$$

Apparent error rate = $1 - .733 = .267$

Table 9.4 Classification Table for the Football Data of Table 8.3 Using Quadratic Classification Functions

Actual Group	Number of Observations	Predicted Group 1	2	3
1	30	27	1	2
2	30	2	21	7
3	30	1	4	25

Apparent correct classification rate $= \dfrac{27 + 21 + 25}{90} = .811$

Apparent error rate $= 1 - .811 = .189$

3 are harder to separate than 1 and 2 or 1 and 3. This pattern is reflected here in the misclassifications. Only 4 of the observation vectors in group 1 are misclassified, while 10 observations in each of groups 2 and 3 are misclassified.

Using the quadratic classification functions $Q_i(\mathbf{y}), i = 1, 2, 3$, in (9.13) and assuming $p_1 = p_2 = p_3$, we obtain the classification results in Table 9.4. There is some improvement in the apparent error rate using quadratic classification functions.

9.5 IMPROVED ESTIMATES OF ERROR RATES

For large samples the apparent error rate has only a small amount of bias for estimating the actual error rate and can be used with little concern. For small samples, however, it is overly optimistic (biased downward) as noted above. We discuss two techniques for improving the apparent error rate, that is, increasing it to a more realistic level.

9.5.1 Partitioning the Sample

One way to avoid bias is to split the sample into two parts, a *training* sample used to construct the classification rule and a *validation* sample used to evaluate it. With the training sample, we calculate linear or quadratic classification functions. We then submit each observation vector in the validation sample to the classification functions obtained from the training sample. Since these observations are not used in calculating the classification functions, the resulting error rate is unbiased. To increase the information gained, we could also reverse the roles of the two samples so that the classification functions are obtained from the validation sample and evaluated on the training sample. The two estimates of error could then be averaged.

Partitioning the sample has at least two disadvantages:

1. It requires large samples that may not be available.
2. It does not evaluate the classification function we will use in practice.

The estimate of error based on half the sample may vary considerably from that based on the entire sample. We prefer to use all or almost all of the data to construct the classification functions so as to minimize the variance of our error rate estimate.

9.5.2 Holdout Method

The *holdout method* is an improved version of the sample splitting procedure in Section 9.5.1. In the holdout procedure, all but one observation is used to compute the classification rule, and this rule is then used to classify the omitted observation. We repeat this procedure for each observation, so that, in a sample of size $N = \Sigma_i n_i$, each observation is classified by a function based on the other $N - 1$ observations. The computation load is increased because N distinct classification procedures have to be constructed. The holdout procedure is also referred to as the *leaving-one-out method* or as *cross validation*. Note that this procedure is used to estimate error rates. The actual classification rule for future observations would be based on all N observations.

Example 9.5.2. We illustrate the holdout method of estimating the error rate for the football data of Table 8.3. Each of the 90 observations is classified by linear classification functions based on the other 89 observations. To begin the procedure, the first observation in group 1 ($\mathbf{y}_{11}$) is held out and the linear classification functions $L_i(\mathbf{y}), i = 1, 2, 3$, in (9.10) are calculated using the remaining 29 observations in group 1 and the 60 observations in groups 2 and 3. The observation $\mathbf{y}_{11}$ is then classified using $L_1(\mathbf{y}), L_2(\mathbf{y})$, and $L_3(\mathbf{y})$. Then $\mathbf{y}_{11}$ is reinserted in group 1 and $\mathbf{y}_{12}$ is held out. The functions $L_1(\mathbf{y}), L_2(\mathbf{y})$, and $L_3(\mathbf{y})$ are recomputed and $\mathbf{y}_{12}$ is then classified. This procedure is followed for each of the 90 observations and the results are in Table 9.5.

As expected, the holdout error rate has increased somewhat from the apparent error rate based on resubstitution in Tables 9.3 and 9.4 in Example 9.4(b).

Table 9.5 Classification Table for the Football Data of Table 8.3 Using the Holdout Method Based on Linear Classification Functions

Actual Group	Number of Observations	Predicted Group 1	2	3
1	30	26	1	3
2	30	1	18	11
3	30	2	9	19

Correct classification rate = $\frac{26 + 18 + 19}{90} = .700$

Error rate = $1 - .700 = .300$

An error rate of .300 is a less optimistic (more realistic) estimate of what the classification functions can do with future samples.

9.6 SUBSET SELECTION

The experimenter often measures a large number of variables so as not to overlook any that might aid in predicting group membership. However, this "scattergun" approach is likely to include some superfluous variables that do not contribute to allocation. For simplicity and parsimony, we may wish to search for redundant variables and delete them. A reduction in the number of variables may in fact lead to improved error rates. An additional consideration is the increase in robustness to nonnormality of linear and quadratic classification functions as p (the number of variables) decreases.

The majority of selection schemes for classification analysis are based on stepwise discriminant analysis or a similar approach (Section 8.9). One finds the subset that best separates groups using Wilks' Λ, for example, and then uses these variables to construct classification functions. Most of the major statistical software packages offer this method. When the "best" subset is selected at each step, an optimistic bias in error rates is introduced. For a discussion of this bias, see Rencher (1992a, 1996, Section 6.7).

Another link between separation and classification is the use of error rates in an informal stopping rule in a stepwise discriminant analysis. Thus, for example, if a subset of 5 variables out of 10 gives a misclassification rate of 33% compared to 30% for the full set of variables, we may decide that the 5 variables are adequate for separating the groups. We could try several subsets of different sizes to see when the error rate begins to escalate noticeably.

Example 9.6(a). In Example 8.9, a stepwise discriminant analysis based on a partial Wilks Λ (or partial F) was carried out for the football data of Table 8.3. Four variables were selected: EYEHD, WDIM, JAW, and EARHD. These same four variables are indicated by the linear classification functions in Example 9.3.1. We now use these four variables to classify the observations using the method of resubstitution to obtain the apparent error rate.

The linear classification functions (9.10) are

$$\begin{aligned}
\text{Group 1}\quad L_1(\mathbf{y}) &= \bar{\mathbf{y}}_1'\mathbf{S}_{\text{pl}}^{-1}\mathbf{y} - \tfrac{1}{2}\bar{\mathbf{y}}_1'\mathbf{S}_{\text{pl}}^{-1}\bar{\mathbf{y}}_1 \\
&= 18.67y_1 + 4.13y_2 + 17.67y_3 + 20.44y_4 - 425.50 \\
\text{Group 2}\quad L_2(\mathbf{y}) &= 21.13y_1 + 1.96y_2 + 16.24y_3 + 18.36y_4 - 392.75 \\
\text{Group 3}\quad L_3(\mathbf{y}) &= 21.87y_1 + 2.67y_2 + 16.13y_3 + 17.46y_4 - 399.63.
\end{aligned}$$

When each observation vector is classified using these linear functions, we obtain the classification results in Table 9.6.

Table 9.6 Classification Table for the Football Data of Table 8.3 Using Linear Classification Functions Based on Four Variables Chosen by Stepwise Selection

Actual Group	Number of Observations	Predicted Group 1	2	3
1	30	26	1	3
2	30	1	20	9
3	30	2	8	20

Table 9.6 is identical to Table 9.3 in Example 9.4(b) where all six variables were used. Thus the four selected variables can classify the sample as well as all six variables.

Example 9.6(b). We illustrate the use of error rates as an informal stopping rule in a stepwise discriminant analysis. Fifteen teacher and pupil behaviors were observed during 5-minute intervals of reading instruction in elementary school classrooms (Rencher, Wadham, and Young 1978). The observations were recorded in rate of occurrences per minute for each variable. The variables were

Teacher Behaviors

1. *Explains*—Explains task to learner.
2. *Models*—Models the task response for the learner.
3. *Questions*—Asks a question to elicit a task response.
4. *Directs*—Gives a direct signal to elicit a task response.
5. *Controls*—Controls management behavior with direction statements or gestures.
6. *Positive*—Gives a positive (affirmative) statement or gesture.
7. *Negative*—Gives a negative statement or gesture.

Pupil Behaviors

8. *Overt delayed*—An overt learner response to task signals that cannot be judged correct or incorrect until later.
9. *Correct*—A correct learner response with relationship to task signals.
10. *Incorrect*—An incorrect learner response with relationship to task signals.
11. *No response*—Learner gives no response with relationship to task signals.
12. *Asks*—Learner asks a question about the task.
13. *Statement*—Learner gives a positive statement or gestures not related to the task.
14. *Inappropriate*—Learner gives inappropriate management behavior.

15. *Other*—Other learner than one being observed gives responses as teacher directs task signals.

The teachers were grouped into four categories:

Group 1: Outstanding teachers
Group 2: Poor teachers
Group 3: First-year teachers
Group 4: Teacher aides

The sample sizes in groups 1–4 were 62, 61, 57, and 41, respectively. Due to the large number of observations, the data are not given here.

The stepwise discriminant analysis was run several times with different threshold F-to-enter values so as to select subsets with different sizes. A classification analysis based on resubstitution was carried out with each of the resulting subsets of variables. In Table 9.7, we compare the overall Wilks' Λ and the apparent correct classification rate.

According to the correct classification rate, we would choose to stop at five variables because of the abrupt change from 5 to 4. On the other hand, the changes in Wilks' Λ are more gradual and no clear stopping point is indicated.

9.7 NONPARAMETRIC PROCEDURES

We have previously discussed both parametric and nonparametric procedures. Welch's optional rule in (9.7) is parametric, whereas Fisher's linear classification rule for two groups as given in (9.5) and (9.6) is essentially nonparametric, since no distributional assumptions were involved in its derivation. However, Fisher's procedure also turns out to be equivalent to the optimal normal-based approach in (9.8). Nonparametric procedures for estimating error rate include

Table 9.7 Stepwise Selection Statistics for the Teacher Data

Number of Variables	Overall Wilks' Λ	Percentage of Correct Classification
15	.132	77.4
10	.159	72.4
9	.170	73.3
8	.182	70.6
7	.195	72.9
6	.211	70.1
5	.231	70.6
4	.256	65.6

the resubstitution and holdout methods. In the next three sections we discuss three additional nonparametric classification procedures.

9.7.1 Multinomial Data

Survey data may consist of observations on several categorical variables. The various combinations of categories constitute the possible outcomes of a multinomial random variable. For example, consider the following four categorical variables: gender (male or female), political party (Republican, Democrat, other), size of city of residence (under 10,000, between 10,000 and 100,000, over 100,000), and education (less than high school graduation, high school graduate, college graduate, advanced degree). The total number of categories in this multinomial distribution is the product of the number of possible states of the individual variables: $2 \times 3 \times 3 \times 4 = 72$. We will use this example to illustrate classification procedures for multinomial data. Suppose we are attempting to predict whether or not a person will vote. Then there are two groups, G_1 and G_2, and we assign a person to one of the groups after observing in which of the 72 categories he or she falls.

Welch's (1939) optimum rule given in (9.7) can be written as: Assign $\mathbf{y}$ to G_1 if

$$\frac{f(\mathbf{y}|G_1)}{f(\mathbf{y}|G_2)} > \frac{p_2}{p_1} \tag{9.16}$$

and to G_2 otherwise. In our categorical example, $f(\mathbf{y}|G_1)$ is represented by q_{1i}, $i = 1, 2, \ldots, 72$, and $f(\mathbf{y}|G_2)$ becomes q_{2i}, $i = 1, 2, \ldots, 72$, where q_{1i} is the probability that a person in Group 1 will fall in the ith category, with an analogous definition for q_{2i}. In terms of these multinomial probabilities, the classification rule in (9.16) becomes: If a person falls in the ith category, assign him or her to G_1 if

$$\frac{q_{1i}}{q_{2i}} > \frac{p_2}{p_1} \tag{9.17}$$

and to G_2 otherwise. If the probabilities q_{1i} and q_{2i} were known, it would be easy to check (9.17) for each i and partition the 72 categories into two subsets, those for which the person would be assigned to G_1 and those corresponding to G_2.

The values of q_{1i} and q_{2i} are usually unknown and must be estimated from a sample. Let n_{1i} and n_{2i} be the numbers of persons in groups 1 and 2 who fall into the ith category, $i = 1, 2, \ldots, 72$. Then we estimate q_{1i} and q_{2i} by

$$\hat{q}_{1i} = \frac{n_{1i}}{N_1} \quad \text{and} \quad \hat{q}_{2i} = \frac{n_{2i}}{N_2}, \tag{9.18}$$

where $N_1 = \Sigma_i n_{1i}$ and $N_2 = \Sigma_i n_{2i}$. However, a large sample size would be required for stable estimates; in any given example, some of the n's may be zero.

Multinomial data can also be classified by ordinary linear classification functions. We must distinguish between ordered and unordered categories. If all of the variables have ordered categories, the data can be submitted directly to an ordinary classification program. In the above example, city size and education are variables of this type. It is customary to assign the categories ranked values such as 1, 2, 3, 4. It has been shown that linear classification functions perform reasonably well on (ordered) discrete data of this type [see Lachenbruch (1975, p.45), Titterington et al. (1981), and Gilbert (1968)].

Unordered categorical variables cannot be handled this same way. Thus the political party variable in the example above should not be coded 1, 2, 3 and entered into the computation of the classification functions. However, an unordered categorical variable with k categories can be replaced by $k - 1$ *dummy* variables (see Sections 6.1.8 and 11.6.2) for use with linear classification functions. For example, the political preference variable with three categories can be converted to two dummy variables as follows:

$$y_1 = \begin{cases} 1 & \text{if Republican} \\ 0 & \text{otherwise} \end{cases} \qquad y_2 = \begin{cases} 1 & \text{if Democrat} \\ 0 & \text{otherwise.} \end{cases}$$

Thus the (y_1, y_2) pair takes the value (1, 0) for a Republican, (0, 1) for a Democrat, and (0, 0) for other. Many software programs will create dummy variables automatically. Note that if a subset selection program is used, the dummy variables for a given categorical variable must be kept together; that is, they must all be included in the chosen subset or all excluded, because all are necessary to describe the categorical variable.

In some cases, such as in medical data collection, there is a mixture of continuous and categorical variables. Various approaches to classification with such data have been discussed by Krzanowski (1975, 1976, 1977, 1979, 1980), Lachenbruch and Goldstein (1979), Tu and Han (1982), and Bayne et al. (1983). See Rencher (1996, Section 6.8) for a discussion of logistic and probit classification, which techniques are useful for certain types of continuous and discrete data that are not normally distributed.

9.7.2 Classification Based on Density Estimators

In (9.8), (9.11), and (9.13) we have linear and quadratic classification rules based on the multivariate normal density and prior probabilities. These normal-based rules arose from Welch's optimal rule that assigns $\mathbf{y}$ to the group for which $p_i f(\mathbf{y}|G_i)$ is maximum. If the form of $f(\mathbf{y}|G_i)$ is nonnormal or unknown, the density can be estimated directly from the data. The approach we describe is known as the *kernel* estimator.

We first describe the kernel method for a univariate continuous random variable y. Suppose y has density $f(y)$, which we wish to estimate using a sample $y_1, y_2, \ldots, y_n$. A simple estimate of $f(y_0)$ for an arbitrary point y_0 is based on the proportion of points in the interval $(y_0 - h, y_0 + h)$. If the number of points in the interval is denoted by $N(y_0)$, then the proportion $N(y_0)/n$ is an estimate of $P(y_0 - h < y < y_0 + h)$, which is approximately equal to $2hf(y_0)$. Thus we estimate $f(y_0)$ by

$$\hat{f}(y_0) = \frac{N(y_0)}{2hn}. \tag{9.19}$$

We can express $\hat{f}(y_0)$ as a function of all y_i in the sample by defining

$$K(u) = \begin{cases} \frac{1}{2} & \text{for } |u| \leq 1 \\ 0 & \text{for } |u| > 1 \end{cases}, \tag{9.20}$$

so that $N(y_0) = 2\sum_{i=1}^{n} K[(y_0 - y_i)/h]$, and (9.19) becomes

$$\hat{f}(y_0) = \frac{1}{hn}\sum_{i=1}^{n} K\left(\frac{y_0 - y_i}{h}\right). \tag{9.21}$$

The function $K(u)$ is called the *kernel.* Kernel estimators were first proposed by Rosenblatt (1956) and Parzen (1962). A good review of nonparametric density estimation including kernel estimators has been given by Silverman (1986), who noted that classification analysis provided the initial motivation for the development of density estimation.

In (9.21), $K[(y_0 - y_i)/h]$ is $\frac{1}{2}$ for any point y_i in the interval $(y_0 - h, y_0 + h)$ and is zero for points outside the interval. Points in the interval add $1/2hn$ to the density and points outside the interval contribute nothing. Thus the kernel defined by (9.20) is rectangular and the graph of $\hat{f}(y_0)$ plotted as a function of y_0 will be a step function, since there will be a jump (or drop) whenever y_0 is a distance h from one of the y_i's. (A moving average has a similar property.)

To obtain a smooth estimator of $f(y)$, we must choose a smooth kernel. Two possibilities are

$$K(u) = \frac{1}{\pi}\,\frac{\sin^2 u}{u^2} \tag{9.22}$$

and

$$K(u) = \frac{1}{\sqrt{2\pi}}\,e^{-u^2/2}, \tag{9.23}$$

which have the property that all n sample points $y_1, y_2, \ldots, y_n$ contribute to $\hat{f}(y_0)$ with the closest points weighted heavier than the more distant points. Even though $K(u)$ in (9.23) has the form of the normal distribution, this does not imply any assumption about the density $f(y)$. We have used the normal density function because it is symmetric and unimodel. Other density functions could be used as kernels.

Cacoullos (1966) provided kernel estimates for multivariate density functions. If $\mathbf{y}_0' = (y_{01}, y_{02}, \ldots, y_{0p})$ is an arbitrary point whose density we wish to estimate, then the extension of (9.21) is

$$\hat{f}(\mathbf{y}_0) = \frac{1}{nh_1h_2, \ldots, h_p} \sum_{i=1}^{n} K\left(\frac{y_{01} - y_{i1}}{h_1}, \ldots, \frac{y_{0p} - y_{ip}}{h_p} \right). \tag{9.24}$$

An estimate $\hat{f}(\mathbf{y}_0)$ based on a multivariate normal kernel is given by

$$\hat{f}(\mathbf{y}_0) = \frac{1}{nh^p|\mathbf{S}_{\text{pl}}|^{1/2}} \sum_{i=1}^{n} e^{-(\mathbf{y}_0-\mathbf{y}_i)'\mathbf{S}_{\text{pl}}^{-1}(\mathbf{y}_0-\mathbf{y}_i)/2h^2}, \tag{9.25}$$

where the h_i's are equal and $\mathbf{S}_{\text{pl}}$ is the pooled covariance matrix from the k groups in the sample. The covariance matrix $h^2\mathbf{S}_{\text{pl}}$ could be replaced by other forms. Two examples are (1) $h^2\mathbf{S}_i$ for the ith group and (2) a diagonal matrix. A diagonal matrix would reduce (9.25) to a sum of products

The choice of the smoothing parameter h is critical in using kernel density estimators. The size of h determines how much each y_i contributes to $\hat{f}(\mathbf{y}_0)$. If h is too small, $\hat{f}(\mathbf{y}_0)$ has a peak at each $\mathbf{y}_i$, and if h is too large, $\hat{f}(\mathbf{y}_0)$ is almost uniform (overly smoothed). Therefore, the value chosen for h must depend on the sample size n to avoid too much or too little smoothing; the larger the sample size, the smaller h should be. In practice, we could try several values of h and check the resulting error rates from the classification analysis.

To use the kernel method of density estimation in classification, we can apply it to each group to obtain $\hat{f}(\mathbf{y}_0|G_1), \hat{f}(\mathbf{y}_0|G_2), \ldots, \hat{f}(\mathbf{y}_0|G_k)$, where $\mathbf{y}_0$ is the vector of measurements for an individual of unknown group membership. The classification rule then becomes: Assign $\mathbf{y}_0$ to the group G_i for which

$$p_i\hat{f}(\mathbf{y}_0|G_i) \text{ is maximum.} \tag{9.26}$$

Habbema et al. (1974) proposed a forward selection method for classification based on density estimation. Wegman (1972) and Habbema et al. (1978) found that the size of the h_i's is more important than the shape of the kernel. The choice of h was investigated by Pfeiffer (1985) in a stepwise mode. Remme et al. (1980) compared linear, quadratic, and kernel classification methods for

Table 9.8 Classification Table for the Football Data of Table 8.3 Using the Density Estimation Method of Classification with Multivariate Normal Kernel

Actual Group	Number of Observations	Predicted Group 1	2	3
1	30	25	1	4
2	30	0	12	18
3	30	0	3	27

Apparent correct classification rate = $\frac{25 + 12 + 27}{90} = .711$

Apparent error rate = 1 − .711 = .289

two groups and reported that for multivariate normal data with equal covariance matrices, the linear classifications were clearly superior. For some cases with departures from these assumptions, the kernel methods gave better results.

Example 9.7.2. We illustrate the density estimation method of classification for the football data of Table 8.3. We use the multivariate normal kernel in (9.25) with $h = 2$ to obtain $f(\mathbf{y}_0|G_i), i = 1, 2, 3$, for the three groups. Using $p_1 = p_2 = p_3$, the rule in (9.26) becomes: Assign $\mathbf{y}_0$ to the group for which $\hat{f}(\mathbf{y}_0|G_i)$ is greatest. To obtain an apparent error rate, we follow this procedure for each of the 90 observations and obtain the classification results in Table 9.8.

Applying a holdout method in which the observation $\mathbf{y}_{ij}$ being classified is excluded from computation of $\hat{f}(\mathbf{y}_{ij}|G_1)$, $\hat{f}(\mathbf{y}_{ij}|G_2)$, and $\hat{f}(\mathbf{y}_{ij}|G_3)$, we obtain the classification results in Table 9.9. As expected, the holdout error rate has increased somewhat from the apparent error rate in Table 9.8.

Table 9.9 Classification Table for the Football Data of Table 8.3 Using the Holdout Method Based on Density Estimation

Actual Group	Number of Observations	Predicted Group 1	2	3
1	30	24	1	5
2	30	0	10	20
3	30	1	3	26

Correct classification rate = $\frac{24 + 10 + 26}{90} = .667$

Error rate = 1 − .667 = .333

9.7.3 Nearest Neighbor Classification Rule

The earliest nonparametric classification method was the *nearest neighbor rule* of Fix and Hodges (1951), also known as the k nearest neighbor rule. The procedure is conceptually simple. We compute the distance from an observation $\mathbf{y}_i$ to all other points $\mathbf{y}_j$ using the distance function

$$(\mathbf{y}_i - \mathbf{y}_j)'\mathbf{S}_{\text{pl}}^{-1}(\mathbf{y}_i - \mathbf{y}_j) \qquad j \neq i.$$

To classify $\mathbf{y}_i$ into one of two groups, the k points nearest to $\mathbf{y}_i$ are examined, and if the majority of the k points belong to G_1, assign $\mathbf{y}_i$ to G_1; otherwise assign $\mathbf{y}_i$ to G_2. If we denote the number of points from G_1 as k_1, with the remaining k_2 points from G_2, where $k = k_1 + k_2$, then the rule can be expressed as: Assign $\mathbf{y}_i$ to G_1 if

$$k_1 > k_2 \tag{9.27}$$

and to G_2 otherwise. If the sample sizes n_1 and n_2 differ, we may wish to use proportions in place of counts: Assign $\mathbf{y}_i$ to G_1 if

$$\frac{k_1}{n_1} > \frac{k_2}{n_2}. \tag{9.28}$$

A further refinement can be made by taking into account prior probabilities: Assign $\mathbf{y}_i$ to G_1 if

$$\frac{k_1/n_1}{k_2/n_2} > \frac{p_2}{p_1}. \tag{9.29}$$

These rules are easily extended to more than two groups. For example, (9.28) becomes: Assign the observation to the group that has the highest proportion k_i/n_i, where k_i is the number of observations from G_i among the k nearest neighbors of the observation in question.

A decision must be made as to the value of k. Loftsgaarden and Quesenberry (1965) suggest choosing k near $\sqrt{n_i}$ for a typical n_i. In practice, one could try several values of k and use the one with the best error rate.

Reviews and extensions of the nearest neighbor method have been given by Hart (1968), Gates (1972), Hand and Batchelor (1978), Chidananda Gowda and Krishna (1979), Rogers and Wagner (1978), and Brown and Koplowitz (1979).

Example 9.7.3. We use the football data of Table 8.3 to illustrate the k nearest neighbor method of estimating error rate, with $k = 5$. Since $n_1 = n_2 = n_3 = 30$ and the p_i's are also assumed to be equal, we simply examine the five points closest to a point $\mathbf{y}$ and classify $\mathbf{y}$ into the group that has the most

Table 9.10 Classification Table for the Football Data of Table 8.3 Using the k Nearest Neighbor Method with $k = 5$

Actual Group	Number of Observations	Predicted Group 1	2	3
1	30	26	0	1
2	30	1	19	9
3	30	1	4	22

Correct classification rate $= \dfrac{26 + 19 + 22}{83} = .807$

Error rate $= 1 - .807 = .193$

points among the five points. If there is a tie, we do not classify the point. For example, if the numbers from G_1, G_2, and G_3 were 1, 2, and 2, respectively, then we do not assign $\mathbf{y}$ to either G_2 or G_3.

For each point $\mathbf{y}_{ij}, i = 1, 2, 3; j = 1, 2, \ldots, 30$, we find the five nearest neighbors and classify the point accordingly. Table 9.10 gives the classification results. As can be seen, there were 3 observations in group 1 that were not classified because of ties, 1 in group 2, and 3 in group 3. This left a total of 83 observations classified.

PROBLEMS

9.1 Show that if $z_{1j} = \mathbf{a}'\mathbf{y}_{1j}, j = 1, 2, \ldots, n_1$, and $z_{2j} = \mathbf{a}'\mathbf{y}_{2j}, j = 1, 2, \ldots, n_2$, where z is the discriminant function defined in (9.1), then $\bar{z}_1 - \bar{z}_2 = (\bar{\mathbf{y}}_1 - \bar{\mathbf{y}}_2)'\mathbf{S}_{\text{pl}}^{-1}(\bar{\mathbf{y}}_1 - \bar{\mathbf{y}}_2)$, as in (9.3).

9.2 With z defined in (9.1), show that $\frac{1}{2}(\bar{z}_1 + \bar{z}_2) = \frac{1}{2}(\bar{\mathbf{y}}_1 - \bar{\mathbf{y}}_2)'\mathbf{S}_{\text{pl}}^{-1}(\bar{\mathbf{y}}_1 + \bar{\mathbf{y}}_2)$, as in (9.4).

9.3 Obtain the normal-based classification rule in (9.8).

9.4 Derive the classification rule in (9.11).

9.5 Derive the quadratic classification function in (9.13).

9.6 Do a classification analysis on the beetle data in Table 5.5 as follows:

(a) Find the classification function $z = (\bar{\mathbf{y}}_1 - \bar{\mathbf{y}}_2)'\mathbf{S}_{\text{pl}}^{-1}\mathbf{y}$ and the cutoff point $\frac{1}{2}(\bar{z}_1 + \bar{z}_2)$.

(b) Find the classification table using the linear classification function in (a).

(c) Find the classification table using the nearest neighbor method.

9.7 Do a classification analysis on the dystrophy data of Table 5.7 as follows:

(a) Find the classification function $z = (\bar{\mathbf{y}}_1 - \bar{\mathbf{y}}_2)'\mathbf{S}_{\text{pl}}^{-1}\mathbf{y}$ and the cutoff point $\frac{1}{2}(\bar{z}_1 + \bar{z}_2)$.

(b) Find the classification table using the linear classification function in (a).

(c) Repeat part (b) using p_1 and p_2 proportional to sample sizes.

9.8 Do a classification analysis on the cyclical data of Table 5.8 as follows:

(a) Find the classification function $z = (\bar{\mathbf{y}}_1 - \bar{\mathbf{y}}_2)'\mathbf{S}_{\text{pl}}^{-1}\mathbf{y}$ and the cutoff point $\frac{1}{2}(\bar{z}_1 + \bar{z}_2)$.

(b) Find the classification table using the linear classification function in (a).

(c) Find the classification table using the holdout method.

(d) Find the classification table using a kernel density estimator method.

9.9 Using the engineer data of Table 5.6, carry out a classification analysis as follows:

(a) Find the classification table using the linear classification function.

(b) Carry out a stepwise discriminant selection of variables (see problem 8.14).

(c) Find the classification table for the variables selected in part (b).

9.10 Do a classification analysis on the fish data in Table 6.17 as follows. Assume $p_1 = p_2 = p_3$.

(a) Find the linear classification functions.

(b) Find the classification table using the linear classification functions in (a) (assuming $\boldsymbol{\Sigma}_1 = \boldsymbol{\Sigma}_2 = \boldsymbol{\Sigma}_3$).

(c) Find the classification table using quadratic classification functions (assuming population covariance matrices are not equal).

(d) Find the classification table using the holdout method.

(e) Find the classification table using the nearest neighbor method.

9.11 Do a classification analysis on the rootstock data of Table 6.2 as follows:

(a) Find the linear classification functions.

(b) Find the classification table using the linear classification functions in (a) (assuming $\boldsymbol{\Sigma}_1 = \boldsymbol{\Sigma}_2 = \boldsymbol{\Sigma}_3$).

(c) Find the classification table using quadratic classification functions (assuming population covariance matrices are not equal).

(d) Find the classification table using the nearest neighbor method.

(e) Find the classification table using a kernel density estimator method.

CHAPTER 10

Multivariate Regression

10.1 INTRODUCTION

In this chapter, we consider the linear relationship between one or more y's (the *dependent* or *response* variables) and one or more x's (the *independent* variables). We will use a linear model to relate the y's to the x's and will be concerned with estimation and testing of the parameters in the model. One aspect of interest will be choosing which variables to include in the model if this is not already known.

We can distinguish three cases according to the number of variables:

1. Simple linear regression: one y and one x. For example, suppose we wish to predict college grade point average (GPA) based on an applicant's high school GPA.
2. Multiple linear regression: one y and several x's. We could attempt to improve our prediction of college GPA by using more than one independent variable, for example, high school GPA, standardized test scores (such as ACT or SAT), or rating of high school.
3. Multivariate multiple linear regression: several y's and several x's. In the above illustration, we may wish to predict several y's (such as number of years of college the person will complete or GPA in the sciences, arts, and humanities). As another example, suppose the Air Force wishes to predict several measures of pilot efficiency. These response variables could be regressed against independent variables (such as math and science skills, reaction time, eyesight acuity, and manual dexterity).

To further distinguish case 2 from case 3, we could designate case 2 as *univariate* multiple regression because there is only one y. Thus in case 3, "multivariate" indicates that there are several y's and "multiple" implies several x's. The term *multivariate regression* usually refers to case 3.

There are two basic types of independent variables, *fixed* and *random*. In the above illustrations, all x's are random variables and are therefore not under the control of the researcher. A person is chosen at random, and all of the y's and x's are measured, or observed, for that person. In some experimental situations, the x's are fixed, that is, under the control of the experimenter. For example, a researcher may wish to relate yield per acre and nutritional value to level of application of various chemical fertilizers. The experimenter can choose the amount of chemicals to be applied and then observe the changes in the yield and nutritional responses.

In order to provide a solid base for multivariate multiple regression, we review several aspects of multiple regression with fixed x's in Section 10.2. The random-x case for multiple regression is discussed briefly in Section 10.3.

10.2 MULTIPLE REGRESSION: FIXED x's

10.2.1 Model for Fixed x's

In the fixed-x regression model, we express each y in a sample of n observations as a linear function of the x's plus a random error, ϵ:

$$\begin{aligned} y_1 &= \beta_0 + \beta_1 x_{11} + \beta_2 x_{12} + \cdots + \beta_q x_{1q} + \epsilon_1 \\ y_2 &= \beta_0 + \beta_1 x_{21} + \beta_2 x_{22} + \cdots + \beta_q x_{2q} + \epsilon_2 \\ &\vdots \\ y_n &= \beta_0 + \beta_1 x_{n1} + \beta_2 x_{n2} + \cdots + \beta_q x_{nq} + \epsilon_n. \end{aligned} \tag{10.1}$$

Note that the number of x's is denoted by q. The β's in (10.1) are called *regression coefficients*. Additional assumptions that accompany the equations of the model are as follows:

1. $E(\epsilon_i) = 0$ for all $i = 1, 2, \ldots, n$.
2. $\text{var}(\epsilon_i) = \sigma^2$ for all $i = 1, 2, \ldots, n$.
3. $\text{cov}(\epsilon_i, \epsilon_j) = 0$ for all $i \neq j$.

Assumption 1 states that the model is linear and that it is correct in the sense that no additional terms are needed to predict y; all remaining variation in y is purely random and unpredictable. Thus if $E(\epsilon_i) = 0$ and the x's are fixed, then $E(y_i) = \beta_0 + \beta_1 x_{i1} + \beta_2 x_{i2} + \cdots + \beta_q x_{iq}$, and the mean of y is expressible in terms of these q x's with no others needed. In assumption 2, the variance of each ϵ_i is the same, which also implies that $\text{var}(y_i) = \sigma^2$, since the x's are fixed. Assumption 3 imposes the condition that the error terms be uncorrelated, from which it follows that the y's are also uncorrelated, that is, $\text{cov}(y_i, y_j) = 0$.

Using matrix notation, the models for the n observations in (10.1) can be written much more concisely in the form

$$\begin{bmatrix} y_1 \\ y_2 \\ \vdots \\ y_n \end{bmatrix} = \begin{bmatrix} 1 & x_{11} & x_{12} & \dots & x_{1q} \\ 1 & x_{21} & x_{22} & \dots & x_{2q} \\ \vdots & \vdots & \vdots & & \vdots \\ 1 & x_{n1} & x_{n2} & \dots & x_{nq} \end{bmatrix} \begin{bmatrix} \beta_0 \\ \beta_1 \\ \vdots \\ \beta_q \end{bmatrix} + \begin{bmatrix} \epsilon_1 \\ \epsilon_2 \\ \vdots \\ \epsilon_n \end{bmatrix}$$

or

$$\mathbf{y} = \mathbf{X}\boldsymbol{\beta} + \boldsymbol{\epsilon}. \qquad (10.2)$$

With this notation, the above three assumptions become

1. $E(\boldsymbol{\epsilon}) = \mathbf{0}$
2. $\text{cov}(\boldsymbol{\epsilon}) = \sigma^2\mathbf{I}$,

which can be rewritten in terms of $\mathbf{y}$ as

1. $E(\mathbf{y}) = \mathbf{X}\boldsymbol{\beta}$
2. $\text{cov}(\mathbf{y}) = \sigma^2\mathbf{I}$.

Note that the second assumption in matrix form incorporates both the second and third assumptions in univariate form; that is, $\text{cov}(\mathbf{y}) = \sigma^2\mathbf{I}$ implies $\text{var}(y_i) = \sigma^2$ and $\text{cov}(y_i, y_j) = 0$.

For estimation and testing purposes, we need to have $n > q + 1$. Therefore, the matrix expression (10.2) has the following typical pattern:

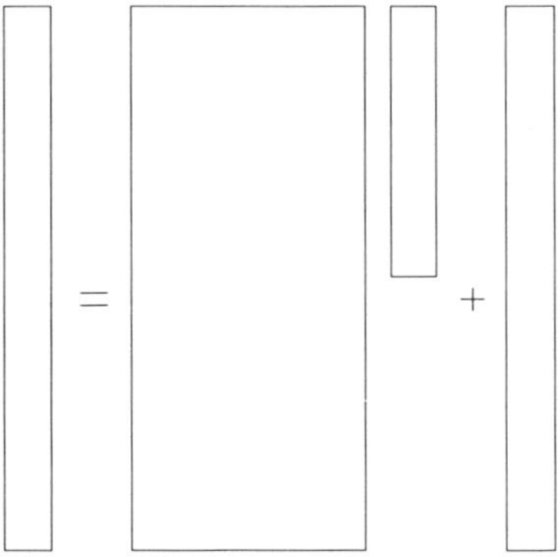

10.2.2 Least Squares Estimation in the Fixed-x Model

As noted above, the first assumption implies $E(y_i) = \beta_0 + \beta_1 x_{i1} + \beta_2 x_{i2} + \cdots + \beta_q x_{iq}$. We seek to estimate the β's and thereby estimate $E(y_i)$. If the estimates are denoted $\hat{\beta}_0, \hat{\beta}_1, \ldots, \hat{\beta}_q$, then $\hat{E}(y_i) = \hat{\beta}_0 + \hat{\beta}_1 x_{i1} + \hat{\beta}_2 x_{i2} + \cdots + \hat{\beta}_q x_{iq}$. However, $\hat{E}(y_i)$ is usually designated $\hat{y}_i$. We now consider the least squares estimates of the β's.

The *least squares* estimates of $\beta_0, \beta_1, \ldots, \beta_q$ minimize the sum of squares of deviations of the n observed y's from their "modeled" values, that is, from their values $\hat{y}_i$ predicted by the model. Thus we seek $\hat{\beta}_0, \hat{\beta}_1, \ldots, \hat{\beta}_q$ that minimize

$$\begin{aligned} \text{SSE} &= \sum_{i=1}^{n} \hat{\epsilon}_i^2 = \sum_{i=1}^{n} (y_i - \hat{y}_i)^2 \\ &= \sum_{i=1}^{n} (y_i - \hat{\beta}_0 - \hat{\beta}_1 x_{i1} - \hat{\beta}_2 x_{i2} - \cdots - \hat{\beta}_q x_{iq})^2. \end{aligned} \tag{10.3}$$

The values of $\hat{\beta}_0, \hat{\beta}_1, \ldots, \hat{\beta}_q$ that minimize SSE in (10.3) can be found in vector form by

$$\hat{\boldsymbol{\beta}} = (\mathbf{X}'\mathbf{X})^{-1}\mathbf{X}'\mathbf{y}, \tag{10.4}$$

where $\hat{\boldsymbol{\beta}}' = (\hat{\beta}_0, \hat{\beta}_1, \ldots, \hat{\beta}_q)$. In (10.4), we assume that $\mathbf{X}'\mathbf{X}$ is nonsingular. This will ordinarily hold if $n > q + 1$ and no x_i is a linear combination of other x's.

In expression (10.4), we see a characteristic pattern. The product $\mathbf{X}'\mathbf{y}$ can be used to compute the covariances of the x's with y. The product $\mathbf{X}'\mathbf{X}$ can be used to obtain the covariance matrix of the x's, which includes the variances of the x's. If $\mathbf{X}'\mathbf{X}$ were diagonal, then $\hat{\beta}_i$ would be large if the (sample) covariance of y and x_i were large or if the (sample) variance of x_i were small. Since $\mathbf{X}'\mathbf{X}$ is typically not diagonal, the $\hat{\beta}_i$'s are also affected by the relationships among the x's.

The least squares estimate of $\boldsymbol{\beta}$ in (10.4) can be derived with calculus, but we demonstrate algebraically that $\hat{\boldsymbol{\beta}} = (\mathbf{X}'\mathbf{X})^{-1}\mathbf{X}'\mathbf{y}$ minimizes SSE. If we designate the ith row of $\mathbf{X}$ as $\mathbf{x}_i' = (1, x_{i1}, x_{i2}, \ldots, x_{iq})$, we can write (10.3) as

$$\text{SSE} = \sum_{i=1}^{n} (y_i - \mathbf{x}_i'\hat{\boldsymbol{\beta}})^2.$$

The quantity $y_i - \mathbf{x}_i'\hat{\boldsymbol{\beta}}$ is the ith element of the vector $\mathbf{y} - \mathbf{X}\hat{\boldsymbol{\beta}}$. Hence, by (2.33),

$$\text{SSE} = (\mathbf{y} - \mathbf{X}\hat{\boldsymbol{\beta}})'(\mathbf{y} - \mathbf{X}\hat{\boldsymbol{\beta}}).$$

Let $\mathbf{b}$ be an alternative estimate that may lead to a smaller value of SSE than $\hat{\boldsymbol{\beta}}$. We add $\mathbf{X}(\hat{\boldsymbol{\beta}} - \mathbf{b})$ to see if this reduces SSE:

$$\text{SSE} = [(\mathbf{y} - \mathbf{X}\hat{\boldsymbol{\beta}}) + \mathbf{X}(\hat{\boldsymbol{\beta}} - \mathbf{b})]'[(\mathbf{y} - \mathbf{X}\hat{\boldsymbol{\beta}}) + \mathbf{X}(\hat{\boldsymbol{\beta}} - \mathbf{b})].$$

We now expand this using the two terms $\mathbf{y} - \mathbf{X}\hat{\boldsymbol{\beta}}$ and $\mathbf{X}(\hat{\boldsymbol{\beta}} - \mathbf{b})$ to obtain

$$\begin{aligned}\text{SSE} &= (\mathbf{y} - \mathbf{X}\hat{\boldsymbol{\beta}})'(\mathbf{y} - \mathbf{X}\hat{\boldsymbol{\beta}}) + [\mathbf{X}(\hat{\boldsymbol{\beta}} - \mathbf{b})]'\mathbf{X}(\hat{\boldsymbol{\beta}} - \mathbf{b}) \\ &\quad + 2[\mathbf{X}(\hat{\boldsymbol{\beta}} - \mathbf{b})]'(\mathbf{y} - \mathbf{X}\hat{\boldsymbol{\beta}}) \\ &= (\mathbf{y} - \mathbf{X}\hat{\boldsymbol{\beta}})'(\mathbf{y} - \mathbf{X}\hat{\boldsymbol{\beta}}) + (\hat{\boldsymbol{\beta}} - \mathbf{b})'\mathbf{X}'\mathbf{X}(\hat{\boldsymbol{\beta}} - \mathbf{b}) \\ &\quad + 2(\hat{\boldsymbol{\beta}} - \mathbf{b})'\mathbf{X}'(\mathbf{y} - \mathbf{X}\hat{\boldsymbol{\beta}}) \\ &= (\mathbf{y} - \mathbf{X}\hat{\boldsymbol{\beta}})'(\mathbf{y} - \mathbf{X}\hat{\boldsymbol{\beta}}) + (\hat{\boldsymbol{\beta}} - \mathbf{b})'\mathbf{X}'\mathbf{X}(\hat{\boldsymbol{\beta}} - \mathbf{b}) \\ &\quad + 2(\hat{\boldsymbol{\beta}} - \mathbf{b})'(\mathbf{X}'\mathbf{y} - \mathbf{X}'\mathbf{X}\hat{\boldsymbol{\beta}}).\end{aligned}$$

The third term vanishes if we substitute $\hat{\boldsymbol{\beta}} = (\mathbf{X}'\mathbf{X})^{-1}\mathbf{X}'\mathbf{y}$ into $\mathbf{X}'\mathbf{X}\hat{\boldsymbol{\beta}}$. The second term is a positive definite quadratic form, and SSE is therefore minimized when $\mathbf{b} = \hat{\boldsymbol{\beta}}$. Thus no value of $\mathbf{b}$ can reduce SSE from the value given by $\hat{\boldsymbol{\beta}}$. For a review of properties of $\hat{\boldsymbol{\beta}}$ and an alternative approach based on the assumption that $\mathbf{y}$ is normally distributed, see Rencher (1996, Chapter 7).

10.2.3 An Estimator for σ^2

It can be shown that

$$E(\text{SSE}) = \sigma^2[n - (q + 1)] = \sigma^2(n - q - 1). \tag{10.5}$$

We can therefore obtain an unbiased estimator of σ^2 as

$$s^2 = \frac{\text{SSE}}{n - q - 1} = \frac{1}{n - q - 1}(\mathbf{y} - \mathbf{X}\hat{\boldsymbol{\beta}})'(\mathbf{y} - \mathbf{X}\hat{\boldsymbol{\beta}}). \tag{10.6}$$

We can also express SSE in the form

$$\text{SSE} = \mathbf{y}'\mathbf{y} - \hat{\boldsymbol{\beta}}'\mathbf{X}'\mathbf{y}, \tag{10.7}$$

and we note that there are n terms in $\mathbf{y}'\mathbf{y}$ and $q + 1$ terms in $\hat{\boldsymbol{\beta}}'\mathbf{X}'\mathbf{y}$. This corresponds to the denominator of SSE in (10.6).

It is not surprising that the degrees of freedom (denominator) for SSE is reduced by $q + 1$, the number of parameters in $\boldsymbol{\beta}$ to be estimated. The need for such an adjustment can be illustrated with a simple random sample of a random variable y from a population with mean μ and variance σ^2, in which case $\sum_i(y_i - \mu)^2$ has n degrees of freedom, while $\sum_i(y_i - \bar{y})^2$ has $n - 1$. It is intuitively clear that

$$E\left[\sum_{i=1}^{n}(y_i - \mu)^2\right] > E\left[\sum_{i=1}^{n}(y_i - \bar{y})^2\right]$$

because $\bar{y}$ fits the sample better than μ, which is the mean of the population but not of the sample. Thus (squared) deviations from $\bar{y}$ will tend to be smaller than deviations from μ. In fact, it is easily shown that

$$\begin{aligned}\sum_{i=1}^{n}(y_i - \mu)^2 &= \sum_{i=1}^{n}(y_i - \bar{y} + \bar{y} - \mu)^2 \\ &= \sum_{i}(y_i - \bar{y})^2 + n(\bar{y} - \mu)^2,\end{aligned} \tag{10.8}$$

whence

$$\sum_{i}(y_i - \bar{y})^2 = \sum_{i}(y_i - \mu)^2 - n(\bar{y} - \mu)^2.$$

Thus $\sum_i (y_i - \bar{y})^2$ is expressible as a sum of n squares minus one square, which corresponds to $n - 1$ degrees of freedom. More formally, we have

$$E\left[\sum_{i}(y_i - \bar{y})^2\right] = n\sigma^2 - \frac{n\sigma^2}{n} = (n - 1)\sigma^2.$$

10.2.4 The Model Corrected for Means

It is sometimes convenient to "center" the x's by subtracting their means, $\bar{x}_1 = \sum_{i=1}^{n} x_{i1}/n, \bar{x}_2 = \sum_{i=1}^{n} x_{i2}/n$, and so on. In terms of centered x's, the model for each y_i in (10.1) becomes

$$y_i = \alpha + \beta_1(x_{i1} - \bar{x}_1) + \beta_2(x_{i2} - \bar{x}_2) + \cdots + \beta_q(x_{iq} - \bar{x}_q) + \epsilon_i. \tag{10.9}$$

If α is defined as

$$\alpha = \beta_0 + \beta_1\bar{x}_1 + \beta_2\bar{x}_2 + \cdots + \beta_q\bar{x}_q, \tag{10.10}$$

then (10.9) is equivalent to (10.1). To estimate

$$\boldsymbol{\beta}_1 = \begin{bmatrix} \beta_1 \\ \beta_2 \\ \vdots \\ \beta_q \end{bmatrix},$$

we use the centered x's in the matrix

$$\mathbf{X}_c = \begin{bmatrix} x_{11} - \bar{x}_1 & x_{12} - \bar{x}_2 & \dots & x_{1q} - \bar{x}_q \\ x_{21} - \bar{x}_1 & x_{22} - \bar{x}_2 & \dots & x_{2q} - \bar{x}_q \\ \vdots & \vdots & & \vdots \\ x_{n1} - \bar{x}_1 & x_{n2} - \bar{x}_2 & \dots & x_{nq} - \bar{x}_q \end{bmatrix}$$

and obtain

$$\hat{\boldsymbol{\beta}}_1 = (\mathbf{X}_c'\mathbf{X}_c)^{-1}\mathbf{X}_c'\mathbf{y}. \tag{10.11}$$

In (10.10), α is seen to be the expected value of y corresponding to the sampie means of the x's. It is not surprising that the estimate of α turns out to be $\bar{y}$:

$$\hat{\alpha} = \bar{y}.$$

In other words, if the origin of the x's is shifted to $(\bar{x}_1, \bar{x}_2, \dots, \bar{x}_q)$, then the intercept of the fitted model is $\bar{y}$. With $\hat{\alpha} = \bar{y}$, we obtain

$$\hat{\beta}_0 = \hat{\alpha} - \hat{\beta}_1\bar{x}_1 - \hat{\beta}_2\bar{x}_2 - \cdots - \hat{\beta}_q\bar{x}_q = \bar{y} - \hat{\boldsymbol{\beta}}_1'\bar{\mathbf{x}} \tag{10.12}$$

as an estimate of β_0 in (10.10). The estimators $\hat{\beta}_0$ and $\hat{\boldsymbol{\beta}}_1$ in (10.12) and (10.11) give the same values as the usual least squares estimators $\hat{\boldsymbol{\beta}}$ in (10.4).

We can express $\hat{\boldsymbol{\beta}}_1$ in (10.11) in terms of sample variances and covariances. The overall sample covariance matrix of y and the x's is

$$\mathbf{S} = \left[\begin{array}{c|cccc} s_{yy} & s_{y1} & s_{y2} & \cdots & s_{yq} \\ \hline s_{1y} & s_{11} & s_{12} & \cdots & s_{1q} \\ \vdots & \vdots & \vdots & & \vdots \\ s_{qy} & s_{q1} & s_{q2} & \cdots & s_{qq} \end{array}\right] = \begin{pmatrix} s_{yy} & \mathbf{s}_{yx}' \\ \mathbf{s}_{yx} & \mathbf{S}_{xx} \end{pmatrix},$$

where s_{yy} is the variance of y, s_{yj} is the covariance of y and x_j, s_{ii} is the variance of x_i, s_{ij} is the covariance of x_i and x_j, and $\mathbf{s}_{yx}' = (s_{y1}, s_{y2}, \dots, s_{yq})$. These sample variances and covariances are mathematically equivalent to analogous formulas (3.21) and (3.23) for random variables, where the sample variances and covariances were estimates of population variances and covariances. However, here the x's are considered to be constants that remain fixed from sample to sample, and a formula such as $s_{11} = \sum_{i=1}^{n} (x_{i1} - \bar{x}_1)^2/(n-1)$ summarizes the spread in the n values of x_1 but does not estimate a population variance.

To express $\hat{\boldsymbol{\beta}}_1$ in terms of $\mathbf{S}_{xx}$ and $\mathbf{s}_{yx}$, we note first that the diagonal elements of $\mathbf{X}_c'\mathbf{X}_c$ are corrected sums of squares. For example, in the second diagonal position, we have

$$\sum_{i=1}^{n} (x_{i2} - \bar{x}_2)^2 = (n-1)s_{22}.$$

The off-diagonal elements of $\mathbf{X}_c'\mathbf{X}_c$ are analogous corrected sums of products; for example, the element in the (1, 2) position is

$$\sum_{i=1}^{n} (x_{i1} - \bar{x}_1)(x_{i2} - \bar{x}_2) = (n-1)s_{12}.$$

Thus

$$\frac{1}{n-1} \mathbf{X}_c'\mathbf{X}_c = \mathbf{S}_{xx}. \tag{10.13}$$

Similarly,

$$\frac{1}{n-1} \mathbf{X}_c'\mathbf{y} = \mathbf{s}_{yx}, \tag{10.14}$$

although this needs some additional justification, because $\mathbf{y}$ has not been centered. The second element of $\mathbf{X}_c'\mathbf{y}$, for example, is $\sum_i (x_{i2} - \bar{x}_2)y_i$, which is equal to $(n-1)s_{2y}$:

$$\begin{aligned}
(n-1)s_{2y} &= \sum_{i=1}^{n} (x_{i2} - \bar{x}_2)(y_i - \bar{y}) \\
&= \sum_{i} (x_{i2} - \bar{x}_2)y_i - \sum_{i} (x_{i2} - \bar{x}_2)\bar{y} \\
&= \sum_{i} (x_{i2} - \bar{x}_2)y_i,
\end{aligned}$$

since $\sum_i (x_{i2} - \bar{x}_2) = 0$. Now, multiplying and dividing by $n - 1$ in (10.11), we obtain

$$\hat{\boldsymbol{\beta}}_1 = (n - 1)(\mathbf{X}_c'\mathbf{X}_c)^{-1} \frac{\mathbf{X}_c'\mathbf{y}}{n-1} = \left(\frac{\mathbf{X}_c'\mathbf{X}_c}{n-1} \right)^{-1} \frac{\mathbf{X}_c'\mathbf{y}}{n-1}$$

$$= \mathbf{S}_{xx}^{-1}\mathbf{s}_{yx} \qquad \text{[by (10.13) and (10.14)]}, \tag{10.15}$$

and substituting this in (10.12) gives

$$\hat{\beta}_0 = \hat{\alpha} - \hat{\boldsymbol{\beta}}_1'\bar{\mathbf{x}} = \bar{y} - \mathbf{s}_{yx}'\mathbf{S}_{xx}^{-1}\bar{\mathbf{x}}. \tag{10.16}$$

10.2.5 Hypothesis Tests

In this section, we review two basic tests on the β's. We assume that $\mathbf{y}$ is $N_n(\mathbf{X}\boldsymbol{\beta}, \sigma^2\mathbf{I})$. For other tests and confidence intervals, see Rencher (1996, Section 7.2.7).

10.2.5a Test of Overall Regression

The overall regression hypothesis that none of the x's predict y can be expressed as $H_0\colon \boldsymbol{\beta}_1 = \mathbf{0}$, since $\boldsymbol{\beta}_1' = (\beta_1, \beta_2, \ldots, \beta_q)$. We do not include β_0, because $\beta_0 = 0$ would restrict y to have a mean of zero (assuming $H_0\colon \boldsymbol{\beta}_1 = \mathbf{0}$ is true). The form of SSE given in (10.7), SSE $= \mathbf{y}'\mathbf{y} - \hat{\boldsymbol{\beta}}'\mathbf{X}'\mathbf{y}$, leads to a natural partitioning of the total sum of squares $\mathbf{y}'\mathbf{y}$,

$$\mathbf{y}'\mathbf{y} = (\mathbf{y}'\mathbf{y} - \hat{\boldsymbol{\beta}}'\mathbf{X}'\mathbf{y}) + \hat{\boldsymbol{\beta}}'\mathbf{X}'\mathbf{y}. \tag{10.17}$$

To correct $\mathbf{y}$ for its mean and thereby avoid inclusion of $\beta_0 = 0$, we subtract $n\bar{y}^2$ from both sides to obtain

$$\mathbf{y}'\mathbf{y} - n\bar{y}^2 = (\mathbf{y}'\mathbf{y} - \hat{\boldsymbol{\beta}}'\mathbf{X}'\mathbf{y}) + (\hat{\boldsymbol{\beta}}'\mathbf{X}'\mathbf{y} - n\bar{y}^2) \tag{10.18}$$

$$= \text{SSE} + \text{SSR},$$

where $\mathbf{y}'\mathbf{y} - n\bar{y}^2 = \sum_i (y_i - \bar{y})^2$ is the total sum of squares adjusted for the mean and SSR $= \hat{\boldsymbol{\beta}}'\mathbf{X}'\mathbf{y} - n\bar{y}^2$ is the overall regression sum of squares adjusted for the intercept.

We can test $H_0\colon \boldsymbol{\beta}_1 = \mathbf{0}$ by means of

$$F = \frac{\text{SSR}/q}{\text{SSE}/(n - q - 1)}, \tag{10.19}$$

which is distributed as $F_{q,n-q-1}$ when $H_0\colon \boldsymbol{\beta}_1 = \mathbf{0}$ is true. We reject H_0 if $F > F_\alpha$.

10.2.5b Test on a Subset of the $\boldsymbol{\beta}$'s

In an attempt to simplify the model, we may be interested in testing the hypothesis that some of the β's are zero. For example, in the model

$$y = \beta_0 + \beta_1 x_1 + \beta_2 x_2 + \beta_3 x_1^2 + \beta_4 x_2^2 + \beta_5 x_1 x_2 + \epsilon,$$

we may wish to test the hypothesis $H_0: \beta_3 = \beta_4 = \beta_5 = 0$. If this is true, the model is linear in x_1 and x_2. In other cases, we may want to ascertain whether a single β_i can be deleted.

For convenience of exposition, let the β's that are candidates for deletion be rearranged to appear last in $\boldsymbol{\beta}$ and denote this subset of β's by $\boldsymbol{\beta}_d$, where d reminds us that these β's are to be *deleted* if $H_0: \boldsymbol{\beta}_d = \mathbf{0}$ is accepted. Let the subset to be *retained* in the *reduced* model be denoted by $\boldsymbol{\beta}_r$. Thus $\boldsymbol{\beta}$ is partitioned into

$$\boldsymbol{\beta} = \begin{pmatrix} \boldsymbol{\beta}_r \\ \boldsymbol{\beta}_d \end{pmatrix}.$$

Let h designate the number of parameters in $\boldsymbol{\beta}_d$. Then there are $q + 1 - h$ parameters in $\boldsymbol{\beta}_r$.

To test the hypothesis $H_0: \boldsymbol{\beta}_d = \mathbf{0}$, we fit the full model containing all the β's in $\boldsymbol{\beta}$ and then fit the reduced model containing only the β's in $\boldsymbol{\beta}_r$. Let $\mathbf{X}_r$ be the columns of $\mathbf{X}$ corresponding to $\boldsymbol{\beta}_r$. Then the reduced model can be written as

$$\mathbf{y} = \mathbf{X}_r \boldsymbol{\beta}_r + \boldsymbol{\epsilon}. \tag{10.20}$$

To compare the fit of the full model and the reduced model, we use

$$\hat{\boldsymbol{\beta}}'\mathbf{X}'\mathbf{y} - \hat{\boldsymbol{\beta}}_r'\mathbf{X}_r'\mathbf{y}, \tag{10.21}$$

where $\hat{\boldsymbol{\beta}}'\mathbf{X}'\mathbf{y}$ is the regression sum of squares from the full model and $\hat{\boldsymbol{\beta}}_r'\mathbf{X}_r'\mathbf{y}$ is the regression sum of squares for the reduced model. The difference $\hat{\boldsymbol{\beta}}'\mathbf{X}'\mathbf{y} - \hat{\boldsymbol{\beta}}_r'\mathbf{X}_r'\mathbf{y}$ is also called the "extra" sum of squares due to $\boldsymbol{\beta}_d$ after adjusting for $\boldsymbol{\beta}_r$. We wish to see what $\boldsymbol{\beta}_d$ contributes "above and beyond" $\boldsymbol{\beta}_r$. We can test $H_0: \boldsymbol{\beta}_d = \mathbf{0}$ with an F-test:

$$F = \frac{(\hat{\boldsymbol{\beta}}'\mathbf{X}'\mathbf{y} - \hat{\boldsymbol{\beta}}_r'\mathbf{X}_r'\mathbf{y})/h}{(\mathbf{y}'\mathbf{y} - \hat{\boldsymbol{\beta}}'\mathbf{X}'\mathbf{y})/(n-q-1)} \tag{10.22}$$

$$= \frac{(\text{SSR}_f - \text{SSR}_r)/h}{\text{SSE}_f/(n-q-1)},$$

where $\text{SSR}_f = \hat{\boldsymbol{\beta}}'\mathbf{X}'\mathbf{y}$ is the regression sum of squares for the full model and $\text{SSR}_r = \hat{\boldsymbol{\beta}}_r\mathbf{X}_r'\mathbf{y}$ is similarly defined for the reduced model. The F-statistic in (10.22) is distributed as $F_{h,n-q-1}$ if H_0 is true, and we reject H_0 if $F > F_\alpha$.

The test in (10.22) is easy to carry out in practice. We fit the full model and obtain the regression and error sums of squares $\hat{\boldsymbol{\beta}}'\mathbf{X}'\mathbf{y}$ and $\mathbf{y}'\mathbf{y} - \hat{\boldsymbol{\beta}}'\mathbf{X}'\mathbf{y}$, respectively. We then fit the reduced model and obtain its regression sum of squares $\hat{\boldsymbol{\beta}}_r'\mathbf{X}_r'\mathbf{y}$ to be subtracted from $\hat{\boldsymbol{\beta}}'\mathbf{X}'\mathbf{y}$. If a software package gives the regression sum of squares in corrected form, this can readily be used to obtain $\hat{\boldsymbol{\beta}}'\mathbf{X}'\mathbf{y} - \hat{\boldsymbol{\beta}}\mathbf{X}_r'\mathbf{y}$, since

$$\hat{\boldsymbol{\beta}}'\mathbf{X}'\mathbf{y} - n\bar{y}^2 - (\hat{\boldsymbol{\beta}}_r'\mathbf{X}_r'\mathbf{y} - n\bar{y}^2) = \hat{\boldsymbol{\beta}}'\mathbf{X}'\mathbf{y} - \hat{\boldsymbol{\beta}}_r'\mathbf{X}_r'\mathbf{y}.$$

Alternatively, we can obtain $\hat{\boldsymbol{\beta}}'\mathbf{X}'\mathbf{y} - \hat{\boldsymbol{\beta}}_r'\mathbf{X}_r'\mathbf{y}$ as the difference between error sums of squares for the two models:

$$\begin{aligned} \text{SSE}_r - \text{SSE}_f &= \mathbf{y}'\mathbf{y} - \hat{\boldsymbol{\beta}}_r'\mathbf{X}_r'\mathbf{y} - (\mathbf{y}'\mathbf{y} - \hat{\boldsymbol{\beta}}'\mathbf{X}'\mathbf{y}) \\ &= \hat{\boldsymbol{\beta}}'\mathbf{X}'\mathbf{y} - \hat{\boldsymbol{\beta}}_r'\mathbf{X}_r'\mathbf{y}. \end{aligned}$$

A test for an individual β_i above and beyond the other β's is readily obtained using (10.22). To test $H_0\colon \beta_i = 0$, we arrange β_i last in $\boldsymbol{\beta}$,

$$\boldsymbol{\beta} = \begin{pmatrix} \boldsymbol{\beta}_r \\ \beta_i \end{pmatrix},$$

where $\boldsymbol{\beta}_r' = (\beta_0, \beta_1, \ldots, \beta_{q-1})$ contains all the β's except β_i. By (10.22), the test statistic is

$$F = \frac{\hat{\boldsymbol{\beta}}'\mathbf{X}'\mathbf{y} - \hat{\boldsymbol{\beta}}_r'\mathbf{X}_r'\mathbf{y}}{\text{SSE}_f/(n-q-1)}, \tag{10.23}$$

which is $F_{1,n-q-1}$. Note that $h = 1$. The test of $H_0\colon \beta_i = 0$ made by the F-statistic in (10.23) is called a *partial F-test*. A detailed breakdown of the effect of each variable in the presence of the others is given by Rencher (1993, 1996, Section 7.3.5).

10.2.6 R^2 in Fixed-x Regression

The proportion of the total variation in the y's that can be attributed to regression on the x's is denoted by R^2:

$$R^2 = \frac{\text{regression sum of squares}}{\text{total sum of squares}} = \frac{\hat{\boldsymbol{\beta}}'\mathbf{X}'\mathbf{y} - n\bar{y}^2}{\mathbf{y}'\mathbf{y} - n\bar{y}^2}. \tag{10.24}$$

The ratio R^2 is called the *coefficient of multiple determination*, or more commonly the *squared multiple correlation*. The *multiple correlation* R is defined as the positive square root of R^2.

The F-test for overall regression in (10.19) can be expressed in terms of R^2 as

$$F = \frac{n-q-1}{q}\,\frac{R^2}{1-R^2}. \tag{10.25}$$

For the reduced model (10.20), R^2 can be written as

$$R_r^2 = \frac{\hat{\boldsymbol{\beta}}_r'\mathbf{X}_r'\mathbf{y} - n\bar{y}^2}{\mathbf{y}'\mathbf{y} - n\bar{y}^2}. \tag{10.26}$$

Then in terms of R^2 and R_r^2, the full and reduced model test in (10.22) for $H_0\colon \boldsymbol{\beta}_d = \mathbf{0}$ becomes

$$F = \frac{(R^2 - R_r^2)/h}{(1-R^2)/(n-q-1)}. \tag{10.27}$$

We can express R^2 in terms of sample variances, covariances, and correlations:

$$R^2 = \frac{\mathbf{s}_{yx}'\mathbf{S}_{xx}^{-1}\mathbf{s}_{yx}}{s_{yy}} = \mathbf{r}_{yx}'\mathbf{R}_{xx}^{-1}\mathbf{r}_{yx}, \tag{10.28}$$

where s_{yy}, $\mathbf{s}_{yx}$, and $\mathbf{S}_{xx}$ are defined following (10.12) and $\mathbf{r}_{yx}$ and $\mathbf{R}_{xx}$ are from an analogous partitioning of the sample correlation matrix of y and the x's:

$$\mathbf{R} = \left[\begin{array}{c|cccc} 1 & r_{y1} & r_{y2} & \cdots & r_{yq} \\ \hline r_{1y} & 1 & r_{12} & \cdots & r_{1q} \\ \vdots & \vdots & \vdots & & \vdots \\ r_{qy} & r_{q1} & r_{q2} & \cdots & 1 \end{array}\right] = \begin{pmatrix} 1 & \mathbf{r}'_{yx} \\ \mathbf{r}_{yx} & \mathbf{R}_{xx} \end{pmatrix}.$$

10.2.7 Subset Selection

In practice, one often has more x's than can be conveniently or economically used in predicting y. Many of them may be redundant and could be discarded. In addition to logistical motivations for deleting variables, there are statistical incentives; for example, if an x is deleted from the fitted model, the variances of the $\hat{\beta}_i$'s and of the $\hat{y}_i$'s are reduced. Certain other aspects of model validation are reviewed by Rencher (1996, Section 7.2.9).

The two most popular approaches to subset selection are to (1) examine all possible subsets and (2) use a stepwise technique. We discuss these in the next two sections.

10.2.7a All Possible Subsets

The optimal approach to subset selection is to use a program that examines all possible subsets of the x's. This may not be computationally feasible if the number of variables is large. In many cases, we can take advantage of algorithms that find the optimum subset of each size without examining all of the subsets. The "leaps-and-bounds" technique of Furnival and Wilson (1974) is an example of this technique.

We discuss three criteria for comparing subsets in a search for the best subset. To conform with established notation in the literature, the number of variables in a subset is denoted by $p-1$, so that with the inclusion of an intercept, there are p parameters in the model. The corresponding total number of available variables from which a subset is to be selected is denoted by $k-1$, with k parameters in the model.

1. R_p^2. The subscript p is an index of the subset size, since it indicates the number of parameters in the model, including an intercept. By its definition in (10.24) as the proportion of total (corrected) sum of squares accounted for by regression, R^2 is clearly a measure of model fit. However, R_p^2 does not reach a maximum for any value of p less than k because it cannot decrease when a variable is added to the model. The usual procedure is to find the subset with largest R_p^2 for each of $p = 2, 3, \ldots, k$ and then choose a value of p beyond which the increases in R^2 appear to be unimportant. This judgment is, of course, subjective.

2. s_p^2. Another useful criterion is the variance estimator for each subset as defined in (10.6):

$$s_p^2 = \text{MSE}_p = \frac{\text{SSE}_p}{n-p}. \tag{10.29}$$

For each of $p = 2, 3, \ldots, k$, we find the subset with smallest s_p^2. If k is fairly large, a typical pattern as p approaches k is for the minimal s_p^2 to decrease to an overall minimum less than s_k^2 and then increase. The minimum value of s_p^2 can be less than s_k^2 if the decrease in SSE_p with an additional variable does not offset the loss of a degree of freedom. It is often suggested that the researcher choose the subset with absolute minimum s_p^2. However, as Hocking (1976, p. 19) notes, this procedure may fit some noise unique to the sample and thereby include one or more extraneous predictor variables. An alternative suggestion is to choose p such that $\min_p s_p^2 = s_k^2$, or more precisely, choose the smallest value of p such that $\min_p s_p^2 < s_k^2$, since there will not be a $p < k$ such that $\min_p s_p^2$ is exactly equal to s_k^2.

3. C_p. The C_p criterion is due to Mallows (1964, 1973). In the following development, we follow Myers (1990, pp. 180–182). The goal is to find a model that achieves a good balance between the bias and variance of the fitted values, $\hat{y}_i$. Bias arises when the $\hat{y}_i$ values are based on an incorrect model, in which $E(\hat{y}_i) \neq E(y_i)$. The *expected squared error*, $E[\hat{y}_i - E(y_i)]^2$, is used in formulating the C_p criterion because it incorporates a variance component and a bias component. If $\hat{y}_i$ were based on the correct model, so that $E(\hat{y}_i) = E(y_i)$, then $E[\hat{y}_i - E(y_i)]^2$ would be equal to $\text{var}(\hat{y}_i)$. In general, however, as we examine many competing models, $\hat{y}_i$ is not based on the correct model, and we have (see problem 10.4)

$$\begin{aligned} E[\hat{y}_i - E(y_i)]^2 &= E[\hat{y}_i - E(\hat{y}_i) + E(\hat{y}_i) - E(y_i)]^2 \\ &= E[\hat{y}_i - E(\hat{y}_i)]^2 + [E(\hat{y}_i) - E(y_i)]^2 \qquad (10.30) \\ &= \text{var}(\hat{y}_i) + (\text{bias in } \hat{y}_i)^2. \qquad (10.31) \end{aligned}$$

The total expected squared error for the n observations in the sample, standardized by dividing by σ^2, becomes

$$\frac{1}{\sigma^2}\sum_{i=1}^{n}[\hat{y}_i - E(y_i)]^2 = \frac{1}{\sigma^2}\sum_{i=1}^{n}\text{var}(\hat{y}_i) + \frac{1}{\sigma^2}\sum_{i=1}^{n}(\text{bias in } \hat{y}_i)^2. \tag{10.32}$$

Before defining C_p as an estimate of (10.32), we can achieve some simplification. We first show that $\sum_i \text{var}(\hat{y}_i)/\sigma^2$ is equal to p. Let the model for all n observations be designated by

$$\mathbf{y} = \mathbf{X}_p\boldsymbol{\beta}_p + \boldsymbol{\epsilon}.$$

We assume that, in general, this prospective model is underspecified and that the true model (which produces σ^2) contains additional β's and additional columns of the $\mathbf{X}$ matrix. If we designate the ith row of $\mathbf{X}_p$ by $\mathbf{x}'_{pi}$, then the first term on the right side of (10.32) becomes (see also problem 10.5)

$$\begin{aligned}
\frac{1}{\sigma^2}\sum_{i=1}^{n} \mathrm{var}(\hat{y}_i) &= \frac{1}{\sigma^2}\sum_{i=1}^{n} \mathrm{var}(\mathbf{x}'_{pi}\hat{\boldsymbol{\beta}}_p) \\
&= \frac{1}{\sigma^2}\sum_{i} \mathbf{x}'_{pi}[\sigma^2(\mathbf{X}'_p\mathbf{X}_p)^{-1}]\mathbf{x}_{pi} && \text{[by (3.65)]} \\
&= \mathrm{tr}\,[\mathbf{X}_p(\mathbf{X}'_p\mathbf{X}_p)^{-1}\mathbf{X}'_p] && \text{[by (3.79)]} \\
&= \mathrm{tr}\,[(\mathbf{X}'_p\mathbf{X}_p)^{-1}\mathbf{X}'_p\mathbf{X}_p] && \text{[by (2.89)]} \\
&= \mathrm{tr}\,\mathbf{I}_p = p. && (10.33)
\end{aligned}$$

It can be shown (Myers 1990, pp. 178–179) that the second term on the right side of (10.32) can be expressed as $\sum_i (\text{bias in } \hat{y}_i)^2 = (n-p)E(s_p^2 - \sigma^2)$. Using this and (10.33), the final simplified form of the total expected squared error in (10.32) is

$$\frac{1}{\sigma^2}\sum_{i=1}^{n}[\hat{y}_i - E(y_i)]^2 = p + \frac{n-p}{\sigma^2}E(s_p^2 - \sigma^2). \tag{10.34}$$

In practice, σ^2 is usually estimated by s_k^2, the MSE from the full model. We thus estimate (10.34) by

$$C_p = p + (n-p)\frac{s_p^2 - s_k^2}{s_k^2}. \tag{10.35}$$

An alternative form is

$$C_p = \frac{\mathrm{SSE}_p}{s_k^2} - (n - 2p). \tag{10.36}$$

In (10.35), we see that if the bias is small for a particular model, C_p will be close to p. For this reason, the line $C_p = p$ is commonly plotted along with the C_p values of several candidate models. We look for small values of C_p that are near this line.

In a Monte Carlo study, Hilton (1983) compared several subset selection criteria based on MSE_p and C_p. The three best procedures were to choose (1) the subset with the smallest p such that $C_p < p$, (2) the subset with the smallest p such that $s_p^2 < s_k^2$, and (3) the subset with minimum s_p^2. The first of these was found to give best results overall, with the second method close behind. The third method performed best in some cases where k was small.

10.2.7b Stepwise Selection

For many data sets, it may be impractical to examine all possible subsets, even with an efficient algorithm such as that of Furnival and Wilson (1974). In such cases, we can use the familiar stepwise approach, which is widely available and has virtually no limit as to the number of variables or observations. A related stepwise technique was discussed in Sections 6.11.2 and 8.9 in connection with selection of dependent variables to separate groups in a MANOVA or discriminant analysis setting. In the present chapter, we are concerned with selecting the independent variables that best predict the dependent variable in regression.

We first review the *forward* selection procedure, which typically uses an F-test at each step. At the first step, y is regressed on each x_i alone and the x with the largest F-value is "entered" into the model. At the second step, we search for the variable with the largest *partial* F-value for testing the significance of each variable in the presence of the variable first entered. Thus, if we denote the first variable to enter as x_1, then at the second step we calculate the partial F-statistic

$$F = \frac{\text{MSR}(x_i|x_1)}{\text{MSE}(x_i, x_1)}$$

for each $i \neq 1$ and choose the variable that maximizes F, where MSR $=$ $(\text{SSR}_f - \text{SSR}_r)/h$ and MSE $= \text{SSE}/(n-q-1)$ are the mean squares for regression and error, respectively, as in (10.22) and the expression following (10.22). In this case, $\text{SSR}_f = \text{SSR}(x_1, x_i)$ and $\text{SSR}_r = \text{SSR}(x_1)$. Note also that $h = 1$ because only one variable is being added, and MSE is calculated using only the variables already entered plus the candidate variable. This procedure continues at each step until the largest partial F for an entering variable falls below a preselected threshold F-value or until the corresponding p-value exceeds some predetermined level.

The *stepwise* selection procedure similarly seeks the best variable to enter at each step. Then after a variable has entered, each of the variables previously included is examined by a partial F-test to see if it is no longer significant and can be dropped from the model.

The *backward elimination* procedure begins with all x's in the model and deletes one at a time. The partial F-statistic for each variable in the presence of the others is calculated, and the variable with smallest F is eliminated. This continues until the smallest F at some step exceeds a preselected threshold value.

Since these sequential methods do not examine all subsets, they will often fail to find the optimum subset, especially if k is large. However, R_p^2, s_p^2, or C_p may not differ substantially between the optimum subset and the one found by stepwise selection. These sequential methods have been popular for a generation, and it is very likely they will continue to be used, even though increased computing power has put the optimal methods within reach for larger data sets.

There are some possible risks in the use of stepwise methods. The stepwise procedure may fail to detect a true predictor (an x_i for which $\beta_i \neq 0$) because s_p^2 is biased upward in an underspecified model, thus artificially reducing the partial F-value. On the other hand, a variable that is not a true predictor of y (an x_i for which $\beta_i = 0$) may enter because of chance correlations in a particular sample. In the presence of such "noise" variables, it is clear that the partial F-statistic for the entering variable does not have an F-distribution because it is maximized at each step. The calculated p-values become optimistic. This problem intensifies when the sample size is relatively small compared to the number of variables. Rencher and Pun (1980) found that in such cases some surprisingly large values of R^2 can occur, even when there is no relationship between y and the x's in the population. In a related study, Flack and Chang (1987) included x's that were authentic contributors as well as noise variables. They found that "for most samples, a large percentage of the selected variables is noise, particularly when the number of candidate variables is large relative to the number of observations. The adjusted R^2 of the selected variables is highly inflated" (p. 84).

10.3 MULTIPLE REGRESSION: RANDOM x's

In Section 10.2, it was assumed that the x's were fixed and would have the same values if another sample were taken; that is, the same $\mathbf{X}$ matrix would be used each time a $\mathbf{y}$ vector was observed. However, many regression applications involve x's that are random variables.

Thus in the random-x case, the values of $x_1, x_2, \ldots, x_q$ are not under the control of the experimenter. They occur randomly along with y. On each subject we observe $y, x_1, x_2, \ldots, x_q$.

If we assume that y and the x's have a multivariate normal distribution, then $\hat{\boldsymbol{\beta}}$, R^2, and the F-tests have the same formulation as in the fixed-x case (for details, see Rencher 1996, Section 7.3). Thus with the multivariate normal assumption, we can proceed with estimation and testing the same way in the random-x case as with fixed x's.

10.4 MULTIVARIATE MULTIPLE REGRESSION: ESTIMATION

10.4.1 The Multivariate Linear Model

We turn now to the *multivariate multiple regression model*, where *multivariate* refers to the dependent variables and *multiple* pertains to the independent variables. In this case, several y's are measured corresponding to each set of

x's. In particular, we have $y_1, y_2, \ldots, y_p$ to be predicted by $x_1, x_2, \ldots, x_q$. Thus $x_1, x_2, \ldots, x_q$ will predict y_1 as well as y_2 and each of the other y's.

In the multivariate model, we have estimation and testing results analogous to those in Section 10.2. In the present section, we assume the x's are fixed.

The n observed values of the vector of y's can be listed as rows in the following matrix:

$$\mathbf{Y} = \begin{bmatrix} y_{11} & y_{12} & \cdots & y_{1p} \\ y_{21} & y_{22} & \cdots & y_{2p} \\ \vdots & \vdots & & \vdots \\ y_{n1} & y_{n2} & \cdots & y_{np} \end{bmatrix} = \begin{bmatrix} \mathbf{y}_1' \\ \mathbf{y}_2' \\ \vdots \\ \mathbf{y}_n' \end{bmatrix}.$$

Thus each row of $\mathbf{Y}$ contains the values of the p variables measured on a subject. Each column of $\mathbf{Y}$ consists of the n observations on one of the p variables and therefore corresponds to the $\mathbf{y}$ vector in the usual (univariate) regression model (10.2).

The n values of $x_1, x_2, \ldots, x_q$ can be placed in a matrix that turns out to be the same as in the multiple regression formulation in Section 10.2.1:

$$\mathbf{X} = \begin{bmatrix} 1 & x_{11} & x_{12} & \cdots & x_{1q} \\ 1 & x_{21} & x_{22} & \cdots & x_{2q} \\ \vdots & \vdots & \vdots & & \vdots \\ 1 & x_{n1} & x_{n2} & \cdots & x_{nq} \end{bmatrix}.$$

We assume that $\mathbf{X}$ is fixed from sample to sample.

Since each of the p y's will depend on the x's in its own way, each column of $\mathbf{Y}$ will need different β's. Thus we have a column of β's for each column of $\mathbf{Y}$. These columns of β's form a matrix of β's that we will denote by $\mathbf{B}$. Our multivariate model is therefore

$$\mathbf{Y} = \mathbf{XB} + \boldsymbol{\Xi}.$$

The notation $\boldsymbol{\Xi}$ (the uppercase version of $\boldsymbol{\xi}$) is adopted here because of its resemblance to ϵ.

We illustrate the multivariate model with $p = 2, q = 3$:

$$\begin{bmatrix} y_{11} & y_{12} \\ y_{21} & y_{22} \\ \vdots & \vdots \\ y_{n1} & y_{n2} \end{bmatrix} = \begin{bmatrix} 1 & x_{11} & x_{12} & x_{13} \\ 1 & x_{21} & x_{22} & x_{23} \\ \vdots & \vdots & \vdots & \vdots \\ 1 & x_{n1} & x_{n2} & x_{n3} \end{bmatrix} \begin{bmatrix} \beta_{01} & \beta_{02} \\ \beta_{11} & \beta_{12} \\ \beta_{21} & \beta_{22} \\ \beta_{31} & \beta_{32} \end{bmatrix} + \begin{bmatrix} \epsilon_{11} & \epsilon_{12} \\ \epsilon_{21} & \epsilon_{22} \\ \vdots & \vdots \\ \epsilon_{n1} & \epsilon_{n2} \end{bmatrix}.$$

The model for the first column of $\mathbf{Y}$ is

$$\begin{bmatrix} y_{11} \\ y_{21} \\ \vdots \\ y_{n1} \end{bmatrix} = \begin{bmatrix} 1 & x_{11} & x_{12} & x_{13} \\ 1 & x_{21} & x_{22} & x_{23} \\ \vdots & \vdots & \vdots & \vdots \\ 1 & x_{n1} & x_{n2} & x_{n3} \end{bmatrix} \begin{bmatrix} \beta_{01} \\ \beta_{11} \\ \beta_{21} \\ \beta_{31} \end{bmatrix} + \begin{bmatrix} \epsilon_{11} \\ \epsilon_{21} \\ \vdots \\ \epsilon_{n1} \end{bmatrix},$$

and for the second column, we have

$$\begin{bmatrix} y_{12} \\ y_{22} \\ \vdots \\ y_{n2} \end{bmatrix} = \begin{bmatrix} 1 & x_{11} & x_{12} & x_{13} \\ 1 & x_{21} & x_{22} & x_{23} \\ \vdots & \vdots & \vdots & \vdots \\ 1 & x_{n1} & x_{n2} & x_{n3} \end{bmatrix} \begin{bmatrix} \beta_{02} \\ \beta_{12} \\ \beta_{22} \\ \beta_{32} \end{bmatrix} + \begin{bmatrix} \epsilon_{12} \\ \epsilon_{22} \\ \vdots \\ \epsilon_{n2} \end{bmatrix}.$$

By analogy with the univariate case in Section 10.2.1, additional assumptions that lead to good estimates are as follows:

1. $E(\mathbf{Y}) = \mathbf{XB}$ or $E(\boldsymbol{\Xi}) = \mathbf{O}$.
2. $\text{cov}(\mathbf{y}_i) = \boldsymbol{\Sigma}$ for all $i = 1, 2, \ldots, n$, where $\mathbf{y}_i'$ is the ith row of $\mathbf{Y}$.
3. $\text{cov}(\mathbf{y}_i, \mathbf{y}_j) = \mathbf{O}$ for all $i \neq j$.

Assumption 1 states that the linear model is correct and that no additional x's are needed to predict the y's. Assumption 2 asserts that each of the n observation vectors (rows) in $\mathbf{Y}$ has the same covariance matrix. Assumption 3 declares that observation vectors (rows of $\mathbf{Y}$) are uncorrelated with each other. Thus observation vectors are independent and have the same covariance matrix; that is, we assume that the y's within an observation vector (row of $\mathbf{Y}$) are correlated with each other but independent of the y's in any other observation vector.

The covariance matrix $\boldsymbol{\Sigma}$ in assumption 2 contains the variances and covariances of $y_{i1}, y_{i2}, \ldots, y_{ip}$ in any $\mathbf{y}_i$:

$$\text{cov}(\mathbf{y}_i) = \boldsymbol{\Sigma} = \begin{bmatrix} \sigma_{11} & \sigma_{12} & \cdots & \sigma_{1p} \\ \sigma_{21} & \sigma_{22} & \cdots & \sigma_{2p} \\ \vdots & \vdots & & \vdots \\ \sigma_{p1} & \sigma_{p2} & \cdots & \sigma_{pp} \end{bmatrix}.$$

The covariance matrix $\mathbf{O}$ in assumption 3 contains the covariances of each of $y_{i1}, y_{i2}, \ldots, y_{ip}$ with each of $y_{j1}, y_{j2}, \ldots, y_{jp}$:

$$\begin{bmatrix} \text{cov}(y_{i1}, y_{j1}) & \text{cov}(y_{i1}, y_{j2}) & \cdots & \text{cov}(y_{i1}, y_{jp}) \\ \text{cov}(y_{i2}, y_{j1}) & \text{cov}(y_{i2}, y_{j2}) & \cdots & \text{cov}(y_{i2}, y_{jp}) \\ \vdots & \vdots & & \vdots \\ \text{cov}(y_{ip}, y_{j1}) & \text{cov}(y_{ip}, y_{j2}) & \cdots & \text{cov}(y_{ip}, y_{jp}) \end{bmatrix} = \begin{bmatrix} 0 & 0 & \cdots & 0 \\ 0 & 0 & \cdots & 0 \\ \vdots & \vdots & & \vdots \\ 0 & 0 & \cdots & 0 \end{bmatrix}.$$

10.4.2 Least Squares Estimation in the Multivariate Model

By analogy with the univariate case in (10.4), we estimate $\mathbf{B}$ with

$$\hat{\mathbf{B}} = (\mathbf{X}'\mathbf{X})^{-1}\mathbf{X}'\mathbf{Y}. \tag{10.37}$$

We call $\hat{\mathbf{B}}$ the *least squares estimator* for $\mathbf{B}$ because it "minimizes" $\mathbf{E} = \hat{\boldsymbol{\Xi}}'\hat{\boldsymbol{\Xi}}$, a matrix analogous to SSE:

$$\mathbf{E} = \hat{\boldsymbol{\Xi}}'\hat{\boldsymbol{\Xi}} = (\mathbf{Y} - \mathbf{X}\hat{\mathbf{B}})'(\mathbf{Y} - \mathbf{X}\hat{\mathbf{B}}).$$

The matrix $\hat{\mathbf{B}}$ minimizes $\mathbf{E}$ in the following sense. If we let $\mathbf{B}_0$ be an estimate that may possibly do better than $\hat{\mathbf{B}}$ and add $\mathbf{X}\hat{\mathbf{B}} - \mathbf{X}\mathbf{B}_0$ to $\mathbf{Y} - \mathbf{X}\hat{\mathbf{B}}$, we find that this adds a positive definite matrix to $\mathbf{E} = (\mathbf{Y} - \mathbf{X}\hat{\mathbf{B}})'(\mathbf{Y} - \mathbf{X}\hat{\mathbf{B}})$. Thus we cannot improve on $\hat{\mathbf{B}}$. The least squares estimate $\hat{\mathbf{B}}$ also minimizes the scalar quantities $\text{tr}\,(\mathbf{Y} - \mathbf{X}\hat{\mathbf{B}})'(\mathbf{Y} - \mathbf{X}\hat{\mathbf{B}})$ and $|(\mathbf{Y} - \mathbf{X}\hat{\mathbf{B}})'(\mathbf{Y} - \mathbf{X}\hat{\mathbf{B}})|$.

We noted earlier that in the model $\mathbf{Y} = \mathbf{X}\mathbf{B} + \boldsymbol{\Xi}$, there is a column of $\mathbf{B}$ corresponding to each column of $\mathbf{Y}$; that is, each y_i, $i = 1, 2, \ldots, p$, is predicted differently by $x_1, x_2, \ldots, x_q$. (This is illustrated in Section 10.4.1 for $p = 2$.) In the estimate $\hat{\mathbf{B}} = (\mathbf{X}'\mathbf{X})^{-1}\mathbf{X}'\mathbf{Y}$, we have a similar pattern. The matrix product $(\mathbf{X}'\mathbf{X})^{-1}\mathbf{X}'$ is multiplied by each column of $\mathbf{Y}$. Thus the ith column of $\hat{\mathbf{B}}$ is the usual least squares estimate $\hat{\boldsymbol{\beta}}$ for the ith dependent variable y_i. To give this a more precise expression, let us denote the p columns of $\mathbf{Y}$ by $\mathbf{y}_{(1)}, \mathbf{y}_{(2)}, \ldots, \mathbf{y}_{(p)}$ to distinguish them from the n rows $\mathbf{y}_i'$, $i = 1, 2, \ldots, n$. Then

$$\begin{aligned} \hat{\mathbf{B}} &= (\mathbf{X}'\mathbf{X})^{-1}\mathbf{X}'\mathbf{Y} = (\mathbf{X}'\mathbf{X})^{-1}\mathbf{X}'(\mathbf{y}_{(1)}, \mathbf{y}_{(2)}, \ldots, \mathbf{y}_{(p)}) \\ &= [(\mathbf{X}'\mathbf{X})^{-1}\mathbf{X}'\mathbf{y}_{(1)}, (\mathbf{X}'\mathbf{X})^{-1}\mathbf{X}'\mathbf{y}_{(2)}, \ldots, (\mathbf{X}'\mathbf{X})^{-1}\mathbf{X}'\mathbf{y}_{(p)}] \\ &= [\hat{\boldsymbol{\beta}}_{(1)}, \hat{\boldsymbol{\beta}}_{(2)}, \ldots, \hat{\boldsymbol{\beta}}_{(p)}]. \end{aligned}$$

Example 10.4.2. The results of a planned experiment involving a chemical reaction are given in Table 10.1 (Box and Youle 1955).

The input (independent) variables are

Table 10.1 Chemical Reaction Data

Experiment Number	Yield Variables y_1	y_2	y_3	Input Variables x_1	x_2	x_3
1	41.5	45.9	11.2	162	23	3
2	33.8	53.3	11.2	162	23	8
3	27.7	57.5	12.7	162	30	5
4	21.7	58.8	16.0	162	30	8
5	19.9	60.6	16.2	172	25	5
6	15.0	58.0	22.6	172	25	8
7	12.2	58.6	24.5	172	30	5
8	4.3	52.4	38.0	172	30	8
9	19.3	56.9	21.3	167	27.5	6.5
10	6.4	55.4	30.8	177	27.5	6.5
11	37.6	46.9	14.7	157	27.5	6.5
12	18.0	57.3	22.2	167	32.5	6.5
13	26.3	55.0	18.3	167	22.5	6.5
14	9.9	58.9	28.0	167	27.5	9.5
15	25.0	50.3	22.1	167	27.5	3.5
16	14.1	61.1	23.0	177	20	6.5
17	15.2	62.9	20.7	177	20	6.5
18	15.9	60.0	22.1	160	34	7.5
19	19.6	60.6	19.3	160	34	7.5

$$
\begin{aligned}
x_1 &= \text{temperature} \\
x_2 &= \text{concentration} \\
x_3 &= \text{time.}
\end{aligned}
$$

The yield (dependent) variables are

$$
\begin{aligned}
y_1 &= \text{percentage of unchanged starting material} \\
y_2 &= \text{percentage converted to the desired product} \\
y_3 &= \text{percentage of unwanted by-product.}
\end{aligned}
$$

Using (10.37), the least squares estimator $\hat{\mathbf{B}}$ for the regression of (y_1, y_2, y_3) on (x_1, x_2, x_3) is given by

$$
\begin{aligned}
\hat{\mathbf{B}} &= (\mathbf{X}'\mathbf{X})^{-1}\mathbf{X}'\mathbf{Y} \\
&= \begin{bmatrix} 332.11 & -26.04 & -164.08 \\ -1.55 & .40 & .91 \\ -1.42 & .29 & .90 \\ -2.24 & 1.03 & 1.15 \end{bmatrix}.
\end{aligned}
$$

Note that the first column of $\hat{\mathbf{B}}$ gives $\hat{\beta}_0, \hat{\beta}_1, \hat{\beta}_2$, and $\hat{\beta}_3$ for regression of y_1 on x_1, x_2, and x_3; the second column of $\hat{\mathbf{B}}$ gives $\hat{\beta}_0, \hat{\beta}_1, \hat{\beta}_2$, and $\hat{\beta}_3$ for regression of y_2 on x_1, x_2, and x_3, and so on.

10.4.3 Properties of Least Squares Estimators $\hat{\mathbf{B}}$

The least squares estimator $\hat{\mathbf{B}}$ can be obtained without imposing the assumptions $E(\mathbf{Y}) = \mathbf{XB}$, $\text{cov}(\mathbf{y}_i) = \mathbf{\Sigma}$, and $\text{cov}(\mathbf{y}_i, \mathbf{y}_j) = \mathbf{O}$. However, when these assumptions hold, $\hat{\mathbf{B}}$ has the following properties:

1. The estimator $\hat{\mathbf{B}}$ is unbiased, that is, $E(\hat{\mathbf{B}}) = \mathbf{B}$. This means that if we sampled repeatedly from the same population, the average value of $\hat{\mathbf{B}}$ would be $\mathbf{B}$.
2. The least squares estimates $\hat{\beta}_{ij}$ in $\hat{\mathbf{B}}$ have minimum variance among all possible linear unbiased estimators. This result is known as the Gauss–Markov theorem. The restriction to unbiased estimators is necessary to exclude trivial estimators such as a constant, which has variance equal to zero, but is of no interest. This minimum variance property of least squares estimators is remarkable for its distributional generality. Normality of the y's is not required.
3. All $\hat{\beta}_{ij}$ in $\hat{\mathbf{B}}$ are intercorrelated with each other. This is due to the intercorrelations among the x's and among the y's. The $\hat{\beta}$'s within a given column of $\hat{\mathbf{B}}$ are correlated because $x_1, x_2, \ldots, x_q$ are correlated. If $x_1, x_2, \ldots, x_q$ were orthogonal to each other, the $\hat{\beta}$'s within each column of $\hat{\mathbf{B}}$ would be uncorrelated. Thus the relationship of the x's to each other affects the relationship of the $\hat{\beta}$'s within each column to each other. On the other hand, the relationship of a $\hat{\beta}$ in one column to $\hat{\beta}$'s in other columns of $\hat{\mathbf{B}}$ is affected by the correlations among $y_1, y_2, \ldots, y_p$. There is a column of $\hat{\mathbf{B}}$ for each y_i, and since the y's are typically correlated, the $\hat{\beta}$'s in different columns are correlated.

   Because of the correlations among the columns of $\hat{\mathbf{B}}$, we need multivariate tests for hypotheses about $\mathbf{B}$. We cannot use an F-test on each column of $\mathbf{B}$ from Section 10.2.5 because these F-tests would not take into account the correlations or preserve the α level. Some appropriate multivariate tests are given in Section 10.5.

10.4.4 An Estimator for $\mathbf{\Sigma}$

By analogy with (10.6) and (10.7), an unbiased estimator of $\text{cov}(\mathbf{y}_i) = \mathbf{\Sigma}$ is given by

$$\frac{\mathbf{E}}{n-q-1} = \frac{(\mathbf{Y}-\mathbf{X}\hat{\mathbf{B}})'(\mathbf{Y}-\mathbf{X}\hat{\mathbf{B}})}{n-q-1} \tag{10.38}$$

$$= \frac{\mathbf{Y}'\mathbf{Y}-\hat{\mathbf{B}}'\mathbf{X}'\mathbf{Y}}{n-q-1}. \tag{10.39}$$

The denominator $n - q - 1$ is necessary to make $\mathbf{E}/(n - q - 1)$ an unbiased estimator of $\mathbf{\Sigma}$.

10.4.5 Model Corrected for Means

If the x's are centered by subtracting their means, we have the centered $\mathbf{X}$ matrix as in Section 10.2.4:

$$\mathbf{X}_c = \begin{bmatrix} x_{11}-\bar{x}_1 & x_{12}-\bar{x}_2 & \dots & x_{1q}-\bar{x}_q \\ x_{21}-\bar{x}_1 & x_{22}-\bar{x}_2 & \dots & x_{2q}-\bar{x}_q \\ \vdots & \vdots & & \vdots \\ x_{n1}-\bar{x}_1 & x_{n2}-\bar{x}_2 & \dots & x_{nq}-\bar{x}_q \end{bmatrix}.$$

The $\mathbf{B}$ matrix can be partitioned as

$$\mathbf{B} = \begin{pmatrix} \boldsymbol{\beta}_0' \\ \mathbf{B}_1 \end{pmatrix} = \left[\begin{array}{cccc} \beta_{01} & \beta_{02} & \cdots & \beta_{0p} \\ \hline \beta_{11} & \beta_{12} & \cdots & \beta_{1p} \\ \vdots & \vdots & & \vdots \\ \beta_{q1} & \beta_{q2} & \cdots & \beta_{qp} \end{array}\right].$$

By analogy with (10.11) and (10.12), the estimates are

$$\hat{\mathbf{B}}_1 = (\mathbf{X}_c'\mathbf{X}_c)^{-1}\mathbf{X}_c'\mathbf{Y} \tag{10.40}$$

$$\hat{\boldsymbol{\beta}}_0' = \bar{\mathbf{y}}' - \bar{\mathbf{x}}'\hat{\mathbf{B}}_1, \tag{10.41}$$

where $\bar{\mathbf{y}}' = (\bar{y}_1, \bar{y}_2, \dots, \bar{y}_p)$ and $\bar{\mathbf{x}}' = (\bar{x}_1, \bar{x}_2, \dots, \bar{x}_q)$. These estimates give the same results as $\hat{\mathbf{B}} = (\mathbf{X}'\mathbf{X})^{-1}\mathbf{X}'\mathbf{Y}$ in (10.37).

As in (10.15), the estimate $\hat{\mathbf{B}}_1$ in (10.40) can be expressed in terms of sample covariance matrices. We multiply and divide (10.40) by $n - 1$ to obtain

$$
\begin{aligned}
\hat{\mathbf{B}}_1 &= (n-1)(\mathbf{X}_c'\mathbf{X}_c)^{-1}\frac{\mathbf{X}_c'\mathbf{Y}}{n-1} = \left(\frac{\mathbf{X}_c'\mathbf{X}_c}{n-1}\right)^{-1}\frac{\mathbf{X}_c'\mathbf{Y}}{n-1} \\
&= \mathbf{S}_{xx}^{-1}\mathbf{S}_{xy},
\end{aligned}
\tag{10.42}
$$

where $\mathbf{S}_{xx}$ and $\mathbf{S}_{xy}$ are blocks from the overall sample covariance matrix of $y_1, y_2, \ldots, y_p, x_1, x_2, \ldots, x_q$:

$$
\mathbf{S} = \begin{pmatrix} \mathbf{S}_{yy} & \mathbf{S}_{yx} \\ \mathbf{S}_{xy} & \mathbf{S}_{xx} \end{pmatrix}.
$$

10.5 MULTIVARIATE MULTIPLE REGRESSION: HYPOTHESIS TESTS

In this section we extend the two tests of Section 10.2.5 to the multivariate y case. We assume the x's are fixed and the y's are multivariate normal. For other tests and confidence intervals, see Rencher (1996, Chapter 7).

10.5.1 Test of Overall Regression

We first consider the hypothesis that none of the x's predict any of the y's, which can be expressed as $H_0\colon \mathbf{B}_1 = \mathbf{O}$, where

$$
\mathbf{B} = \begin{pmatrix} \boldsymbol{\beta}_0' \\ \mathbf{B}_1 \end{pmatrix} = \left[\begin{array}{cccc} \beta_{01} & \beta_{02} & \cdots & \beta_{0p} \\ \hline \beta_{11} & \beta_{12} & \cdots & \beta_{1p} \\ \vdots & \vdots & & \vdots \\ \beta_{q1} & \beta_{q2} & \cdots & \beta_{qp} \end{array}\right].
$$

We do not wish to include $\boldsymbol{\beta}_0' = \mathbf{0}'$ in the hypothesis, because this would restrict all y's to have means equal to zero (assuming $\mathbf{B}_1 = \mathbf{O}$). The alternative hypothesis is $H_1\colon \mathbf{B}_1 \neq \mathbf{O}$, which implies that we want to know if even one $\beta_{ij} \neq 0$, $i = 1, 2, \ldots, q$, $j = 1, 2, \ldots, p$.

The numerator of (10.39) suggests a partitioning of the total sum of squares and products matrix $\mathbf{Y}'\mathbf{Y}$,

$$
\mathbf{Y}'\mathbf{Y} = (\mathbf{Y}'\mathbf{Y} - \hat{\mathbf{B}}'\mathbf{X}'\mathbf{Y}) + \hat{\mathbf{B}}'\mathbf{X}'\mathbf{Y}.
$$

To avoid inclusion of $\boldsymbol{\beta}_0' = \mathbf{0}'$, we subtract $n\overline{\mathbf{y}}\overline{\mathbf{y}}'$ from both sides to obtain

$$\begin{aligned}\mathbf{Y}'\mathbf{Y} - n\overline{\mathbf{y}}\overline{\mathbf{y}}' &= (\mathbf{Y}'\mathbf{Y} - \hat{\mathbf{B}}'\mathbf{X}'\mathbf{Y}) + (\hat{\mathbf{B}}'\mathbf{X}'\mathbf{Y} - n\overline{\mathbf{y}}\overline{\mathbf{y}}') \\ &= \mathbf{E} + \mathbf{H}. \end{aligned} \tag{10.43}$$

The overall regression sum of squares and products matrix $\mathbf{H} = \hat{\mathbf{B}}'\mathbf{X}'\mathbf{Y} - n\overline{\mathbf{y}}\overline{\mathbf{y}}'$ is adjusted for the intercepts $\beta_{01}, \beta_{02}, \ldots, \beta_{0p}$ and can, therefore, be used to test $H_0\colon \mathbf{B}_1 = \mathbf{O}$. The notation $\mathbf{E} = \mathbf{Y}'\mathbf{Y} - \hat{\mathbf{B}}'\mathbf{X}'\mathbf{Y}$ and $\mathbf{H} = \hat{\mathbf{B}}'\mathbf{X}\mathbf{Y} - n\overline{\mathbf{y}}\overline{\mathbf{y}}'$ conforms with usage of $\mathbf{E}$ and $\mathbf{H}$ in Chapter 6.

As in Chapter 6, we can test the significance of $\mathbf{H}$ by Wilks' Λ:

$$\Lambda = \frac{|\mathbf{E}|}{|\mathbf{E}+\mathbf{H}|} = \frac{|\mathbf{Y}'\mathbf{Y} - \hat{\mathbf{B}}'\mathbf{X}'\mathbf{Y}|}{|\mathbf{Y}'\mathbf{Y} - n\overline{\mathbf{y}}\overline{\mathbf{y}}'|}, \tag{10.44}$$

which is distributed as $\Lambda_{p,q,n-q-1}$ when $H_0\colon \mathbf{B}_1 = \mathbf{O}$ is true, where p is the number of y's and q is the number of x's. We reject H_0 if $\Lambda \le \Lambda_{\alpha,p,q,n-q-1}$. The likelihood ratio approach leads to the same test statistic. If $\mathbf{H}$ is "large" due to large values of the $\hat{\beta}_{ij}$'s, then $|\mathbf{E}+\mathbf{H}|$ would be expected to be sufficiently greater than $|\mathbf{E}|$ so that Λ would lead to rejection. By $\mathbf{H}$ "large," we mean that the regression sums of squares on the diagonal are large. Critical values for Λ are available in Table A.9 using $\nu_H = q$ and $\nu_E = n - q - 1$. Note that these degrees of freedom are the same as the univariate test for regression of y on x_1, $x_2, \ldots, x_q$ in (10.19). The F and χ^2 approximations for Λ in (6.13) and (6.14) can also be used.

There are two alternative expressions for Wilks' Λ in (10.44). We can express Λ in terms of the eigenvalues $\lambda_1, \lambda_2, \ldots, \lambda_s$ of $\mathbf{E}^{-1}\mathbf{H}$:

$$\Lambda = \prod_{i=1}^{s} \frac{1}{1+\lambda_i}, \tag{10.45}$$

where $s = \min(p, q)$. Using the partitioned covariance matrix in Section 10.4.5,

$$\mathbf{S} = \begin{pmatrix} \mathbf{S}_{yy} & \mathbf{S}_{yx} \\ \mathbf{S}_{xy} & \mathbf{S}_{xx} \end{pmatrix},$$

the Wilks' Λ statistic in (10.44) or (10.45) can be written in the form

$$\Lambda = \frac{|\mathbf{S}|}{|\mathbf{S}_{xx}||\mathbf{S}_{yy}|}. \tag{10.46}$$

This is the same form of Λ as the test for independence of $\mathbf{y}$ and $\mathbf{x}$ given in (7.28), where $\mathbf{y}$ and $\mathbf{x}$ are both random vectors. In this section, where we are regressing **y** on **x**, the y's are random variables and the x's are fixed. Thus $\mathbf{S}_{yy}$ is the sample covariance matrix of the y's in the usual sense, while $\mathbf{S}_{xx}$ consists of an analogous mathematical expression involving the constant x's (see comments about $\mathbf{S}_{xx}$ in Section 10.2.4).

The union–intersection test of $H_0: \mathbf{B}_1 = \mathbf{O}$ is Roy's test statistic analogous to (6.15),

$$\theta = \frac{\lambda_1}{1 + \lambda_1}, \tag{10.47}$$

where λ_1 is the largest eigenvalue of $\mathbf{E}^{-1}\mathbf{H}$. Upper percentage points θ_α are given in Table A.10. The accompanying parameters are

$$s = \min(p, q) \qquad m = \tfrac{1}{2}(|q - p| - 1) \qquad N = \tfrac{1}{2}(n - q - p - 2).$$

The hypothesis is rejected if $\theta > \theta_\alpha$.

As in Section 6.1.5, Pillai's test statistic is defined as

$$V^{(s)} = \sum_{i=1}^{s} \frac{\lambda_i}{1 + \lambda_i}, \tag{10.48}$$

and the Lawley–Hotelling test statistic is given by

$$U^{(s)} = \sum_{i=1}^{s} \lambda_i, \tag{10.49}$$

where $\lambda_1, \lambda_2, \ldots, \lambda_s$ are the eigenvalues of $\mathbf{E}^{-1}\mathbf{H}$. For $V^{(s)}$, upper percentage points are found in Table A.11, indexed by s, m, and N as defined above in connection with Roy's test. Upper percentage points for $\nu_E U^{(s)}/\nu_H$ (see Section 6.1.5) are provided in Table A.12, where $\nu_H = q$ and $\nu_E = n - q - 1$. Alternatively, we can use the F approximations for $V^{(s)}$ and $U^{(s)}$ in Section 6.1.5.

When H_0 is true, all four test statistics have probability α of rejecting; that is, they all have the same probability of a type I error. When H_0 is false, the power ranking of the tests depends on the configuration of the population eigenvalues, as was noted in Section 6.2. (The sample eigenvalues $\lambda_1, \lambda_2, \ldots, \lambda_s$ from $\mathbf{E}^{-1}\mathbf{H}$ are estimates of the population eigenvalues.) If the population eigenvalues are equal or nearly equal, the power ranking of the tests is $V^{(s)} \geq \Lambda \geq U^{(s)} \geq \theta$. If only one population eigenvalue is nonzero, the powers are reversed: $\theta \geq U^{(s)} \geq \Lambda \geq V^{(s)}$.

In the case of a single nonzero population eigenvalue, the rank of $\mathbf{B}_1$ is 1.

There are various ways this could occur; for example, $\mathbf{B}_1$ could have only one nonzero row, which would indicate that only one of the x's predicts the y's. On the other hand, a single nonzero column implies that only one of the y's is predicted by the x's. Other ways $\mathbf{B}_1$ could have rank 1 would include all rows equal or linear combinations of each other, manifesting that all x's act alike in predicting the y's. Similarly, all columns equal to each other or linear functions of each other would signify only one dimension in the y's as they relate to the x's.

Example 10.5.1. For the chemical data of Table 10.1, we test the overall regression hypothesis $H_0: \mathbf{B}_1 = \mathbf{O}$. The $\mathbf{E}$ and $\mathbf{H}$ matrices are given by

$$\mathbf{E} = \begin{pmatrix} 80.174 & -21.704 & -65.923 \\ -21.704 & 249.462 & -179.496 \\ -65.923 & -179.496 & 231.197 \end{pmatrix}$$
$$\mathbf{H} = \begin{pmatrix} 1707.158 & -492.532 & -996.584 \\ -492.532 & 151.002 & 283.607 \\ -996.584 & 283.607 & 583.688 \end{pmatrix}.$$

The eigenvalues of $\mathbf{E}^{-1}\mathbf{H}$ are 26.3183, .1004, and .0033. The parameters for use in obtaining critical values of the four test statistics are

$$\nu_H = q = 3 \qquad \nu_E = n - q - 1 = 19 - 3 - 1 = 15$$
$$s = \min(3,3) = 3 \qquad m = \tfrac{1}{2}(|q - p| - 1) = -\tfrac{1}{2}$$
$$N = \tfrac{1}{2}(n - q - p - 2) = 5.5.$$

Using the eigenvalues, we obtain the test statistics

$$\Lambda = \prod_{i=1}^{3} \frac{1}{1+\lambda_i} = \frac{1}{1+26.3183}\,\frac{1}{1+.1004}\,\frac{1}{1+.0033}$$
$$= .0332 < \Lambda_{.05,3,3,15} = .3090$$
$$\theta = \frac{\lambda_1}{1+\lambda_1} = .963 > \theta(.05, 3, 0, 5) = .669$$
$$V^{(s)} = \sum_{i=1}^{3} \frac{\lambda_i}{1+\lambda_i} = 1.058 > V^{(s)}_{.05,3,0,15} = .525$$
$$U^{(s)} = \sum_{i=1}^{3} \lambda_i = 26.422 \qquad \frac{\nu_E U^{(s)}}{\nu_H} = 132.11,$$

which exceeds the .05 critical value, 8.936 (interpolated), from Table A.11 (see

Section 6.1.5). Thus all four tests reject H_0. Note that the critical values given for θ and $V^{(s)}$ are conservative, since 0 was used for m in place of $-.5$.

In this case, the first eigenvalue, 26.3183, completely dominates the other two. In Example 10.4.2, we obtained

$$\hat{\mathbf{B}}_1 = \begin{pmatrix} -1.55 & .40 & .91 \\ -1.42 & .29 & .90 \\ -2.24 & 1.03 & 1.15 \end{pmatrix}.$$

The columns are approximately proportional to each other, indicating that there is essentially only one dimension in the y's as they are predicted by the x's. A similar statement can be made about the rows and the dimensionality of the x's as they predict the y's.

10.5.2 Test on a Subset of the x's

We consider the hypothesis that the y's do not depend on the last h of the x's, $x_{q-h+1}, x_{q-h+2}, \ldots, x_q$. By this we mean that none of the p y's is predicted by any of these h x's. To express this hypothesis, write the $\mathbf{B}$ matrix in partitioned form,

$$\mathbf{B} = \begin{pmatrix} \mathbf{B}_r \\ \mathbf{B}_d \end{pmatrix},$$

where, as in Section 10.2.5b, the subscript r denotes the subset of β_{ij}'s to be *retained* in the *reduced* model and d represents the subset of β_{ij}'s to be *deleted* if they are not significant predictors of the y's. Thus $\mathbf{B}_d$ has h rows. The hypothesis can be expressed as

$$H_0\colon \mathbf{B}_d = \mathbf{O}.$$

If $\mathbf{X}_r$ contains the columns of $\mathbf{X}$ corresponding to $\mathbf{B}_r$, then the reduced model is

$$\mathbf{Y} = \mathbf{X}_r\mathbf{B}_r + \mathbf{\Xi}. \qquad (10.50)$$

To compare the fit of the full model and the reduced model, we use the difference between the regression sum of squares and products matrix for the full model, $\hat{\mathbf{B}}'\mathbf{X}'\mathbf{Y}$, and regression sum of squares and products matrix for the reduced model, $\hat{\mathbf{B}}_r'\mathbf{X}_r'\mathbf{Y}$. The difference becomes our $\mathbf{H}$ matrix,

$$\mathbf{H} = \hat{\mathbf{B}}'\mathbf{X}'\mathbf{Y} - \hat{\mathbf{B}}_r'\mathbf{X}_r'\mathbf{Y}. \qquad (10.51)$$

Thus the test of $H_0\colon \mathbf{B}_d = \mathbf{O}$ is a full and reduced model test of the significance of $x_{q-h+1}, x_{q-h+2}, \ldots, x_q$ above and beyond $x_1, x_2, \ldots, x_{q-h}$.

To make the test, we use the $\mathbf{E}$ matrix based on the full model,

$$\mathbf{E} = \mathbf{Y}'\mathbf{Y} - \hat{\mathbf{B}}'\mathbf{X}'\mathbf{Y}.$$

A test statistic based on Wilks' Λ is given by

$$\begin{aligned} \Lambda(x_{q-h+1}, \ldots, x_q | x_1, \ldots, x_{q-h}) &= \frac{|\mathbf{E}|}{|\mathbf{E}+\mathbf{H}|} \\ &= \frac{|\mathbf{Y}'\mathbf{Y} - \hat{\mathbf{B}}'\mathbf{X}'\mathbf{Y}|}{|\mathbf{Y}'\mathbf{Y} - \hat{\mathbf{B}}_r'\mathbf{X}_r'\mathbf{Y}|}, \end{aligned} \tag{10.52}$$

which is distributed as $\Lambda_{p,h,n-q-1}$ when $H_0\colon \mathbf{B}_d = \mathbf{O}$ is true. Critical values are available in Table A.9 with $\nu_H = h$ and $\nu_E = n - q - 1$. Note that these degrees of freedom for the multivariate y case are the same as for the univariate y case (multiple regression) in Section 10.2.5b. The F and χ^2 approximations for Λ in (6.13) and (6.14) can also be used.

As noted following (10.51), Wilks' Λ in (10.52) is a full and reduced model test. We can express it in terms of Λ for the full model and a similar Λ for the reduced model. In the denominator of (10.52), we have

$$\begin{aligned} \mathbf{E} + \mathbf{H} &= (\mathbf{Y}'\mathbf{Y} - \hat{\mathbf{B}}'\mathbf{X}'\mathbf{Y}) + (\hat{\mathbf{B}}'\mathbf{X}'\mathbf{Y} - \hat{\mathbf{B}}_r'\mathbf{X}_r'\mathbf{Y}) \\ &= \mathbf{Y}'\mathbf{Y} - \hat{\mathbf{B}}_r'\mathbf{X}_r'\mathbf{Y} = \mathbf{E}_r, \end{aligned}$$

which is the error matrix for the reduced model (10.50). This error matrix could be used in a test for the significance of overall regression in the reduced model,

$$\Lambda_r = \frac{|\mathbf{Y}'\mathbf{Y} - \hat{\mathbf{B}}_r'\mathbf{X}_r'\mathbf{Y}|}{|\mathbf{Y}'\mathbf{Y} - n\bar{\mathbf{y}}\bar{\mathbf{y}}'|}. \tag{10.53}$$

Since Λ_r in (10.53) has the same denominator as Λ in (10.44), we recognize (10.52) as the ratio of Wilks' Λ for the overall regression test in the full model to Wilks' Λ for the overall regression test in the reduced model,

$$\Lambda(x_{q-h+1}, \ldots, x_q | x_1, \ldots, x_{q-h}) = \frac{|\mathbf{Y}'\mathbf{Y} - \hat{\mathbf{B}}'\mathbf{X}'\mathbf{Y}|}{|\mathbf{Y}'\mathbf{Y} - \hat{\mathbf{B}}_r'\mathbf{X}_r'\mathbf{Y}|}$$

$$= \frac{\dfrac{|\mathbf{Y}'\mathbf{Y} - \hat{\mathbf{B}}'\mathbf{X}'\mathbf{Y}|}{|\mathbf{Y}'\mathbf{Y} - n\bar{\mathbf{y}}\bar{\mathbf{y}}'|}}{\dfrac{|\mathbf{Y}'\mathbf{Y} - \hat{\mathbf{B}}_r'\mathbf{X}_r'\mathbf{Y}|}{|\mathbf{Y}'\mathbf{Y} - n\bar{\mathbf{y}}\bar{\mathbf{y}}'|}}$$

$$= \frac{\Lambda_f}{\Lambda_r}, \qquad (10.54)$$

where Λ_f is given by (10.44). In (10.54), we have a convenient computational device. We run the overall regression test for the full model and again for the reduced model and take the ratio of the resulting Λ values.

The Wilks' Λ in (10.54), comparing the full and reduced models, is similar in appearance to (6.101). However, in (6.101), the full and reduced models involve the dependent variables $y_1, y_2, \ldots, y_p$ in MANOVA, whereas in (10.54), the reduced model is obtained by deleting a subset of the independent variables $x_1, x_2, \ldots, x_q$ in regression. The parameters of the Wilks' Λ distribution are different in the two cases. Note that in (6.101), some of the dependent variables were denoted by $x_1, \ldots, x_q$ for convenience.

Test statistics due to Roy, Pillai, and Lawley–Hotelling can be obtained from the eigenvalues of $\mathbf{E}^{-1}\mathbf{H} = (\mathbf{Y}'\mathbf{Y} - \hat{\mathbf{B}}'\mathbf{X}'\mathbf{Y})^{-1}(\hat{\mathbf{B}}'\mathbf{X}'\mathbf{Y} - \hat{\mathbf{B}}_r'\mathbf{X}_r'\mathbf{Y})$. Critical values or approximate tests for these three test statistics are based on $\nu_H = h$, $\nu_E = n - q - 1$, and

$$s = \min(p, h) \qquad m = \tfrac{1}{2}(|h - p| - 1) \qquad N = \tfrac{1}{2}(n - p - h - 2)$$

Example 10.5.2. The chemical data in Table 10.1 originated from a response surface experiment seeking to locate optimum operating conditions. Therefore, a second-order model is of interest, and we add x_1^2, x_2^2, x_3^2, x_1x_2, x_1x_3, x_2x_3 to the variables x_1, x_2, x_3 considered in Example 10.5.1. With the resulting nine independent variables, we obtain an overall Wilks' Λ of

$$\Lambda = .00145 < \Lambda_{.05,3,9,9} = .0240,$$

where $\nu_H = q = 9$ and $\nu_E = n - q - 1 = 19 - 9 - 1 = 9$. To see whether the six second-order variables add a significant amount to x_1, x_2, x_3 for predicting the y's, we calculate

$$\Lambda = \frac{\Lambda_f}{\Lambda_r} = \frac{.00145}{.0332} = .0438 < \Lambda_{.05,3,6,9} = .0492,$$

where $\nu_H = h = 6$ and $\nu_E = n - q - 1 = 19 - 9 - 1 = 9$. In this case, $\Lambda_r = .0332$ is from Example 10.5.1, in which we considered the model with x_1, x_2, and x_3. Thus we reject $H_0\colon \mathbf{B}_1 = \mathbf{O}$ and conclude that the second-order terms add significant predictability to the model.

10.6 MEASURES OF ASSOCIATION BETWEEN THE y's AND THE x's

The most widely used measures of association between two sets of variables are the *canonical correlations*, which are treated in detail in Chapter 11. In this section, we review other measures of association that have been proposed. A general comparison of these has apparently not been made.

In (10.28), we have $R^2 = \mathbf{s}'_{yx}\mathbf{S}_{xx}^{-1}\mathbf{s}_{yx}/s_{yy}$ for the univariate y case. By analogy, we define an R^2-like measure of association between $y_1, y_2, \ldots, y_p$ and $x_1, x_2, \ldots, x_q$ as

$$R_M^2 = \frac{|\mathbf{S}_{yx}\mathbf{S}_{xx}^{-1}\mathbf{S}_{xy}|}{|\mathbf{S}_{yy}|}, \tag{10.55}$$

where $\mathbf{S}_{yx}$, $\mathbf{S}_{xy}$, $\mathbf{S}_{xx}$, and $\mathbf{S}_{yy}$ are defined following (10.42) and the subscript M indicates multivariate y's.

Another R^2-like measure of association was proposed by Robert and Escoufier (1976):

$$\text{RV} = \frac{\text{tr}\,(\mathbf{S}_{xy}\mathbf{S}_{yx})}{\sqrt{\text{tr}\,(\mathbf{S}_{xx}^2)\,\text{tr}\,(\mathbf{S}_{yy}^2)}}. \tag{10.56}$$

Kabe (1985) discussed the generalized correlation determinant

$$\text{GCD} = \frac{|\mathbf{L}'\mathbf{S}_{xy}\mathbf{M}\mathbf{M}'\mathbf{S}_{yx}\mathbf{L}|}{|\mathbf{L}'\mathbf{S}_{xx}\mathbf{L}||\mathbf{M}'\mathbf{S}_{yy}\mathbf{M}|}$$

for various choices of the transformation matrices $\mathbf{L}$ and $\mathbf{M}$.

In Section 6.1.8, we introduced several measures of association that quantify the amount of relationship between the y's and the dummy grouping variables in a MANOVA context. These are even more appropriate here in the multivariate regression setting, where both the x's and the y's are continuous variables. The R^2-like indices given in (6.32), (6.33), (6.34), (6.35), (6.37), and (6.39) range between 0 and 1 and will be briefly reviewed in the remainder of this section. For more complete commentary, see Section 6.1.8.

The two measures based on Wilks' Λ are

$$\eta_\Lambda^2 = 1 - \Lambda$$

and

$$A_\Lambda = 1 - \Lambda^{1/s},$$

where $s = \min(p, q)$. A measure based on Roy's θ is provided by θ itself,

$$\eta_\theta^2 = \frac{\lambda_1}{1 + \lambda_1} = \theta,$$

where λ_1 is the largest eigenvalue of $\mathbf{E}^{-1}\mathbf{H}$. This was identified in Section 6.1.8 as the square of the first canonical correlation (see Chapter 11) between the y's and the grouping variables in MANOVA. In the multivariate regression setting, θ is the square of the first canonical correlation, r_1^2, between the y's and the x's.

Measures of association based on the Lawley–Hotelling and Pillai statistics are given by

$$A_{\text{LH}} = \frac{U^{(s)}/s}{1 + U^{(s)}/s}$$

and

$$A_P = \frac{V^{(s)}}{s}.$$

The latter is the average of the s canonical correlations, $r_1^2, r_2^2, \ldots, r_3^2$.

Example 10.6. We use the chemical data of Table 10.1 to illustrate some measures of association. For the three dependent variables y_1, y_2, and y_3 and the three independent variables x_1, x_2, and x_3, the partitioned covariance matrix is

$$\mathbf{S} = \begin{pmatrix} \mathbf{S}_{yy} & \mathbf{S}_{yx} \\ \mathbf{S}_{xy} & \mathbf{S}_{xx} \end{pmatrix}$$

$$= \left[\begin{array}{rrr|rrr} 99.30 & -28.57 & -59.03 & -41.95 & -9.49 & -7.37 \\ -28.57 & 22.25 & 5.78 & 11.85 & 1.60 & 3.03 \\ -59.03 & 5.78 & 45.27 & 24.14 & 6.43 & 3.97 \\ \hline -41.95 & 11.85 & 24.14 & 38.67 & -12.17 & -.22 \\ -9.49 & 1.60 & 6.43 & -12.17 & 17.95 & 1.22 \\ -7.36 & 3.03 & 3.97 & -.22 & 1.22 & 2.67 \end{array}\right],$$

from which we obtain R_M^2 and RV directly using (10.55) and (10.56),

$$R_M^2 = \frac{|\mathbf{S}_{yx}\mathbf{S}_{xx}^{-1}\mathbf{S}_{xy}|}{|\mathbf{S}_{yy}|} = .00029$$

$$\text{RV} = \frac{\text{tr}\,(\mathbf{S}_{xy}\mathbf{S}_{yx})}{\sqrt{\text{tr}\,(\mathbf{S}_{xx}^2)\,\text{tr}\,(\mathbf{S}_{yy}^2)}} = .403.$$

Using the results in Example 10.5.1, we obtain

$$\eta_\Lambda^2 = 1 - \Lambda = 1 - .0332 = .967$$

$$A_\Lambda = 1 - \Lambda^{1/s} = 1 - \Lambda^{1/3} = .679$$

$$\eta_\theta^2 = \frac{\lambda_1}{1+\lambda_1} = .963$$

$$A_{\text{LH}} = \frac{U^{(s)}/s}{1+U^{(s)}/s} = \frac{26.422/3}{1+26.422/3} = .898$$

$$A_P = \frac{V^{(s)}}{s} = \frac{1.058}{3} = .352.$$

We have general agreement among η_Λ^2, A_Λ, η_θ^2, and A_{LH}. But R_M^2, RV, and A_P do not appear to be measuring the same level of association, especially R_M^2. It appears that more study is needed before one or more of these measures can be universally recommended.

10.7 SUBSET SELECTION

As in the univariate y case in Section 10.2.7, we may wish to choose a subset of the x's. There are $p+q$ variables in the model, $y_1, y_2, \ldots, y_p$, $x_1, x_2, \ldots, x_q$. But, as in the univariate y case, there may be more potential predictor variables (x's) than are useful in a given situation. Some of the x's may be only slightly correlated with the y's or they may be redundant because of high correlations with other x's.

We may also be interested in deleting some of the y's if they are not related linearly to any of the x's. This would lead to further simplification, yielding a model with no redundant dependent or independent variables.

We present three approaches to subset selection: stepwise procedures, methods involving all possible subsets, and a simultaneous technique.

10.7.1 Stepwise Procedures

Subset selection among the x's is discussed in Section 10.7.1a, followed by selection among the y's in Section 10.7.1b.

10.7.1a Finding a Subset of the x's

We begin with the *forward selection* procedure based on Wilks' Λ. At the first step, we test the regression of the p y's on each x_i. There will be two rows in the $\hat{\mathbf{B}}$ matrix, a row containing the intercepts and a row corresponding to x_i:

$$\hat{\mathbf{B}}_i = \begin{pmatrix} \hat{\beta}_{01} & \hat{\beta}_{02} & \cdots & \hat{\beta}_{0p} \\ \hat{\beta}_{i1} & \hat{\beta}_{i2} & \cdots & \hat{\beta}_{ip} \end{pmatrix}.$$

We use the overall regression test (10.44),

$$\Lambda(x_i) = \frac{|\mathbf{Y}'\mathbf{Y} - \hat{\mathbf{B}}_i'\mathbf{X}'\mathbf{Y}|}{|\mathbf{Y}'\mathbf{Y} - n\bar{\mathbf{y}}\bar{\mathbf{y}}'|},$$

which is distributed as $\Lambda_{p,1,n-2}$, since $\hat{\mathbf{B}}_i$ has two rows. After calculating $\Lambda(x_i)$ for each i, we choose the variable with minimum $\Lambda(x_i)$. Note that at the first step, we are not testing each variable in the presence of the others. We want to know which x_i best predicts the p y's by itself, not above and beyond the other x's.

At the second step, we seek the variable yielding the smallest *partial* Λ for each x adjusted for the variable first entered. The partial Λ-statistic is based on (10.54), the full and reduced model test for a subset of x's:

$$\Lambda(x_{q-h+1}, \ldots, x_q | x_1, \ldots, x_{q-h}) = \frac{\Lambda_f}{\Lambda_r}.$$

After one variable has entered, this becomes

$$\Lambda(x_i | x_1) = \frac{\Lambda(x_1, x_i)}{\Lambda(x_1)}, \qquad (10.57)$$

where x_1 denotes the variable entered at the first step. We calculate (10.57) for each $x_i \neq x_1$ and choose the variable that minimizes $\Lambda(x_i|x_1)$.

If we denote the second variable to enter by x_2, then at the third step we seek the x_i that minimizes

$$\Lambda(x_i|x_1, x_2) = \frac{\Lambda(x_1, x_2, x_i)}{\Lambda(x_1, x_2)}. \tag{10.58}$$

By property 7 in Section 6.1.3, the partial Wilks' Λ-statistic transforms to an exact F since $\nu_H = h = 1$ at each step.

After m variables have been selected, the partial Λ would have the following form at the next step:

$$\Lambda(x_i|x_1, x_2, \ldots, x_m) = \frac{\Lambda(x_1, x_2, \ldots, x_m, x_i)}{\Lambda(x_1, x_2, \ldots, x_m)}, \tag{10.59}$$

where the first m variables to enter are denoted $x_1, x_2, \ldots, x_m$, and x_i is a candidate variable from among the $q - m$ remaining variables. At this step, we would choose the x_i that minimizes (10.59). The partial Wilks' Λ in (10.59) is distributed as $\Lambda_{p,1,n-m-1}$ and, by Table 6.1, transforms to the partial F-statistic $F_{p,n-m-p}$.

The procedure continues, bringing in the "best" variable at each step, until a step is reached at which the minimum partial Λ exceeds a predetermined threshold value or, equivalently, the associated partial F falls below a preselected value. Alternatively, the stopping rule can be cast in terms of the p-value of the partial Λ or F. If the smallest p-value at some step exceeds a predetermined value, the procedure stops.

For each x_i, there corresponds an entire row of the $\hat{\mathbf{B}}$ matrix because x_i has a coefficient for each of the p y's. Thus if a certain x significantly predicts even one of the y's, it should be retained.

The *stepwise* procedure is an extension of forward selection. Each time a variable enters, all the variables that have entered previously are checked by a partial Λ or F to see if the least "significant" one is now redundant and can be deleted.

The *backward elimination* procedure begins with all x's (all rows of $\hat{\mathbf{B}}$) included in the model and deletes one at a time using a partial Λ or F. At the first step, the partial Λ for each x_i is

$$\Lambda(x_i|x_1, \ldots, x_{i-1}, x_{i+1}, \ldots, x_q) = \frac{\Lambda(x_1, \ldots, x_q)}{\Lambda(x_1, \ldots, x_{i-1}, x_{i+1}, \ldots, x_q)}, \tag{10.60}$$

which is distributed as $\Lambda_{p,1,n-q-1}$ and can be converted to $F_{p,n-q-p}$ by Table 6.1. The variable with largest Λ or smallest F is deleted. At the second step, a partial Λ or F is calculated for each of the $q - 1$ remaining variables, and again the least important variable in the presence of the others is eliminated. This process continues until a step is reached at which the largest Λ or smallest F is "significant," indicating that the corresponding variable is apparently not redundant in the presence of its fellows. Some preselected p-value or threshold value of Λ or F is used to determine a stopping rule.

If no automated program is available for subset selection in the multivariate case, a forward selection or backward elimination procedure could easily be carried out by means of a rather simple set of commands based on (10.59) or (10.60).

A sequential procedure such as stepwise selection will often fail to find the optimum subset, especially if a large pool of predictor variables is involved. However, the value of Wilks' Λ found by stepwise selection may not be far from that for the optimum subset.

The remarks in the final paragraph of Section 10.2.7b are pertinent in the multivariate context as well. True predictors of the y's in the population may be overlooked because of inflated error variances or, on the other hand, x's that are not true predictors may appear to be so in the sample. The latter problem can be severe for small sample sizes (Rencher and Pun 1980).

10.7.1b Finding a Subset of the y's

After a subset of x's has been found, the researcher may wish to know if these x's predict all p of the y's. If some of the y's do not relate to any of the x's, they could be deleted from the model to achieve a further simplification. The y's can be checked for redundancy in a manner analogous to the stepwise discriminant approach presented in Sections 6.11.2 and 8.9, which finds subsets of *dependent* variables using a full and reduced model Wilks' Λ for the y's. The partial Λ-statistic for adding or deleting a y is similar to (10.57), (10.58), or (10.59), except that dependent variables are involved rather than independent variables. Thus to add a y at the third step of a *forward selection procedure*, for example, where the first two variables already entered are denoted as y_1 and y_2, we use (6.102) to obtain

$$\Lambda(y_i|y_1, y_2) = \frac{\Lambda(y_1, y_2, y_i)}{\Lambda(y_1, y_2)} \tag{10.61}$$

for each $y_i \neq y_1$ or y_2, and we choose the y_i that minimizes $\Lambda(y_i|y_1, y_2)$. [In (6.102) the dependent variable of interest was denoted by x instead of y_i as here.]

Similarly, if three y's, designated as y_1, y_2, and y_3, were "in the model" and we were checking the feasibility of adding y_i, the partial Λ-statistic would be

$$\Lambda(y_i|y_1, y_2, y_3) = \frac{\Lambda(y_1, y_2, y_3, y_i)}{\Lambda(y_1, y_2, y_3)}, \tag{10.62}$$

which is distributed as $\Lambda_{1,q,n-q-4}$, where q is the number of x's *presently* in the model and 4 is the number of y's *presently* in the model. The two Wilks Λ values in the numerator and denominator of the right side of (10.62), $\Lambda(y_1, y_2, y_3, y_i)$ and $\Lambda(y_1, y_2, y_3)$, are obtained from (10.44). Since $p = 1$, $\Lambda_{1,q,n-q-4}$ in (10.62) transforms to $F_{q,n-q-4}$ (see Table 6.1).

In the first step of a *backward elimination procedure*, we would delete the y_i that maximizes

$$\Lambda(y_i|y_1, \ldots, y_{i-1}, y_{i+1}, \ldots, y_p) = \frac{\Lambda(y_1, \ldots, y_p)}{\Lambda(y_1, \ldots, y_{i-1}, y_{i+1}, \ldots, y_p)}, \qquad (10.63)$$

which is distributed as $\Lambda_{1,\nu_H,\nu_E-p+1}$. In this case, $\nu_H = q$ and $\nu_E = n - q - 1$ so that the distribution of (10.63) is $\Lambda_{1,q,n-q-p}$, which can be transformed to an exact F. Note that q, the number of x's, may have been reduced in a subset selection on the x's as in Section 10.7.1a. Similarly, p is the number of y's and will decrease in subsequent steps.

Stopping rules for either the forward or backward approach could be defined in terms of p-values or threshold values of Λ or the equivalent F. A *stepwise* procedure could be devised as a modification of the forward approach.

If a software program is available that tests the significance of one x as in (10.60), it can be adapted to test one y as in (10.63) by use of property 3 of Section 6.1.3: The distribution of Λ_{p,ν_H,ν_E} is the same as that of $\Lambda_{\nu_H,p,\nu_E+\nu_H-p}$. Thus we can reverse the y's and x's. This can also be seen from the symmetry of Λ expressed as in (10.46):

$$\Lambda = \frac{|\mathbf{S}|}{|\mathbf{S}_{xx}||\mathbf{S}_{yy}|},$$

which is distributed as $\Lambda_{p,q,n-q-1}$ or equivalently as $\Lambda_{q,p,n-p-1}$. Hence we list the x's as dependent variables in the program and the y's as independent variables. Then the test of a single y in (10.62) or (10.63) can be carried out using (10.60) without any adjustment. The partial Λ-statistic in (10.60) is distributed as $\Lambda_{p,1,n-q-1}$. If we interchange p and q, because the y's and x's are interchanged as dependent and independent variables, this becomes $\Lambda_{q,1,n-p-1}$. By property 3 of Section 6.1.3 (repeated above), this is equivalent to $\Lambda_{1,q,n-p-1+1-q} = \Lambda_{1,q,n-p-q}$, which is the distribution of (10.63).

10.7.2 All Possible Subsets

In Section 10.2.7a, we discussed the criteria R_p^2, s_p^2, and C_p for comparing all possible subsets of x's to predict a univariate y in multiple regression, where $p - 1$ denotes the number of x's in a subset selected from a pool of $k - 1$ available independent variables. We now discuss some matrix analogues of these criteria for the multivariate y case, as suggested by Mallows (1973) and Sparks et al. (1983).

In this section and the next, in order to conform with notation in the literature, we will use p for the number of columns in $\mathbf{X}$ (or the number of rows in $\mathbf{B}$), rather than for the number of y's. The number of y's will be indicated by m.

We now extend the three criteria R_p^2, s_p^2, and C_p to analogous matrix expressions $\mathbf{R}_p^2, \mathbf{S}_p$, and $\mathbf{C}_p$. These can be reduced to scalar form using trace or determinant.

1. $\mathbf{R}_p^2$. In the univariate y case, R_p^2 for a $(p-1)$-variable subset of x's is defined by (10.26) as

$$R_p^2 = \frac{\hat{\boldsymbol{\beta}}_p'\mathbf{X}_p'\mathbf{y} - n\overline{y}^2}{\mathbf{y}'\mathbf{y} - n\overline{y}^2}.$$

A direct extension of R_p^2 for the multivariate y case is given by the matrix

$$\mathbf{R}_p^2 = (\mathbf{Y}'\mathbf{Y} - n\overline{\mathbf{y}}\overline{\mathbf{y}}')^{-1}(\hat{\mathbf{B}}_p'\mathbf{X}_p'\mathbf{Y} - n\overline{\mathbf{y}}\overline{\mathbf{y}}'), \tag{10.64}$$

where $p-1$ is the number of x's. One way to convert to scalar form is to use $\text{tr}\,\mathbf{R}_p^2/m$, in which we divide by m, the number of y's, so that $0 \le \text{tr}\,\mathbf{R}_p^2/m \le 1$. As in the univariate case, we identify the subset that maximizes $\text{tr}\,\mathbf{R}_p^2/m$ for each value of $p = 2, 3, \ldots, k$. The criterion $\text{tr}\,\mathbf{R}_p^2/m$ does not attain its maximum until p reaches k, but we look for the value of p at which further increases are deemed unimportant. We could also use $|\mathbf{R}_p^2|$ in place of $\text{tr}\,\mathbf{R}_p^2/m$.

2. $\mathbf{S}_p$. A direct extension of the univariate criterion $s_p^2 = \text{MSE}_p = \text{SSE}_p/(n-p)$ is provided by

$$\mathbf{S}_p = \frac{\mathbf{E}_p}{n-p}, \tag{10.65}$$

where $\mathbf{E}_p = \mathbf{Y}'\mathbf{Y} - \hat{\mathbf{B}}_p'\mathbf{X}_p'\mathbf{Y}$. To convert to a scalar, we can use $\text{tr}\,\mathbf{S}_p$ or $|\mathbf{S}_p|$, either of which will behave in an analogous fashion to s_p^2 in the univariate case. The remarks in Section 10.2.7a apply here as well; one may wish to select the subset with minimum value of $\text{tr}\,\mathbf{S}_p$ or perhaps the subset with smallest p such that $\text{tr}\,\mathbf{S}_p < \text{tr}\,\mathbf{S}_k$. A similar application could be made with $|\mathbf{S}_p|$.

3. $\mathbf{C}_p$. To extend the C_p criterion to the multivariate y case, we write the model under consideration as

$$\mathbf{Y} = \mathbf{X}_p\mathbf{B}_p + \boldsymbol{\Xi},$$

where $p-1$ is the number of x's in the subset and $k-1$ is the number of x's in the "full model." The predicted values of the y's are given by

$$\hat{\mathbf{Y}} = \mathbf{X}_p\hat{\mathbf{B}}_p.$$

We are interested in predicted values of the observation vectors, $\hat{\mathbf{y}}_1, \hat{\mathbf{y}}_2, \ldots, \hat{\mathbf{y}}_n$, which are given by the rows of $\hat{\mathbf{Y}}$:

$$\begin{bmatrix} \hat{\mathbf{y}}_1' \\ \hat{\mathbf{y}}_2' \\ \vdots \\ \hat{\mathbf{y}}_n' \end{bmatrix} = \begin{bmatrix} \mathbf{x}_{p1}' \\ \mathbf{x}_{p2}' \\ \vdots \\ \mathbf{x}_{pn}' \end{bmatrix} \hat{\mathbf{B}}_p = \begin{bmatrix} \mathbf{x}_{p1}'\hat{\mathbf{B}}_p \\ \mathbf{x}_{p2}'\hat{\mathbf{B}}_p \\ \vdots \\ \mathbf{x}_{pn}'\hat{\mathbf{B}}_p \end{bmatrix}.$$

In general, the predicted vectors $\hat{\mathbf{y}}_i$ are biased estimates of $E(\mathbf{y}_i)$ in the correct model, because we are examining many competing models, so that $E(\hat{\mathbf{y}}_i) \neq E(\mathbf{y}_i)$. In place of the univariate expected squared error in (10.30) and (10.31), we define a matrix of expected squares and products of errors, $E[\hat{\mathbf{y}}_i - E(\mathbf{y}_i)][\hat{\mathbf{y}}_i - E(\mathbf{y}_i)]'$. We then add and subtract $E(\hat{\mathbf{y}}_i)$ to obtain (see problem 10.7)

$$\begin{aligned} E[\hat{\mathbf{y}}_i &- E(\mathbf{y}_i)][\hat{\mathbf{y}}_i - E(\mathbf{y}_i)]' \\ &= E[\hat{\mathbf{y}}_i - E(\hat{\mathbf{y}}_i) + E(\hat{\mathbf{y}}_i) - E(\mathbf{y}_i)][\hat{\mathbf{y}}_i - E(\hat{\mathbf{y}}_i) + E(\hat{\mathbf{y}}_i) - E(\mathbf{y}_i)]' \\ &= E[\hat{\mathbf{y}}_i - E(\hat{\mathbf{y}}_i)][\hat{\mathbf{y}}_i - E(\hat{\mathbf{y}}_i)]' + [E(\hat{\mathbf{y}}_i) - E(\mathbf{y}_i)][E(\hat{\mathbf{y}}_i) - E(\mathbf{y}_i)]' \\ &= \text{cov}(\hat{\mathbf{y}}_i) + (\text{bias in } \hat{\mathbf{y}}_i)(\text{bias in } \hat{\mathbf{y}}_i)'. \end{aligned} \tag{10.66}$$

We first find an expression for $\text{cov}(\hat{\mathbf{y}}_i)$, which for convenience we write in row form,

$$\text{cov}(\hat{\mathbf{y}}_i') = \text{cov}(\mathbf{x}_{pi}'\hat{\mathbf{B}}_p) = \text{cov}(\mathbf{x}_{pi}'\hat{\boldsymbol{\beta}}_{p(1)}, \mathbf{x}_{pi}'\hat{\boldsymbol{\beta}}_{p(2)}, \ldots, \mathbf{x}_{pi}'\hat{\boldsymbol{\beta}}_{p(m)}).$$

where $\hat{\mathbf{B}} = (\hat{\boldsymbol{\beta}}_{(1)}, \hat{\boldsymbol{\beta}}_{(2)}, \ldots, \hat{\boldsymbol{\beta}}_{(p)})$, as noted at the end of Section 10.4.2. This can be written as

$$\begin{aligned} \text{cov}(\hat{\mathbf{y}}_i') &= \begin{bmatrix} \sigma_{11}\mathbf{x}_{pi}'(\mathbf{X}_p'\mathbf{X}_p)^{-1}\mathbf{x}_{pi} & \cdots & \sigma_{1m}\mathbf{x}_{pi}'(\mathbf{X}_p'\mathbf{X}_p)^{-1}\mathbf{x}_{pi} \\ \vdots & & \vdots \\ \sigma_{m1}\mathbf{x}_{pi}'(\mathbf{X}_p'\mathbf{X}_p)^{-1}\mathbf{x}_{pi} & \cdots & \sigma_{mm}\mathbf{x}_{pi}'(\mathbf{X}_p'\mathbf{X}_p)^{-1}\mathbf{x}_{pi} \end{bmatrix} \\ &= \mathbf{x}_{pi}'(\mathbf{X}_p'\mathbf{X}_p)^{-1}\mathbf{x}_{pi}\boldsymbol{\Sigma}, \end{aligned} \tag{10.67}$$

where $\boldsymbol{\Sigma} = \text{cov}(\mathbf{y}_i)$ (see problem 10.8). As in (10.33) (see also problem 10.5), we can sum over the n observations and use (3.79) to obtain

$$\sum_{i=1}^{n} \text{cov}(\hat{\mathbf{y}}_i) = \sum_{i=1}^{n} \mathbf{x}'_{pi}(\mathbf{X}'_p\mathbf{X}_p)^{-1}\mathbf{x}_{pi}\boldsymbol{\Sigma}$$
$$= \boldsymbol{\Sigma}\sum_{i=1}^{n} \mathbf{x}'_{pi}(\mathbf{X}'_p\mathbf{X}_p)^{-1}\mathbf{x}_{pi} = p\boldsymbol{\Sigma}. \quad (10.68)$$

To sum the second term on the right of (10.66), we have, by analogy to a remark preceding (10.34),

$$\sum_{i=1}^{n} (\text{bias in } \hat{\mathbf{y}}_i)(\text{bias in } \hat{\mathbf{y}}_i)' = (n - p)E(\mathbf{S}_p - \boldsymbol{\Sigma}), \quad (10.69)$$

where $\mathbf{S}_p$ is given by (10.65).

Now by (10.68) and (10.69), we can sum (10.66) and multiply by $\boldsymbol{\Sigma}^{-1}$ to obtain the matrix of total expected squares and products of error standardized by $\boldsymbol{\Sigma}^{-1}$,

$$\boldsymbol{\Sigma}^{-1}[p\boldsymbol{\Sigma} + (n - p)E(\mathbf{S}_p - \boldsymbol{\Sigma})] = p\mathbf{I} + (n - p)\boldsymbol{\Sigma}^{-1}E(\mathbf{S}_p - \boldsymbol{\Sigma}). \quad (10.70)$$

Using $\mathbf{S}_k = \mathbf{E}_k/(n-k)$, the sample covariance matrix based on all $k-1$ variables, as an estimate of $\boldsymbol{\Sigma}$, we can estimate (10.70) by

$$\mathbf{C}_p = p\mathbf{I} + (n - p)\mathbf{S}_k^{-1}(\mathbf{S}_p - \mathbf{S}_k) \quad (10.71)$$
$$= \mathbf{S}_k^{-1}\mathbf{E}_p + (2p - n)\mathbf{I} \quad [\text{by } (10.65)], \quad (10.72)$$

which is the form suggested by Mallows (1973). We can use $\text{tr}(\mathbf{C}_p)$ or $|\mathbf{C}_p|$ to reduce this to a scalar. But if $2p - n$ is negative, $|\mathbf{C}_p|$ can be negative, and Sparks et al. (1983) suggested a modification of $|\mathbf{C}_p|$,

$$|\mathbf{E}_k^{-1}\mathbf{E}_p|, \quad (10.73)$$

which is always positive.

To find the optimal subset of x's for each value of p, we could examine all possible subsets [or use a computational scheme such as that of Furnival and Wilson (1974)] and look for the "smallest" $\mathbf{C}_p$ matrix for each p. In (10.70), we see that when the bias is $\mathbf{O}$, the "population $\mathbf{C}_p$" is equal to $p\mathbf{I}$. Thus we seek a $\mathbf{C}_p$ that is "small" and near $p\mathbf{I}$. In terms of trace, we seek $\text{tr}\,\mathbf{C}_p$ close to pm, where m is the number of y's in the vector of measurements; that is, $\text{tr}\,\mathbf{I} = m$.

To find a "target" value for (10.73), we write $\mathbf{E}_k^{-1}\mathbf{E}_p$ in terms of $\mathbf{C}_p$ from (10.72),

$$\mathbf{E}_k^{-1}\mathbf{E}_p = \frac{\mathbf{C}_p + (n-2p)\mathbf{I}}{n-k}. \tag{10.74}$$

When the bias is $\mathbf{O}$, we have $\mathbf{C}_p = p\mathbf{I}$, and (10.74) becomes

$$\mathbf{E}_k^{-1}\mathbf{E}_p = \frac{n-p}{n-k}\,\mathbf{I}, \tag{10.75}$$

whence, by (2.79),

$$|\mathbf{E}_k^{-1}\mathbf{E}_p| = \left(\frac{n-p}{n-k}\right)^m. \tag{10.76}$$

Thus we seek subsets such that

$$\operatorname{tr}\mathbf{C}_p \leq pm$$

or

$$|\mathbf{E}_k^{-1}\mathbf{E}_p| \leq \left(\frac{n-p}{n-k}\right)^m.$$

In summary, when examining all possible subsets, any or all of the following criteria may be useful in finding the single best subset or the best subset for each p:

$$\operatorname{tr}\mathbf{R}_p^2/m \qquad |\mathbf{R}_p^2| \qquad \operatorname{tr}\mathbf{S}_p \qquad |\mathbf{S}_p| \qquad \operatorname{tr}\mathbf{C}_p \qquad |\mathbf{E}_k^{-1}\mathbf{E}_p|.$$

10.7.3 Simultaneous Methods

Assuming that $H_0\colon \mathbf{B}_d = \mathbf{O}$ in Section 10.5.2 is true, we have the reduced model in (10.50),

$$\mathbf{Y} = \mathbf{X}_r\mathbf{B}_r + \boldsymbol{\Xi}.$$

If the reduced model holds, then the subset of x's in $\mathbf{X}_r$ is characterized by McKay (1977, 1979) as *adequate*. Equivalently, a subset of x's is adequate if its canonical correlations with the y's are the same as those for the full set of x's.

To find adequate subsets, McKay proposed the use of simultaneous test procedures similar to those of Gabriel (1968, 1969). In this approach (using the notation of Section 10.7.2), a test statistic such as Wilks' Λ for a subset of variables is compared to the critical value for the full set, $\Lambda_{\alpha,m,k-1,n-k}$, where m is the number of y's. (The critical value of Λ decreases as the number of variables increases.) Any and all subsets can be tested with this critical value, and the experimentwise error rate will not exceed α. As with any simultaneous procedure, this test is conservative, especially for smaller subsets. Consequently, a subset may be falsely declared adequate, because the test on its complementary subset fails to reject H_0 when it should reject. Thus the number of adequate sets will tend to be too large. McKay discussed some approaches to reducing this list of adequate subsets. Smith et al. (1985) and Hammond et al. (1981) provided algorithms based on the software package SAS to implement McKay's procedures.

10.8 MULTIVARIATE REGRESSION: RANDOM x's

In Sections 10.4 and 10.5, it was assumed that the x's were fixed and would have the same values in repeated sampling. In many applications, the x's are random variables. In such a case, the values of $x_1, x_2, \ldots, x_q$ are not under the control of the experimenter but occur randomly along with $y_1, y_2, \ldots, y_p$. On each subject, we observe $p + q$ values in the vector $(y_1, y_2, \ldots, y_p, x_1, x_2, \ldots, x_q)$.

If we assume that $(y_1, y_2, \cdots, y_p, x_1, x_2, \cdots, x_q)$ has a multivariate normal distribution, then all estimates and tests have the same formulation as in the fixed-x case (for details, see Rencher 1996, Section 7.7). Thus there is no essential difference in our procedures between the fixed-x case and the random-x case.

PROBLEMS

10.1 Show that $\sum_{i=1}^n (y_i - \mathbf{x}_i'\hat{\boldsymbol{\beta}})^2 = (\mathbf{y} - \mathbf{X}\hat{\boldsymbol{\beta}})'(\mathbf{y} - \mathbf{X}\hat{\boldsymbol{\beta}})$, as in Section 10.2.2.

10.2 Show that $\sum_{i=1}^n (y_i - \mu)^2 = \sum_{i=1}^n (y_i - \bar{y})^2 + n(\bar{y} - \mu)^2$, as in (10.8).

10.3 Show that $\sum_{i=1}^n (x_{i2} - \bar{x}_2)\bar{y} = 0$, as noted below (10.14).

10.4 Show that $E[\hat{y}_i - E(y_i)]^2 = E[\hat{y}_i - E(\hat{y}_i)]^2 + [E(\hat{y}_i) - E(y_i)]^2$, as in (10.30).

10.5 Show that $\sum_{i=1}^n \text{var}(\hat{y}_i)/\sigma^2 = \text{tr}[\mathbf{X}_p(\mathbf{X}_p'\mathbf{X}_p)^{-1}\mathbf{X}_p']$, as used to prove that $\sum_i \text{var}(\hat{y}_i)/\sigma^2 = p$ in (10.33).

10.6 Show that the alternative form of C_p in (10.36) is equal to the original form in (10.35).

10.7 Show that

$$E[\hat{\mathbf{y}}_i - E(\mathbf{y}_i)][\hat{\mathbf{y}}_i - E(\mathbf{y}_i)]'$$
$$= E[\hat{\mathbf{y}}_i - E(\hat{\mathbf{y}}_i)][\hat{\mathbf{y}}_i - E(\hat{\mathbf{y}}_i)]' + [E(\hat{\mathbf{y}}_i) - E(\mathbf{y}_i)][E(\hat{\mathbf{y}}_i) - E(\mathbf{y}_i)]',$$

thus verifying (10.66).

10.8 Show that cov $(\hat{\mathbf{y}}_i')$ has the form given in (10.67).

10.9 Show that the two forms of $\mathbf{C}_p$ in (10.71) and (10.72) are equal.

10.10 Explain why $|\mathbf{E}_k^{-1}\mathbf{E}_p| > 0$, as claimed following (10.73).

10.11 Show that $\mathbf{E}_k^{-1}\mathbf{E}_p = [\mathbf{C}_p + (n - 2p)\mathbf{I}]/(n - k)$, as in (10.74), where $\mathbf{C}_p$ is given in (10.71).

10.12 Show that if $\mathbf{C}_p = p\mathbf{I}$, then $\mathbf{E}_k^{-1}\mathbf{E}_p = [(n - p)/(n - k)]\mathbf{I}$, as in (10.75).

10.13 Use the diabetes data of Table 3.6.

(a) Find the least squares estimate $\hat{\mathbf{B}}$ for the regression of (y_1, y_2) on (x_1, x_2, x_3).

(b) Test the significance of overall regression using all four test statistics.

(c) Determine what the eigenvalues of $\mathbf{E}^{-1}\mathbf{H}$ reveal about the essential rank of $\hat{\mathbf{B}}_1$ and the implications of this rank, such as the relative power of the four tests.

(d) Test the significance of each of x_1, x_2, and x_3 adjusted for the other two x's.

10.14 Use the sons data of Table 3.9.

(a) Find the least squares estimate $\hat{\mathbf{B}}$ for the regression of (y_1, y_2) on (x_1, x_2).

(b) Test the significance of overall regression using all four test statistics.

(c) Determine what the eigenvalues of $\mathbf{E}^{-1}\mathbf{H}$ reveal about the essential rank of $\hat{\mathbf{B}}_1$ and the implications of this rank, such as the relative power of the four tests.

(d) Test the significance of x_1 adjusted for x_2 and of x_2 adjusted for x_1.

10.15 Use the glucose data of Table 3.10.

(a) Find the least squares estimate $\hat{\mathbf{B}}$ for the regression of (y_1, y_2, y_3) on (x_1, x_2, x_3).

(b) Test the significance of overall regression using all four test statistics.

(c) Determine what the eigenvalues of $\mathbf{E}^{-1}\mathbf{H}$ reveal about the essential rank of $\hat{\mathbf{B}}_1$ and the implications of this rank, such as the relative power of the four tests.

(d) Test the significance of each of x_1, x_2, and x_3 adjusted for the other two x's.

(e) Test each y to see if one of them can be deleted by using (10.63).

10.16 Use the seishu data of Table 7.1.

(a) Find the least squares estimate $\hat{\mathbf{B}}$ for the regression of (y_1, y_2) on $(x_1, x_2, \ldots, x_8)$ and test for significance.

(b) Test the significance of (x_7, x_8) adjusted for the other x's.

(c) Test the significance of (x_4, x_5, x_6) adjusted for the other x's.

(d) Test the significance of (x_1, x_2, x_3) adjusted for the other x's.

10.17 Use the seishu data of Table 7.1.

(a) Do a stepwise regression to select a subset of $x_1, x_2, \ldots, x_8$ that adequately predicts (y_1, y_2).

(b) After selecting a subset of x's, use the methods of Section 10.7.1b to check if either of the y's can be deleted.

10.18 Use the temperature data of Table 7.2.

(a) Find the least squares estimate $\hat{\mathbf{B}}$ for the regression of (y_4, y_5, y_6) on (y_1, y_2, y_3) and test for significance.

(b) Find the least squares estimate $\hat{\mathbf{B}}$ for the regression of (y_7, y_8, y_9) on $(y_1, \ldots, y_6)$ and test for significance.

(c) Find the least squares estimate $\hat{\mathbf{B}}$ for the regression of (y_{10}, y_{11}) on $(y_1, \ldots, y_9)$ and test for significance.

10.19 Using the temperature data of Table 7.2, carry out a stepwise regression to select a subset of $y_1, y_2, \ldots, y_9$ that adequately predicts (y_{10}, y_{11}).

CHAPTER 11

Canonical Correlation

11.1 INTRODUCTION

Canonical correlation analysis is concerned with the amount of (linear) relationship between two sets of variables. We often measure two types of variables on each research unit, for example, a set of aptitude variables and a set of achievement variables, a set of personality variables and a set of ability measures, a set of price indices and a set of production indices, a set of student behaviors and a set of teacher behaviors, a set of psychological attributes and a set of physiological attributes, a set of ecological variables and a set of environmental variables, a set of academic achievement variables and a set of measures of job success, a set of closed-book exam scores and a set of open-book exam scores, and a set of personality variables of freshmen students and the same variables on the same students as seniors. In Section 7.4.1, we discussed the hypothesis that two sets of variables were independent. In this chapter, we consider a measure of overall correlation between two sets of variables.

11.2 CANONICAL CORRELATIONS AND CANONICAL VARIATES

We assume that two sets of variables $\mathbf{y}' = (y_1, y_2, \ldots, y_p)$ and $\mathbf{x}' = (x_1, x_2, \ldots, x_q)$ are measured on the same sampling unit. We denote the two sets of variables as $\mathbf{y}$ and $\mathbf{x}$ to conform to notation in Chapters 3, 7, and 10. However, it is not necessary to designate the y's as dependent variables and the x's as independent variables. The two sets are treated symmetrically; it does not matter which is listed first.

The *canonical correlation* is defined later in this section as a measure of correlation between the y's and the x's. It is an extension of multiple correlation, which is the correlation between one y and several x's. Canonical correlation analysis is often a useful complement to a multivariate regression analysis.

We first review multiple correlation. The sample covariances and correlations among $y, x_1, x_2, \ldots, x_q$ can be summarized in the matrices

$$\mathbf{S} = \begin{pmatrix} s_y^2 & \mathbf{s}'_{yx} \\ \mathbf{s}_{yx} & \mathbf{S}_{xx} \end{pmatrix} \qquad (11.1)$$

$$\mathbf{R} = \begin{pmatrix} 1 & \mathbf{r}'_{yx} \\ \mathbf{r}_{yx} & \mathbf{R}_{xx} \end{pmatrix}, \qquad (11.2)$$

where $\mathbf{s}'_{yx} = (s_{y1}, s_{y2}, \ldots, s_{yq})$ contains the sample covariances of y with $x_1, x_2, \ldots, x_q$ and $\mathbf{S}_{xx}$ is the sample covariance matrix of the x's [see the expressions for $\mathbf{S}$ and $\mathbf{S}_{xx}$ following (10.12)]. The partitioned matrix $\mathbf{R}$ is defined analogously; $\mathbf{r}'_{yx} = (r_{y1}, r_{y2}, \ldots, r_{yq})$ contains the sample correlations of y with $x_1, x_2, \ldots, x_q$, and $\mathbf{R}_{xx}$ is the sample correlation matrix of the x's,

$$\mathbf{R}_{xx} = \begin{bmatrix} 1 & r_{12} & \cdots & r_{1q} \\ r_{21} & 1 & \cdots & r_{2q} \\ \vdots & \vdots & & \vdots \\ r_{q1} & r_{q2} & \cdots & 1 \end{bmatrix},$$

where r_{ij} is the sample correlation of x_i and x_j.

By (10.28), the squared multiple correlation between y and the x's can be computed from the partitioned covariance or correlation matrices (11.1) or (11.2) as follows:

$$R^2 = \frac{\mathbf{s}'_{yx}\mathbf{S}_{xx}^{-1}\mathbf{s}_{yx}}{s_y^2} = \mathbf{r}'_{yx}\mathbf{R}_{xx}^{-1}\mathbf{r}_{yx}.$$

In R^2, the q covariances between y and the x's in $\mathbf{s}_{yx}$ or the q correlations between y and the x's in $\mathbf{r}_{yx}$ are channeled into a single measure of relationship between y and the x's. The multiple correlation R can be defined alternatively as the maximum correlation between y and a linear combination of the x's.

We now return to the case of several y's and several x's. The covariance structure associated with two subvectors $\mathbf{y}$ and $\mathbf{x}$ was first discussed in Section 3.8.1. By (3.40), the overall sample covariance matrix can be partitioned as

$$\mathbf{S} = \begin{pmatrix} \mathbf{S}_{yy} & \mathbf{S}_{yx} \\ \mathbf{S}_{xy} & \mathbf{S}_{xx} \end{pmatrix},$$

where $\mathbf{S}_{yy}$ is the $p \times p$ sample covariance matrix of the y's, $\mathbf{S}_{yx}$ is the $p \times q$ matrix of sample covariances between the y's and the x's, and $\mathbf{S}_{xx}$ is the $q \times q$ sample covariance matrix of the x's.

In Section 10.6, we discussed several measures of association between the y's and the x's. The first of these is defined in (10.55) as $R_M^2 = |\mathbf{S}_{yx}\mathbf{S}_{xx}^{-1}\mathbf{S}_{xy}|/|\mathbf{S}_{yy}|$.

By (2.83) and (2.84), this can be rewritten as $R_M^2 = |\mathbf{S}_{yy}^{-1}\mathbf{S}_{yx}\mathbf{S}_{xx}^{-1}\mathbf{S}_{xy}|$. Note that $\mathbf{S}_{yy}^{-1}\mathbf{S}_{yx}\mathbf{S}_{xx}^{-1}\mathbf{S}_{xy}$ is a matrix analogue of $R^2 = \mathbf{s}_{yx}'\mathbf{S}_{xx}^{-1}\mathbf{s}_{yx}/s_y^2$. By (2.99), R_M^2 can be expressed as

$$R_M^2 = |\mathbf{S}_{yy}^{-1}\mathbf{S}_{yx}\mathbf{S}_{xx}^{-1}\mathbf{S}_{xy}| = \prod_{i=1}^{s} r_i^2,$$

where $s = \min(p, q)$ and $r_1^2, r_2^2, \ldots, r_s^2$ are the eigenvalues of $\mathbf{S}_{yy}^{-1}\mathbf{S}_{yx}\mathbf{S}_{xx}^{-1}\mathbf{S}_{xy}$. When written in this form, R_M^2 is seen to be a poor measure of association because $0 \leq r_i^2 \leq 1$ for all i, and $\prod_{i=1}^{s} r_i^2$ will usually be too small to meaningfully reflect the amount of association. (In Example 10.6, $R_M^2 = .00029$ was clearly meaningless compared with the other measures of association illustrated.) We therefore work with the eigenvalues themselves, rather than with their product. The square roots of the eigenvalues, $r_1, r_2, \ldots, r_s$, are called *canonical correlations*.

The largest squared canonical correlation (maximum eigenvalue) r_1^2 of $\mathbf{S}_{yy}^{-1}\mathbf{S}_{yx}\mathbf{S}_{xx}^{-1}\mathbf{S}_{xy}$ is the best overall measure of association, but the other eigenvalues (squared canonical correlations) of $\mathbf{S}_{yy}^{-1}\mathbf{S}_{yx}\mathbf{S}_{xx}^{-1}\mathbf{S}_{xy}$ provide measures of supplemental dimensions of (linear) relationship between $\mathbf{y}$ and $\mathbf{x}$. It can be shown that r_1^2 is the maximum squared correlation between a linear combination of the y's, $u = \mathbf{a}'\mathbf{y}$, and a linear combination of the x's, $v = \mathbf{b}'\mathbf{x}$. We denote the coefficient vectors that yield the maximum correlation as $\mathbf{a}_1$ and $\mathbf{b}_1$. Thus r_1 (the positive square root of r_1^2) is the correlation between $u_1 = \mathbf{a}_1'\mathbf{y}$ and $v_1 = \mathbf{b}_1'\mathbf{x}$. The linear functions u_1 and v_1 are called the first *canonical variates*. There are additional canonical variates corresponding to $r_2, r_3, \ldots, r_s$.

It was noted in Section 2.11.4 that the (nonzero) eigenvalues of $\mathbf{AB}$ are the same as those of $\mathbf{BA}$ as long as $\mathbf{AB}$ and $\mathbf{BA}$ are square, but that the eigenvectors of $\mathbf{AB}$ and $\mathbf{BA}$ are not the same. If we let $\mathbf{A} = \mathbf{S}_{yy}^{-1}\mathbf{S}_{yx}$ and $\mathbf{B} = \mathbf{S}_{xx}^{-1}\mathbf{S}_{xy}$, then $r_1^2, r_2^2, \ldots, r_s^2$ can also be obtained from $\mathbf{BA} = \mathbf{S}_{xx}^{-1}\mathbf{S}_{xy}\mathbf{S}_{yy}^{-1}\mathbf{S}_{yx}$. Thus the eigenvalues can be obtained from either of the characteristic equations

$$|\mathbf{S}_{yy}^{-1}\mathbf{S}_{yx}\mathbf{S}_{xx}^{-1}\mathbf{S}_{xy} - r^2\mathbf{I}| = 0 \qquad (11.3)$$

$$|\mathbf{S}_{xx}^{-1}\mathbf{S}_{xy}\mathbf{S}_{yy}^{-1}\mathbf{S}_{yx} - r^2\mathbf{I}| = 0. \qquad (11.4)$$

The coefficient vectors $\mathbf{a}_i$ and $\mathbf{b}_i$ in the canonical variates $u_i = \mathbf{a}_i'\mathbf{y}$ and $v_i = \mathbf{b}_i'\mathbf{x}$ are the eigenvectors of these same two matrices:

$$(\mathbf{S}_{yy}^{-1}\mathbf{S}_{yx}\mathbf{S}_{xx}^{-1}\mathbf{S}_{xy} - r^2\mathbf{I})\mathbf{a} = \mathbf{0} \qquad (11.5)$$

$$(\mathbf{S}_{xx}^{-1}\mathbf{S}_{xy}\mathbf{S}_{yy}^{-1}\mathbf{S}_{yx} - r^2\mathbf{I})\mathbf{b} = \mathbf{0}. \qquad (11.6)$$

Thus the two matrices $\mathbf{S}_{yy}^{-1}\mathbf{S}_{yx}\mathbf{S}_{xx}^{-1}\mathbf{S}_{xy}$ and $\mathbf{S}_{xx}^{-1}\mathbf{S}_{xy}\mathbf{S}_{yy}^{-1}\mathbf{S}_{yx}$ have the same (nonzero) eigenvalues, as indicated in (11.3) and (11.4), but different eigenvectors, as in (11.5) and (11.6). The eigenvectors $\mathbf{a}$ and $\mathbf{b}$ are ordinarily different in size, since $\mathbf{S}_{yy}^{-1}\mathbf{S}_{yx}\mathbf{S}_{xx}^{-1}\mathbf{S}_{xy}$ is $p \times p$ and $\mathbf{S}_{xx}^{-1}\mathbf{S}_{xy}\mathbf{S}_{yy}^{-1}\mathbf{S}_{yx}$ is $q \times q$, and p is typically not equal to q. In fact, the matrix that is larger in size will be singular, and the smaller one will be nonsingular. We illustrate for $p < q$. In this case $\mathbf{S}_{xx}^{-1}\mathbf{S}_{xy}\mathbf{S}_{yy}^{-1}\mathbf{S}_{yx}$ has the form

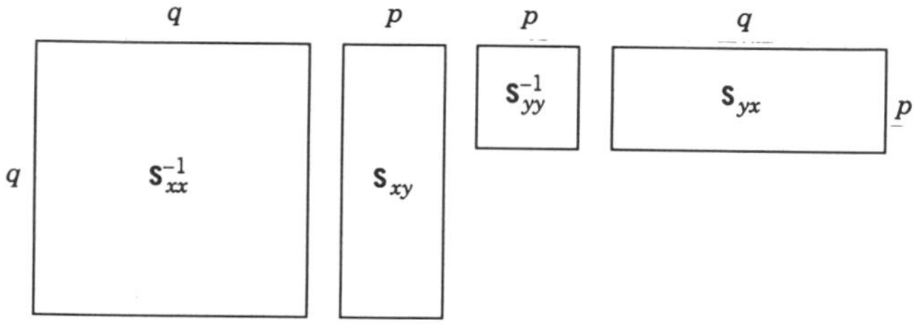

When $p < q$, the rank of $\mathbf{S}_{xx}^{-1}\mathbf{S}_{xy}\mathbf{S}_{yy}^{-1}\mathbf{S}_{yx}$ is p, because $\mathbf{S}_{xx}^{-1}$ has rank q and $\mathbf{S}_{xy}\mathbf{S}_{yy}^{-1}\mathbf{S}_{yx}$ has rank p. In this case p eigenvalues are nonzero and the remaining $q - p$ eigenvalues are equal to zero. In general, there are $s = \min(p, q)$ values of the squared canonical correlation r_i^2 with s corresponding pairs of canonical variates $u_i = \mathbf{a}_i'\mathbf{y}$ and $v_i = \mathbf{b}_i'\mathbf{x}$. For example, if $p = 3$ and $q = 7$, there will be three canonical correlations, r_1, r_2, and r_3.

The canonical correlations $r_1, r_2, \ldots, r_s$ correspond to the s pairs of canonical variates

$$\begin{aligned} u_1 &= \mathbf{a}_1'\mathbf{y} & v_1 &= \mathbf{b}_1'\mathbf{x} \\ u_2 &= \mathbf{a}_2'\mathbf{y} & v_2 &= \mathbf{b}_2'\mathbf{x} \\ &\vdots & &\vdots \\ u_s &= \mathbf{a}_s'\mathbf{y} & v_s &= \mathbf{b}_s'\mathbf{x}. \end{aligned}$$

For $i = 1, 2, \ldots, s$, r_i is the correlation between u_i and v_i. The pairs $(u_i, v_i), i = 1, 2, \ldots, s$, provide the s dimensions of relationship. For simplicity, we would prefer only one dimension of relationship, but this occurs only when $s = 1$, that is, when $p = 1$ or $q = 1$. We examine the elements of the coefficient vectors $\mathbf{a}_i$ and $\mathbf{b}_i$ for the information they provide about the contribution of the y's and x's to r_i. These coefficients can be standardized, as noted in the last paragraph in the present section and in Section 11.5.1.

The s dimensions of relationships $(u_i, v_i), i = 1, 2, \ldots, s$, are nonredundant; the information each pair provides is totally unavailable in the other pairs. The $u_1, u_2, \ldots, u_s$ are, in fact, *uncorrelated*. They are not orthogonal because $\mathbf{a}_1, \mathbf{a}_2, \ldots, \mathbf{a}_s$ are eigenvectors of $\mathbf{S}_{yy}^{-1}\mathbf{S}_{yx}\mathbf{S}_{xx}^{-1}\mathbf{S}_{xy}$, which is nonsymmetric. Similarly, each u_i is uncorrelated with all $v_j, j \neq i$, except, of course, v_i.

As noted, the matrix $\mathbf{S}_{yy}^{-1}\mathbf{S}_{yx}\mathbf{S}_{xx}^{-1}\mathbf{S}_{xy}$ is not symmetric. Many algorithms for computation of eigenvalues and eigenvectors accept only symmetric matrices. Since $\mathbf{S}_{yy}^{-1}\mathbf{S}_{yx}\mathbf{S}_{xx}^{-1}\mathbf{S}_{xy}$ is the product of the two symmetric matrices $\mathbf{S}_{yy}^{-1}$ and $\mathbf{S}_{yx}\mathbf{S}_{xx}^{-1}\mathbf{S}_{xy}$, we can proceed as in (6.18) and work with $(\mathbf{U}')^{-1}\mathbf{S}_{yx}\mathbf{S}_{xx}^{-1}\mathbf{S}_{xy}\mathbf{U}^{-1}$, where $\mathbf{U}'\mathbf{U} = \mathbf{S}_{yy}$ is the Cholesky factorization of $\mathbf{S}_{yy}$. The symmetric matrix $(\mathbf{U}')^{-1}\mathbf{S}_{yx}\mathbf{S}_{xx}^{-1}\mathbf{S}_{xy}\mathbf{U}^{-1}$ has the same eigenvalues as $\mathbf{S}_{yy}^{-1}\mathbf{S}_{yx}\mathbf{S}_{xx}^{-1}\mathbf{S}_{xy}$ but has eigenvectors $\mathbf{U}\mathbf{a}_i$, where $\mathbf{a}_i$ is given in (11.5).

In effect, the pq covariances between the y's and x's in $\mathbf{S}_{yx}$ have been replaced by $s = \min(p, q)$ canonical correlations. These describe the relationships between $\mathbf{y}$ and $\mathbf{x}$ much more succinctly than the pq covariances. In fact, in a typical study, we do not need all s canonical correlations. The smallest eigenvalues can be disregarded to achieve even more simplification. As in (8.12) for discriminant functions, we can judge the importance of each eigenvalue by its relative size:

$$\frac{r_i^2}{\sum_{j=1}^{s} r_j^2}. \tag{11.7}$$

The canonical correlations can also be obtained from the correlation matrix of the y's and x's,

$$\mathbf{R} = \begin{pmatrix} \mathbf{R}_{yy} & \mathbf{R}_{yx} \\ \mathbf{R}_{xy} & \mathbf{R}_{xx} \end{pmatrix},$$

where $\mathbf{R}_{yy}$ is the $p \times p$ sample correlation matrix of the y's, $\mathbf{R}_{yx}$ is the $p \times q$ matrix of sample correlations between the y's and the x's, and $\mathbf{R}_{xx}$ is the $q \times q$ sample correlation matrix of the x's. If we use the partitioned correlation matrix in place of the covariance matrix, we obtain the same canonical correlations as in (11.5) and (11.6) but different eigenvectors:

$$(\mathbf{R}_{yy}^{-1}\mathbf{R}_{yx}\mathbf{R}_{xx}^{-1}\mathbf{R}_{xy} - r^2\mathbf{I})\mathbf{c} = \mathbf{0} \tag{11.8}$$

$$(\mathbf{R}_{xx}^{-1}\mathbf{R}_{xy}\mathbf{R}_{yy}^{-1}\mathbf{R}_{yx} - r^2\mathbf{I})\mathbf{d} = \mathbf{0}. \tag{11.9}$$

Note that the matrix $\mathbf{R}_{yy}^{-1}\mathbf{R}_{yx}\mathbf{R}_{xx}^{-1}\mathbf{R}_{xy}$ is analogous to $R^2 = \mathbf{r}'_{yx}\mathbf{R}_{xx}^{-1}\mathbf{r}_{yx}$ in the univariate y case. The relationship between the eigenvectors $\mathbf{c}$ and $\mathbf{d}$ in (11.8) and (11.9) and the eigenvectors $\mathbf{a}$ and $\mathbf{b}$ in (11.5) and (11.6) is

$$\mathbf{c} = \mathbf{D}_y\mathbf{a} \quad \text{and} \quad \mathbf{d} = \mathbf{D}_x\mathbf{b}, \tag{11.10}$$

where $\mathbf{D}_y = \text{diag}(s_{y_1}, s_{y_2}, \ldots, s_{y_p})$ and $\mathbf{D}_x = \text{diag}(s_{x_1}, s_{x_2}, \ldots, s_{x_q})$.

The characteristic equations corresponding to (11.3) and (11.4),

$$|\mathbf{R}_{yy}^{-1}\mathbf{R}_{yx}\mathbf{R}_{xx}^{-1}\mathbf{R}_{xy} - r^2\mathbf{I}| = 0 \tag{11.11}$$

$$|\mathbf{R}_{xx}^{-1}\mathbf{R}_{xy}\mathbf{R}_{yy}^{-1}\mathbf{R}_{yx} - r^2\mathbf{I}| = 0, \tag{11.12}$$

yield the same eigenvalues $r_1^2, r_2^2, \ldots, r_s^2$ as (11.3) and (11.4).

The eigenvectors **c** and **d** in (11.8), (11.9), and (11.10) are *standardized coefficient vectors*. By analogy to (8.15), they would be applied to standardized variables. To show this, note that in terms of centered variables $\mathbf{y} - \overline{\mathbf{y}}$, we have

$$\begin{aligned}
u &= \mathbf{a}'(\mathbf{y} - \overline{\mathbf{y}}) = \mathbf{a}'\mathbf{D}_y\mathbf{D}_y^{-1}(\mathbf{y} - \overline{\mathbf{y}}) \\
&= \mathbf{c}'\mathbf{D}_y^{-1}(\mathbf{y} - \overline{\mathbf{y}}) \quad [\text{by (11.10)}] \\
&= c_1\frac{y_1 - \overline{y}_1}{s_{y_1}} + c_2\frac{y_2 - \overline{y}_2}{s_{y_2}} + \cdots + c_p\frac{y_p - \overline{y}_p}{s_{y_p}}.
\end{aligned}$$

Hence **c** and **d** are preferred to **a** and **b** for interpretation of the canonical variates u_i and v_i.

11.3 PROPERTIES OF CANONICAL CORRELATIONS

Two interesting properties of canonical correlations are the following (for other properties, see Rencher 1996, Section 8.3):

1. Canonical correlations are invariant to changes of scale on either the y's or the x's. For example, if the measurement scale is changed from inches to centimeters, the canonical correlations will not change (the corresponding eigenvectors will change). This property holds for simple and multiple correlations as well.
2. The first canonical correlation r_1 is the maximum correlation between linear functions of **y** and **x**. Therefore, r_1 exceeds the (absolute value of the) simple correlation between any y and any x or the multiple correlation between any y and all the x's or between any x and all the y's.

Example 11.3. For the chemical data of Table 10.1, we obtain the canonical correlations and illustrate property 2 above. We consider the extended set of nine x's, as in Example 10.5.2. The matrix $\mathbf{R}_{yx}$ of correlations between the y's and the x's is

	x_1	x_2	x_3	x_1x_2	x_1x_3	x_2x_3	x_1^2	x_2^2	x_3^2
y_1	−.68	−.22	−.45	−.41	−.55	−.45	−.68	−.23	−.42
y_2	.40	.08	.39	.16	.44	.33	.40	.12	.33
y_3	.58	.23	.36	.40	.45	.39	.58	.22	.36

The three canonical correlations and their squares are

$$\begin{aligned} r_1 &= .9899 & r_1^2 &= .9800 \\ r_2 &= .9528 & r_2^2 &= .9078 \\ r_3 &= .4625 & r_3^2 &= .2139. \end{aligned}$$

From the relative sizes of the squared canonical correlations, we would consider only the first two to be important. A hypothesis test for the significance of each is carried out in Example 11.4.2.

To confirm that property 2 holds in this case, we compare $r_1 = .9899$ to the individual correlations and the multiple correlations. We first note that .9899 is greater than individual correlations, since the (absolute value of the) largest correlation in $\mathbf{R}_{yx}$ is .68. The multiple correlation $R_{y_i|\mathbf{x}}$ of each y_i with the x's is given by

$$R_{y_1|\mathbf{x}} = .987 \qquad R_{y_2|\mathbf{x}} = .921 \qquad R_{y_3|\mathbf{x}} = .906,$$

and for the multiple correlation of each x with the y's we have

$$\begin{aligned} R_{x_1|\mathbf{y}} &= .691 & R_{x_2|\mathbf{y}} &= .237 & R_{x_3|\mathbf{y}} &= .507 \\ R_{x_1x_2|\mathbf{y}} &= .432 & R_{x_1x_3|\mathbf{y}} &= .585 & R_{x_2x_3|\mathbf{y}} &= .482 \\ R_{x_1^2|\mathbf{y}} &= .690 & R_{x_2^2|\mathbf{y}} &= .234 & R_{x_3^2|\mathbf{y}} &= .466. \end{aligned}$$

The first canonical correlation $r_1 = .9899$ exceeds all multiple correlations. Thus property 2 is satisfied.

11.4 TESTS OF SIGNIFICANCE

In the following two sections we discuss basic tests of significance associated with canonical correlations. For other aspects of model validation, see Rencher (1996, Section 8.5).

11.4.1 Tests of No Relationship between the y's and the x's

In Section 7.4.1, we considered the hypothesis of independence, H_0: $\mathbf{\Sigma}_{yx} = \mathbf{O}$. If $\mathbf{\Sigma}_{yx} = \mathbf{O}$, the covariance of every y_i with every x_j is zero, and all corresponding correlations are likewise zero. Hence under H_0, there is no (linear) relationship between the y's and the x's; and H_0 is equivalent to the statement that all canonical correlations $r_1, r_2, \ldots, r_s$ are nonsignificant. Furthermore, H_0 is equivalent to the overall regression hypothesis in Section 10.5.1, H_0: $\mathbf{B}_1 = \mathbf{O}$, which also relates all the y's to all the x's. Thus by (7.28) or (10.46), the significance of $r_1, r_2, \ldots, r_s$ can be tested by

$$\Lambda = \frac{|\mathbf{S}|}{|\mathbf{S}_{yy}||\mathbf{S}_{xx}|} = \frac{|\mathbf{R}|}{|\mathbf{R}_{yy}||\mathbf{R}_{xx}|}, \tag{11.13}$$

which is distributed as $\Lambda_{p,q,n-1-q}$ or equivalently as $\Lambda_{q,p,n-1-p}$. We reject H_0 if $\Lambda \leq \Lambda_\alpha$. As in (7.29), Λ is expressible in terms of the squared canonical correlations:

$$\Lambda = \prod_{i=1}^{s} (1 - r_i^2). \tag{11.14}$$

If the parameters exceed the range of critical values for Wilks' Λ in Table A.9, we can use the χ^2 approximation in (6.14),

$$\chi^2 = -[n - \tfrac{1}{2}(p + q + 3)] \ln \Lambda, \tag{11.15}$$

which is approximately distributed as χ^2 with pq degrees of freedom. We reject H_0 if $\chi^2 \geq \chi^2_\alpha$. Alternatively, we can use the F approximation given in (6.13):

$$F = \frac{1 - \Lambda^{1/t}}{\Lambda^{1/t}} \frac{\text{df}_2}{\text{df}_1}, \tag{11.16}$$

which has an approximate F-distribution with df_1 and df_2 degrees of freedom, where

$$\text{df}_1 = pq \qquad \text{df}_2 = wt - \tfrac{1}{2}pq + 1$$

$$w = n - \tfrac{1}{2}(p + q + 3) \qquad t = \sqrt{\frac{p^2q^2 - 4}{p^2 + q^2 - 5}}.$$

We reject H_0 if $F > F_\alpha$. When $pq = 2$, t is set equal to 1. If $s = \min(p, q)$ is equal to either 1 or 2, then the F approximation in (11.16) has an exact F-distribution.

For example, if one of the two sets consists of only two variables, an exact test is afforded by the F procedure in (11.16). In contrast, the χ^2 approximation in (11.15) does not reduce to an exact test for any parameter values.

The other three multivariate test statistics in Sections 6.1.4, 6.1.5, and 10.5.1 can also be used. Pillai's test statistic for the significance of canonical correlations is

$$V^{(s)} = \sum_{i=1}^{s} r_i^2. \tag{11.17}$$

Upper percentage points of $V^{(s)}$ are found in Table A.11, indexed by

$$s = \min(p, q) \qquad m = \tfrac{1}{2}(|q - p| - 1) \qquad N = \tfrac{1}{2}(n - q - p - 2).$$

For F approximations for $V^{(s)}$, see Section 6.1.5.

The Lawley–Hotelling statistic for canonical correlations is

$$U^{(s)} = \sum_{i=1}^{s} \frac{r_i^2}{1 - r_i^2}. \tag{11.18}$$

Upper percentage points for $\nu_E U^{(s)}/\nu_H$ (see Section 6.1.5) are given in Table A.12, which is entered with p, $\nu_H = q$, and $\nu_E = n - q - 1$. For F approximations, see Section 6.1.5.

Roy's largest root statistic is given by

$$\theta = r_1^2. \tag{11.19}$$

Upper percentage points are found in Table A.10, with s, m, and N defined as above for Pillai's test. An "upper bound" on F for Roy's test is given in (6.16).

As noted at the beginning of this section, the following three tests are equivalent:

1. Test of H_0: $\boldsymbol{\Sigma}_{yx} = \mathbf{O}$, independence of two sets of variables.
2. Test of H_0: $\mathbf{B}_1 = \mathbf{O}$, significance of overall multivariate multiple regression.
3. Test of significance of the canonical correlations.

Even though these tests are equivalent, we have discussed them separately because each has an extension that is different from the others. The respective extensions are

1. Test of independence of three or more sets of variables (Section 7.4.2).
2. Test of full vs. reduced model in multivariate multiple regression (Section 10.5.2).
3. Test of significance of succeeding canonical correlations after the first (Section 11.4.2).

Example 11.4.1. For the chemical data of Table 10.1, with the extended set of nine x's, we obtained canonical correlations .9899, .9528, and .4625 in Example 11.3. To test the significance of these, we calculate the following four test statistics and associated approximate F's.

Statistic	Approximate F	df_1	df_2	p-Value for F
Wilks' $\Lambda = .00145$	6.537	27	21.09	$< .0001$
Pillai's $V^{(s)} = 2.10$	2.340	27	27	.0155
Lawley-Hotelling $U^{(s)} = 59.03$	12.388	27	17	$< .0001$
Roy's $\theta = .980$	48.908	9	9	$< .0001$

The F approximation for Roy's test is, of course, an "upper bound." Rejection of H_0 in these tests implies that at least r_1^2 is significantly different from zero. The question of how many r_i^2 are significant is treated in the next section.

11.4.2 Test of Significance of Succeeding Canonical Correlations after the First

If the test in (11.14) based on all s canonical correlations rejects H_0, we are not sure if the canonical correlations beyond the first are significant. To test the significance of $r_2, \ldots, r_s$, we delete r_1^2 from Λ to obtain

$$\Lambda_2 = \prod_{i=2}^{s} (1 - r_i^2).$$

If this test rejects the hypothesis, we conclude that at least r_2 is significantly different from zero. We can continue in this manner, testing each r_i in turn, until a test fails to reject the hypothesis. At the kth step, the test statistic is

$$\Lambda_k = \prod_{i=k}^{s} (1 - r_i^2), \tag{11.20}$$

which is distributed as $\Lambda_{p-k+1,q-k+1,n-k-q}$ and tests the significance of $r_k, r_{k+1}, \ldots, r_s$.

The usual χ^2 and F approximations can also be applied to Λ_k. The χ^2 approximation analogous to (11.15) is given by

$$\chi^2 = -[n - \tfrac{1}{2}(p + q + 3)] \ln \Lambda_k, \qquad (11.21)$$

which has $(p - k + 1)(q - k + 1)$ degrees of freedom. The F approximation for Λ_k is a simple modification of (11.16) and the accompanying parameter definitions. In place of p, q, and n, we use $p - k + 1$, $q - k + 1$, and $n - k + 1$ to obtain

$$F = \frac{1 - \Lambda_k^{1/t}}{\Lambda_k^{1/t}} \frac{\text{df}_2}{\text{df}_1},$$

where

$$\begin{aligned}
\text{df}_1 &= (p - k + 1)(q - k + 1) \\
\text{df}_2 &= wt - \tfrac{1}{2}[(p - k + 1)(q - k + 1)] + 1 \\
w &= n - \tfrac{1}{2}(p + q + 3) \\
t &= \sqrt{\frac{(p - k + 1)^2(q - k + 1)^2 - 4}{(p - k + 1)^2 + (q - k + 1)^2 - 5}}.
\end{aligned}$$

Example 11.4.2. We continue our analysis of the canonical correlations for the chemical data in Table 10.1 with three y's and nine x's. The tests are summarized in Table 11.1.

In the case of Λ_2, we have a discrepancy between the exact Wilks Λ-test and the approximate F-test. The test based on Λ is not significant, while the F-test does reach significance. We conclude therefore that only $r_1 = .9899$ is significant. The relative sizes of the squared canonical correlations, .980, .908, and .214, would indicate two dimensions of relationship, but this is not confirmed by the Wilks' test, perhaps because of the small sample size relative to the number of variables ($p + q = 12$ and $n = 19$).

Table 11.1 Tests of Three Canonical Correlations of the Chemical Data

k	Λ_k	$\Lambda_{.05}$	Approximate F	df_1	df_2	p-Value for F
1	.00145	.024	6.537	27	21.1	.0001
2	.0725	.069	2.714	16	16	.0269
3	.786	.209	.350	7	9	.91

To illustrate the computations, we obtain the values in Table 11.1 for $k = 2$. The computation for Λ_2 is

$$\Lambda_2 = \prod_{i=2}^{3}(1 - r_i^2) = (1 - .908)(1 - .214) = .0725.$$

With $k = 2$, $p = 3$, $q = 9$, and $n = 19$, the critical value for Λ_2 is obtained from Table A.9 as

$$\Lambda_{.05,p-k+1,q-k+1,n-k-q} = \Lambda_{.05,2,8,8} = .069.$$

For the approximate F for Λ_2, we have

$$\begin{aligned}
t &= \sqrt{\frac{(3-2+1)^2(9-2+1)^2-4}{(3-2+1)^2+(9-2+1)^2-5}} = 2 \\
w &= 19 - \tfrac{1}{2}(3 + 9 + 3) = 11.5 \\
\text{df}_1 &= (3 - 2 + 1)(9 - 2 + 1) = 16 \\
\text{df}_2 &= (11.5)(2) - \tfrac{1}{2}[(3 - 2 + 1)(9 - 2 + 1)] + 1 = 16 \\
F &= \frac{1 - (.0725)^{1/2}}{(.0725)^{1/2}}\,\frac{16}{16} = 2.714.
\end{aligned}$$

11.5 INTERPRETATION

We now turn to an assessment of the information contained in the canonical correlations and canonical variates. As in the discussion of interpretation of discriminant functions in Section 8.7, a distinction can be made between interpretation of the canonical variates and assessing the contribution of each variable. In the former, the signs of the coefficients are considered; in the latter, the signs are ignored and the coefficients are ranked in order of absolute value.

In the next three sections, we discuss three common tools for interpretation of canonical variates: (1) standardized coefficients, (2) correlation between each variable and the canonical variate, and (3) rotation of the canonical variate coefficients. The second of these is the most widely recommended, but we note in Section 11.5.2 that it is the least useful. In fact, for reasons to be outlined, we recommend only the first, standardized coefficients. In Section 11.5.4, we describe redundancy analysis and discuss its shortcomings as a measure of association between two sets of variables.

11.5.1 Standardized Coefficients

The coefficients in $u_i = \mathbf{a}_i'\mathbf{y}$ and $v_i = \mathbf{b}_i'\mathbf{x}$ reflect differences in scaling of the variables as well as differences in contribution of the variables to canonical correlation. To remove the effect of scaling, $\mathbf{a}_i$ and $\mathbf{b}_i$ can be standardized by multiplying by the standard deviations of the corresponding variables as in (11.10):

$$\mathbf{c}_i = \mathbf{D}_y\mathbf{a}_i \qquad \mathbf{d}_i = \mathbf{D}_x\mathbf{b}_i.$$

Alternatively, the $\mathbf{c}_i$ and $\mathbf{d}_i$ can be obtained directly from (11.8) and (11.9) as eigenvectors of $\mathbf{R}_{yy}^{-1}\mathbf{R}_{yx}\mathbf{R}_{xx}^{-1}\mathbf{R}_{xy}$ and $\mathbf{R}_{xx}^{-1}\mathbf{R}_{xy}\mathbf{R}_{yy}^{-1}\mathbf{R}_{yx}$, respectively. It was noted at the end of Section 11.2 that the coefficients in $\mathbf{c}_i$ are applied to standardized variables. Thus the effect of differences in size or scaling of the variables is removed, and the coefficients $c_{i1}, c_{i2}, \ldots, c_{ip}$ in $\mathbf{c}_i$ reflect the relative contribution of each of $y_1, y_2, \ldots, y_p$ to u_i. A similar statement can be made about $\mathbf{d}_i$.

The standardized coefficients show the contribution of the variables in the presence of each other. Thus if some of the variables are deleted and others added, the coefficients will change. This is precisely the behavior we desire from the coefficients in a multivariate setting. They provide a pertinent multivariate approach to interpretation of the contribution of the variables acting in combination.

Example 11.5.1. For the chemical data in Table 10.1 with the extended set of nine x's, we obtain the following standardized coefficients for the three canonical variates:

	$\mathbf{c}_1$	$\mathbf{c}_2$	$\mathbf{c}_3$
y_1	1.5360	4.4704	5.7961
y_2	0.2108	2.8291	2.2280
y_3	0.4676	3.1309	5.1442

	$\mathbf{d}_1$	$\mathbf{d}_2$	$\mathbf{d}_3$
x_1	5.0125	−38.3053	−12.5072
x_2	5.8551	−17.7390	−24.2290
x_3	1.6500	−7.9699	−32.7392
x_1x_2	−3.9209	19.2937	11.6420
x_1x_3	−2.2968	6.4001	31.2189
x_2x_3	0.5316	0.8096	1.2988
x_1^2	−2.6655	32.7933	4.8454
x_2^2	−1.2346	−3.3641	10.7979
x_3^2	0.5703	0.8733	0.9706

Thus

$$u_1 = 1.54 \frac{y_1 - \overline{y}_1}{s_{y_1}} + .21 \frac{y_2 - \overline{y}_2}{s_{y_2}} + .47 \frac{y_3 - \overline{y}_3}{s_{y_3}}$$

and

$$v_1 = 5.01 \frac{x_1 - \overline{x}_1}{s_{x_1}} + 5.86 \frac{x_2 - \overline{x}_2}{s_{x_2}} + \cdots + .57 \frac{x_3^2 - \overline{x_3^2}}{s_{x_3^2}}.$$

The variables that contribute most to the correlation between u_1 and v_1 are y_1 and $x_1, x_2, x_1x_2, x_1x_3, x_1^2$. The correlation between u_2 and v_2 is due largely to all three y's and x_1, x_2, x_1x_2, x_1^2.

11.5.2 Correlations between Variables and Canonical Variates

Many writers recommend the additional step of converting the standardized coefficients to correlations. Thus, for example, in $\mathbf{c}_1' = (c_{11}, c_{12}, \ldots, c_{1p})$, instead of the second coefficient c_{12} we could examine $r_{y_2u_1}$, the correlation between y_2 and the first canonical variate u_1. Such correlations are sometimes referred to as *loadings* or *structure coefficients*, and it is widely claimed that they provide a more valid interpretation of the canonical variates. Rencher (1988, 1992b) has shown, however, that these correlations yield no information about the joint contribution of the variables to canonical correlation. Instead, they measure only univariate relationships between the two sets of variables and thus become useless in gauging the importance of a given variable in the context of the others. The researcher who uses these correlations for interpretation is unknowingly reducing the multivariate setting to a univariate one.

As noted in Section 8.7.3, if we attempt to improve on the standardized coefficients by converting to correlations, we end up back in the univariate realm.

11.5.3 Rotation

In an attempt to improve interpretability, the canonical variate coefficients can be rotated (see Section 13.5) to increase the number of high and low coefficients and reduce the number of intermediate ones.

We do not recommend rotation of the canonical variate coefficients for two reasons (for proof and further discussion, see Rencher 1992b):

1. Rotation destroys the optimality of the original canonical correlations so that the first canonical correlation is reduced. The sum of the squared

canonical correlations remains invariant with rotation, but they are more evenly distributed.

2. More important, rotation introduces correlations among succeeding canonical variates. Thus, for example, u_1 and u_2 are correlated after rotation. Hence even if the resulting coefficients offer a subjectively more interpretable pattern, this gain is offset by the increased complexity due to interrelationships among the canonical variates. For example, u_2 and v_2 no longer offer a new dimension of relationship uncorrelated with u_1 and v_1. The dimensions now overlap, and some of the information in u_2 and v_2 is already available in u_1 and v_1.

11.5.4 Redundancy Analysis

The *redundancy* is a measure of association between the y's and the x's based on the correlations between variables and canonical variates discussed in Section 11.5.2. But since these correlations do not show the multivariate contribution of the variables, the redundancy turns out to be a univariate rather than a multivariate measure of relationship. In fact, it can be expressed as an average squared multiple correlation. If the squared multiple correlation of y_i regressed on the x's is denoted by $R^2_{y_i|\mathbf{x}}$, then the redundancy of the y's given the v's is

$$\text{Rd}(\mathbf{y}|\mathbf{v}) = \frac{\sum_{i=1}^{p} R^2_{y_i|\mathbf{x}}}{p}. \tag{11.22}$$

Similarly, the redundancy of the x's given the u's is

$$\text{Rd}(\mathbf{x}|\mathbf{u}) = \frac{\sum_{i=1}^{q} R^2_{x_i|\mathbf{y}}}{q}, \tag{11.23}$$

where $R^2_{x_i|\mathbf{y}}$ is the squared multiple correlation of x_i regressed on the y's. Since $\text{Rd}(\mathbf{y}|\mathbf{v})$ is the average multiple correlation squared of each y_i regressed on the x's, it does not take into account the correlations among the y's and is thus an average univariate measure of relationship between the y's and the x's, not a multivariate measure at all.

Thus the so-called redundancy does not really quantify the redundancy among the y's and x's and is, therefore, not a useful measure of association between two sets of variables. For a measure of association we recommend r_1^2 itself.

11.6 RELATIONSHIPS OF CANONICAL CORRELATION ANALYSIS TO OTHER MULTIVARIATE TECHNIQUES

In Section 11.4.1, we noted the equivalence of the test for significance of the canonical correlations and the test for significance of overall regression, H_0: $\mathbf{B}_1 = \mathbf{O}$. Additional relationships between canonical correlation and multivariate regression are developed in Section 11.6.1. Canonical correlation analysis is also closely related to MANOVA and discriminant analysis, as shown in Section 11.6.2.

11.6.1 Regression

There is a direct link between canonical variate coefficients and multivariate multiple regression coefficients. The matrix of regression coefficients of the y's regressed on the x's (corrected for their means) is given in (10.42) as $\hat{\mathbf{B}}_1 = \mathbf{S}_{xx}^{-1}\mathbf{S}_{xy}$. This matrix can be used to relate $\mathbf{a}_i$ and $\mathbf{b}_i$:

$$\mathbf{b}_i = \hat{\mathbf{B}}_1\mathbf{a}_i. \tag{11.24}$$

Hence by (2.62), the canonical variate coefficient vector $\mathbf{b}_i$ is expressible as a linear combination of the columns of $\hat{\mathbf{B}}_1$. A similar expression for $\mathbf{a}_i$ can be obtained from the regression of $\mathbf{x}$ on $\mathbf{y}$.

In Section 11.2, canonical correlation was defined as an extension of multiple correlation. We can also view multiple correlation as a special case of canonical correlation when one of the two sets of variables has only one variable. When $p = 1$, $\mathbf{R}_{yy}$ reduces to 1, and by (11.11), the single squared canonical correlation reduces to $r^2 = \mathbf{r}'_{yx}\mathbf{R}_{xx}^{-1}\mathbf{r}_{yx}$, which we recognize from (10.28) as R^2.

The two Wilks test statistics in multivariate regression in Sections 10.5.1 and 10.5.2, namely, the test for overall regression and the test on a subset of the x's, can both be expressed in terms of canonical correlations. By (10.44) and (11.14), the test for the overall regression hypothesis H_0: $\mathbf{B}_1 = \mathbf{O}$ can be written as

$$\Lambda_f = \frac{|\mathbf{Y}'\mathbf{Y} - \hat{\mathbf{B}}'\mathbf{X}'\mathbf{Y}|}{|\mathbf{Y}'\mathbf{Y} - n\bar{\mathbf{y}}\bar{\mathbf{y}}'|} \tag{11.25}$$

$$= \prod_{i=1}^{s}(1 - r_i^2), \tag{11.26}$$

where r_i^2 is the ith squared canonical correlation.

A test statistic for H_0: $\mathbf{B}_d = \mathbf{O}$, the hypothesis that the y's do not depend on the last h of the x's, is given by (10.54) as

$$\Lambda(x_{q-h+1}, \ldots, x_q | x_1, \ldots, x_{q-h}) = \frac{\Lambda_f}{\Lambda_r}, \tag{11.27}$$

where Λ_f is given in (11.25) and Λ_r is given in (10.53) as

$$\Lambda_r = \frac{|\mathbf{Y}'\mathbf{Y} - \hat{\mathbf{B}}_r'\mathbf{X}_r'\mathbf{Y}|}{|\mathbf{Y}'\mathbf{Y} - n\bar{\mathbf{y}}\bar{\mathbf{y}}'|}. \tag{11.28}$$

By analogy with (11.26), Λ_r can be expressed in terms of the squared canonical correlations $c_1^2, c_2^2, \ldots, c_t^2$ between $y_1, y_2, \ldots, y_p$ and $x_1, x_2, \ldots, x_{q-h}$:

$$\Lambda_r = \prod_{i=1}^{t} (1 - c_i^2), \tag{11.29}$$

where $t = \min(p, q - h)$. We have used the notation c_i^2 instead of r_i^2 to emphasize that the canonical correlations in the reduced model differ from those in the full model. By (11.26) and (11.29), the full and reduced model test of H_0: $\mathbf{B}_d = \mathbf{O}$ in (11.27) can now be expressed in terms of canonical correlations as

$$\Lambda(x_{q-h+1}, \ldots, x_q | x_1, \ldots, x_{q-h}) = \frac{\prod_{i=1}^{s}(1 - r_i^2)}{\prod_{i=1}^{t}(1 - c_i^2)}. \tag{11.30}$$

If $p = 1$, as in multiple regression, then we also have $s = t = 1$ and (11.30) reduces to

$$\Lambda = \frac{1 - R_f^2}{1 - R_r^2}, \tag{11.31}$$

where R_f^2 and R_r^2 are the squared multiple correlations for the full and reduced models. The distribution of Λ in (11.31) is $\Lambda_{1,h,n-q-1}$ when H_0 is true. Since $p = 1$, there is an exact F transformation from Table 6.1,

$$F = \frac{(1 - \Lambda)(n - q - 1)}{\Lambda h},$$

which is distributed as $F_{h,n-q-1}$ when H_0 is true. Substitution of $\Lambda = (1 - R_f^2)/(1$

$- R_r^2$) from (11.31) yields the F-test expressed in terms of R^2,

$$F = \frac{(R_f^2 - R_r^2)(n - q - 1)}{(1 - R_f^2)h}, \tag{11.32}$$

as given in (10.27).

Subset selection in canonical correlation analysis can be handled by the methods for multivariate regression given in Section 10.7. A subset of x's can be found by the procedure of Section 10.7.1a. After a subset of x's is found, the approach in Section 10.7.1b can be used to select a subset of y's, so that x's and y's that do not relate to each other may be deleted.

Muller (1982) discussed the relationship of canonical correlation analysis with multivariate regression and principal components. (Principal components are treated in Chapter 12.)

11.6.2 MANOVA and Discriminant Analysis

In Sections 6.1.8 and 8.4.2, it was noted that in a one-way MANOVA or discriminant analysis setting, $\lambda_i/(1 + \lambda_i)$ is equal to r_i^2, where λ_i is the ith eigenvalue of $\mathbf{E}^{-1}\mathbf{H}$ and r_i^2 is the ith squared canonical correlation between the p dependent variables and the $k - 1$ grouping variables. We now give a justification of this assertion.

Let the dependent variables be denoted by $y_1, y_2, \ldots, y_p$ as usual. We represent the k groups by $k - 1$ *dummy* variables, $x_1, x_2, \ldots, x_{k-1}$, defined for each member of the ith group, $i \leq k - 1$, as $x_1 = 0, \ldots, x_{i-1} = 0$, $x_i = 1$, $x_{i+1} = 0, \ldots, x_{k-1} = 0$. For the kth group, all x's are zero. (See Section 6.1.8 for an introduction to dummy variables.) To illustrate with $k = 4$, the x's are defined as follows in each group:

Group	x_1	x_2	x_3
1	1	0	0
2	0	1	0
3	0	0	1
4	0	0	0

The MANOVA model is equivalent to multivariate regression of $y_1, y_2, \ldots, y_p$ on the dummy grouping variables $x_1, x_2, \ldots, x_{k-1}$. The MANOVA test of H_0: $\boldsymbol{\mu}_1 = \boldsymbol{\mu}_2 = \cdots = \boldsymbol{\mu}_k$ is equivalent to the multivariate regression test of H_0: $\mathbf{B}_1 = \mathbf{O}$, as given by (11.14),

$$\Lambda = \prod_{i=1}^{s} (1 - r_i^2). \tag{11.33}$$

When we compare this form of Λ to the MANOVA test statistic (6.12),

$$\Lambda = \prod_{i=1}^{s} \frac{1}{1+\lambda_i}, \tag{11.34}$$

we obtain the relationships

$$r_i^2 = \frac{\lambda_i}{1+\lambda_i} \tag{11.35}$$

$$\lambda_i = \frac{r_i^2}{1-r_i^2}. \tag{11.36}$$

To establish this relationship more formally, consider the definition for λ in (6.17),

$$(\mathbf{H} - \lambda\mathbf{E})\mathbf{a} = \mathbf{0}, \tag{11.37}$$

and for r^2 in (11.5),

$$(\mathbf{S}_{yy}^{-1}\mathbf{S}_{yx}\mathbf{S}_{xx}^{-1}\mathbf{S}_{xy} - r^2\mathbf{I})\mathbf{a} = \mathbf{0}.$$

We multiply the latter by $\mathbf{S}_{yy}$ to obtain

$$(\mathbf{S}_{yx}\mathbf{S}_{xx}^{-1}\mathbf{S}_{xy} - r^2\mathbf{S}_{yy})\mathbf{a} = \mathbf{0}. \tag{11.38}$$

Using the centered matrix $\mathbf{X}_c$ in Section 10.4.5, with an analogous definition for $\mathbf{Y}_c$, we can write $\mathbf{B}_1$ in the form [see (10.42)]

$$\hat{\mathbf{B}}_1 = \left(\frac{\mathbf{X}_c'\mathbf{X}_c}{n-1}\right)^{-1}\frac{\mathbf{X}_c'\mathbf{Y}_c}{n-1} = \mathbf{S}_{xx}^{-1}\mathbf{S}_{xy}.$$

Then in terms of centered matrices, we have

$$\begin{aligned}\frac{\mathbf{E}}{n-1} &= \frac{\mathbf{Y}_c'\mathbf{Y}_c}{n-1} - \hat{\mathbf{B}}_1'\frac{\mathbf{X}_c'\mathbf{Y}_c}{n-1} \\ &= \mathbf{S}_{yy} - \mathbf{S}_{xy}'\mathbf{S}_{xx}^{-1}\mathbf{S}_{xy} = \mathbf{S}_{yy} - \mathbf{S}_{yx}\mathbf{S}_{xx}^{-1}\mathbf{S}_{xy},\end{aligned}$$

since $\mathbf{S}_{xy}' = \mathbf{S}_{yx}$. Similarly,

$$\frac{\mathbf{H}}{n-1} = \frac{\hat{\mathbf{B}}_1'\mathbf{X}_c'\mathbf{Y}_c}{n-1} = \mathbf{S}_{yx}\mathbf{S}_{xx}^{-1}\mathbf{S}_{xy}.$$

Since MANOVA is equivalent to multivariate regression on dummy grouping variables, we can use these values of **E** and **H** in (11.37) to obtain

$$\mathbf{S}_{yx}\mathbf{S}_{xx}^{-1}\mathbf{S}_{xy}\mathbf{a} = \lambda(\mathbf{S}_{yy} - \mathbf{S}_{yx}\mathbf{S}_{xx}^{-1}\mathbf{S}_{xy})\mathbf{a}. \tag{11.39}$$

Subtracting $r^2\mathbf{S}_{yx}\mathbf{S}_{xx}^{-1}\mathbf{S}_{xy}\mathbf{a}$ from both sides of (11.38) gives

$$\mathbf{S}_{yx}\mathbf{S}_{xx}^{-1}\mathbf{S}_{xy}\mathbf{a} = \frac{r^2}{1-r^2}(\mathbf{S}_{yy} - \mathbf{S}_{yx}\mathbf{S}_{xx}^{-1}\mathbf{S}_{xy})\mathbf{a}. \tag{11.40}$$

A comparison of (11.39) and (11.40) shows that

$$\lambda = \frac{r^2}{1-r^2},$$

as in (11.36). Lindsey et al. (1985) discussed some advantages of using canonical correlation analysis in place of discriminant analysis in the several-group case.

PROBLEMS

11.1 Show that the expression for canonical correlations in (11.8) can be obtained from the analogous expression in terms of variances and covariances in (11.5).

11.2 Verify (11.24), $\mathbf{b}_i = \hat{\mathbf{B}}_1\mathbf{a}_i$.

11.3 Verify (11.31) for Λ when $p = s = t = 1$.

11.4 Verify the expression in (11.32) for F in terms of R_f^2 and R_r^2.

11.5 Solve (11.35), $r_i^2 = \lambda_i/(1+\lambda_i)$, for λ_i to obtain (11.36).

11.6 Verify (11.39), $\mathbf{S}_{yx}\mathbf{S}_{xx}^{-1}\mathbf{S}_{xy}\mathbf{a} = \lambda(\mathbf{S}_{yy} - \mathbf{S}_{yx}\mathbf{S}_{xx}^{-1}\mathbf{S}_{xy})\mathbf{a}$.

11.7 Show that (11.40) can be obtained by subtracting $r^2\mathbf{S}_{yx}\mathbf{S}_{xx}^{-1}\mathbf{S}_{xy}\mathbf{a}$ from both sides of (11.38).

11.8 Use the diabetes data of Table 3.6.
- **(a)** Find the canonical correlations between (y_1, y_2) and (x_1, x_2, x_3).
- **(b)** Find the standardized coefficients for the canonical variates.
- **(c)** Test the significance of each canonical correlation.

11.9 Use the sons data of Table 3.9.
- **(a)** Find the canonical correlations between (y_1, y_2) and (x_1, x_2).
- **(b)** Find the standardized coefficients for the canonical variates.
- **(c)** Test the significance of each canonical correlation.

11.10 Use the glucose data of Table 3.10.
- **(a)** Find the canonical correlations between (y_1, y_2, y_3) and (x_1, x_2, x_3).
- **(b)** Find the standardized coefficients for the canonical variates.
- **(c)** Test the significance of each canonical correlation.

11.11 Use the seishu data of Table 7.1.
- **(a)** Find the canonical correlations between (y_1, y_2) and $(x_1, x_2, \ldots, x_8)$.
- **(b)** Find the standardized coefficients for the canonical variates.
- **(c)** Test the significance of each canonical correlation.

11.12 For parts (b), (c), and (d) of problem 10.16, using the seishu data, find the canonical correlations between (y_1, y_2) and the x's in the indicated reduced models and use (11.30) to carry out the tests.

11.13 Using the temperature data of Table 7.2, find the canonical correlations and the standardized coefficients and carry out significance tests for
- **(a)** (y_1, y_2, y_3) and (y_4, y_5, y_6)
- **(b)** $(y_1, y_2, \ldots, y_6)$ and (y_7, y_8, y_9)
- **(c)** $(y_1, y_2, \ldots, y_9)$ and (y_{10}, y_{11})
- **(d)** $(y_1, y_2, \ldots, y_6)$ and $(y_7, x_8, \ldots, y_{11})$.

CHAPTER 12

Principal Component Analysis

12.1 INTRODUCTION

In principal component analysis, we seek to maximize the variance of a linear combination of the variables. For example, we might want to rank students on the basis of their scores on achievement tests in English, mathematics, reading, and so on. An average score would provide a single scale on which to compare the students, but with unequal weights we can spread the students out further and obtain a better ranking.

Essentially, principal component analysis is a one-sample technique applied to data with no groupings among the observations as in Chapter 8 and no partitioning of the variables into subsets $\mathbf{y}$ and $\mathbf{x}$ as in Chapters 10 and 11. All the linear combinations that we have considered previously were related to other variables or to the data structure. In regression, we have linear combinations of the independent variables that best predict the dependent variable(s); in canonical correlation, we have linear combinations of a subset of variables that maximally correlate with linear combinations of another subset of variables; and discriminant analysis involves linear combinations that maximally separate groups of observations. Principal components, on the other hand, are concerned only with the core structure of a single sample of observations on p variables. None of the variables is designated as dependent, and no grouping of observations is assumed. In seeking a linear combination with maximal variance, we are essentially searching for a dimension along which the observations are maximally separated or spread out. In general, the principal components define different dimensions from those defined by discriminant functions or canonical variates. [For a discussion of the use of principal components with data consisting of several samples or groups, see Rencher (1996, Section 9.9)].

In some applications, the principal components are an end in themselves and may be amenable to interpretation. More often they are obtained for use as input to another analysis. For example, two situations in regression where principal components may be useful are (1) too many independent variables relative to the number of observations may render a test ineffective or even impossible and (2) highly correlated independent variables may produce unstable estimates. In such cases, the independent variables can be reduced to a smaller number of principal components that will yield a better test or more stable estimates of the regression coefficients. For details of this application, see Rencher (1996, Section 9.8).

As another illustration, suppose that in a MANOVA application p is close to ν_E, so that we have more dependent variables than can be tested with reasonable power, or that $p > \nu_E$, in which case we have so many dependent variables that a test cannot be made. In such cases, we can reduce the dependent variables to a smaller set of principal components and then carry out the test.

In these illustrations, principal components are used to reduce the number of dimensions. Another use involving dimension reduction is to evaluate the first two principal components for each observation vector and construct a scatter plot to check for multivariate normality, outliers, and so on.

Finally, we note that in the term *principal components*, we use the adjective *principal*, describing what kind of components—main, primary, fundamental, major, and so on. We do not use the spelling *principle* because principle is a noun and cannot serve as a modifier for *components*.

12.2 GEOMETRIC AND ALGEBRAIC BASIS OF PRINCIPAL COMPONENTS

As noted in Section 12.1, principal components analysis deals with a single sample of observations with no structure in the observations or among the variables within an observation vector. We have a sample of n observation vectors $\mathbf{y}_1, \mathbf{y}_2, \ldots, \mathbf{y}_n$, from which we calculate $\overline{\mathbf{y}}$ and $\mathbf{S}$. The n observation vectors form a swarm of points in a p-dimensional space. Principal components analysis can be applied to any distribution of $\mathbf{y}$, but it will be easier to visualize geometrically if the swarm of points is ellipsoidal.

If the variables $y_1, y_2, \ldots, y_p$ in each $\mathbf{y}_i$ are correlated, the ellipsoidal swarm of points is not orientated parallel to any of the axes represented by $y_1, y_2, \ldots, y_p$. We wish to find the natural axes of the swarm of points with origin at $\overline{\mathbf{y}}$, that is, the axes of the ellipsoid. This is done by translating the origin to $\overline{\mathbf{y}}$ and then rotating the axes. After rotation so that the axes become the natural axes of the ellipsoid, the new variables (principal components) will be uncorrelated.

We could indicate the translation of the origin to $\overline{\mathbf{y}}$ by writing $\mathbf{y}_i - \overline{\mathbf{y}}$ but will not usually do so for economy of notation. We will write $\mathbf{y}_i - \overline{\mathbf{y}}$ when there is an explicit need; otherwise we assume that $\mathbf{y}_i$ has been centered.

The axes can be rotated by multiplying each $\mathbf{y}_i$ by an orthogonal matrix $\mathbf{A}$ [see (2.94) where the orthogonal matrix was denoted by $\mathbf{C}$]:

$$\mathbf{z}_i = \mathbf{A}\mathbf{y}_i. \tag{12.1}$$

Since $\mathbf{A}$ is orthogonal, $\mathbf{A}'\mathbf{A} = \mathbf{I}$, and the distance to the origin is unchanged:

$$\mathbf{z}_i'\mathbf{z}_i = (\mathbf{A}\mathbf{y}_i)'(\mathbf{A}\mathbf{y}_i) = \mathbf{y}_i'\mathbf{A}'\mathbf{A}\mathbf{y}_i = \mathbf{y}_i'\mathbf{y}_i.$$

Thus an orthogonal matrix transforms $\mathbf{y}_i$ to a point $\mathbf{z}_i$ that is the same distance from the origin, and the axes are rotated.

Finding the axes of the ellipsoid is equivalent to finding the orthogonal matrix $\mathbf{A}$ that rotates the axes to line up with the natural extensions of the swarm of points so that the new variables (principal components) $z_1, z_2, \ldots, z_p$ in $\mathbf{z} = \mathbf{A}\mathbf{y}$ are uncorrelated. Thus we want the sample covariance matrix of $\mathbf{z}$ to be of the form

$$\mathbf{S}_z = \begin{bmatrix} s_{z_1}^2 & 0 & \cdots & 0 \\ 0 & s_{z_2}^2 & \cdots & 0 \\ \vdots & \vdots & & \vdots \\ 0 & 0 & \cdots & s_{z_p}^2 \end{bmatrix}. \tag{12.2}$$

By (3.60), if $\mathbf{z} = \mathbf{A}\mathbf{y}$, then $\mathbf{S}_z = \mathbf{A}\mathbf{S}\mathbf{A}'$, and thus

$$\mathbf{A}\mathbf{S}\mathbf{A}' = \begin{bmatrix} s_{z_1}^2 & 0 & \cdots & 0 \\ 0 & s_{z_2}^2 & \cdots & 0 \\ \vdots & \vdots & & \vdots \\ 0 & 0 & \cdots & s_{z_p}^2 \end{bmatrix}, \tag{12.3}$$

where $\mathbf{S}$ is the sample covariance matrix of $\mathbf{y}$. By (2.102), $\mathbf{C}'\mathbf{S}\mathbf{C} = \mathbf{D}$ and the orthogonal matrix $\mathbf{A}$ that diagonalizes $\mathbf{S}$ is the transpose of the matrix $\mathbf{C}$ whose columns are normalized eigenvectors of $\mathbf{S}$. Thus $\mathbf{A}$ can be written as

$$\mathbf{A} = \begin{bmatrix} \mathbf{a}_1' \\ \mathbf{a}_2' \\ \vdots \\ \mathbf{a}_p' \end{bmatrix},$$

where $\mathbf{a}_i$ is the ith normalized ($\mathbf{a}_i'\mathbf{a}_i = 1$) eigenvector of $\mathbf{S}$. The *principal components* are the transformed variables $z_1 = \mathbf{a}_1'\mathbf{y}, z_2 = \mathbf{a}_2'\mathbf{y}, \ldots, z_p = \mathbf{a}_p'\mathbf{y}$ in $\mathbf{z} = \mathbf{A}\mathbf{y}$. For example, $z_1 = a_{11}y_1 + a_{12}y_2 + \cdots + a_{1p}y_p$.

By (2.102), the diagonal elements on the right side of (12.3) are eigenvalues of $\mathbf{S}$. Hence the eigenvalues $\lambda_1, \lambda_2, \ldots, \lambda_p$ of $\mathbf{S}$ are the variances of the principal components $z_i = \mathbf{a}_i'\mathbf{y}$:

$$s_{z_i}^2 = \lambda_i. \tag{12.4}$$

Since the rotation lines up with the natural extensions of the swarm of points, $z_1 = \mathbf{a}_1'\mathbf{y}$ has the largest (sample) variance and $z_p = \mathbf{a}_p'\mathbf{y}$ has the smallest variance. This also follows from (12.4), because the variance of z_1 is λ_1, the largest eigenvalue, and the variance of z_p is λ_p, the smallest eigenvalue. If some of the eigenvalues are small, we can neglect them and represent the points with fewer than p dimensions. For example, if $p = 3$ and λ_3 is small, then the swarm of points is an "elliptical pancake" and a two-dimensional representation will adequately portray the configuration of points.

Because the eigenvalues are variances of the principal components, we can speak of "the proportion of variance explained" by the first k components:

$$\begin{aligned} \text{Proportion of variance} &= \frac{\lambda_1 + \lambda_2 + \cdots + \lambda_k}{\lambda_1 + \lambda_2 + \cdots + \lambda_p} \\ &= \frac{\lambda_1 + \lambda_2 + \cdots + \lambda_k}{\sum_{i=1}^{p} s_{ii}}, \end{aligned} \tag{12.5}$$

since $\sum_{i=1}^{p} \lambda_i = \text{tr}(\mathbf{S})$ by (2.98). Thus we try to represent the p-dimensional points $(y_{i1}, y_{i2}, \ldots, y_{ip})$ with a few principal components $(z_{i1}, z_{i2}, \ldots, z_{ik})$ that account for a large proportion of the total variance. If a few variables have relatively large variances, they will figure disproportionately in $\sum_i s_{ii}$ and in the principal components. For example, if s_{22} is strikingly larger than the other variances, then in $z_1 = a_{11}y_1 + a_{12}y_2 + \cdots + a_{1p}y_p$, the coefficient a_{12} will be large and all other a_{1j} will be small.

When a ratio analogous to (12.5) is used for discriminant functions and canonical variates, it is frequently referred to as *percent of variance*. However, in the case of discriminant functions and canonical variates, the eigenvalues are not variances, as they are in principal components.

If the variables are highly correlated, the essential dimensionality is much smaller than p; that is, the first few eigenvalues will be large, and (12.5) will be close to 1 for a small value of k. On the other hand, if the correlations among the variables are all small, the dimensionality is close to p and the eigenvalues will be nearly equal. In this case, no useful reduction in dimension is achieved, because the principal components essentially duplicate the variables.

Since the principal components represent a rotation of axes, the components $z_i = \mathbf{a}_i'\mathbf{y}$ and $z_j = \mathbf{a}_j'\mathbf{y}$ are orthogonal for $i \neq j$, that is, $\mathbf{a}_i'\mathbf{a}_j = 0$. This orthogonality is also confirmed by the fact that $\mathbf{a}_i$ and $\mathbf{a}_j$ are eigenvectors of the symmetric matrix $\mathbf{S}$ (see Section 2.11.5). Principal components also have the secondary property of being uncorrelated in the sample [see (12.2) and the expression for $\mathbf{S}_z$ following (3.59)]; that is, the covariance of z_i and z_j is zero:

$$s_{z_i z_j} = \mathbf{a}_i'\mathbf{S}\mathbf{a}_j = 0 \qquad \text{for} \quad i \neq j. \tag{12.6}$$

Discriminant functions and canonical variates, on the other hand, have the weaker property of being uncorrelated but not the stronger property of orthogonality. Thus when we plot the first two discriminant functions or canonical variates on perpendicular coordinate axes, there is some distortion of their true relationship because the actual angle between their axes is not 90°.

If we change the scale on one or more of the y's, the shape of the swarm of points will change, and we will need different components to represent the new points. Hence the principal components are not scale invariant. Because of this lack of scale invariance, we need to be concerned with the units in which the variables are measured. If possible, all variables should be expressed in the same units. If the variables have widely disparate variances, we could standardize them before extracting eigenvalues and eigenvectors. This is equivalent to finding principal components of the correlation matrix $\mathbf{R}$ and is treated in Section 12.5.

If one variable has a much greater variance than the other variables, the swarm of points will be elongated and will be nearly parallel to the axis corresponding to the variable with large variance. The first principal component will largely represent that variable, and the other principal components will have negligibly small variances. Such principal components (based on $\mathbf{S}$) do not involve the other $p - 1$ variables, and we may prefer to analyze the correlation matrix $\mathbf{R}$.

An algebraic approach to principal components can be briefly described as follows. As noted in Section 12.1, we seek a linear combination with maximal variance. By (3.52), the variance of $z = \mathbf{a}'\mathbf{y}$ is $\mathbf{a}'\mathbf{S}\mathbf{a}$. Since $\mathbf{a}'\mathbf{S}\mathbf{a}$ has no maximum for arbitrary $\mathbf{a}$, we seek its maximum relative to $\mathbf{a}'\mathbf{a}$. By an argument similar to that used in (8.8)–(8.11), the maximum value of

$$\lambda = \frac{\mathbf{a}'\mathbf{S}\mathbf{a}}{\mathbf{a}'\mathbf{a}}$$

is given by the largest eigenvalue in the expression

$$(\mathbf{S} - \lambda\mathbf{I})\mathbf{a} = \mathbf{0}.$$

The eigenvector $\mathbf{a}_1$ corresponding to the largest eigenvalue λ_1 is the coefficient vector in $z_1 = \mathbf{a}_1'\mathbf{y}$, the linear combination with maximum variance.

Unlike discriminant analysis or canonical correlation, there is no inverse involved before obtaining eigenvectors for principal components. Therefore, a singular matrix is permissible. If $\mathbf{S}$ is singular, some of the eigenvalues are zero and can be ignored.

This tolerance of principal component analysis for a singular $\mathbf{S}$ is important in certain research situations. For example, suppose that one has a one-way MANOVA with 10 observations in each of three groups and that $p = 50$, so that there are 50 variables in each of these 30 observation vectors. A MANOVA test involving $\mathbf{E}^{-1}\mathbf{H}$ cannot be carried out directly in this case because $\mathbf{E}$ is singular, but we could reduce the 50 variables to a small number of principal components and then do a MANOVA test on the components. Thus for entry into the MANOVA program, we would evaluate the principal components for each observation vector. If we are retaining k components, we calculate

$$\begin{aligned} z_{1i} &= \mathbf{a}_1'\mathbf{y}_i \\ z_{2i} &= \mathbf{a}_2'\mathbf{y}_i \\ &\vdots \\ z_{ki} &= \mathbf{a}_k'\mathbf{y}_i \end{aligned} \tag{12.7}$$

for $i = 1, 2, \ldots, 30$. These are sometimes referred to as *component scores.* In vector form, (12.7) can be rewritten as

$$\mathbf{z}_i = \mathbf{A}_k\mathbf{y}_i, \tag{12.8}$$

where

$$\mathbf{z}_i = \begin{bmatrix} z_{1i} \\ z_{2i} \\ \vdots \\ z_{ki} \end{bmatrix} \quad \text{and} \quad \mathbf{A}_k = \begin{bmatrix} \mathbf{a}_1' \\ \mathbf{a}_2' \\ \vdots \\ \mathbf{a}_k' \end{bmatrix}.$$

We then use $\mathbf{z}_1, \mathbf{z}_2, \ldots, \mathbf{z}_{30}$ as input to the MANOVA program.

Note that in this case with $p > n$, the k components will not likely be stable; that is, they would be different in a new sample. However, this is of no concern here because we are using the components only to extract information from the sample at hand in order to compare the three groups.

Example 12.2(a). To illustrate principal components as a rotation when $p = 2$, we use two variables from the sons data of Table 3.9: y_1 is head length and y_2 is head width for the first son. The mean vector and covariance matrix are

$$\overline{\mathbf{y}} = \begin{pmatrix} 185.7 \\ 151.1 \end{pmatrix} \qquad \mathbf{S} = \begin{pmatrix} 95.29 & 52.87 \\ 52.87 & 54.36 \end{pmatrix}.$$

The eigenvalues and eigenvectors of $\mathbf{S}$ are

$$\begin{aligned} \lambda_1 &= 131.52 \qquad \lambda_2 = 18.14 \\ \mathbf{a}_1' &= (a_{11}, a_{12}) = (.825, .565) \qquad \mathbf{a}_2' = (a_{21}, a_{22}) = (-.565, .825). \end{aligned}$$

The symmetric pattern in the eigenvectors is due to their orthogonality: $\mathbf{a}_1'\mathbf{a}_2 = a_{11}a_{21} + a_{12}a_{22} = 0$.

The observations are plotted in Figure 12.1, along with the (translated and) rotated axes. The major axis is the line passing through $\overline{\mathbf{y}}' = (185.7, 151.1)$ in the direction determined by $\mathbf{a}_1' = (.825, .565)$; the slope is $a_{12}/a_{11} = .565/.825$. Alternatively, the equation of the major axis can be obtained by setting $z_2 = 0$:

$$\begin{aligned} z_2 = 0 &= a_{21}(y_1 - \overline{y}_1) + a_{22}(y_2 - \overline{y}_2) \\ &= -.565(y_1 - 185.7) + .825(y_2 - 151.1). \end{aligned}$$

Note that the line formed by the major axis can be considered to be a regression line. It is fit to the points so that the perpendicular distance of the points to the line is minimized, rather than the usual vertical distance (see Section 12.3).

Example 12.2(b). Consider the football data of Table 8.3. In Example 8.8, we saw that high school football players (group 1) differed from the other two groups, college football players and college-age nonfootball players. Therefore, to obtain a homogeneous group of observations, we delete group 1 and use groups 2 and 3 combined. The covariance matrix is as follows:

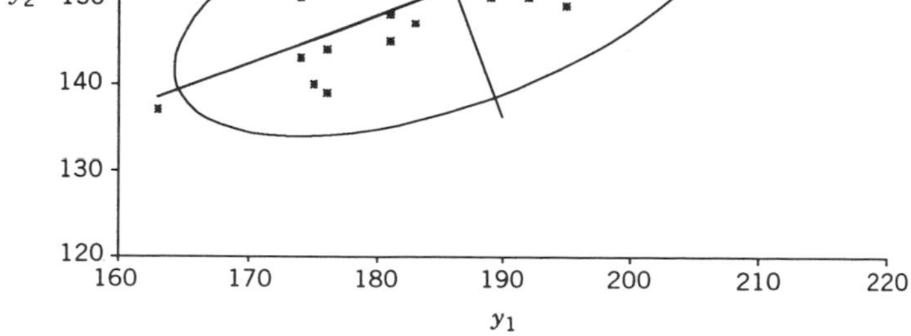

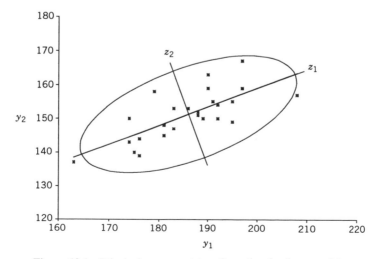

Figure 12.1 Principal component transformation for the sons data.

$$\mathbf{S} = \begin{bmatrix} .370 & .602 & .149 & .044 & .107 & .209 \\ .602 & 2.629 & .801 & .666 & .103 & .377 \\ .149 & .801 & .458 & .012 & -.013 & .120 \\ .044 & .666 & .011 & 1.474 & .252 & -.054 \\ .107 & .103 & -.013 & .252 & .488 & -.036 \\ .209 & .377 & .120 & -.054 & -.036 & .324 \end{bmatrix}.$$

The total variance is

$$\sum_{i=1}^{6} s_{ii} = \sum_{i=1}^{6} \lambda_i = 5.743.$$

The eigenvalues of $\mathbf{S}$ are as follows:

Eigenvalue	Proportion of Variance	Cumulative Proportion
3.323	.579	.579
1.374	.239	.818
.476	.083	.901
.325	.057	.957
.157	.027	.985
.088	.015	1.000

The first two principal components account for 81.8% of the total variance. The corresponding eigenvectors are as follows:

	$\mathbf{a}_1$	$\mathbf{a}_2$
WDIM	.207	−.142
CIRCUM	.873	−.219
FBEYE	.261	−.231
EYEHD	.326	.891
EARHD	.066	.222
JAW	.128	−.187

Thus the first two principal components are

$$z_1 = \mathbf{a}_1'\mathbf{y} = .207y_1 + .873y_2 + .261y_3 + .326y_4 + .066y_5 + .128y_6$$
$$z_2 = \mathbf{a}_2'\mathbf{y} = -.142y_1 - .219y_2 - .231y_3 + .891y_4 + .222y_5 - .187y_6.$$

Notice that the large coefficient in z_1 and the large coefficient in z_2, .873 and .891, respectively, correspond to the two largest variances on the diagonal of **S**. The two variables with large variances, y_2 and y_4, have a notable influence on the first two principal components. However, z_1 and z_2 are still meaningful linear functions. If the six variances were closer in size, the six variables would enter more evenly into the first two principal components. On the other hand, if the variances of y_2 and y_4 were substantially larger, z_1 and z_2 would be essentially equal to y_2 and y_4, respectively.

12.3 PRINCIPAL COMPONENTS AND PERPENDICULAR REGRESSION

It was noted in Section 12.2 that principal components constitute a rotation of axes. Another geometric property of the line formed by the first principal component is that it minimizes the total sum of squared perpendicular distances from the points to the line. This is easily demonstrated in the bivariate case. The first principal component line is plotted in Figure 12.2 for the first two variables of the sons data, as in Example 12.2(a). The perpendicular distance from each point to the line is simply z_2, the second coordinate in the transformed coordinates (z_1, z_2). Hence the sum of squares of perpendicular distances is

$$\sum_{i=1}^{n} z_{2i}^2 = \sum_{i=1}^{n} \left[\mathbf{a}_2'(\mathbf{y}_i - \overline{\mathbf{y}})\right]^2, \tag{12.9}$$

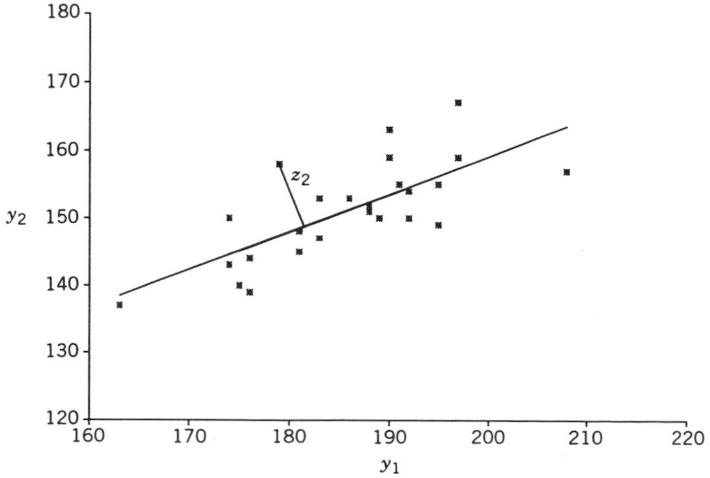

Figure 12.2 The first principal component as a perpendicular regression line.

where $\mathbf{a}_2$ is the second eigenvector of $\mathbf{S}$ and we use $\mathbf{y}_i - \overline{\mathbf{y}}$ because the axes have been translated to the new origin $\overline{\mathbf{y}}$. Since $\mathbf{a}_2'(\mathbf{y}_i - \overline{\mathbf{y}}) = (\mathbf{y}_i - \overline{\mathbf{y}})'\mathbf{a}_2$, we can write (12.9) in the form

$$\begin{aligned} \sum_i z_{2i}^2 &= \sum_i \mathbf{a}_2'(\mathbf{y}_i - \overline{\mathbf{y}})(\mathbf{y}_i - \overline{\mathbf{y}})'\mathbf{a}_2 \\ &= \mathbf{a}_2'\left[\sum_i (\mathbf{y}_i - \overline{\mathbf{y}})(\mathbf{y}_i - \overline{\mathbf{y}})'\right]\mathbf{a}_2 \quad [\text{by } (2.44)] \\ &= (n-1)\mathbf{a}_2'\mathbf{S}\mathbf{a}_2 = (n-1)\lambda_2 \quad [\text{by } (3.25)], \end{aligned} \tag{12.10}$$

which is a minimum by a remark following (12.4).

For the two variables y_1 and y_2, as plotted in Figure 12.2, the ordinary regression line of y_2 on y_1 minimizes the sum of squares of vertical distances from the points to the line. Similarly, the regression of y_1 on y_2 minimizes the sum of squares of horizontal distances from the points to the line. The first principal component line represents a "perpendicular" regression line that lies between the other two. The three lines are compared in Figure 12.3 for the partial sons data. The equation of the first principal component line is easily obtained by setting $z_2 = 0$,

$$\begin{aligned} z_2 = \mathbf{a}_2'(\mathbf{y} - \overline{\mathbf{y}}) &= 0 \\ a_{21}(y_1 - \overline{y}_1) + a_{22}(y_2 - \overline{y}_2) &= 0 \\ -.565(y_1 - \overline{y}_1) + .825(y_2 - \overline{y}_2) &= 0. \end{aligned}$$

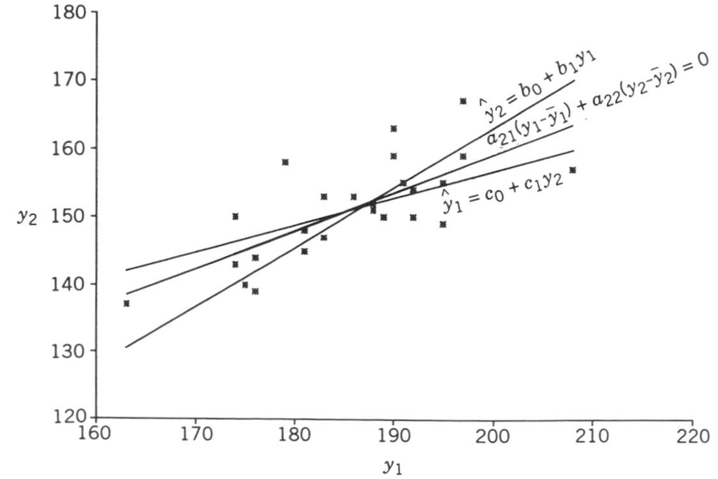

Figure 12.3 Regression lines compared with first principal component line.

12.4 PLOTTING OF PRINCIPAL COMPONENTS

The plots in Figures 12.1 and 12.2 were illustrations of principal components as a rotation of axes when $p = 2$. When $p > 2$, we can plot the first two components as a dimension reduction device. We simply evaluate the first two components (z_1, z_2) for each observation vector and plot these n points. The plot is equivalent to a projection of the p-dimensional data swarm onto the plane that shows the greatest spread of the points. The resulting two-dimensional representation captures more of the overall configuration of the data than does a plot of any two of the original variables.

The plot of the first two components may reveal some important features of the data set. In Example 12.4(a), we show a principal component plot that exhibits a pattern typical of a sample from a multivariate normal distribution. One of the objectives of plotting is to check for departures from normality, such as outliers or nonlinearity. In Examples 12.4(b) and (c), we illustrate principal component plots showing a nonnormal pattern and the presence of outliers. Jackson (1980) provided a test for adequacy of representation of observation vectors in terms of principal components. Rejection often indicates an outlier or some other anomaly in the data.

Gnanadesikan (1977) pointed out that, in general, the first few principal components are sensitive to outliers that inflate variances or distort covariances, and the last few are sensitive to outliers that introduce artificial dimensions or mask singularities. We could examine the bivariate plots of at least the first two and the last two principal components in a search for outliers that may exert undue influence.

Devlin et al. (1981) recommended the extraction of principal components from robust estimates of **S** or **R**. Robust estimation is an approach to estimation that reduces the influence of outliers. Campbell (1980) and Ruymgaart (1981) discussed direct robust estimation of principal components. Critchley (1985) developed methods for detection of influential observations in principal component analysis.

Another feature of the data that a plot of the first two components may reveal is a tendency of the points to cluster. The plot may reveal groupings of points; this is illustrated in Example 12.4(d).

Example 12.4(a). For the modified football data in Example 12.2(b), the first two principal components were given as follows:

$$z_1 = \mathbf{a}_1'\mathbf{y} = .207y_1 + .873y_2 + .261y_3 + .326y_4 + .066y_5 + .128y_6$$
$$z_2 = \mathbf{a}_2'\mathbf{y} = -.142y_1 - .219y_2 - .231y_3 + .891y_4 + .222y_5 - .187y_6.$$

These are evaluated for each observation vector and plotted in Figure 12.4. (For convenience in scaling, $\mathbf{y} - \bar{\mathbf{y}}$ was used in the computations.) The pat-

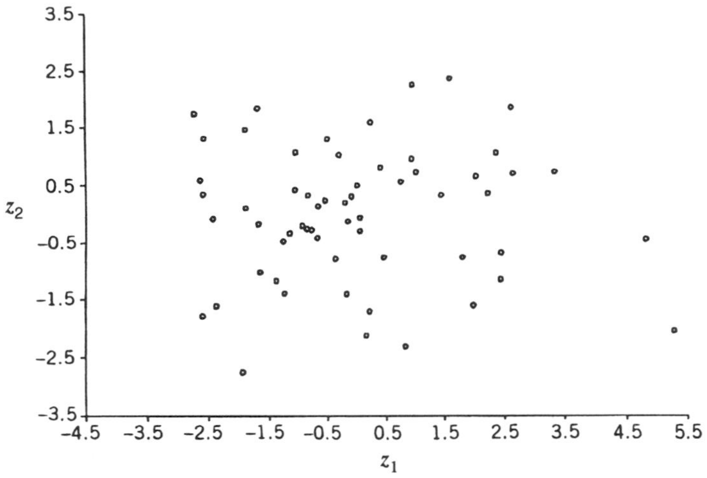

Figure 12.4 Plot of first two components for the modified football data.

tern is typical of that from a multivariate normal distribution. Note that the variance along the z_1 axis is greater than the variance in the z_2 direction, as expected.

Example 12.4(b). In Figures 4.9 and 4.10, the *Q–Q* plot and bivariate scatter plots for the ramus bone data of Table 3.8 exhibit a nonnormal pattern. A principal component analysis using the covariance matrix for the same data is given in Table 12.1, and the first two principal components are plotted in Figure 12.5. The presence of three outliers that cause a nonnormal pattern is evident.

Example 12.4(c). A rather extreme example of the effect of an outlier is given by Devlin et al. (1981). The data set involved $p = 14$ economical variables for $n = 29$ chemical companies. The first two principal components are plotted in Figure 12.6. The sample correlation $r_{z_1 z_2}$ is indeed zero for all 29 points, but

Table 12.1 Principal Components for the Ramus Bone Data of Table 3.8

Eigenvalues		First Two Eigenvectors		
Number	Value	Variable	$\mathbf{a}_1$	$\mathbf{a}_2$
1	25.05	AGE 8	.474	.592
2	1.74	AGE 8.5	.492	.406
3	.22	AGE 9	.515	−.304
4	.11	AGE 9.5	.517	−.627

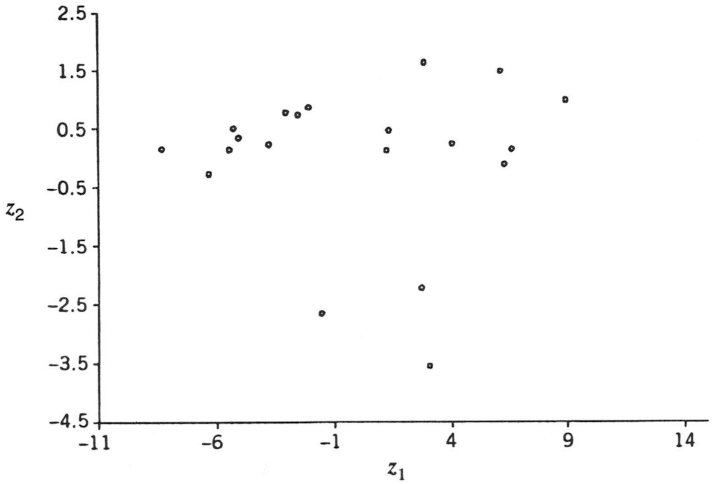

Figure 12.5 First two principal components for the ramus bone data in Table 3.8.

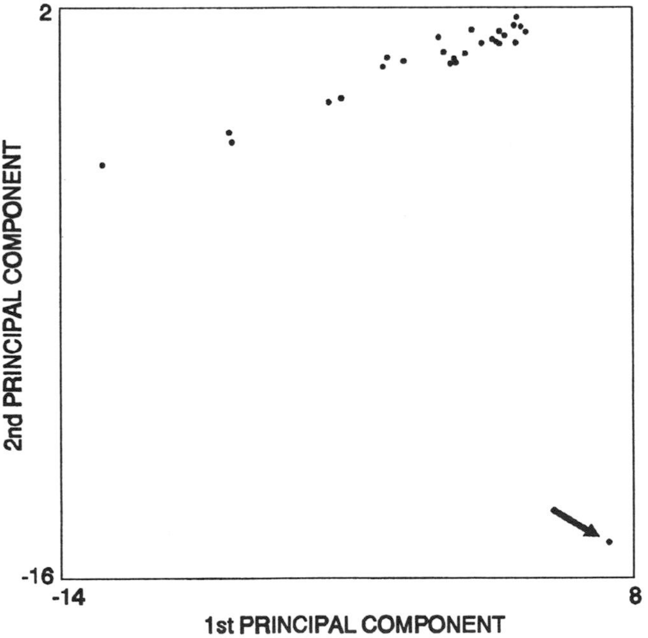

Figure 12.6 First two principal components for economics data.

if the apparent outlier is excluded, then $r_{z_1 z_2} = .99$ for the remaining 28 points. If the outlier were deleted, the axes of the principal components would pass through the natural extensions of the data swarm.

Example 12.4(d). Jeffers (1967) applied principal component analysis to a sample of 40 alate adelges (winged aphids) on which 19 variables had been measured:

LENGTH	body length
WIDTH	body width
FORWING	forewing length
HINWING	hind-wing length
SPIRAC	number of spiracles
ANTSEG 1	length of antennal segment I
ANTSEG 2	length of antennal segment II
ANTSEG 3	length of antennal segment III
ANTSEG 4	length of antennal segment IV
ANTSEG 5	length of antennal segment V
ANTSPIN	number of antennal spines
TARSUS 3	leg length, tarsus III
TIBIA 3	leg length, tibia III
FEMUR 3	leg length, femur III
ROSTRUM	rostrum
OVIPOS	ovipositor
OVSPIN	number of ovipositor spines
FOLD	anal fold
HOOKS	number of hind-wing hooks

An objective in the study was to determine the number of distinct taxa present in the habitat where the sample was taken. Adelges are difficult to identify by the usual taxonomic methods; principal component analysis was used to search for groupings among the 40 individuals in the sample.

The correlation matrix is given in Table 12.2, and the eigenvalues and first four eigenvectors are in Tables 12.3 and 12.4, respectively. The eigenvectors are scaled so that the largest value in each is 1. The first principal component is largely an index of size. The second component is associated with SPIRAC, OVIPOS, OVSPIN, and FOLD.

The first two components were computed for each of the 40 individuals and plotted in Figure 12.7. Since the first two components account for 85% of the total variance, the plot represents the data with very little distortion. There are four major groups, apparently corresponding to species. The groupings form an interesting S shape.

Table 12.2 Correlation Matrix for Winged Aphid Variables (Lower Triangle)

	y_1	y_2	y_3	y_4	y_5	y_6	y_7	y_8	y_9	y_{10}
y_1	1.000									
y_2	.934	1.000								
y_3	.927	.941	1.000							
y_4	.909	.944	.933	1.000						
y_5	.524	.487	.543	.499	1.000					
y_6	.799	.821	.856	.833	.703	1.000				
y_7	.854	.865	.886	.889	.719	.923	1.000			
y_8	.789	.834	.846	.885	.253	.699	.751	1.000		
y_9	.835	.863	.862	.850	.462	.752	.793	.745	1.000	
y_{10}	.845	.878	.863	.881	.567	.836	.913	.787	.805	1.000
y_{11}	−.458	−.496	−.522	−.488	−.174	−.317	−.383	−.497	−.356	−.371
y_{12}	.917	.942	.940	.945	.516	.846	.907	.861	.848	.902
y_{13}	.939	.961	.956	.952	.494	.849	.914	.876	.877	.901
y_{14}	.953	.954	.946	.949	.452	.823	.886	.878	.883	.891
y_{15}	.895	.899	.882	.908	.551	.831	.891	.794	.818	.848
y_{16}	.691	.652	.694	.623	.815	.812	.855	.410	.620	.712
y_{17}	.327	.305	.356	.272	.746	.553	.567	.067	.300	.384
y_{18}	−.676	−.712	−.667	−.736	−.233	−.504	−.502	−.758	−.666	−.629
y_{19}	.702	.729	.746	.777	.285	.499	.592	.793	.671	.668

	y_{11}	y_{12}	y_{13}	y_{14}	y_{15}	y_{16}	y_{17}	y_{18}	y_{19}
y_{11}	1.000								
y_{12}	−.465	1.000							
y_{13}	−.447	.981	1.000						
y_{14}	−.439	.971	.991	1.000					
y_{15}	−.405	.908	.920	.921	1.000				
y_{16}	−.198	.725	.714	.676	.720	1.000			
y_{17}	−.032	.396	.360	.298	.378	.781	1.000		
y_{18}	.492	−.657	−.655	−.678	−.633	−.186	.169	1.000	
y_{19}	−.425	.696	.724	.731	.694	.287	−.026	−.775	1.000

Table 12.3 Eigenvalues of the Correlation Matrix of the Winged Aphid Data

Component	Eigenvalue	Percent of Variance	Cumulative Percent
1	13.861	73.0	73.0
2	2.370	12.5	85.4
3	.748	3.9	89.4
4	.502	2.6	92.0
5	.278	1.4	93.5
6	.266	1.4	94.9
7	.193	1.0	95.9
8	.157	.8	96.7
9	.140	.7	97.4
10	.123	.6	98.1
11	.092	.4	98.6
12	.074	.4	99.0

Table 12.3 (*Continued*)

Component	Eigenvalue	Percent of Variance	Cumulative Percent
13	.060	.3	99.3
14	.042	.2	99.5
15	.036	.2	99.7
16	.024	.1	99.8
17	.020	.1	99.9
18	.011	.1	100.0
19	.003	.0	100.0
	19.000		

Table 12.4 Eigenvectors for the First Four Components of the Winged Aphid Data

	Eigenvectors			
Variable	1	2	3	4
LENGTH	.96	−.06	.03	−.12
WIDTH	.98	−.12	.01	−.16
FORWING	.99	−.06	−.06	−.11
HINWING	.98	−.16	.03	−.00
SPIRAC	.61	.74	−.20	1.00
ANTSEG 1	.91	.33	.04	.02
ANTSEG 2	.96	.30	.00	−.04
ANTSEG 3	.88	−.43	.06	−.18
ANTSEG 4	.90	−.08	.18	−.01
ANTSEG 5	.94	.05	.11	.03
ANTSPIN	−.49	.37	1.00	.27
TARSUS 3	.99	−.02	.03	−.29
TIBIA 3	1.00	−.05	.09	−.31
FEMUR 3	.99	−.12	.12	−.31
ROSTRUM	.96	.02	.08	−.06
OVIPOS	.76	.73	−.03	−.09
OVSPIN	.41	1.00	−.16	−.06
FOLD	−.71	.64	.04	−.80
HOOKS	.76	−.52	.06	.72

12.5 PRINCIPAL COMPONENTS FROM THE CORRELATION MATRIX

Generally, extracting components from **S** rather than **R** remains closer to the spirit and intent of principal component analysis, especially if the components are to be used in further computations. However, in some cases, the principal components will be more interpretable if **R** is used. For example, if the

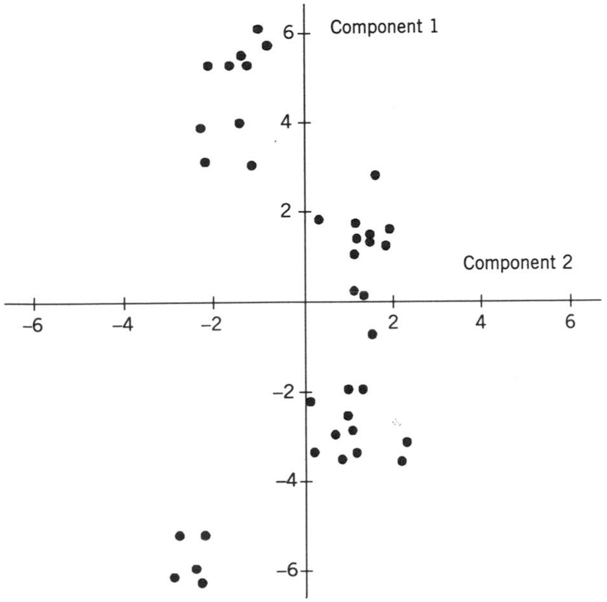

Figure 12.7 Plotted values of the first two components for individual insects.

variances differ widely or if the measurement units are not commensurate, the components of **S** will be dominated by the variables with large variances. The other variables will contribute very little. For a more balanced representation in such cases, components of **R** may be used.

As with any change of scale, when the variables are standardized in transforming to **R**, the shape of the swarm of points will change. (Note, however, that after transforming to **R**, any further changes of scale on the variables would not affect the components because changes of scale do not change **R**.)

To illustrate how the eigenvalues and eigenvectors change when converting from **S** to **R**, we use a simple bivariate example in which one variance is substantially larger than the other. Suppose that **S** and the corresponding **R** have the values

$$\mathbf{S} = \begin{pmatrix} 1 & 4 \\ 4 & 25 \end{pmatrix} \qquad \mathbf{R} = \begin{pmatrix} 1 & .8 \\ .8 & 1 \end{pmatrix}.$$

The eigenvalues and eigenvectors from **S** are

$$\begin{aligned} \lambda_1 &= 25.65 \qquad & \mathbf{a}_1' &= (.160, .987) \\ \lambda_2 &= .35 \qquad & \mathbf{a}_2' &= (.987, -.160). \end{aligned}$$

The patterns we see in $\lambda_1, \lambda_2, \mathbf{a}_1$, and $\mathbf{a}_2$ are quite predictable. The symmetry in $\mathbf{a}_1$ and $\mathbf{a}_2$ is due to their orthogonality, $\mathbf{a}_1'\mathbf{a}_2 = 0$. The large variance of y_2 in $\mathbf{S}$ ensures that the first principal component $z_1 = .160y_1 + .987y_2$ weights y_2 heavily. Thus the first principal component z_1 essentially duplicates y_2 and does not reflect the mutual effect of y_1 and y_2. It is expected also that z_1 would account for virtually all of the total variance:

$$\frac{\lambda_1}{\lambda_1 + \lambda_2} = \frac{25.65}{26} = .9865.$$

The eigenvalues and eigenvectors of $\mathbf{R}$ are

$$\begin{aligned} \lambda_1 &= 1.8 \qquad & \mathbf{a}_1' &= (.707, .707) \\ \lambda_2 &= .2 \qquad & \mathbf{a}_2' &= (.707, -.707). \end{aligned}$$

The first principal component of $\mathbf{R}$,

$$z_1 = .707 \frac{y_1 - \bar{y}_1}{1} + .707 \frac{y_2 - \bar{y}_2}{5},$$

accounts for a high proportion of variance,

$$\frac{\lambda_1}{\lambda_1 + \lambda_2} = \frac{1.8}{2} = .9,$$

because the variables are fairly highly correlated ($r = .8$). But the standardized variables $(y_1 - \bar{y}_1)/1$ and $(y_2 - \bar{y}_2)/5$ are equally weighted in z_1, due to the equality of the diagonal elements ("variances") of $\mathbf{R}$.

We now list some general comparisons of principal components from $\mathbf{R}$ with those from $\mathbf{S}$:

1. The percent of variance in (12.5) accounted for by the components of $\mathbf{R}$ will differ from the percent for $\mathbf{S}$, as illustrated above.
2. The coefficients of the principal components from $\mathbf{R}$ differ from those obtained from $\mathbf{S}$, as illustrated above.
3. If we express the components from $\mathbf{R}$ in terms of the original variables, they still will not agree with the components from $\mathbf{S}$. By transforming the

standardized variables back to the original variables in the above illustration, the components of **R** become

$$\begin{aligned} z_1 &= .707 \frac{y_1 - \bar{y}_1}{1} + .707 \frac{y_2 - \bar{y}_2}{5} \\ &= .707 y_1 + .141 y_2 + \text{const} \\ z_2 &= .707 \frac{y_1 - \bar{y}_1}{1} - .707 \frac{y_2 - \bar{y}_2}{5} \\ &= .707 y_1 - .141 y_2 + \text{const}. \end{aligned}$$

As expected, these are very different from the components extracted directly from **S**. This problem arises, of course, because of the lack of scale invariance of the components of **S**.

4. The components from a given matrix **R** are not unique to that **R**. For example, in the bivariate case, the eigenvalues of

$$\mathbf{R} = \begin{pmatrix} 1 & r \\ r & 1 \end{pmatrix}$$

are given by

$$\lambda_1 = 1 + r \qquad \lambda_2 = 1 - r, \tag{12.11}$$

but the components remain the same for all values of r:

$$\begin{aligned} z_1 &= .707 \frac{y_1 - \bar{y}_1}{s_1} + .707 \frac{y_2 - \bar{y}_2}{s_2} \\ z_2 &= .707 \frac{y_1 - \bar{y}_1}{s_1} - .707 \frac{y_2 - \bar{y}_2}{s_2}. \end{aligned} \tag{12.12}$$

The components in (12.12) do not depend on r. For example, they serve equally well for $r = .01$ and for $r = .99$. For $r = .01$, the proportion of variance explained by z_1 is $\lambda_1/(\lambda_1 + \lambda_2) = (1 + .01)/(1 + .01 + 1 - .01) = 1.01/2 = .505$. For $r = .99$, the ratio is $1.99/2 = .995$. Thus the statement that the first component from a correlation matrix accounts for, say, 90% of the variance is not very meaningful. In general, for $p > 2$, the components from **R** depend only on the ratios (relative values) of the correlations, not on their actual values, and components of a given **R** matrix will serve for other **R** matrices.

12.6 DECIDING HOW MANY COMPONENTS TO RETAIN

In every application, a decision must be made on how many principal components should be retained in order to effectively summarize the data. The following guidelines have been proposed:

1. Retain sufficient components to account for a specified percentage of the total variance, say 80%.
2. Exclude components whose eigenvalues are less than the average of the eigenvalues, $\sum_{i=1}^{p} \lambda_i/p$. For a correlation matrix, this average is 1.
3. Use the *scree graph*, a plot of λ_i versus i. This represents a visual attempt to find a natural break between the "large" eigenvalues and the "small" eigenvalues.
4. Test the significance of the "larger" components. By larger components, we mean the components corresponding to the larger eigenvalues.

We now discuss the above four criteria for choosing the components to keep. Note, however, that the smallest components may carry valuable information that should not be routinely ignored (see Section 12.7).

In method 1, the challenge lies in selecting an appropriate threshold percentage. If we aim too high, we run the risk of including components that are either *sample specific* or *variable specific*. By *sample specific* we mean that a component may not generalize to the population or to other samples. A "variable specific" component is dominated by a single variable and does not represent a composite summary of several variables.

Method 2 is widely used and is the default in many software packages. By (2.98), $\sum_i \lambda_i = \text{tr}\,\mathbf{S}$, and the average eigenvalue is also the average variance of the individual variables. Thus method 2 retains those components that account for more variance than the average for the variables. In cases where the data can be successfully summarized in a relatively small number of dimensions, there is often a wide gap in eigenvalues on both sides of the average. In Example 12.2(a), the average eigenvalue (of $\mathbf{S}$) for the football data is .957, which is amply bracketed by $\lambda_2 = 1.37$ and $\lambda_3 = .48$. In the winged aphid data in Example 12.4(d), the second and third eigenvalues (of $\mathbf{R}$) are 2.370 and .748, leaving a comfortable margin on both sides of 1. In some cases, one may wish to move the cutoff point slightly to accommodate a visible gap in eigenvalues.

The scree graph or scree test in method 3 was named for its similarity in appearance to a cliff with rocky debris at its bottom. The scree graph for the modified football data of Example 12.2(b) exhibits an ideal pattern, as shown in Figure 12.8. The first two eigenvalues form a steep curve followed by a bend and then a straight-line trend. The recommendation is to retain those eigenvalues in the steep curve *before* the first one on the straight line. Thus in Figure

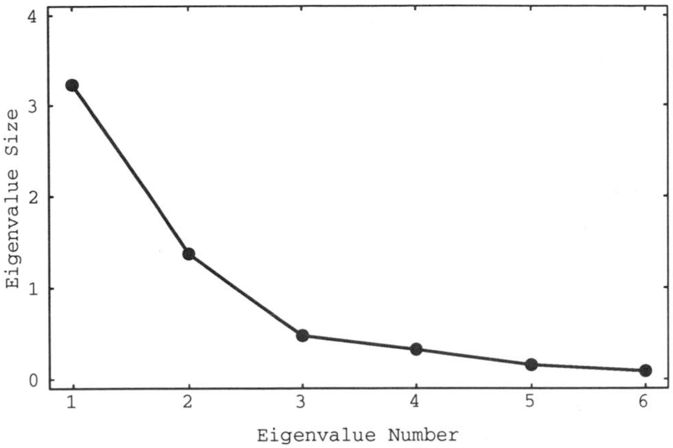

Figure 12.8 Scree graph for eigenvalues of modified football data.

12.8, two components would be retained. In practice, the turning point between the steep curve and the straight line may not be as distinct as this or there may be more than one discernible bend. In such cases, this approach is not as conclusive. The scree graph for the winged aphid data in Example 12.4(d) is plotted in Figure 12.9. The plot would suggest that either two or four components be retained.

The remainder of this section is devoted to method 4, tests of significance. The tests assume multivariate normality, which is not required for estimation of principal components.

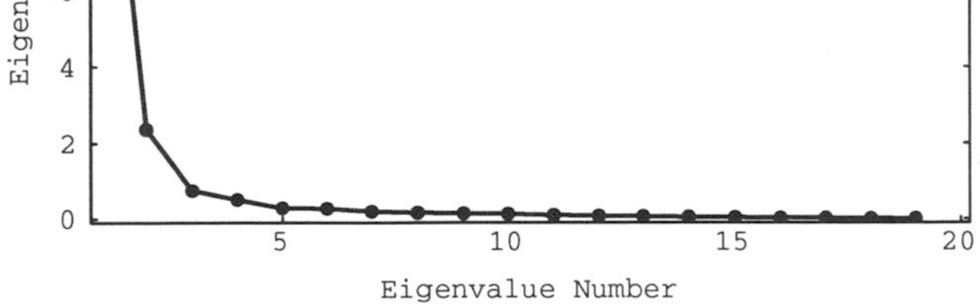

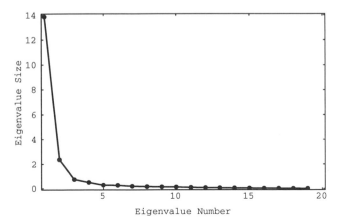

Figure 12.9 Scree graph for eigenvalues of winged aphid data.

It may be useful to make a preliminary test of complete independence of the variables, as in Section 7.4.3: H_0: $\mathbf{\Sigma} = \text{diag}(\sigma_{11}, \sigma_{22}, \ldots, \sigma_{pp})$, or equivalently, H_0: $\mathbf{P}_\rho = \mathbf{I}$. The test statistic is given in (7.35) and (7.36). If the results indicate that the variables are independent, there is no point in extracting principal components, since (except for sampling fluctuation) the variables themselves already form the principal components.

To test the significance of the "larger" components, we test the hypothesis that the last k population eigenvalues are small and equal, H_{0k}: $\gamma_{p-k+1} = \gamma_{p-k+2} = \cdots = \gamma_p$, where $\gamma_1, \gamma_2, \ldots, \gamma_p$ denote the population eigenvalues, namely, the eigenvalues of $\mathbf{\Sigma}$. The implication is that the first sample components capture all the essential dimensions, while the last components reflect noise. If H_0 is true, the last k sample eigenvalues will tend to have the pattern shown by the straight line in the ideal scree graph, such as in Figure 12.8 or 12.9.

To test H_{0k}: $\gamma_{p-k+1} = \cdots = \gamma_p$, we calculate the average of the last k eigenvalues of $\mathbf{S}$,

$$\overline{\lambda} = \sum_{i=p-k+1}^{p} \frac{\lambda_i}{k}$$

and use the test statistic

$$u = \left(n - \frac{2p+11}{6} \right) \left(k \ln \overline{\lambda} - \sum_{i=p-k+1}^{p} \ln \lambda_i \right), \tag{12.13}$$

which has an approximate χ^2-distribution. We reject H_0 if $u \geq \chi^2_{\alpha,\nu}$, where $\nu = \frac{1}{2}(k-1)(k+2)$.

To carry out this procedure, we could begin by testing H_{02}: $\gamma_{p-1} = \gamma_p$. If this is accepted, we could then test H_{03}: $\gamma_{p-2} = \gamma_{p-1} = \gamma_p$ and continue testing in this fashion until H_{0k} is rejected for some value of k. The major disadvantage of this sequential procedure is that, in practice, it tends to retain more principal components than are useful.

Example 12.6. We apply the above four criteria to the modified football data of Example 12.2(b).

For method 1, we simply examine the eigenvalues and their proportion of variance explained, as obtained in Example 12.2(b):

Eigenvalue	Proportion of Variance	Cumulative Proportion
3.323	.579	.579
1.374	.239	.818
.476	.083	.901
.325	.057	.957
.157	.027	.985
.088	.015	1.000

To account for 82% of the variance, we would keep two components. This percent of variance is high enough for most descriptive purposes. For certain other applications, such as input to another analysis, we might wish to retain three components, which would account for 90% of the variance.

To apply method 2, we find the average eigenvalue to be

$$\bar{\lambda} = \sum_{i=1}^{6} \frac{\lambda_i}{6} = \frac{5.742824}{6} = .957.$$

Since only λ_1 and λ_2 exceed .957, we would retain two components.

For method 3, the scree graph in Figure 12.8 indicates conclusively that two components should be retained.

To implement method 4, we carry out the significance tests in (12.13). The values of the test statistic u for $k = 2, 3, \ldots, 6$ are as follows:

Eigenvalue	k	u	df	$\chi^2_{.05}$
3.32341	6	245.57	20	31.41
1.37431	5	123.93	14	23.68
.47607	4	44.10	9	16.92
.32468	3	23.84	5	11.07
.15650	2	4.62	2	5.99
.08785	1			

The tests indicate that only the last two (population) eigenvalues are equal and we should retain the first four. This differs from the results of the other three criteria, which are in close agreement that two components should be retained.

12.7 INFORMATION IN THE LAST FEW PRINCIPAL COMPONENTS

Up to this point, we have focused on using the first few principal components to summarize and simplify the data. However, the last few components may carry information that is useful in some applications.

Since the eigenvalues serve as variances of the principal components, the last few principal components have smaller variances. If the variance of a component is zero or close to zero, the component represents a linear relationship among the variables that is essentially constant; that is, the relationship holds for all $\mathbf{y}_i$ in the sample. Thus if the last eigenvalue is near zero, it signifies the presence of a collinearity that may provide new information for the researcher. Suppose, for example, that there are five variables and $y_5 = \sum_{i=1}^{4} y_i/4$. Then $\mathbf{S}$ is singular, and barring round-off error, λ_5 will be zero. Thus $s_{z_5}^2 = 0$, and z_5 is constant. As noted early in Section 12.2, the $\mathbf{y}_i$'s are centered, because the origin of the principal components is translated to $\overline{\mathbf{y}}$. Hence the constant value of z_5 is its mean, zero:

$$z_5 = \mathbf{a}_5'\mathbf{y} = a_{51}y_1 + a_{52}y_2 + \cdots + a_{55}y_5 = 0.$$

Since this must reflect the dependency of y_5 on y_1, y_2, y_3, and $y_4, \mathbf{a}_5'$ will be proportional to $(1, 1, 1, 1, -4)$.

12.8 INTERPRETATION OF PRINCIPAL COMPONENTS

In Section 12.5, we noted that principal components obtained from **R** are not compatible with those obtained from **S**. Because of this lack of scale invariance of principal components, the coefficients cannot be converted to standardized form, as can be done with coefficients in discriminant functions or canonical variates. Hence interpretation of principal components is not as clear-cut as with previous linear functions that we have discussed. We must choose between components of **S** or **R**, knowing they will have a different interpretation. If the variables have widely disparate variances, we can use **R** instead of **S** to improve interpretation.

For certain patterns of elements in **S** or **R**, the form of the principal components can be predicted. This aid to interpretation is discussed in Section 12.8.1. As with discriminant functions and canonical variates, some writers have advocated rotation and the use of correlations between the variables and the principal components. We argue against the use of these two approaches to interpretation in Sections 12.8.2 and 12.8.3.

12.8.1 Special Patterns in S or R

In the covariance or correlation matrix, we may recognize a distinguishing pattern from which the structure of the principal components can be deduced. For example, we noted in Section 12.2 that if one variable has a much larger variance than the other variables, this variable will account for most of the first component. Another case in which a component will duplicate a variable occurs when the variable is uncorrelated with the other variables. We now demonstrate this by showing that if all p variables are uncorrelated, the variables themselves are the principal components. If the variables were uncorrelated (orthogonal), $\mathbf{S}$ would have the form

$$\mathbf{S} = \begin{bmatrix} s_{11} & 0 & \cdots & 0 \\ 0 & s_{22} & \cdots & 0 \\ \vdots & \vdots & & \vdots \\ 0 & 0 & \cdots & s_{pp} \end{bmatrix}, \tag{12.14}$$

and the characteristic equation would be

$$0 = |\mathbf{S} - \lambda\mathbf{I}| = \prod_{i=1}^{p} (s_{ii} - \lambda) \quad \text{[by (2.76)]},$$

which has solutions

$$\lambda_i = s_{ii} \qquad i = 1, 2, \ldots, p. \tag{12.15}$$

The corresponding normalized eigenvectors have a 1 in the ith position and 0s elsewhere:

$$\mathbf{a}_i' = (0, \ldots, 0, 1, 0, \ldots, 0). \tag{12.16}$$

Thus the ith component is

$$z_i = \mathbf{a}_i'\mathbf{y} = y_i.$$

In practice, the sample correlations (of continuous random variables) will not be zero, but if the correlations are all small, the principal components will largely duplicate the variables.

It was noted in Section 2.11.8 that when all correlations or covariances are positive, all elements of the first eigenvector $\mathbf{a}_1$ are positive. Since the remaining eigenvectors $\mathbf{a}_2, \mathbf{a}_3, \ldots, \mathbf{a}_p$ are orthogonal to $\mathbf{a}_1$, they must have both positive and negative elements. When all elements of $\mathbf{a}_1$ are positive, the first component is a weighted average of the variables and is sometimes referred to as a measure of *size*. Likewise, the positive and negative coefficients in subsequent components may be regarded as defining *shape*. This pattern is often seen when the variables are various measurements of an organism.

Example 12.8.1. In the modified football data of Example 12.2(b), there are a few negative covariances in $\mathbf{S}$, but they are small, and all elements of the first eigenvector remain positive. The second eigenvector therefore has positive and negative elements:

First Two Eigenvectors

	$\mathbf{a}_1$	$\mathbf{a}_2$
WDIM	.207	−.142
CIRCUM	.873	−.219
FBEYE	.261	−.231
EYEHD	.326	.891
EARHD	.066	.222
JAW	.128	−.187

With all positive coefficients, the first component is an overall measure of head size. The second is a shape component that contrasts the vertical measurements EYEHD and EARHD with the three lateral measurements and CIRCUM.

12.8.2 Rotation

The principal components are initially obtained by rotating axes to line up with the natural extensions of the system, so that the new variables become uncorrelated and reflect the directions of maximum variance. If the resulting components do not have a satisfactory interpretation, they can be further rotated, seeking dimensions where many of the coefficients of the linear combinations are near zero to simplify interpretation.

However, the new rotated components are correlated, and they do not successively account for maximum variance. They are therefore no longer principal components in the usual sense, and their routine use is questionable. For improved interpretation, one may wish to try factor analysis (Chapter 13), in which rotation does not destroy any properties. (In factor analysis, the rotation does not involve the axes of the space of the observations $\mathbf{y}_1, \mathbf{y}_2, \ldots, \mathbf{y}_n$, but the axes of another space, that of the factor loadings.)

12.8.3 Correlations between Variables and Principal Components

The use of correlations between variables and principal components is widely recommended as an aid to interpretation. It was noted in Sections 8.7.3 and 11.5.2 that analogous correlations for discriminant functions and canonical variates are not useful in a multivariate context because they provide only univariate information about how each variable operates by itself, ignoring the other variables. Rencher (1992b) obtained a similar result for principal components.

We denote the correlation between the ith variable y_i and the jth principal component z_j by $r_{y_i z_j}$. Because of the orthogonality of the z_j's, we have the simple relationship

$$r^2_{y_i z_1} + r^2_{y_i z_2} + \cdots + r^2_{y_i z_k} = R^2_{y_i | z_1, \ldots, z_k}, \tag{12.17}$$

where k is the number of components retained and $R^2_{y_i|z_1,\ldots,z_k}$ is the squared multiple correlation of y_i with the z_j's. Thus $r^2_{y_i z_j}$ forms part of $R^2_{y_i|z_1,\ldots,z_k}$, which shows how y_i relates to the z's by itself, not what it contributes in the presence of the other y's. The correlations are therefore not informative about the joint contribution of the y's in a principal component.

Since we do not recommend rotation or correlations for interpretation, we are left with the coefficients themselves, obtained from the eigenvectors of either **S** or **R**.

Example 12.8.3. In Example 12.8.1, the eigenvectors of **S** from the modified football data gave a satisfactory interpretation of the first two principal components as head size and shape. We give these in Table 12.5, along with the correlations between variables $y_1, y_2, \ldots, y_6$ and the first two principal components z_1 and z_2. For comparison we also give $R^2_{y_i|z_1,z_2}$ for each variable.

The correlations rank the variables somewhat differently in their contribution to the components, since they form part of the univariate information provided by R^2 for each variable by itself. For example, for the first component, the correlations rank the variables in the order 2, 3, 1, 4, 6, 5, whereas the coefficients (eigenvectors) from **S** rank them in the order 2, 4, 3, 1, 6, 5.

12.9 SELECTION OF VARIABLES

We have previously discussed subset selection in connection with Wilks' Λ (Section 6.11.2), discriminant analysis (Section 8.9), classification analysis (Section 9.6), and regression (Sections 10.2.7 and 10.7). In each case the criterion for selection of variables was the relationship of the variables to some external factor, such as dependent variable(s), separation of groups, or correct classification rates. In the context of principal components, we have no dependent variable, as in regression, or no groupings among the observations, as in

Table 12.5 Eigenvectors Obtained from S, Correlations between Variables and Principal Components, and R^2 for First Two Principal Components

	Eigenvectors from **S**		Correlations		
Variable	$\mathbf{a}_1$	$\mathbf{a}_2$	$r_{y_i z_1}$	$r_{y_i z_2}$	R^2
1	.21	−.14	.62	−.27	.46
2	.87	−.22	.98	−.16	.99
3	.26	−.23	.70	−.40	.66
4	.33	.89	.49	.86	.98
5	.07	.22	.17	.37	.17
6	.13	−.19	.41	−.39	.32

discriminant analysis. With no external influence, we simply wish to find the subset that best captures the internal variation (and covariation) of the variables.

Jolliffe (1972, 1973) discussed eight selection methods and referred to the process as *discarding variables*. The eight methods were based on three basic approaches: multiple correlation, clustering of variables, and principal components. One of the correlation methods, for example, proceeds in a stepwise fashion, deleting at each step the variable with the largest multiple correlation with the other variables. The clustering methods partition the variables into groups or clusters and select a variable from each cluster.

We describe Jolliffe's principal component methods in the context of selecting a subset of 10 variables out of 50. One of his techniques associates a variable with each of the first 10 principal components and retains these 10 variables. Another approach is to associate a variable with each of the last 40 principal components and delete the 40 variables. To associate a variable with a principal component, we choose the variable corresponding to the largest (absolute) coefficient in the component, if the variable has not previously been selected. We can use components extracted from either **S** or **R**. For example, in the two principal components for the football data in Example 12.2(b), we would choose variables 2 and 4, which clearly have the largest coefficients in the two components. Jolliffe's methods could also be applied iteratively, with the principal components being recomputed after a variable is retained or deleted.

Jolliffe (1972) compared the eight methods using both real and simulated data and found that the methods based on principal components performed well in comparison to the regression and cluster-based methods. But he concluded that no single method was uniformly best.

McCabe (1984) suggested several criteria for selection, most of which are based on the conditional covariance matrix of the variables not selected, given those selected. He denoted the selected variables as *principal variables*. Let $\mathbf{y}$ be partitioned as

$$\mathbf{y} = \begin{pmatrix} \mathbf{y}_1 \\ \mathbf{y}_2 \end{pmatrix},$$

where $\mathbf{y}_1$ contains the selected variables and $\mathbf{y}_2$ consists of the variables not selected. The corresponding covariance matrix is

$$\text{cov}(\mathbf{y}) = \mathbf{\Sigma} = \begin{pmatrix} \mathbf{\Sigma}_{11} & \mathbf{\Sigma}_{12} \\ \mathbf{\Sigma}_{21} & \mathbf{\Sigma}_{22} \end{pmatrix}.$$

By (4.8), the conditional covariance matrix is given by (assuming normality)

$$\text{cov}(\mathbf{y}_2|\mathbf{y}_1) = \mathbf{\Sigma}_{22} - \mathbf{\Sigma}_{21}\mathbf{\Sigma}_{11}^{-1}\mathbf{\Sigma}_{12},$$

which is estimated by $\mathbf{S}_{22} - \mathbf{S}_{21}\mathbf{S}_{11}^{-1}\mathbf{S}_{12}$. For a subset of size m, two of McCabe's criteria are to choose the subset $\mathbf{y}_1$ that

1. minimizes $|\mathbf{S}_{22} - \mathbf{S}_{21}\mathbf{S}_{11}^{-1}\mathbf{S}_{12}|$ and
2. maximizes $\sum_{i=1}^{m^*} r_i^2$, where r_i, $i = 1, 2, \ldots, m^* = \min(m, p - m)$ are the canonical correlations between the m selected variables in $\mathbf{y}_1$ and the $p-m$ deleted variables in $\mathbf{y}_2$.

Ideally, these criteria would be evaluated for all possible subsets so as to obtain the best subset of each size. McCabe suggested a regression approach for obtaining a percent of variance explained by a subset of variables to be compared with the percent of variance accounted for by the same number of principal components.

PROBLEMS

12.1 Show that the eigenvalues of

$$\mathbf{R} = \begin{pmatrix} 1 & r \\ r & 1 \end{pmatrix}$$

are $1 \pm r$, as in (12.11), and that the eigenvectors are as given in (12.12).

12.2 Show that when $\mathbf{S}$ is diagonal as in (12.14), the eigenvectors have the form $\mathbf{a}_i' = (0, \ldots, 0, 1, 0, \ldots, 0)$, as given in (12.16).

12.3 Show that $r_{y_iz_1}^2 + r_{y_iz_2}^2 + \cdots + r_{y_iz_k}^2 = R_{y_i|z_1,\ldots,z_k}^2$, as in (12.17). Note that this simple partitioning of R^2 into the sum of squares of simple correlations does not happen in practice because the independent variables are correlated. However, here the z's are principal components and are therefore orthogonal.

12.4 Carry out a principal components analysis of the diabetes data of Table 3.6. Use all five variables, including y's and x's. Use both $\mathbf{S}$ and $\mathbf{R}$. Which do you think is more appropriate here? Show the percent of variance explained. Based on the average eigenvalue or a scree plot, decide how many components to retain. Can you interpret the components of either $\mathbf{S}$ or $\mathbf{R}$?

12.5 Do a principal components analysis of the probe word data of Table 3.7. Use both $\mathbf{S}$ and $\mathbf{R}$. Which do you think is more appropriate here? Show the percent of variance explained. Based on the average eigenvalue or a scree plot, decide how many components to retain. Can you interpret the components of either $\mathbf{S}$ or $\mathbf{R}$?

12.6 Carry out a principal components analysis on all six variables of the glucose data of Table 3.10. Use both **S** and **R**. Which do you think is more appropriate here? Show the percent of variance explained. Based on the average eigenvalue or a scree plot, decide how many components to retain. Can you interpret the components of either **S** or **R**?

12.7 Carry out a principal components analysis on the hematology data of Table 4.3. Use both **S** and **R**. Which do you think is more appropriate here? Show the percent of variance explained. Based on the average eigenvalue or a scree plot, decide how many components to retain. Can you interpret the components of either **S** or **R**? Does the large variance of x_3 affect the pattern of the components of **S**?

12.8 Carry out a principal components analysis separately for males and females in the psychological data of Table 5.1. Compare the results for the two groups. Use **S**.

12.9 Carry out a principal components analysis separately for the two species in the beetle data of Table 5.5. Compare the results for the two groups. Use **S**.

12.10 Carry out a principal components analysis on the engineer data of Table 5.6 as follows:

(a) Use the pooled covariance matrix.

(b) Ignore groups and use a covariance matrix based on all 40 observations.

(c) Which of the approaches in (a) or (b) appears to be more successful?

12.11 Repeat the previous problem for the dystrophy data of Table 5.7.

12.12 Carry out a principal components analysis on all 10 variables of the seishu data of Table 7.1. Use both **S** and **R**. Which do you think is more appropriate here? Show the percent of variance explained. Based on the average eigenvalue or a scree plot, decide how many components to retain. Can you interpret the components of either **S** or **R**?

12.13 Carry out a principal components analysis on the temperature data of Table 7.2. Use both **S** and **R**. Which do you think is more appropriate here? Show the percent of variance explained. Based on the average eigenvalue or a scree plot, decide how many components to retain. Can you interpret the components of either **S** or **R**?

CHAPTER 13

Factor Analysis

13.1 INTRODUCTION

In factor analysis we attempt to represent the variables $y_1, y_2, \ldots, y_p$ as linear combinations of a few random variables $f_1, f_2, \ldots, f_m$ $(m < p)$ called *factors*. The factors are underlying *constructs* or *latent* variables that "generate" the y's. Like the original variables, the factors vary from individual to individual; but unlike the variables, the factors cannot be measured or observed. The existence of these hypothetical variables is therefore open to question.

If the original variables $y_1, y_2, \ldots, y_p$ are at least moderately correlated, the basic dimensionality of the system is less than p. The goal of factor analysis is to characterize the redundancy among the variables by means of a smaller number of factors. Suppose the pattern of the high and low correlations in the correlation matrix is such that the variables in a particular subset have high correlations among themselves but low correlations with all the other variables. Then there may be a single underlying factor that gave rise to the variables in the subset. If the other variables can be similarly grouped into subsets with a like pattern of correlations, then a few factors can represent these groups of variables. Thus the pattern in the correlation matrix may correspond directly to the factors. For example, suppose the correlation matrix has the form

$$\begin{bmatrix} 1 & .9 & .05 & .05 & .05 \\ .9 & 1 & .05 & .05 & .05 \\ .05 & .05 & 1 & .9 & .9 \\ .05 & .05 & .9 & 1 & .9 \\ .05 & .05 & .9 & .9 & 1 \end{bmatrix}$$

Then variables 1 and 2 correspond to a factor and variables 3, 4, and 5 correspond to another factor. In some cases where the correlation matrix does not have such a simple pattern, factor analysis will still partition the variables into clusters.

Factor analysis is related to principal component analysis in that both seek a simpler structure in a set of variables but they differ in many respects (see Section 13.8). For example, two differences in basic approach are as follows:

1. Principal components are defined as linear combinations of the original variables. In factor analysis, the original variables are expressed as linear combinations of the factors.
2. In principal component analysis, we explain a large part of the total variance of the variables, $\sum_i s_{ii}$. In factor analysis, we seek to account for the covariances or correlations among the variables.

In practice, the factor analysis model does not provide a satisfactory fit for many data sets. This is one of the reasons factor analysis is considered controversial by some statisticians. In the past, this skepticism may be owed to the fact that several of the computational methods in use often gave conflicting results. With the advent of present-day computing capability, many early techniques have been abandoned, and present approaches are more consistent. However, factor analysis remains somewhat subjective in many applications. Sometimes a few easily interpretable factors emerge, but for other data sets, neither the number of factors nor the interpretation is clear. Some possible reasons for these failures are discussed in Section 13.7.

13.2 ORTHOGONAL FACTOR MODEL

13.2.1 Model Definition and Assumptions

Factor analysis is basically a one-sample procedure [for possible applications to data with groups, see Rencher (1996, Section 10.8)]. We have a random sample $\mathbf{y}_1, \mathbf{y}_2, \ldots, \mathbf{y}_n$ from a homogeneous population with mean vector $\boldsymbol{\mu}$ and covariance matrix $\boldsymbol{\Sigma}$.

The factor analysis model expresses each variable as a linear combination of underlying *common factors* $f_1, f_2, \ldots, f_m$, with an accompanying error term to account for that part of the variable that is unique (not in common with the other variables). For $y_1, y_2, \ldots, y_p$ in any $\mathbf{y}_i$ the model is as follows:

$$
\begin{aligned}
y_1 - \mu_1 &= \lambda_{11}f_1 + \lambda_{12}f_2 + \cdots + \lambda_{1m}f_m + \epsilon_1 \\
y_2 - \mu_2 &= \lambda_{21}f_1 + \lambda_{22}f_2 + \cdots + \lambda_{2m}f_m + \epsilon_2 \\
&\;\;\vdots \\
y_p - \mu_p &= \lambda_{p1}f_1 + \lambda_{p2}f_2 + \cdots + \lambda_{pm}f_m + \epsilon_p.
\end{aligned}
\tag{13.1}
$$

Ideally, m should be substantially smaller than p, or we have not achieved a parsimonious description of the variables as functions of a few underlying factors. We might regard the f's in (13.1) as random variables that engender the y's. The coefficients λ_{ij} are called *loadings* and serve as weights, showing how each y_i depends on the f's. (In this chapter, we defer to common usage in the factor analysis literature and use the notation λ_{ij} for loadings rather than eigenvalues.) With appropriate assumptions, λ_{ij} indicates the importance of f_j to y_i and can be used in interpretation of f_j. We describe or interpret f_2, for example, by examining its coefficients, $\lambda_{12}, \lambda_{22}, \ldots, \lambda_{p2}$. The larger loadings relate f_2 to the corresponding y's. From these y's, we infer a meaning or description of f_2.

The system of equations (13.1) bears a superficial resemblance to the multiple regression model (10.1), but there are fundamental differences. For example, (1) the f's are unobserved and (2) the model in (13.1) represents only one observation vector, while (10.1) depicts all n observations.

It is assumed that for $j = 1, 2, \ldots, m$, $E(f_j) = 0$, $\text{var}(f_j) = 1$, and $\text{cov}(f_i, f_j) = 0$, $i \neq j$. The assumptions for ϵ_i, $i = 1, 2, \ldots, p$, are similar, except that we must allow each ϵ_i to have a different variance, since it shows the residual part of y_i that is not in common with the other variables. Thus we assume that $E(\epsilon_i) = 0$, $\text{var}(\epsilon_i) = \psi_i$, and $\text{cov}(\epsilon_i, \epsilon_j) = 0$, $i \neq j$. In addition, we assume that $\text{cov}(\epsilon_i, f_j) = 0$ for all i and j. We refer to ψ_i as the *specific variance*.

These assumptions are natural consequences of the basic model (13.1) and the goals of factor analysis. Since $E(y_i - \mu_i) = 0$, we need $E(f_j) = 0$, $j = 1, 2, \ldots, m$. The assumption $\text{cov}(f_i, f_j) = 0$ is made for parsimony in expressing the y's as functions of as few factors as possible. The assumptions $\text{var}(f_j) = 1$, $\text{var}(\epsilon_i) = \psi_i$, $\text{cov}(f_i, f_j) = 0$, and $\text{cov}(\epsilon_i, f_j) = 0$ yield a simple expression for the variance of y_i,

$$\text{var}(y_i) = \lambda_{i1}^2 + \lambda_{i2}^2 + \cdots + \lambda_{im}^2 + \psi_i, \tag{13.2}$$

which plays an important role in our development. Note that the assumption $\text{cov}(\epsilon_i, \epsilon_j) = 0$ implies that the factors account for all the correlations among the y's, that is, all that the y's have in common. Thus the emphasis in factor analysis is on modeling the covariances or correlations among the y's.

Model (13.1) can be written in matrix notation as

$$\mathbf{y} - \boldsymbol{\mu} = \boldsymbol{\Lambda}\mathbf{f} + \boldsymbol{\epsilon}, \tag{13.3}$$

where $\mathbf{y} = (y_1, y_2, \ldots, y_p)'$, $\boldsymbol{\mu} = (\mu_1, \mu_2, \ldots, \mu_p)'$, $\mathbf{f} = (f_1, f_2, \ldots, f_m)'$, $\boldsymbol{\epsilon} = (\epsilon_1, \epsilon_2, \ldots, \epsilon_p)'$, and

$$\boldsymbol{\Lambda} = \begin{bmatrix} \lambda_{11} & \lambda_{12} & \cdots & \lambda_{1m} \\ \lambda_{21} & \lambda_{22} & \cdots & \lambda_{2m} \\ \vdots & \vdots & & \vdots \\ \lambda_{p1} & \lambda_{p2} & \cdots & \lambda_{pm} \end{bmatrix}. \tag{13.4}$$

We illustrate the model in (13.1) and (13.3) with $p = 5$ and $m = 2$. The model for each variable in (13.1) becomes

$$
\begin{aligned}
y_1 - \mu_1 &= \lambda_{11} f_1 + \lambda_{12} f_2 + \epsilon_1 \\
y_2 - \mu_2 &= \lambda_{21} f_1 + \lambda_{22} f_2 + \epsilon_2 \\
y_3 - \mu_3 &= \lambda_{31} f_1 + \lambda_{32} f_2 + \epsilon_3 \\
y_4 - \mu_4 &= \lambda_{41} f_1 + \lambda_{42} f_2 + \epsilon_4 \\
y_5 - \mu_5 &= \lambda_{51} f_1 + \lambda_{52} f_2 + \epsilon_5.
\end{aligned}
$$

In matrix notation as in (13.3), this becomes

$$
\begin{bmatrix} y_1 - \mu_1 \\ y_2 - \mu_2 \\ y_3 - \mu_3 \\ y_4 - \mu_4 \\ y_5 - \mu_5 \end{bmatrix} = \begin{bmatrix} \lambda_{11} & \lambda_{12} \\ \lambda_{21} & \lambda_{22} \\ \lambda_{31} & \lambda_{32} \\ \lambda_{41} & \lambda_{42} \\ \lambda_{51} & \lambda_{52} \end{bmatrix} \begin{pmatrix} f_1 \\ f_2 \end{pmatrix} + \begin{bmatrix} \epsilon_1 \\ \epsilon_2 \\ \epsilon_3 \\ \epsilon_4 \\ \epsilon_5 \end{bmatrix}
$$

or $\mathbf{y} - \boldsymbol{\mu} = \boldsymbol{\Lambda}\mathbf{f} + \boldsymbol{\epsilon}$.

The assumptions listed above can be expressed concisely using the notation in (13.3):

$E(f_j) = 0$, $j = 1, 2, \ldots, m$, becomes

$$E(\mathbf{f}) = \mathbf{0}, \tag{13.5}$$

$\text{var}(f_j) = 1$, $j = 1, 2, \ldots, m$, and $\text{cov}(f_i, f_j) = 0$, $i \neq j$, become

$$\text{cov}(\mathbf{f}) = \mathbf{I}, \tag{13.6}$$

$E(\epsilon_i) = 0$, $i = 1, 2, \ldots, p$, becomes

$$E(\boldsymbol{\epsilon}) = \mathbf{0}, \tag{13.7}$$

$\text{var}(\epsilon_i) = \psi_i$, $i = 1, 2, \ldots, p$, and $\text{cov}(\epsilon_i, \epsilon_j) = 0$, $i \neq j$, become

$$\text{cov}(\boldsymbol{\epsilon}) = \boldsymbol{\Psi} = \begin{bmatrix} \psi_1 & 0 & \cdots & 0 \\ 0 & \psi_2 & \cdots & 0 \\ \vdots & \vdots & & \vdots \\ 0 & 0 & \cdots & \psi_p \end{bmatrix}, \tag{13.8}$$

and $\text{cov}(\epsilon_i, f_j) = 0$ for all i and j becomes

$$\text{cov}(\mathbf{f}, \boldsymbol{\epsilon}) = \mathbf{O}. \tag{13.9}$$

The notation $\text{cov}(\mathbf{f}, \boldsymbol{\epsilon})$ indicates a rectangular matrix containing the covariances of the f's with the ϵ's:

$$\text{cov}(\mathbf{f}, \boldsymbol{\epsilon}) = \begin{bmatrix} \sigma_{f_1\epsilon_1} & \sigma_{f_1\epsilon_2} & \cdots & \sigma_{f_1\epsilon_p} \\ \sigma_{f_2\epsilon_1} & \sigma_{f_2\epsilon_2} & \cdots & \sigma_{f_2\epsilon_p} \\ \vdots & \vdots & & \vdots \\ \sigma_{f_m\epsilon_1} & \sigma_{f_m\epsilon_2} & \cdots & \sigma_{f_m\epsilon_p} \end{bmatrix}.$$

It was noted following (13.2) that the emphasis in factor analysis is on modeling the covariances among the y's. Thus one goal of factor analysis is to express the $\frac{1}{2}p(p-1)$ covariances (and the p variances) of the variables $y_1, y_2, \ldots, y_p$ in terms of a simplified structure involving the pm loadings λ_{ij} and the p specific variances ψ_i. That is, we wish to express $\boldsymbol{\Sigma}$ in terms of $\boldsymbol{\Lambda}$ and $\boldsymbol{\Psi}$. We can do this using the model (13.3) and the assumptions (13.6), (13.8), and (13.9). Since $\boldsymbol{\mu}$ does not affect variances and covariances of $\mathbf{y}$, we have from (13.3)

$$\boldsymbol{\Sigma} = \text{cov}(\mathbf{y}) = \text{cov}(\boldsymbol{\Lambda}\mathbf{f} + \boldsymbol{\epsilon}).$$

By (13.9), $\boldsymbol{\Lambda}\mathbf{f}$ and $\boldsymbol{\epsilon}$ are uncorrelated; therefore, the covariance matrix of their sum is the sum of their covariance matrices:

$$\begin{aligned} \boldsymbol{\Sigma} &= \text{cov}(\boldsymbol{\Lambda}\mathbf{f} + \boldsymbol{\epsilon}) \\ &= \text{cov}(\boldsymbol{\Lambda}\mathbf{f}) + \text{cov}(\boldsymbol{\epsilon}) \\ &= \boldsymbol{\Lambda}\,\text{cov}(\mathbf{f})\boldsymbol{\Lambda}' + \boldsymbol{\Psi} \qquad \text{[by (3.69) and (13.8)]} \\ &= \boldsymbol{\Lambda}\mathbf{I}\boldsymbol{\Lambda}' + \boldsymbol{\Psi} \qquad \text{[by (13.6)]} \\ &= \boldsymbol{\Lambda}\boldsymbol{\Lambda}' + \boldsymbol{\Psi}. \end{aligned} \tag{13.10}$$

If $\boldsymbol{\Lambda}$ has only a few columns, say two or three, then $\boldsymbol{\Sigma} = \boldsymbol{\Lambda}\boldsymbol{\Lambda}' + \boldsymbol{\Psi}$ represents a simplified structure for $\boldsymbol{\Sigma}$. The covariances in $\boldsymbol{\Sigma}$ are modeled by the λ_{ij}'s alone since $\boldsymbol{\Psi}$ is diagonal. For example, in the above illustration with $m = 2$ factors, σ_{12} would be the product of the first two rows of $\boldsymbol{\Lambda}$, that is,

$$\sigma_{12} = \text{cov}(y_1, y_2) = \lambda_{11}\lambda_{21} + \lambda_{12}\lambda_{22},$$

where $(\lambda_{11}, \lambda_{12})$ is the first row of $\boldsymbol{\Lambda}$ and $(\lambda_{21}, \lambda_{22})$ is the second row of $\boldsymbol{\Lambda}$. If y_1 and y_2 have a great deal in common, they will have similar loadings on the common factors f_1 and f_2; that is, $(\lambda_{11}, \lambda_{12})$ will be similar to $(\lambda_{21}, \lambda_{22})$. In this case either $\lambda_{11}\lambda_{21}$ or $\lambda_{12}\lambda_{22}$ is likely to be high. On the other hand, if y_1 and y_2 have little in common, then their loadings λ_{11} and λ_{21} on f_1 will be different and their loadings λ_{12} and λ_{22} on f_2 will likewise differ. With high and low loadings of this type, the products $\lambda_{11}\lambda_{21}$ and $\lambda_{12}\lambda_{22}$ will tend to be small.

We can also find the covariances of the y's with the f's in terms of the λ's. Consider, for example, $\text{cov}(y_1, f_2)$. By (13.1), $y_1 - \mu_1 = \lambda_{11}f_1 + \lambda_{12}f_2 + \cdots + \lambda_{1m}f_m + \epsilon_1$. From (13.6), f_2 is uncorrelated with all other f_j, and by (13.9), f_2 is uncorrelated with ϵ_1. Thus

$$\text{cov}(y_1, f_2) = \text{cov}(\lambda_{12} f_2, f_2) = \lambda_{12}\,\text{var}(f_2) = \lambda_{12},$$

since $\text{var}(f_j) = 1$. Hence the loadings themselves represent covariances of the variables with the factors. In general,

$$\text{cov}(y_i, f_j) = \lambda_{ij} \qquad i = 1, 2, \ldots, p, \quad j = 1, 2, \ldots, m. \tag{13.11}$$

Since λ_{ij} is the (ij)th element of $\mathbf{\Lambda}$, we can write (13.11) in the form

$$\text{cov}(\mathbf{y}, \mathbf{f}) = \mathbf{\Lambda}. \tag{13.12}$$

If standardized variables are used, (13.10) is replaced by $\mathbf{P}_\rho = \mathbf{\Lambda\Lambda}' + \mathbf{\Psi}$, and the loadings become correlations:

$$\text{corr}(y_i, f_j) = \lambda_{ij}. \tag{13.13}$$

In (13.2), we have a breakdown of the variance of y_i into a component due to the common factors, called the *communality*, and a component unique to y_i, called the *specific variance*:

$$\begin{aligned} \sigma_{ii} = \text{var}(y_i) &= (\lambda_{i1}^2 + \lambda_{i2}^2 + \cdots + \lambda_{im}^2) + \psi_i \\ &= h_i^2 + \psi_i \\ &= \text{communality} + \text{specific variance}, \end{aligned}$$

where

$$\begin{aligned} \text{Communality} &= h_i^2 = \lambda_{i1}^2 + \lambda_{i2}^2 + \cdots + \lambda_{im}^2 \\ \text{Specific variance} &= \psi_i. \end{aligned} \tag{13.14}$$

The communality h_i^2 is also referred to as *common variance*, and the specific variance ψ_i has been called *specificity*, *unique variance*, or *residual variance*.

Assumptions (13.5)–(13.9) lead to the simple covariance structure of (13.10). This basic pattern for $\mathbf{\Sigma}$ is an essential part of the factor analysis model. In schematic form, $\mathbf{\Sigma} = \mathbf{\Lambda\Lambda}' + \mathbf{\Psi}$ has the following appearance:

The diagonal elements of $\boldsymbol{\Sigma}$ can be obtained easily by adjusting the diagonal elements of $\boldsymbol{\Psi}$, but $\boldsymbol{\Lambda}\boldsymbol{\Lambda}'$ is a simplified configuration for the off-diagonal elements. Hence the most critical aspect of the model involves the covariances, and this is the major emphasis of factor analysis, as noted in Section 13.1 and in comments following (13.2).

It is a rare population covariance matrix $\boldsymbol{\Sigma}$ that can be expressed exactly as $\boldsymbol{\Sigma} = \boldsymbol{\Lambda}\boldsymbol{\Lambda}' + \boldsymbol{\Psi}$, where $\boldsymbol{\Psi}$ is diagonal and $\boldsymbol{\Lambda}$ is $p \times m$, with m relatively small. In practice, many sample covariance matrices do not come satisfactorily close to this ideal pattern. However, we do not relax the assumptions because the structure $\boldsymbol{\Sigma} = \boldsymbol{\Lambda}\boldsymbol{\Lambda}' + \boldsymbol{\Psi}$ is essential for estimation of $\boldsymbol{\Lambda}$.

One advantage to the factor analysis model is that when it does not fit the data, the estimate of $\boldsymbol{\Lambda}$ clearly reflects this failure. In such cases, there are two problems in the estimates: (1) we cannot tell how many factors there should be and (2) we cannot tell what the factors are. In other statistical estimation procedures, failure of assumptions may not lead to such obvious consequences in the estimates. In factor analysis, the assumptions are essentially self-checking, whereas in other procedures, we typically have to check the assumptions with residual plots, tests, and so on.

13.2.2 Nonuniqueness of Factor Loadings

The loadings in the model (13.3) can be multiplied by an orthogonal matrix without impairing their ability to reproduce the covariance matrix in $\boldsymbol{\Sigma} = \boldsymbol{\Lambda}\boldsymbol{\Lambda}' + \boldsymbol{\Psi}$. To see this, let $\mathbf{T}$ be an arbitrary orthogonal matrix. Then by (2.93), $\mathbf{T}\mathbf{T}' = \mathbf{I}$, and we can insert $\mathbf{T}\mathbf{T}'$ into the basic model (13.3) to obtain

$$\mathbf{y} - \boldsymbol{\mu} = \boldsymbol{\Lambda}\mathbf{T}\mathbf{T}'\mathbf{f} + \boldsymbol{\epsilon}.$$

We then associate $\mathbf{T}$ with $\boldsymbol{\Lambda}$ and $\mathbf{T}'$ with $\mathbf{f}$ so that the model becomes

$$= \boldsymbol{\Lambda}^*\mathbf{f}^* + \boldsymbol{\epsilon}, \tag{13.15}$$

where

$$\boldsymbol{\Lambda}^* = \boldsymbol{\Lambda}\mathbf{T} \tag{13.16}$$
$$\mathbf{f}^* = \mathbf{T}'\mathbf{f}. \tag{13.17}$$

If $\boldsymbol{\Lambda}$ in $\boldsymbol{\Sigma} = \boldsymbol{\Lambda}\boldsymbol{\Lambda}' + \boldsymbol{\Psi}$ is replaced by $\boldsymbol{\Lambda}^* = \boldsymbol{\Lambda}\mathbf{T}$, we have

$$\begin{aligned}\boldsymbol{\Sigma} &= \boldsymbol{\Lambda}^*\boldsymbol{\Lambda}^{*\prime} + \boldsymbol{\Psi} = \boldsymbol{\Lambda}\mathbf{T}(\boldsymbol{\Lambda}\mathbf{T})' + \boldsymbol{\Psi} \\ &= \boldsymbol{\Lambda}\mathbf{T}\mathbf{T}'\boldsymbol{\Lambda}' + \boldsymbol{\Psi} = \boldsymbol{\Lambda}\boldsymbol{\Lambda}' + \boldsymbol{\Psi},\end{aligned}$$

since $\mathbf{TT}' = \mathbf{I}$. Thus the new loadings $\mathbf{\Lambda}^*$ in (13.16) reproduce the covariance matrix, just as $\mathbf{\Lambda}$ does in (13.10):

$$\mathbf{\Sigma} = \mathbf{\Lambda}^*\mathbf{\Lambda}^{*\prime} + \mathbf{\Psi}. \tag{13.18}$$

The new factors $\mathbf{f}^*$ in (13.17) satisfy the assumptions (13.5), (13.6), and (13.9); that is, $E(\mathbf{f}^*) = \mathbf{0}$, $\text{cov}(\mathbf{f}^*) = \mathbf{I}$, and $\text{cov}(\mathbf{f}^*, \boldsymbol{\epsilon}) = \mathbf{O}$.

The communalitites $h_i^2 = \lambda_{i1}^2 + \lambda_{i2}^2 + \cdots + \lambda_{im}^2$, $i = 1, 2, \ldots, p$, as defined in (13.14), are also unaffected by the transformation $\mathbf{\Lambda}^* = \mathbf{\Lambda T}$. This can be seen as follows. The communality h_i^2 is the sum of squares of the ith row of $\mathbf{\Lambda}$. If we denote the ith row of $\mathbf{\Lambda}$ by $\boldsymbol{\lambda}_i'$, then the sum of squares in vector notation is $h_i^2 = \boldsymbol{\lambda}_i'\boldsymbol{\lambda}_i$. The ith row of $\mathbf{\Lambda}^* = \mathbf{\Lambda T}$ is $\boldsymbol{\lambda}_i^{*\prime} = \boldsymbol{\lambda}_i'\mathbf{T}$, and $\boldsymbol{\lambda}_i^*$ becomes $\boldsymbol{\lambda}_i^* = (\boldsymbol{\lambda}_i'\mathbf{T})' = \mathbf{T}'\boldsymbol{\lambda}_i$. In terms of $\boldsymbol{\lambda}_i^*$, the communality is

$$h_i^{*2} = \boldsymbol{\lambda}_i^{*\prime}\boldsymbol{\lambda}_i^* = \boldsymbol{\lambda}_i'\mathbf{TT}'\boldsymbol{\lambda}_i = \boldsymbol{\lambda}_i'\boldsymbol{\lambda}_i = h_i^2.$$

Thus the communalities remain the same for the new loadings. Note that $h_i^2 = \lambda_{i1}^2 + \lambda_{i2}^2 + \cdots + \lambda_{im}^2 = \boldsymbol{\lambda}_i'\boldsymbol{\lambda}_i$ is the distance from the origin to the point $\boldsymbol{\lambda}_i' = (\lambda_{i1}, \lambda_{i2}, \ldots, \lambda_{im})$ in the m-dimensional space of the factor loadings. Since the distance $\boldsymbol{\lambda}_i'\boldsymbol{\lambda}_i$ is the same as that of $\boldsymbol{\lambda}_i^{*\prime}\boldsymbol{\lambda}_i^*$, the points $\boldsymbol{\lambda}_i^*$ are rotated from the points $\boldsymbol{\lambda}_i$. [This also follows because $\boldsymbol{\lambda}_i^{*\prime} = \boldsymbol{\lambda}_i'\mathbf{T}$, where $\mathbf{T}$ is orthogonal. Multiplication of a vector by an orthogonal matrix is equivalent to a rotation of axes; see (2.94).]

The inherent potential to rotate the loadings to a new frame of reference without affecting any assumptions or properties is very useful in interpretation of the factors and will be exploited in Section 13.5.

Note that the coefficients (loadings) in (13.1) are applied to the factors, not to the variables, as they are in discriminant functions and principal components. Thus in factor analysis, the observed variables are not affected by rotation of the coefficients, as happens in discriminant functions and principal components.

13.3 ESTIMATION OF LOADINGS AND COMMUNALITIES

In the Sections 13.3.1–13.3.4, we discuss four approaches to estimation of the loadings.

13.3.1 Principal Component Method

The first technique we consider is commonly called the *principal component* method. This name is perhaps unfortunate in that it adds to the confusion between factor analysis and principal component analysis. In the principal component method of estimation of loadings, we do not actually calculate any principal components. The reason for the name is given later.

From a random sample $\mathbf{y}_1, \mathbf{y}_2, \ldots, \mathbf{y}_n$, we obtain the sample covariance matrix $\mathbf{S}$ and then attempt to find an estimator $\hat{\mathbf{\Lambda}}$ that will approximate the fundamental expression (13.10) with $\mathbf{S}$ in place of $\mathbf{\Sigma}$:

$$\mathbf{S} \cong \hat{\mathbf{\Lambda}}\hat{\mathbf{\Lambda}}' + \hat{\mathbf{\Psi}}. \tag{13.19}$$

In the principal component approach, we neglect $\hat{\mathbf{\Psi}}$ and factor $\mathbf{S}$ into $\mathbf{S} = \hat{\mathbf{\Lambda}}\hat{\mathbf{\Lambda}}'$.

In order to factor $\mathbf{S}$, we use the spectral decomposition in (2.100),

$$\mathbf{S} = \mathbf{CDC}', \tag{13.20}$$

where $\mathbf{C}$ is an orthogonal matrix constructed with normalized eigenvectors ($\mathbf{c}_i'\mathbf{c}_i = 1$) of $\mathbf{S}$ as columns and

$$\mathbf{D} = \begin{bmatrix} \theta_1 & 0 & \cdots & 0 \\ 0 & \theta_2 & \cdots & 0 \\ \vdots & \vdots & & \vdots \\ 0 & 0 & \cdots & \theta_p \end{bmatrix}, \tag{13.21}$$

where $\theta_1, \theta_2, \ldots, \theta_p$ are the eigenvalues of $\mathbf{S}$. We use the notation θ_i for eigenvalues instead of the usual λ_i in order to avoid confusion with the notation λ_{ij} used for the loadings.

To finish factoring $\mathbf{S} = \mathbf{CDC}'$ into the form $\hat{\mathbf{\Lambda}}\hat{\mathbf{\Lambda}}'$, we observe that since the eigenvalues θ_i of the positive definite matrix $\mathbf{S}$ are all positive, we can factor $\mathbf{D}$ into

$$\mathbf{D} = \mathbf{D}^{1/2}\mathbf{D}^{1/2},$$

where

$$\mathbf{D}^{1/2} = \begin{bmatrix} \sqrt{\theta_1} & 0 & \cdots & 0 \\ 0 & \sqrt{\theta_2} & \cdots & 0 \\ \vdots & \vdots & & \vdots \\ 0 & 0 & \cdots & \sqrt{\theta_p} \end{bmatrix}.$$

With this factoring of $\mathbf{D}$, (13.20) becomes,

$$\begin{aligned}\mathbf{S} &= \mathbf{CDC}' = \mathbf{CD}^{1/2}\mathbf{D}^{1/2}\mathbf{C}' \\ &= (\mathbf{CD}^{1/2})(\mathbf{CD}^{1/2})'. \end{aligned} \tag{13.22}$$

This is of the form $\mathbf{S} = \hat{\mathbf{\Lambda}}\hat{\mathbf{\Lambda}}'$, but we do not define $\hat{\mathbf{\Lambda}}$ to be $\mathbf{CD}^{1/2}$ because $\mathbf{CD}^{1/2}$ is $p \times p$, and we are seeking a $\hat{\mathbf{\Lambda}}$ that is $p \times m$ with $m < p$. We therefore define $\mathbf{D}_1$ to contain the m largest eigenvalues $\theta_1 > \theta_2 > \cdots > \theta_m$ and $\mathbf{C}_1$ to contain the corresponding eigenvectors $\mathbf{c}_1, \mathbf{c}_2, \ldots, \mathbf{c}_m$. We then estimate $\mathbf{\Lambda}$ by

$$\hat{\mathbf{\Lambda}} = \mathbf{C}_1\mathbf{D}_1^{1/2}, \tag{13.23}$$

where $\hat{\mathbf{\Lambda}}$ is $p \times m$, $\mathbf{C}_1$ is $p \times m$, and $\mathbf{D}_1^{1/2}$ is $m \times m$.

We illustrate the structure of the $\hat{\lambda}_{ij}$ in (13.23) for $p = 5$ and $m = 2$:

$$\begin{aligned}\begin{bmatrix} \hat{\lambda}_{11} & \hat{\lambda}_{12} \\ \hat{\lambda}_{21} & \hat{\lambda}_{22} \\ \hat{\lambda}_{31} & \hat{\lambda}_{32} \\ \hat{\lambda}_{41} & \hat{\lambda}_{42} \\ \hat{\lambda}_{51} & \hat{\lambda}_{52} \end{bmatrix} &= \begin{bmatrix} c_{11} & c_{12} \\ c_{21} & c_{22} \\ c_{31} & c_{32} \\ c_{41} & c_{42} \\ c_{51} & c_{52} \end{bmatrix} \begin{bmatrix} \sqrt{\theta_1} & 0 \\ 0 & \sqrt{\theta_2} \end{bmatrix} \\ &= \begin{bmatrix} \sqrt{\theta_1}c_{11} & \sqrt{\theta_2}c_{12} \\ \sqrt{\theta_1}c_{21} & \sqrt{\theta_2}c_{22} \\ \sqrt{\theta_1}c_{31} & \sqrt{\theta_2}c_{32} \\ \sqrt{\theta_1}c_{41} & \sqrt{\theta_2}c_{42} \\ \sqrt{\theta_1}c_{51} & \sqrt{\theta_2}c_{52} \end{bmatrix} \quad \text{[by (2.51)]}. \end{aligned} \tag{13.24}$$

We can see in (13.24) the source of the term *principal component* solution. The columns of $\hat{\mathbf{\Lambda}}$ are proportional to the eigenvectors of $\mathbf{S}$, so that the loadings on the jth factor are proportional to coefficients in the jth principal component. The factors are thus related to the first m principal components, and it would seem that interpretation would be the same as for principal components. But after rotation of the loadings, the interpretation of the factors will be different. The researcher will usually prefer the rotated factors for reasons to be treated in Section 13.5.

By (2.48), the ith diagonal element of $\hat{\mathbf{\Lambda}}\hat{\mathbf{\Lambda}}'$ is the sum of squares of the ith row of $\hat{\mathbf{\Lambda}}$, or $\hat{\boldsymbol{\lambda}}_i'\hat{\boldsymbol{\lambda}}_i = \sum_{j=1}^m \hat{\lambda}_{ij}^2$. Hence to complete the approximation of $\mathbf{S}$ in (13.19), we define

$$\hat{\psi}_i = s_{ii} - \sum_{j=1}^{m} \hat{\lambda}_{ij}^2 \tag{13.25}$$

and write

$$\mathbf{S} \cong \hat{\mathbf{\Lambda}}\hat{\mathbf{\Lambda}}' + \hat{\mathbf{\Psi}}, \tag{13.26}$$

where $\hat{\mathbf{\Psi}} = \text{diag}(\hat{\psi}_1, \hat{\psi}_2, \ldots, \hat{\psi}_p)$. Note that in (13.26) the variances on the diagonal of $\mathbf{S}$ are modeled exactly, but the off-diagonal covariances are not. Again, this is the challenge of factor analysis.

In this method of estimation, the sums of squares of the rows and columns of $\hat{\mathbf{\Lambda}}$ are equal to communalities and eigenvalues, respectively. This is easily shown. By analogy with (13.14), the ith communality is estimated by

$$\hat{h}_i^2 = \sum_{j=1}^{m} \hat{\lambda}_{ij}^2, \tag{13.27}$$

which is the sum of squares of the ith row of $\hat{\mathbf{\Lambda}}$. The sum of squares of the jth column of $\hat{\mathbf{\Lambda}}$ is the jth eigenvalue of $\mathbf{S}$, θ_j:

$$\begin{aligned} \sum_{i=1}^{p} \hat{\lambda}_{ij}^2 &= \sum_{i=1}^{p} (\sqrt{\theta_j} c_{ij})^2 \quad \text{[by (13.24)]} \\ &= \theta_j \sum_{i=1}^{p} c_{ij}^2 \\ &= \theta_j, \end{aligned} \tag{13.28}$$

since the normalized eigenvectors (columns of $\mathbf{C}$) have length 1.

If the variables are not commensurate, we can use standardized variables and work with the correlation matrix $\mathbf{R}$. The eigenvalues and eigenvectors of $\mathbf{R}$ are then used in place of those of $\mathbf{S}$ in (13.23) to obtain the loadings. In practice, $\mathbf{R}$ is used more often than $\mathbf{S}$ and is the default in most software packages. Since the emphasis in factor analysis is on reproducing the covariances or correlations rather than the variances, use of $\mathbf{R}$ is more appropriate in factor analysis than in principal components. In applications, $\mathbf{R}$ often gives better results than $\mathbf{S}$.

By (13.25) and (13.27), the variance of the ith variable is partitioned into a part due to the factors and a part due uniquely to the variable:

$$s_{ii} = \hat{h}_i^2 + \hat{\psi}_i = \hat{\lambda}_{i1}^2 + \hat{\lambda}_{i2}^2 + \cdots + \hat{\lambda}_{im}^2 + \hat{\psi}_i.$$

Thus the jth factor contributes $\hat{\lambda}_{ij}^2$ to s_{ii}. The contribution of the jth factor to the total sample variance, $\text{tr}(\mathbf{S}) = s_{11} + s_{22} + \cdots + s_{pp}$, is therefore $\sum_{i=1}^{p} \hat{\lambda}_{ij}^2$, which is the sum of squares of loadings in the jth column of $\hat{\mathbf{\Lambda}}$,

$$\hat{\lambda}_{1j}^2 + \hat{\lambda}_{2j}^2 + \cdots + \hat{\lambda}_{pj}^2. \tag{13.29}$$

By (13.28), this is equal to the jth eigenvalue, θ_j. The proportion of total sample variance due to the jth factor is

$$\frac{\hat{\lambda}_{1j}^2 + \hat{\lambda}_{2j}^2 + \cdots + \hat{\lambda}_{pj}^2}{\text{tr}(\mathbf{S})} = \frac{\theta_j}{\text{tr}(\mathbf{S})}. \tag{13.30}$$

If we are factoring $\mathbf{R}$, the corresponding proportion is

$$\frac{\hat{\lambda}_{1j}^2 + \hat{\lambda}_{2j}^2 + \cdots + \hat{\lambda}_{pj}^2}{\text{tr}(\mathbf{R})} = \frac{\theta_j}{p}, \tag{13.31}$$

where p is the number of variables. Hence if the communalities are small, the proportion of variance in (13.30) or (13.31) will be small.

Example 13.3.1. To illustrate the principal components method of estimation, we use a simple data set collected by Brown et al. (1984). A 12-year-old girl made five ratings on a nine-point semantic differential scale for each of seven of her acquaintances. The ratings were based on the five adjectives "kind," "intelligent," "happy," "likeable," and "just." Her ratings are given in Table 13.1.

The correlation matrix for the five variables (adjectives) is as follows, with the larger values bolded:

$$\mathbf{R} = \begin{bmatrix} 1.000 & .296 & \mathbf{.881} & \mathbf{.995} & .545 \\ .296 & 1.000 & -.022 & .326 & \mathbf{.837} \\ \mathbf{.881} & -.022 & 1.000 & \mathbf{.867} & .130 \\ \mathbf{.995} & .326 & \mathbf{.867} & 1.000 & .544 \\ .545 & \mathbf{.837} & .130 & .544 & 1.000 \end{bmatrix}.$$

Table 13.1 Perception Data: Ratings on Five Adjectives for Seven People

People	Kind	Intelligent	Happy	Likeable	Just
FSM1[a]	1	5	5	1	1
SISTER	8	9	7	9	8
FSM2	9	8	9	9	8
FATHER	9	9	9	9	9
TEACHER	1	9	1	1	9
MSM[b]	9	7	7	9	9
FSM3	9	7	9	9	7

[a]Female schoolmate 1.
[b]Male schoolmate.

The boldface values indicate two groups of variables: $\{1, 3, 4\}$ and $\{2, 5\}$. We would therefore expect that the correlations among the variables can be explained fairly well by two factors.

The eigenvalues of **R** are 3.263, 1.538, .168, .031, and 0. Thus **R** is singular, probably due to having only seven observations on five variables recorded in a single-digit scale. The multicollinearity among the variables induced by the fifth eigenvalue, 0, could be ascertained from the corresponding eigenvector, as noted in Section 12.7 (see problem 13.5).

By (13.31), the first two factors account for (3.263 + 1.538)/ 5 = .96 of the total sample variance. We therefore extract two factors.

The first two eigenvectors are

$$\mathbf{c}_1 = \begin{bmatrix} .537 \\ .288 \\ .434 \\ .537 \\ .390 \end{bmatrix} \quad \text{and} \quad \mathbf{c}_2 = \begin{bmatrix} -.186 \\ .651 \\ -.473 \\ -.169 \\ .538 \end{bmatrix}.$$

When these are multiplied by the square roots of the respective eigenvalues as in (13.24), we obtain the loadings in Table 13.2.

The communalities in Table 13.2 are obtained from the sum of squares of the rows of the loadings, as in (13.27). The first one, for example, is $(.969)^2 + (-.231)^2 = .993$. The specific variances are obtained from (13.25) as $\hat{\psi}_i = 1 - \hat{h}_i^2$ using 1 in place of s_{ii} because we are factoring **R** rather than **S**. The variance

Table 13.2 Factor Loadings by Principal Component Method for Perception Data of Table 13.1

Variables	Loadings $\hat{\lambda}_{1j}$	Loadings $\hat{\lambda}_{2j}$	Communalities, $\hat{h}_i^2$	Specific Variances, $\hat{\psi}_i$
Kind	.969	−.231	.993	.007
Intelligent	.519	.807	.921	.079
Happy	.785	−.587	.960	.040
Likeable	.971	−.210	.987	.013
Just	.704	.667	.940	.060
Variance accounted for	3.263	1.538	4.802	
Proportion of total variance	.653	.308	.960	
Cumulative proportion	.653	.960	.960	

accounted for by each factor is the sum of squares of the corresponding column of the loadings, as in (13.29). By (13.28), the variance accounted for is also equal to the eigenvalue in each case. Notice that the variance accounted for by the two factors adds to the sum of the communalities, since the latter is the sum of all squared loadings. By (13.31), the proportion of total variance for each factor is the variance accounted for divided by 5.

The two factors account for 96% of the total variance and therefore represent the five variables very well. To see how well the two-factor model reproduces the correlation matrix, we examine

$$\hat{\mathbf{\Lambda}}\hat{\mathbf{\Lambda}}' + \hat{\mathbf{\Psi}} = \begin{bmatrix} .969 & -.231 \\ .519 & .807 \\ .785 & -.587 \\ .971 & -.210 \\ .704 & .667 \end{bmatrix} \begin{bmatrix} .969 & .519 & .785 & .971 & .704 \\ -.231 & .807 & -.587 & -.210 & .667 \end{bmatrix}$$

$$+ \begin{bmatrix} .007 & 0 & 0 & 0 & 0 \\ 0 & .079 & 0 & 0 & 0 \\ 0 & 0 & .040 & 0 & 0 \\ 0 & 0 & 0 & .013 & 0 \\ 0 & 0 & 0 & 0 & .060 \end{bmatrix}$$

$$= \begin{bmatrix} 1.000 & .317 & .896 & .990 & .528 \\ .317 & 1.000 & -.066 & .335 & .904 \\ .896 & -.066 & 1.000 & .885 & .161 \\ .990 & .335 & .885 & 1.000 & .543 \\ .528 & .904 & .161 & .543 & 1.000 \end{bmatrix},$$

which is very close to the original $\mathbf{R}$. We will not attempt to interpret the factors at this point but will wait until they have been rotated in Section 13.5.2.

13.3.2 Principal Factor Method

In the principal component approach to estimation of the loadings, we neglected $\mathbf{\Psi}$ and factored $\mathbf{S}$ or $\mathbf{R}$. The *principal factor* method (also called the *principal axis* method) uses an initial estimate $\hat{\mathbf{\Psi}}$ and factors $\mathbf{S} - \hat{\mathbf{\Psi}}$ or $\mathbf{R} - \hat{\mathbf{\Psi}}$ to obtain

$$\mathbf{S} - \hat{\mathbf{\Psi}} \cong \hat{\mathbf{\Lambda}}\hat{\mathbf{\Lambda}}' \tag{13.32}$$

or

$$\mathbf{R} - \hat{\mathbf{\Psi}} \cong \hat{\mathbf{\Lambda}}\hat{\mathbf{\Lambda}}', \tag{13.33}$$

where $\hat{\mathbf{\Lambda}}$ is $p \times m$ and is calculated as in (13.23) using eigenvalues and eigenvectors of $\mathbf{S} - \hat{\mathbf{\Psi}}$ or $\mathbf{R} - \hat{\mathbf{\Psi}}$.

By definition, the diagonal elements of $\mathbf{S} - \hat{\mathbf{\Psi}}$ are the communalities $\hat{h}_i^2 = s_{ii} - \hat{\psi}_i$ and the diagonal elements of $\mathbf{R} - \hat{\mathbf{\Psi}}$ are the communalities $\hat{h}_i^2 = 1 - \hat{\psi}_i$. (Obviously, both $\hat{\psi}_i$ and $\hat{h}_i^2$ have different values for $\mathbf{S}$ than for $\mathbf{R}$.) With these diagonal values, $\mathbf{S} - \hat{\mathbf{\Psi}}$ and $\mathbf{R} - \hat{\mathbf{\Psi}}$ have the form

$$\mathbf{S} - \hat{\mathbf{\Psi}} = \begin{bmatrix} \hat{h}_1^2 & s_{12} & \cdots & s_{1p} \\ s_{21} & \hat{h}_2^2 & \cdots & s_{2p} \\ \vdots & \vdots & & \vdots \\ s_{p1} & s_{p2} & \cdots & \hat{h}_p^2 \end{bmatrix} \tag{13.34}$$

$$\mathbf{R} - \hat{\mathbf{\Psi}} = \begin{bmatrix} \hat{h}_1^2 & r_{12} & \cdots & r_{1p} \\ r_{21} & \hat{h}_2^2 & \cdots & r_{2p} \\ \vdots & \vdots & & \vdots \\ r_{p1} & r_{p2} & \cdots & \hat{h}_p^2 \end{bmatrix}. \tag{13.35}$$

A popular initial estimate for a communality in $\mathbf{R} - \hat{\mathbf{\Psi}}$ is

$$\hat{h}_i^2 = 1 - \frac{1}{r^{ii}}, \tag{13.36}$$

where r^{ii} is the ith diagonal element of $\mathbf{R}^{-1}$. It can be shown that $1 - 1/r^{ii} = R_i^2$, the squared multiple correlation between y_i and the other $p - 1$ variables. Note that if a factor is associated with only one variable, say y_i, the use of $\hat{h}_i^2 = R_i^2$ will result in small loadings for y_i on all factors, including the factor associated with y_i. This is because $\hat{h}_i^2 = R_i^2 = \hat{\lambda}_{i1}^2 + \hat{\lambda}_{i2}^2 + \cdots + \hat{\lambda}_{im}^2$ and R_i^2 will be small due to y_i having little in common with the other $p - 1$ variables.

For $\mathbf{S} - \hat{\mathbf{\Psi}}$, an initial estimate of communality analogous to (13.36) is

$$\hat{h}_i^2 = s_{ii} - \frac{1}{s^{ii}}, \tag{13.37}$$

where s_{ii} is the ith diagonal element of $\mathbf{S}$ and s^{ii} is the ith diagonal element of $\mathbf{S}^{-1}$. It can be shown that (13.37) is equivalent to

$$\hat{h}_i^2 = s_{ii} - \frac{1}{s^{ii}} = s_{ii} R_i^2, \tag{13.38}$$

which is a reasonable estimate of the amount of variance that y_i has in common with the other y's.

To use (13.36) or (13.37), $\mathbf{R}$ or $\mathbf{S}$ must be nonsingular. If $\mathbf{R}$ is singular we can use the absolute value or the square of the largest correlation in the ith row of $\mathbf{R}$ as an estimate of communality.

After obtaining communality estimates, we calculate eigenvalues and eigenvectors of $\mathbf{S}-\hat{\mathbf{\Psi}}$ or $\mathbf{R}-\hat{\mathbf{\Psi}}$ and use (13.23) to obtain estimates of factor loadings, $\hat{\mathbf{\Lambda}}$. Then the columns and rows of $\hat{\mathbf{\Lambda}}$ can be used to obtain new eigenvalues (variance explained) and communalities, respectively. The sum of squares of the jth column of $\hat{\mathbf{\Lambda}}$ is the jth eigenvalue of $\mathbf{S}-\hat{\mathbf{\Psi}}$ or $\mathbf{R}-\hat{\mathbf{\Psi}}$, and the sum of squares of the ith row of $\hat{\mathbf{\Lambda}}$ is the communality of y_i. The proportion of variance explained by the jth factor is

$$\frac{\theta_j}{\mathrm{tr}\,(\mathbf{S}-\hat{\mathbf{\Psi}})} = \frac{\theta_j}{\sum_{i=1}^{p}\theta_i}$$

or

$$\frac{\theta_j}{\mathrm{tr}\,(\mathbf{R}-\hat{\mathbf{\Psi}})} = \frac{\theta_j}{\sum_{i=1}^{p}\theta_i},$$

where θ_j is the jth eigenvalue of $\mathbf{S}-\hat{\mathbf{\Psi}}$ or $\mathbf{R}-\hat{\mathbf{\Psi}}$. The matrices $\mathbf{S}-\hat{\mathbf{\Psi}}$ and $\mathbf{R}-\hat{\mathbf{\Psi}}$ will often have some negative eigenvalues. In such a case, the cumulative proportion of variance will exceed 1 and then decline to 1 as the negative eigenvalues are added.

Example 13.3.2. To illustrate the principal factor method, we use the perception data from Table 13.1. The correlation matrix as given in Example 13.3.1 is singular. Hence in place of multiple correlations as communality estimates, we use the (absolute value of) the largest correlation in each row of $\mathbf{R}$. (The multiple correlation of y with several variables is greater than the simple correlation of y with any of the individual variables.) The diagonal elements of $\mathbf{R}-\hat{\mathbf{\Psi}}$ as given by (13.35) are therefore .995, .837, .881, .995, and .837. The eigenvalues of $\mathbf{R}-\hat{\mathbf{\Psi}}$ are 3.202, 1.395, .030, −.0002, and −.080, whose sum is 4.546. The first two eigenvectors of $\mathbf{R}-\hat{\mathbf{\Psi}}$ are

$$\mathbf{c}_1 = \begin{bmatrix} .548 \\ .272 \\ .431 \\ .549 \\ .373 \end{bmatrix} \quad \text{and} \quad \mathbf{c}_2 = \begin{bmatrix} -.178 \\ .656 \\ -.460 \\ -.159 \\ .549 \end{bmatrix}.$$

When these are multiplied by the square roots of the respective eigenvalues,

Table 13.3 Loadings Obtained by Two Different Methods for Perception Data of Table 13.1

	Principal Component Loadings		Principal Factor Loadings		
Variables	f_1	f_2	f_1	f_2	Communalities
Kind	.969	−.231	.981	−.210	.995
Intelligent	.519	.807	.487	.774	.837
Happy	.785	−.587	.771	−.544	.881
Likeable	.971	−.210	.982	−.188	.995
Just	.704	.667	.667	.648	.837
Variance accounted for	3.263	1.538	3.202	1.395	
Proportion of total variance	.653	.308	.704	.307	
Cumulative proportion	.653	.960	.704	1.01	

we obtain the principal factor loadings. In Table 13.3, these are compared with the loadings obtained by the principal component method in Example 13.3.1. The two sets of loadings are very similar, as we would have expected because of the large size of the communalities. The communalities are for the principal factor loadings, as noted above. The proportion of variance in each case for the principal factor loadings is obtained by dividing the variance accounted for (eigenvalue) by the sum of the eigenvalues, 4.546; for example, 3.202/4.546 = .704.

13.3.3 Iterated Principal Factor Method

The principal factor method can easily be iterated to improve the estimates of communality. After obtaining $\hat{\mathbf{\Lambda}}$ from (13.32) or (13.33) using initial communality estimates, we can obtain new communality estimates from the loadings in $\hat{\mathbf{\Lambda}}$ using (13.27),

$$\hat{h}_i^2 = \sum_{j=1}^{m} \hat{\lambda}_{ij}^2.$$

These values of $\hat{h}_i^2$ are substituted into the diagonal of $\mathbf{S} - \hat{\mathbf{\Psi}}$ or $\mathbf{R} - \hat{\mathbf{\Psi}}$, from which we obtain a new value of $\hat{\mathbf{\Lambda}}$. This process is continued until the communality estimates converge. (For some data sets, the iterative procedure does not converge.) Then the eigenvalues and eigenvectors of the final version of $\mathbf{S} - \hat{\mathbf{\Psi}}$ or $\mathbf{R} - \hat{\mathbf{\Psi}}$ are used in (13.23) to obtain the loadings.

The principal factor method and iterated principal factor method will typically yield results very close to those from the principal component method when *either* of the following is true.

1. The correlations are fairly large, with a resulting small value of m.
2. The number of variables, p, is large.

A shortcoming of the iterative approach is that frequently it leads to a communality estimate exceeding 1 (when factoring **R**). If $\hat{h}_i^2 > 1$, then $\hat{\psi}_i < 0$ by (13.25) and (13.27). Such a result is known as a *Heywood case* (Heywood 1931) and is clearly improper, since we cannot have a negative specific variance. Thus if a communality exceeds 1, the iterative process should stop, with the program reporting that a solution cannot be reached. Some software programs have an option of continuing the iterative procedure by setting the communality equal to 1 in all subsequent iterations. The resulting solution is somewhat questionable because it implies exact dependence of a variable on the factors, a possible but unlikely outcome.

Example 13.3.3. We illustrate the iterated principal factor method using the seishu data in Table 7.1. The data were introduced in Example 7.4.2. The correlation matrix is given below:

$$\mathbf{R} = \begin{bmatrix} 1.00 & .56 & .22 & .10 & .20 & -.04 & .13 & .03 & -.07 & .09 \\ .56 & 1.00 & -.09 & .13 & .20 & -.17 & .17 & .24 & .16 & -.06 \\ .22 & -.09 & 1.00 & .16 & .70 & -.31 & -.45 & -.34 & -.11 & .68 \\ .10 & .13 & .16 & 1.00 & .49 & -.03 & -.16 & .01 & .42 & .37 \\ .20 & .20 & .70 & .49 & 1.00 & -.32 & -.34 & -.19 & .30 & .87 \\ -.04 & -.17 & -.31 & -.03 & -.32 & 1.00 & -.42 & -.57 & -.11 & -.26 \\ .13 & .17 & -.45 & -.16 & -.34 & -.42 & 1.00 & .82 & .23 & -.30 \\ .03 & .24 & -.34 & .01 & -.19 & -.57 & .82 & 1.00 & .45 & -.17 \\ -.07 & .16 & -.11 & .42 & .30 & -.11 & .23 & .45 & 1.00 & .29 \\ .09 & .06 & .68 & .37 & .87 & -.26 & -.30 & -.17 & .29 & 1.00 \end{bmatrix}.$$

The eigenvalues of **R** are 3.17, 2.56, 1.43, 1.28, .54, .47, .25, .12, .10, and .06. There is a notable gap between 1.28 and .54, and we therefore extract four factors (see Section 13.4). The first four eigenvalues account for a proportion

$$\frac{3.17 + 2.56 + 1.43 + 1.28}{10} = .84$$

of tr(**R**).

For initial communality estimates, we use the squared multiple correlation between each variable and the other nine variables. These are given in Table

Table 13.4 Iterated Principal Factor Loadings and Communalities for Seishu Data

Variable	Loadings f_1	f_2	f_3	f_4	Initial Communalities	Final Communalities
Taste	.22	.31	.92	.12	.57	1.00
Odor	.07	.40	.43	−.20	.54	.38
pH	.80	.04	.05	−.40	.78	.79
Acidity 1	.41	.22	−.11	.37	.40	.36
Acidity 2	.14	.28	−.07	.05	.88	.98
Sake-meter	−.13	−.67	.10	.56	.77	.79
Reducing sugar	−.55	.66	.03	−.11	.79	.75
Total sugar	−.45	.88	−.14	−.07	.87	.99
Alcohol	.13	.54	−.37	.54	.66	.74
Formyl-nitrogen	.84	.21	−.17	−.02	.80	.78
Variance accounted for	3.00	2.37	1.25	.96	7.57	7.57

13.4, along with the final communalities after iteration. We multiply the first four eigenvectors of the final iterated version of $\mathbf{R} - \hat{\boldsymbol{\Psi}}$ by the square roots of the respective eigenvalues, as in (13.23) or (13.24), to obtain the factor loadings given in Table 13.4. We will not attempt to interpret the factors until after they have been rotated in Section 13.5.2.

13.3.4 Maximum Likelihood Method

If we assume that the observations $\mathbf{y}_1, \mathbf{y}_2, \ldots, \mathbf{y}_n$ are from $N_p(\boldsymbol{\mu}, \boldsymbol{\Sigma})$, then $\boldsymbol{\Lambda}$ and $\boldsymbol{\Psi}$ can be estimated by the method of maximum likelihood. It can be shown that the estimates $\hat{\boldsymbol{\Lambda}}$ and $\hat{\boldsymbol{\Psi}}$ satisfy the following:

$$\mathbf{S}\hat{\boldsymbol{\Psi}}\hat{\boldsymbol{\Lambda}} = \hat{\boldsymbol{\Lambda}}(\mathbf{I} + \hat{\boldsymbol{\Lambda}}'\hat{\boldsymbol{\Psi}}^{-1}\hat{\boldsymbol{\Lambda}}) \tag{13.39}$$

$$\hat{\boldsymbol{\Psi}} = \operatorname{diag}(\mathbf{S} - \hat{\boldsymbol{\Lambda}}\hat{\boldsymbol{\Lambda}}') \tag{13.40}$$

$$\hat{\boldsymbol{\Lambda}}'\hat{\boldsymbol{\Psi}}^{-1}\hat{\boldsymbol{\Lambda}} \quad \text{is diagonal} \tag{13.41}$$

These equations must be solved iteratively; and in practice the procedure may fail to converge or may yield a Heywood case (Section 13.3.3).

We note that the proportion of variance accounted for by the factors, as given by (13.30) or (13.31), will not necessarily be in descending order for maximum likelihood factors, as it is for factors obtained from the principal component or principal factor method.

Table 13.5 Maximum Likelihood Loadings and Communalities for Seishu Data

Variables	Loadings				Communalities
	f_1	f_2	f_3	f_4	
Taste	1.00	0	0	0	1.00
Odor	.45	−.05	.22	.19	.29
pH	.22	.68	−.20	−.40	.71
Acidity 1	.10	.47	.10	.37	.38
Acidity 2	.20	.98	.02	.00	1.00
Sake-meter	−.04	−.31	−.68	.55	.86
Reducing sugar	.13	−.39	.76	−.02	.75
Total sugar	.03	−.22	.96	.02	.98
Alcohol	−.07	.31	.52	.60	.72
Formyl-nitrogen	.02	.79	−.05	−.10	.63
Variance accounted for	1.33	2.66	2.34	1.00	7.32

Example 13.3.4. We illustrate the maximum likelihood method with the seishu data of Table 7.1. The correlation matrix and its eigenvalues were given in Example 13.3.3. We extract four factors, as in Example 13.3.3. The iterative solution of (13.39), (13.40), and (13.41) yielded the loadings and communalities given in Table 13.5.

The pattern of the loadings is different from that obtained using the iterated principal factor method in Example 13.3.3, but we will not compare them until after rotation in Section 13.5.2. Note that the four values of variance accounted for are not in descending order.

13.4 CHOOSING THE NUMBER OF FACTORS, m

Several criteria have been proposed for choosing m, the number of factors. Some of these methods are similar to those given in Section 12.6 for choosing the number of principal components to retain. We list four criteria, followed by a discussion of each.

1. Choose m equal to the number of factors necessary for the variance accounted for to achieve a predetermined percentage, say 80%, of the total variance tr(**S**) or tr(**R**).
2. Choose m equal to the number of eigenvalues greater than the average eigenvalue. For **R** the average is 1; for **S** it is $\sum_{i=1}^{p} \theta_i/p$.
3. Use the scree test based on a plot of the eigenvalues of **S** or **R**. If the graph drops sharply, followed by a straight line with much smaller slope, choose m equal to the number of eigenvalues before the straight line begins.

4. Test the hypothesis that m is the correct number of factors, namely, $H_0: \boldsymbol{\Sigma} = \boldsymbol{\Lambda}\boldsymbol{\Lambda}' + \boldsymbol{\Psi}$, where $\boldsymbol{\Lambda}$ is $p \times m$.

Method 1 applies particularly to the principal component method. By (13.30), the proportion of total sample variance (variance accounted for) due to the jth factor from $\mathbf{S}$ is $(\hat{\lambda}_{1j}^2 + \hat{\lambda}_{2j}^2 + \cdots + \hat{\lambda}_{pj}^2)/\text{tr}(\mathbf{S})$. The corresponding proportion from $\mathbf{R}$ is $(\hat{\lambda}_{1j}^2 + \hat{\lambda}_{2j}^2 + \cdots + \hat{\lambda}_{pj}^2)/p$, as in (13.31). The contribution of all m factors to $\text{tr}(\mathbf{S})$ or p is therefore $\sum_{i=1}^p \sum_{j=1}^m \hat{\lambda}_{ij}^2$, which is the sum of squares of all elements of $\hat{\boldsymbol{\Lambda}}$. For the principal component method, we see by (13.27) and (13.28) that this sum is also equal to the sum of the first m eigenvalues or to the sum of all p communalities:

$$\sum_{i=1}^{p} \sum_{j=1}^{m} \hat{\lambda}_{ij}^2 = \sum_{i=1}^{p} \hat{h}_i^2 = \sum_{j=1}^{m} \theta_j. \tag{13.42}$$

Thus we choose m sufficiently large so that the sum of the communalities or the sum of the eigenvalues (variance accounted for) constitutes a relatively large portion of tr($\mathbf{S}$).

Method 1 can be extended to the principal factor method, where prior estimates of communalities are used to form $\mathbf{S} - \hat{\boldsymbol{\Psi}}$ or $\mathbf{R} - \hat{\boldsymbol{\Psi}}$. However, $\mathbf{S} - \hat{\boldsymbol{\Psi}}$ or $\mathbf{R} - \hat{\boldsymbol{\Psi}}$ will usually have some negative eigenvalues. Therefore, as values of m range from 1 to p, the cumulative proportion of eigenvalues, $\sum_{j=1}^m \theta_j / \sum_{j=1}^p \theta_j$, will exceed 1.0 and then reduce to 1.0 as the negative eigenvalues are added. Hence a percentage such as 80% will be reached for a lower value of m than would be the case for $\mathbf{S}$ or $\mathbf{R}$, and a better strategy would be to choose m equal to the value for which the percentage first exceeds 100%.

In the iterated principal factor method, m is specified before iteration, and $\sum_i \hat{h}_i^2$ is obtained after iteration as $\sum_i \hat{h}_i^2 = \text{tr}(\mathbf{S} - \hat{\boldsymbol{\Psi}})$. Thus before iterating, one would choose m based on a priori considerations or on the eigenvalues of $\mathbf{S}$ or $\mathbf{R}$, as in the principal component method.

Although method 2 is heuristically based, it often works well in practice. This method is a popular criterion of long standing and is the default in many software packages. A variation to method 2 that has been suggested for use with $\mathbf{R} - \hat{\boldsymbol{\Psi}}$ is to let m equal the number of positive eigenvalues. (There will usually be some negative eigenvalues of $\mathbf{R} - \hat{\boldsymbol{\Psi}}$.) However, this criterion will typically result in too many factors, since the sum of the positive eigenvalues will exceed the sum of the communalities.

The scree test in method 3 was named after the geological term *scree*, referring to the debris at the bottom of a rocky cliff. It also performs well in practice.

In method 4 we wish to test

$$H_0: \boldsymbol{\Sigma} = \boldsymbol{\Lambda}\boldsymbol{\Lambda}' + \boldsymbol{\Psi} \quad \text{vs.} \quad H_1: \boldsymbol{\Sigma} \neq \boldsymbol{\Lambda}\boldsymbol{\Lambda}' + \boldsymbol{\Psi},$$

where $\mathbf{\Lambda}$ is $p \times m$. The test statistic, a function of the likelihood ratio, is

$$\left(n - \frac{2p + 4m + 11}{6}\right) \ln\left(\frac{|\hat{\mathbf{\Lambda}}\hat{\mathbf{\Lambda}}' + \hat{\mathbf{\Psi}}|}{|\mathbf{S}|}\right), \qquad (13.43)$$

which is approximately χ^2_ν when H_0 is true, where $\nu = \frac{1}{2}[(p-m)^2 - p - m]$ and $\hat{\mathbf{\Lambda}}$ and $\hat{\mathbf{\Psi}}$ are the maximum likelihood estimators. If H_0 is rejected, then m is too small and more factors are needed.

In practice, when n is large, the test in method 4 often shows more factors to be significant than do the other three methods. We may therefore consider the value of m indicated by the test to be an upper bound on the number of factors with practical importance.

For many data sets, the choice of m will not be obvious. This indeterminacy leaves many statisticians skeptical as to the validity of factor analysis. A researcher may begin with one of the methods (say method 2) for an initial choice of m, will inspect the resulting percent of tr($\mathbf{R}$) or tr($\mathbf{S}$), and then examine the rotated loadings for interpretability. If the percent of variance or interpretation does not seem satisfactory, the experimenter will try other values of m in a search for an acceptable compromise between percent of tr($\mathbf{R}$) and interpretability of the factors. Admittedly, this is a subjective procedure, and for such data sets one could well question the outcome (see Section 13.7).

When a data set is successfully fitted by a factor analysis model, the first three methods will almost always give the same value of m, and there will be little question of what this value should be. Thus for a "good" data set, the entire procedure becomes much more objective.

Example 13.4(a). We compare the four methods of choosing m for the perception data used in Examples 13.3.1 and 13.3.2.

Method 1 gives $m = 2$, because one eigenvalue accounts for 65% of tr($\mathbf{R}$), while two eigenvalues account for 96%.

Method 2 gives $m = 2$, since $\lambda_2 = 1.54$ and $\lambda_3 = .17$.

For method 3, we examine the scree plot in Figure 13.1. It is clear that $m = 2$ is indicated.

Method 4 is not available for the perception data because $\mathbf{R}$ is singular (fifth eigenvalue is zero), and the test involves $|\mathbf{R}|$.

Hence for the perception data, all three available methods agree on $m = 2$.

Example 13.4(b). We compare the four methods of choosing m for the seishu data used in Examples 13.3.3 and 13.3.4.

Method 1 gives $m = 4$ for the principal component method, because four eigenvalues of $\mathbf{R}$ account for 82% of tr($\mathbf{R}$). For the principal factor method with initial communality estimates R_i^2, the eigenvalues of $\mathbf{R} - \hat{\mathbf{\Psi}}$ and corresponding proportions are as follows:

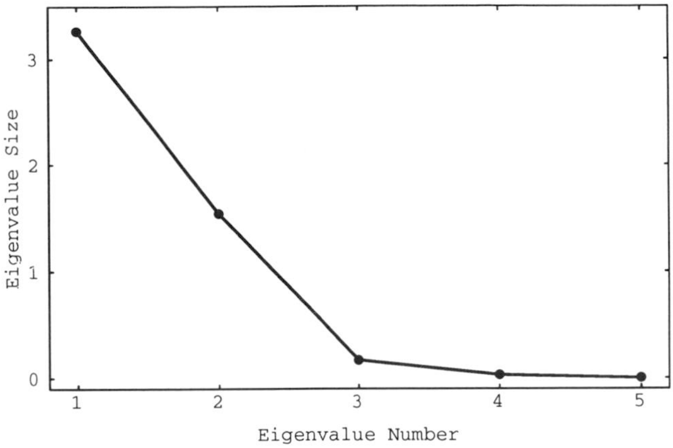

Figure 13.1 Scree graph for the perception data.

Eigenvalues	2.86	2.17	.94	.88	.12	.08	.01	−.06	−.13	−.22
Proportions	.43	.33	.14	.16	.02	.01	.00	−.01	−.02	−.03
Cumulative proportions	.43	.76	.90	1.03	1.05	1.06	1.06	1.06	1.03	1.00

The proportions are obtained by dividing the eigenvalues by their sum, 6.63. Thus the cumulative proportion first exceeds 1.00 for $m = 4$.

Method 2 gives $m = 4$, since $\lambda_4 = 1.31$ and $\lambda_5 = .61$, where λ_4 and λ_5 are eigenvalues of **R**.

For method 3, we examine the scree plot in Figure 13.2. There is clearly a discernible bend in slope at the fifth eigenvalue.

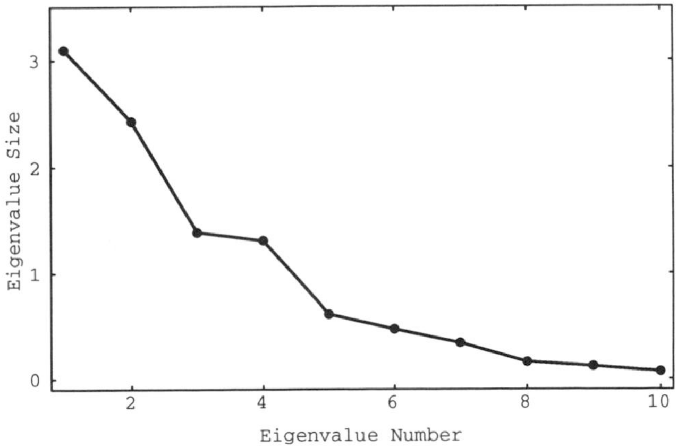

Figure 13.2 Scree graph for the seishu data.

For method 4, we use $m = 4$ in the significance test in (13.43) and obtain $\chi^2 = 9.039$, with degrees of freedom

$$\nu = \tfrac{1}{2}[(p - m)^2 - p - m] = \tfrac{1}{2}[(10 - 4)^2 - 10 - 4] = 11.$$

Since $9.039 < \chi^2_{.05,11} = 19.68$, we do not reject the hypothesis that four factors are adequate.

Thus for the seishu data, all four methods agree on $m = 4$.

13.5 ROTATION

13.5.1 Introduction

As noted in Section 13.2.2, the factor loadings (rows of $\mathbf{\Lambda}$) in the population model are unique only up to multiplication by an orthogonal matrix that rotates the loadings. The rotated loadings preserve the essential properties of the original loadings, in that they reproduce the covariance matrix and satisfy all basic assumptions. The estimated loading matrix $\hat{\mathbf{\Lambda}}$ can likewise be rotated to obtain $\hat{\mathbf{\Lambda}}^* = \hat{\mathbf{\Lambda}}\mathbf{T}$, where $\mathbf{T}$ is orthogonal. Since $\mathbf{TT}' = \mathbf{I}$ by (2.93), the rotated loadings provide the same estimate of the covariance matrix as before:

$$\mathbf{S} \cong \hat{\mathbf{\Lambda}}^*\hat{\mathbf{\Lambda}}^{*\prime} + \hat{\mathbf{\Psi}} = \hat{\mathbf{\Lambda}}\mathbf{TT}'\hat{\mathbf{\Lambda}}' + \hat{\mathbf{\Psi}} = \hat{\mathbf{\Lambda}}\hat{\mathbf{\Lambda}}' + \hat{\mathbf{\Psi}}. \tag{13.44}$$

Geometrically, the loadings in the ith row of $\hat{\mathbf{\Lambda}}$ constitute the coordinates of a point in the loading space corresponding to y_i. Rotation of the p points gives their coordinates with respect to new axes (factors) but otherwise leaves their basic geometric configuration intact. We hope to find a new frame of reference in which the factors are more interpretable. To this end, the goal of rotation is to place the axes close to as many points as possible. If there are clusters of points (corresponding to groupings of y's), we seek to move the axes so as to pass through or near these clusters. This would associate each group of variables with a factor (axis) and make interpretation more objective.

If we can achieve a rotation in which every point is close to an axis, then each variable loads highly on the factor corresponding to the axis and has small loadings on the remaining factors. If each variable loads highly on only one factor, there is no ambiguity. Such a happy state of affairs is called *simple structure*, and interpretation is greatly simplified. We merely observe which variables are associated with each factor, and the factor is defined or named accordingly.

The illustration in Section 13.1 showed a correlation matrix manifesting two clearly discernible groups of variables. The rotation to an interpretable pattern for the loadings (where the variables load highly on only one factor) would associate each of these two groups with a factor. Thus with rotation we are trying to uncover the natural groupings in the variables. Geometrically, these groups of variables plot as clusters of points in the loading space, and we attempt to move the axes as close to these clusters of points as possible. The resulting axes then represent the natural factors.

The number of factors on which a variable has moderate or high loadings is called the *complexity* of the variable. In the ideal situation referred to above as simple structure, the variables all have a complexity of 1.

Two basic types of rotation are used, *orthogonal* and *oblique*. The rotation discussed above in (13.44) is orthogonal; the original perpendicular axes are rotated rigidly and remain perpendicular. In an orthogonal rotation, angles and distances are preserved, communalities are unchanged, and the basic configuration of the points is the same. Only the reference axes differ. In an oblique "rotation" (transformation), the axes are not required to remain perpendicular and are thus free to pass closer to clusters of points.

In the next two sections, we discuss orthogonal and oblique rotations, followed by some guidelines for interpretation in Section 13.5.4.

13.5.2 Orthogonal Rotation

It was noted above that orthogonal rotations preserve communalities. However, since the loadings change, the variance accounted for by each factor as given in (13.29) will change, as will the corresponding proportion in (13.30) or (13.31). The proportions due to the rotated loadings will not necessarily be in descending order.

13.5.2a Graphical Approach

If there are only two factors ($m = 2$), we can use a *graphical* rotation based on a visual inspection of a plot of factor loadings. In this case, the rows of $\hat{\mathbf{\Lambda}}$ are pairs of loadings, $(\hat{\lambda}_{i1}, \hat{\lambda}_{i2})$, $i = 1, 2, \ldots, p$, corresponding to $y_1, y_2, \ldots, y_p$. We choose an angle ϕ through which the axes can be rotated to move them closer to more points. The new rotated loadings $(\hat{\lambda}^*_{i1}, \hat{\lambda}^*_{i2})$ can be measured directly on the graph or calculated from $\hat{\mathbf{\Lambda}}^* = \hat{\mathbf{\Lambda}}\mathbf{T}$, where, for $m = 2$, a convenient representation of an orthogonal matrix in terms of the angle of rotation is

$$\mathbf{T} = \begin{pmatrix} \cos\phi & -\sin\phi \\ \sin\phi & \cos\phi \end{pmatrix}. \tag{13.45}$$

Example 13.5.2a. In Example 13.3.1, the initial factor loadings for the perception data did not provide an interpretation consistent with the two groupings of variables apparent in the pattern of correlations in **R**. The loading pairs $(\hat{\lambda}_{i1}, \hat{\lambda}_{i2})$ corresponding to the five variables are plotted in Figure 13.3. An orthogonal rotation through $-35°$ would bring the axes (factors) closer to the two clusters of points (variables) identified in Example 13.3.1. With the rotation, each cluster of variables corresponds much more closely to a factor. Using

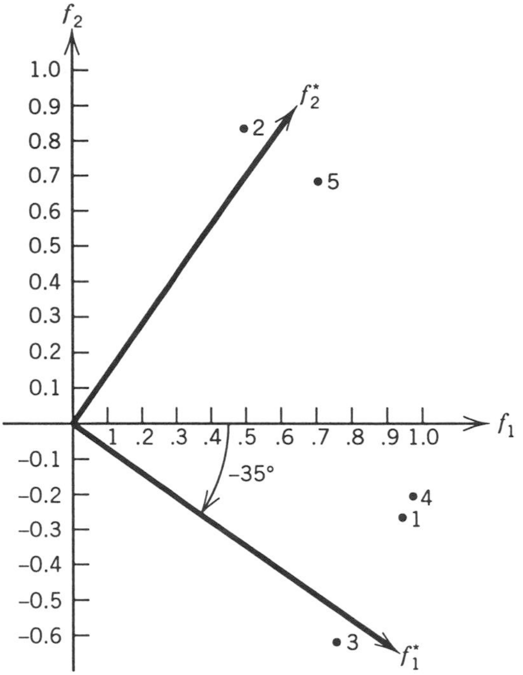

Figure 13.3 Plot of the two loadings for each of the five variables in the perception data of Table 13.1.

$\hat{\mathbf{\Lambda}}$ from Example 13.3.1 and $-35°$ in $\mathbf{T}$ as given in (13.45), we obtain the following rotated loadings:

$$\hat{\mathbf{\Lambda}}^* = \hat{\mathbf{\Lambda}}\mathbf{T} = \begin{bmatrix} .969 & -.231 \\ .519 & .807 \\ .785 & -.587 \\ .971 & -.210 \\ .704 & .667 \end{bmatrix} \begin{pmatrix} .819 & .574 \\ -.574 & .819 \end{pmatrix}$$
$$= \begin{bmatrix} .927 & .367 \\ -.037 & .959 \\ .980 & -.031 \\ .916 & .385 \\ .194 & .950 \end{bmatrix}.$$

In Table 13.6, we compare these rotated loadings with the original loadings.

The interpretation of the rotated loadings is clear. As indicated by the boldface loadings in Table 13.6, the first factor is associated with variables 1, 3, and 4: kind, happy, and likeable. The second factor is associated with the other two

Table 13.6 Graphically Rotated Loadings for Perception Data of Table 13.1

Variables	Principal Component Loadings		Graphically Rotated Loadings		Communalities, $\hat{h}_i^2$
	f_1	f_2	f_1	f_2	
Kind	.969	−.231	**.927**	.367	.993
Intelligent	.519	.807	−.037	**.959**	.921
Happy	.785	−.587	**.980**	−.031	.960
Likeable	.971	−.210	**.916**	.385	.987
Just	.704	.667	.194	**.950**	.940
Variance accounted for	3.263	1.538	2.696	2.106	4.802
Proportion of total variance	.653	.308	.539	.421	.960
Cumulative proportion	.653	.960	.539	.960	.960

variables: intelligent and just. This same grouping of variables was indicated by the pattern in the correlation matrix, and we would also arrive at the same interpretation by a visual examination of the two clusters of points in Figure 13.3. The first factor might be described as representing a person's perceived humanity or amiability, while the second involves more rational practices.

Note that if the rotated axes were allowed to have an angle between them of less than 90°, the lower one representing f_1^* could come closer to the points corresponding to variables 1 and 4 so that the coordinates on f_2^*, .367 and .385, could be reduced. However, the basic interpretation would not change; variables 1 and 4 would still be associated with f_1^*.

13.5.2b Varimax Rotation

The graphical approach to rotation is generally limited to $m = 2$. For $m > 2$, various analytical methods have been proposed. The most popular of these is the *varimax* technique, which seeks rotated loadings that maximize the variance of the squared loadings in each column of $\hat{\mathbf{\Lambda}}^*$. If the loadings in a column were nearly equal, the variance would be close to 0. As the squared loadings approach 0 and 1 (for factoring $\mathbf{R}$), the variance will increase. Thus the varimax method attempts to make the loadings either large or small to facilitate interpretation.

The varimax procedure cannot guarantee that all variables will load highly on only one factor. In fact, no procedure could possibly assure that every variable has a complexity of 1 in all situations. The configuration of the points in the loading space remains fixed; we merely rotate the axes to be as close to as many points as possible. In many cases, the points are not well clustered, and the axes simply cannot be rotated so as to be near all of them. This problem is compounded by having to choose m. If m is changed, the coordinates $(\hat{\lambda}_{i1}, \hat{\lambda}_{i2}, \ldots, \hat{\lambda}_{im})$ change, and the relative position of the points is altered.

The varimax rotation is available in virtually all factor analysis software programs. The output typically includes the rotated loading matrix $\hat{\mathbf{\Lambda}}^*$, the variance accounted for (sum of squares of each column of $\hat{\mathbf{\Lambda}}^*$), the communalities (sum of squares of each row of $\hat{\mathbf{\Lambda}}^*$), and the orthogonal matrix $\mathbf{T}$ used to obtain $\hat{\mathbf{\Lambda}}^* = \hat{\mathbf{\Lambda}}\mathbf{T}$.

Example 13.5.2b(a). In Example 13.5.2a, a graphical rotation was devised visually to achieve interpretable loadings for the Perception data. As we would expect, the varimax method yields a similar result. The varimax rotated loadings are given in Table 13.7. For comparison, we have included the original unrotated loadings from Table 13.3 and the graphically rotated loadings from Table 13.6.

The orthogonal matrix $\mathbf{T}$ for the varimax rotation is

$$\mathbf{T} = \begin{pmatrix} .859 & .512 \\ -.512 & .859 \end{pmatrix}.$$

By (13.45), $-\sin\phi = .512$, and the angle of rotation is given by $\phi = -\sin^{-1}(.512) = -30.8°$. Thus the varimax rotation chose an angle of rotation of $-30.8°$ as compared to the $-35°$ we selected visually, but the results are very close and the interpretation is exactly the same.

Table 13.7 Varimax Rotated Factor Loadings for Perception Data of Table 13.1

	Principal Component Loadings		Graphically Rotated Loadings		Varimax Rotated Loadings		Communalities
Variables	f_1	f_2	f_1	f_2	f_1	f_2	$\hat{h}_i^2$
Kind	.969	−.231	**.927**	.367	**.951**	.298	.993
Intelligent	.519	.807	−.037	**.959**	.033	**.959**	.921
Happy	.785	−.587	**.980**	−.031	**.975**	−.103	.960
Likeable	.971	−.210	**.916**	.385	**.941**	.317	.987
Just	.704	.667	.194	**.950**	.263	**.933**	.940
Variance accounted for	3.263	1.538	2.696	2.106	2.811	1.991	4.802
Proportion of total variance	.653	.308	.539	.421	.562	.398	.960
Cumulative proportion	.653	.960	.539	.960	.562	.960	.960

Table 13.8 Varimax Rotated Loadings for Seishu Data

	Iterated Principal Factor Rotated Loadings				Maximum Likelihood Rotated Loadings			
Variables	f_1	f_2	f_3	f_4	f_1	f_2	f_3	f_4
Taste	.16	−.01	**.99**	−.09	.16	−.00	**.98**	−.10
Odor	−.11	.14	**.48**	.14	−.07	.14	**.49**	.17
pH	**.88**	−.12	.02	−.13	**.82**	−.10	.08	−.15
Acidity 1	.26	−.09	.09	**.54**	.29	−.08	.11	**.53**
Acidity 2	**.89**	−.06	.10	.43	**.91**	−.06	.10	.39
Sake-meter	−.43	**−.76**	.01	.07	−.46	**−.80**	.04	.10
Reducing sugar	−.37	**.76**	.18	.03	−.37	**.75**	.20	.08
Total sugar	−.26	**.92**	.10	.25	−.27	**.91**	.11	.26
Alcohol	−.01	.25	.00	**.80**	−.00	.25	.01	**.81**
Formyl-nitrogen	**.74**	−.07	−.08	.20	**.76**	−.07	−.08	.22
Variance accounted for	2.62	2.12	1.27	1.27	2.61	2.14	1.29	1.28

Example 13.5.2b(b). In Examples 13.3.3 and 13.3.4, we obtained the iterated principal factor loadings and maximum likelihood loadings for the seishu data. In Table 13.8, we show the varimax rotation of these two sets of loadings.

The rotation in each case has achieved a satisfactory simple structure and most variables show a complexity of 1. The boldface loadings indicate the variables associated with each factor for interpretation purposes. These may be meaningful to the researcher. For example, factor 2 is associated with sake-meter, reducing sugar, and total sugar, while factor 3 is aligned with taste and odor.

The similarities in the two sets of rotated loadings are striking. The interpretation in each case is the same. The variances accounted for are virtually identical.

13.5.3 Oblique Rotation

The term *oblique rotation* refers to a transformation in which the axes do not remain perpendicular. Technically, the term oblique rotation is a misnomer, since rotation implies an orthogonal transformation that preserves distances. Thus we should speak of *oblique transformation*, but the term oblique rotation is well established in the literature.

Instead of the orthogonal transformation matrix $\mathbf{T}$ used in (13.15), (13.16), and (13.17), an oblique rotation uses a general nonsingular transformation matrix $\mathbf{Q}$ so that $\mathbf{f}^* = \mathbf{Q}'\mathbf{f}$, and by (3.69), $\text{cov}(\mathbf{f}^*) = \mathbf{Q}'\mathbf{I}\mathbf{Q} = \mathbf{Q}'\mathbf{Q} \neq \mathbf{I}$. Thus the new factors are correlated. Since distances and angles are not preserved, the communalities are changed. Some program packages report communalities obtained from the original loadings, rather than the oblique loadings.

When the axes are not required to be perpendicular, they can more easily pass through the major clusters of points in the loading space (assuming there are clusters). For example, in Figure 13.4, we have plotted the varimax rotated loadings for two factors extracted from the sons data of Table 3.9 (see Example 13.5.3 at the end of this section). Oblique axes with an angle between them of 38° would pass much closer to the points, and the resulting loadings would be very close to 0 and 1. However, the interpretation would not change, since the same points (variables) would be associated with the oblique axes as with the orthogonal axes.

Various analytical methods of achieving oblique rotations have been proposed and are available in program packages. Typically, the output of one of these procedures includes a *pattern matrix*, a *structure matrix*, and a matrix of correlations among the oblique factors. For interpretation, we would usually prefer to use the pattern matrix rather than the structure matrix. The loadings in a row of the pattern matrix are the natural coordinates of the point (variable) on the oblique axes and serve as coefficients in the model relating the variable to the factors.

One use for an oblique rotation is to check on the orthogonality of the factors. The orthogonality in the original factors is imposed by the model and maintained by an orthogonal rotation. If an oblique rotation produces a correlation matrix that is nearly diagonal, we can be more confident that the factors are indeed orthogonal.

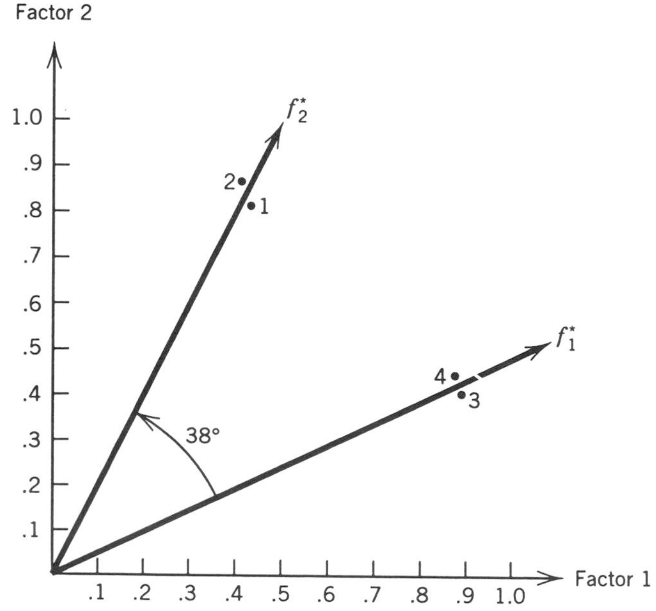

Figure 13.4 Orthogonal and oblique rotations for the sons data.

Example 13.5.3. The correlation matrix for the sons data of Table 3.9 is given below:

$$\mathbf{R} = \begin{bmatrix} 1.000 & .735 & .711 & .704 \\ .735 & 1.000 & .693 & .709 \\ .711 & .693 & 1.000 & .839 \\ .704 & .709 & .839 & 1.000 \end{bmatrix}.$$

The varimax rotated loadings for two factors obtained by the principal component method are given in Table 13.9 and plotted in Figure 13.4. An analytical oblique rotation (Harris–Kaiser orthoblique method in SAS) produced oblique axes with an angle between them of 38°, the same as obtained by a graphical approach. The correlation between the two factors is .79 (obtained from $\mathbf{Q}'\mathbf{Q}$), which is related to the angle by (3.14), $.79 = \cos 38°$. The pattern loadings are given in Table 13.9.

The oblique loadings give a much cleaner simple structure than the varimax loadings, but the interpretation is essentially the same if we neglect loadings below .45 on the varimax rotation.

It is also evident from Figure 13.4 that a single factor would be adequate since the loadings are closer to the 45° line than to the orthogonal axes. The suggestion to let $m = 1$ is also supported by the eigenvalues of $\mathbf{R}$: 3.20, .38, .27, and .16. The first accounts for 80%, the second for an additional 9%. The large correlation, .79, between the two oblique factors constitutes additional evidence that a single factor model would suffice here. In fact, the pattern in $\mathbf{R}$ itself indicates the presence of only one factor. The four variables form only one cluster, since all are highly correlated. There are no small correlations between groupings of variables.

13.5.4 Interpretation

In Sections 13.5.1, 13.5.2, and 13.5.3, we have discussed the usefulness of rotation as an aid to interpretation. Our goal is to achieve a simple structure in which each variable loads highly on only one factor, with small loadings on all other factors. In practice, we often fail to achieve this goal, but rotation usually produces loadings that are closer to the desired simple structure.

We now suggest general guidelines for interpreting the factors by examining the matrix of rotated factor loadings. Starting with the first variable and moving horizontally from left to right across the m loadings, identify the highest loading (in absolute value). If the highest loading is of a significant size (a subjective determination), circle or underline it. This is done for each of the p variables. There may be other significant loadings in a row besides the one circled. If these are considered, the interpretation is less simple. On the other hand, there may be variables with small communalities and no significant loading on any factor. In this case, the researcher may wish to increase the number of factors

Table 13.9 Varimax and Orthoblique Loadings for Sons Data

	Varimax Loadings		Orthoblique Pattern matrix	
Variable	f_1	f_2	f_1	f_2
1	.42	**.82**	.03	**.90**
2	.40	**.85**	−.03	**.96**
3	**.87**	.41	**.97**	−.01
4	**.86**	.43	**.95**	.01

and run the program again so that these variables might associate with a new factor.

To assess significance of factor loadings $\hat{\lambda}_{ij}$ obtained from **R**, a general threshold value of .3 has been advocated by many writers. For most successful applications, however, a critical value of .3 is too low and will result in variables of complexity greater than 1. A target value of .5 or .6 is more typical. The .3 criterion is loosely based on the critical value for significance of an ordinary correlation coefficient, r. However, the distribution of the sample loadings is not the same as that of r arising from the bivariate normal. In addition to distributional considerations, the critical value should be increased because mp values of $\hat{\lambda}_{ij}$ are being tested. On the other hand, the critical value might possibly be reduced for a large number of factors. Since $\hat{h}_i^2 = \sum_{j=1}^m \hat{\lambda}_{ij}^2$ is bounded by 1, an increase in m reduces the average squared loading in a row.

After identifying potentially significant loadings, the experimenter then attempts to discover some meaning in the factors and, ideally, to label or name them. This can readily be done if the group of variables associated with each factor makes sense to the researcher. But in many situations, the groupings are not so logical, and a revision can be tried, such as adjusting the size of loading deemed to be important, changing m, using a different method of estimating the loadings, or employing another type of rotation.

13.6 FACTOR SCORES

In many applications, the researcher wishes only to ascertain whether a factor analysis model fits the data and to identify the factors. In other applications, the experimenter wishes to obtain *factor scores*, which are defined as estimates of the underlying factor values for each observation, $\hat{\mathbf{f}}_i = (\hat{f}_{i1}, \hat{f}_{i2}, \ldots, \hat{f}_{im})'$, $i = 1, 2, \ldots, n$. There are two uses for such scores: (1) the behavior of the observations in terms of the factors may be of interest and (2) we may wish to use the factor scores as input to another analysis, such as MANOVA. The latter usage resembles a similar application of principal components.

Since the f's are not observed, we must estimate them as functions of the observed y's. The most popular approach to estimating the factors is based on regression (Thompson 1951). We will discuss this method and also briefly describe an informal technique that can be used when **R** (or **S**) is singular. For other approaches see Harman (1976, Chapter 16).

Since $E(f_i) = 0$, we relate the f's to the y's by a centered regression model

$$\begin{aligned}
f_1 &= \beta_{11}(y_1 - \bar{y}_1) + \beta_{12}(y_2 - \bar{y}_2) + \cdots + \beta_{1p}(y_p - \bar{y}_p) + \varepsilon_1 \\
f_2 &= \beta_{21}(y_1 - \bar{y}_1) + \beta_{22}(y_2 - \bar{y}_2) + \cdots + \beta_{2p}(y_p - \bar{y}_p) + \varepsilon_2 \\
&\vdots \\
f_m &= \beta_{m1}(y_1 - \bar{y}_1) + \beta_{m2}(y_2 - \bar{y}_2) + \cdots + \beta_{mp}(y_p - \bar{y}_p) + \varepsilon_m,
\end{aligned} \tag{13.46}$$

which can be written in matrix form as

$$\mathbf{f} = \mathbf{B}_1'(\mathbf{y} - \bar{\mathbf{y}}) + \boldsymbol{\varepsilon}. \tag{13.47}$$

We have used the notation $\boldsymbol{\varepsilon}$ to distinguish this error from $\boldsymbol{\epsilon}$ in the original factor model $\mathbf{y} - \boldsymbol{\mu} = \boldsymbol{\Lambda}\mathbf{f} + \boldsymbol{\epsilon}$ given in (13.3). Our approach is to estimate $\mathbf{B}_1$ and use the predicted value $\hat{\mathbf{f}} = \hat{\mathbf{B}}_1'(\mathbf{y} - \bar{\mathbf{y}})$ to estimate $\mathbf{f}$.

The model (13.47) holds for each observation:

$$\mathbf{f}_i = \mathbf{B}_1'(\mathbf{y}_i - \bar{\mathbf{y}}) + \boldsymbol{\varepsilon}_i \qquad i = 1, 2, \ldots, n.$$

In transposed form, the model becomes

$$\mathbf{f}_i' = (\mathbf{y}_i - \bar{\mathbf{y}})'\mathbf{B}_1 + \boldsymbol{\varepsilon}_i' \qquad i = 1, 2, \ldots, n,$$

and these n equations can be combined into a single model,

$$\begin{aligned}
\mathbf{F} = \begin{bmatrix} \mathbf{f}_1' \\ \mathbf{f}_2' \\ \vdots \\ \mathbf{f}_n' \end{bmatrix} &= \begin{bmatrix} (\mathbf{y}_1 - \bar{\mathbf{y}})'\mathbf{B}_1 \\ (\mathbf{y}_2 - \bar{\mathbf{y}})'\mathbf{B}_1 \\ \vdots \\ (\mathbf{y}_n - \bar{\mathbf{y}})'\mathbf{B}_1 \end{bmatrix} + \begin{bmatrix} \boldsymbol{\varepsilon}_1' \\ \boldsymbol{\varepsilon}_2' \\ \vdots \\ \boldsymbol{\varepsilon}_n' \end{bmatrix} \\
&= \begin{bmatrix} (\mathbf{y}_1 - \bar{\mathbf{y}})' \\ (\mathbf{y}_2 - \bar{\mathbf{y}})' \\ \vdots \\ (\mathbf{y}_n - \bar{\mathbf{y}})' \end{bmatrix} \mathbf{B}_1 + \boldsymbol{\Xi} \\
&= \mathbf{Y}_c\mathbf{B}_1 + \boldsymbol{\Xi}.
\end{aligned} \tag{13.48}$$

The model (13.48) has the appearance of a centered multivariate multiple regression model as in Section 10.4.5, with $\mathbf{Y}_c$ in place of $\mathbf{X}_c$. By (10.40), the estimate for $\mathbf{B}_1$ would be

$$\hat{\mathbf{B}}_1 = (\mathbf{Y}_c'\mathbf{Y}_c)^{-1}\mathbf{Y}_c'\mathbf{F}. \tag{13.49}$$

However, $\mathbf{F}$ is unobserved. To evaluate $\hat{\mathbf{B}}_1$ in spite of this, we first use (10.42) to rewrite (13.49) in terms of covariance matrices,

$$\hat{\mathbf{B}}_1 = \mathbf{S}_{yy}^{-1}\mathbf{S}_{yf}. \tag{13.50}$$

In the notation of the present chapter, $\mathbf{S}_{yy}$ is represented by $\mathbf{S}$; for $\mathbf{S}_{yf}$ we use $\hat{\mathbf{\Lambda}}$, since $\hat{\mathbf{\Lambda}}$ estimates $\text{cov}(\mathbf{y}, \mathbf{f}) = \mathbf{\Lambda}$, as given by (13.12). Thus we can write (13.50) as

$$\hat{\mathbf{B}}_1 = \mathbf{S}^{-1}\hat{\mathbf{\Lambda}}. \tag{13.51}$$

Then from model (13.48), the estimated (predicted) $\mathbf{f}_i$ values are given by

$$\hat{\mathbf{F}} = \begin{bmatrix} \hat{\mathbf{f}}_1' \\ \hat{\mathbf{f}}_2' \\ \vdots \\ \hat{\mathbf{f}}_n' \end{bmatrix} = \mathbf{Y}_c\hat{\mathbf{B}}_1$$

$$= \mathbf{Y}_c\mathbf{S}^{-1}\hat{\mathbf{\Lambda}}. \tag{13.52}$$

If $\mathbf{R}$ is factored instead of $\mathbf{S}$, (13.52) becomes

$$\hat{\mathbf{F}} = \mathbf{Y}_s\mathbf{R}^{-1}\hat{\mathbf{\Lambda}}, \tag{13.53}$$

where $\mathbf{Y}_s$ is the observed matrix of standardized variables, $(y_{ij} - \bar{y}_j)/s_j$.

We would ordinarily obtain factor scores for the rotated factors rather than the original factors. Thus $\hat{\mathbf{\Lambda}}$ in (13.52) or (13.53) would be replaced by $\hat{\mathbf{\Lambda}}^*$.

In order to obtain factor scores by the regression method, $\mathbf{S}$ or $\mathbf{R}$ must be nonsingular. When $\mathbf{R}$ (or $\mathbf{S}$) is singular, we can obtain factor scores by a simple method based directly on the rotated loadings. We cluster the variables into groups (factors) according to the loadings and find a score for each factor by averaging the variables associated with the factor. If the variables are not commensurate, the variables should be standardized before averaging. An alternative approach would be to weight the variables by their loadings when averaging.

Example 13.6. The speaking rate of four voices was artificially manipulated by means of a rate changer without altering the pitch (Brown, Strong, and Rencher 1973). There were five rates for each voice:

$$\begin{aligned} \text{FF} &= 45\% \text{ faster} \\ \text{F} &= 25\% \text{ faster} \\ \text{N} &= \text{normal rate} \\ \text{S} &= 22\% \text{ slower} \\ \text{SS} &= 42\% \text{ slower.} \end{aligned}$$

The resulting 20 voices were played to 30 judges who rated them on 15 paired-opposite adjectives with a 14-point scale between poles. The following adjectives were used: intelligent, ambitious, polite, active, confident, happy, just, likeable, kind, sincere, dependable, religious, good-looking, sociable, and strong. The results were averaged over the 30 judges to produce 20 observation vectors of 15 variables each. The averaging produced very reliable data so that even though there were only 20 observations on 15 variables, the factor analysis model fit very well. The correlation matrix is as follows:

$$\mathbf{R} = \begin{bmatrix} 1.00 & .90 & -.17 & .88 & .92 & .88 & .15 & .39 & -.02 & -.16 & .52 & -.15 & -.79 & -.78 & .73 \\ .90 & 1.00 & -.46 & .93 & .87 & .79 & -.16 & .10 & -.35 & -.42 & .25 & -.40 & .68 & -.60 & .62 \\ -.17 & -.46 & 1.00 & -.56 & -.13 & .07 & .85 & .75 & .88 & .91 & .68 & .88 & .21 & .31 & .25 \\ .88 & .93 & -.56 & 1.00 & .85 & .73 & -.25 & -.02 & -.45 & -.57 & .10 & -.53 & .58 & .84 & .50 \\ .92 & .87 & -.13 & .85 & 1.00 & .91 & .20 & .39 & -.09 & -.16 & .49 & -.10 & .85 & .80 & .81 \\ .88 & .79 & .07 & .73 & .91 & 1.00 & .27 & .53 & .12 & .06 & .66 & .08 & .90 & .85 & .78 \\ .15 & -.16 & .85 & -.25 & .20 & .27 & 1.00 & .85 & .81 & .79 & .79 & .81 & .43 & .54 & .53 \\ .39 & .10 & .75 & -.02 & .39 & .53 & .85 & 1.00 & .84 & .79 & .93 & .77 & .71 & .69 & .76 \\ -.02 & -.35 & .88 & -.45 & -.09 & .12 & .81 & .84 & 1.00 & .91 & .76 & .85 & .28 & .36 & .35 \\ -.16 & -.42 & .91 & -.57 & -.16 & .06 & .79 & .79 & .91 & 1.00 & .72 & .96 & .26 & .28 & .29 \\ .52 & .25 & .67 & .10 & .49 & .66 & .79 & .93 & .76 & .72 & 1.00 & .72 & .75 & .77 & .78 \\ -.15 & -.40 & .88 & -.53 & -.10 & .08 & .81 & .77 & .85 & .96 & .72 & 1.00 & .33 & .32 & .34 \\ .79 & .68 & .21 & .58 & .85 & .90 & .43 & .71 & .28 & .26 & .75 & .33 & 1.00 & .86 & .92 \\ .78 & .60 & .31 & .54 & .80 & .85 & .54 & .69 & .36 & .28 & .77 & .32 & .86 & 1.00 & .82 \\ .73 & .62 & .25 & .50 & .81 & .78 & .53 & .76 & .35 & .29 & .78 & .34 & .92 & .82 & 1.00 \end{bmatrix}$$

The eigenvalues of **R** are 7.91, 5.85, .31, .26, ... , .002, with the scree plot in Figure 13.5. Clearly, by any criterion for choosing m, there are two factors.

All four major methods of factor extraction discussed in Section 13.3 produced nearly identical results (after rotation). We give the initial and rotated loadings obtained from the principal component method in Table 13.10.

The two rotated factors were labeled "competence" and "benevolence." The same two factors emerged consistently in similar studies with different voices and different judges.

The factor scores were of primary interest in this study. The goal was to ascertain the effect of the rate manipulations on the two factors. What is the

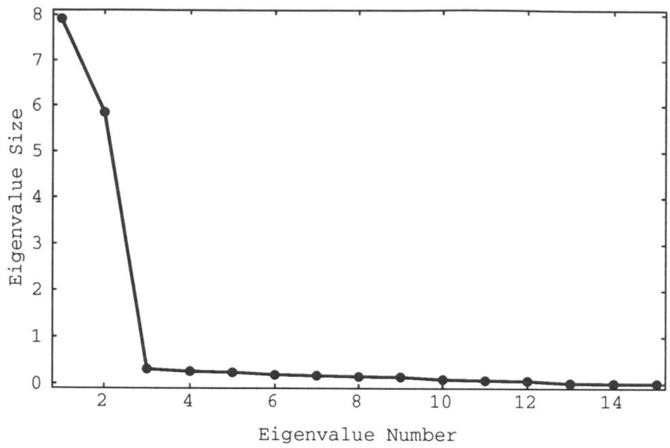

Figure 13.5 Scree graph for voice data.

Table 13.10 Initial and Varimax Rotated Loadings for Voice Data

	Initial Loadings		Rotated Loadings		
Variable	f_1	f_2	f_1	f_2	Communalities
Intelligent	.71	−.65	**.96**	−.06	.93
Ambitious	.48	−.84	**.90**	−.36	.94
Polite	.50	.81	−.12	**.95**	.92
Active	.37	−.91	**.86**	−.48	.97
Confident	.73	−.64	**.97**	−.04	.95
Happy	.83	−.47	**.94**	.15	.91
Just	.71	.58	.20	**.89**	.84
Likeable	.89	.39	.45	**.87**	.95
Kind	.58	.75	−.02	**.95**	.89
Sincere	.52	.82	−.11	**.97**	.95
Dependable	.93	.27	.56	**.79**	.94
Religious	.55	.79	−.07	**.96**	.92
Good Looking	.91	−.29	**.89**	.35	.91
Sociable	.91	−.22	**.84**	.40	.87
Strong	.91	−.21	**.84**	.41	.86
Variance accounted for	7.91	5.85	7.11	6.65	13.76
Proportion of total variance	.53	.39	.47	.44	.92
Cumulative proportion	.53	.92	.47	.92	.92

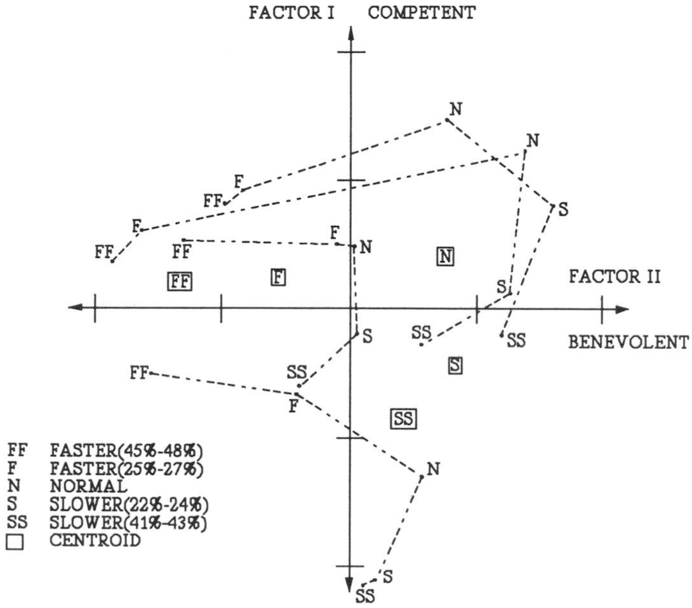

Figure 13.6 Factor scores of adjective rating of voices with five levels of manipulated rate.

perceived change in competence and benevolence when the speaking rate is increased or decreased?

The two factor scores were obtained for each of the 20 voices; these are plotted in Figure 13.6, where the effect of the manipulation of speaking rate can clearly be seen. Decreasing the speaking rate causes the speaker to be rated less competent; increasing the rate causes the speaker to be rated less benevolent. The mean vectors (centroids) are also given for the four speakers in Figure 13.6.

13.7 VALIDITY OF THE FACTOR ANALYSIS MODEL

For many statisticians, factor analysis is controversial and does not belong in a tool kit of legitimate multivariate techniques. The reasons for this mistrust include the following: the difficulty in choosing m, the many methods of extracting factors, the many rotation techniques, the subjectivity in interpretation, and a confusion in many texts and papers between factor analysis and principal components analysis. Some statisticians also criticize factor analysis because of the indeterminacy of the factor loading matrix $\mathbf{\Lambda}$ or $\hat{\mathbf{\Lambda}}$, first noted in Section 13.2.2. However, it is the ability to rotate that gives factor analysis its utility, if not its charm.

The basic question is whether the factors really exist. The model is often used without knowing beforehand if any factors are actually present. The model (13.10) for the covariance matrix is $\mathbf{\Sigma} = \mathbf{\Lambda}\mathbf{\Lambda}' + \mathbf{\Psi}$ or $\mathbf{\Sigma} - \mathbf{\Psi} = \mathbf{\Lambda}\mathbf{\Lambda}'$, where $\mathbf{\Lambda}\mathbf{\Lambda}'$ is of rank m. Many populations have covariance matrices that do not approach this pattern unless m is large. Thus the model will not fit data from such a population when we try to impose a small value of m. On the other hand, for a population in which $\mathbf{\Sigma}$ is reasonably close to $\mathbf{\Lambda\Lambda}' + \mathbf{\Psi}$ for small m, the sampling procedure leading to $\mathbf{S}$ may obscure this pattern. The researcher may believe there are underlying factors but has difficulty collecting data that will reveal them. In many cases, the basic problem is that $\mathbf{S}$ (or $\mathbf{R}$) contains both structure and error and the methods of factor analysis cannot separate the two.

A statistical consultant in a university setting or elsewhere all too often sees the following scenario. A researcher designs a long questionnaire, with answers to be given in, say, a five-point semantic differential scale or Likert scale. The respondents, who vary in attitude from uninterested to resentful, hurriedly mark answers that in many cases are not even good subjective responses to the questions. Then the researcher submits the results to a handy factor analysis program. Being disappointed in the results, he or she appeals to the statistician for help. They attempt to improve the results by trying different methods of extraction, different rotations, different values of m, and so on. But it is all to no avail. The scree plot looks more like the foothills than a steep cliff with gently sloping debris at the bottom. There is no clear value of m. They have to extract 10 or 12 factors to account for, say, 60% of the variance, and interpretation of this large number of factors is hopeless. If a few underlying dimensions exist, they are totally obscured by both systematic and random errors in marking the questionnaire. New techniques and faster computers will not help such a data set. A factor analysis model simply does not fit the data, unless a value of m nearly as large as p is used, which gives useless results.

It is not necessarily the "discreteness" of the data that causes the problem, but the "noisiness" of the data. The specified variables are not measured accurately. In some cases, discrete variables yield satisfactory results, such as in Examples 13.3.1, 13.3.2, 13.5.2a, and 13.5.2b(a), where a 12-year-old girl, responding carefully to a semantic differential scale, produced data leading to an unambiguous factor analysis. On the other hand, continuous variables do not guarantee good results [see Example 13.7(a) below].

In cases in which some factors are found that provide a satisfactory fit to the data, we should still be tentative in interpretation until we can independently establish the existence of the factors. If the same factors emerge in repeated sampling from the same population, then we can have confidence that application of the model has uncovered some real factors. Thus it is good practice to repeat the experiment to check the stability of the factors. If the data set is large enough, it could be split in half and a factor analysis performed on each half. The two solutions could be compared with each other and with that for the complete set.

If there is replication in the data set, it may be helpful to average over the replications. This was done to great advantage in Example 13.6, where several judges rated the same voices. Averaging over the judges produced variables that apparently possessed very low noise. Similar experimentation with different judges always produced the same factors. Unfortunately, replication of this type is unavailable in most situations.

As with other techniques in this book, factor analysis assumes that the variables are approximately linearly related to each other. We could make bivariate scatter plots to check this assumption.

A basic prerequisite for a factor analysis application is that the variables not be independent. To check this requirement, we could test $H_0 : \mathbf{P}_\rho = \mathbf{I}$ by using the test in Section 7.4.3.

Some writers have suggested that $\mathbf{R}^{-1}$ should be a near-diagonal matrix in order to successfully fit a factor analysis model. To assess how close $\mathbf{R}^{-1}$ is to a diagonal matrix, Kaiser (1970) proposed a *measure of sampling adequacy*,

$$\text{MSA} = \frac{\sum_{i \neq j} r_{ij}^2}{\sum_{i \neq j} r_{ij}^2 + \sum_{i \neq j} q_{ij}^2}, \tag{13.54}$$

where r_{ij}^2 is the square of an element from $\mathbf{R}$ and q_{ij}^2 is the square of an element from $\mathbf{Q} = \mathbf{D}\mathbf{R}^{-1}\mathbf{D}$, with $\mathbf{D} = [(\text{diag}\,\mathbf{R}^{-1})^{1/2}]^{-1}$. As $\mathbf{R}$ approaches a diagonal matrix, MSA approaches 1. Kaiser and Rice (1974) suggest that MSA should exceed .8 for satisfactory results to be expected. We show some results for MSA in Example 13.7(b).

In summary, there are many data sets to which factor analysis should not be applied. One indication that $\mathbf{R}$ is inappropriate for factoring is the failure of the methods in Section 13.4 to clearly and rather objectively choose a value for m. If the scree plot does not have a pronounced bend or the eigenvalues do not show a large gap around 1, then $\mathbf{R}$ is likely to be unsuitable for factoring. In addition, the communality estimates should be fairly large.

To balance the "good" examples in this chapter, we now give an example involving a data set that cannot be successfully modeled by factor analysis. Likewise, the problems at the end of the chapter include both "good" and "bad" data sets.

Example 13.7(a). As an illustration of an application of factor analysis that is less successful than previous examples in this chapter, we consider the diabetes data of Table 3.6. The correlation matrix for the five variables is as follows:

$$\mathbf{R} = \begin{bmatrix} 1.00 & .05 & -.13 & .07 & .21 \\ .05 & 1.00 & -.01 & .01 & -.10 \\ -.13 & -.01 & 1.00 & .29 & .05 \\ .07 & .01 & .29 & 1.00 & .21 \\ .21 & -.10 & .05 & .21 & 1.00 \end{bmatrix}.$$

The correlations are all small and the variables do not appear to have much in common. The MSA value is .49. The eigenvalues are 1.40, 1.21, 1.04, .71, and .65. Three factors would be required to account for 73% of the variance and four factors to reach 87%. This is not a useful reduction in dimensionality. The eigenvalues are plotted in a scree graph in Figure 13.7. The lack of a clear value of m is apparent.

It is evident from the small correlations in **R** that the communalities of the variables will not be large. The principal component method, which essentially estimates the initial communalities as 1, gave very different final communality estimates from the iterated principal factor method:

	Communalities				
Principal component method	.71	.91	.71	.67	.64
Iterated principal factor method	.31	.16	.35	.37	.33

The communalities obtained by the iterated approach reflect more accurately the small correlations among the variables.

The varimax rotated factor loadings for three factors extracted by the iterated principal factor method are given in Table 13.11.

The first factor is associated with variables 3 and 4, the second factor with variables 1 and 5, and the third with variable 2. This clustering of variables can be seen in **R**, where variables 1 and 5 have a correlation of .21, variables 3 and 4 have a correlation of .29, and variable 2 has very low correlations with all other variables. However, these correlations (.21 and .29) are small, and this collapsing of five variables to three factors is not a useful reduction in dimensionality, especially since the first three eigenvalues account for only 73% of tr(**R**). The 73% is not significantly greater than 60%, which we would expect from three original variables picked at random. This conclusion is borne

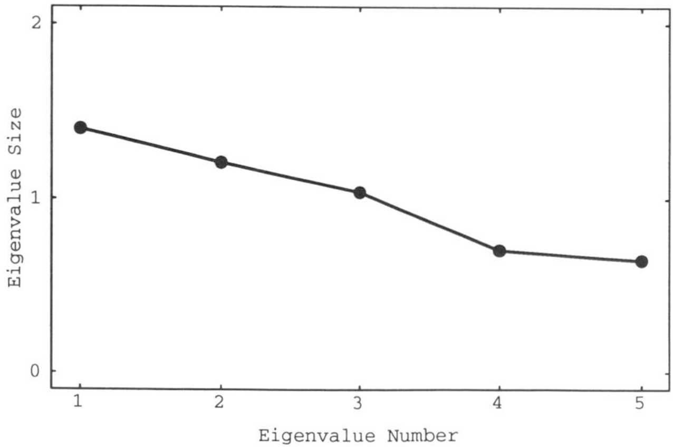

Figure 13.7 Scree graph for diabetes data.

Table 13.11 Varimax Rotated Factor Loadings for Iterated Principal Factors from the Diabetes Data

Variable	Rotated Loadings f_1	f_2	f_3	Communalities
1	−.08	**.54**	.12	.31
2	.01	.01	**.40**	.16
3	**.57**	−.15	−.03	.35
4	**.57**	.22	.02	.37
5	.19	**.47**	−.27	.33
Variance accounted for	.69	.59	.24	1.52

out by a test of $H_0: \mathbf{P}_\rho = \mathbf{I}$. Using (7.35) and (7.36), we obtain

$$u = |\mathbf{R}| = .80276 \qquad \nu = 20 - 1 = 19 \qquad p = 5$$

$$u' = -[\nu - \tfrac{1}{6}(2p+5)] \ln u = -\left(19 - \frac{15}{6}\right)(-.2197) = 3.625.$$

With $\frac{1}{2}p(p-1) = 10$ degrees of freedom, the .05 critical value for this approximate χ^2 test is 18.31, and we have no basis to question the independence of the five variables. Thus the three factors we obtained are very likely an artifact of the present sample and would not reappear in subsequent samples.

Example 13.7(b). For data sets used in previous examples in this chapter, the values of MSA from (13.54) are calculated as follows:

$$\begin{aligned} \text{Seishu data MSA} &= .53 \\ \text{Sons data MSA} &= .82 \\ \text{Voice data MSA} &= .73 \\ \text{Diabetes data MSA} &= .49. \end{aligned}$$

The MSA value cannot be computed for the perception data because **R** is singular.

These results do not suggest great confidence in the MSA index as a sole guide to the suitability of **R** for factoring. We see a wide disparity in the MSA values for the first three data sets. Yet all three yielded successful factor analyses. These three MSA values seem to be inversely related to the number of factors. In the sons data, there were indications that one factor would suffice. The voice data clearly had two factors, and for the seishu data, there were four factors.

The MSA for the diabetes data was close to that of the seishu data. Yet the diabetes data is totally unsuitable for factor analysis, while the factor analysis of the seishu data was very convincing.

13.8 RELATIONSHIP OF FACTOR ANALYSIS TO PRINCIPAL COMPONENT ANALYSIS

Both factor analysis and principal components analysis have the goal of reducing dimensionality. Because the objectives are similar, many authors include principal components analysis as another type of factor analysis. This can be very confusing, and we wish to underscore the distinguishing characteristics of the two techniques.

Two of the differences between factor analysis and principal component analysis were mentioned in Section 13.1: (1) In factor analysis, the variables are expressed as linear combinations of the factors, whereas the principal components are linear functions of the variables, and (2) in principal component analysis, the emphasis is on explaining the total variance $\Sigma_i s_{ii}$, as contrasted with the attempt to explain the covariances in factor analysis.

Additional differences are that (3) principal component analysis requires essentially no assumptions, while factor analysis makes several key assumptions; (4) the principal components are unique (assuming distinct eigenvalues of $\mathbf{S}$), whereas the factors are subject to an arbitrary rotation; and (5) if we change the number of factors, the (estimated) factors change. This does not happen in principal components.

The ability to rotate to improve interpretability is one of the advantages of factor analysis over principal components. If finding and describing some underlying factors is the goal, factor analysis may prove more useful than principal components; we would prefer factor analysis if the factor model fits the data well and we like the interpretation of the rotated factors. On the other hand, if we wish to define a smaller number of variables for input into another analysis, we would ordinarily prefer principal components, although this can sometimes be accomplished with factor scores.

PROBLEMS

13.1 Show that the assumptions lead to the expression in (13.2) for the variance of y_i, $\text{var}(y_i) = \lambda_{i1}^2 + \lambda_{i2}^2 + \cdots + \lambda_{im}^2 + \psi_i$.

13.2 **(a)** Show that $\text{cov}(\lambda_{12} f_2, f_2) = \lambda_{12} \text{var}(f_2)$, as in the illustration preceding (13.11).

(b) Verify (13.12) directly: $\text{cov}(\mathbf{y}, \mathbf{f}) = \mathbf{\Lambda}$.

13.3 Show that $\mathbf{f}^* = \mathbf{T}'\mathbf{f}$ in (13.17) satisfies the assumptions (13.5) and (13.6), $E(\mathbf{f}^*) = \mathbf{0}$ and $\text{cov}(\mathbf{f}^*) = \mathbf{I}$.

13.4 Show that $\sum_{i=1}^{p} \sum_{j=1}^{m} \hat{\lambda}_{ij}^2$ is equal to the sum of the first m eigenvalues and also equal to the sum of all p communalities, as in (13.42).

13.5 In Example 13.3.2, the correlation matrix for the perception data was shown to have an eigenvalue of 0. Find the multicollinearity among the five variables that this implies.

13.6 Use the words data of Table 5.9.

(a) Obtain principal component loadings for two factors.

(b) Do a graphical rotation of the two factors.

(c) Do a varimax rotation and compare with the results in part (b).

13.7 Use the ramus bone data of Table 3.8.

(a) Extract loadings by the principal component method and do a varimax rotation. Use two factors.

(b) Do all variables have a complexity of 1? Carry out an oblique rotation to improve the loadings.

(c) What is the angle between the oblique axes? Would a single factor ($m = 1$) be more appropriate here?

13.8 Carry out a factor analysis of the rootstock data of Table 6.2. Combine the groups into a single sample.

(a) Estimate the loadings for two factors by the principal component method and do a varimax rotation.

(b) Did the rotation improve the loadings?

13.9 Use the fish data of Table 6.17. Combine the groups into a single sample.

(a) Obtain loadings on two factors by the principal component method and do a varimax rotation.

(b) Notice the similarity of loadings for y_1 and y_2. Is there any indication in the correlation matrix as to why this is so?

(c) Compute factor scores.

(d) Carry out a MANOVA on the factor scores comparing the three groups.

13.10 Carry out a factor analysis of the flea data in Table 5.5. Combine the groups into a single sample.

(a) From an examination of the eigenvalues greater than 1, the scree plot, and the percentages, is there a clear choice of m?

(b) Extract two factors by the principal component method and carry out a varimax rotation.

(c) Is the rotation an improvement? Try an oblique rotation.

13.11 Use the engineer data of Table 5.6. Combine the groups into a single sample.

(a) Using a scree plot, the number of eigenvalues greater than 1, and the percentages, is there a clear choice of m?

(b) Extract three factors by the principal component method and carry out a varimax rotation.

(c) Extract three factors by the principal factor method and carry out a varimax rotation.

(d) Compare the results of (b) and (c).

13.12 Use the probe word data of Table 3.7.

(a) Obtain loadings for two factors by the principal component method and carry out a varimax rotation.

(b) Notice the near duplication of loadings for y_2 and y_4. Is there any indication in the correlation matrix as to why this is so?

(c) Is the rotation satisfactory? Try an oblique rotation.

APPENDIX A

Table A.1 Upper Percentiles for $\sqrt{b_1}$

$$\sqrt{b_1} = \frac{\sqrt{n}\sum_{i=1}^{n}(y_i - \bar{y})^3}{\left[\sum_{i=1}^{n}(y_i - \bar{y})^2\right]^{3/2}}$$

The sampling distribution of $\sqrt{b_1}$ is symmetric about zero, and the lower percentage points corresponding to negative skewness are given by the negative of the table values. Reject the hypothesis of normality if $\sqrt{b_1}$ is greater than the table value or less than the negative of the table value.

	Upper Percentiles					
n	10	5	2.5	1	0.5	0.1
4	0.831	0.987	1.070	1.120	1.137	1.151
5	0.821	1.049	1.207	1.337	1.396	1.464
6	0.795	1.042	1.239	1.429	1.531	1.671
7	0.782	1.018	1.230	1.457	1.589	1.797
8	0.765	0.998	1.208	1.452	1.605	1.866
9	0.746	0.977	1.184	1.433	1.598	1.898
10	0.728	0.954	1.159	1.407	1.578	1.906
11	0.710	0.931	1.134	1.381	1.553	1.899
12	0.693	0.910	1.109	1.353	1.526	1.882
13	0.677	0.890	1.085	1.325	1.497	1.859
14	0.662	0.870	1.061	1.298	1.468	1.832
15	0.648	0.851	1.039	1.272	1.440	1.803
16	0.635	0.834	1.018	1.247	1.412	1.773
17	0.622	0.817	0.997	1.222	1.385	1.744
18	0.610	0.801	0.978	1.199	1.359	1.714
19	0.599	0.786	0.960	1.176	1.334	1.685
20	0.588	0.772	0.942	1.155	1.310	1.657
21	0.578	0.758	0.925	1.134	1.287	1.628
22	0.568	0.746	0.909	1.114	1.265	1.602
23	0.559	0.733	0.894	1.096	1.243	1.575
24	0.550	0.722	0.880	1.078	1.223	1.550
25	0.542	0.710	0.866	1.060	1.203	1.526

Table A.2 Coefficients for Transforming $\sqrt{b_1}$ to a Standard Normal

n	δ	$1/\lambda$	n	δ	$1/\lambda$
			62	3.389	1.0400
			64	3.420	1.0449
8	5.563	0.3030	66	3.450	1.0495
9	4.260	0.4080	68	3.480	1.0540
10	3.734	0.4794	70	3.510	1.0581
11	3.447	0.5339	72	3.540	1.0621
12	3.270	0.5781	74	3.569	1.0659
13	3.151	0.6153	76	3.599	1.0695
14	3.069	0.6473	78	3.628	1.0730
15	3.010	0.6753	80	3.657	1.0763
16	2.968	0.7001	82	3.686	1.0795
17	2.937	0.7224	84	3.715	1.0825
18	2.915	0.7426	86	3.744	1.0854
19	2.900	0.7610	88	3.772	1.0882
20	2.890	0.7779	90	3.801	1.0909
21	2.884	0.7934	92	3.829	1.0934
22	2.882	0.8078	94	3.857	1.0959
23	2.882	0.8211	86	3.885	1.0983
24	2.884	0.8336	98	3.913	1.1006
25	2.889	0.8452	100	3.940	1.1028
26	2.895	0.8561	105	4.009	1.1080
27	2.902	0.8664	110	4.076	1.1128
28	2.910	0.8760	115	4.142	1.1172
29	2.920	0.8851	120	4.207	1.1212
30	2.930	0.8938	125	4.272	1.1250
31	2.941	0.9020	130	4.336	1.1285
32	2.952	0.9097	135	4.398	1.1318
33	2.964	0.9171	140	4.460	1.1348
34	2.977	0.9241	145	4.521	1.1377
35	2.990	0.9308	150	4.582	1.1403
36	3.003	0.9372	155	4.641	1.1428
37	3.016	0.9433	160	4.700	1.1452
38	3.030	0.9492	165	4.758	1.1474
39	3.044	0.9548	170	4.816	1.1496
40	3.058	0.9601	175	4.873	1.1516
41	3.073	0.9653	180	4.929	1.1535
42	3.087	0.9702	185	1.985	1.1553
43	3.102	0.9750	190	5.040	1.1570
44	3.117	0.9795	195	5.094	1.1586
45	3.131	0.9840	200	5.148	1.1602
46	3.146	0.9882	205	5.202	1.1616
47	3.161	0.9923	210	5.255	1.1631
48	3.176	0.9963	215	5.307	1.1644

Table A.2 ***(Continued)***

n	δ	$1/\lambda$	n	δ	$1/\lambda$
49	3.192	1.0001	220	5.359	1.1657
50	3.207	1.0038	225	5.410	1.1669
52	3.237	1.0108	230	5.461	1.1681
54	3.268	1.0174	235	5.511	1.1693
56	3.298	1.0235	240	5.561	1.1704
58	3.329	1.0293	245	5.611	1.1714
60	3.359	1.0348	250	5.660	1.1724

Values of δ and $1/\lambda$ are such that $g(\sqrt{b_1}) = \delta \sinh^{-1}(\sqrt{b_1}/\lambda)$ is approximately $N(0, 1)$.

Table A.3 Percentiles for b_2

Upper and lower percentiles for

$$b_2 = \frac{n \sum_{i=1}^{n} (y_i - \bar{y})^4}{\left[\sum_{i=1}^{n} (y_i - \bar{y})^2\right]^2},$$

the sample coefficient of kurtosis. Reject the hypothesis of normality if b_2 is greater than an upper percentile or less than a lower percentile.

Sample size	Percentiles 1	2	2.5	5	10	20	80	90	95	97.5	98	99
7	1.25	1.30	1.34	1.41	1.53	1.70	2.78	3.20	3.55	3.85	3.93	4.23
8	1.31	1.37	1.40	1.46	1.58	1.75	2.84	3.31	3.70	4.09	4.20	4.53
9	1.35	1.42	1.45	1.53	1.63	1.80	2.98	3.43	3.86	4.28	4.41	4.82
10	1.39	1.45	1.49	1.56	1.68	1.85	3.01	3.53	3.95	4.40	4.55	5.00
12	1.46	1.52	1.56	1.64	1.76	1.93	3.06	3.55	4.05	4.56	4.73	5.20
15	1.55	1.61	1.64	1.72	1.84	2.01	3.13	3.62	4.13	4.66	4.85	5.30
20	1.65	1.71	1.74	1.82	1.95	2.13	3.21	3.68	4.17	4.68	4.87	5.36
25	1.72	1.79	1.83	1.91	2.03	2.20	3.23	3.68	4.16	4.65	4.82	5.30
30	1.79	1.86	1.90	1.98	2.10	2.26	3.25	3.68	4.11	4.59	4.75	5.21
35	1.84	1.91	1.95	2.03	2.14	2.31	3.27	3.68	4.10	4.53	4.68	5.13
40	1.89	1.96	1.98	2.07	2.19	2.34	3.28	3.67	4.06	4.46	4.61	5.04
45	1.93	2.00	2.03	2.11	2.22	2.37	3.28	3.65	4.00	4.39	4.52	4.94
50	1.95	2.03	2.06	2.15	2.25	2.41	3.28	3.62	3.99	4.33	4.45	4.88

Table A.4 Percentiles for D'Agostino's Test for Normality

Upper and lower percentiles for the statistic

$$Y = \frac{\sqrt{n}[D - (2\sqrt{\pi})^{-1}]}{.02998598},$$

where

$$D = \frac{\sum_{i=1}^{n} \left[i - \frac{1}{2}(n+1)\right] y_{(i)}}{\sqrt{n^3} \sum_{i=1}^{n} (y_i - \bar{y})^2}$$

and the observations $y_1, y_2, \ldots, y_n$ are ordered as $y_{(1)} \leq y_{(2)} \leq \cdots \leq y_{(n)}$. Reject the hypothesis of normality if Y is greater than an upper percentile or less than a lower percentile.

	Percentiles of Y									
n	0.5	1.0	2.5	5	10	90	95	97.5	99	99.5
10	−4.66	−4.06	−3.25	−2.62	−1.99	0.149	0.235	0.299	0.356	0.385
12	−4.63	−4.02	−3.20	−2.58	−1.94	0.237	0.329	0.381	0.440	0.479
14	−4.57	−3.97	−3.16	−2.53	−1.90	0.308	0.399	0.460	0.515	0.555
16	−4.52	−3.92	−3.12	−2.50	−1.87	0.367	0.459	0.526	0.587	0.613
18	−4.47	−3.87	−3.08	−2.47	−1.85	0.417	0.515	0.574	0.636	0.667
20	−4.41	−3.83	−3.04	−2.44	−1.82	0.460	0.565	0.628	0.690	0.720
22	−4.36	−3.78	−3.01	−2.41	−1.81	0.497	0.609	0.677	0.744	0.775
24	−4.32	−3.75	−2.98	−2.39	−1.79	0.530	0.648	0.720	0.783	0.822
26	−4.27	−3.71	−2.96	−2.37	−1.77	0.559	0.682	0.760	0.827	0.867
28	−4.23	−3.68	−2.93	−2.35	−1.76	0.586	0.714	0.797	0.868	0.910
30	−4.19	−3.64	−2.91	−2.33	−1.75	0.610	0.743	0.830	0.906	0.941
32	−4.16	−3.61	−2.88	−2.32	−1.73	0.631	0.770	0.862	0.942	0.983
34	−4.12	−3.59	−2.86	−2.30	−1.72	0.651	0.794	0.891	0.975	1.02
36	−4.09	−3.56	−2.85	−2.29	−1.71	0.669	0.816	0.917	1.00	1.05
38	−4.06	−3.54	−2.83	−2.28	−1.70	0.686	0.837	0.941	1.03	1.08
40	−4.03	−3.51	−2.81	−2.26	−1.70	0.702	0.857	0.964	1.06	1.11
42	−4.00	−3.49	−2.80	−2.25	−1.69	0.716	0.875	0.986	1.09	1.14
44	−3.98	−3.47	−2.78	−2.24	−1.68	0.730	0.892	1.01	1.11	1.17
46	−3.95	−3.45	−2.77	−2.23	−1.67	0.742	0.908	1.02	1.13	1.19
48	−3.93	−3.43	−2.75	−2.22	−1.67	0.754	0.923	1.04	1.15	1.22
50	−3.91	−3.41	−2.74	−2.21	−1.66	0.765	0.937	1.06	1.18	1.24
60	−3.81	−3.34	−2.68	−2.17	−1.64	0.812	0.997	1.13	1.26	1.34
70	−3.73	−3.27	−2.64	−2.14	−1.61	0.849	1.05	1.19	1.33	1.42
80	−3.67	−3.22	−2.60	−2.11	−1.59	0.878	1.08	1.24	1.39	1.48
90	−3.61	−3.17	−2.57	−2.09	−1.58	0.902	1.12	1.28	1.44	1.54
100	−3.57	−3.14	−2.54	−2.07	−1.57	0.923	1.14	1.31	1.48	1.59
150	−3.409	−3.009	−2.452	−2.004	−1.520	0.990	1.233	1.423	1.623	1.746
200	−3.302	−2.922	−2.391	−1.960	−1.491	1.032	1.290	1.496	1.715	1.853
250	−3.227	−2.861	−2.348	−1.926	−1.471	1.060	1.328	1.545	1.779	1.927

Table A.5 Upper Percentiles for $b_{1,p}$ and Upper and Lower Percentiles for $b_{2,p}$

Reject the hypothesis of multivariate normality if $b_{1,p}$ is greater than table value. Reject if $b_{2,p}$ is greater than upper percentile or if $b_{2,p}$ is less than lower percentile. The statistics $b_{1,p}$ and $b_{2,p}$ are defined in Section 4.4.2.

$p = 2$, Upper Percentiles for $b_{1,p}$

n	Percentiles 90	92.5	95	97.5	99	99.9
10	2.994	3.263	3.694	4.294	5.194	6.994
12	2.681	2.944	3.319	3.931	4.938	6.744
14	2.419	2.669	3.031	3.619	4.581	6.419
16	2.219	2.444	2.775	3.337	4.231	6.062
18	2.050	2.256	2.556	3.100	3.962	5.737
20	1.894	2.081	2.356	2.881	3.669	5.425
25	1.581	1.744	1.969	2.438	3.106	4.719
30	1.363	1.513	1.687	2.094	2.681	4.238
40	1.050	1.181	1.319	1.606	2.087	3.369
50	0.862	0.969	1.069	1.306	1.744	2.706
60	0.731	0.819	0.906	1.094	1.444	2.200
70	0.631	0.725	0.794	0.937	1.244	1.863
80	0.544	0.637	0.694	0.812	1.056	1.587
90	0.487	0.569	0.638	0.725	0.919	1.400
100	0.438	0.506	0.581	0.656	0.831	1.231
150	0.281	0.344	0.400	0.444	0.531	0.794
200	0.219	0.269	0.300	0.331	0.394	0.569
300	0.144	0.169	0.209	0.225	0.256	0.369
400	0.116	0.129	0.141	0.166	0.197	0.275
600	0.077	0.085	0.094	0.110	0.131	0.183
800	0.058	0.064	0.071	0.083	0.099	0.137
1000	0.046	0.051	0.057	0.066	0.079	0.110
1500	0.031	0.034	0.038	0.044	0.053	0.074
2500	0.019	0.021	0.023	0.027	0.032	0.044
3000	0.016	0.017	0.019	0.022	0.027	0.037
4000	0.012	0.013	0.014	0.017	0.020	0.028
5000	0.009	0.010	0.011	0.013	0.016	0.022

$p = 2$, Upper and Lower Percentiles for $b_{2,p}$

n	Percentiles .5	1.25	2.5	5	95	97.5	98.75	99.5
10	4.580	4.722	4.887	5.057	8.606	9.203	9.781	10.378
12	4.732	4.899	5.053	5.232	8.947	9.593	10.150	10.881
14	4.842	5.015	5.179	5.358	9.162	9.769	10.375	11.159
16	4.977	5.149	5.318	5.482	9.331	9.941	10.562	11.387
18	5.045	5.219	5.382	5.555	9.403	10.005	10.628	11.478
20	5.175	5.262	5.533	5.717	9.469	10.114	10.691	11.609
25	5.351	5.525	5.689	5.871	9.503	10.159	10.584	11.628
30	5.518	5.692	5.855	6.038	9.516	10.156	10.556	11.594
40	5.703	5.871	6.139	6.229	9.497	10.109	10.563	11.453
50	5.909	6.083	6.239	6.403	9.453	9.987	10.372	11.181
60	6.015	6.189	6.335	6.505	9.401	9.889	10.250	10.994
70	6.139	6.290	6.437	6.602	9.356	9.781	10.106	10.753
80	6.223	6.372	6.539	6.683	9.309	9.694	9.981	10.537
90	6.332	6.475	6.622	6.749	9.256	9.688	9.885	10.325
100	6.389	6.521	6.665	6.793	9.210	9.556	9.806	10.188
150	6.615	6.749	6.858	6.972	9.027	9.300	9.475	10.253
200	6.761	6.889	6.979	7.083	8.919	9.141	9.269	9.506
300	6.949	7.052	7.142	7.245	8.776	8.916	9.031	9.219
400	7.079	7.171	7.252	7.342	8.664	8.787	8.917	9.061
600	7.232	7.295	7.369	7.464	8.547	8.647	8.749	8.874
800	7.304	7.372	7.451	7.536	8.472	8.562	8.641	8.747
1000	7.367	7.433	7.504	7.585	8.419	8.497	8.569	8.656
1500	7.460	7.537	7.595	7.661	8.339	8.405	8.463	8.532
2000	7.535	7.599	7.649	7.707	8.293	8.351	8.401	8.461
2500	7.588	7.641	7.686	7.738	8.262	8.314	8.359	8.412
3000	7.624	7.673	7.714	7.760	8.240	8.286	8.327	8.376
4000	7.674	7.716	7.752	7.793	8.207	8.248	8.284	8.326
5000	7.709	7.746	7.778	7.714	8.186	8.222	8.254	8.291

Table A.5 (*Continued*)

$p = 3$, Upper Percentiles for $b_{1,p}$

n	Percentiles 90	92.5	95	97.5	99	99.9
10	6.0	6.5	6.9	7.7	8.8	11.5
12	5.5	5.9	6.4	7.1	8.1	10.5
14	5.0	5.4	5.9	6.5	7.4	9.7
16	4.6	4.9	5.4	6.1	6.8	8.9
18	4.2	4.6	5.1	5.6	6.4	8.3
20	3.9	4.2	4.7	5.3	6.0	7.7
25	3.3	3.5	3.9	4.5	5.2	6.5
30	2.8	3.0	3.3	3.9	4.4	5.6
40	2.2	2.4	2.7	3.0	3.5	4.2
50	1.7	1.9	2.2	2.4	2.8	3.4
60	1.5	1.6	1.8	2.0	2.4	2.9
70	1.3	1.4	1.5	1.7	2.0	2.5
80	1.13	1.2	1.3	1.5	1.7	2.2
90	1.01	1.08	1.16	1.3	1.5	1.9
100	0.92	0.97	1.05	1.18	1.3	1.7
150	0.62	0.66	0.71	0.80	0.90	1.15
200	0.47	0.50	0.54	0.60	0.68	0.87
300	0.32	0.33	0.36	0.40	0.46	0.58
400	0.237	0.252	0.272	0.30	0.34	0.44
600	0.159	0.168	0.182	0.203	0.230	0.294
800	0.119	0.127	0.137	0.153	0.173	0.221
1000	0.095	0.010	0.109	0.122	0.139	0.177
1500	0.064	0.068	0.073	0.082	0.093	0.118
2000	0.048	0.051	0.055	0.061	0.069	0.089
3000	0.032	0.034	0.037	0.041	0.046	0.059
4000	0.024	0.025	0.027	0.031	0.035	0.044
5000	0.019	0.020	0.022	0.025	0.028	0.035

$p = 3$, Upper and Lower Percentiles for $b_{2,p}$

n	Percentiles .5	1.25	2.5	5	95	97.5	98.75	99.5
10	10.0	10.2	10.4	10.7	14.0	14.4	15.0	15.6
12	10.2	10.4	10.7	11.0	14.7	15.2	15.9	16.4
14	10.4	10.6	10.9	11.3	15.1	15.8	16.5	17.1
16	10.5	10.8	11.1	11.5	15.4	16.1	16.8	17.5
18	10.7	11.0	11.3	11.6	15.5	16.4	17.1	17.8
20	10.8	11.1	11.4	11.8	15.7	16.5	17.2	18.0
25	11.1	11.4	11.8	12.1	15.9	16.7	17.4	18.2
30	11.3	11.6	12.0	12.3	16.0	16.7	17.5	18.3
40	11.7	12.0	12.4	12.7	16.1	16.7	17.4	18.2
50	11.9	12.3	12.6	12.9	16.1	16.7	17.3	18.0
60	12.1	12.5	12.8	13.1	16.1	16.6	17.2	17.9
70	12.3	12.6	13.0	13.2	16.1	16.6	17.1	17.7
80	12.4	12.8	13.1	13.3	16.1	16.5	17.0	17.6
90	12.5	12.9	13.2	13.5	16.0	16.5	16.9	17.5
100	12.6	13.0	13.3	13.5	16.0	16.4	16.8	17.4
150	13.0	13.3	13.6	13.8	15.9	16.2	16.5	17.0
200	13.2	13.5	13.8	14.0	15.8	16.1	16.3	16.8
300	13.6	13.8	14.0	14.2	15.7	15.9	16.1	16.5
400	13.7	13.9	14.1	14.3	15.6	15.8	16.0	16.3
600	13.9	14.1	14.3	14.4	15.51	15.67	15.81	15.97
800	14.1	14.2	14.3	14.5	15.45	15.59	15.71	15.85
1000	14.17	14.30	14.41	14.53	15.41	15.53	15.64	15.77
1500	14.33	14.43	14.52	14.62	15.34	15.44	15.53	15.63
2000	14.42	14.51	14.58	14.67	15.30	15.39	15.46	15.55
3000	14.53	14.60	14.66	14.73	15.25	15.32	15.38	15.45
4000	14.59	14.65	14.71	14.77	15.21	15.28	15.33	15.39
5000	14.63	14.69	14.74	14.80	15.19	15.25	15.30	15.35

Table A.5 (*Continued*)

	$p = 4$, Upper Percentiles for $b_{1,p}$							$p = 4$, Upper and Lower Percentiles for $b_{2,p}$							
	Percentiles							Percentiles							
n	90	92.5	95	97.5	99	99.9	n	.5	1.25	2.5	5	95	97.5	98.75	99.5
10	11.1	11.6	12.2	13.3	15.3	17.9	10	17.0	17.3	17.6	17.8	21.5	22.4	23.0	24.0
12	10.1	10.6	11.2	12.2	13.9	16.2	12	17.4	17.7	18.0	18.3	22.3	23.3	24.2	25.4
14	9.2	9.7	10.2	11.2	12.7	14.8	14	17.7	18.0	18.3	18.6	23.0	24.0	25.0	26.1
16	8.4	8.8	9.4	10.3	11.6	13.6	16	18.0	18.2	18.6	18.9	23.4	24.4	25.4	26.6
18	7.7	8.0	8.7	9.5	10.7	12.6	18	18.2	18.4	18.8	19.2	23.8	24.7	25.8	26.9
20	7.0	7.4	8.0	8.8	9.9	11.6	20	18.4	18.6	19.0	19.4	24.0	25.0	26.1	27.1
25	5.9	6.2	6.6	7.1	8.1	9.7	25	18.8	19.1	19.5	19.8	24.5	25.4	26.4	27.3
30	5.0	5.3	5.6	6.0	6.8	8.1	30	19.1	19.4	19.8	20.2	24.7	25.5	26.6	27.4
40	3.9	4.1	4.3	4.6	5.2	6.2	40	19.6	19.9	20.3	21.0	25.0	25.7	26.7	27.4
50	3.1	3.3	3.5	3.8	4.2	5.0	50	20.0	20.3	20.6	21.0	25.1	25.7	26.6	27.3
60	2.7	2.8	2.9	3.2	3.5	4.2	60	20.2	20.5	20.9	21.3	25.14	25.7	26.6	27.2
70	2.3	2.4	2.5	2.8	3.0	3.7	70	20.4	20.7	21.0	21.5	25.15	25.7	26.5	27.0
80	2.0	2.1	2.2	2.4	2.7	3.2	80	20.6	21.0	21.2	21.7	25.15	25.6	26.4	26.9
90	1.81	1.89	2.0	2.2	2.4	2.9	90	20.8	21.1	21.4	21.8	25.14	25.6	26.3	26.8
100	1.64	1.71	1.81	1.97	2.2	2.6	100	20.9	21.2	21.5	21.9	25.12	25.6	26.2	26.7
150	1.11	1.16	1.22	1.33	1.46	1.76	150	21.4	21.7	22.0	22.33	25.03	25.42	25.9	26.3
200	0.84	0.87	0.92	1.00	1.10	1.33	200	21.7	22.0	22.2	22.57	24.95	25.29	25.6	26.0
300	0.56	0.59	0.62	0.67	0.74	0.89	300	22.1	22.33	22.57	22.85	24.83	25.11	25.3	25.7
400	0.42	0.44	0.47	0.51	0.56	0.67	400	22.3	22.56	22.77	23.02	24.75	24.99	25.20	25.46
600	0.282	0.295	0.31	0.34	0.37	0.45	600	22.63	22.83	23.01	23.21	24.63	24.83	25.01	25.21
800	0.212	0.222	0.234	0.255	0.280	0.34	800	22.82	22.99	23.15	23.32	24.56	24.74	24.89	25.06
1000	0.170	0.177	0.188	0.204	0.224	0.271	1000	22.94	23.10	23.24	23.40	24.51	24.67	24.80	24.96
1500	0.113	0.118	0.125	0.136	0.150	0.181	1500	23.14	23.27	23.38	23.51	24.42	24.55	24.66	24.79
2000	0.085	0.089	0.094	0.102	0.112	0.136	2000	23.26	23.37	23.47	23.58	24.37	24.48	24.58	24.69
3000	0.057	0.059	0.063	0.068	0.075	0.091	3000	23.40	23.49	23.57	23.66	24.31	24.40	24.48	24.57
4000	0.043	0.045	0.047	0.051	0.056	0.068	4000	23.48	23.56	23.63	23.71	24.27	24.35	24.42	24.50
5000	0.034	0.039	0.038	0.041	0.045	0.054	5000	23.54	23.61	23.67	23.74	24.24	24.31	24.37	24.45

Table A.6 Upper Percentiles for Test of Single Multivariate Normal Outlier

Upper percentage points for the test statistic

$$D^2_{(n)} = \max_{1 \leq i \leq n} (\mathbf{y}_i - \bar{\mathbf{y}})' \mathbf{S}^{-1} (\mathbf{y}_i - \bar{\mathbf{y}}).$$

This tests for a single outlier in a sample of size n from a multivariate normal distribution. Reject and conclude that the outlier is significant if $D^2_{(n)}$ exceeds the table value.

	$p = 2$		$p = 3$		$p = 4$		$p = 5$	
n	$\alpha = 0.05$	$\alpha = 0.01$	$\alpha = 0.05$	$\alpha = 0.01$	$\alpha = 0.05$	$\alpha = 0.01$	$\alpha = 0.05$	$\alpha = 0.01$
5	3.17	3.19						
6	4.00	4.11	4.14	4.16				
7	4.71	4.95	5.01	5.10	5.12	5.14		
8	5.32	5.70	5.77	5.97	6.01	6.09	6.11	6.12
9	5.85	6.37	6.43	6.76	6.80	6.97	7.01	7.08
10	6.32	6.97	7.01	7.47	7.50	7.79	7.82	7.98
12	7.10	8.00	7.99	8.70	8.67	9.20	9.19	9.57
14	7.74	8.84	8.78	9.71	9.61	10.37	10.29	10.90
16	8.27	9.54	9.44	10.56	10.39	11.36	11.20	12.02
18	8.73	10.15	10.00	11.28	11.06	12.20	11.96	12.98
20	9.13	10.67	10.49	11.91	11.63	12.93	12.62	13.81
25	9.94	11.73	11.48	13.18	12.78	14.40	13.94	15.47
30	10.58	12.54	12.24	14.14	13.67	15.51	14.95	16.73
35	11.10	13.20	12.85	14.92	14.37	16.40	15.75	17.73
40	11.53	13.74	13.36	15.56	14.96	17.13	16.41	18.55
45	11.90	14.20	13.80	16.10	15.46	17.74	16.97	19.24
50	12.23	14.60	14.18	16.56	15.89	18.27	17.45	19.83
100	14.22	16.95	16.45	19.26	18.43	21.30	20.26	23.17
200	15.99	18.94	18.42	21.47	20.59	23.72	22.59	25.82
500	18.12	21.22	20.75	23.95	23.06	26.37	25.21	28.62

Table A.7 Upper Percentage Points of Hotelling's T^2 Distribution

Degrees of Freedom, ν	$p = 1$	$p = 2$	$p = 3$	$p = 4$	$p = 5$	$p = 6$	$p = 7$	$p = 8$	$p = 9$	$p = 10$
				$\alpha = 0.05$						
2	18.513									
3	10.128	57.000								
4	7.709	25.472	114.986							
5	6.608	17.361	46.383	192.468						
6	5.987	13.887	29.661	72.937	289.446					
7	5.591	12.001	22.720	44.718	105.157	405.920				
8	5.318	10.828	19.028	33.230	62.561	143.050	541.890			
9	5.117	10.033	16.766	27.202	45.453	83.202	186.622	697.356		
10	4.965	9.459	15.248	23.545	36.561	59.403	106.649	235.873	872.317	
11	4.844	9.026	14.163	21.108	31.205	47.123	75.088	132.903	290.806	1066.774
12	4.747	8.689	13.350	19.376	27.656	39.764	58.893	92.512	161.967	351.421
13	4.667	8.418	12.719	18.086	25.145	34.911	49.232	71.878	111.676	193.842
14	4.600	8.197	12.216	17.089	23.281	31.488	42.881	59.612	86.079	132.582
15	4.543	8.012	11.806	16.296	21.845	28.955	38.415	51.572	70.907	101.499
16	4.494	7.856	11.465	15.651	20.706	27.008	35.117	45.932	60.986	83.121
17	4.451	7.722	11.177	15.117	19.782	25.467	32.588	41.775	54.041	71.127
18	4.414	7.606	10.931	14.667	19.017	24.219	30.590	38.592	48.930	62.746
19	4.381	7.504	10.719	14.283	18.375	23.189	28.975	36.082	45.023	56.587
20	4.351	7.415	10.533	13.952	17.828	22.324	27.642	34.054	41.946	51.884
21	4.325	7.335	10.370	13.663	17.356	21.588	26.525	32.384	39.463	48.184
22	4.301	7.264	10.225	13.409	16.945	20.954	25.576	30.985	37.419	45.202
23	4.279	7.200	10.095	13.184	16.585	20.403	24.759	29.798	35.709	42.750
24	4.260	7.142	9.979	12.983	16.265	19.920	24.049	28.777	34.258	40.699
25	4.242	7.089	9.874	12.803	15.981	19.492	23.427	27.891	33.013	38.961
26	4.225	7.041	9.779	12.641	15.726	19.112	22.878	27.114	31.932	37.469
27	4.210	6.997	9.692	12.493	15.496	18.770	22.388	26.428	30.985	36.176
28	4.196	6.957	9.612	12.359	15.287	18.463	21.950	25.818	30.149	35.043
29	4.183	6.919	9.539	12.236	15.097	18.184	21.555	25.272	29.407	34.044
30	4.171	6.885	9.471	12.123	14.924	17.931	21.198	24.781	28.742	33.156
35	4.121	6.744	9.200	11.674	14.240	16.944	19.823	22.913	26.252	29.881
40	4.085	6.642	9.005	11.356	13.762	16.264	18.890	21.668	24.624	27.783
45	4.057	6.564	8.859	11.118	13.409	15.767	18.217	20.781	23.477	26.326
50	4.034	6.503	8.744	10.934	13.138	15.388	17.709	20.117	22.627	25.256
55	4.016	6.454	8.652	10.787	12.923	15.090	17.311	19.600	21.972	24.437
60	4.001	6.413	8.577	10.668	12.748	14.850	16.992	19.188	21.451	23.790
70	3.978	6.350	8.460	10.484	12.482	14.485	16.510	18.571	20.676	22.834
80	3.960	6.303	8.375	10.350	12.289	14.222	16.165	18.130	20.127	22.162
90	3.947	6.267	8.309	10.248	12.142	14.022	15.905	17.801	19.718	21.663
100	3.936	6.239	8.257	10.167	12.027	13.867	15.702	17.544	19.401	21.279
110	3.927	6.216	8.215	10.102	11.934	13.741	15.540	17.340	19.149	20.973
120	3.920	6.196	8.181	10.048	11.858	13.639	15.407	17.172	18.943	20.725
150	3.904	6.155	8.105	9.931	11.693	13.417	15.121	16.814	18.504	20.196
200	3.888	6.113	8.031	9.817	11.531	13.202	14.845	16.469	18.083	19.692
400	3.865	6.052	7.922	9.650	11.297	12.890	14.447	15.975	17.484	18.976
1000	3.851	6.015	7.857	9.552	11.160	12.710	14.217	15.692	17.141	18.570
∞	3.841	5.991	7.815	9.488	11.070	12.592	14.067	15.507	16.919	18.307

Table A.7 (*Continued*)

Degrees of Freedom, ν	$p = 1$	$p = 2$	$p = 3$	$p = 4$	$p = 5$	$p = 6$	$p = 7$	$p = 8$	$p = 9$	$p = 10$
					$\alpha = 0.01$					
2	98.503									
3	34.116	297.000								
4	21.198	82.177	594.997							
5	16.258	45.000	147.283	992.494						
6	13.745	31.857	75.125	229.679	1489.489					
7	12.246	25.491	50.652	111.839	329.433	2085.984				
8	11.259	21.821	39.118	72.908	155.219	446.571	2781.978			
9	10.561	19.460	32.598	54.890	98.703	205.293	581.106	3577.472		
10	10.044	17.826	28.466	44.838	72.882	128.067	262.076	733.045	4472.464	
11	9.646	16.631	25.637	38.533	58.618	93.127	161.015	325.576	902.392	5466.956
12	9.330	15.722	23.588	34.251	49.739	73.969	115.640	197.555	395.797	1089.149
13	9.074	15.008	22.041	31.171	43.745	62.114	90.907	140.429	237.692	472.742
14	8.862	14.433	20.834	28.857	39.454	54.150	75.676	109.441	167.499	281.428
15	8.683	13.960	19.867	27.060	36.246	48.472	65.483	90.433	129.576	196.853
16	8.531	13.566	19.076	25.626	33.672	44.240	58.241	77.755	106.391	151.316
17	8.400	13.231	18.418	24.458	31.788	40.975	52.858	68.771	90.969	123.554
18	8.285	12.943	17.861	23.487	30.182	38.385	48.715	62.109	80.067	105.131
19	8.185	12.694	17.385	22.670	28.852	36.283	45.435	56.992	71.999	92.134
20	8.096	12.476	16.973	21.972	27.734	34.546	42.779	52.948	65.813	82.532
21	8.017	12.283	16.613	21.369	26.781	33.088	40.587	49.679	60.932	75.181
22	7.945	12.111	16.296	20.843	25.959	31.847	38.750	46.986	56.991	69.389
23	7.881	11.958	16.015	20.381	25.244	30.779	37.188	44.730	53.748	64.719
24	7.823	11.820	15.763	19.972	24.616	29.850	35.846	42.816	51.036	60.879
25	7.770	11.695	15.538	19.606	24.060	29.036	34.680	41.171	48.736	57.671
26	7.721	11.581	15.334	19.279	23.565	28.316	33.659	39.745	46.762	54.953
27	7.677	11.478	15.149	18.983	23.121	27.675	32.756	38.496	45.051	52.622
28	7.636	11.383	14.980	18.715	22.721	27.101	31.954	37.393	43.554	50.604
29	7.598	11.295	14.825	18.471	22.359	26.584	31.236	36.414	42.234	48.839
30	7.562	11.215	14.683	18.247	22.029	26.116	30.589	35.538	41.062	47.283
35	7.419	10.890	14.117	17.366	20.743	24.314	28.135	32.259	36.743	41.651
40	7.314	10.655	13.715	16.750	19.858	23.094	26.502	30.120	33.984	38.135
45	7.234	10.478	13.414	16.295	19.211	22.214	25.340	28.617	32.073	35.737
50	7.171	10.340	13.181	15.945	18.718	21.550	24.470	27.504	30.673	33.998
55	7.119	10.228	12.995	15.667	18.331	21.030	23.795	26.647	29.603	32.682
60	7.077	10.137	12.843	15.442	18.018	20.613	23.257	25.967	28.760	31.650
70	7.011	9.996	12.611	15.098	17.543	19.986	22.451	24.957	27.515	30.139
80	6.963	9.892	12.440	14.849	17.201	19.536	21.877	24.242	26.642	29.085
90	6.925	9.813	12.310	14.660	16.942	19.197	21.448	23.710	25.995	28.310
100	6.895	9.750	12.208	14.511	16.740	18.934	21.115	23.299	25.496	27.714
110	6.871	9.699	12.125	14.391	16.577	18.722	20.849	22.972	25.101	27.243
120	6.851	9.657	12.057	14.292	16.444	18.549	20.632	22.705	24.779	26.862
150	6.807	9.565	11.909	14.079	16.156	18.178	20.167	22.137	24.096	26.054
200	6.763	9.474	11.764	13.871	15.877	17.819	19.720	21.592	23.446	25.287
400	6.699	9.341	11.551	13.569	15.473	17.303	19.080	20.818	22.525	24.209
1000	6.660	9.262	11.426	13.392	15.239	17.006	18.743	20.376	22.003	23.600
∞	6.635	9.210	11.345	13.277	15.086	16.812	18.475	20.090	21.666	23.209

Note: p = number of variables.

Table A.8 Bonferonni t-Values, $t_{\alpha/2k,\nu}$, $\alpha = 0.05$

	(1)	(2)	(3)	(4)	(5)	(6)	(7)	(8)	(9)	(10)
	100α/k									
ν	5.0000	2.5000	1.6667	1.2500	1.0000	0.8333	0.7143	0.6250	0.5556	0.5000
2	4.3027	6.2053	7.6488	8.8602	9.9248	10.8859	11.7687	12.5897	13.3604	14.0890
3	3.1824	4.1765	4.8567	5.3919	5.8409	6.2315	6.5797	6.8952	7.1849	7.4533
4	2.7764	3.4954	3.9608	4.3147	4.6041	4.8510	5.0675	5.2611	5.4366	5.5976
5	2.5706	3.1634	3.5341	3.8100	4.0321	4.2193	4.3818	4.5257	4.6553	4.7733
6	2.4469	2.9687	3.2875	3.5212	3.7074	2.8630	3.9971	4.1152	4.2209	4.3168
7	2.3646	2.8412	3.1276	3.3353	3.4995	3.6358	3.7527	3.8552	3.9467	4.0293
8	2.3060	2.7515	3.0158	3.2060	3.3554	3.4789	3.5844	3.6766	3.7586	3.8325
9	2.2622	2.6850	2.9333	3.1109	3.2498	3.3642	3.4616	3.5465	3.6219	3.6897
10	2.2281	2.6338	2.8701	3.0382	3.1693	3.2768	3.3682	3.4477	3.5182	3.5814
11	2.2010	2.5931	2.8200	2.9809	3.1058	3.2081	3.2949	3.3702	3.4368	3.4966
12	2.1788	2.5600	2.7795	2.9345	3.0545	3.1527	3.2357	3.3078	3.3714	3.4284
13	2.1604	2.5326	2.7459	2.8961	3.0123	3.1070	3.1871	3.2565	3.3177	3.3725
14	2.1448	2.5096	2.7178	2.8640	2.9768	3.0688	3.1464	3.2135	3.2727	3.3257
15	2.1314	2.4899	2.6937	2.8366	2.9467	3.0363	3.1118	3.1771	3.2346	3.2860
16	2.1199	2.4729	2.6730	2.8131	2.9208	3.0083	3.0821	3.1458	3.2019	3.2520
17	2.1098	2.4581	2.6550	2.7925	2.8982	2.9840	3.0563	3.1186	3.1735	3.2224
18	2.1009	2.4450	2.6391	2.7745	2.8784	2.9627	3.0336	3.0948	3.1486	3.1966
19	2.0930	2.4334	2.6251	2.7586	2.8609	2.9439	3.0136	3.0738	3.1266	3.1737
20	2.0860	2.4231	2.6126	2.7444	2.8453	2.9271	2.9958	3.0550	3.1070	3.1534
21	2.0796	2.4138	2.6013	2.7316	2.8314	2.9121	2.9799	3.0382	3.0895	3.1352
22	2.0739	2.4055	2.5912	2.7201	2.8188	2.8985	2.9655	3.0231	3.0737	3.1188
23	2.0687	2.3979	2.5820	2.7097	2.8073	2.8863	2.9525	3.0095	3.0595	3.1040
24	2.0639	2.3909	2.5736	2.7002	2.7969	2.8751	2.9406	2.9970	3.0465	3.0905
25	2.0595	2.3846	2.5660	2.6916	2.7874	2.8649	2.9298	2.9856	3.0346	3.0782
26	2.0555	2.3788	2.5589	2.6836	2.7787	2.8555	2.9199	2.9752	3.0237	3.0669
27	2.0518	2.3734	2.5525	2.6763	2.7707	2.8469	2.9107	2.9656	3.0137	3.0565
28	2.0484	2.3685	2.5465	2.6695	2.7633	2.8389	2.9023	2.9567	3.0045	3.0469
29	2.0452	2.3638	2.5409	2.6632	2.7564	2.8316	2.8945	2.9485	2.9959	3.0380
30	2.0423	2.3596	2.5357	2.6574	2.7500	2.8247	2.8872	2.9409	2.9880	3.0298
35	2.0301	2.3420	2.5145	2.6334	2.7238	2.7966	2.8575	2.9097	2.9554	2.9960
40	2.0211	2.3289	2.4989	2.6157	2.7045	2.7759	2.8355	2.8867	2.9314	2.9712
45	2.0141	2.3189	2.4868	2.6021	2.6896	2.7599	2.8187	2.8690	2.9130	2.9521
50	2.0086	2.3109	2.4772	2.5913	2.6778	2.7473	2.8053	2.8550	2.8984	2.9370
55	2.0040	2.3044	2.4694	2.5825	2.6682	2.7370	2.7944	2.8436	2.8866	2.9247
60	2.0003	2.2990	2.4630	2.5752	2.6603	2.7286	2.7855	2.8342	2.8768	2.9146
70	1.9944	2.2906	2.4529	2.5639	2.6479	2.7153	2.7715	2.8195	2.8615	2.8987
80	1.9901	2.2844	2.4454	2.5554	2.6387	2.7054	2.7610	2.8086	2.8502	2.8870
90	1.9867	2.2795	2.4395	2.5489	2.6316	2.6978	2.7530	2.8002	2.8414	2.8779
100	1.9840	2.2757	2.4349	2.5437	2.6259	2.6918	2.7466	2.7935	2.8344	2.8707
110	1.9818	2.2725	2.4311	2.5394	2.6213	2.6868	2.7414	2.7880	2.8287	2.8648
120	1.9799	2.2699	2.4280	2.5359	2.6174	2.6827	2.7370	2.7835	2.8240	2.8599
250	1.9695	2.2550	2.4102	2.5159	2.5956	2.6594	2.7124	2.7577	2.7972	2.8322
500	1.9647	2.2482	2.4021	2.5068	2.5857	2.6488	2.7012	2.7460	2.7850	2.8195
1000	1.9623	2.2448	2.3980	2.5022	2.5808	2.6435	2.6957	2.7402	2.7790	2.8133
∞	1.9600	2.2414	2.3940	2.4977	2.5758	2.6383	2.6901	2.7344	2.7729	2.8070

Note: Column numbers in parentheses are k-values.

Table A.8 (*Continued*)

	(11)	(12)	(13)	(14)	(15) $100\alpha/k$	(16)	(17)	(18)	(19)
ν	0.4545	0.4167	0.3846	0.3571	0.3333	0.3125	0.2941	0.2778	0.2632
2	14.7818	15.4435	16.0780	16.6883	17.2772	17.8466	18.3984	18.9341	19.4551
3	7.7041	7.9398	8.1625	8.3738	8.5752	8.7676	8.9521	9.1294	9.3001
4	5.7465	5.8853	6.0154	6.1380	6.2541	6.3643	6.4693	6.5697	6.6659
5	4.8819	4.9825	5.0764	5.1644	5.2474	5.3259	5.4005	5.4715	5.5393
6	4.4047	4.4858	4.5612	4.6317	4.6979	4.7604	4.8196	4.8759	4.9295
7	4.1048	4.1743	4.2388	4.2989	4.3553	4.4084	4.4586	4.5062	4.5514
8	3.8999	3.9618	4.0191	4.0724	4.1224	4.1693	4.2137	4.2556	4.2955
9	3.7513	3.8079	3.8602	3.9088	3.9542	3.9969	4.0371	4.0752	4.1114
10	3.6388	3.6915	3.7401	3.7852	3.8273	3.8669	3.9041	3.9394	3.9728
11	3.5508	3.6004	3.6462	3.6887	3.7283	3.7654	3.8004	3.8335	3.8648
12	3.4801	3.5274	3.5709	3.6112	3.6489	3.6842	3.7173	3.7487	3.7783
13	3.4221	3.4674	3.5091	3.5478	3.5838	3.6176	3.6493	3.6793	3.7076
14	3.3736	3.4173	3.4576	3.4949	3.5296	3.5621	3.5926	3.6214	3.6487
15	3.3325	3.3749	3.4139	3.4501	3.4837	3.5151	3.5447	3.5725	3.5989
16	3.2973	3.3386	3.3765	3.4116	3.4443	3.4749	3.5036	3.5306	3.5562
17	3.2667	3.3070	3.3440	3.3783	3.4102	3.4400	3.4680	3.4944	3.5193
18	3.2399	3.2794	3.3156	3.3492	3.3804	3.4095	3.4369	3.4626	3.4870
19	3.2163	3.2550	3.2906	3.3235	3.3540	3.3826	3.4094	3.4347	3.4585
20	3.1952	3.2333	3.2683	3.3006	3.3306	3.3587	3.3850	3.4098	3.4332
21	3.1764	3.2139	3.2483	3.2802	3.3097	3.3373	3.3632	3.3876	3.4106
22	3.1595	3.1965	3.2304	3.2618	3.2909	3.3181	3.3436	3.3676	3.3903
23	3.1441	3.1807	3.2142	3.2451	3.2739	3.3007	3.3259	3.3495	3.3719
24	3.1302	3.1663	3.1994	3.2300	3.2584	3.2849	3.3097	3.3331	3.3552
25	3.1175	3.1532	3.1859	3.2162	3.2443	3.2705	3.2950	3.3181	3.3400
26	3.1058	3.1412	3.1736	3.2035	3.2313	3.2572	3.2815	3.3044	3.3260
27	3.0951	3.1301	3.1622	3.1919	3.2194	3.2451	3.2691	3.2918	3.3132
28	3.0852	3.1199	3.1517	3.1811	3.2084	3.2339	3.2577	3.2801	3.3013
29	3.0760	3.1105	3.1420	3.1712	3.1982	3.2235	3.2471	3.2694	3.2904
30	3.0675	3.1017	3.1330	3.1620	3.1888	3.2138	3.2373	3.2594	3.2802
35	3.0326	3.0658	3.0962	3.1242	3.1502	3.1744	3.1971	3.2185	3.2386
40	3.0069	3.0393	3.0690	3.0964	3.1218	3.1455	3.1676	3.1884	3.2081
45	2.9872	3.0191	3.0482	3.0751	3.1000	3.1232	3.1450	3.1654	3.1846
50	2.9716	3.0030	3.0318	3.0582	3.0828	3.1057	3.1271	3.1472	3.1661
55	2.9589	2.9900	3.0184	3.0446	3.0688	3.0914	3.1125	3.1324	3.1511
60	2.9485	2.9792	3.0074	3.0333	3.0573	3.0796	3.1005	3.1202	3.1387
70	2.9321	2.9624	2.9901	3.0156	3.0393	3.0613	3.0818	3.1012	3.1194
80	2.9200	2.9500	2.9773	3.0026	3.0259	3.0476	3.0679	3.0870	3.1050
90	2.9106	2.9403	2.9675	2.9924	3.0156	3.0371	3.0572	3.0761	3.0939
100	2.9032	2.9327	2.9596	2.9844	3.0073	3.0287	3.0487	3.0674	3.0851
110	2.8971	2.9264	2.9532	2.9778	3.0007	3.0219	3.0417	3.0604	3.0779
120	2.8921	2.9212	2.9479	2.9724	2.9951	3.0162	3.0360	3.0545	3.0720
250	2.8635	2.8919	2.9178	2.9416	2.9637	2.9842	3.0034	3.0213	3.0383
500	2.8505	2.8785	2.9041	2.9276	2.9494	2.9696	2.9885	3.0063	3.0230
1000	2.8440	2.8719	2.8973	2.9207	2.9423	2.9624	2.9812	2.9988	3.0154
∞	2.8376	2.8653	2.8905	2.9137	2.9352	2.9552	2.9738	2.9913	3.0078

Table A.9 Lower Critical Values of Wilks Λ, α = 0.05

$$\Lambda = \frac{|\mathbf{E}|}{|\mathbf{E}+\mathbf{H}|} = \prod_{i=1}^{s} \frac{1}{1+\lambda_i}$$

where $\lambda_1, \lambda_2, \ldots, \lambda_s$ are eigenvalues of $\mathbf{E}^{-1}\mathbf{H}$. Reject H_0 if $\Lambda \le$ table value.

ν_E	$\nu_H = 1$	$\nu_H = 2$	$\nu_H = 3$	$\nu_H = 4$	$\nu_H = 5$	$\nu_H = 6$	$\nu_H = 7$	$\nu_H = 8$	$\nu_H = 9$	$\nu_H = 10$	$\nu_H = 11$	$\nu_H = 12$	ν_E
							$p = 1$						
1	.006157	.002501	.001543	.001112	.000868	.000712	.000603	.000523	.000462	.000413	.000374	.000341	1
2	.097504	.050003	.033615	.025322	.020309	.016953	.014549	.012741	.011333	.010208	.009281	.008512	2
3	.228516	.135712	.097321	.076019	.062408	.052963	.046005	.040672	.036446	.033020	.030182	.027794	3
4	.341614	.223602	.168243	.135345	.113373	.097610	.085724	.076447	.068985	.062851	.057724	.053375	4
5	.430725	.301697	.235535	.194031	.165283	.144073	.127777	.114822	.104279	.095505	.088120	.081787	5
6	.500549	.368408	.295990	.248596	.214783	.189255	.169266	.153168	.139893	.128754	.119278	.111115	6
7	.555908	.424896	.349304	.298096	.260620	.231812	.208893	.190186	.174606	.161423	.150116	.140289	7
8	.600708	.472870	.396057	.342590	.302612	.271332	.246124	.225311	.207825	.192902	.180008	.168747	8
9	.637512	.513916	.437164	.382446	.340790	.307770	.280823	.258362	.239288	.222931	.208679	.196182	9
10	.668243	.549286	.473389	.418213	.375519	.341248	.313019	.289246	.268936	.251373	.235992	.222443	10
11	.694275	.580017	.505463	.450317	.407104	.372040	.342834	.318054	.296768	.278229	.261932	.247467	11
12	.716553	.606964	.534027	.479309	.435913	.400299	.370453	.344940	.322876	.303528	.286469	.271240	12
13	.735840	.630737	.559570	.505524	.462189	.426361	.396057	.369995	.347321	.327362	.309662	.293823	13
14	.752686	.651825	.582581	.529327	.486267	.450348	.419800	.393372	.370239	.349823	.331589	.315247	14
15	.767548	.670715	.603333	.551025	.508362	.472534	.441864	.415222	.391754	.370941	.352325	.335541	15
16	.780701	.687653	.622162	.570862	.528717	.493103	.462433	.435638	.411957	.390869	.371918	.354797	16
17	.792480	.702972	.639343	.589081	.547516	.512177	.481598	.454742	.430939	.409637	.390472	.373077	17
18	.803070	.716858	.655029	.605835	.564911	.529907	.499481	.472687	.448807	.427368	.408020	.390411	18
19	.812622	.729553	.669434	.621307	.581024	.546448	.516235	.489502	.465637	.444138	.424652	.406891	19
20	.821320	.741135	.682709	.635651	.596039	.561890	.531952	.505341	.481506	.459991	.440430	.422546	20

Table A.9 (*Continued*)

ν_E	$\nu_H = 1$	$\nu_H = 2$	$\nu_H = 3$	$\nu_H = 4$	$\nu_H = 5$	$\nu_H = 6$	$\nu_H = 7$	$\nu_H = 8$	$\nu_H = 9$	$\nu_H = 10$	$\nu_H = 11$	$\nu_H = 12$	ν_E
21	.829224	.751770	.694977	.648941	.610046	.576355	.546692	.520264	.496521	.475006	.455414	.437469	21
22	.836472	.761597	.706329	.661316	.623108	.589905	.560562	.534332	.510712	.489258	.469635	.451660	22
23	.843140	.770660	.716858	.672867	.635361	.602631	.573639	.547638	.524139	.502762	.483185	.465179	23
24	.849274	.779083	.726685	.683655	.646851	.614609	.585968	.560211	.536896	.515594	.496078	.478088	24
25	.854950	.786896	.735870	.693771	.657639	.625900	.597626	.572128	.548981	.527817	.508362	.490402	25
26	.860199	.794189	.744446	.703278	.667786	.636566	.608643	.583435	.560486	.539459	.520081	.502167	26
27	.865112	.800995	.752487	.712189	.677383	.646637	.619080	.594147	.571411	.550537	.531281	.513428	27
28	.869675	.807373	.760040	.720612	.686432	.656174	.628998	.604370	.581833	.561127	.541962	.524200	28
29	.873947	.813339	.767151	.728546	.694992	.665222	.638428	.614075	.591766	.571228	.552200	.534515	29
30	.877945	.818970	.773865	.736053	.703110	.673798	.647385	.623322	.601242	.580872	.561996	.544418	30
40	.907349	.860886	.824463	.793274	.765594	.740540	.717575	.696365	.676636	.658188	.640884	.624603	40
60	.937485	.904968	.878807	.855911	.835175	.816055	.798233	.781494	.765686	.750702	.736420	.722809	60
80	.952827	.927841	.907471	.889450	.872940	.857590	.843124	.829437	.816391	.803925	.791962	.780464	80
100	.962128	.941845	.925179	.910324	.896637	.883835	.871696	.860153	.849083	.838455	.828201	.818314	100
120	.968363	.951297	.937200	.924578	.912894	.901916	.891475	.881501	.871901	.862660	.853706	.845045	120
140	.972836	.958107	.945890	.934921	.924731	.915131	.905971	.897200	.888734	.880563	.872625	.864929	140
170	.977588	.965370	.955195	.946025	.937478	.929401	.921669	.914245	.907057	.900101	.893324	.886738	170
200	.980926	.970487	.961768	.953893	.946532	.939564	.932877	.926443	.920200	.914149	.908239	.902486	200
240	.984086	.975345	.968024	.961396	.955187	.949296	.943631	.938171	.932861	.927705	.922660	.917740	240
320	.988046	.981451	.975907	.970876	.966145	.961649	.957311	.953121	.949035	.945058	.941155	.937344	320
440	.991295	.986475	.982411	.978715	.975232	.971914	.968704	.965599	.962561	.959605	.956692	.953846	440
600	.993610	.990064	.987067	.984337	.981759	.979301	.976917	.974611	.972349	.970144	.967969	.965842	600
800	.995204	.992539	.990282	.988225	.986279	.984422	.982619	.980873	.979158	.977487	.975834	.974218	800
1000	.996161	.994026	.992216	.990566	.989003	.987512	.986062	.984658	.983276	.981931	.980598	.979296	1000

Table A.9 *(Continued)*

ν_E	$\nu_H = 1$	$\nu_H = 2$	$\nu_H = 3$	$\nu_H = 4$	$\nu_H = 5$	$\nu_H = 6$	$\nu_H = 7$	$\nu_H = 8$	$\nu_H = 9$	$\nu_H = 10$	$\nu_H = 11$	$\nu_H = 12$	ν_E
						$p = 2$							
1	.000000	.000000	.000000	.000000	.000000	.000000	.000000	.000000	.000000	.000000	.000000	.000000	1
2	.002500	.000641	.000287	.000162	.000104	.000072	.000053	.000041	.000032	.000026	.000022	.000018	2
3	.049998	.018318	.009528	.005844	.003950	.002849	.002152	.001683	.001352	.001110	.000928	.000787	3
4	.135725	.061800	.035817	.023460	.016578	.012346	.009555	.007615	.006212	.005165	.004362	.003734	4
5	.223606	.117368	.073621	.050765	.037211	.028476	.022507	.018244	.015092	.012695	.010826	.009343	5
6	.301715	.174902	.116450	.083663	.063188	.049481	.039834	.032772	.027440	.023320	.020068	.017453	6
7	.368405	.229737	.160239	.118984	.092129	.073571	.060172	.050155	.042465	.036426	.031600	.027678	7
8	.424876	.280187	.202813	.154741	.122376	.099380	.082397	.069475	.059404	.051386	.044908	.039579	8
9	.472866	.325883	.243151	.189781	.152779	.125881	.105643	.089993	.077615	.067661	.059515	.052772	9
10	.513885	.367036	.280802	.223433	.182644	.152421	.129282	.111138	.096610	.084797	.075044	.066901	10
11	.549281	.404052	.315720	.255369	.211592	.178545	.152898	.132506	.116013	.102453	.091178	.081680	11
12	.580029	.437339	.347988	.285511	.239373	.203997	.176155	.153782	.135511	.120356	.107656	.096885	12
13	.606971	.467384	.377744	.313837	.265838	.228568	.198874	.174774	.154909	.138311	.124284	.112321	13
14	.630737	.494599	.405216	.340396	.291016	.252171	.220930	.195325	.174061	.156149	.140923	.127849	14
15	.651851	.519281	.430564	.365263	.314863	.274786	.242249	.215357	.192837	.173755	.157442	.143350	15
16	.670711	.541775	.454003	.388530	.337412	.296391	.262763	.234782	.211185	.191059	.173755	.158740	16
17	.687662	.562317	.475724	.410322	.358763	.316990	.282502	.253583	.229036	.208000	.189807	.173946	17
18	.702982	.581146	.495888	.430784	.378964	.336632	.301430	.271723	.246366	.224530	.205530	.188918	18
19	.716866	.598489	.514629	.449961	.398041	.355335	.319573	.289225	.263169	.240614	.220915	.203611	19
20	.729531	.614483	.532092	.467968	.416109	.373163	.336951	.306072	.279429	.256249	.235937	.218013	20

Table A.9 (*Continued*)

ν_E	$\nu_H = 1$	$\nu_H = 2$	$\nu_H = 3$	$\nu_H = 4$	$\nu_H = 5$	$\nu_H = 6$	$\nu_H = 7$	$\nu_H = 8$	$\nu_H = 9$	$\nu_H = 10$	$\nu_H = 11$	$\nu_H = 12$	ν_E
21	.741124	.629283	.548399	.484925	.433211	.390129	.353609	.322287	.295147	.271437	.250565	.232083	21
22	.751776	.643011	.563622	.500886	.449429	.406286	.369555	.337873	.310325	.286147	.264800	.245821	22
23	.761598	.655775	.577893	.515922	.464800	.421699	.384810	.352883	.324978	.300409	.278639	.259224	23
24	.770680	.667666	.591286	.530135	.479373	.436391	.399429	.367295	.339116	.314213	.292087	.272280	24
25	.779088	.678783	.603884	.543551	.493227	.450412	.413436	.381165	.352775	.327593	.305127	.285006	25
26	.786893	.689182	.615752	.556269	.506409	.463802	.426867	.394506	.365946	.340539	.317798	.297372	26
27	.794192	.698945	.626937	.568306	.518951	.476588	.439744	.407337	.378645	.353047	.330095	.309407	27
28	.800992	.708108	.637517	.579727	.530891	.488822	.452093	.419700	.390911	.365171	.342019	.321110	28
29	.807354	.716737	.647497	.590582	.542291	.500519	.463948	.431586	.402753	.376900	.353591	.332484	29
30	.813343	.724899	.656962	.600899	.553155	.511722	.475325	.443028	.414182	.388244	.364802	.343537	30
40	.857594	.786433	.729818	.681627	.639419	.601870	.568076	.537426	.509476	.483873	.460296	.438550	40
60	.903437	.852599	.810662	.773804	.740586	.710190	.682157	.656096	.631804	.609029	.587643	.567501	60
80	.926967	.887496	.854347	.824736	.797636	.772490	.748974	.726849	.705927	.686107	.667279	.649328	80
100	.941272	.909051	.881684	.856993	.834186	.812834	.792697	.773596	.755405	.738034	.721395	.705440	100
120	.950898	.923673	.900382	.879233	.859569	.841056	.823491	.806739	.790700	.775302	.760485	.746201	120
140	.957812	.934247	.913983	.895493	.878224	.861896	.846339	.831442	.817125	.803326	.789999	.777105	140
170	.965169	.945562	.928606	.913057	.898465	.884603	.871338	.858581	.846267	.834352	.822797	.811574	170
200	.970341	.953554	.938982	.925569	.912940	.900904	.889349	.878202	.867412	.856939	.846755	.836834	200
240	.975243	.961158	.948887	.937554	.926848	.916613	.906758	.897224	.887968	.878959	.870174	.861593	240
320	.981393	.970741	.961415	.952766	.944563	.936691	.929082	.921692	.914493	.907461	.900579	.893835	320
440	.986445	.978644	.971788	.965408	.959337	.953491	.947824	.942303	.936908	.931623	.926435	.921337	440
600	.990047	.984298	.979233	.974507	.969998	.965648	.961420	.957293	.953251	.949283	.945380	.941537	600
800	.992529	.988203	.984384	.980814	.977404	.974108	.970900	.967763	.964687	.961662	.958683	.955744	800
1000	.994021	.990552	.987487	.984620	.981877	.979224	.976640	.974110	.971627	.969184	.966775	.964397	1000

Table A.9 ***(Continued)***

ν_E	$\nu_H = 1$	$\nu_H = 2$	$\nu_H = 3$	$\nu_H = 4$	$\nu_H = 5$	$\nu_H = 6$	$\nu_H = 7$	$\nu_H = 8$	$\nu_H = 9$	$\nu_H = 10$	$\nu_H = 11$	$\nu_H = 12$	ν_E
						$p = 3$							
1	.000000	.000000	.000000	.000000	.000000	.000000	.000000	.000000	.000000	.000000	.000000	.000000	1
2	.000000	.000000	.000000	.000000	.000000	.000001	.000002	.000004	.000005	.000008	.000010	.000013	2
3	.001698	.000354	.000179	.000127	.000105	.000095	.000091	.000090	.000091	.000092	.000095	.000098	3
4	.033740	.009612	.004205	.002314	.001479	.001052	.000809	.000659	.000562	.000496	.000449	.000416	4
5	.097355	.035855	.017521	.010010	.006357	.004369	.003195	.002458	.001971	.001636	.001397	.001222	5
6	.168271	.073634	.039672	.024047	.015792	.011018	.008067	.006148	.004849	.003939	.003281	.002793	6
7	.235525	.116476	.067711	.043226	.029433	.021043	.015642	.012012	.009485	.007674	.006345	.005347	7
8	.295976	.160244	.098932	.065947	.046378	.033966	.025706	.019990	.015911	.012927	.010697	.008997	8
9	.349277	.202814	.131378	.090794	.065660	.049161	.037855	.029838	.023995	.019637	.016323	.013763	9
10	.396084	.243139	.163846	.116701	.086448	.066012	.051643	.041238	.033514	.027654	.023135	.019593	10
11	.437147	.280808	.195556	.142927	.108110	.083979	.066659	.053876	.044225	.036801	.030993	.026391	11
12	.473377	.315719	.226090	.168939	.130131	.102644	.082534	.067443	.055894	.046882	.039757	.034049	12
13	.505452	.347981	.255220	.194414	.152160	.121656	.098973	.081704	.068298	.057724	.049278	.042437	13
14	.534018	.377735	.282849	.219113	.173959	.140775	.115736	.096413	.081246	.069166	.059407	.051442	14
15	.559570	.405221	.308951	.242944	.195322	.159796	.132619	.111416	.094593	.081052	.070029	.060954	15
16	.582577	.430566	.333588	.265812	.216138	.178574	.149493	.126564	.108178	.093264	.081026	.070875	16
17	.603338	.454006	.356777	.287689	.236338	.197017	.166236	.141728	.121917	.105704	.092299	.081109	17
18	.622168	.475728	.378631	.308599	.255858	.215044	.182762	.156827	.135694	.118273	.103768	.091588	18
19	.639337	.495908	.399223	.328552	.274710	.232604	.199009	.171789	.149446	.130904	.115361	.102241	19
20	.655028	.514622	.418629	.347546	.292843	.249666	.214918	.186544	.163097	.143521	.127018	.113012	20

Table A.9 ***(Continued)***

ν_E	$\nu_H = 1$	$\nu_H = 2$	$\nu_H = 3$	$\nu_H = 4$	$\nu_H = 5$	$\nu_H = 6$	$\nu_H = 7$	$\nu_H = 8$	$\nu_H = 9$	$\nu_H = 10$	$\nu_H = 11$	$\nu_H = 12$	ν_E
21	.669437	.532101	.436898	.365676	.310304	.266216	.230467	.201077	.176620	.156088	.138689	.123835	21
22	.682712	.548393	.454182	.382934	.327083	.282253	.245626	.215325	.189969	.168561	.150321	.134680	22
23	.694960	.563637	.470473	.399402	.343191	.297740	.260397	.229291	.203123	.180907	.161896	.145521	23
24	.706310	.577895	.485889	.415077	.358665	.312738	.274743	.242939	.216044	.193091	.173370	.156313	24
25	.716875	.591311	.500491	.430041	.373523	.327222	.288709	.256276	.228718	.205103	.184720	.167023	25
26	.726681	.603899	.514336	.444332	.387790	.341199	.302238	.269280	.241137	.216929	.195944	.177651	26
27	.735837	.615757	.527453	.457946	.401488	.354711	.315386	.281968	.253300	.228535	.206998	.188160	27
28	.744404	.626944	.539914	.470981	.414658	.367742	.328131	.294313	.265188	.239935	.217899	.198546	28
29	.752437	.637514	.551741	.483431	.427307	.380334	.340477	.306326	.276805	.251110	.228615	.208809	29
30	.759984	.647501	.563023	.495347	.439475	.392490	.352461	.318033	.288158	.262062	.239155	.218912	30
40	.816139	.723938	.651356	.590773	.538846	.493686	.453976	.418785	.387401	.359271	.333940	.311045	40
60	.874843	.807778	.752424	.704238	.661334	.622640	.587440	.555224	.525598	.498272	.472957	.449477	60
80	.905160	.852653	.808266	.768805	.732964	.700027	.669520	.641124	.614572	.589678	.566281	.544236	80
100	.923660	.880557	.843610	.810333	.779746	.751296	.724666	.699598	.675935	.653520	.632235	.611999	100
120	.936178	.899588	.867973	.839253	.812632	.787686	.764150	.741841	.720623	.700389	.681054	.662546	120
140	.945137	.913391	.885776	.860534	.836998	.814820	.793780	.773732	.754565	.736197	.718557	.701592	140
170	.954680	.928199	.904999	.883652	.863624	.844636	.826518	.809156	.792465	.776383	.760857	.745847	170
200	.961395	.938685	.918687	.900202	.882782	.866197	.850307	.835018	.820262	.805990	.792160	.778739	200
240	.967765	.948679	.931793	.916116	.901281	.887100	.873459	.860284	.847521	.835131	.823081	.811346	240
320	.975762	.961296	.948422	.936405	.924972	.913987	.903369	.893064	.883033	.873250	.863692	.854341	320
440	.982336	.971725	.962235	.953337	.944835	.936632	.928671	.920913	.913333	.905910	.898630	.891482	440
600	.987028	.979198	.972173	.965563	.959229	.953099	.947133	.941302	.935589	.929978	.924461	.919029	600
800	.990261	.984364	.979060	.974060	.969257	.964600	.960057	.955610	.951243	.946947	.942713	.938538	800
1000	.992204	.987475	.983215	.979193	.975326	.971571	.967905	.964310	.960776	.957296	.953863	.950473	1000

Table A.9 (*Continued*)

ν_E	$\nu_H = 1$	$\nu_H = 2$	$\nu_H = 3$	$\nu_H = 4$	$\nu_H = 5$	$\nu_H = 6$	$\nu_H = 7$	$\nu_H = 8$	$\nu_H = 9$	$\nu_H = 10$	$\nu_H = 11$	$\nu_H = 12$	ν_E
							$p = 4$						
1	.000000	.000000	.000000	.000000	.000000	.000000	.000000	.000000	.000000	.000000	.000000	.000000	1
2	.000000	.000000	.000000	.000000	.000000	.000000	.000000	.000000	.000000	.000000	.000000	.000000	2
3	.000000	.000000	.000000	.000000	.000000	.000001	.000001	.000001	.000002	.000002	.000002	.000003	3
4	.001378	.000292	.000127	.000075	.000052	.000040	.000033	.000029	.000026	.000025	.000023	.000022	4
5	.025529	.006091	.002314	.001128	.000647	.000416	.000292	.000218	.000172	.000141	.000120	.000105	5
6	.076071	.023604	.010010	.005073	.002903	.001818	.001223	.000872	.000652	.000508	.000409	.000338	6
7	.135374	.050839	.024047	.013014	.007737	.004938	.003338	.002365	.001745	.001333	.001050	.000848	7
8	.194043	.083695	.043226	.024857	.015415	.010129	.006975	.004994	.003698	.002819	.002206	.001766	8
9	.248619	.118995	.065947	.039919	.025729	.017408	.012249	.008907	.006664	.005112	.004009	.003208	9
10	.298130	.154758	.090794	.057378	.038260	.026586	.019107	.014130	.010706	.008288	.006542	.005254	10
11	.342593	.189778	.116701	.076502	.052524	.037385	.027402	.020589	.015806	.012365	.009839	.007948	11
12	.382448	.223411	.142927	.096664	.068077	.049495	.036933	.028170	.021899	.017314	.013895	.011302	12
13	.418181	.255376	.168939	.117377	.084546	.062632	.047493	.036731	.028895	.023075	.018675	.015303	13
14	.450335	.285511	.194414	.138286	.101586	.076537	.058886	.046115	.036676	.029572	.024133	.019917	14
15	.479286	.313829	.219113	.159131	.118954	.090983	.070925	.056188	.045140	.036722	.030208	.025101	15
16	.505512	.340400	.242944	.179688	.136434	.105779	.083443	.066806	.054181	.044440	.036830	.030804	16
17	.529312	.365253	.265812	.199832	.153891	.120780	.096316	.077856	.063688	.052645	.043936	.036980	17
18	.551035	.388530	.287689	.219490	.171171	.135856	.109411	.089236	.073577	.061263	.051456	.043568	18
19	.570858	.410325	.308599	.238570	.188209	.150905	.122643	.100843	.083764	.070213	.059338	.050514	19
20	.589077	.430766	.328552	.257052	.204926	.165853	.135926	.112607	.094180	.079441	.067513	.057782	20

Table A.9 (*Continued*)

ν_E	$\nu_H = 1$	$\nu_H = 2$	$\nu_H = 3$	$\nu_H = 4$	$\nu_H = 5$	$\nu_H = 6$	$\nu_H = 7$	$\nu_H = 8$	$\nu_H = 9$	$\nu_H = 10$	$\nu_H = 11$	$\nu_H = 12$	ν_E
21	.605832	.449947	.347546	.274909	.221288	.180626	.149180	.124462	.104757	.088877	.075938	.065315	21
22	.621318	.467988	.365676	.292142	.237242	.195197	.162364	.136342	.115440	.098474	.084565	.073068	22
23	.635634	.484922	.382934	.308765	.252783	.209511	.175434	.148204	.126185	.108191	.093352	.081008	23
24	.648934	.500883	.399402	.324767	.267896	.223535	.188341	.160009	.136950	.117977	.102254	.089100	24
25	.661320	.515918	.415077	.340175	.282568	.237277	.201067	.171726	.147695	.127818	.111240	.097305	25
26	.672864	.530124	.430041	.355004	.296810	.250710	.213597	.183333	.158399	.137656	.120274	.105608	26
27	.683663	.543561	.444332	.369254	.310608	.263809	.225900	.194794	.169017	.147483	.129346	.113968	27
28	.693769	.556262	.457946	.382979	.323980	.276602	.237971	.206105	.179569	.157274	.138418	.122368	28
29	.703259	.568303	.470981	.396197	.336947	.289051	.249798	.217241	.189991	.167006	.147478	.130785	29
30	.712188	.579734	.483431	.408914	.349488	.301188	.261373	.228198	.200311	.176673	.156516	.139205	30
40	.778877	.668158	.582817	.513297	.455181	.405867	.363565	.326959	.295085	.267163	.242600	.220888	40
60	.849044	.767047	.700066	.642556	.592126	.547349	.507256	.471148	.438462	.408771	.381699	.356960	60
80	.885442	.820705	.766251	.718260	.675124	.635912	.600023	.566986	.536460	.508176	.481887	.457414	80
100	.907714	.854312	.808614	.767700	.730354	.695928	.663968	.634166	.606280	.580112	.555487	.532298	100
120	.922736	.877325	.838018	.802443	.769650	.739118	.710513	.683595	.658183	.634132	.611324	.589657	120
140	.933554	.894066	.859605	.828176	.798994	.771635	.745829	.721386	.698162	.676045	.654943	.634778	140
170	.945088	.912072	.883006	.856283	.831279	.807662	.785224	.763821	.743347	.723717	.704865	.686733	170
200	.953211	.924848	.899727	.876499	.854647	.833900	.814087	.795095	.776838	.759251	.742281	.725885	200
240	.960919	.937047	.915781	.896012	.877319	.859482	.842366	.825881	.809961	.794554	.779622	.765130	240
320	.970605	.952477	.936212	.920990	.906503	.892593	.879164	.866153	.853513	.841211	.829220	.817517	320
440	.978571	.965253	.953233	.941922	.931100	.920655	.910522	.900654	.891022	.881602	.872376	.863331	440
600	.984259	.974422	.965507	.957084	.948995	.941160	.933530	.926075	.918772	.911606	.904563	.897634	600
800	.988181	.980767	.974028	.967644	.961498	.955529	.949702	.943994	.938390	.932877	.927446	.922092	800
1000	.990538	.984589	.979173	.974034	.969078	.964257	.959545	.954922	.950376	.945898	.941481	.937120	1000

Table A.9 (*Continued*)

ν_E	$\nu_H = 1$	$\nu_H = 2$	$\nu_H = 3$	$\nu_H = 4$	$\nu_H = 5$	$\nu_H = 6$	$\nu_H = 7$	$\nu_H = 8$	$\nu_H = 9$	$\nu_H = 10$	$\nu_H = 11$	$\nu_H = 12$	ν_E
							$p = 5$						
1	.000000	.000000	.000000	.000000	.000000	.000000	.000000	.000000	.000000	.000000	.000000	.000000	1
2	.000000	.000000	.000000	.000000	.000000	.000000	.000000	.000000	.000000	.000000	.000000	.000000	2
3	.000000	.000000	.000000	.000000	.000000	.000000	.000000	.000000	.000000	.000000	.000000	.000000	3
4	.000000	.000000	.000000	.000000	.000001	.000001	.000001	.000001	.000001	.000001	.000001	.000001	4
5	.001598	.000291	.000105	.000052	.000031	.000021	.000015	.000012	.000010	.000008	.000007	.000007	5
6	.021145	.004391	.001479	.000647	.000335	.000197	.000126	.000087	.000064	.000049	.000039	.000032	6
7	.062771	.016898	.006357	.002903	.001514	.000872	.000544	.000361	.000253	.000185	.000141	.000110	7
8	.113526	.037390	.015792	.007737	.004208	.002479	.001557	.001032	.000716	.000516	.000385	.000296	8
9	.165351	.063279	.029433	.015415	.008787	.005348	.003433	.002304	.001607	.001159	.000861	.000657	9
10	.214794	.092191	.046378	.025729	.015321	.009639	.006343	.004335	.003062	.002225	.001660	.001267	10
11	.260635	.122403	.065660	.038260	.023674	.015360	.010358	.007216	.005173	.003802	.002858	.002192	11
12	.302608	.152793	.086448	.052524	.033618	.022418	.015467	.010980	.007991	.005946	.004512	.003486	12
13	.340813	.182662	.108110	.068077	.044878	.030680	.021607	.015611	.011530	.008685	.006659	.005187	13
14	.375528	.211602	.130131	.084546	.057198	.039965	.028683	.021061	.015774	.012024	.009313	.007317	14
15	.407128	.239373	.152160	.101586	.070324	.050117	.036584	.027266	.020687	.015949	.012475	.009885	15
16	.435899	.265851	.173959	.118954	.084048	.060965	.045199	.034145	.026219	.020428	.016129	.012885	16
17	.462173	.291015	.195322	.136434	.098187	.072367	.054409	.041618	.032312	.025427	.020252	.016307	17
18	.486266	.314859	.216138	.153891	.112582	.084178	.064111	.049602	.038909	.030904	.024819	.020133	18
19	.508362	.337418	.236338	.171171	.127108	.096308	.074209	.058024	.045951	.036810	.029790	.024339	19
20	.528714	.358776	.255858	.188209	.141662	.108634	.084619	.066805	.053373	.043100	.035137	.028896	20

Table A.9 (*Continued*)

ν_E	$\nu_H = 1$	$\nu_H = 2$	$\nu_H = 3$	$\nu_H = 4$	$\nu_H = 5$	$\nu_H = 6$	$\nu_H = 7$	$\nu_H = 8$	$\nu_H = 9$	$\nu_H = 10$	$\nu_H = 11$	$\nu_H = 12$	ν_E
21	.547516	.378956	.274710	.204926	.156176	.121083	.095254	.075885	.061122	.049724	.040817	.033782	21
22	.564905	.398038	.292843	.221288	.170563	.133590	.106063	.085203	.069149	.056652	.046803	.038962	22
23	.581036	.416105	.310304	.237242	.184782	.146095	.116974	.094699	.077408	.063832	.053052	.044411	23
24	.596032	.433216	.327083	.252783	.198795	.158544	.127948	.104337	.085849	.071231	.059537	.050103	24
25	.610030	.449429	.343191	.267896	.212568	.170898	.138945	.114058	.094444	.078809	.066222	.056005	25
26	.623126	.464800	.358665	.282568	.226071	.183129	.149909	.123843	.103144	.086536	.073084	.062103	26
27	.635368	.479382	.373523	.296810	.239294	.195207	.160826	.133657	.111931	.094385	.080093	.068358	27
28	.646832	.493247	.387790	.310608	.252224	.207116	.171667	.143454	.120766	.102328	.087220	.074761	28
29	.657645	.506421	.401488	.323980	.264873	.218828	.182408	.153240	.129630	.110336	.094455	.081283	29
30	.667803	.518945	.414658	.336947	.277200	.230347	.193043	.162971	.138499	.118393	.101767	.087901	30
40	.744010	.617178	.521747	.446045	.384424	.333492	.290896	.254963	.224433	.198322	.175874	.156480	40
60	.824764	.729155	.652037	.586878	.530670	.481578	.438367	.400085	.365997	.335520	.308193	.283593	60
80	.866847	.790730	.727186	.671775	.622536	.578316	.538319	.501966	.468774	.438392	.410497	.384827	80
100	.892643	.829563	.775817	.728040	.684827	.645343	.609037	.575509	.544420	.515540	.488629	.463515	100
120	.910071	.856268	.809790	.767957	.729656	.694256	.661341	.630608	.601822	.574793	.549362	.525395	120
140	.922634	.875748	.834850	.797705	.763400	.731431	.701466	.673268	.646653	.621477	.597616	.574968	140
170	.936039	.896748	.862122	.830370	.800777	.772953	.746649	.721687	.697934	.675284	.653648	.632953	170
200	.945486	.911680	.881674	.853973	.827989	.803406	.780024	.757705	.736343	.715856	.696177	.677251	200
240	.954455	.925960	.900496	.876838	.854512	.833264	.812938	.793426	.774647	.756540	.739054	.722148	240
320	.965732	.944055	.924519	.906224	.888827	.872146	.856074	.840535	.825476	.810855	.796641	.782805	320
440	.975013	.959064	.944590	.930949	.917894	.905302	.893096	.881226	.869655	.858357	.847311	.836500	440
600	.981642	.969850	.959096	.948913	.939124	.929642	.920411	.911396	.902572	.893921	.885429	.877084	600
800	.986214	.977320	.969181	.961450	.953996	.946753	.939682	.932756	.925957	.919273	.912693	.906209	800
1000	.988963	.981823	.975277	.969047	.963029	.957171	.951441	.945820	.940292	.934848	.929480	.924182	1000

Table A.9 (Continued)

ν_E	$\nu_H = 1$	$\nu_H = 2$	$\nu_H = 3$	$\nu_H = 4$	$\nu_H = 5$	$\nu_H = 6$	$\nu_H = 7$	$\nu_H = 8$	$\nu_H = 9$	$\nu_H = 10$	$\nu_H = 11$	$\nu_H = 12$	ν_E
							$p = 6$						
1	.000000	.000000	.000000	.000000	.000000	.000000	.000000	.000000	.000000	.000000	.000000	.000000	1
2	.000000	.000000	.000000	.000000	.000000	.000000	.000000	.000000	.000000	.000000	.000000	.000000	2
3	.000000	.000000	.000000	.000000	.000000	.000000	.000000	.000000	.000000	.000000	.000000	.000000	3
4	.000000	.000000	.000000	.000000	.000000	.000000	.000000	.000000	.000000	.000000	.000000	.000000	4
5	.000007	.000002	.000001	.000001	.000001	.000000	.000000	.000000	.000000	.000000	.000000	.000000	5
6	.002045	.000315	.000095	.000040	.000021	.000012	.000008	.000006	.000004	.000003	.000003	.000002	6
7	.018804	.003479	.001052	.000416	.000197	.000106	.000063	.000040	.000027	.000020	.000015	.000011	7
8	.053911	.012883	.004369	.001818	.000872	.000465	.000270	.000168	.000111	.000076	.000055	.000041	8
9	.098038	.028824	.011018	.004938	.002479	.001358	.000798	.000497	.000325	.000222	.000157	.000115	9
10	.144274	.049685	.021043	.010129	.005348	.003035	.001826	.001155	.000762	.000521	.000369	.000269	10
11	.189355	.073697	.033966	.017408	.009639	.005672	.003507	.002263	.001514	.001046	.000744	.000543	11
12	.231866	.099450	.049161	.026586	.015360	.009348	.005940	.003915	.002664	.001865	.001338	.000983	12
13	.271356	.125933	.066012	.037385	.022418	.014071	.009172	.006173	.004273	.003033	.002200	.001630	13
14	.307797	.152453	.083979	.049495	.030680	.019795	.013205	.009066	.006381	.004592	.003370	.002520	14
15	.341285	.178581	.102644	.062632	.039965	.026433	.018012	.012593	.009005	.006568	.004877	.003682	15
16	.372033	.204010	.121656	.076537	.050117	.033893	.023544	.016741	.012147	.008974	.006740	.005137	16
17	.400304	.228568	.140775	.090983	.060965	.042061	.029737	.021472	.015794	.011811	.008966	.006898	17
18	.426364	.252176	.159796	.105779	.072367	.050834	.036522	.026746	.019924	.015070	.011554	.008971	18
19	.450349	.274785	.178574	.120780	.084178	.060119	.043825	.032520	.024510	.018734	.014503	.011356	19
20	.472562	.296393	.197017	.135856	.096308	.069818	.051576	.038739	.029518	.022785	.017796	.014049	20

Table A.9 (*Continued*)

ν_E	$\nu_H = 1$	$\nu_H = 2$	$\nu_H = 3$	$\nu_H = 4$	$\nu_H = 5$	$\nu_H = 6$	$\nu_H = 7$	$\nu_H = 8$	$\nu_H = 9$	$\nu_H = 10$	$\nu_H = 11$	$\nu_H = 12$	ν_E
21	.493091	.316990	.215044	.150905	.108634	.079840	.059715	.045350	.034906	.027193	.021418	.017040	21
22	.512182	.336628	.232604	.165853	.121083	.090122	.068178	.052311	.040646	.031936	.025354	.020317	22
23	.529913	.355328	.249666	.180626	.133590	.100596	.076899	.059574	.046695	.036988	.029582	.023864	23
24	.546452	.373143	.266216	.195197	.146095	.111189	.085836	.067090	.053016	.042316	.034078	.027670	24
25	.561889	.390109	.282253	.209511	.158544	.121873	.094944	.074824	.059586	.047895	.038825	.031716	25
26	.576348	.406285	.297740	.223535	.170898	.132587	.104168	.082735	.066362	.053696	.043795	.035986	26
27	.589899	.421688	.312738	.237277	.183129	.143309	.113485	.090793	.073318	.059697	.048977	.040460	27
28	.602633	.436379	.327222	.250710	.195207	.153998	.122849	.098970	.080420	.065867	.054339	.045123	28
29	.614602	.450416	.341199	.263809	.207116	.164629	.132250	.107224	.087654	.072196	.059866	.049957	29
30	.625896	.463794	.354711	.276602	.218828	.175171	.141648	.115539	.094994	.078649	.065542	.054951	30
40	.710937	.569976	.466792	.387183	.324162	.273470	.232192	.198251	.170132	.146678	.126985	.110367	40
60	.801604	.693451	.607528	.536153	.475641	.423707	.378774	.339636	.305361	.275238	.248638	.225098	60
80	.849063	.762264	.690479	.628610	.574313	.526153	.483144	.444543	.409736	.378269	.349725	.323787	80
100	.878218	.805945	.744748	.690824	.642495	.598763	.558956	.522538	.489125	.458377	.430004	.403784	100
120	.897944	.836112	.782919	.735354	.692128	.652489	.615927	.582063	.550602	.521300	.493955	.468392	120
140	.912172	.858176	.811198	.768751	.729786	.693709	.660119	.628724	.599296	.571649	.545628	.521100	140
170	.927365	.882016	.842092	.805615	.771776	.740119	.710350	.682254	.655667	.630455	.606507	.583730	170
200	.938078	.899001	.864314	.832375	.802523	.774395	.747758	.722444	.698328	.675308	.653300	.632233	200
240	.948255	.915270	.885761	.858391	.832628	.808187	.784886	.762599	.741229	.720701	.700953	.681935	240
320	.961056	.935919	.913212	.891956	.871772	.852459	.833892	.815985	.798676	.781916	.765666	.749894	320
440	.971597	.953076	.936212	.920308	.905097	.890438	.876249	.862471	.849063	.835995	.823242	.810784	440
600	.979129	.965422	.952870	.940969	.929529	.918448	.907669	.897152	.886868	.876798	.866924	.857233	600
800	.984325	.973979	.964469	.955420	.946689	.938203	.929921	.921812	.913858	.906042	.898354	.890785	800
1000	.987450	.979142	.971487	.964187	.957129	.950256	.943532	.936937	.930455	.924073	.917783	.911578	1000

Table A.9 (*Continued*)

ν_E	$\nu_H = 1$	$\nu_H = 2$	$\nu_H = 3$	$\nu_H = 4$	$\nu_H = 5$	$\nu_H = 6$	$\nu_H = 7$	$\nu_H = 8$	$\nu_H = 9$	$\nu_H = 10$	$\nu_H = 11$	$\nu_H = 12$	ν_E
							$p = 7$						
1	.000000	.000000	.000000	.000000	.000000	.000000	.000000	.000000	.000000	.000000	.000000	.000000	1
2	.000000	.000000	.000000	.000000	.000000	.000000	.000000	.000000	.000000	.000000	.000000	.000000	2
3	.000000	.000000	.000000	.000000	.000000	.000000	.000000	.000000	.000000	.000000	.000000	.000000	3
4	.000000	.000000	.000000	.000000	.000000	.000000	.000000	.000000	.000000	.000000	.000000	.000000	4
5	.000000	.000000	.000000	.000000	.000000	.000000	.000000	.000000	.000000	.000000	.000000	.000000	5
6	.000043	.000006	.000002	.000001	.000001	.000000	.000000	.000000	.000000	.000000	.000000	.000000	6
7	.002625	.000350	.000091	.000033	.000015	.000008	.000005	.000003	.000002	.000002	.000001	.000001	7
8	.017612	.002953	.000809	.000292	.000126	.000063	.000034	.000020	.000013	.000009	.000006	.000005	8
9	.047835	.010329	.003195	.001223	.000543	.000270	.000147	.000086	.000053	.000035	.000024	.000017	9
10	.086645	.023060	.008067	.003338	.001558	.000798	.000440	.000259	.000160	.000104	.000070	.000049	10
11	.128234	.040186	.015642	.006974	.003433	.001826	.001035	.000619	.000387	.000252	.000170	.000119	11
12	.169506	.060396	.025707	.012249	.006343	.003508	.002048	.001252	.000796	.000525	.000357	.000249	12
13	.209026	.082538	.037857	.019109	.010357	.005940	.003571	.002234	.001448	.000967	.000665	.000468	13
14	.246203	.105734	.051646	.027402	.015466	.009172	.005668	.003628	.002395	.001625	.001131	.000804	14
15	.280861	.129346	.066659	.036933	.021607	.013206	.008371	.005476	.003682	.002537	.001787	.001285	15
16	.313032	.152929	.082533	.047494	.028684	.018013	.011688	.007801	.005337	.003733	.002664	.001936	16
17	.342842	.176179	.098971	.058884	.036586	.023544	.015606	.010611	.007379	.005235	.003782	.002778	17
18	.370455	.198894	.115731	.070921	.045199	.029736	.020096	.013900	.009814	.007057	.005159	.003829	18
19	.396050	.220944	.132623	.083445	.054409	.036520	.025122	.017653	.012640	.009204	.006805	.005102	19
20	.419802	.242252	.149498	.096315	.064111	.043824	.030640	.021845	.015847	.011676	.008725	.006605	20

Table A.9 (*Continued*)

ν_E	$\nu_H = 1$	$\nu_H = 2$	$\nu_H = 3$	$\nu_H = 4$	$\nu_H = 5$	$\nu_H = 6$	$\nu_H = 7$	$\nu_H = 8$	$\nu_H = 9$	$\nu_H = 10$	$\nu_H = 11$	$\nu_H = 12$	ν_E
21	.441876	.262777	.166240	.109415	.074209	.051579	.036603	.026450	.019422	.014469	.010921	.008342	21
22	.462425	.282503	.182765	.122645	.084616	.059717	.042965	.031435	.023345	.017571	.013387	.010314	22
23	.481587	.301432	.199007	.135923	.095257	.068177	.049678	.036769	.027595	.020971	.016120	.012521	23
24	.499486	.319577	.214919	.149181	.106063	.076901	.056697	.042416	.032148	.024653	.019108	.014956	24
25	.516238	.336959	.230467	.162364	.116978	.085838	.063980	.048346	.036980	.028599	.022341	.017614	25
26	.531942	.353606	.245631	.175429	.127951	.094941	.071488	.054525	.042067	.032794	.025807	.020487	26
27	.546689	.369546	.260395	.188340	.138940	.104168	.079183	.060924	.047385	.037217	.029493	.023565	27
28	.560561	.384810	.274752	.201068	.149909	.113482	.087032	.067514	.052911	.041851	.033384	.026838	28
29	.573629	.399430	.288701	.213591	.160826	.122851	.095005	.074268	.058622	.046678	.037467	.030296	29
30	.585961	.413438	.302243	.225894	.171667	.132247	.103073	.081161	.064496	.051680	.041727	.033928	30
40	.679228	.525996	.417050	.335433	.272668	.223571	.184671	.153533	.128393	.107941	.091192	.077392	40
60	.779306	.659576	.566032	.489695	.426135	.372561	.327012	.288026	.254476	.225471	.200293	.178361	60
80	.831906	.735024	.655779	.588321	.529875	.478709	.433602	.393626	.358051	.326284	.297833	.272287	80
100	.864288	.783251	.715144	.655689	.602930	.555673	.513081	.474521	.439488	.407570	.378421	.351744	100
120	.886219	.816680	.757179	.704361	.656738	.613420	.573796	.537400	.503866	.472893	.444226	.417647	120
140	.902052	.841199	.788462	.741086	.697881	.658148	.621410	.587314	.555578	.525974	.498306	.472408	140
170	.918970	.867751	.822764	.781839	.744063	.708913	.676042	.645194	.616167	.588800	.562955	.538514	170
200	.930906	.886705	.847518	.811553	.778074	.746666	.717058	.689053	.662499	.637274	.613274	.590412	200
240	.942249	.904887	.871471	.840546	.811527	.784091	.758031	.733198	.709478	.686784	.665038	.644178	240
320	.956525	.928004	.902213	.878097	.855239	.833417	.812491	.792362	.772959	.754224	.736112	.718583	320
440	.968286	.947243	.928043	.909937	.892635	.875985	.859892	.844294	.829142	.814403	.800046	.786051	440
600	.976693	.961103	.946788	.933208	.920155	.907522	.895244	.883276	.871588	.860157	.848964	.837994	600
800	.982494	.970720	.959861	.949517	.939535	.929836	.920373	.911114	.902038	.893128	.884371	.875758	800
1000	.985983	.976524	.967778	.959426	.951346	.943478	.935782	.928236	.920822	.913527	.906342	.899259	1000

Table A.9 (*Continued*)

ν_E	$\nu_H = 1$	$\nu_H = 2$	$\nu_H = 3$	$\nu_H = 4$	$\nu_H = 5$	$\nu_H = 6$	$\nu_H = 7$	$\nu_H = 8$	$\nu_H = 9$	$\nu_H = 10$	$\nu_H = 11$	$\nu_H = 12$	ν_E
							$p = 8$						
1	.000000	.000000	.000000	.000000	.000000	.000000	.000000	.000000	.000000	.000000	.000000	.000000	1
2	.000000	.000000	.000000	.000000	.000000	.000000	.000000	.000000	.000000	.000000	.000000	.000000	2
3	.000000	.000000	.000000	.000000	.000000	.000000	.000000	.000000	.000000	.000000	.000000	.000000	3
4	.000000	.000000	.000000	.000000	.000000	.000000	.000000	.000000	.000000	.000000	.000000	.000000	4
5	.000000	.000000	.000000	.000000	.000000	.000000	.000000	.000000	.000000	.000000	.000000	.000000	5
6	.000000	.000000	.000000	.000000	.000000	.000000	.000000	.000000	.000000	.000000	.000000	.000000	6
7	.000138	.000015	.000004	.000001	.000001	.000000	.000000	.000000	.000000	.000000	.000000	.000000	7
8	.003295	.000393	.000090	.000029	.000012	.000006	.000003	.000002	.000001	.000001	.000001	.000000	8
9	.017079	.002632	.000659	.000218	.000087	.000040	.000020	.000011	.000007	.000004	.000003	.000002	9
10	.043574	.008626	.002458	.000872	.000361	.000168	.000086	.000047	.000028	.000017	.000011	.000008	10
11	.078039	.019031	.006148	.002365	.001032	.000497	.000259	.000144	.000085	.000052	.000034	.000023	11
12	.115676	.033314	.012011	.004993	.002304	.001155	.000619	.000351	.000209	.000130	.000084	.000056	12
13	.153630	.050518	.019990	.008908	.004335	.002263	.001252	.000727	.000441	.000278	.000181	.000122	13
14	.190453	.069716	.029839	.014129	.007216	.003915	.002234	.001331	.000824	.000527	.000347	.000235	14
15	.225477	.090151	.041241	.020590	.010980	.006173	.003628	.002215	.001399	.000910	.000608	.000416	15
16	.258443	.111245	.053875	.028171	.015610	.009065	.005476	.003422	.002203	.001457	.000987	.000683	16
17	.289300	.132575	.067447	.036729	.021061	.012594	.007801	.004982	.003269	.002197	.001509	.001057	17
18	.318105	.153836	.081699	.046115	.027265	.016740	.010611	.006915	.004617	.003151	.002194	.001555	18
19	.344966	.174814	.096415	.056185	.034144	.021472	.013900	.009228	.006265	.004339	.003060	.002194	19
20	.370015	.195359	.111416	.066805	.041616	.026747	.017653	.011923	.008219	.005771	.004120	.002987	20

Table A.9 ***(Continued)***

ν_E	$\nu_H = 1$	$\nu_H = 2$	$\nu_H = 3$	$\nu_H = 4$	$\nu_H = 5$	$\nu_H = 6$	$\nu_H = 7$	$\nu_H = 8$	$\nu_H = 9$	$\nu_H = 10$	$\nu_H = 11$	$\nu_H = 12$	ν_E
21	.393387	.215374	.126559	.077857	.049601	.032519	.021845	.014991	.010483	.007456	.005386	.003946	21
22	.415217	.234796	.141726	.089233	.058021	.038737	.026450	.018419	.013053	.009397	.006863	.005078	22
23	.435632	.253588	.156826	.100843	.066804	.045350	.031435	.022192	.015923	.011593	.008555	.006390	23
24	.454749	.271732	.171785	.112606	.075884	.052311	.036769	.026287	.019081	.014041	.010462	.007885	24
25	.472677	.289225	.186549	.124457	.085199	.059573	.042416	.030685	.022515	.016733	.012583	.009565	25
26	.489514	.306072	.201075	.136338	.094698	.067091	.048346	.035361	.026210	.019663	.014914	.011428	26
27	.505352	.322285	.215331	.148203	.104332	.074826	.054525	.040293	.030150	.022818	.017449	.013472	27
28	.520271	.337880	.229293	.160010	.114060	.082739	.060924	.045457	.034319	.026189	.020182	.015694	28
29	.534345	.352879	.242945	.171728	.123844	.090796	.067514	.050831	.038700	.029764	.023104	.018089	29
30	.547639	.367302	.256277	.183330	.133653	.098967	.074268	.056394	.043276	.033529	.026207	.020651	30
40	.648630	.484826	.371902	.289857	.228618	.182082	.146235	.118316	.096365	.078964	.065068	.053897	40
60	.757690	.627279	.527185	.447009	.381482	.327255	.281978	.243910	.211718	.184362	.161015	.141011	60
80	.815243	.708843	.622840	.550577	.488795	.435425	.388992	.348380	.312704	.281253	.253441	.228779	80
100	.850742	.761330	.686819	.622411	.565838	.515687	.470954	.430871	.394827	.362322	.332935	.306310	100
120	.874811	.797857	.732425	.674791	.623251	.576764	.534599	.496197	.461114	.428982	.399491	.372376	120
140	.892201	.824719	.766516	.714559	.667497	.624521	.585067	.548712	.515117	.484002	.455129	.428296	140
170	.910793	.853874	.804039	.758920	.717494	.679163	.643522	.610267	.579158	.549999	.522621	.496881	170
200	.923918	.874725	.831204	.791410	.754525	.720081	.687764	.657345	.628642	.601508	.575820	.551470	200
240	.936396	.894758	.857556	.823223	.791114	.760867	.732246	.705079	.679234	.654605	.631100	.608645	240
320	.952108	.920269	.891472	.864586	.839159	.814944	.791784	.769570	.748216	.727659	.707843	.688723	320
440	.965057	.941534	.920045	.899793	.880463	.861889	.843968	.826629	.809821	.793502	.777641	.762209	440
600	.974316	.956873	.940825	.925599	.910972	.896826	.883093	.869724	.856684	.843948	.831494	.819306	600
800	.980707	.967524	.955338	.943721	.932512	.921624	.911008	.900630	.890464	.880494	.870704	.861085	800
1000	.984551	.973956	.964134	.954746	.945661	.936815	.928167	.919691	.911367	.903183	.895128	.887192	1000

Table A.10 Upper Critical Values for Roy's Test, $\alpha = .05$

Roy's test statistic is given by

$$\theta = \frac{\lambda_1}{1+\lambda_1},$$

where λ_1 is the largest eigenvalue of $\mathbf{E}^{-1}\mathbf{H}$. The parameters are

$$s = \min(\nu_H, p)$$

$$m = \frac{|\nu_H - p| - 1}{2}$$

$$N = \frac{\nu_E - p - 1}{2}.$$

Reject if $\theta >$ table value.

	m								
N	0	1	2	3	4	5	7	10	15
					$s = 2$				
5	.565	.651	.706	.746	.776	.799	.834	.868	.901
10	.374	.455	.514	.561	.598	.629	.679	.732	.789
15	.278	.348	.402	.446	.483	.515	.567	.627	.696
20	.221	.281	.329	.369	.404	.434	.486	.546	.620
25	.184	.236	.278	.314	.346	.375	.424	.484	.558
30	.157	.203	.241	.274	.303	.330	.376	.433	.507
40	.122	.159	.190	.218	.243	.266	.306	.359	.428
50	.099	.130	.157	.180	.202	.222	.259	.306	.370
60	.084	.110	.133	.154	.173	.191	.223	.266	.326
80	.064	.085	.103	.119	.135	.149	.176	.211	.263
120	.043	.058	.070	.082	.093	.104	.123	.150	.190
240	.022	.030	.036	.042	.048	.054	.065	.080	.103
					$s = 3$				
5	.669	.729	.770	.800	.822	.840	.867	.894	.920
10	.472	.537	.586	.625	.656	.683	.725	.770	.819
15	.362	.422	.469	.508	.541	.569	.616	.669	.730
20	.293	.346	.390	.427	.458	.486	.533	.589	.656
25	.246	.294	.333	.367	.397	.424	.470	.525	.594
30	.212	.255	.291	.322	.350	.375	.419	.473	.543
40	.166	.201	.232	.259	.283	.305	.345	.395	.462
50	.136	.167	.192	.216	.237	.257	.292	.339	.402
60	.116	.142	.164	.185	.204	.221	.254	.296	.355
80	.089	.109	.127	.144	.160	.174	.201	.237	.288
120	.061	.075	.088	.100	.111	.122	.142	.169	.209
240	.031	.039	.046	.052	.058	.064	.075	.090	.114

Table A.10 (*Continued*)

N	m = 0	1	2	3	4	5	7	10	15
					$s = 4$				
5	.739	.782	.813	.836	.854	.868	.889	.911	.933
10	.547	.601	.641	.674	.700	.723	.759	.798	.840
15	.431	.482	.523	.558	.587	.612	.654	.701	.756
20	.354	.402	.441	.474	.503	.529	.572	.623	.684
25	.301	.344	.380	.412	.440	.464	.507	.559	.624
30	.261	.301	.334	.364	.390	.414	.455	.507	.572
40	.207	.240	.269	.294	.318	.339	.377	.426	.490
50	.171	.199	.224	.247	.268	.287	.322	.367	.428
60	.145	.170	.193	.213	.232	.249	.280	.322	.380
80	.112	.132	.150	.167	.182	.196	.223	.259	.309
120	.077	.091	.104	.116	.127	.138	.158	.185	.226
240	.040	.047	.054	.061	.067	.073	.084	.100	.124
					$s = 5$				
5	.788	.821	.845	.863	.877	.888	.906	.924	.942
10	.607	.651	.685	.713	.735	.755	.786	.820	.857
15	.488	.533	.569	.599	.625	.648	.685	.728	.777
20	.407	.449	.485	.515	.542	.565	.604	.651	.708
25	.349	.388	.422	.451	.477	.500	.540	.588	.648
30	.305	.341	.373	.400	.425	.448	.487	.535	.597
40	.243	.275	.302	.327	.349	.370	.406	.453	.514
50	.202	.230	.254	.276	.296	.315	.348	.392	.451
60	.173	.197	.219	.238	.257	.274	.304	.345	.401
80	.134	.154	.171	.188	.203	.217	.243	.278	.329
120	.093	.107	.120	.132	.143	.154	.174	.201	.241
240	.048	.056	.063	.069	.076	.082	.093	.109	.134

Table A.10 (*Continued*)

N	m = 0	1	2	3	4	5	7	10	15
				$s = 6$					
5	.825	.850	.869	.883	.895	.904	.918	.934	.949
10	.655	.692	.721	.744	.764	.781	.808	.838	.871
15	.537	.576	.608	.635	.658	.678	.711	.750	.795
20	.454	.491	.523	.551	.575	.596	.632	.676	.728
25	.392	.428	.458	.485	.509	.531	.568	.613	.669
30	.345	.378	.407	.433	.457	.478	.514	.560	.618
40	.278	.307	.333	.356	.378	.397	.432	.477	.536
50	.232	.258	.281	.302	.322	.340	.372	.414	.472
60	.200	.223	.243	.262	.280	.297	.327	.366	.421
80	.156	.174	.192	.208	.222	.236	.262	.297	.346
120	.108	.122	.134	.146	.157	.168	.188	.215	.255
240	.056	.064	.071	.078	.084	.090	.101	.118	.142
				$s = 7$					
5	.852	.872	.887	.899	.908	.917	.929	.941	.955
10	.695	.726	.750	.771	.788	.802	.826	.853	.882
15	.579	.613	.641	.665	.686	.704	.734	.769	.810
20	.494	.528	.557	.582	.604	.624	.657	.697	.745
25	.431	.463	.491	.516	.538	.558	.593	.635	.688
30	.381	.412	.439	.463	.485	.505	.540	.583	.638
40	.309	.337	.362	.384	.404	.423	.456	.499	.555
60	.224	.246	.266	.285	.302	.318	.347	.386	.439
80	.176	.194	.211	.226	.241	.255	.280	.314	.363
100	.145	.160	.175	.188	.200	.212	.235	.265	.310
200	.077	.085	.093	.101	.109	.116	.129	.148	.175
300	.052	.058	.064	.069	.074	.079	.089	.103	.125
500	.032	.036	.039	.042	.046	.049	.055	.064	.078
1000	.016	.018	.020	.022	.023	.025	.028	.033	.041

Table A.10 (*Continued*)

N	m = 0	1	2	3	4	5	7	10	15
					s = 8				
5	.874	.890	.902	.912	.920	.927	.937	.948	.959
10	.728	.754	.775	.793	.808	.821	.842	.865	.892
15	.615	.645	.670	.692	.710	.727	.754	.786	.824
20	.531	.561	.587	.610	.630	.648	.679	.716	.761
25	.466	.495	.521	.544	.565	.583	.616	.655	.705
30	.414	.443	.468	.491	.511	.530	.563	.603	.655
40	.339	.365	.388	.409	.428	.446	.478	.519	.573
60	.248	.269	.288	.306	.323	.338	.367	.404	.456
80	.195	.213	.229	.244	.259	.272	.297	.330	.378
100	.161	.176	.190	.203	.216	.228	.250	.279	.323
200	.086	.094	.103	.110	.118	.125	.138	.157	.185
300	.058	.065	.070	.076	.081	.086	.096	.109	.130
500	.036	.040	.043	.047	.050	.053	.059	.068	.081
1000	.018	.020	.022	.024	.025	.027	.030	.035	.042
					s = 9				
5	.891	.904	.914	.922	.929	.935	.944	.953	.963
10	.756	.778	.797	.812	.825	.837	.855	.876	.901
15	.647	.674	.696	.715	.732	.747	.771	.801	.835
20	.563	.591	.614	.635	.654	.670	.698	.733	.775
25	.497	.525	.549	.570	.589	.606	.636	.673	.720
30	.445	.471	.495	.516	.535	.552	.583	.622	.671
40	.366	.391	.413	.433	.451	.468	.499	.538	.590
60	.270	.291	.309	.326	.343	.358	.385	.421	.472
80	.214	.231	.247	.262	.276	.289	.313	.346	.392
100	.177	.192	.206	.219	.231	.242	.264	.293	.336
200	.095	.104	.112	.119	.127	.134	.147	.166	.194
300	.065	.071	.077	.082	.087	.092	.102	.115	.136
500	.040	.043	.047	.051	.054	.057	.063	.072	.086
1000	.020	.022	.024	.026	.028	.029	.032	.037	.044
					s = 10				
5	.905	.916	.924	.931	.937	.941	.949	.958	.967
10	.780	.799	.815	.829	.840	.851	.867	.886	.908
15	.675	.699	.719	.736	.751	.764	.787	.814	.846
20	.592	.617	.639	.658	.675	.690	.716	.748	.787
25	.526	.551	.573	.593	.611	.627	.655	.690	.734
30	.473	.497	.519	.539	.557	.573	.603	.639	.686
40	.392	.415	.436	.455	.473	.489	.518	.555	.605
60	.292	.311	.329	.346	.361	.376	.402	.438	.487
80	.232	.249	.264	.278	.292	.305	.329	.361	.406
100	.193	.207	.220	.233	.245	.256	.278	.306	.348
200	.104	.112	.120	.128	.135	.142	.156	.174	.202
300	.071	.077	.083	.088	.093	.098	.108	.122	.143
500	.044	.047	.051	.054	.058	.061	.067	.076	.090
1000	.022	.024	.026	.028	.030	.031	.034	.039	.047

Table A.11 Upper Critical Values of Pillai's Statistic $V^{(s)}$, $\alpha = 0.05$

$$V^{(s)} = \sum_{i=1}^{s} \frac{\lambda_i}{1+\lambda_i}$$

where $\lambda_1, \lambda_2, \ldots, \lambda_s$ are eigenvalues of $\mathbf{E}^{-1}\mathbf{H}$. Reject H_0 if $V^{(s)}$ exceeds table value. The parameters s, m, and N are defined in Table A.10.

	N													
m	0	1	2	3	4	5	6	7	8	9	10	15	20	25
							$s = 2$							
0	1.536	1.232	1.031	0.890	0.782	0.698	0.629	0.573	0.526	0.485	0.451	0.333	0.263	0.218
1	1.706	1.452	1.258	1.109	0.991	0.896	0.817	0.751	0.694	0.646	0.604	0.455	0.364	0.304
2	1.784	1.573	1.397	1.254	1.137	1.039	0.956	0.886	0.825	0.772	0.725	0.556	0.451	0.379
3	1.829	1.649	1.492	1.358	1.245	1.149	1.065	0.993	0.930	0.875	0.825	0.643	0.526	0.445
4	1.859	1.703	1.560	1.436	1.329	1.235	1.153	1.081	1.018	0.961	0.910	0.719	1.594	0.506
5	1.880	1.742	1.613	1.497	1.395	1.305	1.226	1.155	1.091	1.034	0.983	0.786	0.655	0.561
6	1.895	1.772	1.654	1.546	1.450	1.364	1.286	1.217	1.154	1.098	1.046	0.846	0.710	0.612
7	1.907	1.796	1.687	1.586	1.495	1.413	1.338	1.270	1.209	1.153	1.102	0.901	0.761	0.658
8	1.917	1.815	1.714	1.620	1.534	1.455	1.383	1.317	1.257	1.202	1.151	0.950	0.808	0.702
9	1.924	1.831	1.737	1.649	1.567	1.491	1.422	1.358	1.299	1.245	1.195	0.995	0.851	0.743
10	1.931	1.844	1.757	1.673	1.595	1.523	1.456	1.394	1.337	1.284	1.235	1.036	0.891	0.781
15	1.951	1.888	1.822	1.758	1.695	1.636	1.580	1.527	1.477	1.430	1.386			
20	1.963	1.913	1.860	1.807	1.756	1.706	1.658	1.612	1.568	1.527	1.487			
25	1.969	1.929	1.885	1.840	1.796	1.753	1.711	1.671	1.632	1.595	1.559			

Table A.11 *(Continued)*

	N													
m	0	1	2	3	4	5	6	7	8	9	10	15	20	25
							$s = 3$							
0	2.037	1.710	1.473	1.294	1.153	1.040	0.947	0.869	0.803	0.746	0.697	0.524	0.420	0.350
1	2.297	1.988	1.751	1.564	1.412	1.287	1.183	1.094	1.017	0.950	0.892	0.682	0.552	0.453
2	2.447	2.168	1.943	1.759	1.606	1.477	1.367	1.273	1.190	1.117	1.053	0.818	0.668	0.565
3	2.544	2.294	2.084	1.907	1.757	1.628	1.517	1.420	1.334	1.258	1.190	0.937	0.772	0.656
4	2.612	2.386	2.191	2.023	1.878	1.752	1.641	1.543	1.456	1.378	1.308	1.042	0.866	0.740
5	2.662	2.457	2.276	2.117	1.978	1.854	1.745	1.648	1.561	1.482	1.411	1.137	0.952	0.818
6	2.701	2.514	2.345	2.194	2.061	1.941	1.835	1.739	1.652	1.573	1.502	1.222	1.030	0.890
7	2.732	2.559	2.402	2.259	2.131	2.016	1.912	1.818	1.732	1.654	1.582	1.300	1.103	0.957
8	2.757	2.597	2.449	2.314	2.192	2.081	1.979	1.887	1.803	1.726	1.655	1.371	1.170	1.020
9	2.777	2.629	2.490	2.362	2.244	2.137	2.039	1.949	1.866	1.790	1.720	1.436	1.23	
10	2.795	2.656	2.525	2.403	2.291	2.187	2.092	2.004	1.923	1.848	1.779	1.496	1.3	
15	2.853	2.748	2.646	2.549	2.457	2.370	2.288	2.211	21.39	2.071	2.007			
20	2.885	2.802	2.718	2.637	2.560	2.485	2.414	2.347	2.283	2.222	2.163			
25	2.906	2.836	2.766	2.697	2.630	2.565	2.503	2.443	2.385					
							$s = 4$							
0	2.549	2.194	1.926	1.717	1.548	1.410	1.294	1.196	1.112	1.038	0.974	0.744	0.602	
1	2.852	2.510	2.241	2.023	1.844	1.693	1.566	1.456	1.360	1.277	1.203	0.932	0.761	
2	3.052	2.733	2.472	2.256	2.074	1.919	1.786	1.670	1.567	1.477	1.396	1.097	0.903	
3	3.193	2.898	2.650	2.440	2.260	2.104	1.969	1.849	1.743	1.649	1.564	1.243	1.032	
4	3.298	3.025	2.791	2.589	2.413	2.259	2.123	2.002	1.895	1.798	1.710	1.375	1.149	
5	3.378	3.126	2.905	2.711	2.541	2.390	2.255	2.135	2.027	1.929	1.840	1.494		
6	3.442	3.208	2.999	2.814	2.649	2.502	2.370	2.251	2.143	2.044	1.955	1.602		
7	3.494	3.276	3.079	2.902	2.743	2.600	2.470	2.353	2.246	2.148	2.058	1.70		
8	3.537	3.333	3.146	2.977	2.824	2.685	2.559	2.444	2.338	2.241	2.151	1.8		
9	3.574	3.382	3.205	3.043	2.896	2.761	2.638	2.525	2.421	2.325	2.236			
10	3.605	3.424	3.256	3.101	2.959	2.829	2.708	2.598	2.496	2.401	2.313			
15	3.710	3.570	3.436	3.310	3.191	3.079	2.974	2.876	2.783	2.696	2.615			
20	3.771	3.657	3.546	3.440	3.338	3.241	3.149							

Table A.11 (*Continued*)

	N													
m	0	1	2	3	4	5	6	7	8	9	10	15	20	25
							$s = 5$							
0	3.055	2.681	2.389	2.155	1.962	1.801	1.664	1.547	1.445	1.356	1.277			
1	3.390	3.025	2.731	2.488	2.285	2.122	1.964	1.835	1.722	1.622	1.533			
2	3.628	3.281	2.993	2.751	2.545	2.367	2.213	2.077	1.957	1.850	1.754			
3	3.805	3.478	3.201	2.964	2.759	2.580	2.423	2.284	2.160	2.048	1.948			
4	3.941	3.635	3.370	3.140	2.938	2.761	2.604	2.463	2.337	2.222	2.119			
5	4.050	3.762	3.510	3.288	3.091	2.916	2.760	2.619	2.492	2.377	2.271			
6	4.138	3.868	3.627	3.414	3.223	3.052	2.897	2.758	2.630	2.514	2.408			
7	4.212	3.957	3.728	3.522	3.337	3.170	3.018	2.880						
8	4.274	4.033	3.815	3.617	3.438	3.275	3.126							
9	4.327	4.099	3.890	3.700	3.527	3.369								
10	4.372	4.156	3.957	3.774	3.607	3.45								
							$s = 6$							
0	3.559	3.171	2.859	2.604	2.390	2.209	2.053	1.918	1.799	1.694	1.601			
1	3.917	3.535	3.221	2.958	2.734	2.542	2.375	2.229	2.099	1.984	1.881			
2	4.185	3.817	3.508	3.245	3.018	2.821	2.647	2.494	2.358	2.235	2.125			
3	4.391	4.041	3.741	3.482	3.256	3.057	2.881	2.724	2.583	2.456	2.341			
4	4.556	4.223	3.934	3.681	3.458	3.260	3.084	2.925	2.782	2.652	2.534			
5	4.690	4.375	4.097	3.851	3.633	3.438	3.262	3.103	2.959	2.827	2.706			
6	4.802	4.502	4.236	3.998	3.785									
7	4.896	4.611	4.356	4.126	3.919									
8	4.976	4.706	4.461	4.239										
9	5.045	4.788	4.553											
10	5.106	4.860	4.635											

Table A.12 Upper Critical Values for the Lawley–Hotelling Statistic, $\alpha = 0.05$

The test statistic is $\nu_E U^{(s)}/\nu_H$, where $U^{(s)}$ is the Lawley–Hotelling statistic. Reject H_0 if $\nu_E U^{(s)}/\nu_H >$ table value.

	ν_H												
ν_E	2	3	4	5	6	8	10	12	15	20	25	40	60
						$p = 2$							
2^a	9.8591	10.659	11.098	11.373	11.562	11.952	11.804	12.052	12.153	12.254	12.316	12.409	12.461
3	58.428	58.915	59.161	59.308	59.407	59.531	59.606	59.655	59.705	59.755	59.785	59.830	59.855
4	23.999	23.312	22.918	22.663	22.484	22.250	22.104	22.003	21.901	21.797	21.733	21.636	21.582
5	15.639	14.864	14.422	14.135	13.934	13.670	13.504	13.391	13.275	13.156	13.083	12.972	12.909
6	12.175	11.411	10.975	10.691	10.491	10.228	10.063	9.9489	9.8320	9.7118	9.6381	9.5251	9.4610
7	10.334	9.5937	9.1694	8.8927	8.6975	8.4396	8.2765	8.16399	8.0480	7.9285	7.8549	7.7417	7.6773
8	9.2069	8.4881	8.0752	7.8054	7.6145	7.3614	7.2008	7.0896	6.9748	6.8560	6.7826	6.6694	6.6048
10	7.9095	7.2243	6.8294	6.5702	6.3860	6.1405	5.9837	5.8745	5.7612	5.6433	5.5701	5.4564	5.3910
12	7.1902	6.5284	6.1461	5.8942	5.7147	5.4744	5.3200	5.2122	5.0997	4.9820	4.9085	4.7938	4.7274
14	6.7350	6.0902	4.7168	5.4703	5.2941	5.0574	4.9048	4.7977	4.6856	4.5678	4.4939	4.3780	4.3105
16	6.4217	5.7895	5.4230	5.1804	5.0067	4.7727	4.6213	4.5147	4.4028	4.2846	4.2102	4.0930	4.0243
18	6.1932	5.5708	5.2095	4.9700	4.7982	4.5663	4.4157	4.3094	4.1976	4.0791	4.0042	3.8855	3.8158
20	6.0192	5.4046	5.0475	4.8105	4.6402	4.4099	4.2600	4.1539	4.0420	3.9231	3.8477	3.7278	3.6569
25	5.7244	5.1237	4.7741	2.5415	4.3740	4.1465	3.9977	3.8919	3.7798	3.6598	3.5832	3.4605	3.3868
30	5.5401	4.9487	4.6040	4.3743	4.2086	3.9829	3.8347	3.7291	3.6166	3.4957	3.4181	3.2926	3.2168
35	5.4140	4.8291	4.8880	4.2604	4.0959	3.8715	3.7237	3.6181	3.5054	3.3836	3.3051	3.1774	3.1000
40	5.3224	4.7424	4.4039	4.1778	4.0143	3.7908	3.6433	3.5377	3.4247	3.3022	3.2230	3.0933	3.0140
50	5.1981	4.6249	4.2900	4.0661	3.9039	3.6817	3.5346	3.4289	3.3154	3.1919	3.1115	2.9787	2.8965
60	5.1178	4.5490	4.2166	3.9941	3.8328	3.6114	3.4646	3.3588	3.2450	3.1206	3.0392	2.9041	2.8196
70	5.0616	4.4960	4.1653	3.9439	3.7831	3.5624	3.4157	3.3099	3.1957	3.0706	2.9886	2.8516	2.7652
80	5.0200	4.4569	4.1275	3.9068	3.7465	3.5262	3.3796	3.2737	3.1594	3.0338	2.9512	2.8126	2.7247
100	4.9628	4.4030	4.0754	3.8557	3.6961	3.4764	3.3300	3.2240	3.1093	2.9829	2.8994	2.7586	2.6683
200	4.8514	4.2982	3.9742	3.7567	3.5983	3.3798	3.2336	3.1275	3.0120	2.8838	2.7984	2.6520	2.5559
∞	4.7442	4.1973	3.8769	3.6614	3.5044	3.2870	3.1410	3.0346	2.9182	2.7879	2.7002	2.5470	2.4428

Table A.12 (*Continued*)

ν_E	ν_H 3	4	5	6	8	10	12	15	20	25	40	60
						$p = 3$						
3^b	25.930	26.996	27.665	28.125	28.712	29.073	29.316	29.561	29.809	29.959	30.19	30.31
4^b	1.1880	1.1929	1.1959	1.1978	1.2003	1.2018	1.2028	1.2038	1.2048	1.2054	1.2063	1.2068
5	42.474	41.764	1.305	40.983	40.562	40.300	40.120	39.937	39.750	39.635	39.462	39.366
6	25.456	24.715	24.235	23.899	23.458	23.182	22.992	22.799	22.600	22.479	22.294	22.190
7	18.752	18.056	17.605	17.288	16.870	16.608	16.427	16.241	16.051	15.934	15.755	15.653
8	15.308	14.657	14.233	13.934	13.540	13.290	13.118	12.941	12.758	12.646	12.473	12.375
10	11.893	11.306	10.921	10.649	10.287	10.057	9.8974	9.7320	9.5603	9.4541	9.2897	9.1955
12	10.229	9.6825	9.3234	9.0680	8.7271	8.5088	8.3566	8.1982	8.0330	7.9301	7.7700	7.6777
14	9.2550	8.7356	8.3935	8.1495	7.8225	7.6122	7.4649	7.3110	7.1497	7.0488	6.8908	6.7991
16	8.6180	8.1183	7.7884	7.5526	7.2355	7.0307	6.8868	6.7360	6.5772	6.4774	6.3204	6.2287
18	8.1701	7.6851	7.3644	7.1347	6.8251	6.6244	6.4830	6.3343	6.1771	6.0780	5.9212	5.8292
20	7.8384	7.3649	7.0513	6.8263	6.5224	6.3249	6.1853	6.0383	5.8822	5.7834	5.6266	5.5341
25	7.2943	6.8407	6.5394	6.3227	6.0287	5.8365	5.7001	5.5555	5.4010	5.3025	5.1446	5.0503
30	6.9654	6.5245	6.2311	6.0196	5.7319	5.5431	5.4085	5.2654	5.1116	5.0129	4.8535	4.7575
35	6.7453	6.3132	6.0253	5.8175	5.5341	5.3476	5.2143	5.0720	4.9185	4.8195	4.6586	4.5608
40	6.5877	6.1621	5.8783	5.6732	5.3929	5.2081	5.0757	4.9340	4.7806	4.6813	4.5189	4.4195
50	6.3773	5.9606	5.6823	5.4809	5.2050	5.0224	4.8911	4.7502	4.5967	4.4968	4.3319	4.2297
60	6.2433	5.8324	5.5577	5.3587	5.0856	4.9044	4.7739	4.6334	4.4798	4.3793	4.2123	4.1078
70	6.1504	5.7436	5.4715	5.2742	5.0031	4.8229	4.6929	4.5526	4.3988	4.2979	4.1292	4.0227
80	6.0823	5.6786	5.4084	5.2122	4.9426	4.7632	4.6336	4.4935	4.3395	4.2381	4.0680	3.9600
100	5.9891	5.5896	5.3220	5.1276	4.8601	4.6817	4.5525	4.4126	4.2583	4.1563	3.9840	3.8734
200	5.8099	5.4186	5.1562	4.9653	4.7017	4.5252	4.3970	4.2574	4.1023	3.9988	3.8212	3.7042
∞	5.6397	5.2565	4.9992	4.8116	4.5519	4.3773	4.2499	4.1104	3.9541	3.8487	3.6642	3.5384

Table A.12 (Continued)

ν_E	ν_H 4	5	6	8	10	12	15	20	25	40	60
					$p = 4$						
4^b	49.964	51.204	52.054	53.142	53.808	54.258	54.71	55.17	55.46	—	—
5^b	1.9964	2.0013	2.0046	2.0087	2.0112	2.0128	2.0145	2.0171	2.0171	2.019	—
6	65.715	64.999	64.497	63.841	63.432	63.151	62.866	62.573	62.396	62.13	—
7	37.343	36.629	36.129	35.474	35.064	34.782	34.495	34.200	34.019	33.75	—
8	26.516	25.868	25.413	24.814	24.437	24.178	23.912	23.639	23.471	23.214	23.072
10	17.875	17.326	16.938	16.424	16.098	15.872	15.640	15.399	15.250	15.021	14.891
12	14.338	13.848	13.500	13.037	12.741	12.535	12.321	12.099	11.961	11.747	11.624
14	12.455	12.002	11.680	11.248	10.972	10.778	10.577	10.366	10.234	10.029	9.9103
16	11.295	10.868	10.563	10.154	9.8904	9.7054	9.5119	9.3085	9.1810	8.9808	8.8644
18	10.512	10.104	9.8121	9.4190	9.1647	8.9857	8.7978	8.5996	8.4748	8.2778	8.1626
20	9.9500	9.5560	9.2736	8.8926	8.6453	8.4708	8.2871	8.0926	7.9696	7.7748	7.6601
25	9.0585	8.6884	8.4223	8.0616	7.8261	7.6590	7.4821	7.2933	7.1730	6.9805	6.8659
30	8.5377	8.1825	7.9265	7.5784	7.3502	7.1876	7.0147	6.8291	6.7101	6.5181	6.4026
35	8.1968	7.8517	7.6026	7.2631	7.0397	6.8801	6.7099	6.5262	6.4079	6.2156	6.0989
40	7.9566	7.6188	7.3746	7.0413	6.8214	6.6640	6.4955	6.3131	6.1952	6.0023	5.8844
50	7.6404	7.3125	7.0751	6.7501	6.5350	6.3804	6.2143	6.0334	5.9157	5.7214	5.6011
60	7.4417	7.1202	6.8872	6.5676	6.3555	6.2027	6.0381	5.8581	5.7403	5.5446	5.4222
70	7.3054	6.9884	6.7584	6.4426	6.2325	6.0809	5.9173	5.7378	5.6200	5.4230	5.2987
80	7.2061	6.8924	6.6646	6.3515	6.1430	5.9924	5.8294	5.6503	5.5323	5.3343	5.2084
100	7.0711	6.7619	6.5372	6.2279	6.0215	5.8721	5.7101	5.5313	5.4131	5.2133	5.0849
200	6.8143	6.5139	6.2952	5.9933	5.7910	5.6439	5.4836	5.3053	5.1863	4.9819	4.8471
∞	6.5741	6.2821	6.0692	5.7743	5.5758	5.4309	5.2721	5.0940	4.9737	4.7629	4.6190

Table A.12 (*Continued*)

ν_E	ν_H 5	6	8	10	12	15	20	25	40	60
					$p = 5$					
5^a	81.991	83.352	85.093	86.160	86.88	—	—	—	—	—
6^a	3.0093	3.0142	3.0204	3.0241	3.0266	3.0291	3.032	—	—	—
7	93.762	93.042	92.102	91.515	91.113	90.705	90.29	90.04	—	—
8	51.339	50.646	49.739	49.170	48.780	48.382	47.973	47.723	47.35	—
10	27.667	27.115	26.387	25.927	25.610	25.284	24.947	24.740	24.422	—
12	20.169	19.701	19.079	18.683	18.409	18.124	17.830	17.647	17.365	17.20
14	16.643	16.224	15.666	15.309	15.059	14.800	14.530	14.361	14.100	13.95
16	14.624	14.239	13.722	13.389	13.157	12.914	12.659	12.499	12.250	12.105
18	13.326	12.963	12.476	12.161	11.939	11.708	11.463	11.310	11.068	10.928
20	12.424	12.078	11.612	11.310	11.097	10.874	10.637	10.488	10.252	10.113
25	11.046	10.728	10.297	10.016	9.8168	9.6061	9.3814	9.2386	9.0102	8.8745
30	10.270	9.9689	9.5592	9.2907	9.0995	8.8964	8.6785	8.5389	8.3141	8.1790
35	9.7739	9.4836	9.0879	8.8277	8.6419	8.4437	8.2301	8.0926	7.8693	7.7339
40	9.4292	9.1469	8.7613	8.5070	8.3250	8.1303	7.9195	7.7833	7.5607	7.4247
50	8.9825	8.7107	8.3385	8.0921	7.9150	7.7248	7.5177	7.3829	7.1605	7.0229
60	8.7057	8.4406	8.0769	7.8355	7.6615	7.4741	7.2692	7.1351	6.9124	6.7730
70	8.5174	8.2570	7.8991	7.6612	7.4894	7.3039	7.1004	6.9667	6.7434	6.6024
80	8.3811	8.1241	7.7705	7.5351	7.3648	7.1807	6.9782	6.8448	6.6208	6.4785
100	8.1969	7.9446	7.5969	7.3649	7.1968	7.0145	6.8133	6.6801	6.4550	6.3103
200	7.8505	7.6070	7.2706	7.0451	6.8811	6.7023	6.5032	6.3702	6.1416	5.9908
∞	7.5305	7.2955	6.9698	6.7505	6.5902	6.4144	6.2171	6.0838	5.8499	5.6899

Table A.12 *(Continued)*

ν_E	ν_H 6	8	10	12	15	20	25	30	35
					$p = 6$				
10	45.722	44.677	44.019	43.567	43.103	42.626	42.334	42.136	41.993
12	28.959	28.121	27.590	27.223	26.843	26.451	26.209	26.044	25.925
14	22.321	21.600	21.141	20.821	20.489	20.144	19.929	19.783	19.677
16	18.858	18.210	17.795	17.505	17.202	16.886	16.688	16.553	16.455
18	16.755	16.157	15.772	15.501	15.218	14.921	14.735	14.607	14.513
20	15.351	14.788	14.424	14.168	13.899	13.615	13.436	13.313	13.223
25	13.293	12.786	12.456	12.222	11.975	11.711	11.544	11.428	11.343
30	12.180	11.705	11.395	11.173	10.939	10.687	10.526	10.414	10.331
35	11.484	11.031	10.733	10.520	10.293	10.049	9.8921	9.7820	9.7003
40	11.009	10.571	10.282	10.075	9.8535	9.6142	9.4596	9.3508	9.2699
50	10.402	9.9832	9.7060	9.5067	9.2927	9.0598	8.9082	8.8009	8.7207
60	10.031	9.6246	9.3547	9.1602	8.9507	8.7215	8.5717	8.4651	8.3851
70	9.7813	9.3830	9.1182	8.9269	8.7204	8.4938	8.3450	8.2388	8.1589
80	9.6014	9.2093	8.9480	8.7591	8.5548	8.3300	8.1819	8.0759	7.9959
100	9.3598	8.9760	8.7197	8.5340	8.3326	8.1102	7.9629	7.8572	7.7771
200	8.9099	8.5419	8.2950	8.1153	7.9193	7.7011	7.5552	7.4494	7.3685
∞	8.4997	8.1463	7.9082	7.7340	7.5430	7.3284	7.1832	7.0768	6.9945

[a]Multiply entry by 100.
[b]Multiply entry by 10^4.

Table A.13 Orthogonal Polynomial Contrasts

		Variable										
p	Polynomial	1	2	3	4	5	6	7	8	9	10	$\mathbf{c}_i'\mathbf{c}_i$
3	Linear	−1	0	1								2
	Quadratic	1	−2	1								6
4	Linear	−3	−1	1	3							20
	Quadratic	1	−1	−1	1							4
	Cubic	−1	3	−3	1							20
5	Linear	−2	−1	0	1	2						10
	Quadratic	2	−1	−2	−1	2						14
	Cubic	−1	2	0	−2	1						10
	Quartic	1	−4	6	−4	1						70
6	Linear	−5	−3	−1	1	3	5					70
	Quadratic	5	−1	−4	−4	−1	5					84
	Cubic	−5	7	4	−4	−7	5					180
	Quartic	1	−3	2	2	−3	1					28
	Quintic	−1	5	−10	10	−5	1					252
7	Linear	−3	−2	−1	0	1	2	3				28
	Quadratic	5	0	−3	−4	−3	0	5				84
	Cubic	−1	1	1	0	−1	−1	1				6
	Quartic	3	−7	1	6	1	−7	3				154
	Quintic	−1	4	−5	0	5	−4	1				84
	Sextic	1	−6	15	−20	15	−6	1				924
8	Linear	−7	−5	−3	−1	1	3	5	7			168
	Quadratic	7	1	−3	−5	−5	−3	1	7			168
	Cubic	−7	5	7	3	−3	−7	−5	7			264
	Quartic	7	−13	−3	9	9	−3	−13	7			616
	Quintic	−7	23	−17	−15	15	17	−23	7			2,184
	Sextic	1	−5	9	−5	−5	9	−5	1			264
	Septic	−1	7	−21	35	−35	21	−7	1			3,432
9	Linear	−4	−3	−2	−1	0	1	2	3	4		60
	Quadratic	28	7	−8	−17	−20	−17	−8	7	28		2,772
	Cubic	−14	7	13	9	0	−9	−13	−7	14		990
	Quartic	14	−21	−11	9	18	9	−11	−21	14		2,002
	Quintic	−4	11	−4	−9	0	9	4	−11	4		468
	Sextic	4	−17	22	1	−20	1	22	−17	4		1,980
	Septic	−1	6	−14	14	0	−14	14	−6	1		858
	Octic	1	−8	28	−56	70	−56	28	−8	1		12,870
10	Linear	−9	−7	−5	−3	−1	1	3	5	7	9	330
	Quadratic	6	2	−1	−3	−4	−4	−3	−1	2	6	132
	Cubic	−42	14	35	31	12	−12	−31	−35	−14	42	8,580
	Quartic	18	−22	−17	3	18	18	3	−17	−22	18	2,860
	Quintic	−6	14	−1	−11	−6	6	11	1	−14	6	780
	Sextic	3	−11	10	6	−8	−8	6	10	11	3	660
	Septic	−9	47	−86	92	56	−56	−42	86	−47	9	29,172
	Octic	1	−7	20	−28	14	14	−28	20	−7	1	2,860
	Novic	−1	9	−36	84	−126	126	−84	36	−9	1	48,620

Note: Entries are rows $\mathbf{c}_i'$ of the $(p-1)\times p$ matrix $\mathbf{C}$ in Section 6.10.1.

Table A.14 Test for Equal Covariance Matrices, $\alpha = 0.05$

ν	$k = 2$	$k = 3$	$k = 4$	$k = 5$	$k = 6$	$k = 7$	$k = 8$	$k = 9$	$k = 10$
					$p = 2$				
3	12.18	18.70	24.55	30.09	35.45	40.68	45.81	50.87	55.86
4	10.70	16.65	22.00	27.07	31.97	36.75	41.45	46.07	50.64
5	9.97	15.63	20.73	25.57	30.23	34.79	39.26	43.67	48.02
6	9.53	15.02	19.97	24.66	29.19	33.61	37.95	42.22	46.45
7	9.24	14.62	19.46	24.05	28.49	32.83	37.08	41.26	45.40
8	9.04	14.33	19.10	23.62	27.99	32.26	36.44	40.57	44.64
9	8.88	14.11	18.83	23.30	27.62	31.84	35.98	40.05	44.08
10	8.76	13.94	18.61	23.05	27.33	31.51	35.61	39.65	43.64
11	8.67	13.81	18.44	22.85	27.10	31.25	35.32	39.33	43.29
12	8.59	13.70	18.30	22.68	26.90	31.03	35.08	39.07	43.00
13	8.52	13.60	18.19	22.54	26.75	30.85	34.87	38.84	42.76
14	8.47	13.53	18.10	22.42	26.61	30.70	34.71	38.66	42.56
15	8.42	13.46	18.01	22.33	26.50	30.57	34.57	38.50	42.38
16	8.38	13.40	17.94	22.24	26.40	30.45	34.43	38.36	42.23
17	8.35	13.35	17.87	22.17	26.31	30.35	34.32	38.24	42.10
18	8.32	13.30	17.82	22.10	26.23	30.27	34.23	38.13	41.99
19	8.28	13.26	17.77	22.04	26.16	30.19	34.14	38.04	41.88
20	8.26	13.23	17.72	21.98	26.10	30.12	34.07	37.95	41.79
25	8.17	13.10	17.55	21.79	25.87	29.86	33.78	37.63	41.44
30	8.11	13.01	17.44	21.65	25.72	29.69	33.59	37.42	41.21
					$p = 3$				
4	22.41	35.00	46.58	57.68	68.50	79.11	89.60	99.94	110.21
5	19.19	30.52	40.95	50.95	60.69	70.26	79.69	89.03	98.27
6	17.57	28.24	38.06	47.49	56.67	65.69	74.58	83.39	92.09
7	16.59	26.84	36.29	45.37	54.20	62.89	71.44	79.90	88.30
8	15.93	25.90	35.10	43.93	52.54	60.99	69.32	77.57	85.73
9	15.46	25.22	34.24	42.90	51.33	59.62	67.78	75.86	83.87
10	15.11	24.71	33.59	42.11	50.42	58.57	66.62	74.58	82.46
11	14.83	24.31	33.08	41.50	49.71	57.76	65.71	73.57	81.36
12	14.61	23.99	32.67	41.00	49.13	57.11	64.97	72.75	80.45
13	14.43	23.73	32.33	40.60	48.65	56.56	64.36	72.09	79.72
14	14.28	23.50	32.05	40.26	48.26	56.11	63.86	71.53	79.11
15	14.15	23.32	31.81	39.97	47.92	55.73	63.43	71.05	78.60
16	14.04	23.16	31.60	39.72	47.63	55.40	63.06	70.64	78.14
17	13.94	23.02	31.43	39.50	47.38	55.11	62.73	70.27	77.76
18	13.86	22.89	31.26	39.31	47.16	54.86	62.45	69.97	77.41
19	13.79	22.78	31.13	39.15	46.96	54.64	62.21	69.69	77.11
20	13.72	22.69	31.01	39.00	46.79	54.44	61.98	69.45	76.84
25	13.48	22.33	30.55	38.44	46.15	53.70	61.16	68.54	75.84
30	13.32	22.10	30.25	38.09	45.73	53.22	60.62	67.94	75.18

Table A.14 (*Continued*)

ν	$k = 2$	$k = 3$	$k = 4$	$k = 5$	$k = 6$	$k = 7$	$k = 8$	$k = 9$	$k = 10$
					$p = 4$				
5	35.39	56.10	75.36	93.97	112.17	130.11	147.81	165.39	182.80
6	30.06	48.62	65.90	82.60	98.93	115.03	130.94	146.69	162.34
7	27.31	44.69	60.89	76.56	91.88	106.98	121.90	136.71	151.39
8	25.61	42.24	57.77	72.77	87.46	101.94	116.23	130.43	144.50
9	24.45	40.57	55.62	70.17	84.42	98.46	112.32	126.08	139.74
10	23.62	39.34	54.04	68.26	82.19	95.90	109.46	122.91	136.24
11	22.98	38.41	52.84	66.81	80.48	93.95	107.27	120.46	133.57
12	22.48	37.67	51.90	65.66	79.14	92.41	105.54	118.55	131.45
13	22.08	37.08	51.13	64.73	78.04	91.15	104.12	116.98	129.74
14	21.75	36.59	50.50	63.95	77.13	90.12	102.97	115.69	128.32
15	21.47	36.17	49.97	63.30	76.37	89.26	101.99	114.59	127.14
16	21.24	35.82	49.51	62.76	75.73	88.51	101.14	113.67	126.10
17	21.03	35.52	49.12	62.28	75.16	87.87	100.42	112.87	125.22
18	20.86	35.26	48.78	61.86	74.68	87.31	99.80	112.17	124.46
19	20.70	35.02	48.47	61.50	74.25	86.82	99.25	111.56	123.79
20	20.56	34.82	48.21	61.17	73.87	86.38	98.75	111.02	123.18
25	20.06	34.06	47.23	59.98	72.47	84.78	96.95	109.01	120.99
30	19.74	33.59	46.61	59.21	71.58	83.74	95.79	107.71	119.57
					$p = 5$				
6	51.11	81.99	110.92	138.98	166.54	193.71	220.66	247.37	273.88
7	43.40	71.06	97.03	122.22	146.95	171.34	195.49	219.47	243.30
8	39.29	65.15	89.45	113.03	136.18	159.04	181.65	204.14	226.48
9	36.71	61.39	84.62	107.17	129.30	151.17	172.80	194.27	215.64
10	34.93	58.78	81.25	103.06	124.48	145.64	166.56	187.37	208.02
11	33.62	56.85	78.75	100.02	120.92	141.54	161.98	182.24	202.37
12	32.62	55.37	76.83	97.68	118.15	138.38	158.38	178.23	198.03
13	31.83	54.19	75.30	95.82	115.96	135.86	155.54	175.10	194.51
14	31.19	53.23	74.05	94.29	114.16	133.80	153.21	172.49	191.68
15	30.66	52.44	73.01	93.02	112.66	132.07	151.29	170.36	189.38
16	30.22	51.76	72.14	91.94	111.41	130.61	149.66	166.53	187.32
17	29.83	51.19	71.39	91.03	110.34	129.38	148.25	166.99	185.61
18	29.51	50.69	70.74	90.23	109.39	128.29	147.03	165.65	184.10
19	29.22	50.26	70.17	89.54	108.57	127.36	145.97	164.45	182.81
20	28.97	49.88	69.67	88.93	107.85	126.52	145.02	163.38	181.65
25	28.05	48.48	67.86	86.70	105.21	123.51	141.62	159.60	177.49
30	27.48	47.61	66.71	85.29	103.56	121.60	139.47	157.22	174.87

Note: Table contains upper percentage points for

$$-2 \ln M = \nu \left(k \ln |\mathbf{S}| - \sum_{i=1}^{k} \ln |\mathbf{S}_i| \right)$$

for k samples, each with ν degrees of freedom. Reject $H_0: \boldsymbol{\Sigma}_1 = \boldsymbol{\Sigma}_2 = \cdots = \boldsymbol{\Sigma}_k$ if $-2 \ln M >$ table value.

Table A.15 Test for Independence of p Variables

Upper percentage points for

$$u' = -\left(\nu - \frac{2p+5}{6}\right) \ln \left(\frac{|\mathbf{S}|}{s_{11} \cdots s_{pp}}\right) = -\left(\nu - \frac{2p+5}{6}\right) \ln |\mathbf{R}|,$$

where ν is the degrees of freedom of **S** or **R**. Reject if u' is greater than table value. The χ^2_α values are shown for comparison, since u' is approximately χ^2 distributed with $f = \frac{1}{2}p(p-1)$ degrees of freedom.

	n	$p = 3$	$p = 4$	$p = 5$	$p = 6$	$p = 7$	$p = 8$	$p = 9$	$p = 10$
					$\alpha = 0.05$				
	4	8.020							
	5	7.834	15.22						
	6	7.814	13.47	24.01					
	7	7.811	13.03	20.44	34.30				
	8	7.811	12.85	19.45	28.75	46.05			
	9	7.811	12.76	19.02	27.11	38.41	59.25		
	10	7.812	12.71	18.80	26.37	36.03	49.42	73.79	
	11	7.812	12.68	18.67	25.96	34.91	46.22	61.76	89.92
	12	7.813	12.66	18.58	25.71	34.28	44.67	57.68	75.45
	13	7.813	12.65	18.52	25.55	33.89	43.78	55.65	70.43
	14	7.813	12.64	18.48	25.44	33.63	43.21	54.46	67.87
	15	7.813	12.63	18.45	25.36	33.44	42.82	53.69	66.34
	16	7.814	12.62	18.43	25.30	33.31	42.55	53.15	65.33
	17	7.814	12.62	18.41	25.25	33.20	42.34	52.77	64.63
	18	7.814	12.62	18.40	25.21	33.12	42.19	52.48	64.12
	19	7.814	12.61	18.38	25.19	33.06	42.06	52.26	63.73
	20	7.814	12.61	18.37	25.16	33.01	41.97	52.08	63.43
$\chi^2_{0.05}$		7.815	12.59	18.31	25.00	32.67	41.34	51.00	61.66
					$\alpha = 0.01$				
	4	11.79							
	5	11.41	21.18						
	6	11.36	18.27	32.16					
	7	11.34	17.54	26.50	44.65				
	8	11.34	17.24	24.95	36.09	58.61			
	9	11.34	17.10	24.29	33.63	47.05	74.01		
	10	11.34	17.01	23.95	32.54	43.59	59.36	90.87	
	11	11.34	16.96	23.75	31.95	42.00	54.83	73.03	109.53
	12	11.34	16.93	23.62	31.60	41.13	52.70	67.37	88.05
	13	11.34	16.90	23.53	31.36	40.59	51.49	64.64	81.20
	14	11.34	16.89	23.47	31.20	40.23	50.73	63.06	77.83
	15	11.34	16.87	23.42	31.09	39.97	50.22	62.05	75.84
	16	11.34	16.86	23.39	31.00	39.79	49.85	61.36	74.56
	17	11.34	16.86	23.36	30.94	39.65	49.59	60.86	73.66
	18	11.34	16.85	23.34	30.88	39.54	49.38	60.49	73.01
	19	11.34	16.85	23.32	30.84	39.46	49.22	60.21	72.52
	20	11.34	16.84	23.31	30.81	39.39	49.09	59.99	72.15
$\chi^2_{0.01}$		11.34	16.81	23.21	30.58	38.93	48.28	58.57	69.92

APPENDIX B

Answers and Hints to Problems

CHAPTER 2

2.1 **(a)** $\mathbf{A}+\mathbf{B}=\begin{pmatrix} 7 & 0 & 7 \\ 13 & 14 & 3 \end{pmatrix}$, $\quad \mathbf{A}-\mathbf{B}=\begin{pmatrix} 1 & 4 & -1 \\ 1 & -4 & 13 \end{pmatrix}$

(b) $\mathbf{A}'\mathbf{A}=\begin{pmatrix} 65 & 43 & 68 \\ 43 & 29 & 46 \\ 68 & 46 & 73 \end{pmatrix}$, $\quad \mathbf{A}\mathbf{A}'=\begin{pmatrix} 29 & 62 \\ 62 & 138 \end{pmatrix}$

2.2 **(a)** $(\mathbf{A}+\mathbf{B})'=\begin{pmatrix} 7 & 13 \\ 0 & 14 \\ 7 & 3 \end{pmatrix}$, $\quad \mathbf{A}'+\mathbf{B}'=\begin{pmatrix} 7 & 13 \\ 0 & 14 \\ 7 & 3 \end{pmatrix}$

(b) $\mathbf{A}'=\begin{pmatrix} 4 & 7 \\ 2 & 5 \\ 3 & 8 \end{pmatrix}$, $\quad (\mathbf{A}')'=\begin{pmatrix} 4 & 2 & 3 \\ 7 & 5 & 8 \end{pmatrix}=\mathbf{A}$

2.3 **(a)** $\mathbf{AB}=\begin{pmatrix} 5 & 15 \\ 3 & -5 \end{pmatrix}$, $\quad \mathbf{BA}=\begin{pmatrix} 2 & 6 \\ 11 & -2 \end{pmatrix}$

(b) $|\mathbf{AB}|=-70, \quad |\mathbf{A}|=-7, \quad |\mathbf{B}|=10$

2.4 **(a)** $\mathbf{A}+\mathbf{B}=\begin{pmatrix} 3 & 3 \\ 3 & 4 \end{pmatrix}$, $\quad \text{tr}(\mathbf{A}+\mathbf{B})=7$

(b) $\text{tr}(\mathbf{A})=0, \quad \text{tr}(\mathbf{B})=7$

2.5 **(a)** $\mathbf{AB}=\begin{pmatrix} 4 & 1 \\ 3 & -3 \end{pmatrix}$, $\quad \mathbf{BA}=\begin{pmatrix} -1 & 8 & 7 \\ 2 & 4 & 6 \\ 1 & -3 & -2 \end{pmatrix}$

(b) $\text{tr}(\mathbf{AB})=1, \quad \text{tr}(\mathbf{BA})=1$

2.6 **(b)** $\mathbf{x}=(1 \quad 1 \quad -1)'$

2.7 (a) $\mathbf{Bx} = (13, 6, 9)'$ (b) $\mathbf{y}'\mathbf{B} = (25, -1, 17)$ (c) $\mathbf{x}'\mathbf{Ax} = 16$
(d) $\mathbf{x}'\mathbf{Ay} = 43$ (e) $\mathbf{x}'\mathbf{x} = 6$ (f) $\mathbf{x}'\mathbf{y} = 3$

(g) $\mathbf{xx}' = \begin{pmatrix} 1 & -1 & 2 \\ -1 & 1 & -2 \\ 2 & -2 & 4 \end{pmatrix}$ (h) $\mathbf{xy}' = \begin{pmatrix} 3 & 2 & 1 \\ -3 & -2 & -1 \\ 6 & 4 & 2 \end{pmatrix}$

(i) $\mathbf{B}'\mathbf{B} = \begin{pmatrix} 62 & 7 & 22 \\ 7 & 14 & 7 \\ 22 & 7 & 41 \end{pmatrix}$

2.8 (a) $\mathbf{x} + \mathbf{y} = (4, 1, 3)'$, $\mathbf{x} - \mathbf{y} = (-2, -3, 1)'$, (b) $(\mathbf{x} - \mathbf{y})'\mathbf{A}(\mathbf{x} - \mathbf{y}) = -31$

2.9 $$\mathbf{Bx} = \mathbf{b}_1 x_1 + \mathbf{b}_2 x_2 + \mathbf{b}_3 x_3 = (1)\begin{pmatrix} 3 \\ 7 \\ 2 \end{pmatrix} + (-1)\begin{pmatrix} -2 \\ 1 \\ 3 \end{pmatrix} + (2)\begin{pmatrix} 4 \\ 0 \\ 5 \end{pmatrix} = \begin{pmatrix} 13 \\ 6 \\ 9 \end{pmatrix}$$

2.10 (a) $(\mathbf{AB})' = \begin{pmatrix} 7 & 16 \\ 8 & 4 \\ 7 & 11 \end{pmatrix}$, $\mathbf{B}'\mathbf{A}' = \begin{pmatrix} 7 & 16 \\ 8 & 4 \\ 7 & 11 \end{pmatrix}$ (c) $|\mathbf{A}| = 5$

2.11 (a) $\mathbf{a}'\mathbf{b} = 5$, $(\mathbf{a}'\mathbf{b})^2 = 25$ (b) $\mathbf{bb}' = \begin{pmatrix} 4 & 2 & 6 \\ 2 & 1 & 3 \\ 6 & 3 & 9 \end{pmatrix}$, $\mathbf{a}'(\mathbf{bb}')\mathbf{a} = 25$

2.12 $\mathbf{DA} = \begin{pmatrix} a & 2a & 3a \\ 4b & 5b & 6b \\ 7c & 8c & 9c \end{pmatrix}$, $\mathbf{AD} = \begin{pmatrix} a & 2b & 3c \\ 4a & 5b & 6c \\ 7a & 8b & 9c \end{pmatrix}$,

$\mathbf{DAD} = \begin{pmatrix} a^2 & 2ab & 3ac \\ 4ab & 5b^2 & 6bc \\ 7ac & 8bc & 9c^2 \end{pmatrix}$

2.13 $$\mathbf{AB} = \left[\begin{array}{ccc|c} 8 & 9 & 5 & 6 \\ 7 & 5 & 5 & 4 \\ \hline 3 & 4 & 2 & 2 \end{array}\right]$$

2.14 $\mathbf{AB} = \begin{pmatrix} 3 & 5 \\ 1 & 4 \end{pmatrix}$, $\mathbf{CB} = \begin{pmatrix} 3 & 5 \\ 1 & 4 \end{pmatrix}$

2.15 (a) $\text{tr}(\mathbf{A}) = 5$, $\text{tr}(\mathbf{B}) = 5$

(b) $\mathbf{A} + \mathbf{B} = \begin{pmatrix} 6 & 4 & 5 \\ 2 & -2 & 1 \\ 4 & 9 & 6 \end{pmatrix}$, $\text{tr}(\mathbf{A} + \mathbf{B}) = 10$

(c) $|\mathbf{A}| = 0$, $|\mathbf{B}| = 2$

(d) $\mathbf{AB} = \begin{pmatrix} 9 & 12 & 17 \\ 3 & -1 & 5 \\ 6 & 13 & 12 \end{pmatrix}$, $\det(\mathbf{AB}) = 0$

2.16 **(a)** $|\mathbf{A}| = 36$ **(b)** $\mathbf{T} = \begin{pmatrix} 1.7321 & 2.3094 & 1.7321 \\ 0 & 1.6330 & 1.2247 \\ 0 & 0 & 2.1213 \end{pmatrix}$

2.17 **(a)** $\det(\mathbf{A}) = 1$ **(b)** $\mathbf{T} = \begin{pmatrix} 1.7321 & -2.8868 & -.5774 \\ 0 & 2.1602 & -.7715 \\ 0 & 0 & .2673 \end{pmatrix}$

2.18 **(a)** $\mathbf{C} = \begin{pmatrix} .4082 & -.5774 & .7071 \\ .8165 & .5774 & .0000 \\ .4082 & -.5774 & -.7071 \end{pmatrix}$

2.19 **(a)** Eigenvalues: $2, 1, -1$

Eigenvectors: $\begin{pmatrix} .3015 \\ .9045 \\ .3015 \end{pmatrix}, \begin{pmatrix} .7999 \\ .5368 \\ .2684 \end{pmatrix}, \begin{pmatrix} .7071 \\ 0 \\ .7071 \end{pmatrix}$

(b) $\text{tr}(\mathbf{A}) = 2, \quad |\mathbf{A}| = -2$

2.20 **(a)** $\mathbf{C} = \begin{pmatrix} .0000 & .5774 & -.8165 \\ -.7071 & -.5774 & -.4082 \\ .7071 & -.5774 & -.4082 \end{pmatrix}$,

(b) $\mathbf{C}'\mathbf{AC} = \begin{pmatrix} -2 & 0 & 0 \\ 0 & 1 & 0 \\ 0 & 0 & 4 \end{pmatrix}$

(c) $\mathbf{CDC}' = \begin{pmatrix} 3 & 1 & 1 \\ 1 & 0 & 2 \\ 1 & 2 & 0 \end{pmatrix} = \mathbf{A}$

2.21 Eigenvalues: $1, 3$, $\mathbf{C} = \begin{pmatrix} -.7071 & -.7071 \\ -.7071 & .7071 \end{pmatrix}$,

$$\mathbf{A}^{1/2} = \mathbf{CD}^{1/2}\mathbf{C}' = \begin{pmatrix} 1.3660 & -.3660 \\ -.3660 & 1.3660 \end{pmatrix}$$

2.22 **(a)** $\mathbf{j}'\mathbf{a} = (1)a_1 + (1)a_2 + \cdots + (1)a_n = \sum_i a_i = \mathbf{a}'\mathbf{j}$

(b) $$\mathbf{Aj} = \begin{bmatrix} (1)a_{11} + (1)a_{12} + \cdots + (1)a_{1p} \\ (1)a_{21} + (1)a_{22} + \cdots + (1)a_{2p} \\ \vdots \qquad \vdots \qquad \vdots \\ (1)a_{n1} + (1)a_{n2} + \cdots + (1)a_{np} \end{bmatrix} = \begin{bmatrix} \sum_j a_{1j} \\ \sum_j a_{2j} \\ \vdots \\ \sum_j a_{nj} \end{bmatrix}$$

(c) $$\begin{aligned}\mathbf{j}'\mathbf{A} &= [(1)a_{11} + (1)a_{21} + \cdots + (1)a_{n1}, \ldots, (1)a_{1p} \\ &\quad + (1)a_{2p} + \cdots + (1)a_{np}] \\ &= (\textstyle\sum_i a_{i1}, \sum_i a_{i2}, \ldots, \sum_i a_{ip})\end{aligned}$$

2.23 $$\begin{aligned}(\mathbf{x} - \mathbf{y})'(\mathbf{x} - \mathbf{y}) &= (\mathbf{x}' - \mathbf{y}')(\mathbf{x} - \mathbf{y}) \\ &= \mathbf{x}'\mathbf{x} - \mathbf{x}'\mathbf{y} - \mathbf{y}'\mathbf{x} + \mathbf{y}'\mathbf{y} \\ &= \mathbf{x}'\mathbf{x} - 2\mathbf{x}'\mathbf{y} + \mathbf{y}'\mathbf{y}\end{aligned}$$

2.24 By (2.27), $(\mathbf{A}'\mathbf{A})' = \mathbf{A}'(\mathbf{A}')'$. By (2.6), $(\mathbf{A}')' = \mathbf{A}$. Thus, $(\mathbf{A}'\mathbf{A})' = \mathbf{A}'\mathbf{A}$.

2.25 (a) $$\begin{aligned}\textstyle\sum_i \mathbf{a}'\mathbf{x}_i &= \mathbf{a}'\mathbf{x}_1 + \mathbf{a}'\mathbf{x}_2 + \cdots + \mathbf{a}'\mathbf{x}_n \\ &= \mathbf{a}'(\mathbf{x}_1 + \mathbf{x}_2 + \cdots + \mathbf{x}_n) \quad \text{[by (2.21)]} \\ &= \mathbf{a}'\textstyle\sum_i \mathbf{x}_i\end{aligned}$$

(b) $$\begin{aligned}\textstyle\sum_i \mathbf{A}\mathbf{x}_i &= \mathbf{A}\mathbf{x}_1 + \mathbf{A}\mathbf{x}_2 + \cdots + \mathbf{A}\mathbf{x}_n \\ &= \mathbf{A}(\mathbf{x}_1 + \mathbf{x}_2 + \cdots + \mathbf{x}_n) \quad \text{[by (2.21)]} \\ &= \mathbf{A}\textstyle\sum_i \mathbf{x}_i\end{aligned}$$

(c) $$\begin{aligned}\textstyle\sum_i (\mathbf{a}'\mathbf{x}_i)^2 &= \textstyle\sum_i \mathbf{a}'(\mathbf{x}_i\mathbf{x}_i')\mathbf{a} \quad \text{[by (2.40)]} \\ &= \mathbf{a}'(\textstyle\sum_i \mathbf{x}_i\mathbf{x}_i')\mathbf{a} \quad \text{[by (2.29)]}\end{aligned}$$

(d) $\sum_i \mathbf{A}\mathbf{x}_i(\mathbf{A}\mathbf{x}_i)' = \sum_i \mathbf{A}\mathbf{x}_i\mathbf{x}_i'\mathbf{A}' = \mathbf{A}(\sum_i \mathbf{x}_i\mathbf{x}_i')\mathbf{A}'$

2.26 (a) $$\mathbf{Ax} = \begin{pmatrix} \mathbf{a}_1' \\ \mathbf{a}_2' \end{pmatrix} \mathbf{x} = \begin{pmatrix} \mathbf{a}_1'\mathbf{x} \\ \mathbf{a}_2'\mathbf{x} \end{pmatrix}$$

(b) $$\begin{aligned}\mathbf{A}'\mathbf{S}\mathbf{A} &= \begin{pmatrix} \mathbf{a}_1' \\ \mathbf{a}_2' \end{pmatrix} \mathbf{S}(\mathbf{a}_1, \mathbf{a}_2) = \begin{pmatrix} \mathbf{a}_1' \\ \mathbf{a}_2' \end{pmatrix} (\mathbf{S}\mathbf{a}_1, \mathbf{S}\mathbf{a}_2) \quad \text{[by (2.47)]} \\ &= \begin{pmatrix} \mathbf{a}_1'\mathbf{S}\mathbf{a}_1 & \mathbf{a}_1'\mathbf{S}\mathbf{a}_2 \\ \mathbf{a}_2'\mathbf{S}\mathbf{a}_1 & \mathbf{a}_2'\mathbf{S}\mathbf{a}_2 \end{pmatrix}\end{aligned}$$

2.27 If $\mathbf{A} = \begin{bmatrix} \mathbf{a}_1' \\ \mathbf{a}_2' \\ \vdots \\ \mathbf{a}_n' \end{bmatrix}$, then by (2.63), $\mathbf{A}' = (\mathbf{a}_1, \mathbf{a}_2, \ldots, \mathbf{a}_n)$ and

$$\mathbf{A}'\mathbf{A} = (\mathbf{a}_1, \mathbf{a}_2, \ldots, \mathbf{a}_n)\begin{bmatrix} \mathbf{a}_1' \\ \mathbf{a}_2' \\ \vdots \\ \mathbf{a}_n' \end{bmatrix}$$

$$= \mathbf{a}_1\mathbf{a}_1' + \mathbf{a}_2\mathbf{a}_2' + \cdots + \mathbf{a}_n\mathbf{a}_n' \quad [\text{by } (2.60)]$$

2.28 $$\frac{1}{b}\begin{pmatrix} b\mathbf{A}_{11}^{-1} + \mathbf{A}_{11}^{-1}\mathbf{a}_{12}\mathbf{a}_{12}'\mathbf{A}_{11}^{-1} & -\mathbf{A}_{11}^{-1}\mathbf{a}_{12} \\ -\mathbf{a}_{12}'\mathbf{A}_{11}^{-1} & 1 \end{pmatrix}\begin{pmatrix} \mathbf{A}_{11} & \mathbf{a}_{12} \\ \mathbf{a}_{12}' & a_{22} \end{pmatrix}$$

$$= \frac{1}{b}\begin{pmatrix} b\mathbf{I} + \mathbf{A}_{11}^{-1}\mathbf{a}_{12}\mathbf{a}_{12}' - \mathbf{A}_{11}^{-1}\mathbf{a}_{12}\mathbf{a}_{12}' & b\mathbf{A}_{11}^{-1}\mathbf{a}_{12} + \mathbf{A}_{11}^{-1}\mathbf{a}_{12}\mathbf{a}_{12}'\mathbf{A}_{11}^{-1}\mathbf{a}_{12} - \mathbf{A}_{11}^{-1}\mathbf{a}_{12}a_{22} \\ -\mathbf{a}_{12}' + \mathbf{a}_{12}' & -\mathbf{a}_{12}'\mathbf{A}_{11}^{-1}\mathbf{a}_{12} + a_{22} \end{pmatrix}$$

$$= \frac{1}{b}\begin{pmatrix} b\mathbf{I} & \mathbf{0} \\ \mathbf{0}' & b \end{pmatrix} \quad \text{where } b = a_{22} - \mathbf{a}_{12}'\mathbf{A}_{11}^{-1}\mathbf{a}_{12}$$

$$= \begin{pmatrix} \mathbf{I} & \mathbf{0} \\ \mathbf{0}' & 1 \end{pmatrix}$$

2.29 $$(\mathbf{B} + \mathbf{cc}')\left(\mathbf{B}^{-1} - \frac{\mathbf{B}^{-1}\mathbf{cc}'\mathbf{B}^{-1}}{1 + \mathbf{c}'\mathbf{B}^{-1}\mathbf{c}}\right)$$

$$= \mathbf{I} - \frac{\mathbf{cc}'\mathbf{B}^{-1}}{1 + \mathbf{c}'\mathbf{B}^{-1}\mathbf{c}} + \mathbf{cc}'\mathbf{B}^{-1} - \frac{\mathbf{cc}'\mathbf{B}^{-1}\mathbf{cc}'\mathbf{B}^{-1}}{1 + \mathbf{c}'\mathbf{B}^{-1}\mathbf{c}} \quad [\text{by } (2.26)]$$

$$= \mathbf{I} - \mathbf{cc}'\mathbf{B}^{-1}\left(\frac{1 + \mathbf{c}'\mathbf{B}^{-1}\mathbf{c}}{1 + \mathbf{c}'\mathbf{B}^{-1}\mathbf{c}}\right) + \mathbf{cc}'\mathbf{B}^{-1} = \mathbf{I}$$

2.30 $\mathbf{AA}^{-1} = \mathbf{I}$

$|\mathbf{AA}^{-1}| = |\mathbf{I}|$

$|\mathbf{A}||\mathbf{A}^{-1}| = 1 \quad [\text{by } (2.83)]$

$$|\mathbf{A}^{-1}| = \frac{1}{|\mathbf{A}|}$$

2.31 In (2.86) and (2.87), let $\mathbf{A}_{11} = \mathbf{B}, \mathbf{A}_{12} = \mathbf{c}, \mathbf{A}_{21} = -\mathbf{c}'$, and $\mathbf{A}_{22} = 1$. Then equate the right-hand sides of (2.86) and (2.87) to obtain (2.109).

2.32 Show that $|\mathbf{C}| \neq 0$ by taking the determinant of both sides of $\mathbf{C}'\mathbf{C} = \mathbf{I}$. Thus $\mathbf{C}$ is nonsingular and $\mathbf{C}^{-1}$ exists. Multiply $\mathbf{C}'\mathbf{C} = \mathbf{I}$ on the right by $\mathbf{C}^{-1}$ and on the left by $\mathbf{C}$.

2.33 Multiply $\mathbf{ABx} = \lambda\mathbf{x}$ on the left by $\mathbf{B}$. Then λ is an eigenvalue of $\mathbf{BA}$ and $\mathbf{Bx}$ is an eigenvector.

2.34 (a) $(\mathbf{A}^{1/2})^2 = (\mathbf{CD}^{1/2}\mathbf{C}')^2 = \mathbf{CD}^{1/2}\mathbf{C}'\mathbf{CD}^{1/2}\mathbf{C}'$

$= \mathbf{CDC}'$ [by (2.92)]
$= \mathbf{A}$ [by (2.100)]

(b) By (2.105), $\mathbf{A}^{1/2}\mathbf{A}^{1/2} = \mathbf{A}$. By (2.83),

$$\begin{aligned} |\mathbf{A}^{1/2}\mathbf{A}^{1/2}| &= |\mathbf{A}| \\ |\mathbf{A}^{1/2}||\mathbf{A}^{1/2}| &= |\mathbf{A}| \\ |\mathbf{A}^{1/2}|^2 &= |\mathbf{A}| \end{aligned}$$

(c) Since $\mathbf{A}$ is positive definite, we have, from part (b), $|\mathbf{A}^{1/2}| = |\mathbf{A}|^{1/2}$.

2.35 $\mathbf{A}^{-1}\mathbf{A} = \mathbf{I}$
$(\mathbf{A}^{-1}\mathbf{A})' = \mathbf{I}' = \mathbf{I}$
$\mathbf{A}'(\mathbf{A}^{-1})' = \mathbf{I}$
$(\mathbf{A}')^{-1}\mathbf{A}'(\mathbf{A}^{-1})' = (\mathbf{A}')^{-1}\mathbf{I} = (\mathbf{A}')^{-1}$
$(\mathbf{A}^{-1})' = (\mathbf{A}')^{-1}$

CHAPTER 3

3.1 $\bar{z} = \sum_{i=1}^n z_i/n = \sum_i ay_i/n = (ay_1 + \cdots + ay_n)/n$. Now factor a out of the sum.

3.2 The numerator of s_z^2 is $\sum_{i=1}^n (z_i - \bar{z})^2 = \sum_i (ay_i - a\bar{y})^2 = \sum_i [a(y_i - \bar{y})]^2$.

3.3 $\bar{x} = 4, \bar{y} = 4$:

x	y	$x - \bar{x}$	$y - \bar{y}$	$(x - \bar{x})(y - \bar{y})$
2	2	−2	−2	4
2	4	−2	0	0
2	6	−2	2	−4
4	2	0	−2	0
4	4	0	0	0
4	6	0	2	0
6	2	2	−2	−4
6	4	2	0	0
6	6	2	2	4
				Sum = 0

3.4 $\mathbf{x} - \bar{x}\mathbf{j} = \begin{bmatrix} x_1 \\ x_2 \\ \vdots \\ x_n \end{bmatrix} - \bar{x}\begin{bmatrix} 1 \\ 1 \\ \vdots \\ 1 \end{bmatrix} = \begin{bmatrix} x_1 \\ x_2 \\ \vdots \\ x_n \end{bmatrix} - \begin{bmatrix} \bar{x} \\ \bar{x} \\ \vdots \\ \bar{x} \end{bmatrix} = \begin{bmatrix} x_1 - \bar{x} \\ x_2 - \bar{x} \\ \vdots \\ x_n - \bar{x} \end{bmatrix}$

3.5 $\mathbf{y}_i - \bar{\mathbf{y}} = \begin{pmatrix} y_{i1} \\ y_{i2} \\ y_{i3} \end{pmatrix} - \begin{pmatrix} \bar{y}_1 \\ \bar{y}_2 \\ \bar{y}_3 \end{pmatrix} = \begin{pmatrix} y_{i1} - \bar{y}_1 \\ y_{i2} - \bar{y}_2 \\ y_{i3} - \bar{y}_3 \end{pmatrix}$

$$\sum_{i=1}^n (\mathbf{y}_i - \bar{\mathbf{y}})(\mathbf{y}_i - \bar{\mathbf{y}})' = \sum_{i=1}^n \begin{pmatrix} y_{i1} - \bar{y}_1 \\ y_{i2} - \bar{y}_2 \\ y_{i3} - \bar{y}_3 \end{pmatrix} (y_{i1} - \bar{y}_1, y_{i2} - \bar{y}_2, y_{i3} - \bar{y}_3)$$

$$= \sum_{i=1}^n \begin{pmatrix} (y_{i1} - \bar{y}_1)^2 & (y_{i1} - \bar{y}_1)(y_{i2} - \bar{y}_2) & (y_{i1} - \bar{y}_1)(y_{i3} - \bar{y}_3) \\ (y_{i2} - \bar{y}_2)(y_{i1} - \bar{y}_1) & (y_{i2} - \bar{y}_2)^2 & (y_{i2} - \bar{y}_2)(y_{i3} - \bar{y}_3) \\ (y_{i3} - \bar{y}_3)(y_{i1} - \bar{y}_1) & (y_{i3} - \bar{y}_3)(y_{i2} - \bar{y}_2) & (y_{i3} - \bar{y}_3)^2 \end{pmatrix}$$

3.6 $\bar{z} = \sum_{i=1}^n z_i/n = \sum_i \mathbf{a}'\mathbf{y}_i/n = (\mathbf{a}'\mathbf{y}_1 + \cdots + \mathbf{a}'\mathbf{y}_n)/n$. Now factor out $\mathbf{a}'$ on the left. See also (2.42).

3.7 The numerator of s_z^2 is $\sum_{i=1}^n (z_i - \bar{z})^2 = \sum_i (\mathbf{a}'\mathbf{y}_i - \mathbf{a}'\bar{\mathbf{y}})^2 = \sum_i (\mathbf{a}'\mathbf{y}_i - \mathbf{a}'\bar{\mathbf{y}})(\mathbf{a}'\mathbf{y}_i - \mathbf{a}'\bar{\mathbf{y}})$. The scalar $\mathbf{a}'\mathbf{y}_i$ is equal to its transpose, as in (2.39). Thus $\mathbf{a}'\mathbf{y}_i = (\mathbf{a}'\mathbf{y}_i)' = \mathbf{y}_i'\mathbf{a}$, and $\sum_i (\mathbf{a}'\mathbf{y}_i - \mathbf{a}'\bar{\mathbf{y}})(\mathbf{a}'\mathbf{y}_i - \mathbf{a}'\bar{\mathbf{y}}) = \sum_i (\mathbf{a}'\mathbf{y}_i - \mathbf{a}'\bar{\mathbf{y}})(\mathbf{y}_i'\mathbf{a} - \bar{\mathbf{y}}'\mathbf{a})$. By (2.22) and (2.24), this becomes $\sum_i \mathbf{a}'(\mathbf{y}_i - \bar{\mathbf{y}})(\mathbf{y}_i - \bar{\mathbf{y}})'\mathbf{a}$. Now factor out $\mathbf{a}'$ on the left and $\mathbf{a}$ on the right. See also (2.44).

3.8 See problem 2.26(b).

3.9 In the discussion preceding (3.60), we have $\mathbf{ASA}'$ in the form

$$\mathbf{ASA}' = \begin{bmatrix} \mathbf{a}_1'\mathbf{Sa}_1 & \mathbf{a}_1'\mathbf{Sa}_2 & \cdots & \mathbf{a}_1'\mathbf{Sa}_k \\ \mathbf{a}_2'\mathbf{Sa}_1 & \mathbf{a}_2'\mathbf{Sa}_2 & \cdots & \mathbf{a}_2'\mathbf{Sa}_k \\ \vdots & \vdots & & \vdots \\ \mathbf{a}_k'\mathbf{Sa}_1 & \mathbf{a}_k'\mathbf{Sa}_2 & \cdots & \mathbf{a}_k'\mathbf{Sa}_k \end{bmatrix},$$

from which the result follows immediately.

3.10
$$\begin{aligned} \text{cov}(\mathbf{z}) &= \text{cov}\,[(\boldsymbol{\Sigma}^{1/2})^{-1}\bar{\mathbf{y}} - (\boldsymbol{\Sigma}^{1/2})^{-1}\boldsymbol{\mu}] \\ &= (\boldsymbol{\Sigma}^{1/2})^{-1}\,\text{cov}\,(\bar{\mathbf{y}})[(\boldsymbol{\Sigma}^{1/2})^{-1}]' \quad [\text{by (3.71)}] \\ &= (\boldsymbol{\Sigma}^{1/2})^{-1}\left(\frac{\boldsymbol{\Sigma}}{n}\right)(\boldsymbol{\Sigma}^{1/2})^{-1} \end{aligned}$$

$$= \frac{1}{n}(\boldsymbol{\Sigma}^{1/2})^{-1}\boldsymbol{\Sigma}^{1/2}\boldsymbol{\Sigma}^{1/2}(\boldsymbol{\Sigma}^{1/2})^{-1} \quad \text{[by (2.105)]}$$
$$= \frac{1}{n}\mathbf{I}$$

3.11 Answers are given in Examples 3.6 and 3.7.

3.12 **(a)** $|\mathbf{S}| = 459.956$, **(b)** $\text{tr}(\mathbf{S}) = 213.043$

3.13 **(a)** $|\mathbf{S}| = 27{,}236{,}586$, **(b)** $\text{tr}(\mathbf{S}) = 292.891$

3.14
$$\mathbf{R} = \begin{bmatrix} 1.000 & .614 & .757 & .575 & .413 \\ .614 & 1.000 & .547 & .750 & .548 \\ .757 & .547 & 1.000 & .605 & .692 \\ .575 & .750 & .605 & 1.000 & .524 \\ .413 & .548 & .692 & .524 & 1.000 \end{bmatrix}$$

3.15 $\bar{z} = 83.298$, $s_z^2 = 1048.659$

3.16 $r_{zw} = -.6106$

3.17 $y_1 = (1, 0, 0)\mathbf{y} = \mathbf{a}'\mathbf{y}$, $\frac{1}{2}(y_2 + y_3) = \left(0, \frac{1}{2}, \frac{1}{2}\right)\mathbf{y} = \mathbf{b}'\mathbf{y}$.Use (3.54) to obtain $r_{zw} = .4873$.

3.18 **(a)**
$$\bar{\mathbf{z}} = \begin{pmatrix} 38.369 \\ 40.838 \\ -51.727 \end{pmatrix}, \qquad \mathbf{S}_z = \begin{pmatrix} 323.64 & 19.25 & -460.98 \\ 19.25 & 588.67 & 104.07 \\ -460.98 & 104.07 & 686.27 \end{pmatrix}$$

(b)
$$\mathbf{R}_z = \begin{pmatrix} 1.0000 & .0441 & -.9781 \\ .0441 & 1.0000 & .1637 \\ -.9781 & .1637 & 1.0000 \end{pmatrix}$$

3.19 **(a)**
$$\bar{\mathbf{y}} = \begin{bmatrix} 48.655 \\ 49.625 \\ 50.570 \\ 51.445 \end{bmatrix}, \qquad \mathbf{S} = \begin{bmatrix} 6.3300 & 6.1891 & 5.7770 & 5.5348 \\ 6.1891 & 6.4493 & 6.1534 & 5.9057 \\ 5.7770 & 6.1534 & 6.9180 & 6.9267 \\ 5.5348 & 5.9057 & 6.9267 & 7.4331 \end{bmatrix}$$

$$\mathbf{R} = \begin{bmatrix} 1.0000 & .9687 & .8730 & .8069 \\ .9687 & 1.0000 & .9212 & .8530 \\ .8730 & .9212 & 1.0000 & .9659 \\ .8069 & .8530 & .9659 & 1.0000 \end{bmatrix}$$

(b) $|\mathbf{S}| = 1.0865$, $\text{tr}(\mathbf{S}) = 27.1304$

3.20 **(a)** $\bar{z} = 44.1400$, $s_z^2 = 21.2309$, $\overline{w} = 103.8850$, $s_w^2 = 30.8161$
(b) $s_{zw} = 6.5359$, $r_{zw} = .2555$

3.21 $\bar{\mathbf{z}} = \begin{pmatrix} 401.40 \\ -47.55 \\ 150.48 \end{pmatrix}$, $\quad \mathbf{S}_z = \begin{pmatrix} 398.33 & -44.35 & 148.35 \\ -44.35 & 12.36 & -16.90 \\ 148.35 & -16.90 & 59.46 \end{pmatrix}$

$$\mathbf{R}_z = \begin{pmatrix} 1.00 & -.63 & .96 \\ -.63 & 1.00 & -.62 \\ .96 & -.62 & 1.00 \end{pmatrix}$$

3.22 **(a)** $\begin{pmatrix} \bar{\mathbf{y}} \\ \bar{\mathbf{x}} \end{pmatrix} = \begin{bmatrix} 185.72 \\ 151.12 \\ \hline 183.84 \\ 149.24 \end{bmatrix}$

(b) $\mathbf{S} = \left[\begin{array}{cc|cc} 95.29 & 52.87 & 69.66 & 46.11 \\ 52.87 & 54.36 & 51.31 & 35.05 \\ \hline 69.66 & 51.31 & 100.81 & 56.54 \\ 46.11 & 35.05 & 56.54 & 45.02 \end{array}\right]$

3.23 $\begin{pmatrix} \bar{\mathbf{y}} \\ \bar{\mathbf{x}} \end{pmatrix} = \begin{bmatrix} 70.08 \\ 73.54 \\ 75.10 \\ \hline 109.68 \\ 104.24 \\ 109.98 \end{bmatrix}$

$$\mathbf{S} = \left[\begin{array}{ccc|ccc} 95.54 & 17.61 & 12.18 & 60.52 & 23.00 & 62.84 \\ 17.61 & 73.19 & 14.25 & 5.73 & 61.28 & -1.66 \\ 12.18 & 14.25 & 76.17 & 46.75 & 32.77 & 69.84 \\ \hline 60.52 & 5.73 & 46.75 & 808.63 & 320.59 & 227.36 \\ 23.00 & 61.28 & 32.77 & 320.59 & 505.86 & 167.35 \\ 62.84 & -1.66 & 69.84 & 227.36 & 167.35 & 508.71 \end{array}\right]$$

CHAPTER 4

4.1 $|\boldsymbol{\Sigma}_1| = 1$, $\quad \text{tr}(\boldsymbol{\Sigma}_1) = 20$, $\quad |\boldsymbol{\Sigma}_2| = 4$, $\quad \text{tr}(\boldsymbol{\Sigma}_2) = 15$. Thus $\text{tr}(\boldsymbol{\Sigma}_1) > \text{tr}(\boldsymbol{\Sigma}_2)$, but $|\boldsymbol{\Sigma}_1| < |\boldsymbol{\Sigma}_2|$. When converted to correlations, we have

$$\mathbf{P}_{\rho 1} = \begin{pmatrix} 1 & .96 & .80 \\ .96 & 1 & .89 \\ .80 & .89 & 1 \end{pmatrix} \qquad \mathbf{P}_{\rho 2} = \begin{pmatrix} 1 & .87 & .41 \\ .87 & 1 & .71 \\ .41 & .71 & 1 \end{pmatrix}$$

As noted at the end of Section 4.1.3, a decrease in intercorrelations or an increase in the variances will lead to a larger $|\boldsymbol{\Sigma}|$. In this case, the decrease in correlations from $\boldsymbol{\Sigma}_1$ to $\boldsymbol{\Sigma}_2$ outweighed the increase in the variances (the increase in trace).

4.2

$$\begin{aligned} E(\mathbf{z}) &= (\mathbf{T}')^{-1}[E(\mathbf{y}) - \boldsymbol{\mu}] \quad &&[\text{by (3.68)}] \\ &= (\mathbf{T}')^{-1}[\boldsymbol{\mu} - \boldsymbol{\mu}] = \mathbf{0} \\ \text{cov}(\mathbf{z}) &= (\mathbf{T}')^{-1}\boldsymbol{\Sigma}[(\mathbf{T}')^{-1}]' \quad &&[\text{by (3.71)}] \\ &= (\mathbf{T}')^{-1}\mathbf{T}'\mathbf{T}\mathbf{T}^{-1} \quad &&[\text{by (2.110) and (2.73)}] \\ &= \mathbf{I} \end{aligned}$$

4.3 By the last expression in Section 2.3.1,

$$\prod_{i=1}^{n} \frac{1}{(\sqrt{2\pi})^p |\boldsymbol{\Sigma}|^{1/2}} = \frac{1}{(\sqrt{2\pi})^{np} |\boldsymbol{\Sigma}|^{n/2}}.$$

The sum in the exponent of (4.13) follows from the basic algebra of exponents.

4.4 The other two terms are of the form $\frac{1}{2}\sum_{i=1}^{n}(\overline{\mathbf{y}} - \boldsymbol{\mu})'\boldsymbol{\Sigma}^{-1}(\mathbf{y}_i - \overline{\mathbf{y}})$, which is equal to $\frac{1}{2}[(\overline{\mathbf{y}} - \boldsymbol{\mu})'\boldsymbol{\Sigma}^{-1}]\sum_{i=1}^{n}(\overline{\mathbf{y}}_i - \overline{\mathbf{y}})$. This vanishes because $\sum_{i=1}^{n}(\mathbf{y}_i - \overline{\mathbf{y}}) = n\overline{\mathbf{y}} - n\overline{\mathbf{y}} = \mathbf{0}$.

4.5 We replace y_i in $\sqrt{b_1}$ by $z_i = ay_i + b$. By an extension of (3.3), $\overline{z} = a\overline{y} + b$. Then (4.18) becomes

$$\begin{aligned} \frac{\sqrt{n}\sum_{i=1}^{n}(z_i - \overline{z})^3}{\left[\sum_{i=1}^{n}(z_i - \overline{z})^2\right]^{3/2}} &= \frac{\sqrt{n}\sum_i (ay_i + b - a\overline{y} - b)^3}{\left[\sum_i (ay_i + b - a\overline{y} - b)^2\right]^{3/2}} \\ &= \frac{\sqrt{n}a^3\sum_i (y_i - \overline{y})^3}{\left[a^2\sum_i (y_i - \overline{y})^2\right]^{3/2}} \\ &= \frac{\sqrt{n}\sum_i (y_i - \overline{y})^3}{\left[\sum_i (y_i - \overline{y})^2\right]^{3/2}} = \sqrt{b_1}. \end{aligned}$$

Similarly, if (4.19) is expressed in terms of $z_i = ay_i + b$, it reduces to b_2 in terms of y_i.

4.6 $\beta_{2,p} = E[(\mathbf{y} - \boldsymbol{\mu})'\boldsymbol{\Sigma}^{-1}(\mathbf{y} - \boldsymbol{\mu})]^2$ by (4.33). But when $\mathbf{y}$ is $N_p(\boldsymbol{\mu}, \boldsymbol{\Sigma})$, $v = (\mathbf{y} - \boldsymbol{\mu})'\boldsymbol{\Sigma}^{-1}(\mathbf{y} - \boldsymbol{\mu})$ is distributed as $\chi^2(p)$ by property 3 in Section 4.2. Then $E(v^2) = \text{var}(v) + [E(v)]^2$.

4.7 To show that $b_{1,p}$ and $b_{2,p}$ are invariant under the transformation $\mathbf{z} = \mathbf{A}\mathbf{y}_i + \mathbf{b}$, where $\mathbf{A}$ is nonsingular, it is sufficient to show that $g_{ij}(\mathbf{z}) = (\mathbf{y}_i - \overline{\mathbf{y}})'\hat{\boldsymbol{\Sigma}}^{-1}(\mathbf{y}_j - \overline{\mathbf{y}})$. By (3.62) and (3.63), $\overline{\mathbf{z}} = \mathbf{A}\overline{\mathbf{y}} + \mathbf{b}$ and $\hat{\boldsymbol{\Sigma}}_z = \mathbf{A}\hat{\boldsymbol{\Sigma}}\mathbf{A}'$. Then g_{ij} for $\mathbf{z}$ becomes

$$\begin{aligned} g_{ij}(\mathbf{z}) &= (\mathbf{z}_i - \overline{\mathbf{z}})'\hat{\boldsymbol{\Sigma}}_z^{-1}(\mathbf{z}_j - \overline{\mathbf{z}}) \\ &= (\mathbf{A}\mathbf{y}_i + \mathbf{b} - \mathbf{A}\overline{\mathbf{y}} - \mathbf{b})'(\mathbf{A}\hat{\boldsymbol{\Sigma}}\mathbf{A}')^{-1}(\mathbf{A}\mathbf{y}_j + \mathbf{b} - \mathbf{A}\overline{\mathbf{z}} - \mathbf{b}) \\ &= (\mathbf{y}_i - \overline{\mathbf{y}})'\mathbf{A}'(\mathbf{A}')^{-1}\hat{\boldsymbol{\Sigma}}^{-1}\mathbf{A}^{-1}\mathbf{A}(\mathbf{y}_j - \overline{\mathbf{y}}) \\ &= (\mathbf{y}_i - \overline{\mathbf{y}})'\hat{\boldsymbol{\Sigma}}^{-1}(\mathbf{y}_j - \overline{\mathbf{y}}) = g_{ij}(\mathbf{y}) \end{aligned}$$

4.8 Let $i = (n)$ in (4.48); then substitute (4.47) into (4.48) to obtain $F_{(n)}$ in terms of w, as in (4.49).

4.9 **(a)** $\mathbf{a}' = (2, -1, 3)$, $\quad z = \mathbf{a}'\mathbf{y}$ is $N(17, 21)$

(b) $\mathbf{A} = \begin{pmatrix} 1 & 1 & 1 \\ 1 & -1 & 2 \end{pmatrix}$, $\quad \mathbf{z} = \mathbf{A}\mathbf{y}$ is $N_2\left[\begin{pmatrix} 8 \\ 10 \end{pmatrix}, \begin{pmatrix} 29 & -1 \\ -1 & 9 \end{pmatrix}\right]$

(c) By property 4b in Section 4.2, y_2 is $N(1, 13)$.

(d) By property 4a in Section 4.2, $\begin{pmatrix} y_1 \\ y_3 \end{pmatrix}$ is $N_2\left[\begin{pmatrix} 3 \\ 4 \end{pmatrix}, \begin{pmatrix} 6 & -2 \\ -2 & 4 \end{pmatrix}\right]$.

(e) $\mathbf{A} = \begin{pmatrix} 1 & 0 & 0 \\ 0 & 0 & 1 \\ \frac{1}{2} & \frac{1}{2} & 0 \end{pmatrix}$, $\quad \mathbf{A}\mathbf{y}$ is $N_3\left[\begin{pmatrix} 3 \\ 4 \\ 2 \end{pmatrix}, \begin{pmatrix} 6 & -2 & 3.5 \\ -2 & 4 & 1 \\ 3.5 & 1 & 5.25 \end{pmatrix}\right]$

4.10 **(a)** $\mathbf{z} = \begin{pmatrix} .408 & 0 & 0 \\ -.047 & .279 & 0 \\ .285 & -.247 & .731 \end{pmatrix}\begin{pmatrix} y - 3 \\ y - 1 \\ y - 4 \end{pmatrix}$

(b) $\mathbf{z} = \begin{pmatrix} .465 & -.070 & .170 \\ -.070 & .326 & -.166 \\ .170 & -.166 & .692 \end{pmatrix} \begin{pmatrix} y-3 \\ y-1 \\ y-4 \end{pmatrix}$

(c) By (4.6), $(\mathbf{y}-\boldsymbol{\mu})'\boldsymbol{\Sigma}^{-1}(\mathbf{y}-\boldsymbol{\mu})$ is distributed as χ^2_3.

4.11 **(a)** $\mathbf{a}' = (4, -2, 1, -3)$, $\quad z = \mathbf{a}'\mathbf{y}$ is $N(-30, 153)$

(b) $\mathbf{A} = \begin{pmatrix} 1 & 1 & 1 & 1 \\ -2 & 3 & 1 & -2 \end{pmatrix}$, $\quad \mathbf{z} = \mathbf{A}\mathbf{y}$ is $N_2\left[\begin{pmatrix} 5 \\ 2 \end{pmatrix}, \begin{pmatrix} 27 & -79 \\ -79 & 361 \end{pmatrix}\right]$

(c) $\mathbf{A} = \begin{pmatrix} 3 & 1 & -4 & -1 \\ -1 & -3 & 1 & -2 \\ 2 & 2 & 4 & -5 \end{pmatrix}$,

$$\mathbf{z} = \mathbf{A}\mathbf{y} \text{ is } N_3\left[\begin{pmatrix} -4 \\ -18 \\ -27 \end{pmatrix}, \begin{pmatrix} 35 & -18 & -6 \\ -18 & 46 & 14 \\ -6 & 14 & 93 \end{pmatrix}\right]$$

(d) By property 4b in Section 4.2, y_3 is $N(-1, 2)$.

(e) By property 4a in Section 4.2, $\begin{pmatrix} y_2 \\ y_4 \end{pmatrix}$ is $N_2\left[\begin{pmatrix} 3 \\ 5 \end{pmatrix}, \begin{pmatrix} 9 & -6 \\ -6 & 9 \end{pmatrix}\right]$.

(f) $\mathbf{A} = \begin{bmatrix} 1 & 0 & 0 & 0 \\ \frac{1}{2} & \frac{1}{2} & 0 & 0 \\ \frac{1}{3} & \frac{1}{3} & \frac{1}{3} & 0 \\ \frac{1}{4} & \frac{1}{4} & \frac{1}{4} & \frac{1}{4} \end{bmatrix}$,

$$\mathbf{A}\mathbf{y} \text{ is } N_4\left[\begin{bmatrix} -2 \\ .5 \\ 0 \\ 1.25 \end{bmatrix}, \begin{bmatrix} 11 & 1.5 & 2 & 3.75 \\ 1.5 & 1 & .67 & .875 \\ 2 & .67 & .67 & 1 \\ 3.75 & .875 & 1 & 1.688 \end{bmatrix}\right]$$

4.12 **(a)** $\mathbf{z} = \begin{bmatrix} .302 & 0 & 0 & 0 \\ .408 & .561 & 0 & 0 \\ -.087 & .261 & 1.015 & 0 \\ -.858 & -.343 & -.686 & .972 \end{bmatrix} \begin{bmatrix} y+2 \\ y-3 \\ y+1 \\ y-5 \end{bmatrix}$

(b) $\mathbf{z} = \begin{bmatrix} .810 & .305 & .143 & -.479 \\ .305 & .582 & .249 & -.083 \\ .143 & .249 & 1.153 & -.298 \\ -.480 & -.083 & -.298 & .787 \end{bmatrix} \begin{bmatrix} y+2 \\ y-3 \\ y+1 \\ y-5 \end{bmatrix}$

(c) $(\mathbf{y}-\boldsymbol{\mu})'\boldsymbol{\Sigma}^{-1}(\mathbf{y}-\boldsymbol{\mu}) = (\mathbf{y}-\boldsymbol{\mu})'\boldsymbol{\Sigma}^{-1/2}\boldsymbol{\Sigma}^{-1/2}(\mathbf{y}-\boldsymbol{\mu}) = \mathbf{z}'\mathbf{z}$, which is $\chi^2(p) = \chi^2(4)$

4.13 The variables in (b), (c), and (d) are independent.

4.14 The variables in (a), (c), (d), (f), (i), (j), and (n) are independent.

4.15 **(a)** $E(\mathbf{y}|\mathbf{x}) = \boldsymbol{\mu}_y + \boldsymbol{\Sigma}_{yx}\boldsymbol{\Sigma}_{xx}^{-1}(\mathbf{x} - \boldsymbol{\mu}_x)$

$$= \begin{pmatrix} 2 \\ -1 \end{pmatrix} + \begin{pmatrix} -3 & 2 \\ 0 & 4 \end{pmatrix} \begin{pmatrix} 5 & -2 \\ -2 & 4 \end{pmatrix}^{-1} \begin{pmatrix} x_1 - 3 \\ x_2 - 1 \end{pmatrix}$$

$$= \begin{pmatrix} 2 \\ -1 \end{pmatrix} + \begin{pmatrix} -.5 & .25 \\ .5 & 1.25 \end{pmatrix} \begin{pmatrix} x_1 - 3 \\ x_2 - 1 \end{pmatrix}$$

$$= \begin{pmatrix} 3.25 \\ -3.75 \end{pmatrix} + \begin{pmatrix} -.5 & .25 \\ .5 & 1.25 \end{pmatrix} \begin{pmatrix} x_1 \\ x_2 \end{pmatrix}$$

(b) $\text{cov}\,(\mathbf{y}|\mathbf{x}) = \boldsymbol{\Sigma}_{yy} - \boldsymbol{\Sigma}_{yx}\boldsymbol{\Sigma}_{xx}^{-1}\boldsymbol{\Sigma}_{xy}$

$$= \begin{pmatrix} 7 & 3 \\ 3 & 6 \end{pmatrix} - \begin{pmatrix} -3 & 2 \\ 0 & 4 \end{pmatrix} \begin{pmatrix} 5 & -2 \\ -2 & 4 \end{pmatrix}^{-1} \begin{pmatrix} -3 & 0 \\ 2 & 4 \end{pmatrix}$$

$$= \begin{pmatrix} 7 & 3 \\ 3 & 6 \end{pmatrix} - \begin{pmatrix} 2 & 1 \\ 1 & 5 \end{pmatrix} = \begin{pmatrix} 5 & 2 \\ 2 & 1 \end{pmatrix}$$

4.16 **(a)** $E(\mathbf{y}|\mathbf{x}) = \boldsymbol{\mu}_y + \boldsymbol{\Sigma}_{yx}\boldsymbol{\Sigma}_{xx}^{-1}(\mathbf{x} - \boldsymbol{\mu}_x)$

$$= \begin{pmatrix} 3 \\ -2 \end{pmatrix} + \begin{pmatrix} 15 & 0 & 3 \\ 8 & 6 & -2 \end{pmatrix} \begin{pmatrix} 50 & 8 & 5 \\ 8 & 4 & 0 \\ 5 & 0 & 1 \end{pmatrix}^{-1} \begin{pmatrix} x_1 - 4 \\ x_2 + 3 \\ x_3 - 5 \end{pmatrix}$$

$$= \begin{pmatrix} 3 \\ -2 \end{pmatrix} - \begin{pmatrix} 15 \\ -24.5 \end{pmatrix} + \begin{pmatrix} 0 & 0 & 3 \\ .67 & .167 & -5.33 \end{pmatrix} \begin{pmatrix} x_1 \\ x_2 \\ x_3 \end{pmatrix}$$

$$= \begin{pmatrix} -12 \\ 22.5 \end{pmatrix} + \begin{pmatrix} 0 & 0 & 3 \\ .67 & .167 & -5.33 \end{pmatrix} \begin{pmatrix} x_1 \\ x_2 \\ x_3 \end{pmatrix}$$

(b) $\text{cov}\,(\mathbf{y}|\mathbf{x}) = \boldsymbol{\Sigma}_{yy} - \boldsymbol{\Sigma}_{yx}\boldsymbol{\Sigma}_{xx}^{-1}\boldsymbol{\Sigma}_{xy}$

$$= \begin{pmatrix} 14 & -8 \\ -8 & 18 \end{pmatrix} - \begin{pmatrix} 15 & 0 & 3 \\ 8 & 6 & -2 \end{pmatrix} \begin{pmatrix} 50 & 8 & 5 \\ 8 & 4 & 0 \\ 5 & 0 & 1 \end{pmatrix}^{-1} \cdot \begin{pmatrix} 15 & 8 \\ 0 & 6 \\ 3 & -2 \end{pmatrix}$$

$$= \begin{pmatrix} 14 & -8 \\ -8 & 18 \end{pmatrix} - \begin{pmatrix} 9 & -6 \\ -6 & 17 \end{pmatrix} = \begin{pmatrix} 5 & -2 \\ -2 & 1 \end{pmatrix}$$

4.17 **(a)** By the central limit theorem in Section 4.3.2, $\sqrt{n}(\bar{\mathbf{y}} - \boldsymbol{\mu})$ is approximately $N_p(\mathbf{0}, \boldsymbol{\Sigma})$.

(b) $\bar{\mathbf{y}}$ is approximately $N_p(\boldsymbol{\mu}, \boldsymbol{\Sigma}/n)$.

4.18 **(a)** The plots show almost no deviation from normality.

(b)

Variable	y_1	y_2	y_3	y_4
$\sqrt{b_1}$	.3069	.3111	.0645	.0637
b_2	1.932	2.107	1.792	1.570

The values of $\sqrt{b_1}$ show a small amount of positive skewness, but none exceeds the upper 2.5% critical value for $\sqrt{b_1}$ given in Table A.1 as .942. The values of b_2 show negative kurtosis. For y_4, the kurtosis is significant, since $b_2 < 1.74$, the lower 2.5 percentile in Table A.3.

(c)

Variable	y_1	y_2	y_3	y_4
D	.2848	.2841	.2866	.2851
Y	.4021	.2934	.6730	.4491

From Table A.4, the lower 2.5 percentile for Y is -3.04 and the upper 97.5 percentile is .628. We reject the hypothesis of normality only for y_3.

(d) z defined in (4.24) is approximately $N(0, 3/n)$. To obtain a $N(0, 1)$ statistic, we calculate $z^* = z/\sqrt{3/n}$.

Variable	y_1	y_2	y_3	y_4
z^*	−.3366	−.3095	−.0737	−.0856

4.19 **(a)**

i	1	2	3	4	5	6	7	8	9	10
D_i^2	1.06	1.60	7.54	3.54	4.61	.63	.81	2.47	.95	3.78

(b) The .05 critical value from Table A.6 is 7.01. $D^2_{(10)} = 7.54 > 7.01$.

(c)

i	1	2	3	4	5	6	7	8	9	10
$u_{(i)}$	.08	.10	.12	.13	.20	.30	.44	.47	.57	.93
v_i	.08	.18	.27	.37	.46	.56	.66	.75	.85	.94

The plot of $(v_i, u_{(i)})$ shows some evidence of nonlinearity and an outlier.

(d) $b_{1,p} = 7.255$, $b_{2,p} = 14.406$. Both exceed critical values in Table A.5.

4.20 (b)

Variable	y_1	y_2	y_3	y_4	y_5
$\sqrt{b_1}$	.2176	.5857	.7461	−.3327	−.1772
b_2	2.079	1.681	2.583	1.774	2.456

None of the values of $\sqrt{b_1}$ exceeds 1.134 (from Table A.1) or is less than −1.134. None of the values of b_2 is less than 1.53 (from Table A.3). Thus there is no significant departure from normality.

(c)

Variable	y_1	y_2	y_3	y_4	y_5
D	.279	.269	.275	.281	.276
Y	−.305	−1.399	−.805	−.114	−.669

(d) $z^* = z/\sqrt{3/n}$, where z is defined in (4.24).

Variable	y_1	y_2	y_3	y_4	y_5
z^*	−.4848	−1.7183	−1.3627	.8091	.3686

4.21 (a)

i	1	2	3	4	5	6	7	8	9	10	11
D_i^2	5.20	2.15	7.63	5.34	5.54	1.73	5.21	5.90	2.72	6.02	2.56

(c) The plot shows a sharp break from the fourth to the fifth points.

(d) $b_{1,p} = 12.985$, $b_{2,p} = 29.072$

4.22 (a) The Q–Q plots for y_1 and y_5 show little departure from normality. The Q–Q plots for y_2 and y_3 show some evidence of heavier tails than the normal. The Q–Q plots for y_4 and y_6 show some evidence of positive skewness.

(b)

Variable	y_1	y_2	y_3	y_4	y_5	y_6
$\sqrt{b_1}$	.5521	.0302	.7827	1.4627	.2219	.9974
b_2	3.160	3.275	2.772	6.675	2.176	4.528

(c)

Variable	y_1	y_2	y_3	y_4	y_5	y_6
D	.276	.274	.275	.260	.286	.271
Y	−1.469	−1.845	−1.675	−5.249	.889	−2.741

(d)

Variable	y_1	y_2	y_3	y_4	y_5	y_6
z^*	−1.640	−.062	−2.803	−2.961	−.870	−2.456

4.23 (a) $D_i^2 = 7.816, 3.640, 5.730, \ldots, 6.433$

(b) $D^2_{(51)} = 25.628$. By extrapolation in Table A.6, the .05 critical value for $p = 6$ is approximately 19. Thus we reject the hypothesis of multivariate normality.

(c) $(v_i, u_{(i)}) = (.013, .011), (.033, .015), \ldots, (.990, .986)$. The plot shows nonlinearity

(d) $b_{1,p} = 16.287, b_{2,p} = 58.337$. By extrapolation in Table A.5, both appear to exceed their critical values.

CHAPTER 5

5.1 By (5.4), we have

$$(\bar{\mathbf{y}} - \boldsymbol{\mu}_0)' \left(\frac{\mathbf{S}}{n} \right)^{-1} (\bar{\mathbf{y}} - \boldsymbol{\mu}_0) = (\bar{\mathbf{y}} - \boldsymbol{\mu}_0)' \left(\frac{1}{n} \right)^{-1} \mathbf{S}^{-1} (\bar{\mathbf{y}} - \boldsymbol{\mu}_0)$$
$$= n(\bar{\mathbf{y}} - \boldsymbol{\mu}_0)' \mathbf{S}^{-1} (\bar{\mathbf{y}} - \boldsymbol{\mu}_0)$$

5.2 From (5.8), we have

$$\frac{n_1 n_2}{n_1 + n_2} (\bar{\mathbf{y}}_1 - \bar{\mathbf{y}}_2)' \mathbf{S}_{\text{pl}}^{-1} (\bar{\mathbf{y}}_1 - \bar{\mathbf{y}}_2)$$
$$= (\bar{\mathbf{y}}_1 - \bar{\mathbf{y}}_2)' \left(\frac{n_1 + n_2}{n_1 n_2} \right)^{-1} \mathbf{S}_{\text{pl}}^{-1} (\bar{\mathbf{y}}_1 - \bar{\mathbf{y}}_2)$$
$$= (\bar{\mathbf{y}}_1 - \bar{\mathbf{y}}_2)' \left(\frac{n_1 + n_2}{n_1 n_2} \mathbf{S}_{\text{pl}} \right)^{-1} (\bar{\mathbf{y}}_1 - \bar{\mathbf{y}}_2)$$
$$= (\bar{\mathbf{y}}_1 - \bar{\mathbf{y}}_2)' \left[\left(\frac{1}{n_1} + \frac{1}{n_2} \right) \mathbf{S}_{\text{pl}} \right]^{-1} (\bar{\mathbf{y}}_1 - \bar{\mathbf{y}}_2)$$

5.3 $$\bar{d} = \frac{1}{n} \sum_{i=1}^{n} d_i = \frac{1}{n} \sum_{i=1}^{n} (y_i - x_i) = \frac{1}{n} \sum_i y_i - \frac{1}{n} \sum_i x_i = \bar{y} - \bar{x}$$

$$s_d^2 = \frac{1}{n-1} \sum_{i=1}^{n} (d_i - \bar{d})^2 = \frac{1}{n-1} \sum_i (y_i - x_i - \bar{y} + \bar{x})^2$$
$$= \frac{1}{n-1} \sum_i [(y_i - \bar{y}) - (x_i - \bar{x})]^2$$

When this is expanded, we obtain $s_d^2 = s_y^2 + s_x^2 - 2s_{yx}$.

5.4 Similar to Problem 5.1.

5.5 It is assumed that y and x have a bivariate normal distribution. Let $\mathbf{y}_i = \binom{y_i}{x_i}$. Then d_i can be expressed as $d_i = y_i - x_i = \mathbf{a}'\mathbf{y}_i$, where $\mathbf{a}' =$

(1, −1). By property 1b in Section 4.2, d_i is $N(\mathbf{a}'\boldsymbol{\mu}, \mathbf{a}'\boldsymbol{\Sigma}\mathbf{a})$. Show that $\mathbf{a}'\overline{\mathbf{y}} = \overline{y} - \overline{x}$, $\mathbf{a}'\mathbf{S}\mathbf{a} = s_y^2 - 2s_{yx} + s_x^2 = s_d^2$, and that $T^2 = n(\mathbf{a}'\overline{\mathbf{y}})'(\mathbf{a}'\mathbf{S}\mathbf{a})^{-1}(\mathbf{a}'\overline{\mathbf{y}})$ is the square of $t = \overline{d}/(s_d/\sqrt{n})$.

5.7 Under H_{03}, we have $\mathbf{C}\boldsymbol{\mu}_1 = \mathbf{0}$ and $\mathbf{C}\boldsymbol{\mu}_2 = \mathbf{0}$. Then

$$E(\mathbf{C}\overline{\mathbf{y}}) = \mathbf{C}E(\overline{\mathbf{y}}) = \mathbf{C}E\left(\frac{n_1\overline{\mathbf{y}}_1 + n_2\overline{\mathbf{y}}_2}{n_1 + n_2} \right) = \frac{n_1\mathbf{C}\boldsymbol{\mu}_1 + n_2\mathbf{C}\boldsymbol{\mu}_2}{n_1 + n_2} = \mathbf{0}.$$

Since $\overline{\mathbf{y}}_1$ and $\overline{\mathbf{y}}_2$ are independent,

$$\begin{aligned} \mathrm{cov}\,(\overline{\mathbf{y}}) &= \mathrm{cov}\left(\frac{n_1\overline{\mathbf{y}}_1 + n_2\overline{\mathbf{y}}_2}{n_1 + n_2} \right) = \frac{n_1^2\boldsymbol{\Sigma}/n_1 + n_2^2\boldsymbol{\Sigma}/n_2}{(n_1 + n_2)^2} \\ &= \frac{(n_1 + n_2)\boldsymbol{\Sigma}}{(n_1 + n_2)^2}. \end{aligned}$$

5.8 $\mathbf{C}\mathbf{S}_{\mathrm{pl}}\mathbf{C}'/(n_1+n_2)$ is the sample covariance matrix of $\mathbf{C}\overline{\mathbf{y}}$. Hence the equation immediately above (5.39) exhibits the characteristic form of the T^2-distribution.

5.9 $T^2 = .061$

5.10 **(a)** $T^2 = 85.3327$

(b) $t_1 = 2.5039, t_2 = .2665, t_3 = -2.5157, t_4 = .9510, t_5 = .3161$

5.11 $T^2 = 30.2860$

5.12 **(a)** $T^2 = 1.8198$

(b) $t_1 = 1.1643, t_2 = 1.1006, t_3 = .9692, t_4 = .7299$. None of these is significant. In fact, ordinarily they would not have been examined because the T^2-test in part (a) did not reject.

5.13 $T^2 = 79.5510$

5.14 **(a)** $T^2 = 133.4873$

(b) $t_1 = 3.8879, t_2 = -3.8652, t_3 = -5.6911, t_4 = -5.0426$

(c) $\mathbf{a}' = (.345, -.130, -.106, -.143)$

(d) $T^2 = 133.4873$

(e) $R^2 = .782975, T^2 = 133.4873$

(f) By (5.33), $t^2(y_1|y_2, y_3, y_4) = 35.9336, t^2(y_2|y_1, y_3, y_4) = 5.7994,$ $t^2(y_3|y_1, y_2, y_4) = 1.7749, t^2(y_4|y_1, y_2, y_3) = 8.2592$

(g) By (5.31), $T^2(y_3, y_4|y_1, y_2) = 12.5206, F(y_3, y_4|y_1, y_2) = 6.0814$

5.15 By (5.35), the test for parallelism gives $T^2 = 132.6863$. The discriminant function coefficient vector is given by (5.36) as $\mathbf{a}' = (-.362, -.223, -.137)$.

5.16 **(a)** $T^2 = 66.6604$

(b) $t_1 = -.6556, t_2 = 2.6139, t_3 - 3.2884, t_4 = -4.6315, t_5 = 1.8873, t_6 = -3.2205$

(c) By (5.33), $t^2(y_1|y_2, y_3, y_4, y_5, y_6) = .0758, t^2(y_2|y_1, y_3, y_4, y_5, y_6) = 6.4513$, $t^2(y_3|y_1, y_2, y_4, y_5, y_6) = 6.9518, t^2(y_4|y_1, y_2, y_3, y_5, y_6) = 6.0309$, $t^2(y_5|y_1, y_2, y_3, y_4, y_6) = 3.7052, t^2(y_6|y_1, y_2, y_3, y_4, y_5) = 6.2619$

(d) By (5.31), $T^2(y_4, y_5, y_6|y_1, y_2, y_3) = 27.547$

5.17 **(a)** $T^2 = 70.5679$

(b) $T^2(y_5, y_6|y_3, y_4) = 13.1517$

(c) $T^2(y_1, y_2|y_3, y_4, y_5, y_6) = 8.5162$

5.18 **(a)** $T^2 = 18.4625$ **(b)** $\mathbf{a}' = (-.057, -.010, -.242, -.071)$
(c) By (5.33), $t^2(y_1|y_2, y_3, y_4) = 3.3315, t^2(y_2|y_1, y_3, y_4) = .0102,$ $t^2(y_3|y_1, y_2, y_4) = 1.4823, t^2(y_4|y_1, y_2, y_3) = .0013$

5.19 **(a)** $T^2 = 15.1912$ **(b)** $\mathbf{a}' = (-.036, .048)$ **(c)** $t_1 = -3.8371, t_2 = -2.4362$

5.20 $T^2 = 22.3238$

5.21 **(a)** $T^2 = 206.1188$

(b) $t^2(d_1|d_2, d_3) = 59.0020, t^2(d_2|d_1, d_3) = 53.4507$, $t^2(d_3|d_1, d_2) = 80.9349$

CHAPTER 6

6.1 **(a)** Using $\overline{y}_{i.} = y_{i.}/n$, we have

$$\begin{aligned}\sum_{i=1}^{k}\sum_{j=1}^{n}(y_{ij} - \overline{y}_{i.})^2 &= \sum_{ij}(y_{ij}^2 - 2y_{ij}\overline{y}_{i.} + \overline{y}_{i.}^2)\\ &= \sum_{ij} y_{ij}^2 - \sum_i \overline{y}_{i.}\sum_j y_{ij} + n\sum_i \overline{y}_{i.}^2\\ &= \sum_{ij} y_{ij}^2 - 2\sum_i \frac{y_{i.}}{n}y_{i.} + n\sum_i\left(\frac{y_{i.}}{n}\right)^2\\ &= \sum_{ij} y_{ij}^2 - 2\sum_i \frac{y_{i.}^2}{n} + \sum_i \frac{y_{i.}^2}{n}.\end{aligned}$$

6.2 $(\mathbf{E}^{-1}\mathbf{H} - \lambda\mathbf{I})\mathbf{a} = \mathbf{0}$
$[(\mathbf{E}^{1/2}\mathbf{E}^{1/2})^{-1}\mathbf{H} - \lambda\mathbf{I}]\mathbf{a} = \mathbf{0}$
$[(\mathbf{E}^{1/2})^{-1}(\mathbf{E}^{1/2})^{-1}\mathbf{H} - \lambda\mathbf{I}]\mathbf{a} = \mathbf{0}$
$[(\mathbf{E}^{1/2})^{-1}\mathbf{H} - \lambda\mathbf{E}^{1/2}]\mathbf{a} = \mathbf{0}$
$[(\mathbf{E}^{1/2})^{-1}\mathbf{H} - \lambda\mathbf{E}^{1/2}](\mathbf{E}^{1/2})^{-1}\mathbf{E}^{1/2}\mathbf{a} = \mathbf{0}$
$[(\mathbf{E}^{1/2})^{-1}\mathbf{H}(\mathbf{E}^{1/2})^{-1} - \lambda\mathbf{I}]\mathbf{E}^{1/2}\mathbf{a} = \mathbf{0}$

6.3 When $s = 1$, we have $V^{(1)} = \lambda_1/(1+\lambda_1)$, $U^{(1)} = \lambda_1$, $\Lambda = 1/(1+\lambda_1)$, and $\theta = \lambda_1/(1+\lambda_1)$. Solving the last of these for λ_1 gives $\lambda_1 = \theta/(1-\theta)$, and the results in (6.27), (6.28), and (6.29) follow immediately.

6.4 With $T^2 = (n_1 + n_2 - 2)U^{(1)}$ and $U^{(1)} = \theta/(1-\theta)$, we obtain (5.20). We obtain (5.19) from (5.20) by $V^{(1)} = \theta$. A similar argument leads to (6.27).

6.5 With $\overline{\mathbf{y}}_{i.} = \mathbf{y}_{i.}/n_i$ and $\overline{\mathbf{y}}_{..} = \mathbf{y}_{..}/N$, we obtain

$$\begin{aligned}\mathbf{H} &= \sum_{i=1}^{k} n_i(\overline{\mathbf{y}}_{i.} - \overline{\mathbf{y}}_{..})(\overline{\mathbf{y}}_{i.} - \overline{\mathbf{y}}_{..})'\\ &= \sum_i n_i(\overline{\mathbf{y}}_{i.}\overline{\mathbf{y}}_{i.}' - \overline{\mathbf{y}}_{i.}\overline{\mathbf{y}}_{..}' - \overline{\mathbf{y}}_{..}\overline{\mathbf{y}}_{i.}' + \overline{\mathbf{y}}_{..}\overline{\mathbf{y}}_{..}')\end{aligned}$$

$$= \sum_i n_i \overline{\mathbf{y}}_{i.} \overline{\mathbf{y}}_{i.}' - \left(\sum_i n_i \overline{\mathbf{y}}_{i.} \right) \overline{\mathbf{y}}_{..}' - \overline{\mathbf{y}}_{..} \sum_i n_i \overline{\mathbf{y}}_{i.}' + \overline{\mathbf{y}}_{..} \overline{\mathbf{y}}_{..}' \sum_i n_i$$

$$= \sum_i n_i \frac{\mathbf{y}_{i.} \mathbf{y}_{i.}'}{n_i^2} - \frac{\left(\sum_i \mathbf{y}_{i.} \right) \mathbf{y}_{..}}{N} - \frac{\mathbf{y}_{..}}{N} \sum_i \mathbf{y}_{i.} + \frac{N \mathbf{y}_{..} \mathbf{y}_{..}'}{N^2}$$

$$= \sum_i \frac{\mathbf{y}_{i.} \mathbf{y}_{i.}'}{n_i} - \frac{\mathbf{y}_{..} \mathbf{y}_{..}'}{N} - \frac{\mathbf{y}_{..} \mathbf{y}_{..}'}{N} + \frac{\mathbf{y}_{..} \mathbf{y}_{..}'}{N}.$$

6.6 $\overline{\mathbf{y}}_{1.} - \overline{\mathbf{y}}_{..}$ becomes

$$\overline{\mathbf{y}}_{1.} - \frac{n_1 \overline{\mathbf{y}}_{1.} + n_2 \overline{\mathbf{y}}_{2.}}{n_1 + n_2} = \frac{n_1 \overline{\mathbf{y}}_{1.} + n_2 \overline{\mathbf{y}}_{1.} - n_1 \overline{\mathbf{y}}_{1.} - n_2 \overline{\mathbf{y}}_{2.}}{n_1 + n_2} = \frac{n_2 (\overline{\mathbf{y}}_{1.} - \overline{\mathbf{y}}_{2.})}{n_1 + n_2}.$$

The first term in the sum is

$$\frac{n_1 n_2^2}{(n_1 + n_2)^2} (\overline{\mathbf{y}}_{1.} - \overline{\mathbf{y}}_{2.})(\overline{\mathbf{y}}_{1.} - \overline{\mathbf{y}}_{2.})'.$$

The second term in the sum is

$$\frac{n_1^2 n_2}{(n_1 + n_2)^2} (\overline{\mathbf{y}}_{1.} - \overline{\mathbf{y}}_{2.})(\overline{\mathbf{y}}_{1.} - \overline{\mathbf{y}}_{2.})'.$$

6.7 $\theta = \dfrac{\lambda_1}{1 + \lambda_1} = \dfrac{\text{SSH}(z)/\text{SSE}(z)}{1 + \text{SSH}(z)/\text{SSE}(z)} = \dfrac{\text{SSE}(z)}{\text{SSE}(z) + \text{SSH}(z)}$

6.8 From $r_i^2 = \lambda_i/(1 + \lambda_i)$, obtain $\lambda_i = r_i^2/(1 - r_i^2)$. Substitute this into $1/(1 + \lambda_i)$ to obtain the result.

6.9 Substitute $A_p = V^{(s)}/s$ into (6.38) to obtain (6.21).

6.10 When $s = 1$, (6.39) becomes

$$A_{\text{LH}} = \frac{U^{(1)}}{1 + U^{(1)}}.$$

By (6.27), $U^{(1)} = \lambda_1$.

6.11 Substitute $A_{\text{LH}} = U^{(s)}/(s + U^{(s)})$ from (6.39) into (6.40) to obtain F_3 at the end of Section 6.1.5.

6.12 To show $\text{cov}(c_i\bar{\mathbf{y}}_{i.}) = c_i^2\boldsymbol{\Sigma}$, use (3.69), $\text{cov}(\mathbf{Ay}) = \mathbf{A\Sigma A}'$, with $\mathbf{A} = c_i\mathbf{I}$.

6.13 By (6.7),

$$\begin{aligned}
\mathbf{H}_z &= n\sum_{i=1}^{k}(\bar{\mathbf{z}}_{i.} - \bar{\mathbf{z}}_{..})(\bar{\mathbf{z}}_{i.} - \bar{\mathbf{z}}_{..})' \\
&= n\sum_{i}(\mathbf{C}\bar{\mathbf{y}}_{i.} - \mathbf{C}\bar{\mathbf{y}}_{..})(\mathbf{C}\bar{\mathbf{y}}_{i.} - \mathbf{C}\bar{\mathbf{y}}_{..})' \\
&= n\sum_{i}[\mathbf{C}(\bar{\mathbf{y}}_{i.} - \bar{\mathbf{y}}_{..})][\mathbf{C}(\bar{\mathbf{y}}_{i.} - \bar{\mathbf{y}}_{..})]' \\
&= n\mathbf{C}\left[\sum_{i}(\bar{\mathbf{y}}_{i.} - \bar{\mathbf{y}}_{..})(\bar{\mathbf{y}}_{i.} - \bar{\mathbf{y}}_{..})'\right]\mathbf{C}' \quad \text{[by (2.45)]}
\end{aligned}$$

6.14 $\mathbf{C}$ is not square.

6.15 $E(\mathbf{C}\bar{\mathbf{y}}_{..}) = \mathbf{C}E(\bar{\mathbf{y}}_{..}) = \mathbf{C}E(\sum_{i=1}^{k}\bar{\mathbf{y}}_{i.}/k)$
$= \mathbf{C}\sum_i E(\bar{\mathbf{y}}_{i.})/k = \mathbf{C}\sum_i \boldsymbol{\mu}_i/k$
$= \mathbf{0}$ [by H_{03} in (6.66)]
$\text{cov}(\mathbf{C}\bar{\mathbf{y}}_{..}) = \mathbf{C\Sigma C}'/kn$ if there are no differences in the group means, $\mathbf{C}\boldsymbol{\mu}_1, \mathbf{C}\boldsymbol{\mu}_2, \ldots, \mathbf{C}\boldsymbol{\mu}_k$. This condition is assured by H_{01} in (6.61).

6.16 For our purposes, it will suffice to show that T^2 has the characteristic form of the T^2-distribution in (5.5).

6.17 The (univariate) expected mean square corresponding to $\bar{\mu}_.$ in a one-way ANOVA is $\sigma^2 + N\mu^2$. Thus the mean square for $\bar{\mu}_.$ is tested with MSE.

6.18 From (6.88) we have

$$\Lambda = \frac{|\mathbf{AEA}'|}{|\mathbf{A}(\mathbf{E}+\mathbf{H}^*)\mathbf{A}'|} = \frac{|\mathbf{AEA}'|}{|\mathbf{AEA}' + \mathbf{AH}^*\mathbf{A}'|}.$$

Substitute $\mathbf{H}^* = kn\bar{\mathbf{y}}_{..}\bar{\mathbf{y}}_{..}'$ to obtain

$$\Lambda = \frac{|\mathbf{AEA}'|}{|\mathbf{AEA}' + \sqrt{kn}\mathbf{A}\bar{\mathbf{y}}_{..}(\sqrt{kn}\mathbf{A}\bar{\mathbf{y}}_{..})'|}.$$

Now use (2.109) in problem 2.31 with $\mathbf{B} = \mathbf{AEA}'$ and $\mathbf{c} = \sqrt{kn}\mathbf{A}\bar{\mathbf{y}}_{..}$ to obtain

$$\Lambda = \frac{1}{1 + kn(\mathbf{A}\bar{\mathbf{y}}_{..})'(\mathbf{AEA}')^{-1}(\mathbf{A}\bar{\mathbf{y}}_{..})}.$$

Multiply and divide by ν_E and use (6.84) to obtain (6.89).

6.19 Solve for T^2 in (6.89).

6.20 In $\mathbf{C}_1\mathbf{A}'$ the rows of $\mathbf{C}_1$ are multiplied by the rows of $\mathbf{A}$. Show that $\mathbf{C}_1\mathbf{A}' = \mathbf{O}$.

6.21 As noted, the function $(\bar{\mathbf{y}} - \mathbf{A}\hat{\boldsymbol{\beta}})'\mathbf{S}^{-1}(\bar{\mathbf{y}} - \mathbf{A}\hat{\boldsymbol{\beta}})$ is similar to SSE = $(\mathbf{y} - \mathbf{X}\hat{\boldsymbol{\beta}})'(\mathbf{y} - \mathbf{X}\hat{\boldsymbol{\beta}})$ in (10.3) and (10.6). By an argument similar to that used in Section 10.2.2 to obtain $\hat{\boldsymbol{\beta}} = (\mathbf{X}'\mathbf{X})^{-1}\mathbf{X}'\mathbf{y}$, it follows that $\hat{\boldsymbol{\beta}} = (\mathbf{A}'\mathbf{S}^{-1}\mathbf{A})^{-1}\mathbf{A}'\mathbf{S}^{-1}\bar{\mathbf{y}}$. An alternative approach (for those familiar with differentiation with respect to a vector) is to expand $(\bar{\mathbf{y}} - \mathbf{A}\hat{\boldsymbol{\beta}})'\mathbf{S}^{-1}(\bar{\mathbf{y}} - \mathbf{A}\hat{\boldsymbol{\beta}})$ to four terms, differentiate with respect to $\hat{\boldsymbol{\beta}}$, and set the result equal to $\mathbf{0}$.

6.22 Expand $n(\bar{\mathbf{y}} - \mathbf{A}\hat{\boldsymbol{\beta}})'\mathbf{S}^{-1}(\bar{\mathbf{y}} - \mathbf{A}\hat{\boldsymbol{\beta}})$ to four terms and substitute $\hat{\boldsymbol{\beta}} = (\mathbf{A}'\mathbf{S}^{-1}\mathbf{A})^{-1}\mathbf{A}'\mathbf{S}^{-1}\bar{\mathbf{y}}$ into the last one.

6.23 **(a)**
$$\mathbf{E} = \begin{bmatrix} 13.41 & 7.72 & 8.68 & 5.86 \\ 7.72 & 8.48 & 7.53 & 6.21 \\ 8.68 & 7.53 & 11.61 & 7.04 \\ 5.86 & 6.21 & 7.04 & 10.57 \end{bmatrix}$$

$$\mathbf{H} = \begin{bmatrix} 1.05 & 2.17 & -1.38 & -.76 \\ 2.17 & 4.88 & -2.37 & -1.26 \\ -1.38 & -2.37 & 2.38 & 1.38 \\ -.76 & -1.26 & 1.38 & .81 \end{bmatrix}$$

$\Lambda = .224$, $V^{(s)} = .860$, $U^{(s)} = 3.08$, and $\theta = .747$ All four are significant.

(b) $\eta^2_\Lambda = 1 - \Lambda = .776$, $\eta^2_\theta = \theta = .747$, $A_\Lambda = 1 - \Lambda^{1/s} = .526$, $A_{\text{LH}} = .606$, $A_p = V^{(s)}/s = .430$

(c) The eigenvalues of $\mathbf{E}^{-1}\mathbf{H}$ are 2.9515 and .1273. The essential dimensionality of the space of the mean vectors is 1.

(d) For 1, 2 vs. 3 we have $\Lambda = .270$, $V^{(s)} = .730$, $U^{(s)} = 2.702$, and $\theta = .730$. All four are significant. For 1 vs. 2 we obtain $\Lambda = .726$, $V^{(s)} = .274$, $U^{(s)} = .377$, and $\theta = .274$. All four are significant.

(e)

Variable	y_1	y_2	y_3	y_4
F	1.29	9.50	3.39	1.27

Only the F for y_2 is significant. The discriminant function coefficient vector from $z = \mathbf{a}'\mathbf{y}$, where $\mathbf{a}$ is the first eigenvector of $\mathbf{E}^{-1}\mathbf{H}$, is $\mathbf{a}' = (.021, .533, -.347, -.135)$. Again y_2 contributes most to separation of groups.

(f) By (6.101), $\Lambda(y_3, y_4 | y_1, y_2) = \Lambda(y_1, y_2, y_3, y_4)/\Lambda(y_1, y_2) = .224/.568 = .395 < \Lambda_{.05} = .725$

(g) By (6.102),

$$\begin{aligned}
\Lambda(y_1 | y_2, y_3, y_4) &= \Lambda(y_1, y_2, y_3, y_4)/\Lambda(y_2, y_3, y_4) \\
&= .224/.240 = .934 > \Lambda_{.05} = .819 \\
\Lambda(y_2 | y_1, y_3, y_4) &= .224/.538 = .417 < .819 \\
\Lambda(y_3 | y_1, y_2, y_4) &= .224/.369 = .609 < .819 \\
\Lambda(y_4 | y_1, y_2, y_3) &= .224/.243 = .924 > .819
\end{aligned}$$

6.24 **(a)** S effect: $\Lambda = .00065, V^{(s)} = 2.357, U^{(s)} = 142.304, \theta = .993$. All are significant.
V effect: $\Lambda = .065, V^{(s)} = 1.107, U^{(s)} = 11.675, \theta = .920$. All are significant.
SV interaction: $\Lambda = .138, V^{(s)} = 1.321, U^{(s)} = 3.450, \theta = .726$. All are significant.

(b) Contrast on V comparing 2 vs. 1, 3: $\Lambda = .0804, V^{(s)} = .920, U^{(s)} = 11.445, \theta = .920$. All are significant.

(c) Linear contrast for S: $\Lambda = .0073, V^{(s)} = .993, U^{(s)} = 135.273, \theta = .993$. All are significant.
Quadratic contrast for S: $\Lambda = .168, V^{(s)} = .832, U^{(s)} = 4.956, \theta = .832$. All are significant.
Cubic contrast for S: $\Lambda = .325, V^{(s)} = .675, U^{(s)} = 2.076, \theta = .675$. All are significant.

(d) ANOVA F-tests for each variable:

Source	y_1	y_2	y_3	y_4
S	980.21	214.24	876.13	73.91
V	251.22	9.47	14.77	27.12
SV	20.37	2.84	3.44	2.08

All F's are significant except the last one, 2.08.

(e) Test of significance of y_3 and y_4 adjusted for y_1 and y_2:

	S	V	SV
$\Lambda(y_3, y_4 \mid y_1, y_2)$	.1226	.9336	.6402

(f) Test of significance of each variable adjusted for the other three:

	S	V	SV
$\Lambda(y_1 \mid y_2, y_3, y_4)$	.1158	.2099	.3082
$\Lambda(y_2 \mid y_1, y_3, y_4)$	.5586	.8134	.7967
$\Lambda(y_3 \mid y_1, y_2, y_4)$	.2271	.9627	.7604
$\Lambda(y_4 \mid y_1, y_2, y_3)$	.6692	.9795	.8683

6.25 V = velocity (fixed), L = lubricant (random)
V effect (using $\mathbf{H}_{VL}$ for error matrix): $\Lambda = .0492, V^{(s)} = .951, U^{(s)} = 19.315, \theta = .951$. With $p = 2, \nu_H = 1$, and $\nu_E = 3, \Lambda_{.05} = .049998, V^{(s)}_{.05} = .950, U^{(s)}_{.05} = T^2_{.05}/\nu_E = 19.00, \theta_{.05} = .950$. Thus all four test statistics reject H_0.
L effect (using $\mathbf{E}$ for error matrix): $\Lambda = .692, V^{(s)} = .314, U^{(s)} = .438, \theta = .295$. None is significant.
VL interaction (using $\mathbf{E}$ for error matrix): $\Lambda = .932, V^{(s)} = .069, U^{(s)} = .073, \theta = .061$. None is significant.

6.26

Source	Λ	$V^{(s)}$	$U^{(s)}$	θ	Significant
(a) Reagent	.0993	1.126	6.911	.868	Yes
(b) Contrast 1 vs. 2, 3, 4	.146	.854	5.871	.854	Yes
Subjects	.00000082	2.847	1091.127	.999	Yes

6.27 P = proportion of filler, T = surface treatment, F = filler:

Source	Λ	$V^{(s)}$	$U^{(s)}$	θ	Significant
P	.138	.977	5.441	.841	Yes
T	.080	.920	11.503	.920	Yes
PT	.712	.295	.396	.271	No
F	.019	.980	51.180	.981	Yes
PF	.179	.958	3.835	.784	Yes
TF	.355	.645	1.815	.645	Yes
PTF	.752	.264	.309	.172	No

6.28 A = period; P, T, and F are defined in problem 6.27:

Source	Λ	$V^{(s)}$	$U^{(s)}$	θ	Significant
A	.021	.979	47.099	.979	Yes
AP	.475	.545	1.063	.505	No
AT	.142	.858	6.049	.858	Yes
APT	.777	.228	.282	.208	No
AF	.095	.905	9.486	.905	Yes
APF	.622	.387	.594	.363	No
ATF	.387	.613	1.586	.613	Yes
$APTF$	.781	.229	.267	.169	No

For the between-subject factors and interactions, we have

Source	df	F	p-Value
P	2	21.79	< .0001
T	1	78.34	< .0001
PT	2	1.28	.3143
F	1	345.04	< .0001
PF	2	15.79	.0004
TF	1	5.36	.0392
PTF	2	.48	.6294
Error	12		

6.29 For parallelism, we use (6.62) to obtain $\Lambda = .2397$. For levels, we use (6.64) and (6.65) to obtain $\Lambda = .9651$ and $F = .597$. For flatness we use (6.67) to obtain $T^2 = 110.521$.

6.30 **(a)** By (6.73), $T^2 = 20.7420$. By (6.88), $\Lambda = .5655$.

(b) For each row $\mathbf{c}_i'$ of $\mathbf{C}$, we use $T_i^2 = n(\mathbf{c}_i'\overline{\mathbf{y}})'(\mathbf{c}_i'\mathbf{S}\mathbf{c}_i)^{-1}\mathbf{c}_i'\overline{\mathbf{y}}$, as in Example 6.9.2:
$T_1^2 = 17.0648, T_2^2 = .3238, T_3^2 = .2714$.
This can also be done by Wilks' Λ using $\Lambda_i = \mathbf{c}_i'\mathbf{E}\mathbf{c}_i/\mathbf{c}_i'(\mathbf{E} + \mathbf{H}^*)\mathbf{c}_i$: $\Lambda_1 = .6127, \Lambda_2 = .9882, \Lambda_3 = .9900$.

6.31 The six variables represent two within-subjects factors: y_1 is A_1B_1, y_2 is A_1B_2, y_3 is A_1B_3, x_1 is A_2B_1, x_2 is A_2B_2, and x_3 is A_2B_2. Using linear and quadratic effects (other orthogonal contrasts could be used), the matrices **A**, **B**, and **G** in (6.81), (6.82), and (6.83) become

$$\mathbf{A} = (1 \quad 1 \quad 1 \quad -1 \quad -1 \quad -1)$$

$$\mathbf{B} = \begin{pmatrix} 1 & 0 & -1 & 1 & 0 & -1 \\ 1 & -2 & 1 & 1 & -2 & 1 \end{pmatrix}$$

$$\mathbf{G} = \begin{pmatrix} 1 & 0 & -1 & -1 & 0 & 1 \\ 1 & -2 & 1 & -1 & 2 & -1 \end{pmatrix}.$$

Using these in T^2 as given by (6.84), (6.85), and (6.86), we obtain $T_A^2 = 193.0901$, $T_B^2 = 2.8000$, and $T_{AB}^2 = 6.8676$. Using MANOVA tests for the same within-subjects factors, we obtain

Source	Λ	$V^{(s)}$	$U^{(s)}$	θ	Significant
A	.202	.798	3.941	.798	Yes
B	.946	.054	.057	.054	No
AB	.877	.123	.140	.123	Yes

6.32 MANOVA tests for the within-subjects effect, time (T), and interactions of time with the between-subjects effects, cancer (C) and gender (G):

Source	Λ	$V^{(s)}$	$U^{(s)}$	θ
T	.258	.742	2.874	.742
TC	.363	.809	1.299	.444
TG	.929	.071	.077	.071
TCG	.809	.201	.225	.130

ANOVA F-tests for between-subjects factors and interactions:

Source	df	F	p-Value
C	5	4.16	.003
G	1	2.69	.107
CG	5	.37	.869

6.33 **(a)** $T^2 = 79.551$

(b) Using $t_i = \mathbf{c}_i'\overline{\mathbf{y}}/\sqrt{\mathbf{c}_i'\mathbf{S}\mathbf{c}_i/n}$, where $\mathbf{c}_i'$ is the ith row of $\mathbf{C}$, we obtain $t_1 = 7.155, t_2 = -.445, t_3 = -.105$.

6.34 **(a)** $T^2 = 1712.2201$

(b) Using $t_i^2 = N(\mathbf{c}_i'\overline{\mathbf{y}})^2/\mathbf{c}_i'\mathbf{S}\mathbf{c}_i$, we obtain $t_1^2 = 18.231, t_2^2 = -7.388, t_3^2 = .237, t_4^2 = 2.763, t_5^2 = 2.084, t_6^2 = 1.403$.

6.35 **(a)** Using $T^2 = N(\mathbf{C}\overline{\mathbf{y}}_{..})'(\mathbf{C}\mathbf{S}_{\text{pl}}\mathbf{C}')^{-1}(\mathbf{C}\overline{\mathbf{y}}_{..})$ in Section 6.10.2, we obtain $T^2 = 17.582 < T^2_{.05,3,9} = 27.202$.

(b) $t_1 = .951, t_2 = 1.606, t_3 = .127$ (Since the T^2-test did not reject, these would ordinarily not be calculated.)

(c) Using $\Lambda = |\mathbf{CEC}'|/|\mathbf{C}(\mathbf{E}+\mathbf{H})\mathbf{C}'|$ in Section 6.10.2, we obtain $\Lambda = .3107$.

(d) Using $\Lambda_i = \mathbf{c}_i'\mathbf{E}\mathbf{c}_i/\mathbf{c}_i(\mathbf{E}+\mathbf{H})\mathbf{c}_i$, we have $\Lambda_1 = .833, \Lambda_2 = .988, \Lambda_3 = .650$. [Since there was nothing significant in part (b), we would ordinarily not have calculated these.]

6.36 **(a)** Using $T^2 = N(\mathbf{C}\bar{\mathbf{y}}_{..})'(\mathbf{CS}_{\text{pl}}\mathbf{C}')^{-1}(\mathbf{C}\bar{\mathbf{y}}_{..})$ in Section 6.10.2, we obtain $T^2 = 33.802 > T^2_{.05,4,24} = 12.983$.

(b) Using $t_i^2 = N(\mathbf{c}_i'\bar{\mathbf{y}})^2/\mathbf{c}_i'\mathbf{S}_{\text{pl}}\mathbf{c}_i$, we obtain $t_1^2 = .675$, $t_2^2 = .393$, $t_3^2 = 32.626$. Only the cubic effect is significant.

(c) For an overall test comparing groups, we have $\Lambda = |\mathbf{CEC}'|/|\mathbf{C}(\mathbf{E}+\mathbf{H})\mathbf{C}'| = .4361$.

(d) To compare groups using each row of $\mathbf{C}$, we have $\Lambda_i = \mathbf{c}_i'\mathbf{E}\mathbf{c}_i/\mathbf{c}_i'(\mathbf{E}+\mathbf{H})\mathbf{c}_i$. $\Lambda_1 = .534, \Lambda_2 = .764, \Lambda_3 = .941$

6.37 **(a)** Using $T^2 = N(\mathbf{C}\bar{\mathbf{y}}_{..})'(\mathbf{CS}_{\text{pl}}\mathbf{C}')^{-1}(\mathbf{C}\bar{\mathbf{y}}_{..})$ in Section 6.10.2, we obtain $T^2 = 45.500$.

(b) Using $t_i^2 = N(\mathbf{c}_i'\bar{\mathbf{y}})^2/\mathbf{c}_i'\mathbf{S}_{\text{pl}}\mathbf{c}_i$, we obtain $t_1^2 = 18.410, t_2^2 = 8.385, t_3^2 = 3.446, t_4^2 = .011, t_5^2 = .098, t_6^2 = 2.900$.

(c) For an overall test comparing groups, we have $\Lambda = |\mathbf{CEC}'|/|\mathbf{C}(\mathbf{E}+\mathbf{H})\mathbf{C}'| = .304$.

(d) To compare groups using each row of $\mathbf{C}$, we have $\Lambda_i = \mathbf{c}_i'\mathbf{E}\mathbf{c}_i/\mathbf{c}_i'(\mathbf{E}+\mathbf{H})\mathbf{c}_i$. $\Lambda_1 = .695, \Lambda_2 = .925, \Lambda_3 = .731, \Lambda_4 = .814, \Lambda_5 = .950, \Lambda_6 = .894$

6.38 **(a)** Combined groups (pooled covariance matrix). Using t = number of minutes − 30, we obtain, by (6.93),

$$\hat{\boldsymbol{\beta}}' = (98.1, .981, .0418, -.00101, -.000048)$$

By (6.94), we obtain $T^2 = .216$. By (6.96), we have

$$\hat{\boldsymbol{\mu}}' = (95.5, 96.7, 95.6, 93.8, 98.1, 99.2)$$

(b) Group 1: $\hat{\boldsymbol{\beta}}_1' = (100.7, .819, .040, -.00085, -.000038)$, $T^2 = .0113$, $\hat{\boldsymbol{\mu}}_1' = (105.2, 104.4, 101.5, 98.6, 100.6, 108.1)$

(c) Groups 2–4: $\hat{\boldsymbol{\beta}}_2' = (97.4, 1.010, .0403, -.00103, -.000049)$, $T^2 = .2554$, $\hat{\boldsymbol{\mu}}_2' = (92.6, 94.4, 93.8, 92.4, 97.4, 96.6)$

6.39 **(a)** For the control group, the overall test is $T^2 = n_1(\mathbf{C}\bar{\mathbf{y}}_{1.})'(\mathbf{CS}_1\mathbf{C}')^{-1}$

$(\mathbf{C}\bar{\mathbf{y}}_{1.}) = 554.749$. For each row of $\mathbf{C}$ (linear, quadratic, etc.), we have $t_i^2 = n_1(\mathbf{c}_i'\bar{\mathbf{y}}_{1.})^2/\mathbf{c}_i'\mathbf{S}_1\mathbf{c}_i$: $t_1^2 = 5.714, t_2^2 = 50.111, t_3^2 = 50.767, t_4^2 = 8.011, t_5^2 = .508$.

(b) For the obese group, we obtain $T^2 = n_2(\mathbf{C}\bar{\mathbf{y}}_{2.})'(\mathbf{C}\mathbf{S}_2\mathbf{C}')^{-1}(\mathbf{C}\bar{\mathbf{y}}_{2.}) = 128.552$. For each row of $\mathbf{C}$, we obtain $t_1^2 = 4.978, t_2^2 = 107.129, t_3^2 = 5.225, t_4^2 = 10.750, t_5^2 = 3.572$.

(c) For the combined groups ($\mathbf{S}$ = pooled covariance matrix), we use $T^2 = N(\mathbf{C}\bar{\mathbf{y}}_{..})'(\mathbf{C}\mathbf{S}\mathbf{C}')^{-1}(\mathbf{C}\bar{\mathbf{y}}_{..})$ to obtain $T^2 = 247.0079$. We test for linear, quadratic, etc., trends using the rows of $\mathbf{C}$: $t_i^2 = N(\mathbf{c}_i'\bar{\mathbf{y}}_{..})^2/\mathbf{c}_i'\mathbf{S}\mathbf{c}_i$. $t_1^2 = 1.162, t_2^2 = 155.017, t_3^2 = 30.540, t_4^2 = 1.319, t_5^2 = .506$. To compare groups, we use $\Lambda = |\mathbf{C}\mathbf{E}\mathbf{C}'|/|\mathbf{C}(\mathbf{E}+\mathbf{H})\mathbf{C}'|$ and $\Lambda_i = \mathbf{c}_i'\mathbf{E}\mathbf{c}_i/\mathbf{c}_i'(\mathbf{E}+\mathbf{H})\mathbf{c}_i$: $\Lambda = .4902, \Lambda_1 = .7947, \Lambda_2 = .9940, \Lambda_3 = .7987, \Lambda_4 = .6228, \Lambda_5 = .9172$.

6.40 Control group: By (6.93), $\hat{\boldsymbol{\beta}}_1' = (3.129, .656, -.283, -.334, .192, .037, -.020)$. By (6.94), $T^2 = .7633$. By (6.96),

$$\hat{\boldsymbol{\mu}}_1' = (\hat{\mu}_{11}, \hat{\mu}_{12}, \ldots, \hat{\mu}_{18}) = (4.11, 3.29, 2.71, 2.71, 3.04, 3.39, 3.54, 3.95).$$

Obese group: $\hat{\boldsymbol{\beta}}_2' = (3.207, -.187, .463, .056, -.102, -.010, .010)$, $T^2 = .3943$, $\hat{\boldsymbol{\mu}}_2' = (4.51, 4.12, 3.81, 3.48, 3.24, 3.37, 3.70, 4.02)$
Combined groups (pooled covariance matrix):
$\hat{\boldsymbol{\beta}}' = (3.15, .162, .183, -.115, .012, .010, -.002), T^2 = .0158$, $\hat{\boldsymbol{\mu}}' = (4.36, 3.80, 3.36, 3.15, 3.13, 3.37, 3.63, 3.98)$

6.41 A = activator, T = time, G = group. In (6.84), (6.85), and (6.86), we use

$$\mathbf{A} = \begin{pmatrix} 2 & 2 & 2 & -1 & -1 & -1 & -1 & -1 & -1 \\ 0 & 0 & 0 & 1 & 1 & 1 & -1 & -1 & -1 \end{pmatrix}$$

$$\mathbf{T} = \begin{pmatrix} -1 & 0 & 1 & -1 & 0 & 1 & -1 & 0 & 1 \\ 1 & -2 & 1 & 1 & -2 & 1 & 1 & -2 & 1 \end{pmatrix}$$

$$\mathbf{C} = \begin{bmatrix} -2 & 0 & 2 & 1 & 0 & -1 & 1 & 0 & -1 \\ 2 & -4 & 2 & -1 & 2 & -1 & -1 & 2 & -1 \\ 0 & 0 & 0 & 1 & 0 & -1 & -1 & 0 & 1 \\ 0 & 0 & 0 & 1 & -2 & 1 & -1 & 2 & -1 \end{bmatrix}$$

$T_A^2 = 5072.579, T_T^2 = 268.185, T_{AT}^2 = 143.491$. The same within-sample factors and interaction can be tested with Wilks' Λ using (6.88) and the other three MANOVA tests:

Source	Λ	$V^{(s)}$	$U^{(s)}$	θ	Significant
A	.003	.997	317.04	.997	Yes
T	.056	.944	16.76	.944	Yes
AT	.100	.900	8.97	.900	Yes

The interactions of the within factors with the between factor G are tested with Wilks' Λ (Section 6.9.5) and with the other three MANOVA tests:

Source	Λ	$V^{(s)}$	$U^{(s)}$	θ	Significant
AG	.884	.116	.131	.116	No
TG	.889	.111	.125	.111	No
ATG	.795	.205	.258	.205	No

The between-subjects factor G is tested with an ANOVA F-test: F = .47, p-value = .504.

CHAPTER 7

7.1 If $\boldsymbol{\Sigma}_0$ is substituted for $\mathbf{S}$ in (7.1), we have

$$u = \nu[\ln|\boldsymbol{\Sigma}_0| - \ln|\boldsymbol{\Sigma}_0| + \text{tr}(\mathbf{I}) - p] = \nu[0 + p - p] = 0$$

7.2
$$\begin{aligned} \ln|\boldsymbol{\Sigma}_0| - \ln|\mathbf{S}| &= -\ln|\boldsymbol{\Sigma}_0|^{-1} - \ln|\mathbf{S}| \\ &= -\ln|\boldsymbol{\Sigma}_0^{-1}| - \ln|\mathbf{S}| \quad \text{[by (2.84)]} \\ &= -(\ln|\mathbf{S}| + \ln|\boldsymbol{\Sigma}_0^{-1}|) \\ &= -\ln|\mathbf{S}\boldsymbol{\Sigma}_0^{-1}| \quad \text{[by (2.83)]} \end{aligned}$$

7.3
$$\begin{aligned} -\ln(\Pi_{i=1}^p \lambda_i) + \textstyle\sum_{i=1}^p \lambda_i &= -\textstyle\sum_{i=1}^p \ln\lambda_i + \sum_{i=1}^p \lambda_i \\ &= \textstyle\sum_{i=1}^p (\lambda_i - \ln\lambda_i) \end{aligned}$$

7.4 As noted in Section 7.1, the likelihood ratio in this case involves the ratio of the determinants of the sample covariance matrices under H_0 and H_1. Under H_1, which is essentially unrestricted, the maximum likelihood estimate of $\boldsymbol{\Sigma}$ (corrected for bias) is given by (4.12) as $\mathbf{S}$. Under H_0 it is assumed that each of the p y_i's in $\mathbf{y}$ has variance σ^2 and that all y_i's are independent. Thus we estimate σ^2 (unbiasedly) in each of the p columns of the $\mathbf{Y}$ matrix [see (3.16)] and pool the p estimates to obtain

$$\hat{\sigma}^2 = \sum_{i=1}^{n}\sum_{j=1}^{p} \frac{(y_{ij} - \bar{y}_j)^2}{(n-1)p}.$$

Show that by (3.21) this is equal to

$$\hat{\sigma}^2 = \sum_{i=1}^{p} \frac{s_{ii}}{p} = \frac{\operatorname{tr} \mathbf{S}}{p}.$$

Thus the likelihood ratio is

$$\text{LR} = \left(\frac{|\mathbf{S}|}{|\hat{\sigma}^2 \mathbf{I}|} \right)^{n/2} = \left(\frac{|\mathbf{S}|}{|\mathbf{I} \operatorname{tr} \mathbf{S}/p|} \right)^{n/2}.$$

Show that by (2.79) this becomes

$$\text{LR} = \left(\frac{|\mathbf{S}|}{(\operatorname{tr} \mathbf{S}/p)^p} \right)^{n/2}.$$

7.5 If $\lambda_1 = \lambda_2 = \cdots = \lambda_p = \lambda$, say, then

$$u = \frac{p^p \prod_{i=1}^{p} \lambda_i}{\left(\sum_{i=1}^{p} \lambda_i \right)^p} = \frac{p^p \lambda^p}{(p\lambda)^p} = 1$$

7.6

$$[(1-\rho)\mathbf{I} + \rho\mathbf{J}] = \begin{bmatrix} 1-\rho & 0 & \cdots & 0 \\ 0 & 1-\rho & \cdots & 0 \\ \vdots & \vdots & & \vdots \\ 0 & 0 & \cdots & 1-\rho \end{bmatrix} + \begin{bmatrix} \rho & \rho & \cdots & \rho \\ \rho & \rho & \cdots & \rho \\ \vdots & \vdots & & \vdots \\ \rho & \rho & \cdots & \rho \end{bmatrix}$$
$$= \begin{bmatrix} 1 & \rho & \cdots & \rho \\ \rho & 1 & \cdots & \rho \\ \vdots & \vdots & & \vdots \\ \rho & \rho & \cdots & 1 \end{bmatrix}$$

7.7 $M = \dfrac{|\mathbf{S}_1|^{\nu_1/2}|\mathbf{S}_2|^{\nu_2/2}\cdots|\mathbf{S}_k|^{\nu_k/2}}{|\mathbf{S}|^{\sum_i \nu_i/2}} = \dfrac{|\mathbf{S}_1|^{\nu_1/2}|\mathbf{S}_2|^{\nu_2/2}\cdots|\mathbf{S}_k|^{\nu_k/2}}{|\mathbf{S}|^{\nu_1/2}|\mathbf{S}|^{\nu_2/2}\cdots|\mathbf{S}|^{\nu_k/2}}$

7.8 **(a)** $M = .7015$, **(b)** $M = .0797$

7.9 When all $p_i = 1$, we have $k = p$, and the submatrices in the denominators of (7.31) and (7.32) reduce to $\mathbf{S}_{ii} = s_{ii}, i = 1, 2, \ldots, p$, and $\mathbf{R}_{ii} = 1, i = 1, 2, \ldots, p$.

7.10 When all $p_i = 1$, we have $k = p$ and

$$a_2 = p^2 - \sum_{i=1}^{p} p_i^2 = p^2 - p, \qquad a_3 = p^3 - p$$

$$\begin{aligned} c &= 1 - \frac{1}{12f\nu}(2a_3 + 3a_2) \\ &= 1 - \frac{1}{6(p^2 - p)\nu}[2(p^3 - p) + 3(p^2 - p)] \\ &= 1 - \frac{1}{6(p-1)\nu}[2(p^2 - 1) + 3(p - 1)] \\ &= 1 - \frac{1}{6(p-1)\nu}[2(p - 1)(p + 1) + 3(p - 1)] \\ &= 1 - \frac{1}{6\nu}[2p + 5] \end{aligned}$$

7.11 By (7.1) and (7.2), $u = 11.094$ and $u' = 10.668$.

7.12 By (7.7), $u = .0000594$. By (7.9), $u' = 23.519$. For H_0: $\mathbf{C\Sigma C}' = \sigma^2\mathbf{I}$, $u = .471$ and $u' = 2.050$.

7.13 For H_0: $\mathbf{\Sigma} = \sigma^2\mathbf{I}$, $u = .00513$ and $u' = 131.922$. For H_0: $\mathbf{C\Sigma C}' = \sigma^2\mathbf{I}$, $u = .129$ and $u' = 36.278$.

7.14 For H_0: $\mathbf{\Sigma} = \sigma^2\mathbf{I}$, $u = .00471$ and $u' = 136.190$. For H_0: $\mathbf{C\Sigma C}' = \sigma^2\mathbf{I}$, $u = .747$ and $u' = 7.486$.

7.15 By (7.15), $u' = 6.3323$ with 13 degrees of freedom. The F approximation is $F = .4802$ with 13 and 1147 degrees of freedom.

7.16 $u' = 21.488$, $F = 2.511$ with 8 and 217 degrees of freedom

7.17 $u' = 35.795, F = 4.466$ with 8 and 4905 degrees of freedom

7.18 $u = 8.7457, F = .8730$ with 10 and 6502 degrees of freedom

7.19 $|\mathbf{S}_1| = 2.620 \times 10^{14}, |\mathbf{S}_2| = 2.410 \times 10^{14}, |\mathbf{S}_{pl}| = 4.368 \times 10^{14}, u = 17.502, F = .829$

7.20 $\ln M = -85.965, u = 156.434, a_1 = 21, a_2 = 17,797, F = 7.4396$

7.21 $\ln M = -7.082, u = 10.565, a_1 = 10, a_2 = 1340, F = 1.046$

7.22 $\ln M = -8.6062, u = 14.222, a_1 = 20, a_2 = 3909, F = .707$

7.23 $\ln M = -28.917, u = 44.018, a_1 = 50, a_2 = 3238, F = .8625$

7.24 $\ln M = -142.435, u = 174.285, a_1 = 110, a_2 = 2084, F = 1.448$

7.25 $|\mathbf{S}| = 1,207,109.5, |\mathbf{S}_{yy}| = 2385.1, |\mathbf{S}_{xx}| = 1341.9, \Lambda = .3772$

7.26 $|\mathbf{S}| = 4.237 \times 10^{13}, |\mathbf{S}_{yy}| = 484,926.6, |\mathbf{S}_{xx}| = 131,406,938, \Lambda = .6650$

7.27 $|\mathbf{S}| = 9.676 \times 10^{-8}, |\mathbf{S}_{yy}| = .02097, |\mathbf{S}_{xx}| = 9.94 \times 10^{-6}, \Lambda = .4642$

7.28 $|\mathbf{S}| = 1.7148 \times 10^{16}, |\mathbf{S}_{11}| = 11,284.967, |\mathbf{S}_{22}| = 11,891.15, |\mathbf{S}_{33}| = 25,951.605, |\mathbf{S}_{44}| = 22,227.158, |\mathbf{S}_{55}| = 214.06, u = .00104, u' = 274.787, \text{df} = 46$

7.29 $|\mathbf{S}| = 459.96, |\mathbf{S}_{11}| = 140.54, |\mathbf{S}_{22}| = 72.25, |\mathbf{S}_{33}| = .250, u = .1811, u' = 12.246, \text{df} = 3$

7.30 $u = .0001379, u' = 16.297$

7.31 $u = .0005176, u' = 127.367$

7.32 $u = .005071, u' = 131.226$

CHAPTER 8

8.1 Using $\mathbf{a} = \mathbf{S}_{\text{pl}}^{-1}(\overline{\mathbf{y}}_1 - \overline{\mathbf{y}}_2)$, we obtain

$$\begin{aligned}\frac{[\mathbf{a}'(\overline{\mathbf{y}}_1 - \overline{\mathbf{y}}_2)]^2}{\mathbf{a}'\mathbf{S}_{\text{pl}}\mathbf{a}} &= \frac{[(\overline{\mathbf{y}}_1 - \overline{\mathbf{y}}_2)'\mathbf{S}_{\text{pl}}^{-1}(\overline{\mathbf{y}}_1 - \overline{\mathbf{y}}_2)]^2}{(\overline{\mathbf{y}}_1 - \overline{\mathbf{y}}_2)'\mathbf{S}_{\text{pl}}^{-1}\mathbf{S}_{\text{pl}}\mathbf{S}_{\text{pl}}^{-1}(\overline{\mathbf{y}}_1 - \overline{\mathbf{y}}_2)} \\ &= \frac{[(\overline{\mathbf{y}}_1 - \overline{\mathbf{y}}_2)'\mathbf{S}_{\text{pl}}^{-1}(\overline{\mathbf{y}}_1 - \overline{\mathbf{y}}_2)]^2}{(\overline{\mathbf{y}}_1 - \overline{\mathbf{y}}_2)'\mathbf{S}_{\text{pl}}^{-1}(\overline{\mathbf{y}}_1 - \overline{\mathbf{y}}_2)}.\end{aligned}$$

8.2 You may wish to use the following steps:

(i) In Section 5.6.2 the grouping variable w is defined as $n_2/(n_1 + n_2)$ for each observation in group 1 and $-n_1/(n_1+n_2)$ for group 2. Show that with this formulation, $\overline{w} = 0$.

(ii) Because $\overline{w} = 0$, there is no intercept and the fitted model becomes

$$\begin{aligned}\hat{w}_i &= b_1(y_{i1} - \overline{y}_1) + b_2(y_{i2} - \overline{y}_2) + \cdots + b_p(y_{ip} - \overline{y}_p), \\ i &= 1, 2, \ldots, n_1 + n_2.\end{aligned}$$

Denote the resulting matrix of y values corrected for their means as $\mathbf{Y}_c$ and the vector of w's as $\mathbf{w}$. Then the least squares estimate $\mathbf{b} = (b_1, b_2, \ldots, b_p)'$ is obtained as

$$\mathbf{b} = (\mathbf{Y}_c'\mathbf{Y}_c)^{-1}\mathbf{Y}_c'\mathbf{w}.$$

Using (2.108) in problem 2.27, show that

$$\begin{aligned}\mathbf{Y}_c'\mathbf{Y}_c &= \sum_{i=1}^{2}\sum_{j=1}^{n_i}(\mathbf{y}_{ij} - \overline{\mathbf{y}})(\mathbf{y}_{ij} - \overline{\mathbf{y}})' \\ &= \sum_{i=1}^{2}\sum_{j=1}^{n_i}(\mathbf{y}_{ij} - \overline{\mathbf{y}}_i)(\mathbf{y}_{ij} - \overline{\mathbf{y}}_i)' + \frac{n_1 n_2}{n_1 + n_2}(\overline{\mathbf{y}}_1 - \overline{\mathbf{y}}_2)(\overline{\mathbf{y}}_1 - \overline{\mathbf{y}}_2)',\end{aligned}$$

where $\overline{\mathbf{y}} = (n_1\overline{\mathbf{y}}_1 + n_2\overline{\mathbf{y}}_2)/(n_1 + n_2)$. It will be helpful to write the first sum above as

$$\sum_{j=1}^{n_1}(\mathbf{y}_{1j} - \overline{\mathbf{y}})(\mathbf{y}_{1j} - \overline{\mathbf{y}})' + \sum_{j=1}^{n_2}(\mathbf{y}_{2j} - \overline{\mathbf{y}})(\mathbf{y}_{2j} - \overline{\mathbf{y}})'$$

and add and subtract $\overline{\mathbf{y}}_1$ in the first term and $\overline{\mathbf{y}}_2$ in the second.

(iii) Show that

$$\mathbf{Y}_c'\mathbf{w} = \sum_{i=1}^{2}\sum_{j=1}^{n_i}(\mathbf{y}_{ij} - \overline{\mathbf{y}})w_{ij} = \frac{n_1 n_2}{n_2 + n_2}(\overline{\mathbf{y}}_1 - \overline{\mathbf{y}}_2).$$

Again it will be helpful to sum separately over the two groups.

(iv) From (ii) and (iii) we have

$$\mathbf{b} = (\nu\mathbf{S} + k\overline{\mathbf{d}}\,\overline{\mathbf{d}}')^{-1}k\overline{\mathbf{d}},$$

where $\mathbf{S} = \sum_{ij}(\mathbf{y}_{ij} - \overline{\mathbf{y}}_i)(\mathbf{y}_{ij} - \overline{\mathbf{y}}_i)'/(n_1 + n_2 - 2)$, $\nu = n_1 + n_2 - 2$, $k = n_1 n_2/(n_1 + n_2)$, and $\overline{\mathbf{d}} = \overline{\mathbf{y}}_1 - \overline{\mathbf{y}}_2$. Use (2.71) for the inverse of a patterned matrix of the type $\nu\mathbf{S} + k\overline{\mathbf{d}}\,\overline{\mathbf{d}}'$ to obtain (8.4).

8.3 You may want to use the following steps:

(i) R^2 is defined as [see (10.24)]

$$R^2 = \frac{\mathbf{b}'\mathbf{Y}_c'\mathbf{w} - n\overline{w}^2}{\mathbf{w}'\mathbf{w} - n\overline{w}^2}.$$

In this case the expression simplifies because $\overline{w} = 0$. Using $\mathbf{Y}_c'\mathbf{w}$ in problem 8.2(iii), show that $R^2 = \mathbf{b}'(\overline{\mathbf{y}}_1 - \overline{\mathbf{y}}_2)$.

(ii) Show that

$$\mathbf{b}'(\overline{\mathbf{y}}_1 - \overline{\mathbf{y}}_2) = \frac{n_1 n_2 D^2}{(n_1 + n_2)(n_1 + n_2 - 2) + n_1 n_2 D^2}.$$

8.4 $[\mathbf{a}'(\overline{\mathbf{y}}_1 - \overline{\mathbf{y}}_2)]^2 = \mathbf{a}'(\overline{\mathbf{y}}_1 - \overline{\mathbf{y}}_2)\mathbf{a}'(\overline{\mathbf{y}}_1 - \overline{\mathbf{y}}_2) = \mathbf{a}'(\overline{\mathbf{y}}_1 - \overline{\mathbf{y}}_2)(\overline{\mathbf{y}}_1 - \overline{\mathbf{y}}_2)'\mathbf{a}$

8.5 $\mathbf{Ha} - \lambda\mathbf{Ea} = \mathbf{0}$
$\mathbf{E}^{-1}(\mathbf{Ha} - \lambda\mathbf{Ea}) = \mathbf{E}^{-1}\mathbf{0}$
$\mathbf{E}^{-1}\mathbf{Ha} - \lambda\mathbf{E}^{-1}\mathbf{Ea} = \mathbf{0}$
$(\mathbf{E}^{-1}\mathbf{H} - \lambda\mathbf{I})\mathbf{a} = \mathbf{0}$

8.6 Using $a_r^* = s_r a_r$, $r = 1, 2, \ldots, p$, we obtain

$$
\begin{aligned}
z_{ij} &= s_1 a_1 \frac{y_{ij1} - \bar{y}_{i1}}{s_1} + s_2 a_2 \frac{y_{ij2} - \bar{y}_{i2}}{s_2} + \cdots + s_p a_p \frac{y_{ijp} - \bar{y}_{ip}}{s_p} \\
&= a_1 y_{ij1} + a_2 y_{ij2} + \cdots + a_p y_{ijp} - a_1 \bar{y}_{i1} - a_2 \bar{y}_{i2} - \cdots - a_p \bar{y}_{ip} \\
&= a_1 y_{ij1} + a_2 y_{ij2} + \cdots + a_p y_{ijp} - \mathbf{a}'\bar{\mathbf{y}}_i
\end{aligned}
$$

8.7 **(a)** $\mathbf{a}*' = (1.366, -.810, 2.525, -1.463)$

(b) $t_1 = 5.417, t_2 = 2.007, t_3 = 7.775, t_4 = .688$

(c) The standardized coefficients rank the variables in the order y_3, y_4, y_1, y_2. The t-tests rank them in the order y_3, y_1, y_2, y_4.

(d) The partial F's calculated by (8.19) are $F(y_1|y_2, y_3, y_4) = 7.844$, $F(y_2|y_1, y_3, y_4) = 2.612, F(y_3|y_1, y_2, y_4) = 40.513$, and $F(y_4|y_1, y_2, y_3) = 9.938$.

8.8 **(a)** $\mathbf{a}' = (.345, -.130, -.106, -.143)$

(b) $\mathbf{a}*' = (4.137, -2.501, -1.158, -2.068)$

(c) $t_1 = 3.888, t_2 = -3.865, t_3 = -5.691, t_4 = -5.043$

(e) $F(y_1|y_2, y_3, y_4) = 35.934, F(y_2|y_1, y_3, y_4) = 5.799$
$F(y_3|y_1, y_2, y_4) = 1.775, F(y_4|y_1, y_2, y_3) = 8.259$

8.9 **(a)** $\mathbf{a}' = (-.145, .052, -.005, -.089, -.007, -.022)$

(b) $\mathbf{a}*' = (-1.016, .147, -.542, -1.035, -.107, -1.200)$

(c) $t_1 = -4.655, t_2 = .592, t_3 = -4.354, t_4 = -5.257, t_5 = -4.032, t_6 = -6.439$

(e) $F(y_1|y_2, y_3, y_4, y_5, y_6) = 8.081$, $F(y_2|y_1, y_3, y_4, y_5, y_6) = .150$,
$F(y_3|y_1, y_2, y_4, y_5, y_6) = .835$, $F(y_4|y_1, y_2, y_3, y_5, y_6) = 8.503$,
$F(y_5|y_1, y_2, y_3, y_4, y_6) = .028, F(y_6|y_1, y_2, y_3, y_4, y_5) = 9.192$

8.10 **(a)** $\mathbf{a}' = (.057, .010, .242, .071)$ (signs reversed)

(b) $\mathbf{a}*' = (1.390, .083, 1.025, .032)$ (signs reversed)

(c) $t_1 = -3.713, t_2 = .549, t_3 = -3.262, t_4 = -.724$

(e) $F(y_1|y_2, y_3, y_4) = 3.332, F(y_2|y_1, y_3, y_4) = .010$
$F(y_3|y_1, y_2, y_4) = 1.482, F(y_4|y_1, y_2, y_3) = .001$

8.11 **(a)** $\mathbf{a}_1' = (.021, .533, -.347, -.135)$, $\mathbf{a}_2' = (-.317, .298, .243, -.026)$

(b) $\lambda_1/(\lambda_1 + \lambda_2) = .958, \lambda_2/(\lambda_1 + \lambda_2) = .042$. Using the methods of Section 8.6.2, we have two tests: the first for significance of λ_1 and λ_2 and the second for significance of λ_2:

Test	Λ	F	p-Value for F
1	.2245	8.3294	<.0001
2	.8871	1.3157	.2869

(c) $\mathbf{a}_1^{*\prime} = (.076, 1.553, -1.182, -.439), \mathbf{a}_2^{*\prime} = (-1.162, .869, .828, -.085)$

(d) $F(y_1|y_2, y_3, y_4) = 1.067, F(y_2|y_1, y_3, y_4) = 20.975$
$F(y_3|y_1, y_2, y_4) = 9.630, F(y_4|y_1, y_2, y_3) = 1.228$

(e) In the plot, the first discriminant function separates groups 1 and 2 from group 3, but the second is ineffective in separating group 1 from group 2.

8.12 **(a)**

λ_i	$\lambda_i / \sum_{j=1}^4 \lambda_j$	Eigenvector
1.8757	.6421	$\mathbf{a}_1' = (.470, -.263, .653, -.074)$
.7907	.2707	$\mathbf{a}_2' = (.176, .188, -1.058, 1.778)$
.2290	.0784	$\mathbf{a}_3' = (-.155, .258, .470, -.850)$
.0260	.0089	$\mathbf{a}_4' = (-3.614, .475, .310, -.479)$

(b) Test of significance of each eigenvalue and those that follow it:

Test	Λ	Approximate F	p-Value for F
1	.1540	4.937	<.0001
2	.4429	3.188	.0006
3	.7931	1.680	.1363
4	.9747	.545	.5839

(c) $\mathbf{a}_1^{*\prime} = (.266, -.915, 1.353, -.097), \mathbf{a}_2^{*\prime} = (.100, .654, -2.291, 2.333)$,
$\mathbf{a}_3^{*\prime} = (-.087, .899, .973, -1.115), \mathbf{a}_4^{*\prime} = (-2.044, 1.654, .643, -.628)$

(d) $F(y_1|y_2, y_3, y_4) = .299, F(y_2|y_1, y_3, y_4) = 1.931$
$F(y_3|y_1, y_2, y_4) = 6.085, F(y_4|y_1, y_2, y_3) = 4.659$

(e) In the plot, the first discriminant function separates groups 1, 4, and 6 from groups 2, 3, and 5. The second function achieves some separation of group 4 from groups 1 and 6 and some separation of group 3 from groups 2 and 5.

8.13 Three variables entered the model in the stepwise selection. We show the summary table:

Step	Variable Entered	Overall Λ	p-Value	Partial Λ	Partial F	p-Value
1	y_4	.4086	<.0001	.4086	12.158	<.0001
2	y_3	.2655	<.0001	.6499	4.418	.0026
3	y_2	.1599	<.0001	.6022	5.284	.0008

8.14 Summary table:

Step	Variable Entered	Overall Λ	p-Value	Partial Λ	Partial F	p-Value
1	y_4	.6392	<.0001	.6392	21.451	<.0001
2	y_3	.5430	<.0001	.8495	6.554	.0147
3	y_6	.4594	<.0001	.8461	6.549	.0148
4	y_2	.4063	<.0001	.8843	4.578	.0394
5	y_5	.3639	<.0001	.8957	3.959	.0547

In this case, the fifth variable to enter, y_5, would not ordinarily be included in the subset. The p-value of .0547 is large in this setting, where several tests are run at each step and the variable with smallest p-value is selected.

8.15 Summary table:

Step	Variable Entered	Overall Λ	p-Value	Partial Λ	Partial F	p-Value
1	y_2	.6347	.0006	.6347	9.495	.0006
2	y_3	.2606	<.0001	.4106	22.975	<.0001

CHAPTER 9

9.1 $\bar{z}_1 - \bar{z}_2 = \mathbf{a}'\bar{\mathbf{y}}_1 - \mathbf{a}'\bar{\mathbf{y}}_2 = \mathbf{a}'(\bar{\mathbf{y}}_1 - \bar{\mathbf{y}}_2) = (\bar{\mathbf{y}}_1 - \bar{\mathbf{y}}_2)'\mathbf{S}_{\text{pl}}^{-1}(\bar{\mathbf{y}}_1 - \bar{\mathbf{y}}_2)$

9.2
$$\begin{aligned}\tfrac{1}{2}(\bar{z}_1 + \bar{z}_2) &= \tfrac{1}{2}(\mathbf{a}'\bar{\mathbf{y}}_1 + \mathbf{a}'\bar{\mathbf{y}}_2) = \tfrac{1}{2}\mathbf{a}'(\bar{\mathbf{y}}_1 + \bar{\mathbf{y}}_2) \\ &= \tfrac{1}{2}(\bar{\mathbf{y}}_1 - \bar{\mathbf{y}}_2)'\mathbf{S}_{\text{pl}}^{-1}(\bar{\mathbf{y}}_1 + \bar{\mathbf{y}}_2)\end{aligned}$$

9.3 Write (9.7) in the form

$$\frac{f(\mathbf{y}|G_1)}{f(\mathbf{y}|G_2)} > \frac{p_2}{p_1}$$

and substitute $f(\mathbf{y}|G_i) = N_p(\boldsymbol{\mu}_i, \boldsymbol{\Sigma})$ from (4.2) to obtain

$$\frac{f(\mathbf{y}|G_1)}{f(\mathbf{y}|G_2)} = e^{(\boldsymbol{\mu}_1-\boldsymbol{\mu}_2)'\boldsymbol{\Sigma}^{-1}\mathbf{y} - (\boldsymbol{\mu}_1-\boldsymbol{\mu}_2)'\boldsymbol{\Sigma}^{-1}(\boldsymbol{\mu}_1+\boldsymbol{\mu}_2)/2} > \frac{p_2}{p_1}.$$

Substitute estimates for $\boldsymbol{\mu}_1$, $\boldsymbol{\mu}_2$, and $\boldsymbol{\Sigma}$, and take the logarithm of both sides to obtain (9.8). Note that if $a > b$, then $\ln a > \ln b$.

9.4 Maximizing $p_i f(\mathbf{y}, G_i)$ is equivalent to maximizing $\ln[p_i f(\mathbf{y}|G_i)]$. Use $f(\mathbf{y}|G_i) = N_p(\boldsymbol{\mu}_i, \boldsymbol{\Sigma})$ from (4.2) and take the logarithm to obtain

$$\ln[p_i f(\mathbf{y}|G_i)] = \ln p_i - \tfrac{1}{2}p \ln(2\pi) - \tfrac{1}{2}|\boldsymbol{\Sigma}| - \tfrac{1}{2}(\mathbf{y} - \boldsymbol{\mu}_i)'\boldsymbol{\Sigma}^{-1}(\mathbf{y} - \boldsymbol{\mu}_i).$$

Expand the last term, delete terms common to all groups (terms that do not involve i), and substitute estimators of $\boldsymbol{\mu}_i$ and $\boldsymbol{\Sigma}$ to obtain (9.11).

9.5 Use $f(\mathbf{y}|G_i) = N_p(\boldsymbol{\mu}_i, \boldsymbol{\Sigma}_i)$ in $\ln[p_i f(\mathbf{y}|G_i)]$, delete $-(p/2)\ln(2\pi)$, and substitute $\overline{\mathbf{y}}_i$ and $\mathbf{S}_i$ for $\boldsymbol{\mu}_i$ and $\boldsymbol{\Sigma}_i$.

9.6 **(a)** $\mathbf{a}' = (\overline{\mathbf{y}}_1 - \overline{\mathbf{y}}_2)'\mathbf{S}_{\text{pl}}^{-1} = (.345, -.130, -.106, -.143)$
$\frac{1}{2}(\overline{z}_1 + \overline{z}_2) = -15.8054$

(b)

Actual Group	Number of Observations	Predicted Group 1	Predicted Group 2
1	19	19	0
2	20	1	19

Error rate $= \frac{1}{39} = .0256$

(c) Using the k nearest neighbor method with $k = 5$, we obtain the same classification table as in part (b). With $k = 4$, two observations are misclassified, and the error rate becomes $2/39 = .0513$.

9.7 **(a)** $\mathbf{a}' = (\overline{\mathbf{y}}_1 - \overline{\mathbf{y}}_2)'\mathbf{S}_{\text{pl}}^{-1} = (-.145, .052, -.005, -.089, -.007, -.022)$
$\frac{1}{2}(\overline{z}_1 + \overline{z}_2) = -17.045$

(b) Linear Classification

Actual Group	Number of Observations	Predicted Group 1	Predicted Group 2
1	39	37	2
2	34	8	26

Error rate $= (2 + 8)/73 = .1370$

(c) p_1 and p_2 Proportional to Sample Sizes

Actual Group	Number of Observations	Predicted Group 1	Predicted Group 2
1	39	37	2
2	34	8	26

Error rate = (2 + 8)/73 = .1370

9.8 **(a)** $\mathbf{a}' = (\overline{\mathbf{y}}_1 - \overline{\mathbf{y}}_2)'\mathbf{S}_{\text{pl}}^{-1} = (-.057, -.010, -.242, -.071)$
$\frac{1}{2}(\overline{z}_1 + \overline{z}_2) = -7.9686$

(b) Linear Classification

Actual Group	Number of Observations	Predicted Group 1	Predicted Group 2
1	9	8	1
2	10	1	9

Error rate = $\frac{2}{19}$ = .1053

(c) Holdout Method

Actual Group	Number of Observations	Predicted Group 1	Predicted Group 2
1	9	6	3
2	10	3	7

Error rate = (3 + 3)/19 = .3158

(d) Kernel Density Estimator with $h = 2$

Actual Group	Number of Observations	Predicted Group 1	Predicted Group 2
1	9	9	0
2	10	1	9

Error rate = $\frac{1}{19}$ = .0526

9.9 **(a)**

Actual Group	Number of Observations	Predicted Group 1	2
1	20	18	2
2	20	2	18

Error rate = (2 + 2)/40 = .100.

(b) Four variables were selected by the stepwise discriminant analysis: y_2, y_3, y_4, and y_6 (see problem 8.14). With these four variables we obtain the classification table in (c).

(c)

Actual Group	Number of Observations	Predicted Group 1	2
1	20	18	2
2	20	2	18

Error rate = (2 + 2)/40 = .100. The four variables classified the sample as well as did all six variables in part (a).

9.10 **(a)** By (9.10), $L_i(\mathbf{y}) = \overline{\mathbf{y}}_i'\mathbf{S}_{\text{pl}}^{-1}\mathbf{y} - \frac{1}{2}\overline{\mathbf{y}}_i'\mathbf{S}_{\text{pl}}^{-1}\overline{\mathbf{y}}_i = \mathbf{c}_i'\mathbf{y} + c_{0i}$. The vectors $\begin{pmatrix} c_{0i} \\ \mathbf{c}_i \end{pmatrix}$, $i =$ 1,2,3, are

Group 1	Group 2	Group 3
−72.77	−65.18	−68.57
.81	2.12	.68
15.15	10.11	2.79
−1.03	−.24	6.54
10.02	11.06	13.09

(b) Linear Classification

Actual Group	Number of Observations	Predicted Group 1	2	3
1	12	9	3	0
2	12	3	7	2
3	12	0	1	11

Error rate = (3 + 3 + 2 + 1)/36 = .250

(c) Quadratic Classification

Actual Group	Number of Observations	Predicted Group 1	2	3
1	12	10	2	0
2	12	2	8	2
3	12	0	1	11

Error rate = $(2 + 2 + 2 + 1)/36 = .194$

(d) Linear Classification–Holdout Method

Actual Group	Number of Observations	Predicted Group 1	2	3
1	12	7	5	0
2	12	4	5	3
3	12	0	1	11

Error rate = $(5 + 4 + 3 + 1)/12 = .361$

(e) k Nearest Neighbor with $k = 5$

Actual Group	Number of Observations	Predicted Group 1	2	3
1	11	9	2	0
2	11	2	7	2
3	12	0	1	11

Error rate = $(2 + 2 + 2 + 1)/34 = .206$

9.11 **(a)** By (9.10), $L_i(\mathbf{y}) = \overline{\mathbf{y}}_i'\mathbf{S}_{\text{pl}}^{-1}\mathbf{y} - \frac{1}{2}\overline{\mathbf{y}}_i'\mathbf{S}_{\text{pl}}^{-1}\overline{\mathbf{y}}_i = \mathbf{c}_i'\mathbf{y} + c_{0i}$. The vectors $\binom{c_{0i}}{\mathbf{c}_i}$, $i = 1, 2, \ldots, 6$, are

Group 1	Group 2	Group 3	Group 4	Group 5	Group 6
−300.0	−353.2	−328.5	−291.8	−347.5	−315.8
314.6	317.1	324.6	307.3	316.8	311.3
−59.4	−64.0	−65.2	−59.4	−65.8	−63.1
149.6	168.2	154.9	147.7	168.2	160.6
−161.2	−172.6	−150.4	−153.4	−172.9	−175.5

(b) Linear Classification

Actual Group	Number of Observations	Predicted Group 1	2	3	4	5	6
1	8	5	0	0	1	0	2
2	8	0	3	2	1	2	0
3	8	0	0	6	1	1	0
4	8	3	0	1	4	0	0
5	8	0	3	1	0	3	1
6	8	2	0	0	0	2	4

Correct classification rate = $(5 + 3 + 6 + 4 + 3 + 4)/48 = .521$
Error rate = $1 - .521 = .479$

(c) Quadratic Classification

Actual Group	Number of Observations	Predicted Group 1	2	3	4	5	6
1	8	8	0	0	0	0	0
2	8	0	7	0	1	0	0
3	8	1	0	6	0	1	0
4	8	0	0	1	7	0	0
5	8	0	3	0	0	4	1
6	8	2	0	0	0	1	5

Correct classification rate = $(8 + 7 + 6 + 7 + 4 + 5)/48 = .771$
Error rate = $1 - .771 = .229$

(d) k Nearest Neighbor with $k = 3$

Actual Group	Number of Observations	Predicted Group 1	2	3	4	5	6	Ties
1	8	5	0	0	2	0	0	1
2	8	0	4	0	0	1	0	3
3	8	1	0	6	0	1	0	0
4	8	0	0	0	5	0	0	3
5	8	0	1	0	0	6	1	0
6	8	2	0	0	0	0	5	1

Correct classification rate = $(5 + 4 + 6 + 5 + 6 + 5)/40 = .775$
Error rate = $1 - .775 = .225$

(e) Normal Kernel with $h = 1$
(For this data set, larger values of h do much worse.)

Actual Group	Number of Observations	Predicted Group 1	2	3	4	5	6
1	8	8	0	0	0	0	0
2	8	0	8	0	0	0	0
3	8	1	0	6	0	1	0
4	8	1	0	0	7	0	0
5	8	0	0	0	0	7	1
6	8	2	0	0	0	0	6

Correct classification rate = $(8 + 8 + 6 + 7 + 7 + 6)/48 = .875$
Error rate = $1 - .875 = .125$

CHAPTER 10

10.1 $$\mathbf{y} - \mathbf{X}\hat{\boldsymbol{\beta}} = \begin{bmatrix} y_1 \\ y_2 \\ \vdots \\ y_n \end{bmatrix} - \begin{bmatrix} \mathbf{x}_1' \\ \mathbf{x}_2' \\ \vdots \\ \mathbf{x}_n' \end{bmatrix} \hat{\boldsymbol{\beta}} = \begin{bmatrix} y_1 \\ y_2 \\ \vdots \\ y_n \end{bmatrix} - \begin{bmatrix} \mathbf{x}_1'\hat{\boldsymbol{\beta}} \\ \mathbf{x}_2'\hat{\boldsymbol{\beta}} \\ \vdots \\ \mathbf{x}_n'\hat{\boldsymbol{\beta}} \end{bmatrix} = \begin{bmatrix} y_1 - \mathbf{x}_1'\hat{\boldsymbol{\beta}} \\ y_2 - \mathbf{x}_2'\hat{\boldsymbol{\beta}} \\ \vdots \\ y_n - \mathbf{x}_n'\hat{\boldsymbol{\beta}} \end{bmatrix}$$

By (2.33), $\sum_{i=1}^n (y_i - \mathbf{x}_i'\hat{\boldsymbol{\beta}})^2 = (\mathbf{y} - \mathbf{X}\hat{\boldsymbol{\beta}})'(\mathbf{y} - \mathbf{X}\hat{\boldsymbol{\beta}})$.

10.2 $$\begin{aligned} \textstyle\sum_{i=1}^n (y_i - \mu)^2 &= \textstyle\sum_{i=1}^n (y_i - \bar{y} + \bar{y} - \mu)^2 \\ &= \textstyle\sum_{i=1}^n (y_i - \bar{y})^2 + 2\sum_{i=1}^n (y_i - \bar{y})(\bar{y} - \mu) + \sum_{i=1}^n (\bar{y} - \mu)^2 \\ &= \textstyle\sum_{i=1}^n (y_i - \bar{y})^2 + (\bar{y} - \mu)\sum_{i=1}^n (y_i - \bar{y}) + n(\bar{y} - \mu)^2 \\ &= \textstyle\sum_i (y_i - \bar{y})^2 + n(\bar{y} - \mu)^2 \quad [\text{since } \sum_{i=1}^n (y_i - \bar{y}) = 0] \end{aligned}$$

10.3 $\sum_{i=1}^n (x_{i2} - \bar{x}_2)\bar{y} = \bar{y} \sum_{i=1}^n (x_{i2} - \bar{x}_2) = \bar{y}(\sum_{i=1}^n x_{i2} - n\bar{x}_2) = \bar{y}(n\bar{x}_2 - n\bar{x}_2)$

10.4 $$\begin{aligned} E[\hat{y}_i - E(y_i)]^2 &= E[\hat{y}_i - E(\hat{y}_i) + E(\hat{y}_i) - E(y_i)]^2 \\ &= E[\hat{y}_i - E(\hat{y}_i)]^2 + 2E[\hat{y}_i - E(\hat{y}_i)][E(\hat{y}_i) - E(y_i)] \\ &\quad + E[E(\hat{y}_i) - E(y_i)]^2 \end{aligned}$$

The second term on the right vanishes because $[E(\hat{y}_i) - E(y_i)]$ is constant and $E[\hat{y}_i - E(\hat{y}_i)] = E(\hat{y}_i) - E(\hat{y}_i) = 0$. For the third term, we have $E[E(\hat{y}_i) - E(y_i)]^2 = [E(\hat{y}_i) - E(y_i)]^2$, because $[E(\hat{y}_i) - E(y_i)]^2$ is constant.

10.5 First show that $\text{cov}(\hat{\boldsymbol{\beta}}_p) = \sigma^2(\mathbf{X}_p'\mathbf{X}_p)^{-1}$. This can be done by noting that $\hat{\boldsymbol{\beta}}_p = (\mathbf{X}_p'\mathbf{X}_p)^{-1}\mathbf{X}_p'\mathbf{y} = \mathbf{A}\mathbf{y}$, say. Then, by (3.69), $\text{cov}(\mathbf{A}\mathbf{y}) = \mathbf{A}\,\text{cov}(\mathbf{y})\mathbf{A}' = \mathbf{A}(\sigma^2\mathbf{I})\mathbf{A}' = \sigma^2\mathbf{A}\mathbf{A}'$. By substituting $\mathbf{A} = (\mathbf{X}_p'\mathbf{X}_p)^{-1}\mathbf{X}_p'$, this becomes $\text{cov}(\hat{\boldsymbol{\beta}}_p) = \sigma^2(\mathbf{X}_p'\mathbf{X}_p)^{-1}$. Then by (3.65), $\text{var}(\mathbf{x}_{pi}'\hat{\boldsymbol{\beta}}_p) = \mathbf{x}_{pi}'\text{cov}(\hat{\boldsymbol{\beta}}_p)\mathbf{x}_{pi}$ and the remaining steps follow as indicated.

10.6 By (10.29), $s_p^2 = \mathrm{SSE}_p/(n-p)$. Then by (10.35),

$$
\begin{aligned}
C_p &= p + (n-p)\frac{s_p^2 - s_k^2}{s_k^2} = p + (n-p)\left(\frac{s_p^2}{s_k^2} - 1\right) \\
&= p + (n-p)\frac{s_p^2}{s_k^2} - (n-p) = (n-p)\frac{\mathrm{SSE}_p/s_k^2}{n-p} - n + 2p \\
&= \frac{\mathrm{SSE}_p}{s_k^2} - (n - 2p).
\end{aligned}
$$

10.7
$$
\begin{aligned}
E[\hat{\mathbf{y}}_i - E(\mathbf{y}_i)][\hat{\mathbf{y}}_i - E(\mathbf{y}_i)]' &= E[\hat{\mathbf{y}}_i - E(\hat{\mathbf{y}}_i) + E(\hat{\mathbf{y}}_i) - E(\mathbf{y}_i)][\hat{\mathbf{y}}_i - E(\hat{\mathbf{y}}_i) \\
&\quad + E(\hat{\mathbf{y}}_i) - E(\mathbf{y}_i)]' \\
&= E[\hat{\mathbf{y}}_i - E(\hat{\mathbf{y}}_i)][\hat{\mathbf{y}}_i - E(\hat{\mathbf{y}}_i)]' \\
&\quad + E[\hat{\mathbf{y}}_i - E(\hat{\mathbf{y}}_i)][E(\hat{\mathbf{y}}_i) - E(\mathbf{y}_i)]' \\
&\quad + E[E(\hat{\mathbf{y}}_i) - E(\mathbf{y}_i)][\hat{\mathbf{y}}_i - E(\hat{\mathbf{y}}_i)]' \\
&\quad + E[E(\hat{\mathbf{y}}_i) - E(\mathbf{y}_i)][E(\hat{\mathbf{y}}_i) - E(\mathbf{y}_i)]'
\end{aligned}
$$

The second and third terms are equal to $\mathbf{O}$ because $[E(\hat{\mathbf{y}}_i) - E(\mathbf{y}_i)]$ is a constant vector and $E[\hat{\mathbf{y}}_i - E(\hat{\mathbf{y}}_i)] = E(\hat{\mathbf{y}}_i) - E(\hat{\mathbf{y}}_i) = \mathbf{0}$. The fourth term is a constant matrix and the first E can be deleted.

10.8 As in problem 10.5, we have $\mathrm{cov}(\hat{\boldsymbol{\beta}}_{p(i)}) = \sigma_{ii}(\mathbf{X}_p'\mathbf{X}_p)^{-1}$, where $\sigma_{ii} = \mathrm{var}(y_i)$ is the ith diagonal element of $\boldsymbol{\Sigma} = \mathrm{cov}(\mathbf{y})$. Similarly, $\mathrm{cov}(\hat{\boldsymbol{\beta}}_{p(i)}, \hat{\boldsymbol{\beta}}_{p(j)}) = \sigma_{ij}(\mathbf{X}_p'\mathbf{X}_p)^{-1}$, where $\sigma_{ij} = \mathrm{cov}(y_i, y_j)$ is the (ij)th element of $\boldsymbol{\Sigma}$. The notation $\mathrm{cov}(\hat{\boldsymbol{\beta}}_{p(i)}, \hat{\boldsymbol{\beta}}_{p(j)})$ indicates a matrix containing the covariance of each element of $\hat{\boldsymbol{\beta}}_{p(i)}$ and each element of $\hat{\boldsymbol{\beta}}_{p(j)}$. Now for the covariance matrix, $\mathrm{cov}(\hat{\mathbf{y}}_i') = \mathrm{cov}(\mathbf{x}_{pi}'\hat{\boldsymbol{\beta}}_{p(1)}, \ldots, \mathbf{x}_{pi}'\hat{\boldsymbol{\beta}}_{p(m)})$, we need the variance of each of the m random variables and the covariance of each pair. By problem 10.5 and (3.65), $\mathrm{var}(\mathbf{x}_{pi}'\hat{\boldsymbol{\beta}}_{p(1)}) = \mathbf{x}_{pi}'\mathrm{cov}(\hat{\boldsymbol{\beta}}_{p(1)})\mathbf{x}_{pi} = \sigma_{11}\mathbf{x}_{pi}'(\mathbf{X}_p'\mathbf{X}_p)^{-1}\mathbf{x}_{pi}$. Similarly, $\mathrm{cov}(\mathbf{x}_{pi}'\hat{\boldsymbol{\beta}}_{p(1)}, \mathbf{x}_{pi}'\hat{\boldsymbol{\beta}}_{p(2)}) = \sigma_{12}\mathbf{x}_{pi}'(\mathbf{X}_p'\mathbf{X}_p)^{-1}\mathbf{x}_{pi}$. The other variances and covariances can be obtained in an analogous manner.

10.9 By (10.65), $\mathbf{S}_p = \mathbf{E}_p/(n-p)$. Then by (10.71),

$$
\begin{aligned}
\mathbf{C}_p &= p\mathbf{I} + (n-p)\mathbf{S}_k^{-1}(\mathbf{S}_p - \mathbf{S}_k) \\
&= p\mathbf{I} + (n-p)\mathbf{S}_k^{-1}\frac{\mathbf{E}_p}{n-p} - (n-p)\mathbf{I} \\
&= \mathbf{S}_k^{-1}\mathbf{E}_p + (2p - n)\mathbf{I}.
\end{aligned}
$$

10.10 $|\mathbf{E}_k^{-1}\mathbf{E}_p| = |\mathbf{E}_k^{-1}||\mathbf{E}_p| > 0$, because both $\mathbf{E}_k^{-1}$ and $\mathbf{E}_p$ are positive definite.

10.11 By (10.72), $\mathbf{C}_p = \mathbf{S}_k^{-1}\mathbf{E}_p + (2p - n)\mathbf{I}$. Using $\mathbf{S}_k = \mathbf{E}_k/(n-k)$, we obtain

$$\left(\frac{\mathbf{E}_k}{n-k}\right)^{-1}\mathbf{E}_p = \mathbf{C}_p - (2p - n)\mathbf{I}$$
$$(n-k)\mathbf{E}_k^{-1}\mathbf{E}_p = \mathbf{C}_p + (n - 2p)\mathbf{I}.$$

10.12 If $\mathbf{C}_p$ is replaced by $p\mathbf{I}$ in (10.74), we obtain

$$\mathbf{E}_k^{-1}\mathbf{E}_p = \frac{\mathbf{C}_p + (n-2p)\mathbf{I}}{n-k} = \frac{p\mathbf{I} + n\mathbf{I} - 2p\mathbf{I}}{n-k}$$
$$= \frac{(n-p)\mathbf{I}}{n-k}.$$

10.13 **(a)** $\hat{\mathbf{B}} = \begin{bmatrix} .6264 & 83.243 \\ .0009 & .029 \\ -.0010 & -.013 \\ .0015 & -.004 \end{bmatrix}$

(b) $\Lambda = .742, V^{(s)} = .280, U^{(s)} = .375, \theta = .283$

(c) $\lambda_1 = .3594, \lambda_2 = .0160$. The essential rank of $\hat{\mathbf{B}}_1$ is 1, and the power ranking is $\theta > U^{(s)} > \Lambda > V^{(s)}$.

(d) The Wilks' Λ test of x_2 adjusted for x_1 and x_3, for example, is given by (10.54) as

$$\Lambda(x_2|x_1, x_3) = \frac{\Lambda(x_1, x_2, x_3)}{\Lambda(x_1, x_3)},$$

which is distributed as $\Lambda_{p,1,n-4}$ and has an exact F transformation. The tests for x_1 and x_3 are similar. For the three tests we obtain the following:

	Λ	F	p-Value
$x_1 \| x_2, x_3$	.931	1.519	.231
$x_2 \| x_1, x_3$	.887	2.606	.086
$x_3 \| x_1, x_2$	.762	6.417	.004

10.14 **(a)** $\hat{\mathbf{B}} = \begin{pmatrix} 34.282 & 35.802 \\ .394 & .245 \\ .529 & .471 \end{pmatrix}$

(b) $\Lambda = .377, V^{(s)} = .625, U^{(s)} = 1.647, \theta = .622$

(c) $\lambda_1 = 1.644, \lambda_2 = .0029$. The essential rank of $\hat{\mathbf{B}}_1$ is 1, and the power ranking is $\theta > U^{(s)} > \Lambda > V^{(s)}$.

(d)

	Λ	F	p-Value
$x_1 \mid x_2$	.888	1.327	.287
$x_2 \mid x_1$	.875	1.506	.245

10.15 **(a)** $\hat{\mathbf{B}} = \begin{bmatrix} 54.870 & 65.679 & 58.106 \\ .054 & -.048 & .018 \\ -.024 & .163 & .012 \\ .107 & -.036 & .125 \end{bmatrix}$

(b) $\Lambda = .665, V^{(s)} = .365, U^{(s)} = .458, \theta = .240$

(c) $\lambda_1 = .3159, \lambda_2 = .1385, \lambda_3 = .0037$. The essential rank of $\hat{\mathbf{B}}_1$ is 2, and the power ranking is $V^{(s)} > \Lambda > U^{(s)} > \theta$.

(d)

	Λ	F	p-Value
$x_1 \mid x_2, x_3$	.942	.903	.447
$x_2 \mid x_1, x_3$	.847	2.653	.060
$x_3 \mid x_1, x_2$	.829	3.020	.040

(e)

	Λ	F	p-Value
$y_1 \mid y_2, y_3$	.890	1.804	.160
$y_2 \mid y_1, y_3$	.833	2.932	.044
$y_3 \mid y_1, y_2$	.872	2.159	.106

10.16 **(a)** $\hat{\mathbf{B}} = \begin{bmatrix} -4.140 & 4.935 \\ 1.103 & -.955 \\ .231 & -.222 \\ 1.171 & 1.773 \\ .111 & .048 \\ .617 & -.058 \\ .267 & .485 \\ -.263 & -.209 \\ -.004 & -.004 \end{bmatrix}$

Test of overall regression of (y_1, y_2) on $(x_1, x_2, \ldots, x_8)$: $\Lambda = .505$ (with

$p = 2$, exact $F = 1.018$, p-value = .459). Tests on subsets (the F's are exact because $p = 2$):

	Λ	F	p-Value
(b) $x_7, x_8 \| x_1, x_2, \ldots, x_6$	.856	.808	.527
(c) $x_4, x_5, x_6 \| x_1, x_2, x_3, x_7, x_8$	.674	1.457	.218
(d) $x_1, x_2, x_3 \| x_4, x_5, \ldots, x_8$	.569	2.170	.066

10.17 **(a)** The overall test of (y_1, y_2) on $(x_1, x_2, \ldots, x_8)$ gives $\Lambda = .5051$, with (exact) $F = 1.018$ (p-value = .459). Even though this test result is not significant, we give the results of a backward elimination for illustrative purposes:

Partial Λ-Test on Each x_i Using (10.60)

Step	x_1	x_2	x_3	x_4	x_5	x_6	x_7	x_8
1	.780	.977	.839	.846	.845	.951	.896	.891
2	.793		.811	.838	.856	.954	.898	.883
3	.791		.841	.820	.755		.914	.889
4	.730		.891	.856	.812			.908
5	.759		.940	.878	.834			
6	.724			.873	.838			
7	.830				.932			
8	.854							

At each step, the variable deleted was not significant. In fact, the variable remaining at the last step, x_1, is not a significant predictor of y_1 and y_2.

(b) There were no significant x's, but to illustrate, we will use the three x's at step 6 above:

	Λ	F	p-Value
$y_1 \| y_2$	.734	3.018	.049
$y_2 \| y_1$	.845	1.526	.232

10.18 **(a)** $\hat{\mathbf{B}} = \begin{bmatrix} 58.927 & 4.493 & 51.502 \\ -.019 & -.032 & -.421 \\ -.168 & 1.074 & -.070 \\ .262 & -.047 & 1.041 \end{bmatrix}$

$\Lambda = .093,\ V^{(s)} = 1.205,\ U^{(s)} = 6.680,\ \theta = .861$

$$\textbf{(b)}\ \hat{\mathbf{B}} = \begin{bmatrix} 99.817 & -29.120 & 121.595 \\ -.008 & -.224 & -.027 \\ .097 & 1.252 & 5.775 \\ -.049 & -.442 & -1.768 \\ -.022 & -.631 & -.488 \\ -.159 & 2.128 & 4.387 \\ .054 & -.037 & -.476 \end{bmatrix}$$

$$\Lambda = .110,\ V^{(s)} = 1.350,\ U^{(s)} = 4.319,\ \theta = .769$$

$$\textbf{(c)}\ \hat{\mathbf{B}} = \begin{bmatrix} 710.236 & 123.403 \\ -1.625 & .055 \\ 24.648 & .094 \\ -8.622 & -.334 \\ -8.224 & .462 \\ 23.626 & -.110 \\ 2.862 & .427 \\ -16.186 & -.267 \\ -.268 & .014 \\ -1.160 & -.336 \end{bmatrix}$$

$$\Lambda = .102,\ V^{(s)} = 1.236,\ U^{(s)} = 5.475,\ \theta = .827$$

10.19 Using a backward elimination based on (10.60), we obtain the following partial Λ-values:

Step	x_1	x_2	x_3	x_4	x_5	x_6	x_7	x_8	x_9
1	.993	.962	.916	.958	.919	.879	.981	.999	.797
2	.994	.962	.916	.956	.909	.874	.980		.626
3		.951	.883	.954	.912	.873	.981		.626
4		.948	.884	.955	.861	.867			.561
5		.953	.862		.840	.803			.561
6			.830		.781	.783			.535

At step 6, we stop and retain all four x's because each has a p-value less than .05.

CHAPTER 11

11.1 By (3.36), $\mathbf{S}_{yy} = \mathbf{D}_y\mathbf{R}_{yy}\mathbf{D}_y$ and $\mathbf{S}_{xx} = \mathbf{D}_x\mathbf{R}_{xx}\mathbf{D}_x$, where $\mathbf{D}_y$ and $\mathbf{D}_x$ are defined below (11.10). Similarly, $\mathbf{S}_{yx} = \mathbf{D}_y\mathbf{R}_{yx}\mathbf{D}_x$ and $\mathbf{S}_{xy} = \mathbf{D}_x\mathbf{R}_{xy}\mathbf{D}_y$.

Substitute these into (11.5), replace $\mathbf{I}$ by $\mathbf{D}_y^{-1}\mathbf{D}_y$, and factor out $\mathbf{D}_y$ on the right.

11.2 Multiply (11.5) by $\mathbf{S}_{xx}^{-1}\mathbf{S}_{xy}$ to obtain $(\mathbf{S}_{xx}^{-1}\mathbf{S}_{xy}\mathbf{S}_{yy}^{-1}\mathbf{S}_{yx}\mathbf{S}_{xx}^{-1}\mathbf{S}_{xy} - r^2\mathbf{S}_{xx}^{-1}\mathbf{S}_{xy})\mathbf{a} = \mathbf{0}$. Factor out $\mathbf{S}_{xx}^{-1}\mathbf{S}_{xy}$ to write this in the form $(\mathbf{S}_{xx}^{-1}\mathbf{S}_{xy}\mathbf{S}_{yy}^{-1}\mathbf{S}_{yx} - r^2\mathbf{I})\mathbf{S}_{xx}^{-1}\mathbf{S}_{xy}\mathbf{a} = \mathbf{0}$. Upon comparing this to (11.6), we see that $\mathbf{b} = \mathbf{S}_{xx}^{-1}\mathbf{S}_{xy}\mathbf{a}$.

11.3 When $p = 1$, s is also 1, and there is only one canonical correlation, which is equal to R^2 from multiple regression [see comments between (11.24) and (11.25)]. Thus

$$\Lambda = \frac{1 - r_1^2}{1 - c_1^2} = \frac{1 - R_f^2}{1 - R_r^2}.$$

11.4
$$F = \frac{(1-\Lambda)(n-q-1)}{\Lambda h} = \frac{[1 - (1-R_f^2)/(1-R_r^2)](n-q-1)}{[(1-R_f^2)/(1-R_r^2)]h}$$
$$= \frac{[1 - R_r^2 - (1-R_f^2)](n-q-1)}{(1-R_f^2)h}$$
$$= \frac{(R_f^2 - R_r^2)(n-q-1)}{(1-R_f^2)h}$$

11.5 By (11.35),

$$\begin{aligned} r_i^2 &= \frac{\lambda_i}{1+\lambda_i} \\ r_i^2 + r_i^2\lambda_i &= \lambda_i \\ \lambda_i(1 - r_i^2) &= r_i^2 \end{aligned}$$

11.6 Substitute $\mathbf{E} = (n-1)(\mathbf{S}_{yy} - \mathbf{S}_{yx}\mathbf{S}_{xx}^{-1}\mathbf{S}_{xy})$ and $\mathbf{H} = (n-1)\mathbf{S}_{yx}\mathbf{S}_{xx}^{-1}\mathbf{S}_{xy}$ into (11.37):

$$\begin{aligned} (\mathbf{H} - \lambda\mathbf{E})\mathbf{a} &= \mathbf{0} \\ (n-1)[\mathbf{S}_{yx}\mathbf{S}_{xx}^{-1}\mathbf{S}_{xy} - \lambda(\mathbf{S}_{yy} - \mathbf{S}_{yx}\mathbf{S}_{xx}^{-1}\mathbf{S}_{xy})]\mathbf{a} &= \mathbf{0} \\ \mathbf{S}_{yx}\mathbf{S}_{xx}^{-1}\mathbf{S}_{xy}\mathbf{a} &= \lambda(\mathbf{S}_{yy} - \mathbf{S}_{yx}\mathbf{S}_{xx}^{-1}\mathbf{S}_{xy})\mathbf{a} \end{aligned}$$

11.7 By (11.38), $(\mathbf{S}_{yx}\mathbf{S}_{xx}^{-1}\mathbf{S}_{xy} - r^2\mathbf{S}_{yy})\mathbf{a} = \mathbf{0}$. Subtracting $r^2\mathbf{S}_{yx}\mathbf{S}_{xx}^{-1}\mathbf{S}_{xy}\mathbf{a}$ from both sides gives

$$\mathbf{S}_{yx}\mathbf{S}_{xx}^{-1}\mathbf{S}_{xy}\mathbf{a} - r^2\mathbf{S}_{yx}\mathbf{S}_{xx}^{-1}\mathbf{S}_{xy}\mathbf{a} = r^2\mathbf{S}_{yy} - r^2\mathbf{S}_{yx}\mathbf{S}_{xx}^{-1}\mathbf{S}_{xy}\mathbf{a}$$
$$(1 - r^2)\mathbf{S}_{yx}\mathbf{S}_{xx}^{-1}\mathbf{S}_{xy}\mathbf{a} = r^2(\mathbf{S}_{yy} - \mathbf{S}_{yx}\mathbf{S}_{xx}^{-1}\mathbf{S}_{xy})\mathbf{a}$$

11.8 **(a)** $r_1 = .5142, \quad r_2 = .1255$

(b)

	$\mathbf{c}_1$	$\mathbf{c}_2$
y_1	1.020	−.048
y_2	−.160	1.009

	$\mathbf{d}_1$	$\mathbf{d}_2$
x_1	.436	.823
x_2	−.704	−.455
x_3	1.081	−.401

(c)

k	Λ	Approximate F	p-Value
1	.7240	2.395	.035
2	.9843	.336	.716

11.9 **(a)** $r_1 = .7885, \quad r_2 = .0537$

(b)

	$\mathbf{c}_1$	$\mathbf{c}_2$
y_1	.5522	−1.3664
y_2	.5215	1.3784

	$\mathbf{d}_1$	$\mathbf{d}_2$
x_1	.5044	−1.7686
x_2	.5383	1.7586

(c)

k	Λ	Approximate F	p-Value
1	.3772	6.5972	.0003
2	.9971	0.0637	.8031

11.10 **(a)** $r_1 = .4900, \quad r_2 = .3488, \quad r_3 = .0609$

(b)

	$\mathbf{c}_1$	$\mathbf{c}_2$	$\mathbf{c}_3$
y_1	.633	.091	.806
y_2	−.624	816	.147
y_3	.643	.400	−.690

	$\mathbf{d}_1$	$\mathbf{d}_2$	$\mathbf{d}_3$
x_1	.482	−.262	1.054
x_2	−.578	1.024	−.059
x_3	.865	.216	−.626

(c)

k	Λ	Approximate F	p-Value
1	.665	2.175	.029
2	.875	1.552	.194
3	.996	.171	.681

11.11 **(a)** $r_1 = .6251, \quad r_2 = .4135$

(b)

	$\mathbf{c}_1$	$\mathbf{c}_2$
y_1	1.120	−.007
y_2	−.498	1.003

	$\mathbf{d}_1$	$\mathbf{d}_2$
x_1	1.091	−.794
x_2	.184	−.288
x_3	.842	1.807
x_4	.944	.641
x_5	1.040	−.154
x_6	.215	1.256
x_7	−.603	−.528
x_8	−.641	−.588

(c)

k	Λ	Approximate F	p-Value
1	.4642	1.1692	.3321
2	.7553	.9718	.4766

11.12 **(b)** By (11.30),

$$\Lambda(x_7, x_8 | x_1, x_2, \ldots, x_6) = \frac{\prod_{i=1}^{2}(1 - r_i^2)}{\prod_{i=1}^{2}(1 - c_i^2)},$$

where r_1^2 and r_2^2 are the squared canonical correlations from the full model and c_1^2 and c_2^2 are the squared canonical correlations from the reduced model:

$$\Lambda(x_7, x_8 | x_1, x_2, \ldots, x_6) = \frac{(1 - .6208^2)(1 - .4947^2)}{(1 - .2650^2)(1 - .0886^2)} = \frac{.4643}{.9225} = .5033$$

(c) $$\Lambda(x_4, x_5, x_6 | x_1, x_2, x_3, x_7, x_8) = \frac{(1 - .6208^2)(1 - .4947^2)}{(1 - .3301^2)(1 - .1707^2)}$$

$$= \frac{.4643}{.8651} = .5367$$

(d) $$\Lambda(x_1, x_2, x_3 | x_4, x_5, \ldots, x_8) = \frac{(1 - .6208^2)(1 - .4947^2)}{(1 - .4831^2)(1 - .2185^2)}$$

$$= \frac{.4643}{.7300} = .6359$$

11.13 **(a)** $r_1 = .9279, r_2 = .5622, r_3 = .1660$:

k	Λ	Approximate F	p-Value
1	.0925	17.9776	< .0001
2	.6651	4.6366	.0020
3	.9725	1.1898	.2816

(b) $r_1 = .8770, r_2 = .6776, r_3 = .3488$:

k	Λ	Approximate F	p-Value
1	.1097	6.919	< .0001
2	.4751	3.427	.001
3	.8783	1.351	.269

(c) $r_1 = .9095, r_2 = .6395$:

k	Λ	Approximate F	p-Value
1	.1022	8.2757	< .0001
2	.5911	3.1129	.0089

(d) $r_1 = .9029, r_2 = .7797, r_3 = .3597, r_4 = .3233, r_5 = .0794$:

k	Λ	Approximate F	p-Value
1	.0561	4.992	< .0001
2	.3037	2.601	.0007
3	.7747	.829	.6210
4	.8898	.761	.6030
5	.9937	.124	.8840

CHAPTER 12

12.1 $|\mathbf{R} - \lambda\mathbf{I}| = 0$

$$\begin{vmatrix} 1-\lambda & r \\ r & 1-\lambda \end{vmatrix} = (1-\lambda)^2 - r^2 = 0$$

$(1 - \lambda + r)(1 - \lambda - r) = 0$
$\lambda = 1 \pm r$
With $\lambda_1 = 1 + r$ in $(\mathbf{R} - \lambda_1\mathbf{I})\mathbf{a}_1 = \mathbf{0}$, we obtain

$$\begin{pmatrix} -r & r \\ r & -r \end{pmatrix} \begin{pmatrix} a_{11} \\ a_{12} \end{pmatrix} = \begin{pmatrix} 0 \\ 0 \end{pmatrix}$$

which gives $a_{11} = a_{12}$ for any r. Normalizing to $\mathbf{a}_1'\mathbf{a}_1 = 1$, yields $a_{11} = 1/\sqrt{2}$.

12.2 If $\mathbf{S}$ is diagonal, then $\lambda_i = s_{ii}$, as in (12.15). Thus

$$\mathbf{S}\mathbf{a}_i = \lambda_i \mathbf{a}_i = s_{ii}\mathbf{a}_i$$

$$\begin{bmatrix} s_{11} & 0 & \cdots & 0 \\ 0 & s_{22} & \cdots & 0 \\ \vdots & \vdots & & \vdots \\ 0 & 0 & \cdots & s_{pp} \end{bmatrix} \begin{bmatrix} a_{i1} \\ a_{i2} \\ \vdots \\ a_{ip} \end{bmatrix} = \begin{bmatrix} s_{11}a_{i1} \\ s_{22}a_{i2} \\ \vdots \\ s_{pp}a_{ip} \end{bmatrix} = \begin{bmatrix} s_{ii}a_{i1} \\ s_{ii}a_{i2} \\ \vdots \\ s_{ii}a_{ip} \end{bmatrix}$$

From the first element, we obtain $s_{11}a_{i1} = s_{ii}a_{i1}$ or $(s_{11} - s_{ii})a_{i1} = 0$. Since $s_{11} - s_{ii} \neq 0$, we must have $a_{i1} = 0$ (unless $i = 1$).

12.3 By (10.28) and (12.2),

$$\begin{aligned} R^2_{y_i|z_1,\ldots,z_k} &= \frac{\mathbf{s}'_{y_iz}\mathbf{S}_{zz}^{-1}\mathbf{s}_{y_iz}}{s^2_{y_i}} \\ &= (s_{y_iz_1}, s_{y_iz_2}, \ldots, s_{y_iz_k}) \begin{bmatrix} s^2_{z_1} & 0 & \cdots & 0 \\ 0 & s^2_{z_2} & \cdots & 0 \\ \vdots & & & \vdots \\ 0 & \cdots & & s^2_{z_k} \end{bmatrix}^{-1} \begin{bmatrix} s_{y_iz_1} \\ s_{y_iz_2} \\ \vdots \\ s_{y_iz_k} \end{bmatrix} \Bigg/ s^2_{y_i}. \end{aligned}$$

Show that this is equal to

$$R^2_{y_i|z_1,\ldots,z_k} = \sum_{j=1}^{k} \frac{s^2_{y_iz_j}}{s^2_{z_j}s^2_{y_i}} = \sum_{j=1}^{k} r^2_{y_iz_j}.$$

12.4 The variances of y_1, y_2, x_1, x_2, and x_3 on the diagonal of $\mathbf{S}$ are .016, 70.6, 1106.4, 2381.9, and 2136.4. The eigenvalues of $\mathbf{S}$ and $\mathbf{R}$ are as follows:

S			R		
λ_i	$\lambda_i/\sum_j \lambda_j$	Cumulative	λ_i	$\lambda_i/\sum_j \lambda_j$	Cumulative
3466.18	.608607	.60861	1.72	.34	.34
1264.47	.222021	.83063	1.23	.25	.59
895.27	.157195	.98782	.96	.19	.78
69.34	.012174	.99999	.79	.16	.94
.01	.000002	1.00000	.30	.06	1.00

Two principal components of **S** account for 83% of the variance, but it requires three principal components of **R** to reach 78%. For most purposes we would use two components of **S**, although with three we could account for 99% of the variance. However, we show all five eigenvectors below because of the interesting pattern they exhibit. The first principal component is largely a weighted average of the last two variables, x_2 and x_3, which have the largest variances. The second and third components represent contrasts in the last three variables and could be described as "shape" components. The fourth and fifth components are associated uniquely with y_2 and y_1, respectively. These components are "variable specific," as described in the discussion of method 1 in Section 12.6.

As expected, the principal components of **R** show an entirely different pattern. All five variables contribute to the first three components of **R**; whereas, in **S**, y_1 and y_2 have small variances and contribute almost nothing to the first three components. The eigenvectors of **S** and **R** are as follows:

	S					R				
	$\mathbf{a}_1$	$\mathbf{a}_2$	$\mathbf{a}_3$	$\mathbf{a}_4$	$\mathbf{a}_5$	$\mathbf{a}_1$	$\mathbf{a}_2$	$\mathbf{a}_3$	$\mathbf{a}_4$	$\mathbf{a}_5$
y_1	.0004	−.0008	.0018	.0029	.9999	.42	.53	−.42	−.40	.46
y_2	−.0080	.0166	.0286	.9994	−.0029	.07	.68	.16	.70	−.10
x_1	.1547	.6382	.7535	−.0309	−.0008	.36	.20	.76	−.44	−.24
x_2	.7430	.4279	−.5145	.0136	.0009	.54	−.43	.25	.39	.56
x_3	.6511	−.6397	.4083	.0042	−.0015	.63	−.18	−.40	.10	−.64

12.5 $\mathbf{S} = \begin{bmatrix} 65.1 & 33.6 & 47.6 & 36.8 & 25.4 \\ 33.6 & 46.1 & 28.9 & 40.3 & 28.4 \\ 47.6 & 28.9 & 60.7 & 37.4 & 41.1 \\ 36.8 & 40.3 & 37.4 & 62.8 & 31.7 \\ 25.4 & 28.4 & 41.1 & 31.7 & 58.2 \end{bmatrix}$

$$\mathbf{R} = \begin{bmatrix} 1.00 & .61 & .76 & .58 & .41 \\ .61 & 1.00 & .55 & .75 & .55 \\ .76 & .55 & 1.00 & .61 & .69 \\ .58 & .75 & .61 & 1.00 & .52 \\ .41 & .55 & .69 & .52 & 1.00 \end{bmatrix}$$

The eigenvalues of **S** and **R** are as follows:

S			**R**		
λ_i	$\lambda_i/\sum_j \lambda_j$	Cumulative	λ_i	$\lambda_i/\sum_j \lambda_j$	Cumulative
200.4	.684	.684	3.42	.683	.683
36.1	.123	.807	.61	.123	.806
34.1	.116	.924	.57	.114	.921
15.0	.051	.975	.27	.054	.975
7.4	.025	1.000	.13	.025	1.000

The first three eigenvectors of **S** and **R** are as follows:

S			**R**		
$\mathbf{a}_1$	$\mathbf{a}_2$	$\mathbf{a}_3$	$\mathbf{a}_1$	$\mathbf{a}_2$	$\mathbf{a}_3$
.47	−.58	−.42	.44	−.20	−.68
.39	−.11	.45	.45	−.43	.35
.49	.10	−.48	.47	.37	−.38
.47	−.12	.62	.45	−.39	.33
.41	.80	−.09	.41	.70	.41

The variances in **S** are nearly identical and the covariances are likewise similar in magnitude. Consequently, the percent of variance explained by the eigenvalues of **S** and **R** are indistinguishable. The interpretation of the second principal component from **S** is slightly different from that of the second one from **R**, but otherwise there is little to choose between them.

12.6 The variances on the diagonal of **S** are 95.5, 73.2, 76.2, 808.6, 505.9, and 508.7. The eigenvalues of **S** and **R** are as follows:

S			R		
λ_i	$\lambda_i/\sum_j \lambda_j$	Cumulative	λ_i	$\lambda_i/\sum_j \lambda_j$	Cumulative
1152.0	.557	.557	2.17	.363	.363
394.1	.191	.748	1.08	.180	.543
310.8	.150	.898	.98	.163	.706
97.8	.047	.945	.87	.144	.850
68.8	.033	.978	.55	.092	.942
44.6	.022	1.000	.35	.058	1.000

We could keep either two or three components from **S**. The first three components of **S** account for a larger percent of variance than do those from **R**. The first three eigenvectors of **S** and **R** are as follows:

S			R		
$\mathbf{a}_1$	$\mathbf{a}_2$	$\mathbf{a}_3$	$\mathbf{a}_1$	$\mathbf{a}_2$	$\mathbf{a}_3$
.080	.092	−.069	.336	.176	.497
.034	−.018	.202	.258	.843	−.093
.076	.122	−.011	.370	.049	.466
.758	−.446	−.469	.475	−.329	−.358
.493	−.081	.844	.486	.079	−.567
.412	.878	−.147	.471	−.376	.278

As expected, the first three principal components from **S** are heavily influenced by the last three variables because of their relatively large variances.

12.7 The variances on the diagonal of **S** are .69; 5.4; 2,006,682.4; 90.3; 56.4; 18.1. With the large variance of y_3, we would expect the first principal component from **S** to account for most of the variance, and y_3 would essentially constitute that single component. This is indeed the pattern that emerges in the eigenvalues and eigenvectors of **S**. The principal components from **R**, on the other hand, are not dominated by y_3. The eigenvalues of **S** and **R** are as follows:

S		R		
λ_i	$\lambda_i/\sum_j \lambda_j$	λ_i	$\lambda_i/\sum_j \lambda_j$	Cumulative
2,006,760	.999954	2.42	.404	.404
65	.000033	1.40	.234	.638
18	.000009	1.03	.171	.809
7	.000003	.92	.153	.963
3	.000001	.20	.033	.996
0	.000000	.02	.004	1.000

Most of the correlations in **R** are small (only three exceed .3), and its first three principal components account for only 72% of the variance. The first three eigenvectors of **S** and **R** are as follows:

S			R		
$\mathbf{a}_1$	$\mathbf{a}_2$	$\mathbf{a}_3$	$\mathbf{a}_1$	$\mathbf{a}_2$	$\mathbf{a}_3$
.00016	.005	−.0136	.424	−.561	−.150
.00051	.017	.0787	.446	−.528	.087
.99998	−.001	−.0002	.563	.387	−.051
.00529	.698	.0174	.454	.267	.166
.00322	−.716	.0195	.303	.425	−.296
.00020	.025	.9965	.073	.069	.923

12.8 Covariance matrix for males:

$$\mathbf{S}_M = \begin{bmatrix} 5.19 & 4.55 & 6.52 & 5.25 \\ 4.55 & 13.18 & 6.76 & 6.27 \\ 6.52 & 6.76 & 28.67 & 14.47 \\ 5.25 & 6.27 & 14.47 & 16.65 \end{bmatrix}$$

Covariance matrix for females:

$$\mathbf{S}_F = \begin{bmatrix} 9.14 & 7.55 & 4.86 & 4.15 \\ 7.55 & 18.60 & 10.22 & 5.45 \\ 4.86 & 10.22 & 30.04 & 13.49 \\ 4.15 & 5.45 & 13.49 & 28.00 \end{bmatrix}$$

The eigenvalues are as follows:

Males			Females		
λ_i	$\lambda_i/\sum_j \lambda_j$	Cumulative	λ_i	$\lambda_i/\sum_j \lambda_j$	Cumulative
43.56	.684	.684	48.96	.571	.571
11.14	.175	.858	18.46	.215	.786
6.47	.102	.960	13.54	.158	.944
2.52	.040	1.000	4.82	.056	1.000

The first two eigenvectors are as follows:

Males		Females	
$\mathbf{a}_1$	$\mathbf{a}_2$	$\mathbf{a}_1$	$\mathbf{a}_2$
.24	.21	.22	.27
.31	.85	.39	.62
.76	−.48	.68	.17
.52	.09	.58	−.72

The variances in $\mathbf{S}_M$ have a slightly wider range (5.19–28.67) than those in $\mathbf{S}_F$ (9.14–30.25), and this is reflected in the eigenvalues. The first two components account for 86% of the variance from $\mathbf{S}_M$, whereas the first two account for 79% from $\mathbf{S}_F$.

12.9 Covariance matrix for species 1:

$$\mathbf{S}_1 = \begin{bmatrix} 187.6 & 176.9 & 48.4 & 113.6 \\ 176.9 & 345.4 & 76.0 & 118.8 \\ 48.4 & 76.0 & 66.4 & 16.2 \\ 113.6 & 118.8 & 16.2 & 239.9 \end{bmatrix}$$

Covariance matrix for species 2:

$$\mathbf{S}_2 = \begin{bmatrix} 101.8 & 128.1 & 37.0 & 32.6 \\ 128.1 & 389.0 & 165.4 & 94.4 \\ 37.0 & 165.4 & 167.5 & 66.5 \\ 32.6 & 94.4 & 66.5 & 177.9 \end{bmatrix}$$

The eigenvalues are as follows:

Species 1			Species 2		
λ_i	$\lambda_i/\sum_j \lambda_j$	Cumulative	λ_i	$\lambda_i/\sum_j \lambda_j$	Cumulative
561.3	.669	.669	555.7	.664	.664
169.0	.201	.870	145.4	.174	.838
65.3	.078	.948	93.5	.112	.950
43.7	.052	1.000	41.7	.050	1.000

The first two eigenvectors are as follows:

Species 1		Species 2	
$\mathbf{a}_1$	$\mathbf{a}_2$	$\mathbf{a}_1$	$\mathbf{a}_2$
.50	.01	.28	−.20
.72	−.48	.81	−.34
.17	−.22	.42	.14
.45	.85	.30	.91

The variances in $\mathbf{S}_1$ have a wider range than those in $\mathbf{S}_2$, and the first two components of $\mathbf{S}_1$ account for a higher percent of variance.

12.10 The variances on the diagonal of **S** in each case are:

(a) Pooled: 536.0, 59.9, 116.0, 896.4, 248.1, 862.0

(b) Unpooled: 528.2, 68.9, 145.2, 1366.4, 264.4, 1069.1

The eigenvalues are as follows:

Pooled			Unpooled		
λ_i	$\lambda_i/\sum_j \lambda_j$	Cumulative	λ_i	$\lambda_i/\sum_j \lambda_j$	Cumulative
1050.6	.386	.386	1722.0	.500	.500
858.3	.316	.702	878.4	.255	.755
398.9	.147	.849	401.4	.117	.872
259.2	.095	.944	261.1	.076	.948
108.1	.040	.984	128.9	.037	.985
43.4	.016	1.000	50.4	.015	1.000

The first three eigenvectors are as follows:

Pooled			Unpooled		
$\mathbf{a}_1$	$\mathbf{a}_2$	$\mathbf{a}_3$	$\mathbf{a}_1$	$\mathbf{a}_2$	$\mathbf{a}_3$
.441	−.190	.864	.212	.389	.888
.041	−.038	.082	−.039	.064	.096
−.039	.031	.143	.080	−.066	.081
.450	.892	−.033	.776	−.608	.081
−.019	−.001	−.054	−.096	.010	.015
.774	−.407	−.471	.580	.686	−.434

(c) The pattern in both eigenvalues and eigenvectors is similar for the pooled and unpooled cases. The first three principal components account for 87.2% of the variance in the unpooled case compared to 84.9% for the pooled case.

12.11 The variances on the diagonal of **S** in each case are:

(a) Pooled: 49.1, 8.1, 12,140.8, 136.2, 210.8, 2983.9

(b) Unpooled: 63.2, 8.0, 15,168.9, 186.6, 255.4, 4660.7

The eigenvalues are as follows:

Pooled			Unpooled		
λ_i	$\lambda_i/\sum_j \lambda_j$	Cumulative	λ_i	$\lambda_i/\sum_j \lambda_j$	Cumulative
12, 809.0	.8249	.8249	17, 087.0	.8400	.8400
2, 455.9	.1582	.9830	2, 958.0	.1454	.9854
137.1	.0088	.9918	168.6	.0083	.9937
77.2	.0055	.9968	77.1	.0038	.9974
42.2	.0027	.9995	44.7	.0022	.9996
7.4	.0005	1.0000	7.3	.0004	1.0000

The eigenvectors are as follows:

Pooled		Unpooled	
$\mathbf{a}_1$	$\mathbf{a}_2$	$\mathbf{a}_1$	$\mathbf{a}_2$
−.004	−.000	.013	.027
−.005	.004	−.004	.004
.968	−.233	.931	−.355
−.002	.023	.028	.069
.103	.041	.103	.021
.228	.971	.350	.932

12.12 The variances on the diagonal of **S** are all less than 1 except $s_{x4}^2 = 5.02$ and $s_{x8}^2 = 1541.08$. We therefore expect the last variable, x_8, to dominate the principal components of **S**. This is the case for **S** but not for **R**. The eigenvalues of **S** and **R** are as follows:

S		R		
λ_i	$\lambda_i/\sum_j \lambda_j$	λ_i	$\lambda_i/\sum_j \lambda_j$	Cumulative
1541.55	.996273	3.174	.317	.317
4.83	.003123	2.565	.256	.574
.44	.000286	1.432	.143	.717
.27	.000174	1.277	.128	.845
.10	.000066	.542	.054	.899
.07	.000043	.473	.047	.946
.02	.000014	.251	.025	.971
.02	.000011	.118	.012	.983
.01	.000005	.104	.010	.994
.00	.000003	.064	.006	1.000

The eigenvectors of **S** and **R** are as follows:

S		**R**			
$\mathbf{a}_1$	$\mathbf{a}_2$	$\mathbf{a}_1$	$\mathbf{a}_2$	$\mathbf{a}_3$	$\mathbf{a}_4$
.0009	−.005	.12	.19	.69	.10
.0007	−.034	.06	.32	.54	.26
.0029	−.007	.46	−.06	.07	−.38
.0014	.004	.29	.17	−.18	.49
.0059	−.009	.52	.14	−.04	−.01
−.0150	.982	−.09	−.42	.07	.55
−.0028	−.092	−.31	.45	−.01	−.14
−.0022	−.158	−.23	.54	−.14	−.10
.0044	−.011	.09	.36	−.38	.44
.9998	.014	.50	.11	−.13	−.09

12.13 The variances in the diagonal of **S** are: 55.7, 10.9, 402.7, 25.7, 13.4, 438.3, 1.5, 106.2, 885.6, 22227.2, 214.1
The eigenvalues of **S** and **R** are as follows:

S			**R**		
λ_i	$\lambda_i/\sum_j \lambda_j$	Cumulative	λ_i	$\lambda_i/\sum_j \lambda_j$	Cumulative
22,303.5	.91479	.91479	6.020	.54730	.54730
1590.7	.06524	.98003	2.119	.19267	.73996
358.0	.01469	.99471	1.130	.10275	.84272
63.4	.00260	.99731	.760	.06909	.91181
29.3	.00120	.99852	.355	.03231	.94411
17.1	.00070	.99922	.259	.02358	.96769
12.7	.00052	.99974	.122	.01110	.97879
2.8	.00012	.99986	.110	.01004	.98883
1.9	.00008	.99994	.060	.00544	.99427
.9	.00004	.99997	.042	.00384	.99810
.7	.00003	1.00000	.021	.00190	1.00000

The eigenvectors of **S** and **R** are as follows:

	S		R			
	$\mathbf{a}_1$	$\mathbf{a}_2$	$\mathbf{a}_1$	$\mathbf{a}_2$	$\mathbf{a}_3$	$\mathbf{a}_4$
y_1	−.0097	.1331	.3304	−.0787	.0880	−.2807
y_2	.0006	.0608	.3542	.1928	.1071	−.2301
y_3	−.0141	.4397	.3923	.0518	.1105	−.1413
y_4	−.0033	.1078	.3820	.0474	.1334	−.0104
y_5	.0101	.0398	.2323	.5303	.0154	−.0710
y_6	.0167	.4290	.3621	.2361	.1198	.1350
y_7	−.0012	−.0072	−.0884	.0213	.7946	.5414
y_8	.0275	−.1844	−.2501	.5023	.0826	−.1506
y_9	.0456	−.6657	−.3111	.3595	.2136	−.2278
y_{10}	.9982	.0346	−.0243	.4685	−.4669	.5001
y_{11}	.0034	.3311	.3357	−.1153	−.1853	.4550

For most purposes, one or two principal components would suffice for **S**, with 91% or 98% of the variance explained. For **R**, on the other hand, three components are required to explain 84% of the variance, and seven components are necessary to reach 98%.

The reduction to one or two components for **S** is due in part to the relatively large variances of y_3, y_6, y_9, and y_{10}. In the eigenvectors of **S**, we see that these four variables figure prominently in the first two principal components.

CHAPTER 13

13.1 $\text{var}(y_i) = \text{var}(y_i - \mu_i) = \text{var}(\lambda_{i1} f_1 + \lambda_{i2} f_2 + \cdots + \lambda_{im} f_m + \epsilon_i)$

$$= \sum_{j=1}^{m} \lambda_{ij}^2 \,\text{var}(f_j) + \text{var}(\epsilon_i) + \sum_{j \neq k} \lambda_{ij}\lambda_{ik} \,\text{cov}\,(f_j, f_k)$$

$$+ \sum_{j=1}^{m} \lambda_{ij} \,\text{cov}\,(f_j, \epsilon_i)$$

$$= \sum_{j=1}^{m} \lambda_{ij}^2 + \psi_i.$$

The last equality follows by the assumptions $\text{var}(f_j) = 1, \text{var}(\epsilon_i) = \psi_i, \text{cov}\,(f_j, f_k) = 0$, and $\text{cov}\,(f_j, \epsilon_i) = 0$.

13.2 **(a)** $\text{cov}\,(\lambda_{12} f_2, f_2) = E[\lambda_{12} f_2 - E(\lambda_{12} f_2)][f_2 - E(f_2)]$
$= E[\lambda_{12} f_2 - \lambda_{12} E(f_2)][f_2 - E(f_2)]$
$= \lambda_{12} E[f_2 - E(f_2)]^2 = \lambda_{12}\text{var}(f_2)$

(b) $\text{cov}\,(\mathbf{y}, \mathbf{f}) = \text{cov}\,(\mathbf{\Lambda f} + \boldsymbol{\epsilon}, \mathbf{f})$ [by (13.3)]
$= \text{cov}\,(\boldsymbol{\lambda}\mathbf{f}, \mathbf{f})$ [by (13.9)]

$$
\begin{aligned}
&= E[\boldsymbol{\Lambda}\mathbf{f} - E(\boldsymbol{\Lambda}\mathbf{f})][\mathbf{f} - E(\mathbf{f})]' && \text{[by analogy to (3.29)]} \\
&= E[\boldsymbol{\Lambda}\mathbf{f} - \boldsymbol{\Lambda}E(\mathbf{f})][\mathbf{f} - E(\mathbf{f})]' \\
&= \boldsymbol{\Lambda}E[\mathbf{f} - E(\mathbf{f})][\mathbf{f} - E(\mathbf{f})]' \\
&= \boldsymbol{\Lambda}\,\text{cov}(\mathbf{f}) = \boldsymbol{\Lambda} && \text{[by (13.6)]}
\end{aligned}
$$

13.3 $E(\mathbf{f}^*) = E(\mathbf{T}'\mathbf{f}) = \mathbf{T}'E(\mathbf{f}) = \mathbf{T}'\mathbf{0} = \mathbf{0}$
$\text{cov}(\mathbf{f}^*) = \text{cov}(\mathbf{T}'\mathbf{f}) = \mathbf{T}'\text{cov}(\mathbf{f})\mathbf{T} = \mathbf{T}'\mathbf{I}\mathbf{T} = \mathbf{I}$

13.4 $\displaystyle\sum_{i=1}^{p}\sum_{j=1}^{m}\hat{\lambda}_{ij}^2 = \sum_{i=1}^{p}\left[\sum_{j=1}^{m}\hat{\lambda}_{ij}^2\right] = \sum_{i=1}^{p}\hat{h}_i^2$ [by (13.27)]

By interchanging the order of summation, we have

$$\sum_{i=1}^{p}\sum_{j=1}^{m}\hat{\lambda}_{ij}^2 = \sum_{j=1}^{m}\sum_{i=1}^{p}\hat{\lambda}_{ij}^2 = \sum_{j=1}^{m}\theta_j \quad \text{[by (13.28)].}$$

13.5 We use the covariance matrix to avoid working with standardized variables. The eigenvalues of **S** are 39.16, 8.78, .66, .30, and 0. The eigenvector corresponding to $\lambda_5 = 0$ is

$$\mathbf{a}_5' = (-.75, -.25, .25, .50, .25).$$

As noted in Section 12.7, $s_{z_5}^2 = 0$ implies $z_5 = 0$. Thus

$$z_5 = \mathbf{a}_5'\mathbf{y} = -.75y_1 - .25y_2 + .25y_3 + .50y_4 + .25y_5 = 0$$

or

$$3y_1 + y_2 = y_3 + 2y_4 + y_5.$$

13.6 Words data of Table 5.9:

	Principal Component Loadings		Varimax Rotated Loadings		Communalities, $\hat{h}_i^2$
	f_1	f_2	f_1	f_2	
Variables					
Informal words	.802	−.535	**.956**	.129	.930
Informal verbs	.856	−.326	**.858**	.321	.839
Formal words	.883	.270	.484	**.786**	.853
Formal verbs	.714	.658	.101	**.966**	.943
Variance	2.666	.899	1.894	1.671	3.565
Proportion	.666	.225	.474	.418	.891

The orthogonal matrix **T** for the varimax rotation as given by (13.45) is

$$\mathbf{T} = \begin{pmatrix} .750 & .661 \\ -.661 & .750 \end{pmatrix}.$$

Thus $\sin \phi = -.661$ and $\phi = -41.4°$. A graphical rotation of $-40°$ would produce results very close to the varimax rotation.

13.7 Ramus bone data of Table 3.8:

	Principal Component Loadings		Varimax Rotated Loadings		Communalities, $\hat{h}_i^2$	Orthoblique Pattern Loadings	
	f_1	f_2	f_1	f_2		f_1	f_2
Variables							
8 years	.949	−.295	**.884**	.455	.988	−.108	**1.087**
$8\frac{1}{2}$ years	.974	−.193	**.830**	.545	.986	.106	**.900**
9 years	.978	.171	.578	**.808**	.986	**.825**	.188
$9\frac{1}{2}$ years	.943	.319	.449	**.888**	.991	**1.099**	−.121
Variance	3.695	.255	2.005	1.946	3.951		
Proportion	.924	.064	.501	.486	.988		

The Harris-Kaiser orthoblique rotation produced loadings for which the variables have a complexity of 1. These oblique loadings provide a much cleaner simple structure than the varimax loadings. For interpretation, we see that one factor represents variables 1 and 2, and the other factor represents variables 3 and 4. This same clustering of variables can be deduced from the varimax loadings if we simply use the higher of the two loadings for each variable.

The correlation between the two oblique factors is .87. The angle between the oblique axes is $\cos^{-1}(.87) = 29.5°$. With such a small angle between the axes and a large correlation between the factors, it is clear that a single factor would better represent the variables. This is also borne out by the eigenvalues of the correlation matrix: 3.695, .255, .033, and .017. The first accounts for 92% of the variance and the second for only 6%.

13.8 Rootstock data of Table 6.2:

	Principal Component Loadings		Varimax Rotated Loadings		Communalities, $\hat{h}_i^2$
	f_1	f_2	f_1	f_2	
Variables					
Trunk 4 years	.787	.575	.167	**.960**	.949
Extension 4 years	.849	.467	.287	**.925**	.939
Trunk 15 years	.875	−.455	**.946**	.280	.973
Weight 15 years	.824	−.547	**.973**	.179	.978
Variance	2.785	1.054	1.951	1.888	3.839
Proportion	.696	.264	.488	.472	.960

The rotation was successful in producing variables with a complexity of 1, that is, partitioning the variables into two groups, each with two variables.

13.9 Fish data of Table 6.17:

	Principal Component Loadings		Varimax Rotated Loadings		Communalities, $\hat{h}_i^2$
	f_1	f_2	f_1	f_2	
Variables					
y_1	.830	−.403	**.874**	.294	.851
y_2	.783	−.504	**.911**	.189	.866
y_3	.803	.432	.270	**.871**	.831
y_4	.769	.497	.200	**.893**	.838
Variance	2.537	.850	1.709	1.678	3.386
Proportion	.634	.213	.427	.420	.847

(b) The loadings for y_1 and y_2 are similar. In **R** below we see some indication of the reason for this; y_1 and y_2 are more highly correlated than any other pair of variables and their correlations with y_3

and y_4 are similar:

$$\mathbf{R} = \begin{bmatrix} 1.00 & .71 & .51 & .40 \\ .71 & 1.00 & .38 & .40 \\ .51 & .38 & 1.00 & .67 \\ .40 & .40 & .67 & 1.00 \end{bmatrix}.$$

(c) By (13.51) and (13.53), the factor score coefficient matrix is

$$\hat{\mathbf{B}}_1 = \mathbf{R}^{-1}\hat{\boldsymbol{\Lambda}} = \begin{bmatrix} .566 & -.109 \\ .636 & -.207 \\ -.130 & .584 \\ -.194 & .630 \end{bmatrix},$$

where $\hat{\boldsymbol{\Lambda}}$ is the matrix of rotated factor loadings given above in part (a). The factor scores are given by (13.53) as follows:

Method 1		Method 2		Method 3	
$\hat{f}_1$	$\hat{f}_2$	$\hat{f}_1$	$\hat{f}_2$	$\hat{f}_1$	$\hat{f}_2$
.544	1.151	−.254	.309	−1.156	2.104
1.250	−.254	−.309	−1.534	−.321	.878
1.017	1.120	−1.865	−1.558	−.671	.947
−.147	−1.583	−.999	−.690	.067	1.130
.219	−.103	.520	−.343	−1.610	.458
1.007	.679	.919	−.111	.557	.491
1.413	−.186	−.443	−.018	−.454	1.157
−.666	−2.279	−.265	.676	−.961	.063
1.057	−1.870	1.449	−.295	−.230	1.721
.388	−.440	1.371	.295	−1.309	.054
1.328	−.298	1.260	−.027	−1.766	−.111
.694	−.033	−.000	−1.452	−1.636	−.048

(d) A one-way MANOVA on the two factor scores comparing the three methods yielded the following values for **E** and **H**:

$$\mathbf{E} = \begin{pmatrix} 21.8606 & 10.3073 \\ 10.3073 & 25.2081 \end{pmatrix} \qquad \mathbf{H} = \begin{pmatrix} 13.1394 & -10.3073 \\ -10.3073 & 9.7919 \end{pmatrix}.$$

The four MANOVA test statistics are $\Lambda = .3631$, $V^{(s)} = .6552$, $U^{(s)} = 1.7035$, and $\theta = .6259$. All are highly significant.

13.10 **(a)** For the flea data of Table 5.5, the eigenvalues of **R** are 2.273, 1.081, .450, and .196. There is a noticeable gap between 1.081 and .450, and the first two factors account for 83.9% of the variance. Thus $m = 2$ factors seem to be indicated for this set of data.

(b)

	Principal Component Loadings		Varimax Rotated Loadings		Communalities, $\hat{h}_i^2$	Orthoblique Pattern Loadings	
	f_1	f_2	f_1	f_2		f_1	f_2
Variables							
y_1	−.038	.989	−.025	**.990**	.980	−.003	**.990**
y_2	.889	.269	**.892**	.256	.862	**.898**	.253
y_3	.893	−.157	**.891**	−.170	.823	**.887**	−.173
y_4	.827	−.073	**.823**	−.084	.689	**.824**	−.087
Variance	2.273	1.081	2.273	1.081	3.354		
Proportion	.568	.270	.568	.270	.839		

(The variance explained by the varimax rotated factors remains the same as for the initial factors when rounded to three decimal places.)

(c) In this case, neither of the rotations changes the initial loadings appreciably. The reason for this unusual outcome can be seen in the correlation matrix:

$$\mathbf{R} = \begin{bmatrix} 1.00 & .18 & -.17 & -.07 \\ .18 & 1.00 & .73 & .59 \\ -.17 & .73 & 1.00 & .59 \\ -.07 & .59 & .59 & 1.00 \end{bmatrix}.$$

There are clearly two clusters of variables: $\{y_1\}$ and $\{y_2, y_3, y_4\}$. We would expect two factors corresponding to these groupings to emerge after rotation. That the same pattern surfaces in the initial factor loadings (based on eigenvectors) is due to their affiliation with principal components. As noted in Section 12.8.1, if a variable has small correlations with all other variables, the variable itself will essentially constitute a principal component. In this case, y_1 has this property and makes up most of the second principal component. The first component is comprised of the other three variables.

13.11 **(a)** For the engineer data of Table 5.6, the number of eigenvalues greater than 1 is 3, but the three account for only 70% of the variance. It requires four eigenvalues to reach 84%. The scree plot also indicates four eigenvalues.

(b)

	Principal Component Loadings			Varimax Rotated Loadings			Communalities
	f_1	f_2	f_3	f_1	f_2	f_3	$\hat{h}_i^2$
Variables							
y_1	.536	.461	.478	−.063	**.834**	.170	.729
y_2	−.129	.870	−.182	−.357	.100	**.818**	.806
y_3	.514	−.254	−.448	**.724**	−.026	.068	.529
y_4	.724	−.366	−.110	**.739**	.295	−.193	.670
y_5	−.416	−.414	.649	−.484	−.013	**−.729**	.766
y_6	.715	.124	.420	.239	**.800**	−.069	.702
Variance	1.775	1.354	1.073	1.493	1.435	1.275	4.202
Proportion	.296	.226	.179	.249	.239	.212	.700

(c) The initial communality estimates for the six variables are given by (13.36) as

$$.215, .225, .113, .255, .161, .248.$$

With these substituted for the diagonal of $\mathbf{R}$, the eigenvalues of $\mathbf{R} - \hat{\mathbf{\Psi}}$ are

Eigenvalue	.994	.569	.255	−.025	−.237	−.339
Proportion	.816	.468	.209	−.020	−.195	−.278
Cumulative	.816	1.284	1.493	1.473	1.278	1.000

The principal factor loadings and varimax rotation are as follows:

Variables	Principal Factor Loadings f_1	f_2	f_3	Varimax Rotated Loadings f_1	f_2	f_3	Communalities $\hat{h}_i^2$
y_1	.403	.312	.227	.030	**.536**	.151	.311
y_2	−.106	.569	−.100	−.288	.083	**.505**	.345
y_3	.343	−.139	−.197	**.413**	.060	.037	.176
y_4	.559	−.247	−.090	**.564**	.233	−.094	.381
y_5	−.286	−.246	.328	−.262	−.088	−**.417**	.250
y_6	.556	.089	.197	.258	**.537**	.003	.356

(d) The pattern of loadings is similar in parts (b) and (c), and the interpretation of the three factors would be the same.

13.12 Probe word data of Table 3.7:

	Principal Component Loadings f_1	f_2	Varimax Rotated Loadings f_1	f_2	Communalities, $\hat{h}_i^2$	Orthoblique Pattern Loadings f_1	f_2
Variables							
y_1	.817	−.157	**.732**	.395	.692	**.737**	.131
y_2	.838	−.336	**.861**	.271	.815	**.963**	−.092
y_3	.874	.288	.494	**.776**	.847	.248	**.734**
y_4	.838	−.308	**.844**	.292	.798	**.931**	−.057
y_5	.762	.547	.244	**.905**	.879	−.134	**1.023**
Variance	3.416	.614	2.294	1.736	4.031		
Proportion	.683	.123	.459	.347	.806		

The loadings for y_2 and y_4 are very similar in all three sets of loadings. The reason for this can be seen in the correlation matrix

$$\mathbf{R} = \begin{bmatrix} 1.00 & .61 & .76 & .58 & .41 \\ .61 & 1.00 & .55 & .75 & .55 \\ .76 & .55 & 1.00 & .61 & .69 \\ .58 & .75 & .61 & 1.00 & .52 \\ .41 & .55 & .69 & .52 & 1.00 \end{bmatrix}.$$

The correlations of y_2 with y_1, y_3, and y_5 are very similar to the correlations of y_4 with y_1, y_3, and y_5.

APPENDIX C

About the Diskette

C.1 DISK CONTENTS

The disk that accompanies this book contains an installation program that installs the following types of files:

1. Most of the data sets in the book
2. SAS command files for most of the examples in the book
3. An annotated example introducing SAS IML (denoted by *example.sas*)

The disk includes three installation files and a README file that contains information about the disk files. The command files include illustrations of the use of the following SAS procedures: IML, CANCORR, CANDISC, DISCRIM, FACTOR, GLM, PRINCOMP, REG, STEPDISC. The command files can be adapted for use in working problems or for analyzing the reader's data sets. The files have the format "ex6-1-7.sas," for example, representing Example 6.1.7. The following examples are specifically covered (in some cases, others are included):

Chapter	Examples Specifically Covered in SAS Command Files
3	3.2 3.5 3.8.1 3.9.1a
4	4.5.2
5	5.2.2 5.3.2 5.4.2 5.5 5.6
6	6.1.7 6.3.2 6.4 6.5.2 6.10.1 6.10.2 6.11.1
7	7.2.2 7.3.2 7.4.1 7.4.2 7.4.3
8	8.2 8.3 8.4.1 8.9
9	9.3.1 9.5.2 9.7.2 9.7.2b 9.7.3 9.7.3a 9.7.3b
10	10.4.2 10.6
11	11.3
12	12.2a 12.2b 12.4b 12.6
13	13.3.1 13.3.3

The data sets are labeled on the diskette as follows:

Chapter	Table in Text	Label on Diskette
3	3.1	heightwt.dat
	3.2	vote.dat
	3.5	calcium.dat
	3.6	diab.dat
	3.7	probe.dat
	3.8	bone.dat
	3.9	sons.dat
	3.10	glucose.dat
4	4.3	hematol.dat
5	5.1	psych.dat
	5.6	pilot.dat
	5.7	muscdys.dat
	5.8	goods.dat
	5.9	essay.dat
	5.10	broncus.dat
	5.15	fbeetles.dat
6	6.2	root.dat
	6.6	barsteel.dat
	6.12	calcspd.dat
	6.14	repmeas.dat
	6.16	dental.dat
	6.17	fish.dat
	6.18	snapbean.dat
	6.19	reagent.dat
	6.20	wear.dat
	6.21	cork.dat
	6.22	survtime.dat
	6.23	mice.dat
	6.24	trout.dat
	6.25	rat.dat
	6.26	corsinp.dat
	6.27	bp.dat
	6.28	plasma.dat
	6.29	mandible.dat
7	7.1	seishu.dat
	7.2	temperat.dat
8	8.1	steel.dat
	8.3	head.dat
	8.3	football.dat
10	10.1	chem.dat

C.2 HARDWARE AND SOFTWARE REQUIREMENTS

The disk can be installed on any PC-compatible computer. All the data sets are provided in an ASCII text format. The SAS command files can be run on any standard platform that supports SAS.

C.3 MAKING A BACKUP COPY

Before using the enclosed diskette, make a backup copy of the original. This backup is for personal use and will only be required in case of damage to the original. Any other use of the diskette violates copyright law. Assuming the floppy drive you will be using is drive A and your target diskette is the same format as the original diskette, please to the following:

1. Insert the original diskette included with the book into drive A.
2. At the A:> prompt, type DISKCOPY A: A: and press Return.

You will be prompted to place the source diskette into drive A.

3. Press Return and wait until you are prompted to place the target diskette in drive A.
4. Remove the original diskette and replace it with your blank backup diskette. Press Return. The backup will be completed.

C.4 INSTALLING THE DISKETTE FILES

To install the files included on the disk you will need approximately 80KB of free disk space on your hard drive. To install, do the following:

1. Assuming you will be using the drive A as the floppy drive for your diskette, at the A:> prompt type INSTALL. You may also type A:INSTALL at the C:> prompt.
2. Follow the instructions displayed by the installation program. The default choice for the installation directory is MULTIVAR and the default drive is C. You may change the settings by using the cursor to move to any of the lines, pressing ENTER, and then either selecting or typing your preferred setting at the prompt. At the end of the process, you will be given the opportunity to review the README file for more information about the diskette files.

References

Anderson, E. (1960), "A Semi-Graphical Method for Analysis of Complex Problems," *Technometrics*, **2,** 287–292.

Andrews, D. F., and Herzberg, A. M. (1985), *Data*, New York: Springer-Verlag.

Bailey, B. J. R. (1977), "Tables of the Bonferroni *t* Statistic," *Journal of American Statistical Association*, **72,** 469–479.

Barcikowski, R. S. (1981), "Statistical Power with Group Means as the Unit of Analysis," *Journal of Educational Statistics*, **6,** 267–285.

Barnett, V., and Lewis, T. (1978), *Outliers in Statistical Data*, New York: Wiley.

Bartlett, M. S. (1937), "Properties of Sufficiency and Statistical Tests," *Proceedings of the Royal Society of London, Ser. A*, **160,** 268–282.

Baten, W. D., Tack, P. I., and Baeder, H. A. (1958), "Testing for Differences between Methods of Preparing Fish by Use of Discriminant Function," *Industrial Quality Control*, **14,** 6–10.

Bayne, C. K., Beauchamp, J. J., Kane, V. E., and McCabe, G. P. (1983), "Assessment of Fisher and Logistic Linear and Quadratic Discrimination Models," *Computational Statistics and Data Analysis*, **1,** 257–273.

Beall, G. (1945), "Approximate Methods in Calculating Discriminant Functions," *Psychometrika*, **10**(3), 205–217.

Beauchamp, J. J., and Hoel, D. G. (1974), "Some Investigation and Simulation Studies of Canonical Analysis," *Journal of Statistical Computation and Simulation*.

Beckman, R. J., and Cook, R. D. (1983), "Outliers" (with comments), *Technometrics*, **25,** 119–163.

Bock, R. D. (1963), "Multivariate Analysis of Variance of Repeated Measurements," in *Problems of Measuring Change*, C. W. Harris (ed.), Madison, WS: University of Wisconsin Press, pp. 85–103.

Bock, R. D. (1975), *Multivariate Statistical Methods in Behavioral Research*, New York: McGraw-Hill.

Boik, R. J. (1981), "A Priori Tests in Repeated Measures Designs: Effects of Nonsphericity," *Psychometrika*, **46,** 241–255.

Bonferroni, C. E. (1936), "Il Calcolo delle Assicurazioni su Gruppi di Teste," in *Studii in Onore del Profesor S. O. Carboni Roma.*

Box, G. E. P. (1949), "A General Distribution Theory for a Class of Likelihood Criteria," *Biometrika*, **36,** 317–346.

Box, G. E. P. (1950), "Problems in the Analysis of Growth and Linear Curves," *Biometrics*, **6,** 362–389.

Box, G. E. P. (1954), "Some Theorems on Quadratic Forms Applied in the Study of Analysis of Variance Problems: II. The Effect of Inequality of Variance and of Correlation between Errors in the Two-Way Classification," *Annals of Mathematical Statistics*, **25,** 484–498.

Box, G. E. P., and Youle, P. V. (1955), "The Exploration of Response Surfaces: An Example of the Link between the Fitted Surface and the Basic Mechanism of the System," *Biometrics*, **11,** 287–323.

Brown, B. L., Strong, W. J., and Rencher, A. C. (1973), "Perceptions of Personality from Speech: Effects of Manipulations of Acoustical Parameters," *The Journal of the Acoustical Society of America*, **54,** 29–35.

Brown, B. L., Williams, R. N., and Barlow, C. D. (1984), "PRIFAC: A Pascal Factor Analysis Program," *Journal of Pascal, Ada, and Modula*, **2,** 18–24.

Brown, T. A., and Koplowitz, J. (1979), "The Weighted Nearest Neighbor Rule for Class Dependent Sample Sizes," *IEEE Transactions on Information Theory*, **IT-25,** 617–619.

Bryce, G. R. (1980), "Some Observations on the Analysis of Growth Curves," Paper No. SD-025-R, Brigham Young University, Department of Statistics.

Buck, S. F. A. (1960), "A Method of Estimation of Missing Values in Multivariate Data Suitable for Use With an Electronic Computer," *Journal of the Royal Statistical Society, Series B*, **22,** 302–307.

Burdick, R. K. (1979), "On the Use of Canonical Analysis to Test MANOVA Hypotheses," Presented at the Annual Meeting of the American Statistical Association, Washington, DC, August 1979.

Cacoullos, T. (1966), "Estimation of a Multivariate Density," *Annals of the Institute of Statistical Mathematics*, **18,** 179–189.

Cameron, E., and Pauling, L. (1978), "Supplemental Ascorbate in the Supportive Treatment of Cancer: Reevaluation of Prolongation of Survival Times in Terminal Human Cancer," *Proceedings of the National Academy of Science, U.S.A.*, **75,** 4538–4552.

Campbell, N. A. (1980), "Robust Procedures in Multivariate Analysis I. Robust Covariance Estimation," *Applied Statistics*, **29,** 231–237.

Chambers, J. M., and Kleiner, B. (1982), "Graphical Techniques for Multivariate Data and for Clustering," in *Handbook of Statistics*, Vol. 2, P. R. Krishnaiah and L. N. Kanal (eds.), New York: North-Holland, pp. 209–244.

Chernoff, H. (1973), "The Use of Faces to Represent Points in *k*-dimensional Space Graphically," *Journal of the American Statistical Association*, **68,** 361–368.

Chidananda Gowda, K., and Krishna, G. (1979), "The Condensed Nearest Neighbor Rule Using the Concept of Mutual Nearest Neighborhood," *IEEE Transactions on Information Theory*, **IT-25,** 488–490.

Cochran, W. G., and Cox, G. M. (1957), *Experimental Designs*, 2nd ed. New York: Wiley.

Collier, R. O., Baker, F. B., Mandeville, G. K., and Hayes, T. F. (1967), "Estimates of Test Size for Several Procedures Based on Conventional Ratios in the Repeated Measure Design," *Psychometrika*, **32,** 339–353.

Cramer, E. M., and Nicewander, W. A. (1979), "Some Symmetric, Invariant Measures of Multivariate Association," *Psychometrika*, **44,** 43–54.

Crepeau, H., Koziol, J., Reid, N., and York, Y. S. (1985), "Analysis of Incomplete Multivariate Data from Repeated Measurement Experiments," *Biometrics*, **41,** 505–514.

Critchley, F. (1985), "Influence in Principal Components Analysis," *Biometrika*, **72,** 627–636.

Crowder, M. J., and Hand, D. J. (1990), *Analysis of Repeated Measures*, New York: Chapman & Hall.

D'Agostino, R. B. (1971), "An Omnibus Test of Normality for Moderate and Large Size Samples," *Biometrika*, **58,** 341–348.

D'Agostino, R. B. (1972), "Small Sample Probability Points for the *D* Test of Normality," *Biometrika*, **59,** 219–221.

D'Agostino, R. B., and Pearson, E. S. (1973), "Tests for Departure from Normality. Empirical Results for the Distributions of b_2 and $\sqrt{b_1}$," *Biometrika*, **60,** 613–622; correction, **61,** 647.

D'Agostino, R. B., and Tietjen, G. L. (1971), "Simulated Probability Points of b_2 for Small Samples," *Biometrika*, **58,** 669–672.

Davidson, M. L. (1972), "Univariate versus Multivariate Tests in Repeated Measures Experiments," *Psychological Bulletin*, **77,** 446–452.

Davis, A. W. (1970a), "Exact Distributions of Hotelling's Generalized T_0^2-Test," *Biometrika*, **57,** 187–191.

Davis, A. W. (1970b), "Further Applications of a Differential Equation for Hotelling's Generalized T_0^2-Test," *Annals of the Institute of Statistical Mathematics*, **22,** 77–87.

Davis, A. W. (1980), "Further Tabulation of Hotelling's Generalized T_0^2-Test," *Communications in Statistics—Part B Simulation and Computation*, **9,** 321–336.

Dempster, A. P., Laird, N. M., and Rubin, D. B. (1977), "Maximum Likelihood from Incomplete Data via the EM Algorithm," *Journal of the Royal Statistical Society, Series B*, **39,** 1–38.

Devlin, S. J., Gnanadesikan, R., and Kettenring, J. R. (1981), "Robust Estimation of Dispersion Matrices and Principal Components," *Journal of the American Statistical Association*, **76,** 354–362.

Ehrenberg, A. S. C. (1977), "Rudiments of Numeracy," *Journal of the Royal Statistical Society, Series A*, **140,** 277–297.

Elston, R. C., and Grizzle, J. E. (1962), "Estimation of Time-response Curves and Their Confidence Bands," *Biometrics*, **18,** 148–159.

Fearn, T. (1975), "A Bayesian Approach to Growth Curves," *Biometrika*, **62,** 89–100.

Fearn, T. (1977), "A Two-Stage Model for Growth Curves which Leads to Rao's Covariance Adjusted Estimators," *Biometrika*, **64,** 141–143.

Federer, W. T. (1986), "On Planning Repeated Measures Experiments," Unit Report BU-909-M, Cornell University, Biometrics Unit.

Fehlberg, W. T. (1980), "Repeated Measures: The Effect of a Preliminary Test for the Structure of the Covariance Matrix upon the Alpha Levels of the Tests for Experimental Design Factors," M. A. Thesis, Brigham Young University, Department of Statistics.

Ferguson, T. S. (1961), "On the Rejection of Outliers," *Proceedings of the Fourth Berkeley Symposium on Mathematical Statistics and Probability*, Vol. 1, Los Angeles, CA: University of California Press, pp. 253–287.

Fisher, R. A. (1936), "The Use of Multiple Measurement in Taxonomic Problems," *Annals of Eugenics*, **7,** 179–188.

Fix, E., and Hodges, J. L. (1951), "Discriminatory Analysis, Nonparametric Discrimination: Consistency Properties," Report No. 4, Project No. 21-49-004, Brooks Air Force Base, USAF School of Aviation Medicine.

Flack, V. F., and Chang, P. C. (1987), "Frequency of Selecting Noise Variables in Subset Regression Analysis: A Simulation Study," *American Statistician*, **41,** 84–86.

Flury, B., and Riedwyl, H. (1981), "Graphical Representation of Multivariate Data by Means of Asymmetric Faces," *Journal of the American Statistical Association*, **76,** 757–765.

Flury, B., and Riedwyl, H. (1985), "T^2 Tests, the Linear Two-Group Discriminant Function, and Their Computation by Linear Regression," *American Statistician*, **39,** 20–25.

Frets, G. P. (1921), "Heredity of Head Form in Man," *Genetica*, **3,** 193–384.

Furnival, G. M., and Wilson, R. W. (1974), "Regression by Leaps and Bounds," *Technometrics*, **16,** 499–511.

Gabriel, K. R. (1968), "Simultaneous Test Procedures in Multivariate Analysis of Variance," *Biometrika*, **55,** 489–504.

Gabriel, K. R. (1969), "Simultaneous Test Procedures—Some Theory of Multiple Comparisons," *Annals of Mathematics and Statistics*, **40,** 224–250.

Gates, G. W. (1972), "The Reduced Nearest Neighbor Rule," *IEEE Transactions on Information Theory*, **IT-18,** 431.

Geisser, S. (1980), "Growth Curve Analysis," In *Handbook of Statistics*, Vol. 1, P. R. Krishnaiah (ed.), Amsterdam: North-Holland, pp. 89–115.

Gilbert, E. S. (1968), "On Discrimination Using Qualitative Variables," *Journal of the American Statistical Association*, **63,** 1399–1412.

Gnanadesikan, R. (1977), *Methods for Statistical Data Analysis of Multivariate Observations*, New York: Wiley.

Gnanadesikan, R., and Kettenring, J. R. (1972), "Robust Estimates, Residuals, and Outlier Detection with Multiresponse Data," *Biometrics*, **28,** 81–124.

Graybill, F. A. (1969), *Introduction to Matrices with Applications in Statistics*, Belmont, CA: Wadsworth.

Greenhouse, S. W., and Geisser, S. (1959), "On Methods in the Analysis of Profile Data," *Psychometrika*, **24,** 95–112.

Grizzle, J. E., and Allen, D. M. (1969), "Analysis of Growth and Dose Response Curves," *Biometrics*, **25,** 357–381.

Guttman, I. (1982), *Linear Models: An Introduction*, New York: Wiley.

Habbema, J. D. F., Hermans, J., and Remme, J. (1978), "Variable Kernel Density Estimation in Discriminant Analysis," In *Compstat 1978: Proceedings in Computational Statistics*, G. Bruckman (ed.), Vienna: Physica-Verlag, pp. 178–185.

Habbema, J. D. F., Hermans, J., and van den Broek, K. (1974), "A Stepwise Discriminant Analysis Program Using Density Estimation," In *Comstat*, G. Bruckmann, F. Ferschl, and L. Schmetterer (eds.), Vienna: Physica-Verlag, pp. 101–110.

Hammond, J., Smith, D., and Gill, D. S. (1981), "Selection of Adequate Subsets in Multivariate Multiple Regression," Presented at the Annual Meeting of the American Statistical Association, August 10–13, Detroit, MI.

Hand, D. J., and Batchelor, B. G. (1978), "An Edited Condensed Nearest Neighbor Rule," *Information Sciences*, **14,** 171–180.

Harman, H. H. (1976), *Modern Factor Analysis*, 3rd rev. ed., Chicago: University of Chicago Press.

Hart, P. E. (1968), "The Condensed Nearest Neighbor Rule," *IEEE Transactions on Information Theory*, **IT-14,** 515–516.

Hartigan, J. A. (1975), "Printer Graphics for Clustering," *Journal of Statistical Computation and Simulation*, **4,** 187–213.

Hawkins, D. M. (1980), *Identification of Outliers*, London: Chapman & Hall.

Healy, M. J. R. (1969), "Rao's Paradox Concerning Multivariate Tests of Significance," *Biometrics*, **25,** 411–413.

Heywood, H. B. (1931), "On Finite Sequences of Real Numbers," *Proceedings of the Royal Society, Series A*, **134,** 486–501.

Hilton, D. K. (1983), "A Simulation and Comparison of Three Criterion Functions Used for Subset Selection in Linear Regression," M.A. Thesis, Brigham Young University, Department of Statistics.

Hocking, R. R. (1976), "The Analysis and Selection of Variables in Linear Regression," *Biometrics*, **32,** 1–51.

Hotelling, H. (1931), "The Generalization of Student's Ratio," *Annals of Mathematical Statistics*, **2,** 360–378.

Hotelling, H. (1951), "A Generalized T Test and Measure of Multivariate Dispersion," *Proceedings of the Second Berkeley Symposium on Mathematical Statistics and Probability*, **1,** 23–41.

Hsu, P. L. (1938), "Notes on Hotelling's Generalized T^2," *Annals of Mathematical Statistics*, **9,** 231–243.

Huberty, C. J. (1975), "The Stability of Three Indices of Relative Variable Contribution in Discriminant Analysis," *Journal of Experimental Education*, **44,** 59–64.

Hummel, T. J., and Sligo, J. (1971), "Empirical Comparison of Univariate and Multivariate Analysis of Variance Procedures," *Psychological Bulletin*, **76,** 49–57.

Huynh, H. (1978), "Some Approximate Tests in Repeated Measures Designs," *Psychometrika*, **43,** 1582–1589.

Huynh, H., and Feldt, L. S. (1970), "Conditions Under Which Mean Square Ratios in Repeated Measurement Designs Have Exact F Distributions," *Journal of the American Statistical Association*, **65,** 1582–1589.

Huynh, H., and Feldt, L. S. (1976), "Estimation of the Box Correction for Degrees of

Freedom from Sample Data in Randomized Block and Split-Plot Designs," *Journal of Educational Statistics*, **1**(1), 69–82.

Izenman, A. J., and Williams, J. S., "A Class of Linear Spectral Models and Analyses for the Study of Longitudinal Data," *Biometrics*, **45,** 831–849.

Jackson, D., and Bryce, G. R. (1981), "A Univariate-Like Method of Analyzing Growth Curve Data with Nonuniform Variance-Covariance Matrices," Paper No. SD-028-R, Brigham Young University, Department of Statistics.

Jackson, J. E. (1980), "Principal Components and Factor Analysis: Part I—Principal Components," *Journal of Quality Technology*, **12,** 201–213.

Jeffers, J. N. R. (1967), "Two Case Studies in the Application of Principal Component Analysis," *Applied Statistics*, **16,** 225–236.

Jensen, D. R. (1982), "Efficiency and Robustness in the Use of Repeated Measurements," *Biometrics*, **38,** 813–825.

Jolliffe, I. T. (1972), "Discarding Variables in a Principal Component Analysis, I: Artificial Data," *Applied Statistics*, **21,** 160–173.

Jolliffe, I. T. (1973), "Discarding Variables in a Principal Component Analysis, II: Real Data," *Applied Statistics*, **22,** 21–31.

Kabe, D. G. (1985), "On Some Multivariate Statistical Methodology with Applications to Statistics, Psychology, and Mathematical Programming," *Journal of the Industrial Mathematics Society*, **35,** 1–18.

Kaiser, H. F. (1970), "A Second Generation Little Jiffy," *Psychometrika*, **35,** 401–415.

Kaiser, H. F., and Rice, J. (1974), "Little Jiffy, mark IV," *Educational and Psychological Measurement*, **34,** 111-117.

Keuls, M., Martakis, G. F. P., and Magid, A. H. A. (1984), "The Relationship between Pod Yield and Specific Leaf Area in Snapbeans; an Example of Stepwise Multivariate Analysis of Variance," *Scientia Horticulturae*, **23,** 231–246.

Kleinbaum, D. G., Kupper, L. L., and Muller, K. E. (1988), *Applied Regression Analysis and Other Multivariable Methods*, Boston: PWS-KENT.

Kleiner, B., and Hartigan, J. A. (1981), "Representing Points in Many Dimensions by Trees and Castles," *Journal of the American Statistical Association*, **76,** 260–269.

Kramer, C. Y. (1972), *A First Course in Multivariate Analysis*, Published by the author, Blacksburg, VA.

Kramer, C. Y., and Jensen, D. R. (1969a), "Fundamentals of Multivariate Analysis, Part I. Inference about Means," *Journal of Quality Technology*, **1**(2), 120–133.

Kramer, C. Y., and Jensen, D. R. (1969b), "Fundamentals of Multivariate Analysis, Part II. Inference about Two Treatments," *Journal of Quality Technology*, **1**(3), 189–204.

Kramer, C. Y., and Jensen, D. R. (1970), "Fundamentals of Multivariate Analysis, Part IV. Analysis of Variance for Balanced Experiments," *Journal of Quality Technology*, **2,** 32–40.

Krzanowski, W. J. (1975), "Discrimination and Classification Using Both Binary and Continuous Variables," *Journal of the American Statistical Association*, **70,** 782–790.

Krzanowski, W. J. (1976), "Canonical Representation of the Location Model for Discrimination or Classification," *Journal of the American Statistical Association*, **71,** 845–848.

Krzanowski, W. J. (1977), "The Performance of Fisher's Linear Discriminant Function Under Non-Optimal Conditions," *Technometrics*, **19,** 191–200.

Krzanowski, W. J. (1979), "Some Linear Transformations for Mixtures of Binary and Continuous Variables, with Particular Reference to Linear Discriminant Analysis," *Biometrika*, **66,** 33–39.

Krzanowski, W. J. (1980), "Mixtures of Continuous and Categorical Variables in Discriminant Analysis," *Biometrics*, **36,** 493–499.

Lachenbruch, P. A. (1975), *Discriminant Analysis*, New York: Hafner.

Lachenbruch, P. A., and Goldstein, M. (1979), "Discriminant Analysis," *Biometrics*, **35,** 69–85.

Lavine, B., and Carlson, D. (1987), "European Bee or Africanized Bee?: Species Identification through Chemical Analysis," *Analytical Chemistry*, **59**(6), 468–470.

Lawley, D. N. (1938), "A Generalization of Fisher's *Z*-Test," *Biometrika*, **30,** 180–187.

Lee, J. C. (1988), "Prediction and Estimation of Growth Curves with Special Covariance Structures," *Journal of the American Statistical Association*, **83,** 432–440.

Lee, J. C., Chang, T. C., and Krishnaiah, P. R. (1977), "Approximations to the Distributions of the Likelihood Ratio Statistics for Testing Certain Structures on the Covariance Matrices of Real Multivariate Normal Populations," In *Multivariate Analysis*, Vol. 4, P. R. Krishnaiah (ed.), Amsterdam: North-Holland, pp. 105–118.

Lin, C. C., and Mudholkar, G. S. (1980), "A Simple Test for Normality Against Asymmetric Alternatives," *Biometrika*, **67,** 455–461.

Lindsey, H., Webster, J. T., and Halpern, H. (1985), "Canonical Correlation as a Discriminant Tool in a Periodontal Problem," *Biometrika*, **27,** 257–264.

Loftsgaarden, D. O., and Quesenberry, C. P. (1965), "A Nonparametric Estimate of a Multivariate Density Function," *Annals of Mathematical Statistics*, **36,** 1049–1051.

Lubischew, A. A. (1962), "On the Use of Discriminant Functions in Taxonomy," *Biometrics*, **18,** 455–477.

Mahalanobis, P. C. (1936), "On the Generalized Distance in Statistics," *Proceedings of the National Institute of Sciences of India*, **12,** 49–55.

Mallows, C. L. (1964), "Choosing Variables in a Linear Regression: A Graphical Aid," Paper presented at the Central Regional Meeting of the Institute of Mathematical Statistics, Manhatten, KS.

Mallows, C. L. (1973), "Some Comments on C_p," *Technometrics*, **15,** 661–675.

Mardia, K. V. (1970), "Measures of Multivariate Skewness and Kurtosis with Applications," *Biometrika*, **57,** 519–530.

Mardia, K. V. (1974), "Applications of Some Measures of Multivariate Skewness and Kurtosis for Testing Normality and Robustness Studies," *Sankhya, Series B*, **36,** 115–128.

Marks, R. G., and Rao, P. V. (1978), "A Modified Tiao–Guttman Rule for Multiple Outliers," *Communications in Statistics*, **A7,** 113–126.

Mathai, A. M., and Katiyar, R. S. (1979), "Exact Percentage Points for Testing Independence," *Biometrika*, **66,** 353–356.

Mauchly, J. W. (1940), "Significance Test for Sphericity of a Normal *N*-Variate Distribution," *Annals of Mathematical Statistics*, **11,** 204–209.

Maxwell, S. E., and Avery, R. D. (1982), "Small Sample Profile Analysis with Many Variables," *Psychological Bulletin*, **92,** 778–785.

McCabe, G. P. (1984), "Principal Variables," *Technometrics*, **26,** 137–144.

McKay, R. J. (1977), "Variable Selection in Multivariate Regression: An Application of Simultaneous Test Procedures," *Journal of the Royal Statistical Society, Series B*, **39,** 371–380.

McKay, R. J. (1979), "The Adequacy of Variable Subsets in Multivariate Regression," *Technometrics*, **21,** 475–480.

Morrison, D. F. (1972), "The Analysis of a Single Sample of Repeated Measurements," *Biometrics*, **28,** 55–71.

Morrison, D. F. (1983), *Applied Linear Statistical Methods*, Englewood Cliffs, NJ: Prentice-Hall.

Morrison, D. F. (1990), *Multivariate Statistical Methods*, 3rd ed, New York: McGraw-Hill.

Mosteller, F., and Wallace, D. L. (1984), *Applied Bayesian and Classicial Inference: The Case of the Federalist Papers*, New York: Springer.

Mulholland, H. P. (1977), "On the Null Distribution of $\sqrt{b_1}$ for Samples of Size at Most 25, with Tables," *Biometrika*, **64,** 401–409.

Muller, K. E. (1982), "Understanding Canonical Correlation through the General Linear Model and Principal Components," *American Statistician*, **36,** 342–354.

Muller, K. E., and Peterson, B. L. (1984), "Practical Methods for Computing Power in Testing the Multivariate General Linear Hypothesis," *Computational Statistics and Data Analysis*, **2,** 143–158.

Myers, R. H. (1990), *Classical and Modern Regression with Applications*, 2nd ed, Boston: PWS-KENT.

Nagarsenker, B. N., and Pillai, K. C. S. (1973), "The Distribution of the Sphericity Test Criterion," *Journal of Multivariate Analysis*, **3,** 226–235.

Olson, C. L. (1974), "Comparative Robustness of Six Tests in Multivariate Analysis of Variance," *Journal of the American Statistical Association*, **69,** 894–908.

O'Sullivan, J. B., and Mahan, C. M. (1966), "Glucose Tolerance Test: Variability in Pregnant and Non-pregnant Women," *American Journal of Clinical Nutrition*, **19,** 345–351.

Parzen, E. (1962), "On Estimation of a Probability Density Function and Mode," *Annals of Mathematical Statistics*, **33,** 1065–1076.

Patel, H. I. (1986), "Analysis of Repeated Measures Designs with Changing Covariates in Clinical Trials," *Biometrika*, **73,** 707–715.

Pearson, E. S., and Hartley, H. O. (eds.) (1972), *Biometrika Tables for Statisticians*, Volume 2, Cambridge University Press; Cambridge.

Pfieffer, K. P. (1985), "Stepwise Variable Selection and Maximum Likelihood Estimation of Smoothing Factors of Kernel Functions for Nonparametric Discriminant Functions Evaluated by Different Criteria," *Computers and Biomedical Research*, **18,** 46–61.

Pillai, K. C. S. (1964), "On the Distribution of the Largest of Seven Roots of a Matrix in Multivariate Analysis," *Biometrika*, **51,** 270–275.

Pillai, K. C. S. (1965), "On the Distribution of the Largest Characteristic Root of a Matrix in Multivariate Analysis," *Biometrika*, **52,** 405–414.

Posten, H. O. (1962), "Analysis of Variance and Analysis of Regression with More than One Response," *Proceedings of a Symposium on Applications of Statistics and Computers to Fuel and Lubricant Problems*, 91–109.

Potthoff, R. F., and Roy, S. N. (1964), "A Generalized Multivariate Analysis of Variance Model Useful Especially for Growth Curve Problems," *Biometrika*, **51,** 313–326.

Rao, C. R. (1948), "Tests of Significance in Multivariate Analysis," *Biometrika*, **35,** 58–79.

Rao, C. R. (1959), "Some Problems Involving Linear Hypotheses in Multivariate Analysis," *Biometrika*, **46,** 49–58.

Rao, C. R. (1966), "Covariance Adjustment and Related Problems in Multivariate Analysis," In *Multivariate Analysis*, P. Krishnaiah (ed.), New York: Academic, pp. 87–103.

Rao, C. R. (1973), *Linear Statistical Inference and Its Applications*, 2nd ed., New York: Wiley.

Rao, C. R. (1984), "Prediction of Future Observations in Polynomial Growth Curve Models," in *Proceedings of the Indian Statistical Institute Golden Jubilee International Conference on Statistics: Applications and New Directions*, Calcutta: Indian Statistical Institute, pp. 512–520.

Rao, C. R. (1987), "Prediction of Future Observations in Growth Curve Models," *Statistical Science*, **2,** 434–471.

Reaven, G. M., and Miller, R. G. (1979), "An Attempt to Define the Nature of Chemical Diabetes Using a Multidimensional Analysis," *Diabetologia*, **16,** 17–24.

Reinsel, G. (1982), "Multivariate Repeated-Measurement or Growth Curve Models with Random-Effects Covariance Structure," *Journal of the American Statistical Association*, **77,** 190–195.

Remme, J., Habbema, J. D. F., and Hermans, J. (1980), "A Simulative Comparison of Linear Quadratic and Kernel Discrimination," *Journal of Statistical Computation and Simulation*, **11,** 87–106.

Rencher, A. C. (1988), "On the Use of Correlations to Interpret Canonical Functions," *Biometrika*, **75,** 363–365.

Rencher, A. C. (1992a), "Bias in Apparent Classification Rates in Stepwise Discriminant Analysis," *Communications in Statistics—Simulation and Computation*, **21,** 373–389.

Rencher, A. C. (1992b), "Interpretation of Canonical Discriminant Functions, Canonical Variates, and Principal Components," *American Statistician*, **46,** 217–225.

Rencher, A. C. (1993), "The Contribution of Individual Variables to Hotelling's T^2, Wilks' Λ, and R^2," *Biometrics*, **49,** 479–489.

Rencher, A. C. (1996), *An Introduction to Multivariate Statistical Inference and Applications*, New York: Wiley.

Rencher, A. C., and Larson, S. F. (1980), "Bias in Wilks' Λ in Stepwise Discriminant Analysis," *Technometrics*, **22,** 349–356.

Rencher, A. C., and Pun, F. C. (1980), "Inflation of R^2 in Best Subset Regression," *Technometrics*, **22,** 49–53.

Rencher, A. C., and Scott, D. T. (1990), "Assessing the Contribution of Individual Variables Following Rejection of a Multivariate Hypothesis," *Communications in Statistics: Simulation and Computation*, **19**(2), 535–553.

Rencher, A. C., Wadham, R. A., and Young, J. R. (1978), "A Discriminant Analysis of Four Levels of Teacher Competence," *Journal of Experimental Education*, **46,** 46–51.

Robert, P., and Escoufier, Y. (1976), "A Unifying Tool for Linear Multivariate Analysis: The RV-Coefficient," *Journal of the Royal Statistical Society, Series C*, **25,** 257–265.

Roebruck, P. (1982), "Canonical Forms and Tests of Hypotheses; Part II: Multivariate Mixed Linear Models," *Statistica Neerlandica*, **36,** 75–80.

Rogan, J. C., Keselman, H. J., and Mendoza, J. L. (1979), "Analysis of Repeated Measurements," *British Journal of Mathematical and Statistical Psychology*, **32,** 269–286.

Rogers, W. H., and Wagner, T. J. (1978), "A Finite Sample Distribution Free Performance Bound for Local Discrimination Rules," *Annals of Statistics*, **6,** 506–514.

Rosenblatt, M. (1956), "Remarks on Some Nonparametric Estimates of a Density Function," *Annals of Mathematical Statistics*, **27,** 832–837.

Royston, J. P. (1983), "Some Techniques for Assessing Multivariate Normality Based on the Shapiro-Wilk *W*," *Applied Statistics*, **32,** 121–133.

Ruymgaart, F. H. (1981), "A Robust Principal Component Analysis," *Journal of Multivariate Analysis*, **11,** 485–497.

Satterthwaite, F. E. (1941), "Synthesis of Variances," *Psychometrika*, **6,** 309–316.

Schott, J. R., and Saw, J. G. (1984), "A Multivariate One-Way Classification Model with Random Effects," *Journal of Multivariate Analysis*, **15,** 1–12.

Schuurmann, F. J., Krishnaiah, P. R., and Chattopadhyay, A. K. (1975), "Exact Percentage Points of the Distribution of the Trace of a Multivariate Beta Matrix," *Journal of Statistical Computation and Simulation*, **3,** 331–343.

Schwager, S. J., and Margolin, B. H. (1982), "Detection of Multivariate Normal Outliers," *Annals of Statistics*, **10,** 943–954.

Searle, S. R., (1982), *Matrix Algebra Useful for Statistics*, New York: Wiley.

Seber, G. A. F. (1984), *Multivariate Observations*, New York: Wiley.

Shapiro, S. S., and Wilk, M. B. (1965), "An Analysis of Variance Test for Normality (Complete Samples)," *Biometrika*, **52,** 591–611.

Silverman, B. W. (1986), *Density Estimation for Statistics and Data Analysis*, New York: Chapman & Hall.

Siotani, M., Hayakawa, T., and Fujikoshi, Y. (1985), *Modern Multivariate Statistical Analysis*, Columbus, OH: American Sciences.

Siotani, M., Yoshida, K., Kawakami, H., Nojiro, K., Kawashima, K., and Sato, M. (1963), "Statistical Research on the Taste Judgement: Analysis of the Preliminary Experiment on Sensory and Chemical Characters of SEISHU," *Proceedings of the Institute of Statistical Mathematics*, **10,** 99–118.

Sinha, B. K. (1984), "Detection of Multivariate Outliers in Elliptically Symmetric Distributions," *Annals of Statistics*, **12,** 1558–1565.

Smith, D. W., Gill, D. S., and Hammond, J. J. (1985), "Variable Selection in Multivariate Multiple Regression," *Journal of Statistical Computation and Simulation*, **22,** 217–227.

Snee, R. D. (1972), "On the Analysis of Response Curve Data," *Technometrics*, **14,** 47–62.

Snee, R. D., Acuff, S. J., and Gibson, J. R. (1979), "A Useful Method for the Analysis of Growth Studies," *Biometrics*, **35,** 835–848.

Sparks, R. S., Coutsourides, D., and Troskie, L. (1983), "The Multivariate C_p," *Communications in Statistics—Part A Theory and Methods*, **12**(15), 1775–1793.

Thomas, D. R. (1983), "Univariate Repeated Measures Techniques Applied to Multivariate Data," *Psychometrika*, **48,** 451–464.

Thomson, G. H. (1951), *The Factorial Analysis of Human Ability*, London: London University Press.

Timm, N. H. (1975), *Multivariate Analysis: With Applications in Education and Psychology*, Monterey, CA: Brooks/Cole.

Timm, N. H. (1980), "Multivariate Analysis of Variance of Repeated Measures," In *Handbook of Statistics*, Vol. 1, P. R. Krishnaiah (ed.), Amsterdam: North-Holland, pp. 41–87.

Tinter, G. (1946), "Some Applications of Multivariate Analysis to Economic Data," *Journal of the American Statistical Association*, **41,** 472–500.

Titterington, D. M., Murray, G. D., Murray, L. S., Speigelhalter, D. J., Skene, A. M., Habbema, J. D. F., and Gelpke, D. J. (1981), "Comparison of Discrimination Techniques Applied to a Complex Data Set of Head Injured Patients," *Journal of the Royal Statistical Society, Series A*, **144,** 145–175.

Travers, R. M. W. (1939), "The Use of a Discriminant Function in the Treatment of Psychological Group Differences," *Psychometrika*, **4**(1), 25–32.

Tu, C. T., and Han, C. P. (1982), "Discriminant Analysis Based on Binary and Continuous Variables," *Journal of the American Statistical Association*, **77,** 447–454.

Venables, W. N. (1976), "Some Implications of the Union Intersection Principle for Test of Sphericity," *Journal of Multivariate Analysis*, **6,** 185–190.

Vonesh, E. F. (1986), "Sample Sizes in the Multivariate Analysis of Repeated Measurements," *Biometrics*, **42,** 601–610.

Wainer, H. (1981), "Comments on a Paper by Kleiner and Hartigan," *Journal of the American Statistical Association*, **76,** 270–272.

Wall, F. J. (1967), *The Generalized Variance Ratio or U-Statistic*, Albuquerque, NM: The Dikewood Corporation.

Wang, C. M. (1983), "On the Analysis of Multivariate Repeated Measures Designs," *Communications in Statistics—Part A Theory and Methods*, **12,** 1647–1659.

Wegman, E. J. (1972), "Nonparametric Probability Density Estimation I: A Summary of Available Methods," *Technometrics*, **14,** 533–546.

Welch, B. L. (1939), "Note on Discriminant Functions," *Biometrika*, **31,** 218–220.

Wilks, S. S. (1932), "Certain Generalizations in the Analysis of Variance," *Biometrika*, **24,** 471–494.

Wilks, S. S. (1946), "Sample Criteria for Testing Equality of Means, Equality of Variances and Equality of Covariances in a Normal Multivariate Distribution," *Annals of Mathematical Statistics*, **17,** 257–281.

Wilks, S. S. (1963), "Multivariate Statistical Outliers," *Sankhya, Series A*, **25,** 407–426.

Williams, E. J. (1970), "Comparing Means of Correlated Variates," *Biometrika*, **57,** 459–461.

Williams, J. S., and Izenman, A. J. (1981), "A Class of Linear Spectral Models and Analysis for the Study of Longitudinal Data," Technical Report, Colorado State University, Department of Statistics.

Yang, S. S., and Lee, Y. (1987), "Identification of a Multivariate Outlier," Presented at the Annual Meeting of the American Statistical Association, San Francisco, CA, August 1987.

Zerbe, G. O. (1979a), "Randomization Analysis of the Completely Randomized Design Extended to Growth Curves," *Journal of the American Statistical Association*, **74,** 215–221.

Zerbe, G. O. (1979b), "Randomization Analysis of Randomized Block Design Extended to Growth and Response Curves," *Communications in Statistics—Part A Theory and Methods*, **8,** 191–205.

Index

WILEY SERIES IN PROBABILITY
AND MATHEMATICAL STATISTICS

ESTABLISHED BY WALTER A. SHEWHART AND SAMUEL S. WILKS
Editors
Vic Barnett, Ralph A. Bradley, Nicholas I. Fisher, J. Stuart Hunter, J. B. Kadane, David G. Kendall, David W. Scott, Adrian F. M. Smith, Jozef L. Teugels, Geoffrey S. Watson

Probability and Mathematical Statistics

ANDERSON · An Introduction to Multivariate Statistical Analysis, *Second Edition*
*ANDERSON · The Statistical Analysis of Time Series
ARNOLD, BALAKRISHNAN, and NAGARAJA · A First Course in Order Statistics
BACCELLI, COHEN, OLSDER, and QUADRAT · Synchronization and Linearity: An Algebra for Discrete Event Systems
BARNETT · Comparative Statistical Inference, *Second Edition*
BERNARDO and SMITH · Bayesian Statistical Concepts and Theory
BHATTACHARYYA and JOHNSON · Statistical Concepts and Methods
BILLINGSLEY · Convergence of Probability Measures
BILLINGSLEY · Probability and Measure, *Second Edition*
BOROVKOV · Asymptotic Methods in Queuing Theory
BRANDT, FRANKEN, and LISEK · Stationary Stochastic Models
CAINES · Linear Stochastic Systems
CHEN · Recursive Estimation and Control for Stochastic Systems
CONSTANTINE · Combinatorial Theory and Statistical Design
COOK and WEISBERG · An Introduction to Regression Graphics
COVER and THOMAS · Elements of Information Theory
*DOOB · Stochastic Processes
DUDEWICZ and MISHRA · Modern Mathematical Statistics
ETHIER and KURTZ · Markov Processes: Characterization and Convergence
FELLER · An Introduction to Probability Theory and Its Applications, Volume I, *Third Edition*, Revised; Volume II, *Second Edition*
FULLER · Introduction to Statistical Time Series
FULLER · Measurement Error Models
GIFI · Nonlinear Multivariate Analysis
GUTTORP · Statistical Inference for Branching Processes
HALD · A History of Probability and Statistics and Their Applications before 1750
HALL · Introduction to the Theory of Coverage Processes
HANNAN and DEISTLER · The Statistical Theory of Linear Systems
HEDAYAT and SINHA · Design and Inference in Finite Population Sampling
HOEL · Introduction to Mathematical Statistics, *Fifth Edition*
HUBER · Robust Statistics
IMAN and CONOVER · A Modern Approach to Statistics
JUREK and MASON · Operator-Limit Distributions in Probability Theory
KAUFMAN and ROUSSEEUW · Finding Groups in Data: An Introduction to Cluster Analysis
LARSON · Introduction to Probability Theory and Statistical Inference, *Third Edition*
LESSLER and KALSBEEK · Nonsampling Error in Surveys
LINDVALL · Lectures on the Coupling Method
MANTON, WOODBURY, and TOLLEY · Statistical Applications Using Fuzzy Sets
MORGENTHALER and TUKEY · Configural Polysampling: A Route to Practical Robustness
MUIRHEAD · Aspects of Multivariate Statistical Theory
OLIVER and SMITH · Influence Diagrams, Belief Nets and Decision Analysis
*PARZEN · Modern Probability Theory and Its Applications

*Now available in a lower priced paperback edition in the Wiley Classics Library.

Probability and Mathematical Statistics (Continued)

PILZ · Bayesian Estimation and Experimental Design in Linear Regression Models
PRESS · Bayesian Statistics: Principles, Models, and Applications
PUKELSHEIM · Optimal Experimental Design
PURI and SEN · Nonparametric Methods in General Linear Models
PURI, VILAPLANA, and WERTZ · New Perspectives in Theoretical and Applied Statistics
RAO · Asymptotic Theory of Statistical Inference
RAO · Linear Statistical Inference and Its Applications, *Second Edition*
ROBERTSON, WRIGHT, and DYKSTRA · Order Restricted Statistical Inference
ROGERS and WILLIAMS · Diffusions, Markov Processes, and Martingales, Volume II: Îto Calculus
ROHATGI · A Introduction to Probability Theory and Mathematical Statistics
ROSS · Stochastic Processes
RUBINSTEIN · Simulation and the Monte Carlo Method
RUZSA and SZEKELY · Algebraic Probability Theory
SCHEFFE · The Analysis of Variance
SEBER · Linear Regression Analysis
SEBER · Multivariate Observations
SEBER and WILD · Nonlinear Regression
SERFLING · Approximation Theorems of Mathematical Statistics
SHORACK and WELLNER · Empirical Processes with Applications to Statistics
SMALL and McLEISH · Hilbert Space Methods in Probability and Statistical Inference
STAUDTE and SHEATHER · Robust Estimation and Testing
STOYANOV · Counterexamples in Probability
STYAN · The Collected Papers of T. W. Anderson: 1943–1985
WHITTAKER · Graphical Models in Applied Multivariate Statistics
YANG · The Construction Theory of Denumerable Markov Processes

Applied Probability and Statistics

ABRAHAM and LEDOLTER · Statistical Methods for Forecasting
AGRESTI · Analysis of Ordinal Categorical Data
AGRESTI · Categorical Data Analysis
ANDERSON and LOYNES · The Teaching of Practical Statistics
ANDERSON, AUQUIER, HAUCK, OAKES, VANDAELE, and WEISBERG · Statistical Methods for Comparative Studies
*ARTHANARI and DODGE · Mathematical Programming in Statistics
ASMUSSEN · Applied Probability and Queues
*BAILEY · The Elements of Stochastic Processes with Applications to the Natural Sciences
BARNETT · Interpreting Multivariate Data
BARNETT and LEWIS · Outliers in Statistical Data, *Second Edition*
BARTHOLOMEW, FORBES, and McLEAN · Statistical Techniques for Manpower Planning, *Second Edition*
BATES and WATTS · Nonlinear Regression Analysis and Its Applications
BELSLEY · Conditioning Diagnostics: Collinearity and Weak Data in Regression
BELSLEY, KUH, and WELSCH · Regression Diagnostics: Identifying Influential Data and Sources of Collinearity
BHAT · Elements of Applied Stochastic Processes, *Second Edition*
BHATTACHARYA and WAYMIRE · Stochastic Processes with Applications
BIEMER, GROVES, LYBERG, MATHIOWETZ, and SUDMAN · Measurement Errors in Surveys
BIRKES and DODGE · Alternative Methods of Regression
BLOOMFIELD · Fourier Analysis of Time Series: An Introduction
BOLLEN · Structural Equations with Latent Variables

*Now available in a lower priced paperback edition in the Wiley Classics Library.

Applied Probability and Statistics (Continued)

BOULEAU · Numerical Methods for Stochastic Processes
BOX · R. A. Fisher, the Life of a Scientist
BOX and DRAPER · Empirical Model-Building and Response Surfaces
BOX and DRAPER · Evolutionary Operation: A Statistical Method for Process Improvement
BOX, HUNTER, and HUNTER · Statistics for Experimenters: An Introduction to Design, Data Analysis, and Model Building
BROWN and HOLLANDER · Statistics: A Biomedical Introduction
BUCKLEW · Large Deviation Techniques in Decision, Simulation, and Estimation
BUNKE and BUNKE · Nonlinear Regression, Functional Relations and Robust Methods: Statistical Methods of Model Building
CHATTERJEE and HADI · Sensitivity Analysis in Linear Regression
CHATTERJEE and PRICE · Regression Analysis by Example, *Second Edition*
CLARKE and DISNEY · Probability and Random Processes: A First Course with Applications, *Second Edition*
COCHRAN · Sampling Techniques, *Third Edition*
*COCHRAN and COX · Experimental Designs, *Second Edition*
CONOVER · Practical Nonparametric Statistics, *Second Edition*
CONOVER and IMAN · Introduction to Modern Business Statistics
CORNELL · Experiments with Mixtures, Designs, Models, and the Analysis of Mixture Data, *Second Edition*
COX · A Handbook of Introductory Statistical Methods
*COX · Planning of Experiments
CRESSIE · Statistics for Spatial Data, *Revised Edition*
DANIEL · Applications of Statistics to Industrial Experimentation
DANIEL · Biostatistics: A Foundation for Analysis in the Health Sciences, *Sixth Edition*
DAVID · Order Statistics, *Second Edition*
*DEGROOT, FIENBERG, and KADANE · Statistics and the Law
*DEMING · Sample Design in Business Research
DILLON and GOLDSTEIN · Multivariate Analysis: Methods and Applications
DODGE and ROMIG · Sampling Inspection Tables, *Second Edition*
DOWDY and WEARDEN · Statistics for Research, *Second Edition*
DRAPER and SMITH · Applied Regression Analysis, *Second Edition*
DUNN · Basic Statistics: A Primer for the Biomedical Sciences, *Second Edition*
DUNN and CLARK · Applied Statistics: Analysis of Variance and Regression, *Second Edition*
ELANDT-JOHNSON and JOHNSON · Survival Models and Data Analysis
EVANS, PEACOCK, and HASTINGS · Statistical Distributions, *Second Edition*
FISHER and VAN BELLE · Biostatistics: A Methodology for the Health Sciences
FLEISS · The Design and Analysis of Clinical Experiments
FLEISS · Statistical Methods for Rates and Proportions, *Second Edition*
FLEMING and HARRINGTON · Counting Processes and Survival Analysis
FLURY · Common Principal Components and Related Multivariate Models
GALLANT · Nonlinear Statistical Models
GLASSERMAN and YAO · Monotone Structure in Discrete-Event Systems
GROSS and HARRIS · Fundamentals of Queueing Theory, *Second Edition*
GROVES · Survey Errors and Survey Costs
GROVES, BIEMER, LYBERG, MASSEY, NICHOLLS, and WAKSBERG · Telephone Survey Methodology
HAHN and MEEKER · Statistical Intervals: A Guide for Practitioners
HAND · Discrimination and Classification
*HANSEN, HURWITZ, and MADOW · Sample Survey Methods and Theory, Volume I: Methods and Applications
*HANSEN, HURWITZ, and MADOW · Sample Survey Methods and Theory, Volume II: Theory

*Now available in a lower priced paperback edition in the Wiley Classics Library.

Applied Probability and Statistics (Continued)

HEIBERGER · Computation for the Analysis of Designed Experiments
HELLER · MACSYMA for Statisticians
HINKELMAN and KEMPTHORNE: Design and Analysis of Experiments, Volume I: Introduction to Experimental Design
HOAGLIN, MOSTELLER, and TUKEY · Exploratory Approach to Analysis of Variance
HOAGLIN, MOSTELLER, and TUKEY · Exploring Data Tables, Trends and Shapes
HOAGLIN, MOSTELLER, and TUKEY · Understanding Robust and Exploratory Data Analysis
HOCHBERG and TAMHANE · Multiple Comparison Procedures
HOEL · Elementary Statistics, *Fifth Edition*
HOGG and KLUGMAN · Loss Distributions
HOLLANDER and WOLFE · Nonparametric Statistical Methods
HOSMER and LEMESHOW · Applied Logistic Regression
HØYLAND and RAUSAND · System Reliability Theory: Models and Statistical MethodsL
HUBERTY · Applied Discriminant Analysis
IMAN and CONOVER · Modern Business Statistics
JACKSON · A User's Guide to Principle Components
JOHN · Statistical Methods in Engineering and Quality Assurance
JOHNSON · Multivariate Statistical Simulation
JOHNSON and KOTZ · Distributions in Statistics
Continuous Univariate Distributions—2
Continuous Multivariate Distributions
JOHNSON, KOTZ, and BALAKRISHNAN · Continuous Univariate Distributions, Volume 1, *Second Edition*
JOHNSON, KOTZ, and KEMP · Univariate Discrete Distributions, *Second Edition*
JUDGE, GRIFFITHS, HILL, LÜTKEPOHL, and LEE · The Theory and Practice of Econometrics, *Second Edition*
JUDGE, HILL, GRIFFITHS, LÜTKEPOHL, and LEE · Introduction to the Theory and Practice of Econometrics, *Second Edition*
KALBFLEISCH and PRENTICE · The Statistical Analysis of Failure Time Data
KASPRZYK, DUNCAN, KALTON, and SINGH · Panel Surveys
KISH · Statistical Design for Research
*KISH · Survey Sampling
LANGE, RYAN, BILLARD, BRILLINGER, CONQUEST, and GREENHOUSE · Case Studies in Biometry
LAWLESS · Statistical Models and Methods for Lifetime Data
LEBART, MORINEAU, and WARWICK · Multivariate Descriptive Statistical Analysis: Correspondence Analysis and Related Techniques for Large Matrices
LEE · Statistical Methods for Survival Data Analysis, *Second Edition*
LePAGE and BILLARD · Exploring the Limits of Bootstrap
LEVY and LEMESHOW · Sampling of Populations: Methods and Applications
LINHART and ZUCCHINI · Model Selection
LITTLE and RUBIN · Statistical Analysis with Missing Data
MAGNUS and NEUDECKER · Matrix Differential Calculus with Applications in Statistics and Econometrics
MAINDONALD · Statistical Computation
MALLOWS · Design, Data, and Analysis by Some Friends of Cuthbert Daniel
MANN, SCHAFER, and SINGPURWALLA · Methods for Statistical Analysis of Reliability and Life Data
MASON, GUNST, and HESS · Statistical Design and Analysis of Experiments with Applications to Engineering and Science
McLACHLAN · Discriminant Analysis and Statistical Pattern Recognition
MILLER · Survival Analysis
MONTGOMERY and PECK · Introduction to Linear Regression Analysis, *Second Edition*
NELSON · Accelerated Testing, Statistical Models, Test Plans, and Data Analyses
NELSON · Applied Life Data Analysis

Applied Probability and Statistics (Continued)

OCHI · Applied Probability and Stochastic Processes in Engineering and Physical Sciences
OKABE, BOOTS, and SUGIHARA · Spatial Tesselations: Concepts and Applications of Voronoi Diagrams
OSBORNE · Finite Algorithms in Optimization and Data Analysis
PANKRATZ · Forecasting with Dynamic Regression Models
PANKRATZ · Forecasting with Univariate Box-Jenkins Models: Concepts and Cases
PORT · Theoretical Probability for Applications
PUTERMAN · Markov Decision Processes: Discrete Stochastic Dynamic Programming
RACHEV · Probability Metrics and the Stability of Stochastic Models
RENCHER · Methods of Multivariate Analysis
RÉNYI · A Diary on Information Theory
RIPLEY · Spatial Statistics
RIPLEY · Stochastic Simulation
ROSS · Introduction to Probability and Statistics for Engineers and Scientists
ROUSSEEUW and LEROY · Robust Regression and Outlier Detection
RUBIN · Multiple Imputation for Nonresponse in Surveys
RYAN · Statistical Methods for Quality Improvement
SCHUSS · Theory and Applications of Stochastic Differential Equations
SCOTT · Multivariate Density Estimation: Theory, Practice, and Visualization
SEARLE · Linear Models
SEARLE · Linear Models for Unbalanced Data
SEARLE · Matrix Algebra Useful for Statistics
SEARLE, CASELLA, and McCULLOCH · Variance Components
SKINNER, HOLT, and SMITH · Analysis of Complex Surveys
STOYAN · Comparison Methods for Queues and Other Stochastic Models
STOYAN, KENDALL, and MECKE · Stochastic Geometry and Its Applications
STOYAN and STOYAN · Fractals, Random Shapes and Point Fields
THOMPSON · Empirical Model Building
THOMPSON · Sampling
TIERNEY · LISP-STAT: An Object-Oriented Environment for Statistical Computing and Dynamic Graphics
TIJMS · Stochastic Modeling and Analysis: A Computational Approach
TITTERINGTON, SMITH, and MAKOV · Statistical Analysis of Finite Mixture Distributions
UPTON and FINGLETON · Spatial Data Analysis by Example, Volume I: Point Pattern and Quantitative Data
UPTON and FINGLETON · Spatial Data Analysis by Example, Volume II: Categorical and Directional Data
VAN RIJCKEVORSEL and DE LEEUW · Component and Correspondence Analysis
WEISBERG · Applied Linear Regression, *Second Edition*
WESTFALL and YOUNG · Resampling-Based Multiple Testing: Examples and Methods for *p*-Value Adjustment
WHITTLE · Optimization Over Time: Dynamic Programming and Stochastic Control, Volume I and Volume II
WHITTLE · Systems in Stochastic Equilibrium
WONNACOTT and WONNACOTT · Econometrics, *Second Edition*
WONNACOTT and WONNACOTT · Introductory Statistics, *Fifth Edition*
WONNACOTT and WONNACOTT · Introductory Statistics for Business and Economics, *Fourth Edition*
WOODING · Planning Pharmaceutical Clinical Trials: Basic Statistical Principles
WOOLSON · Statistical Methods for the Analysis of Biomedical Data

Tracts on Probability and Statistics

BILLINGSLEY · Convergence of Probability Measures
TOUTENBURG · Prior Information in Linear Models